Molecular Biology
of Aging

Molecular Biology of Aging

EDITED BY

Leonard P. Guarente
Massachusetts Institute of Technology, Cambridge

Linda Partridge
University College London

Douglas C. Wallace
University of California, Irvine

COLD SPRING HARBOR LABORATORY PRESS
Cold Spring Harbor, New York • www.cshlpress.com

Molecular Biology of Aging

Monograph 51

Publisher	John Inglis
Acquisition Editor	Alex Gann
Development Director	Jan Argentine
Project Coordinator	Joan Ebert
Production Editors	Patricia Barker and Kaaren Hegquist
Desktop Editor	Lauren Heller
Production Manager	Denise Weiss
Marketing Manager	Ingrid Benirschke
Sales Manager	Elizabeth Powers
Compositor	Stratford Publishing Services, Inc.
Cover Designer	Ed Atkeson

Front cover artwork: Old clock with numbers and a grunge background.

Library of Congress Cataloging-in-Publication Data

Molecular biology of aging / edited by Leonard P. Guarente, Linda Partridge, Douglas C. Wallace.
      p. ; cm. -- (Monograph ; 51)
   Includes bibliographical references and index.
   ISBN 978-0-87969-824-9 (alk. paper)
   1. Aging--Molecular aspects. I. Guarente, Leonard. II. Partridge, Linda.
III. Wallace, Douglas C. IV. Series: Cold Spring Harbor monograph series ;
51.
   [DNLM: 1. Aging--physiology. 2. Aging--genetics. 3. Cell
Aging--physiology. 4. Molecular Biology. WT 104 M71745 2008]

   QP86.M685 2008
   612.6'7--dc22

                                                    2007032831

10 9 8 7 6 5 4 3 2 1

All Cold Spring Harbor Laboratory Press publications may be ordered directly from Cold Spring Harbor Laboratory Press, 500 Sunnyside Boulevard, Woodbury, New York 11797-2924. Phone: 1-800-843-4388 in Continental U.S. and Canada. All other locations: (516) 422-4100. FAX: (516) 422-4097. E-mail: cshpress@cshl.edu. For a complete catalog of Cold Spring Harbor Laboratory Press publications, visit our World Wide Web Site http://www.cshlpress.com/.

Contents

Preface

THE PAST DECADE OR SO HAS WITNESSED AN EXPLOSION in knowledge about the molecular basis of the aging process and its modulation by genes and diet. This book was designed to capture the field of aging research at this exciting moment in its history. The chapters were chosen to represent many of the important threads that are woven together to provide the current framework of our understanding of aging, at the level of molecules, cells, tissues, and the whole organism.

One group of chapters focuses on the model organisms that have been used to dissect the genetics and molecular biology of aging in recent years. These include yeast, *C. elegans*, *Drosophila*, and mice. The strengths and weaknesses of each system are on display, and the major findings that have emerged are described in detail. Several of the pathways identified show evolutionary conservation among these model systems, and are hence candidates for modulation of human aging. The roles of stress resistance and of DNA repair are discussed, as are genomic systems and population genetic approaches currently used to investigate aging in model organisms and humans. Human aging may have unique features that are not shared by the model organisms, and this is also addressed.

Another theme in this book is the interaction between diet, metabolism, and life span. It has been known for about 75 years that a low calorie diet, termed calorie restriction or CR, can extend lifespan in rodents and in an expanding range of organisms, and these physiological and population studies are brought up to date in the book. In recent years, it has become clearer that the physiological and life-extending effects of CR are not passive events but are highly regulated processes. Candidate regulators of CR, including sirtuins and insulin/IGF signaling, are well described in this book.

CR has long been known to delay or ameliorate most forms of aging-related pathology and disease in rodents, and very recently it is starting to appear that single gene mutations that extend life span can have similar effects. A significant consequence of our understanding of CR and other molecular mechanisms that can slow down aging may, therefore, well be protection against the major diseases of aging. There are chapters that focus in detail on research in Alzheimer's disease and cancer.

An important area emphasized in the book is oxidative stress and mitochondrial function. The notion that oxidative damage causes aging is the oldest and still leading contender as an explanation for at least some of the degenerative aspects of aging. Recently, there has been a convergence of some of these themes, which reinforces their place of importance in aging. For example, CR often triggers mitochondrial biogenesis and this may dictate some of the protective effects of this diet by reducing oxidative damage.

Another important topic in aging research relates to the control of cell division and the role of cellular turnover and senescence in aging. There are chapters that focus on cellular senescence, telomeres, DNA damage and repair, stem cells, and cancer, which bring these interacting areas into sharper focus.

Finally, the evolutionary perspective of aging and how it has led the aging processes now observed in different organisms is covered. All told, this book covers a wide swath of the field of aging. Without giving short shrift to classic studies in the field, the emphasis is on the more recent molecular findings that drive our current understanding of aging. It seems likely that pharmacology will soon leverage these findings into important new therapeutics for aging and its associated diseases.

We are greatly indebted to the authors of these chapters for producing a fresh, lively, and authoritative set of reviews that well convey the current excitement and scientific opportunity in aging research. These will provide food for thought for experts in the field, and will also provide excellent and intelligible accounts for people reading for general interest or who are considering starting research on aging. We also extend our warm thanks to the wonderful staff at Cold Spring Harbor Laboratory Press, John Inglis, Alex Gann, Patricia Barker, Kaaren Hegquist, Lauren Heller, and Joan Ebert, for their expert help and encouragement, and for keeping this project moving along with great tact and good humor.

LENNY GUARENTE
LINDA PARTRIDGE
DOUG WALLACE

1

The Human Mitochondrion and Pathophysiology of Aging and Age-related Diseases

Douglas C. Wallace

Director, Center for Molecular and Mitochondrial Medicine and Genetics
University of California, Irvine
Irvine, California 92697

DURING THE PAST DECADE, INTEREST HAS GROWN rapidly in the possibility that mitochondrial dysfunction has a significant role in the etiology of aging and the age-related diseases (Wallace 1992b). However, the potential importance of the mitochondrion in these processes has not been fully explored due in part to the dominance of the anatomical and Mendelian paradigms in Western medicine. Although these two paradigms have been highly successful in addressing organ-specific symptoms and Mendelian inherited diseases, respectively, they have been relatively unsuccessful in clarifying the etiology of multisystem, age-related disorders.

Aging affects a variety of systems, although to different extents in different individuals. Furthermore, in stark contrast to the prediction of Mendelian genetics in which genetic traits are biallelic and thus quantized (+/+, +/−, −/−), age-related symptoms show a gradual decline suggestive of quantitative rather than quantized genetics.

These ambiguities might be explained by adding the mitochondrial energetic and genetics paradigms to the existing anatomical and Mendelian paradigms. The mitochondria generate the energy for the body, although different tissues rely on mitochondrial energy to different extents. Moreover, each cell contains hundreds of mitochondria and thousands of mitochondrial DNAs (mtDNAs), with each mtDNA encoding the same 13 proteins that are critical for mitochondrial energy production. The mtDNA also has a very high mutation rate, such that mtDNA mutations accumulate in tissues over time. This results in a stochastic decline in energy output that ultimately falls below the minimal energetic threshold, resulting in cell loss, tissue dysfunction, and symptoms.

WHY DO WE HAVE A MATERNALLY INHERITED MtDNA?

The life of each cell results from the interplay among structure, energy, and information. Each eukaryotic cell is the sum of two independent life forms that formed a symbiosis about 2 billion years ago. One of the cells is now represented by the nucleus and the cytosol and the information of this organism is contained in the nuclear DNA (nDNA) located in the chromosomes. The other organism is the mitochondrion with a subset of its genes contained on the mtDNA. Originally, both organisms encoded all of the information for an autonomous cell. However, as the symbiosis evolved, the DNAs of the two organisms became specialized, with the nDNA acquiring all of the genes for structure and the mtDNA retaining critical genes for energy production. The structural genes that were originally encoded by the mtDNA have been transferred to the nDNA, where many acquired a positively charged amino-terminal targeting peptide. This mitochondrial targeting peptide now directs these cytosolically synthesized proteins back into the negatively charged mitochondrial matrix to assemble the structure of the mitochondrion. However, the mtDNA has retained 13 polypeptides that appear to be the pivotal electron and proton conductors (wires) necessary for the mitochondrial energy production circuitry. Thus, by analogy, the nDNA contains the architect's blueprints for building a house, but the mtDNA contains the wiring diagram for energizing it (Wallace 2005b, 2007).

Mitochondrial Biochemistry and Physiology

The 13 polypeptides of the mtDNA are integral components of the mitochondrial energy-generating pathway oxidative phosphorylation (OXPHOS). In addition, the human mtDNA codes for the 22S tRNA and the 12S and 16S rRNA components of the mitochondrial protein synthesis system necessary for the synthesis of these proteins (Wallace 1983).

OXPHOS (Fig. 1) burns the calories (hydrogen or reducing equivalents) from the carbohydrates and fats in our diet with the oxygen that we breathe to generate water (H_2O). The energy that is released is used to maintain our body temperature and to generate ATP. The electrons from dietary carbohydrates are extracted by the tricarboxylic acid (TCA)

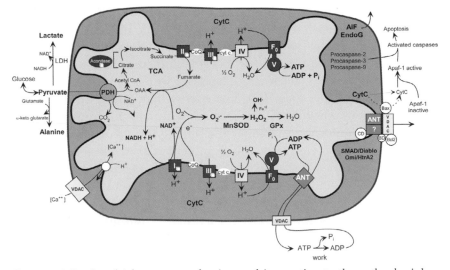

Figure 1. Mitochondrial energy production and its reaction to the pathophysiology of disease: Three features of mitochondrial metabolism are central to the pathophysiology of the common age-related diseases: (1) energy production by oxidative phosphorylation (OXPHOS), (2) reactive oxygen species (ROS) generation as a toxic by-product of OXPHOS, and (3) regulation of apoptosis through activation of the mitochondrial permeability transition pore (mtPTP). (ADP or ATP) Adenosine di- or tri phosphate; (ANT) adenine nucleotide translocator; (Cytc) cytochrome c; (GPx) glutathione peroxidase-1; (LDH) lactate dehydrogenase; (MnSOD) manganese superoxide dismutase or SOD2; (NADH) reduce nicotinamide adenine dinucleotide; (TCA) tricarboxylic acid cycle; (VDAC) voltage-dependent anion channel; (I, II, III, IV, and V) OXPHOS complexes I to V. Complex I is composed of 45 polypeptides, seven (ND1, 2, 3, 4L, 4, 5, 6) encoded by the mammalian mtDNA; complex II consists of four nDNA-encoded polypeptides; complex III by 11 polypeptides, one (Cytb) encoded by the mtDNA; complex IV is composed of 13 polypeptides, three (COI, II, III) encoded by the mtDNA; and complex V is composed of about 17 polypeptides, two (ATP6,8) encoded by the mtDNA. (Reprinted from Wallace 2005b [© Annual Reviews] and MITOMAP 2006 [© mitomap.org].)

cycle, whereas electrons from fats are extracted by β-oxidation. These reducing equivalents are transferred to either NAD^+ to generate NADH + H^+ or FAD to give $FADH_2$. The electrons are then oxidized by the electron transport chain (ETC). The ETC begins with complex I (NADH dehydrogenase), which collects electrons from NADH; complex II (succinate dehydrogenase), which collects electrons from succinate of the TCA cycle; or the electron transfer flavoprotein (ETF) and ETF dehydrogenase, which collect electrons from the fatty acyl–CoA dehydrogenases. All of these initial complexes transfer their electrons to coenzyme Q_{10} (CoQ), reducing CoQ sequentially from ubiquinone (no electrons) to ubisemiquinone (one electron) to ubiquinol (two electrons). From ubiquinol, the two electrons are transferred to complex III (bc_1 complex), then to cytochrome c, and then to complex IV (cytochrome c oxidase) which uses the electrons to reduce one-half O_2 to give H_2O. The energy that is released during the flow of electrons down the ETC is used to pump protons from the matrix out across the mitochondrial inner membrane through complexes I, III, and IV to create an electrochemical gradient ($\Delta P = \Delta \Psi + \Delta \mu^{H+}$). This biological capacitor is then used as a source of potential energy to drive complex V (ATP synthase) to condense ADP + Pi to give ATP. The mitochondrial ATP is exchanged for cytosolic ADP across the mitochondrial inner membrane by the adenine nucleotide translocators (ANTs). The efficiency with which ΔP is converted to ATP is known as the OXPHOS "coupling efficiency." Tightly coupled OXPHOS generates the maximum ATP and minimum heat, whereas loosely coupled OXPHOS generates less ATP but more heat (Wallace 1999, 2001; Wallace et al. 2001; Wallace and Lott 2002).

OXPHOS also generates most of endogenous cellular reactive oxygen species (ROS). This is the result of the redirection of electrons from the early stages of the ETC surrounding CoQ directly to O_2 to generate superoxide anion (O_2^-). Superoxide anion is highly reactive, but it can be detoxified by the mitochondrial manganese superoxide dismutase (MnSOD) to generate hydrogen peroxide (H_2O_2). Hygrogen peroxide is relatively stable. However, in the presence of reduced transition metals, it can be further reduced to hydroxyl radical ($^.OH$) which is the most reactive ROS. H_2O_2 can be reduced to H_2O by glutathione peroxidase (GPx1) or converted to H_2O and O_2 by catalase (Fig. 1). The GPxs derive electrons from the glutathione thiol (G-SH)/disulfide (G-S-S-G) redox cycle. GSH/GSSG is reduced by glutathione reductase using NADPH. NADPH can be generated by the transfer of reducing equivalents from NADH via the nicotinamide nucleoside transhydrogenase (NNT), energized by ΔP (Freeman et al. 2006; Huang et al. 2006). The GSH/GSSG system is also coupled to the cysteine (Cys)/cystine (CySS) redox system and the nuclear-cytosol and mitochondrial thioredoxin-1,2 (Trx1-$[SH]_2$)/thioredoxin-1,2 (Trx1-SS) redox systems (Hansen et al. 2006; Jones 2006). These redox systems regulate cellular growth, quiescence, and death through oxidation reduction of critical thiols in activator protein-1 (AP-1, *Fos/Jun*), NF-κB, apurinic/apyrimidinic endonuclease 1 (APE-1), PAX, *HIF-1α*, p53, and various kinases (Src kinase, protein kinase C, mitogen-activated protein kinase [MAPK], and receptor tyrosine kinases) (Evans et al. 2000; McCord 2000; Kelley and Parsons 2001; Hansen et al. 2006; Jones 2006).

Under normal circumstances, mitochondrial ROS is now thought to have an important role in communication between the mitochondria and the nucleus (Hansen et al. 2006; Jones 2006). However, ROS can also oxidize and damage cellular lipids, proteins, and nucleic acids. This toxic aspect of mitochondrial ROS production becomes accentuated when the ETC becomes excessively reduced. In these instances, the excess electrons accumulate in the electron carriers and stimulate the transfer of electrons to O_2, increasing O_2^- production. Overreduction of the ETC carriers resulting in increased mitochondrial ROS production can result from inhibition of the ETC or the availability of excess dietary calories. Inhibition of the ETC can be caused by mutations in mtDNA or nDNA mitochondrial genes that alter the kinetics of one of the OXPHOS complexes, inhibiting the flux of electrons through the ETC. Ingestion of excess calories can also feed more calories into the ETC than are required for ATP production, which also reduces the electon carriers and increases ROS production. ROS damage of mitochondrial proteins, lipids, and nucleic acids further inhibits OXPHOS and exacerbates ROS production. Ultimately, the mitochondria become damaged sufficiently that they cannot generate adequate energy for the cell, and the malfunctioning cell is removed from the tissue by apoptosis (Wallace 1999, 2001; Wallace et al. 2001; Wallace and Lott 2002).

Cells with faulty mitochondria are destroyed by apoptosis through the activation of the mitochondrial permeability transition pore (mtPTP). Although the structure of the mtPTP is much debated, one popular model envisions it composed of the outer-membrane voltage-dependent anion channel (VDAC) proteins, the inner membrane ANT, the pro- and anti-apoptotic Bcl2-Bax gene family proteins, and cyclophillin D. This complex senses changes in mitochondrial ΔP, adenine nucleotides, ROS, and Ca^{++}, and when these factors become sufficiently out of balance, the mtPTP is activated and opens a channel between the cytosol and the matrix. This depolarizes ΔP and causes the mitochondrion to swell and to release proteins from the mitochondrial intermembrane space into the cytosol. The released proteins include cytochrome c, procaspase 9, apoptosis initiating factor (AIF), and endonucelase G. Cytochrome c interacts with and activates cytosolic Apaf-1 to convert procaspase 9 to an active caspase. Caspase 9 activates procaspases 2 and 3, and they degrade cellular and mitochondrial proteins. AIF and endonuclease G are transported to the nucleus where they degrade the nDNA. The cell is thus digested from within and removed from the tissue (Fig. 1) (Wallace 2005b, 2007).

mtDNA Genetics

The 13 mtDNA polypeptide genes encode seven (ND1, 2, 3, 4L, 4, 5, 6) of the approximately 45 subunits of complex I, one (Cytb) of the 11 subunits of complex III, three (COI, II, III) of the 13 subunits of complex IV, and two (ATP6,8) of the about 17 subunits of complex V. Since the mtDNA is located in the cytosol, it is inherited through the oocyte cytoplasm and thus is exclusively maternally inherited (Giles et al. 1980). In addition, each cell harbors thousands of mtDNAs so that when a mutation arises in an mtDNA, this creates an intracellular mixture of mutant and normal mtDNAs, a state known as heteroplasmy. When a heteroplasmic cell divides, the mutant and normal mtDNAs are distributed into the daughter cells by chance and thus the mtDNA genotype can drift during mitotic replication, ultimately segregating to pure mutant or wild-type mtDNA (homoplasmic) cytoplasms.

Because of its chronic exposure to ROS, the mtDNA has a very high mutation rate and because the mtDNA genes contain no introns or noncoding sequences, a high percentage of the mutations affect gene function. Since mtDNAs replicate in both mitotic and meiotic cells, mutations can arise in and be transmitted through the female germ line. Such maternally inherited variants fall into two major categories: ancient polymorphisms and recent mutations. The mitochondria and mtDNAs also turned over in postmitotic cells. Hence, the mtDNA must continue to replicate with the result that mtDNA mutations accumulate over time in postmitotic tissues.

Mutations in the mtDNA can be deleterious, neutral, or beneficial. Since the mtDNA encodes only energy genes, deleterious mtDNA mutants generate energetic defects that are eliminated by purifying selection as a multisystem degenerative disease. The nature and severity of the disease are affected by the degree of the patient's energetic defect, which depends on the severity of the mutation and the percentage of the mutant mtDNAs in the cell, tissue, or individual. Considering that deleterious mutations are rapidly removed from the population by selection, those deleterious mutations that are found within the population must have arisen recently. These deleterious mutations preferentially affect the central nervous system and heart, muscle, renal, and endocrine systems, the tissues that are most reliant on mitochondrial energy production and are also affected in aging.

Since OXPHOS lies at the interface between calorie availability and environmental demands on cellular energy production to generate ATP and heat, certain mtDNA variants that alter OXPHOS coupling efficiency can be beneficial in appropriate environments. Such adaptive mutations are rare, however. Consequently, most of the adaptive mutations are relatively ancient. Furthermore, as the ancient adaptive mutation is beneficial, it is enriched by natural selection. Consequently, mtDNAs harboring such mutations become enriched in specific environments (Wallace 2005b, 2007).

The mtDNA Encodes the Components of an Integrated Circuit

OXPHOS uses the potential energy stored in the mitochondrial electrochemical gradient for all energy transformations, and thus all of the processes to generate, sustain, and utilize this gradient must be electrically balanced. Otherwise, this biological capacitor would short-circuit. The only way to change the circuit while keeping the components in balance is to add each new mutation sequentially to the preexisting mtDNA component variants and then test the new combination by natural selection. Therefore, only mutations that are beneficial within the environment will be retained. Since different mtDNA sequences become optimized for different environments, recombination between two different mtDNAs would be likely to create incompatible combinations of mitochondrial circuit elements. The resulting short circuits would be highly deleterious. Hence, recombination must be avoided. This is accomplished through uniparental inheritance which blocks the mixing of mtDNAs adapted to different environments (Wallace 2007).

Evidence that the mtDNA encodes the key electrical components for coupling electron transport with proton pumping comes from comparative genomics of all of the available mtDNA sequences. This analysis revealed that all mtDNAs retain the COI and Cytb genes and most retain the COII and COIII genes, the core proteins that couple electron transport to proton pumping in complexes III and IV. Hence, these four proteins are sufficient to generate and maintain a mitochondrial electrochemical gradient. Most mtDNAs also retain the electron and proton wire proteins of complex V, including the ATP6 and ATP8 genes and frequently the ATP9 gene. If the mitochondrion is going to use the NADH dehydrogenase for proton pumping, then the mtDNA also retains the complex I genes ND1, 2, 3, 4L, 4, 5, and 6. Animal mtDNAs retain essentially all of these genes (Wallace 2007).

Mutations in these mtDNA-encoded core OXPHOS proteins can adjust the ratio of protons pumped per electrons oxidized. For example, in complex III, electron transport is coupled to proton pumping through the "Q cycle." This cycle involves two CoQ-binding sites, one on the inside and the other on the outside of the mitochondrial inner membrane. Both CoQ sites are located within the cytochrome b protein, and key adaptive mutations in northern climates inactivate one or the other of these two sites. Mutations in the cytochrome b CoQ-binding sites will disrupt the Q cycle, permitting electrons to pass through complex III without pumping protons. This means that more calories must be hydrolyzed to generate the same amount of ATP; hence, the coupling efficiency is reduced. The increased calorie oxidation results in increased heat production that can be beneficial in colder climates. However, since this mutation reduces the amount of ATP generated per calorie consumed, it would be disadvantageous in a warmer climate. Hence, mtDNA mutations that change the coupling efficiency can be adaptive for different environments. Therefore, the high mutation rate of the maternally inherited mtDNA OXPHOS coupling proteins provides a robust system for permitting animals to adapt to environmental changes (Wallace 2007).

MTDNA VARIATION IN HUMAN ORIGINS AND ADAPTATION

Because of the maternal inheritance and high mutation rate of the mtDNA, mtDNA mutations accumulate along radiating maternal lineages at a rate within the time frame of historical human events. Hence, mtDNA variation has proven to be a powerful tool for reconstructing ancient female origins and migrations. However, the current distribution of mtDNA lineages appears to have been influenced by both chance (genetic drift) and adaptive selection. Hence, the mtDNA story is about both migration and the struggle to survive of our ancestors.

Using mtDNA Variation to Trace Ancient Human Migrations

To a first approximation, the number of nucleotide differences between any two mtDNAs is proportional to the time that the individuals shared a common maternal ancestor. Therefore, by correlating the geographic distribution of the aboriginal peoples with their mtDNA genetic distances, it has been possible to reconstruct a mitochondrial history of women (Fig. 2). Since mtDNA mutations accumulate sequentially, variants that founded particular mtDNA lineages define branches of the

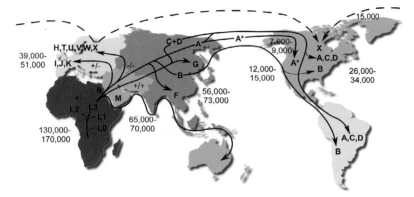

Figure 2. Diagram outlining the migratory history of the human mtDNA haplogroups. *Homo sapiens* mtDNAs arose in Africa about 150,000 to 200,000 years before present (YBP), with the first African-specific haplogroup branch being L0, followed by the appearance in Africa of lineages L1, L2, and L3. In northeastern Africa, L3 gave rise to two new lineages M and N. M and N are distinguished by the restriction site polymorphisms for DdeI at nucleotide 10394 (A to G at nucleotide 10398) and AluI at nucleotide 10397 (C to T at nucleotide 10400): +/+ for M and +/− and −/− for N. Only M and N mtDNAs succeeded in leaving Africa and colonizing all of Eurasia about 65,000 YBP. In Europe, N gave rise to the H, I, J, Uk, T, U, V, W, and X haplogroups. In Asia, M and N gave rise to a diverse range of mtDNA lineages including A, B, and F from N and C, D, and G from M. A, C, and D became enriched in northeastern Siberia and crossed the Bering Land Bridge about 20,000 to 30,000 YBP to found the Paleo-Indians. At 15,000 YBP, haplogroup X came to central Canada either from across the frozen Atlantic or by an Asian route of which there are no clear remnants today. At 12,000 to 15,000 YBP, haplogroup B entered the Americas, bypassing Siberia and the arctic, likely by moving along the Beringian Coast. Next, at 7000 to 9000 YBP, a migration bringing a modified haplogroup A moved from northeastern Siberia into northwestern North America to found the Na-Dene (Athebaskins, Dogrib, Apaches, and Navajos). Finally, relatively recently, derivatives of A and D moved along the Arctic Circle to found the Eskimos. These observations revealed two major latitudinal discontinuities in mtDNA variation: one between the Africa L haplogroups and the Eurasia and N and M derivatives and the other between the plethora of Central Asian mtDNA lineages and the almost exclusive presence of lineages A, C, and D northeastern Siberia, the later spawning the Native American migrations. Since these discontinuities correspond to the transitions from tropical and subtropical to temperate and from temperate to arctic, we have proposed that these discontinuities were the result of climatic selection of specific mtDNA mutations that permitted certain female lineages to prosper in the increasingly colder northern latitudes. (Reprinted from Wallace 2005b [© Annual Reviews] and MITOMAP 2006 [© mitomap.org].)

mtDNA tree and thus are shared by a group of related haplotypes, called a haplogroup (Wallace et al. 1999; Wallace 2005b).

Analysis of mtDNA variation has revealed that women arose in Africa about 200,000 years before present (YBP) and radiated in the African continent for about 150,000 years. This generated four major sub-Saharan African haplogroups: L0, L1, L2, and L3 (Chen et al. 2000). About 65,000 YBP, two new lineages arose from L3 in northeastern Africa: M and N. Only these two lineages left Africa to colonize the rest of the World. One migration carrying lineage N moved north into Europe and gave rise to the European-specific haplogroups H, I, J, K, T, U, V, W, and X (Wallace 2005b). In addition, both N and M radiated into Asia, N giving rise to multiple haplogroups including A, B, and F and M giving rise to a plethora of haplogroups including C, D, E, G, Y, and Z (Schurr and Wallace 2002).

Of the Asian haplogroups, only A, C, and D became established in extreme northeastern Siberia. Consequently, these were the only mtDNAs that were in a position to cross the Bering Land Bridge when it appeared about 20,000 YBP. Thus, these were the mtDNA lineages of the first Native Americans. Subsequently, haplogroup B migrated along the coast of Asia to reach the Americas about 12,000–15,000 YBP and mixed with haplogroups A, C, and D in the temperate and tropical zones. Immigrants from either Europe or Asia carried haplogroup X into the Great Lakes region of North America about 15,000 YBP (Brown et al. 1998). A subsequent migration from the Sea of Okhotsk region about 7000–9000 YBP carried a variant of haplogroup A and gave rise to the Na-Dene speakers of North America. Finally, recent migrations carrying haplogroups A and D from Chukotka gave rise to the Eskimos and Aleuts (Starikovskaya et al. 1998; Wallace et al. 2000; Derbeneva et al. 2002). After the Bering Land Bridge submerged, mtDNA haplogroups G, Y, and Z arose in Asia and moved into the Sea of Okhotsk region (Schurr et al. 1999).

mtDNA Variation and Human Adaptation

From this historical genetic reconstruction, it is clear that human mtDNA variation exhibits two major discontinuities (1) between the great African mtDNA diversity and the colonization of all of Eurasia by only two mtDNA lineages (M and N) and (2) between the plethora of Asian derivatives of lineages M and N and the presence of only haplogroups A, C, D, and G in far northeastern Siberia. The defining geographic difference across these mtDNA discontinuities is latitude and thus climate. Therefore, it has been hypothesized that mtDNA mutations that changed the coupling efficiency and thus the relative level of ATP and heat pro-

duction permitted humans to adapt to the increasingly cold environments encountered as they migrated north from Africa.

This hypothesis has been supported by the analysis of the mtDNA
protein sequence variation from 104 complete mtDNA sequences encompassing mtDNA haplotypes from tropical Africa, temperate Eurasia, and
arctic Siberian and North America. This revealed that the mtDNA ATP6
protein sequence was hypervariable in arctic Siberian and North American
mtDNAs, Cytb was hypervariable in temperate European mtDNAs, and
COI was most variable in tropical African mtDNAs (Mishmar et al. 2003).

Further support for this hypothesis was obtained by assembling 1125
complete mtDNA-coding sequences and using these to reconstruct the
entire human global mtDNA phylogeny encompassing all nucleotide
changes within a sequential mutational tree. This revealed that nodal
amino acid substitution mutations increased in frequency for mitochondrial lineages from tropical Africa, to temperate Eurasia, to arctic Siberia
and North America. Moreover, at the base of many of the temperate and
arctic mtDNA lineages, distinctive, highly conserved, amino acid substitution mutations were found that could affect the efficiency either of ETC
proton pumping or of ATP generation. Thus, mtDNA mutations that
could have affected coupling efficiency did become established as people
moved out of Africa (Ruiz-Pesini et al. 2004).

Various mtDNA haplogroups have also been correlated with a variety of common degenerative disease phenotypes as well as with longevity.
The European haplogroups J and Uk have been found to be protective
for Alzheimer's disease (AD) and Parkinson's disease (PD) and also associated with increased longevity. These associations are particularly striking, since both European haplogroups J1 and Uk independently acquired
the same cytochrome *b* mutation at nucleotide 14798C (Ivanova et al.
1998; Chagnon et al. 1999; De Benedictus et al. 1999; Ross et al. 2001;
Niemi et al. 2003; van der Walt et al. 2003, 2004). This mutation alters
one of the CoQ-binding sites of complex III and is conserved in 79% of
the species analyzed. Furthermore, individuals harboring haplogroup H
have been found to be more resistant to sepsis (Baudouin et al. 2005),
those harboring haplogroup T are enriched among individuals with bipolar disease (McMahon et al. 2000), and those harboring haplogroup J
mtDNAs are more prone to developing blindness upon acquisition of the
milder mutations that cause Leber hereditary optic neuropathy (LHON)
(Brown et al. 1997, 2002; Torroni et al. 1997).

Potentially adaptive mtDNA mutations have been identified in both
the mtDNA polypeptide (Ruiz-Pesini et al. 2004) and the tRNA and rRNA
(Ruiz-Pesini and Wallace 2006) genes. Since uncoupled mitochondria

would remain more oxidized and thus generate fewer ROS, more uncoupled mtDNAs could reduce mitochondrial oxidative stress. This would preserve mitochondrial and cellular functions and permit prolonged survival of postmitotic cells (Ruiz-Pesini et al. 2004). Therefore, the same mtDNA variants that permitted our ancestors to adapt to colder climates may be affecting individual predisposition to modern diseases and aging.

Since mtDNAs do not recombine, neutral variants that happen to be present in the mtDNA when the beneficial mutation occurred were expanded in the population along with the adaptive selection by hitchhiking. This then explains why arrays of mtDNA variants are found together.

MITOCHONDRIAL MUTATIONS IN DEGENERATIVE DISEASE

Deleterious mutations can arise in either the mtDNA or the nDNA genes of the mitochondrial genome. Since the mtDNA contains the blueprints for the electrical circuit but the nDNA encodes the genes that specify the mitochondrial structure, mitochondrial diseases can result from either mtDNA or nDNA gene mutations or a combination of the two (Wallace 2005b, 2007).

mtDNA Mutations in Inherited Disease

Pathogenic mtDNA base-substitution mutations can alter either polypeptide genes or rRNA and tRNA genes. One example of a disease resulting from mtDNA polypeptide gene mutation is the sudden-onset midlife blindness in LHON. LHON can be caused by any one of several complex I gene missense mutations including mutations at nucleotides G11778A (Wallace et al. 1988a), G3460A (Huoponen et al. 1991), T14484C (Johns et al. 1992), T10663C (Brown et al. 2002), and A14459G (Jun et al. 1994).

An example of an mtDNA protein synthesis gene mutation is the tRNA^{Lys} A8344G mutation that causes the MERRF syndrome. MERRF is a multisystem neuromuscular disease that in the extreme form causes myoclonic epilepsy and ragged-red-fiber disease (MERRF). However, variation in the percentage of heteroplasmy can result in a wide spectrum of clinical symptoms from sensorineural hearing loss to epilepsy and muscle disease to cardiomyopathy to progressive dementia (Wallace et al. 1988b; Shoffner et al. 1990).

To date, more than 229 pathogenic base-substitution mutations have been identified that cause a broad spectrum of symptoms affecting the central nervous system causing vision and hearing loss, movement and balance problems, and memory loss; muscle wasting with progressive weakness and myaglia; heart dysfunction leading to hypertrophic and dilated cardiomyopathy; endocrine system disorders resulting in diabetes mellitus; renal system failure; etc. These reported disease mutations are catalogued in the MITOMAP Web site (Wallace 2005b; MITOMAP 2006) and in a recent review (Wallace et al. 2007).

That mtDNA base-substitution mutations can cause degenerative disease was confirmed by introducing into the mouse germ line the mtDNA 16S rRNA chloramphenicol (CAP) resistance (CAP^R) mutation at T2433C and demonstrating that the mutant mice developed mitochondrial myopathy and cardiomyopathy. These mice were generated by enucleating CAP^R mouse cultured cells and fusing the cytoplasmic fragment containing the mutant mtDNA to a female embryonic stem (ES) cell that had been depleted of its resident mtDNAs by treatment with the mitochondrial toxic rhodamine 6G. CAP^R female ES cell cybrids were injected into mouse bastocysts, chimeric animals were identified and phenotypically characterized, and the females were bred with the result that the CAP^R mtDNAs were transferred into progeny in either the homoplasmic or heteroplasmic state. Phenotypic analysis of the chimeric mice revealed that they developed cataracts, retinopathy, and optic nerve head hamartomas. The CAP^R progeny mice died within months after birth in association with cardiomyopathy and mitochondrial myopathy (Sligh et al. 2000). Thus, the T2433C mtDNA base-substitution mutation was both necessary and sufficient to cause progressive degenerative disease.

In addition to base-substitution mutations, the mtDNA is also prone to rearrangement mutations (Holt et al. 1988). Rearrangements can include insertions, deletions, and reciprocal combinations of the two. Since most rearrangements remove one or more tRNAs, rearrangement mutations frequently result in protein synthesis defects. Common clinical presentations of rearrangement mutations include the frequently lethal Pearson marrow/pancreas syndrome (Rötig et al. 1988), the Kearns-Sayre syndrome (KSS), and chronic progressive external opthalmoplegia (CPEO) (Moraes et al. 1989).

The pathogenicity of mtDNA rearrangement mutations has been confirmed by introducing rearranged mtDNAs into the mouse germ line. An mtDNA with a 4696-nucleotide deletion that removed six tRNAs and seven structural genes was recovered from a mouse brain by fusion of synaptosomes to cultured mouse cells. The resulting synaptosome cybrids were enucleated and the cytoplasts were fused to pronucleus-stage embryos, which were then implanted into the oviducts of pseudopregnant females. This resulted in 24 animals having 6–42% deleted mtDNAs in their muscle. Females with 6–13% deleted mtDNA were mated and the rearranged mtDNAs were transmitted through three successive generations, with the percentage of deleted mtDNAs increasing with successive generations to a maximum of 90% in the muscles of some animals. Although mtDNA duplications were not observed in the original synaptosome cybrid cells, they were found in the postmitotic tissues of the animals. This raises the possibility that the maternal transmission of the rearranged mtDNA was through a duplicated mtDNA intermediate, as proposed for the human maternally inherited mtDNA rearrangement pedigree presenting with diabetes mellitus and deafness (Ballinger et al. 1992, 1994). Although RRFs were not observed in these animals, fibers with greater than 85% mutant mtDNAs were COX-negative, and many fibers had aggregates of subsacolemmal mitochondria. The heart tissues of heteroplasmic animals were also a mosaic of COX-positive and COX-negative fibers, and the amount of lactic acid in peripheral blood was proportional to the amount of mutant mtDNA in the muscle tissues. Mice with predominantly mutant mtDNAs in their muscle tissue died within 200 days with systemic ischemia and enlarged kidneys with granulated surfaces and dilation of the proximal and distal renal tubules. These animals also developed high concentrations of blood urea and creatinine (Inoue et al. 2000). Hence, mtDNA deletion mutations can also cause disease in mice, but the phenotypes and inheritance patterns are somewhat different from those seen in most human mtDNA rearrangement patients.

Finally, a wide array of base-substitution and small insertion deletion mutations have been identified in the mtDNA control region (MITOMAP 2006). In one case, a large intracontrol region duplication was reported to be associated with increased predisposition to CPEO and KSS (Brockington et al. 1993); however, the predisposition to this duplication was subsequently linked to a base substitution in European haplogroup I which created a long direct repeat that consistently encompassed the duplication (Torroni et al. 1994). In addition, base-substitution mutations in the control region have been associated with diabetes mellitus and cardiomyopathy (Khogali et al. 2001; Poulton et al. 2002) and endometrial cancer (Liu et al. 2003), and a number of mutations in the control region termination-associated sequence (TAS) have been reported in conjunction with cyclic vomiting syndrome and migraine without aura (Wang et al. 2004). However, considerably more research is required to understand the role of control region mutations in disease.

Pathophysiology of nDNA Mitochondrial Disease

Mutations in nDNA-encoded mitochondrial genes are now also being discovered. These fall into two general classes: those that alter structural proteins involved in mitochondrial energy metabolism (Procaccio and Wallace 2004) and those that alter enzymes that are important in the maintenance of the mtDNA such as mtDNA polymerase γ (POLG) (Naimi et al. 2006). Since the pathophysiology of mutations in nDNA-encoded mitochondrial genes will also affect the same processes as mtDNA mutations, introduction of nDNA mutations can provide insight into the pathophysiology of mitochondrial disease.

To determine the importance of alterations in mitochondrial ROS generation, energy production, and apoptosis in age-related diseases, mice have been engineered with deficiencies in the mitochondrial antioxidant enzymes for MnSOD, glutathione peroxidase-1 (GPx1) and GPx4, and in the energy transduction and mtPTP proteins for the adenine nucleotide translocator isoforms 1 and 2 (Ant1 and Ant2). Although the loss of GPx1 resulted in a relatively mild energetic defect (Esposito et al. 2000), genetic inactivation of GPx4 was lethal. Since GPx1 detoxifies soluble peroxides, whereas GPx4 detoxifies lipid peroxides, these results demonstrate that lipid peroxides in the mitochondrial inner membrane are extremely toxic (Yant et al. 2003; Ran et al. 2004; Liang et al. 2007). Inactivation of the MnSOD gene also proved to be lethal (Li et al. 1995; Melov et al. 1999). Moreover, heterozygous MnSOD animals showed reduced mitochondrial function, early activation of the mtPTP, and increased apopotosis (Kokoszka et al. 2001). The importance of ROS in the pathophysiology of mitochondrial disease has been confirmed by showing that many of the cardiomyopathy and certain neurological symptoms can be reversed by treatment with metalloporphoryin antioxidants such as MnTBAP (Melov et al. 1998) and manganese intercalated galene catalytic antioxidants such as EUK8 (Melov et al. 2001). Thus, ROS toxicity is a major component of the pathophysiology of mitochondrial disease.

To determine the importance of energy deficiency and alterations in the mtPTP in mitochondrial disease, the heart muscle isoform of the mitochondrial ANT (Ant1) was also genetically inactivated. This inhibited the export of mitochondrial ATP into the muscle and heart cells and resulted in severe exercise intolerance and a hypertrophic cardiomyopathy that progressed to dilated cardiomyopathy. The resulting inhibition of the mitochondrial respiratory chain also increased ROS production, which caused the premature accumulation of mtDNA deletions in the heart associated with heart failure. Hence, reductions in mitochondrial ATP production are an important factor in mitochondrial disease, and inhibition of OXPHOS is also tied to increased mitochondrial oxidative damage (Graham et al. 1997; Esposito et al. 1999).

To further explore the importance of energy deficiency in disease, the systemic ANT isoform gene, *Ant2*, located on the X chromosome, was also inactivated. Complete inactivation of the *Ant2* gene proved to be an embryonic-lethal. However, a conditional knockout allele was generated by flanking the *Ant2* gene with recombinogenic *LoxP* sites, resulting in a floxed *Ant2* gene (*Ant2^{floxed}*). Mating the *Ant2^{floxed}* mice, which were also *Ant1$^{-/-}$*, with animals in which the *Cre* recombinase was transcribed from a liver-specific albumin promoter (*alb-Cre*) resulted in mice with liver mitochondria that were unable to exchange ATP or ADP across the mitochondrial inner membrane. However, these liver mitochondria were still able to undergo permeability transition, although the sensitivity of the ANT-deficient mtPTP to Ca^{++} activation was greatly reduced and the mtPTP could no longer be modulated by ADP or other ligands of the ANT. Thus, the ANTs are not required to form the actual mtPTP pore, but they are essential for the regulation of mtPTP (Kokoszka et al. 2004).

These studies then confirm that mitochondrial energy production, ROS toxicity, and regulation of apoptosis are all important factors in the pathophysiology of mitochondrial disease.

Somatic mtDNA Mutations in Aging and Age-related Disease

One of the most distinctive features of mitochondrial diseases is that they have a delayed onset and a progressive course. Therefore, mitochondrial diseases must be modulated by an aging clock. Since mtDNA pathogenic mutations have been shown to cause all of the symptoms associated with aging, it follows that the accumulation of somatic mtDNA mutations may cause the progressive decline in mitochondrial function, ultimately resulting in organ failure and clinical symptoms (Wallace 1992a,b).

Both rearrangement and base-substitution somatic mtDNA mutations have been documented to accumulate in postmitotic tissues with age. Rearrangement mutations have been repeatedly documented to accumulate in the tissues most prone to age-related decline (Corral-Debrinski et al. 1992; Cortopassi and Arnheim 1992; Song et al. 2004; Bender et al. 2006; Kraytsberg et al. 2006). Furthermore, base-substitution mutations in the mtDNA control region have also been reported to accumulate with age, and most surprisingly, specific mutations have been reported to accumulate in particular tissues. A mutation in the mitochondrial transcription-factor-binding site (mtTFA) for the L-strand promoter at nucleotide T414G has been found to accumulate with age in human skin fibroblasts (Michikawa et al. 1999), mutations at A189G and T408A preferentially accumulate in muscle (Wang et al. 2001), and a mutation at T150C accumulates in white blood cells and has been proposed to be protective of aging and stress (Zhang et al. 2003). Furthermore, somatic mtDNA mutations accumulate in all animal tissues, at a rate consistent with the aging rate of that species (Wallace 2005b).

Evidence that the accumulation of somatic mtDNA mutations is an important factor in aging has come from two types of mouse genetic studies. In the first, mice were created in which the normal mtDNA POLG gene was substituted with a POLG in which the proofreading exonuclease function was inactivated by a D257A amino acid substitution. These mice had a shortened life span and developed a premature aging phenotype involving weight loss, reduction of subcutaneous fat, hair loss (alopecia), curvature of the spine (kyphosis), osteoporosis, anemia, reduced fertility, and heart enlargement. This was associated with an age-related decline in respiratory complexes I and IV and in mitochondrial ATP production rates in the heart. Analysis of the mtDNA revealed that POLG mutant mice had a three- to fivefold increase in mtDNA base-substitution mutations in brain, heart, and liver, with the number of mutations being higher in the coding region cytochrome b gene than in the control region (Trifunovic et al. 2004; Kujoth et al. 2005; Vermulst et al. 2007). Alternative assays have provided somewhat different conclusions (Vermulst et al. 2007). Still, these studies indicate that increasing the mtDNA mutation rate increases the aging rate.

In both of the experiments involving POLG mutant mice, the mitochondrial ROS production rate was not observed to be increased (Kujoth et al. 2005; Loeb et al. 2005; Trifunovic et al. 2005). However, this is reasonable since the mutator POLG produces random mutations in the various cellular mtDNAs, each causing a different mitochondrial defect. In contrast, in mtDNA diseases, most or all of the mtDNAs harbor the same mutation leading to the same biochemical defect. The discrete OXPHOS defect resulting from a high percent of a single heteroplasmic mutation would stall the ETC causing electrons to accumulate in the carriers. The excess electrons could then be transferred directly to O_2 to generate ROS. In contrast, in the POLG mutator mouse where each mtDNA has a different mutation, the mitochondria within a cell are constantly undergoing fusion and fission, randomly mixing the various mutant mtDNAs and their defective gene products. The resulting uniform distribution of the different defective gene products across all of the mitochondria permits the complementation of the mutant mtDNAs in *trans* (Oliver and Wallace 1982). Consequently, no discrete block would develop in the ETC even though the overall efficiency of OXPHOS energy production would decline. Hence, apoptosis would increase due to energetic failure, in the absence of excessive ROS production.

Although increasing the mtDNA mutation rate shortens life span, to prove that somatic mtDNA mutations limit life span requires demonstrating that diminishing the mtDNA mutation rate through removing ROS increases life span. To accomplish this, a transgenic mouse was created in which catalase, which is normally peroxisomal, was targeted to the mitochondrial matrix. Catalase rapidly degrades H_2O_2. In this construct, the catalase peroxisomal targeting peptide was removed and the amino-terminal ornithine transcarbamylase (OTC) mitochondrial targeting peptide was added. Mice harboring this "mCAT" transgene not only lived 20% longer, but had reduced mitochondrial and mtDNA oxidative stress, reduced age-related accumulation of mtDNA rearrangement mutations, and protected cardiovascular function (Schriner et al. 2005). Thus, the accumulation of somatic mtDNA mutations is a component of the aging clock.

Mitochondrial diseases and aging are thus the same process, differing only in their time course (Fig. 3). Individuals begin life with a particular energetic capacity determined by their inherited mtDNA and nDNA mitochondrial variants. As individuals age, they accumulate somatic mtDNA mutations that erode their cellular mitochondrial function. This increases the apoptosis rate, causing a progressive decline in cellularity until tissue integrity is insufficient for normal organ function. Eventually, cellular loss results in organ failure and the symptoms called aging. If thresholds are crossed early, these same processes are referred to as disease (Wallace 2005b).

The phenotypic expression of certain mild mtDNA disease mutations can also be influenced by the background mtDNA haplogroup on which the mutations arose. In Europeans, the milder LHON mutations, nucleotides 11778, 14484, and 10366, are enhanced in their expressivity when they arise on European mtDNA haplogroup J (Brown et al. 1997, 2001, 2002). The more biochemically severe mutations cause LHON independent of the background mtDNA lineage (Shoffner et al. 1995). Hence, the diminished ATP production of haplogroup J with its adaptive cytochrome *b* variants must augment the energetic defect of the milder LHON complex I mutations sufficiently to create a clinical phenotype.

THRESHOLD HYPOTHESIS
OXPHOS Capacity vs. mtDNA Damage

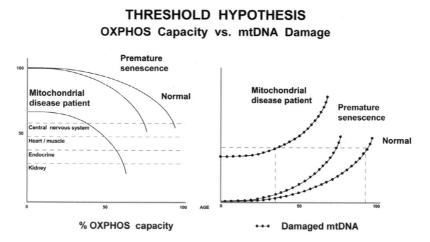

Figure 3. The threshold hypothesis of the clinical expression of mitochondrial energy deficiency. (*Left panel*) Age-related decline in mitochondrial energy-generating capacity showing the starting energetic levels of individuals with initially highly functional mitochondria (*upper lines*) or partially dysfunctional mitochondria (*lower line*). (*Upper line*) Two different kinetics: slow decay resulting in normal aging or more rapid decay, possibly due to increased mitochondrial ROS production, resulting in "premature" development of age-related disease symptoms. Common symptoms associated with mitochondrial decline include type II diabetics, Alzheimer's disease, Parkinson's disease, cardiovascular disease, or ophthalmologic deterioration. (*Right panel*) Relative proportion of mutant mtDNAs that accumulate with age showing the cumulative effects of inherited plus somatic mtDNA mutations. (Reprinted, with modifications, from Wallace 1992a [© Annual Reviews].)

These observations then provide a general paradigm for age-related diseases and aging. Mitochondrial energy production generates ROS as a toxic by-product. This results in the accumulation of mtDNA mutations that erode mitochondrial function. Ultimately, mitochondrial function diminishes to the point that the mtPTP is activated and the cell dies. Since mtDNA mutations that alter a vital function occur by chance, the loss of cells is stochastic. To compensate for this inherent cellular loss rate, humans are born with extra cells in postmitotic tissues so that there will be sufficient cells to accommodate the random cell loss while still retaining cells to function adequately to the end of reproductive age (Fig. 4) (Wallace 2005b).

Evidence that this process is acting in age-related degenerative diseases has been obtained by studying the accumulation of mtDNA control region somatic mutations in AD brains. AD brains have been found to have on average 63% more heteroplasmic somatic mutations than age-matched

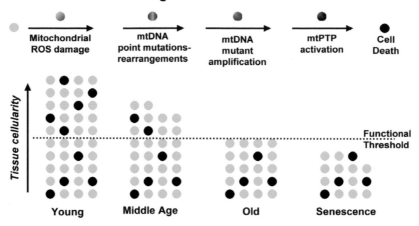

Figure 4. Mitochondrial and cellular model of aging. The upper line of cells diagrams the mitochondrial role in the energetic life and death of a cell. All postmitotic cells have a finite probability of dying due to the accumulation of mitochondrial and mtDNA damage associated with OXPHOS ROS production, followed by mutant mtDNA amplification, activation of the mtPTP, and death by apoptosis. The bottom diagram represents the loss of cells in a tissue over the life of an individual through mitochondrial-mediated death (*black cells*). Each individual is born with sufficient extra cells such that even as cells are lost by mitochondrial-induced apoptosis, sufficient cells remain to maintain tissue function. Ultimately, however, enough cells are lost that the organ begins to malfunction, resulting in the clinical manifestations of old age. (*Dashed line*) Minimum number of cells needed for the tissue to function normally. (Reprinted from Wallace 2005 [© Annual Reviews] and MITOMAP 2006 [© mitomap.org].)

controls. Moreover, the somatic control region mutations in AD brains selectively cluster within the known functional elements that regulate mtDNA transcription and replication, whereas those mutations that are present in the control brain mtDNAs occurred outside of functional elements. For example, the T414G mutation that displaces the light (L)-strand promoter transcription factor was detected in 65% of AD brains but in none of the control brains. That these control region variants are functionally significant was indicated by an observed 50% reduction in the brain mtDNA-to-nDNA ratio and 50% reduction in the level of the L-strand ND6 transcript (Coskun et al. 2004).

The mtDNA somatic mutation rate has also been found to be elevated in the keratoconus of the cornea (Atilano et al. 2005), Parkinson's disease (Bender et al. 2006; Kraytsberg et al. 2006), as well as other "spontaneous" clinical syndromes. Taken together, these observations suggest that the accumulation of somatic mtDNA mutations may be a central factor in the etiology of a variety of age-related clinical syndromes.

mtDNA Mutations in Cancer

Cancer is also an age-related disease that has increasingly been shown to be associated with mitochondrial abnormalities. More than 70 years ago, Otto Warburg observed that cancer cells generated excessive lactate while undergoing aerobic metabolism, a situation that he called "aerobic-glycolysis." Warburg then posited that mitochondrial defects were important in the etiology of cancer. However, the theoretical and experimental tools of mitochondrial medicine have only recently become available to test Warburg's conjecture (Wallace 2005a).

In a study of prostate cancer, increased variation in the mtDNA COI gene was observed, consistent with proteomic studies that had revealed an increased ratio of nDNA complex IV subunits to mtDNA complex IV subunits in prostecomies (Herrmann et al. 2003; Krieg et al. 2004). Sequencing the COI gene from 260 prostecomies revealed that 11% harbored COI mutations, whereas 0% of the prostate-cancer-negative controls harbored COI mutations. Both germ-line and somatic COI mutations were observed. Four of the germ-line mutations changed highly conserved amino acids and were found in multiple independent cases on different mtDNA haplogroups, strongly indicating that these mtDNA mutations have a causal role in prostate cancer (Petros et al. 2005).

Surprisingly, analysis of the mtDNA COI gene variation in the general population revealed that about 5.5% of individuals also harbored COI mutations. Presumably, these mutations can persist in the general population since the mtDNA is maternally inherited but women are not affected by prostate cancer (Petros et al. 2005).

To confirm that deleterious mtDNA mutations can contribute to prostate cancer, the mtDNAs of the prostate cancer cell line PC-3 were substituted for mtDNAs from a patient that was heteroplasmic for the pathogenic ATP6 T8993G mutation that causes Leigh syndrome. PC-3 cells that were homoplasmic for the mutant T8993G mtDNA generated large rapidly growing tumors in nude mice, whereas PC-3 cells that were homoplasmic for the wild-type T8993T mtDNAs did not generate tumors. Biochemical analysis of these tumors revealed that those with the mutant T8993G mtDNA generated high levels of ROS, whereas those with the wild-type T899T mtDNA did not generate much ROS (Petros et al. 2005). Hence, it would appear that deleterious mtDNA mutations can greatly enhance tumorigenicity and that one relevant factor might be increased mitochondrial generation of ROS.

A meta-analysis of all somatic mtDNA mutations that have been reported in cancers revealed that about half of the mtDNA-coding-region mutations fit into the category of deleterious and thus might have contributed to establishing the tumorigenic phenotype. However, the other half of the coding-region mutations and more than 80% of the control region mutations proved to be the same mtDNA sequence variants that had been observed as population variants. These observations suggest that somatic mtDNA mutations in cancer may fall into two classes: (1) those that initially inhibited OXPHOS, resulting in increased ROS production and tumor initiation and promotion, and (2) those that subsequently altered mitochondrial energy metabolism to permit the cancer cells to adapt to the changing environmental availability of substrates and oxygen and to the effects on apoptosis (Brandon et al. 2006).

MITOCHONDRIA AND ENVIRONMENTAL ADAPTATION AND CALORIE RESTRICTION

Physiological Modulation of Energy Metabolism

The calories burned by the mitochondria ultimately come from the collection of the energy of sunlight by plant chloroplasts. Chloroplasts use the sun's energy to split H_2O into hydrogen and oxygen, and the resulting reducing equivalents are combined with carbon derived from CO_2 to generate glucose. Therefore, during the growing season, plants generate excess carbohydrates that animals ingest. The carbohydrates can be metabolized by glycolysis to generate ATP and the excess energy stored as fat. In contrast, in the nongrowing season when carbohydrates are limited, animals must mobilize their fat stores and burn them by OXPHOS to generate the ATP to sustain them.

When glucose is plentiful, high animal serum glucose levels are detected by the pancreatic β cells that then secrete insulin into the circulation. Insulin then signals to the energy-utilizing cells to down-regulate OXPHOS, since there is ample glucose to generate energy by glycolysis. The insulin also instructs the energy storage tissues to store the excess calories as fat in white adipose tissue (WAT).

When glucose availability declines, animal serum glucose levels also decline. This decreases insulin secretion by the pancreatic β cells and increases glucagon secretion by the pancreatic α cells. Glucagon is delivered to the WAT and signals it to mobilize fats. Glucagon also signals the energy-utilizing cells to up-regulate mitochondrial OXPHOS to burn the fat. Thus, animals utilize the energy stored as fat to survive calorie privation during the nongrowing season (Fig. 5).

At the target cells, insulin binds to the insulin receptor on the plasma membrane of the energy-utilizing cells (Fig. 5). This activates the insulin receptor tyrosine kinase that phosphorylates insulin target proteins. These proteins activate the phosphoinositol-3 kinase (PI3K) pathway that activates protein kinase B (PKB or *Akt*). PKB then phosphorylates the FOXO proteins, excluding them from the nucleus, thus inactivating them as transcription factors.

When not phosphorylated and thus active, FOXOs bind to insulin response elements (IREs) in the promoters of nuclear genes including peroxisome-proliferation-activated receptor γ (PPARγ)–coactivator-1α (PGC-1α) whose promoter contains three IREs (Daitoku et al. 2003). PGC-1α, in turn, interacts with a variety of nuclear transcription factors and up-regulates nDNA gene transcription for mitochondrial biogenesis genes, including the mitochondrial transcription factor, mtTFA. mtTFA migrates to the mitochondrion and up-regulates mtDNA transcription and replication (Fig. 5, right side) (Wu et al. 1999; Herzig et al. 2000, 2001; Scarpulla 2002; Kelly and Scarpulla 2004). Activation of FOXOs also up-regulates the transcription of antioxidant genes such as MnSOD and catalase. Because of this regulatory pathway, during the growing season when animal serum glucose is high, insulin secretion is high, and the FOXOs are phosphorylated and inactivated. FOXO inactivation means that they cannot bind to the IREs. Consequently, PGC-1α and antioxidant gene transcription are reduced and mitochondrial OXPHOS and antioxidant defenses are down-regulated. This shifts the animal's energy generation from OXPHOS toward glycolysis.

In contrast, during the nongrowing season, the low animal serum glucose reduces serum insulin levels, inactivating PKB and leaving the FOXOs unphosphorylated and active. The active FOXOs migrate to the nucleus and bind to the IREs, thus up-regulating PGC-1α and mitochondrial OXPHOS and antioxidant defenses. Down-regulation of insulin also mobilizes WAT fat stores to be secreted into the blood stream to serve as fuel for the mitochondria (Fig. 5).

At the same time, reduced serum glucose stimulates the pancreatic α cells to secrete glucagon. Glucagon binds to the plasma membrane glucagon receptor activating adenylcyclase. The resulting rise in cAMP activates protein kinase A (PKA) which phosphorylates the cAMP response element-binding protein (CREP). Phosphorylated CREP migrates into the nucleus where it binds to the cAMP response elements (CRPs). The

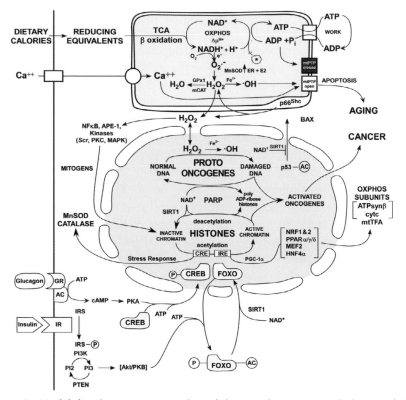

Figure 5. Model for the environmental modulation of energy metabolism and the mitochondrial physiological basis of age-related diseases and aging. The primary interactions are diagrammed between mitochondrial energy production, ROS generation, and initiation of apoptosis through activation of the mtPTP. These components of energy metabolism are modulated by environmental constraints such as calorie availability and thermal stress through the regulation of the FOXO and CREB transcription factors. The FOXO transcription factors are inactivated by phosphorylation by the Akt 1/2 (protein kinase B, PKB) kinase, which is up-regulated by insulin binding to the insulin receptor through activation of phosphatidyl inositol signaling and also inactivated by their acetylation via NAD-dependent SIRT1 that is modulated by the NADH/NAD ratio and thus the metabolic status of the cell. The CREB transcription factor is activated by phosphorylation by PKA when PKA is activated by cAMP. cAMP is generated by adenylylcylase in response to the binding of glucagons the glucagons receptor. The PPARγ-coactivator 1 (PGC-1α) factor is a nuclear transcriptional coactivator that interacts and enhances the action of multiple transcription factors that coordinately up-regulate genes for mitochondrial biogenesis including the mtDNA transcription factor mtTFA. The promoter of PGC-1α has three IREs that bind the dephosphorylated and deacetylated FOXOs and has a CRE that binds phosphorylated CREB. When serum glucose is high, insulin is high, the FOXOs are phosphorylated and inactive, CREP is not phosphorylated and is inactive, and mitochondrial OXPHOS is down-regulated in association with the down-

mammalian PGC-1α gene encompasses a strong cAMP response element (CRE) (Daitoku et al. 2003) and thus is induced. This augments the effects of the FOXOs, further up-regulating mitochondrial OXPHOS and antioxidant defenses.

Studies in *Caenorhabditis elegans, Drosophila,* and mice have shown that the genetic inactivation of the insulin-like growth factor (IGF) receptor-Akt (PKB) signal transduction pathway consistently increases life span. Since inactivation of the IGF-Akt pathway would activate the forkhead (FOXO) transcription factors, this would activate PGC-1α and up-regulate OXPHOS. Activation of the forkheads would also up-regulate the expression of MnSOD and other antioxidant genes. Hence, inactivation of the IGF signal transduction pathway should up-regulate mitochondrial OXPHOS and antioxidant defenses (Furuyama et al. 2003; Accili and Arden 2004; Wallace 2005b, 2007).

FOXO activity is also inhibited by acetylation by GCN5 (Nemoto et al. 2005; Rodgers et al. 2005; Koo and Montminy 2006; Gerhart-Hines et al. 2007) or Cbp and p300 (Accili and Arden 2004), whereas FOXOs are activated by deacetylation mediated by the NAD$^+$-dependent sirtuins, specifically SIRT1 (Brunet et al. 2004). In mammals, SIRT1 binds to FOXO3 under oxidative stress, deacetyating and thus activating the FOXOs. This up-regulates the antioxidant and stress-response genes. Overexpression of the sirtuins in various model organisms also increases life span (Brunet et al. 2004; Motta et al. 2004; Giannakou and Partridge 2004; Wallace 2005b). SIRT1 deacetylation uses NAD$^+$, but not NADH + H$^+$, as a substrate and thus may be sensitive to the cellular redox state (Nemoto et al. 2005; Rodgers et al. 2005; Koo and Montminy 2006;

---

regulation of the antioxidant defenses, MnSOD and catalase. When serum glucose in low, insulin signaling declines, the FOXOs are dephosphorylated and activated and migrate to the nucleus where they activate transcription of PGC-1α and the antioxidant enzymes. Glucagon binds the glucagon receptor that activates adenylcyclase generating cAMP. The cAMP-activated PKA then phosphorylates and activates CREB which binds to CREs and up-regulates PGC-1α and the antioxidant enzymes. The Sir2 system further modulates the apoptosis effectors and transcription and replication mediators with respect to the metabolic status of the cell. (PARP) Poly ADP-ribose polymerase; (SIRT1) mammalian homolog to Sir2; (FOXO3) the most ubiquitous mammalian forkhead transcription factor; (P) a phosphorylated protein; (Ac) an acetylated protein; (ILL) insulin-like ligand; (ILR) insulin-like growth factor receptor; (PI3K) PI3 kinase; (PI2) membrane-bound phosphotidyl-inositol diphosphate; (PI3) membrane-bound phosphatidyl inositol triphosphate; (AKT 1/2) the AKT kinases also known as protein kinase B (PKB). (Reprinted, with modifications, from Wallace 2007 [© Annual Reviews].)

Gerhart-Hines et al. 2007). Sirtuins have also been shown to deacetylate and activate PGC-1α and deacetylate and inactivate p53 (Imai et al. 2000; Luo et al. 2001; Vaziri et al. 2001; Langley et al. 2002; Wallace 2005b [#4430]). Resveratrol, which is an activator of SIRT1 (Howitz et al. 2003), extends life span in a variety of model organisms (Wood et al. 2004; Valenzano et al. 2006) and strongly induces mitochondrial function and performance in mammals through activation of PGC-1α (Baur et al. 2006; Lagouge et al. 2006). Thus, the Sir2s also appears to be modulating energy metabolism and antioxidant defenses in response to caloric availability.

In *Drosophila* studies, the importance of cAMP signal transduction in regulating longevity has also been demonstrated. In flies, the neurofibromatosis-1 (NF1) gene (*NF1*) product, neurofibromin, stimulates adenylylcyclase (AdCy). Inactivation of the *Drosophila NF1* gene reduces fly cAMP, and this is associated with increased sensitivity to oxidative and thermal stress, reduced physical performance, reduced OXPHOS complex I activity, a doubling of mitochondrial ROS production, and a shortened life span. Feeding the *NF1* mutant flies with catalytic antioxidants (MnTBAP or MnTDEIP) increases physical performance and restores longevity, indicating that the important factor in reduced physical robustness and life span is increased mitochondrial ROS (Tong et al. 2007).

In contrast, overexpression of the *NF1* gene in *Drosophila* doubles the cAMP level and increases resistance to oxidative and thermal stress, physical performance, and longevity. This is associated with an approximately 30% increase in OXPHOS complex I activity and about a 60% reduction in mitochondrial ROS production. That cAMP level and consequently mitochondrial ROS production is the major factor in the *NF1* extension of life span has been demonstrated showing that inactivation of the cAMP phosphodiesterase, expression of a constitutive PKA, or simply feeding *Drosophila* dibutyrl-cAMP or 8-bromo-cAMP can all extend *Drosophila* life span (Tong et al. 2007).

Assuming that cAMP-mediated pathways for regulating mitochondrial biogenesis are conserved between flies and mammals, we would predict that the fly PKA should phosphorylate the *Drosophila* equivalent of cAMP response element binding (CREB). Phosphorylated CREB should then bind to CREs in the *Drosophila* equivalent of PGC-1α, CG9809. CG9809 or *dPGC-1α* encodes a 1088-amino-acid protein that has 68% homology with the mammalian PGC-1α carboxy-terminal RNA-binding motif (Gershman et al. 2007). The expression *dPGC-1α* has recently been reported to increase 49% when adult flies are shifted from nonyeast to a yeast-containing diet. Within seven hours of yeast refeeding, the upregulation of *dPGC-1α* co-occurs with a shift in the transcription of genes involved in fuel metabolism and mitochondrial biogenesis. One regulatory system for *dPGC-1α* has been shown to be the *Drosophila* homolog of the FOXOs (dFOXO) (Gershman et al. 2007). Since *Drosophila PGC-1α* is regulated by FOXO (Gershman et al. 2007), as is mammalian *PGC-1α,* it is reasonable to hypothesize the *dPGC-1α* gene will also be regulated by CREB (Daitoku et al. 2003).

In mammalian cells, cAMP has also been shown to drive the post-translational phosphorylation of mitochondrial OXPHOS subunits. Mitochondrially localized cAMP-activated PKA has been reported to mediate serine-threonine phosphorylation of the complex I subunits (Technikova-Dobrova et al. 2001) ESSS and MWFE (Chen et al. 2004) and/or 42 kD and B14.5A (Pocsfalvi et al. 2007). The phosphorylation of the complex I subunits is associated with increased complex I V_{max} and $NADH^{+}$-linked respiration and decreased mitochondrial ROS production (Technikova-Dobrova et al. 2001; Papa et al. 2002; Bellomo et al. 2006; Piccoli et al. 2006). Hence, cAMP modulation of mitochondrial function could be occurring at both the transcriptional and posttranscriptional levels.

In summary, animal mitochondrial energy metabolism is coupled to plant chloroplast photosynthesis through glucose availability. This seasonal quantitative regulation of OXPHOS levels in response to available substrates is superimposed on the constitutive, qualitative changes in OXPHOS that result from the adaptive mtDNA mutations. Thus, longevity is coupled to environmental variation through regulation of energy metabolism and the regulatory and secondary toxic effects of ROS.

Calorie Restriction and Longevity

Calorie restriction (CR) is the most robust experimental procedure currently known to increase life span, having been shown to extend life span in a wide variety of eukaryotes including yeast, *C. elegans, Drosophila,* and rodents. In rodents, CR also suppresses tumor formation (Harrison and Archer 1987; Masoro et al. 1992; McCarter and Palmer 1992; Masoro 1993; Sohal et al. 1994). The effects of CR can be nicely interpreted within the context of mitochondrial function.

When carbohydrates and thus calories are in excess and OXPHOS and the antioxidant defenses are down-regulated, the ETC remains in a chronically reduced state. The resulting excess of electrons in the presence of saturating O_2 favors increased ROS production. Chronically increased ROS then increases the probability of mutagenesis of the mtDNA and activation of the mtPTP, resulting in premature cell death and aging. Furthermore, mitochondrial H_2O_2, which is relatively stable, can diffuse out of the mitochondrion and into the cytosol and the nucleus. In the nucleus, H_2O_2 can be converted to hydroxyl ion that can mutagenize proto-oncogenes, converting them to oncogenes and initiating tumor formation (Fig. 5). Moreover, mitochondrial H_2O_2 can also activate cytosolic and nuclear mitogenic pathways, driving the cell into mitosis, thus promoting tumor growth. As a consequence, animals that ingest excess calories over those needed for ATP and heat production are prone to premature aging and cancer. In contrast, CR reduces the excess electrons, induces OXPHOS and antioxidant defenses, and thus extends life span and reduces cancer risk (Wallace 2005b).

THE ONE CELL–TWO ORGANISM MODEL OF MEDICINE

The inclusion of mitochondrial biology and genetics into medicine provides a more complete explanation of the pathophysiology and genetics of the age-related diseases: aging and cancer. As the amalgamation of the nuclear-cytosol host cell and the mitochondrial symbiont, the eukaryotic cell encompasses two distinct biologies and two distinct genetics: the anatomical biology of the nuclear-cytosol organism with its quantized structural genes and the energetic biology of the mitochondria with their quantitative energy genes (Wallace 2007).

Up until now, modern medicine has focused almost exclusively on the nuclear-cytosol organism, thus being confined to an exclusively structural perspective of human disease and a Mendelian perspective of genetics. But the common age-related diseases, cancer and aging, are generally not tissue-specific and do not show a strictly quantized Mendelian inheritance. Nothing in Mendelian genetics can explain the striking variability in human phenotypes and the progressive age-related decline in function so characteristic of aging and age-related diseases.

This dilemma can be resolved by considering the effects of the mitochondrial organism in medicine. Defects in the mitochondrial energetic system result in systemic biochemical defects that preferentially affect the tissues and organs most reliant on mitochondrial energy production. Furthermore, the thousands of copies of the mtDNAs can accommodate a continuous distribution of ratios of mutant and wild-type mtDNAs and thus a progressive energy decline until functional thresholds are traversed and symptoms ensue (Fig. 6) (Wallace 2007).

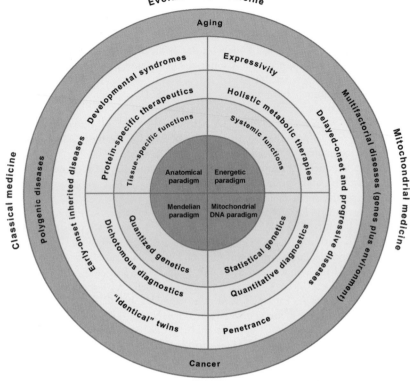

Figure 6. New biomedical paradigm that combines the traditional structural biology and Mendelian genetics of the nuclear-cytosolic organism with the energy biology and mtDNA genetics of the mitochondrial organism. The four sectors indicate the different aspects of human biology and genetics determined by the functional and genetic components of the two different organisms. (Reprinted, with permission, from Wallace 2007 [© Annual Reviews].)

In conclusion, the structural and Mendelian paradigms derived from the nuclear-cytosol organism encompass Classical Western Medicine, whereas the energy and mtDNA genetics paradigms constitute Mitochondrial Medicine. Taken together, the quantized Mendelian genetics and quantitative mtDNA genetics explain the diverse features of human Genetic Medicine, and the combination of the structure and energetic biologies provides a coherent framework in which to integrate genetic and environmental factors to create the new discipline of Environmental Medicine (Fig. 6) (Wallace 2007).

REFERENCES

Accili D. and Arden K.C. 2004. FoxOs at the crossroads of cellular metabolism, differentiation, and transformation. *Cell* **117:** 421–426.

Atilano S.R., Coskun P., Chwa M., Jordan N., Reddy V., Le K., Wallace D.C., and Kenney M.C. 2005. Accumulation of mitochondrial DNA damage in keratoconus corneas. *Invest. Ophthalmol. Vis. Sci.* **46:** 1256–1263.

Ballinger S.W., Shoffner J.M., Gebhart S., Koontz D.A., and Wallace D.C. 1994. Mitochondrial diabetes revisited. *Nat. Genet.* **7:** 458–459.

Ballinger S.W., Shoffner J.M., Hedaya E.V., Trounce I., Polak M.A., Koontz D.A., and Wallace D.C. 1992. Maternally transmitted diabetes and deafness associated with a 10.4 kb mitochondrial DNA deletion. *Nat. Genet.* **1:** 11–15.

Baudouin S.V., Saunders D., Tiangyou W., Elson J.L., Poynter J., Pyle A., Keers S., Turnbull D.M., Howell N., and Chinnery P.F. 2005. Mitochondrial DNA and survival after sepsis: A prospective study. *Lancet* **366:** 2118–2121.

Baur J.A., Pearson K.J., Price N.L., Jamieson H.A., Lerin C., Kalra A., Prabhu V.V., Allard J.S., Lopez-Lluch G., Lewis K., Pistell P.J., Poosala S., Becker K.G., Boss O., Gwinn D., Wang M., Ramaswamy S., Fishbein K.W., Spencer R.G., Lakatta E.G., Le Couteur D., Shaw R.J., Navas P., Puigserver P., and Ingram D.K. 2006. Resveratrol improves health and survival of mice on a high-calorie diet. *Nature* **444:** 337–342.

Bellomo F., Piccoli C., Cocco T., Scacco S., Papa F., Gaballo A., Boffoli D., Signorile A., D'Aprile A., Scrima R., Sardanelli A.M., Capitanio N., and Papa S. 2006. Regulation by the cAMP cascade of oxygen free radical balance in mammalian cells. *Antioxid. Redox Signal.* **8:** 495–502.

Bender A., Krishnan K.J., Morris C.M., Taylor G.A., Reeve A.K., Perry R.H., Jaros E., Hersheson J.S., Betts J., Klopstock T., Taylor R.W., and Turnbull D.M. 2006. High levels of mitochondrial DNA deletions in substantia nigra neurons in aging and Parkinson disease. *Nat. Genet.* **38:** 515–517.

Brandon M., Baldi P., and Wallace D.C. 2006. Mitochondrial mutations in cancer. *Oncogene* **25:** 4647–4662.

Brockington M., Sweeney M.G., Hammans S.R., Morgan-Hughes J.A., and Harding A.E. 1993. A tandem duplication in the D-loop of human mitochondrial DNA is associated with deletions in mitochondrial myopathies. *Nat. Genet.* **4:** 67–71.

Brown M.D., Sun F., and Wallace D.C. 1997. Clustering of caucasian Leber hereditary optic neuropathy patients containing the 11778 or 14484 mutations on an mtDNA lineage. *Am. J. Hum. Genet.* **60:** 381–387.

Brown M.D., Allen J.C., Van Stavern G.P., Newman N.J., and Wallace D.C. 2001. Clinical, genetic, and biochemical characterization of a Leber hereditary optic neuropathy family containing both the 11778 and 14484 primary mutations. *Am. J. Med. Genet.* **104:** 331–338.

Brown M.D., Starikovskaya E., Derbeneva O., Hosseini S., Allen J.C., Mikhailovskaya I.E., Sukernik R.I., and Wallace D.C. 2002. The role of mtDNA background in disease expression: A new primary LHON mutation associated with Western Eurasian haplogroup J. *Hum. Genet.* **110:** 130–138.

Brown M.D., Hosseini S.H., Torroni A., Bandelt H.J., Allen J.C., Schurr T.G., Scozzari R., Cruciani F., and Wallace D.C. 1998. mtDNA haplogroup X: An ancient link between Europe/Western Asia and North America? *Am. J. Hum. Genet.* **63:** 1852–1861.

Brunet A., Sweeney L.B., Sturgill J.F., Chua K.F., Greer P.L., Lin Y., Tran H., Ross S.E., Mostoslavsky R., Cohen H.Y., Hu L.S., Cheng H.L. Jedrychowski M.P., Gygi S.P., Sinclair D.A., Alt F.W., and Greenberg M.E. 2004. Stress-dependent regulation of FOXO transcription factors by the SIRT1 deacetylase. *Science* **303:** 2011–2015.

Chagnon P., Gee M., Filion M., Robitaille Y., Belouchi M., and Gauvreau D. 1999. Phylogenetic analysis of the mitochondrial genome indicates significant differences between patients with Alzheimer disease and controls in a French-Canadian founder population. *Am. J. Med. Genet.* **85:** 20–30.

Chen R., Fearnley I.M., Peak-Chew S.Y., and Walker J.E. 2004. The phosphorylation of subunits of complex I from bovine heart mitochondria. *J. Biol. Chem.* **279:** 26036–26045.

Chen Y.S., Olckers A., Schurr T.G., Kogelnik A.M., Huoponen K., and Wallace D.C. 2000. mtDNA variation in the South African Kung and Khwe and their genetic relationships to other African populations. *Am. J. Hum. Genet.* **66:** 1362–1383.

Corral-Debrinski M., Horton T., Lott M.T., Shoffner J.M., Beal M.F., and Wallace D.C. 1992. Mitochondrial DNA deletions in human brain: Regional variability and increase with advanced age. *Nat. Genet.* **2:** 324–329.

Cortopassi G. and Arnheim N. 1992. Accumulation of mitochondrial DNA mutation in normal aging brain and muscle. In *Mitochondrial DNA in human pathology* (ed. S. DiMauro and D.C. Wallace), pp. 125–136. Raven Press, New York.

Coskun P.E., Beal M.F., and Wallace D.C. 2004. Alzheimer's brains harbor somatic mtDNA control-region mutations that suppress mitochondrial transcription and replication. *Proc. Natl. Acad. Sci.* **101:** 10726–10731.

Daitoku H., Yamagata K., Matsuzaki H., Hatta M., and Fukamizu A. 2003. Regulation of PGC-1 promoter activity by protein kinase B and the forkhead transcription factor FKHR. *Diabetes* **52:** 642–649.

De Benedictis G., Rose G., Carrieri G., De Luca M., Falcone E., Passarino G., Bonafe M., Monti D., Baggio G., Bertolini S., Mari D., Mattace R., and Franceschi C. 1999. Mitochondrial DNA inherited variants are associated with successful aging and longevity in humans. *FASEB J.* **13:** 1532–1536.

Derbeneva O.A., Sukernik R.I., Volodko N.V., Hosseini S.H., Lott M.T., and Wallace D.C. 2002. Analysis of mitochondrial DNA diversity in the Aleuts of the Commander Islands and its implications for the genetic history of Beringia. *Am. J. Hum. Genet.* **71:** 415–421.

Esposito L.A., Melov S., Panov A., Cottrell B.A., and Wallace D.C. 1999. Mitochondrial disease in mouse results in increased oxidative stress. *Proc. Natl. Acad. Sci.* **96:** 4820–4825.

Esposito L.A., Kokoszka J.E., Waymire K.G., Cottrell B., MacGregor G.R., and Wallace D.C. 2000. Mitochondrial oxidative stress in mice lacking the glutathione peroxidase-1 gene. *Free Radical Biol. Med.* **28:** 754–766.

Evans A.R., Limp-Foster M., and Kelley M.R. 2000. Going APE over ref-1. *Mutat. Res.* **461:** 83–108.

Freeman H.C., Hugill A., Dear N.T., Ashcroft F.M., and Cox R.D. 2006. Deletion of nicotinamide nucleotide transhydrogenase: A new quantitive trait locus accounting for glucose intolerance in C57BL/6J mice. *Diabetes* **55:** 2153–2156.

Furuyama T., Kitayama K., Yamashita H., and Mori N. 2003. Forkhead transcription factor FOXO1 (FKHR)-dependent induction of PDK4 gene expression in skeletal muscle during energy deprivation. *Biochem. J.* **375:** 365–371.

Gerhart-Hines Z., Rodgers J.T., Bare O., Lerin C., Kim S.H., Mostoslavsky R., Alt F.W., Wu Z., and Puigserver P. 2007. Metabolic control of muscle mitochondrial function and fatty acid oxidation through SIRT1/PGC-1alpha. *EMBO J.* **26:** 1913–1923.

Gershman B., Puig O., Hang L., Peitzsh R.M., Tatar M., and Garofalo R.S. 2007. High resolution dynamics of the transcriptional response to nutrition in *Drosophila:* A key role for dFOXO. *Physiol. Genomics* **29:** 24–34.

Giannakou M.E. and Partridge L. 2004. The interaction between FOXO and SIRT1: Tipping the balance towards survival. *Trends Cell Biol.* **14:** 408–412.

Giles R.E., Blanc H., Cann H.M., and Wallace D.C. 1980. Maternal inheritance of human mitochondrial DNA. *Proc. Natl. Acad. Sci.* **77:** 6715–6719.

Graham B.H., Waymire K.G., Cottrell B., Trounce I.A., MacGregor G.R., and Wallace D.C. 1997. A mouse model for mitochondrial myopathy and cardiomyopathy resulting from a deficiency in the heart/skeletal muscle isoform of the adenine nucleotide translocator. *Nat. Genet.* **16:** 226–234.

Hansen J.M., Go Y.M., and Jones D.P. 2006. Nuclear and mitochondrial compartmentation of oxidative stress and redox signaling. *Annu. Rev. Pharmacol. Toxicol.* **46:** 215–234.

Harrison D.E. and Archer J.R. 1987. Genetic differences in effects of food restriction on aging in mice. *J. Nutr.* **117:** 376–382.

Herrmann P.C., Gillespie J.W., Charboneau L., Bichsel V.E., Paweletz C.P., Calvert V.S., Kohn E.C., Emmert-Buck M.R., Liotta L.A., and Petricoin E.F., III. 2003. Mitochondrial proteome: Altered cytochrome c oxidase subunit levels in prostate cancer. *Proteomics* **3:** 1801–1810.

Herzig R.P., Scacco S., and Scarpulla R.C. 2000. Sequential serum-dependent activation of CREB and NRF-1 leads to enhanced mitochondrial respiration through the induction of cytochrome c. *J. Biol. Chem.* **275:** 13134–13141.

Herzig S., Long F., Jhala U.S., Hedrick S., Quinn R., Bauer A., Rudolph D., Schutz G., Yoon C., Puigserver P., Spiegelman B., and Montminy M. 2001. CREB regulates hepatic gluconeogenesis through the coactivator PGC-1. *Nature* **413:** 179–183.

Holt I.J., Harding A.E., and Morgan-Hughes J.A. 1988. Deletions of muscle mitochondrial DNA in patients with mitochondrial myopathies. *Nature* **331:** 717–719.

Howitz K.T., Bitterman K.J., Cohen H.Y., Lamming D.W., Lavu S., Wood J.G., Zipkin R.E., Chung P., Kisielewski A., Zhang L.L., Scherer B., and Sinclair D.A. 2003. Small molecule activators of sirtuins extend *Saccharomyces cerevisiae* lifespan. *Nature* **425:** 191–196.

Huang T.T., Naeemuddin M., Elchuri S., Yamaguchi M., Kozy H.M., Carlson E.J., and Epstein C.J. 2006. Genetic modifiers of the phenotype of mice deficient in mitochondrial superoxide dismutase. *Hum. Mol. Genet.* **15:** 1187–1194.

Huoponen K., Vilkki J., Aula P., Nikoskelainen E.K., and Savontaus M.L. 1991. A new mtDNA mutation associated with Leber hereditary optic neuroretinopathy. *Am. J. Hum. Genet.* **48:** 1147–1153.

Imai S., Armstrong C.M., Kaeberlein M., and Guarente L. 2000. Transcriptional silencing and longevity protein Sir2 is an NAD-dependent histone deacetylase. *Nature* **403:** 795–800.

Inoue K., Nakada K., Ogura A., Isobe K., Goto Y., Nonaka I., and Hayashi J.I. 2000. Generation of mice with mitochondrial dysfunction by introducing mouse mtDNA carrying a deletion into zygotes. *Nat. Genet.* **26:** 176–181.

Ivanova R., Lepage V., Charron D., and Schächter F. 1998. Mitochondrial genotype associated with French Caucasian centenarians. *Gerontology* **44:** 349.

Johns D.R., Neufeld M.J., and Park R.D. 1992. An ND-6 mitochondrial DNA mutation associated with Leber hereditary optic neuropathy. *Biochem. Biophys. Res. Commun.* **187:** 1551–1557.

Jones D.P. 2006. Disruption of mitochondrial redox circuitry in oxidative stress. *Chem. Biol. Interact.* **163:** 38–53.

Jun A.S., Brown M.D., and Wallace D.C. 1994. A mitochondrial DNA mutation at np 14459 of the ND6 gene associated with maternally inherited Leber's hereditary optic neuropathy and dystonia. *Proc. Natl. Acad. Sci.* **91:** 6206–6210.

Kelley M.R. and Parsons S.H. 2001. Redox regulation of the DNA repair function of the human AP endonuclease Ape1/ref-1. *Antioxid. Redox Signal.* **3:** 671–683.

Kelly D.P. and Scarpulla R.C. 2004. Transcriptional regulatory circuits controlling mitochondrial biogenesis and function. *Genes Dev.* **18:** 357–368.

Khogali S.S., Mayosi B.M., Beattie J.M., McKenna W.J., Watkins H., and Poulton J. 2001. A common mitochondrial DNA variant associated with susceptibility to dilated cardiomyopathy in two different populations. *Lancet* **357:** 1265–1267.

Kokoszka J.E., Coskun P., Esposito L., and Wallace D.C. 2001. Increased mitochondrial oxidative stress in the Sod2 (+/−) mouse results in the age-related decline of mitochondrial function culminating in increased apoptosis. *Proc. Natl. Acad. Sci.* **98:** 2278–2283.

Kokoszka J.E., Waymire K.G., Levy S.E., Sligh J.E., Cai J., Jones D.P., MacGregor G.R., and Wallace D.C. 2004. The ADP/ATP translocator is not essential for the mitochondrial permeability transition pore. *Nature* **427:** 461–465.

Koo S.H. and Montminy M. 2006. In vino veritas: A tale of two sirt1s? *Cell* **127:** 1091–1093.

Kraytsberg Y., Kudryavtseva E., McKee A.C., Geula C., Kowall N.W., and Khrapko K. 2006. Mitochondrial DNA deletions are abundant and cause functional impairment in aged human substantia nigra neurons. *Nat. Genet.* **38:** 518–520.

Krieg R.C., Knuechel R., Schiffmann E., Liotta L.A., Petricoin E.F., III, and Herrmann P.C. 2004. Mitochondrial proteome: Cancer-altered metabolism associated with cytochrome c oxidase subunit level variation. *Proteomics* **4:** 2789–2795.

Kujoth G.C., Hiona A., Pugh T.D., Someya S., Panzer K., Wohlgemuth S.E., Hofer T., Seo A.Y., Sullivan R., Jobling W.A., Morrow J.D., Van Remmen H., Sedivy J.M., Yamasoba T., Tanokura M., Weindruch R., Leeuwenburgh C., and Prolla T.A. 2005. Mitochondrial DNA mutations, oxidative stress, and apoptosis in mammalian aging. *Science* **309:** 481–484.

Lagouge M., Argmann C., Gerhart-Hines Z., Meziane H., Lerin C., Daussin F., Messadeq N., Milne J., Lambert P., Elliott P., Geny B., Laakso M., Puigserver P., and Auwerx J. 2006. Resveratrol improves mitochondrial function and protects against metabolic disease by activating SIRT1 and PGC-1alpha. *Cell* **127:** 1109–1122.

Langley E., Pearson M., Faretta M., Bauer U.M., Frye R.A., Minucci S., Pelicci P.G., and Kouzarides T. 2002. Human SIR2 deacetylates p53 and antagonizes PML/p53-induced cellular senescence. *EMBO J.* **21:** 2383–2396.

Li Y., Huang T.T., Carlson E.J., Melov S., Ursell P.C., Olson J.L., Noble L.J., Yoshimura M.P., Berger C., Chan P.H., Wallace D.C., and Epstein C.J. 1995. Dilated cardiomyopathy and neonatal lethality in mutant mice lacking manganese superoxide dismutase. *Nat. Genet.* **11:** 376–381.

Liang H., Remmen H.V., Frohlich V., Lechleiter J., Richardson A., and Ran Q. 2007. Gpx4 protects mitochondrial ATP generation against oxidative damage. *Biochem. Biophys. Res. Commun.* **356:** 893–898.

Liu V.W., Wang Y., Yang H.J., Tsang P.C., Ng T.Y., Wong L.C., Nagley P., and Ngan H.Y. 2003. Mitochondrial DNA variant 16189T>C is associated with susceptibility to endometrial cancer. *Hum. Mutat.* **22:** 173–174.

Loeb L.A., Wallace D.C., and Martin G.M. 2005. The mitochondrial theory of aging and its relationship to reactive oxygen species damage and somatic mtDNA mutations. *Proc. Natl. Acad. Sci.* **102:** 18769–18770.

Luo J., Nikolaev A.Y., Imai S., Chen D., Su F., Shiloh A., Guarente L., and Gu W. 2001. Negative control of p53 by Sir2alpha promotes cell survival under stress. *Cell* **107:** 137–148.

Masoro E.J. 1993. Dietary restriction and aging. *J. Am. Geriatr. Soc.* **41:** 994–999.

Masoro E.J., McCarter R.J., Katz M.S., and McMahan C.A. 1992. Dietary restriction alters characteristics of glucose fuel use (erratum in *J. Gerontol.* [1993] **48:** B73). *J. Gerontol.* **47:** B202–B208.

McCarter R.J. and Palmer J. 1992. Energy metabolism and aging: A lifelong study of Fischer 344 rats. *Am. J. Physiol.* **263:** E448–E452.

McCord J.M. 2000. The evolution of free radicals and oxidative stress. *Am. J. Med. Genet.* **108:** 652–659.

McMahon F.J., Chen Y.S., Patel S., Kokoszka J., Brown M.D., Torroni A., DePaulo J.R., and Wallace D.C. 2000. Mitochondrial DNA sequence diversity in bipolar affective disorder. *Am. J. Psychiatry* **157:** 1058–1064.

Melov S., Schneider J.A., Day B.J., Hinerfeld D., Coskun P., Mirra S.S., Crapo J.D., and Wallace D.C. 1998. A novel neurological phenotype in mice lacking mitochondrial manganese superoxide dismutase. *Nat. Genet.* **18:** 159–163.

Melov S., Doctrow S.R., Schneider J.A., Haberson J., Patel M., Coskun P.E., Huffman K., Wallace D.C., and Malfroy B. 2001. Lifespan extension and rescue of spongiform encephalopathy in superoxide dismutase 2 nullizygous mice treated with superoxide dismutase-catalase mimetics. *J. Neurosci.* **21:** 8348–8353.

Melov S., Coskun P., Patel M., Tunistra R., Cottrell B., Jun A.S., Zastawny T.H., Dizdaroglu M., Goodman S.I., Huang T.T., Miziorko H., Epstein C.J., and Wallace D.C. 1999. Mitochondrial disease in superoxide dismutase 2 mutant mice. *Proc. Natl. Acad. Sci.* **96:** 846–851.

Michikawa Y., Mazzucchelli F., Bresolin N., Scarlato G., and Attardi G. 1999. Aging-dependent large accumulation of point mutations in the human mtDNA control region for replication. *Science* **286:** 774–779.

Mishmar D., Ruiz-Pesini E.E., Golik P., Macaulay V., Clark A.G., Hosseini S., Brandon M., Easley K., Chen E., Brown M.D., Sukernik R.I., Olckers A., and Wallace D.C. 2003. Natural selection shaped regional mtDNA variation in humans. *Proc. Natl. Acad. Sci.* **100:** 171–176.

MITOMAP. 2006. MITOMAP: A human mitochondrial genome database. http://www.mitomap.org.

Moraes C.T., DiMauro S., Zeviani M., Lombes A., Shanske S., Miranda A.F., Nakase H., Bonilla E., Werneck L.C., and Servidei S., et al. 1989. Mitochondrial DNA deletions in progressive external ophthalmoplegia and Kearns-Sayre syndrome. *N. Engl. J. Med.* **320:** 1293–1299.

Motta M.C., Divecha N., Lemieux M., Kamel C., Chen D., Gu W., Bultsma Y., McBurney M., and Guarente L. 2004. Mammalian SIRT1 represses forkhead transcription factors. *Cell* **116:** 551–563.

Naimi M., Bannwarth S., Procaccio V., Pouget J., Desnuelle C., Pelissier J.F., Rötig A., Munnich A., Calvas P., Richelme C., Jonveaux P., Castelnovo G., Simon M., Clanet M.,

Wallace D., and Paquis-Flucklinger V. 2006. Molecular analysis of ANT1, TWINKLE and POLG in patients with multiple deletions or depletion of mitochondrial DNA by a dHPLC-based assay. *Eur. J. Hum. Genet.* **14:** 917–922.

Nemoto S., Fergusson M.M., and Finkel T. 2005. SIRT1 functionally interacts with the metabolic regulator and transcriptional coactivator PGC-1{alpha}. *J. Biol. Chem.* **280:** 16456–16460.

Niemi A.K., Hervonen A., Hurme M., Karhunen P.J., Jylha M., and Majamaa K. 2003. Mitochondrial DNA polymorphisms associated with longevity in a Finnish population. *Hum. Genet.* **112:** 29–33.

Oliver N.A. and Wallace D.C. 1982. Assignment of two mitochondrially synthesized polypeptides to human mitochondrial DNA and their use in the study of intracellular mitochondrial interaction. *Mol. Cell Biol.* **2:** 30–41.

Papa S., Scacco S., Sardanelli A.M., Petruzzella V., Vergari R., Signorile A., and Technikova-Dobrova Z. 2002. Complex I and the cAMP cascade in human physiopathology. *Biosci. Rep.* **22:** 3–16.

Petros J.A., Baumann A.K., Ruiz-Pesini E., Amin M.B., Sun C.Q., Hall J., Lim S., Issa M.M., Flanders W.D., Hosseini S.H., Marshall F.F., and Wallace D.C. 2005. mtDNA mutations increase tumorigenicity in prostate cancer. *Proc. Natl. Acad. Sci.* **102:** 719–724.

Piccoli C., Scacco S., Bellomo F., Signorile A., Iuso A., Boffoli D., Scrima R., Capitanio N., and Papa S. 2006. cAMP controls oxygen metabolism in mammalian cells. *FEBS Lett.* **580:** 4539–4543.

Pocsfalvi G., Cuccurullo M., Schlosser G., Scacco S., Papa S., and Malorni A. 2007. Phosphorylation of B14.5a subunit from bovine heart complex I identified by titanium dioxide selective enrichment and shotgun proteomics. *Mol. Cell Proteomics* **6:** 231–237.

Poulton J., Luan J., Macaulay V., Hennings S., Mitchell J., and Wareham N.J. 2002. Type 2 diabetes is associated with a common mitochondrial variant: Evidence from a population-based case-control study. *Hum. Mol. Genet.* **11:** 1581–1583.

Procaccio V. and Wallace D.C. 2004. Late-onset Leigh syndrome in a patient with mitochondrial complex I NDUFS8 mutations. *Neurology* **62:** 1899–1901.

Ran Q., Liang H., Gu M., Qi W., Walter C.A., Roberts L.J., Herman B., Richardson A., and Van Remmen H. 2004. Transgenic mice overexpressing glutathione peroxidase 4 are protected against oxidative stress-induced apoptosis. *J. Biol. Chem.* **279:** 55137–55146.

Rodgers J.T., Lerin C., Haas W., Gygi S.P., Spiegelman B.M., and Puigserver P. 2005. Nutrient control of glucose homeostasis through a complex of PGC-1alpha and SIRT1. *Nature* **434:** 113–118.

Ross O.A., McCormack R., Curran M.D., Duguid R.A., Barnett Y.A., Rea I.M., and Middleton D. 2001. Mitochondrial DNA polymorphism: Its role in longevity of the Irish population. *Exp. Gerontol.* **36:** 1161–1178.

Rötig A., Colonna M., Blanche S., Fischer A., Le Deist F., Frezal J., Saudubray J.M., and Munnich A. 1988. Deletion of blood mitochondrial DNA in pancytopenia. *Lancet* **2:** 567–568.

Ruiz-Pesini E. and Wallace D.C. 2006. Evidence for adaptive selection acting on the tRNA and rRNA genes of the human mitochondrial DNA. *Hum. Mutat.* **27:** 1072–1081.

Ruiz-Pesini E., Mishmar D., Brandon M., Procaccio V., and Wallace D.C. 2004. Effects of purifying and adaptive selection on regional variation in human mtDNA. *Science* **303:** 223–226.

Scarpulla R.C. 2002. Nuclear activators and coactivators in mammalian mitochondrial biogenesis. *Biochim. Biophys. Acta* **1576:** 1–14.

Schriner S.E., Linford N.J., Martin G.M., Treuting P., Ogburn C.E., Emond M., Coskun P.E., Ladiges W., Wolf N., Van Remmen H., Wallace D.C., and Rabinovitch P.S. 2005. Extension of murine life span by overexpression of catalase targeted to mitochondria. *Science* **308:** 1909–1911.

Schurr T.G. and Wallace D.C. 2002. Mitochondrial DNA diversity in Southeast Asian populations. *Hum. Biol.* **74:** 431–452.

Schurr T.G., Sukernik R.I., Starikovskaya Y.B., and Wallace D.C. 1999. Mitochondrial DNA variation in Koryaks and Itel'men: Population replacement in the Okhotsk Sea-Bering Sea region during the Neolithic. *Am. J. Phys. Anthropol.* **108:** 1–39.

Shoffner J.M., Lott M.T., Lezza A.M., Seibel P., Ballinger S.W., and Wallace D.C. 1990. Myoclonic epilepsy and ragged-red fiber disease (MERRF) is associated with a mitochondrial DNA tRNA^{Lys} mutation. *Cell* **61:** 931–937.

Shoffner J.M., Brown M.D., Stugard C., Jun A.S., Pollock S., Haas R.H., Kaufman A., Koontz D., Kim Y., and Graham J.R., et al. 1995. Leber's hereditary optic neuropathy plus dystonia is caused by a mitochondrial DNA point mutation in a complex I subunit. *Ann. Neurol.* **38:** 163–169.

Sligh J.E., Levy S.E., Waymire K.G., Allard P., Dillehay D.L., Nusinowitz S., Heckenlively J.R., MacGregor G.R., and Wallace D.C. 2000. Maternal germ-line transmission of mutant mtDNAs from embryonic stem cell-derived chimeric mice. *Proc. Natl. Acad. Sci.* **97:** 14461–14466.

Sohal R.S., Ku H.H., Agarwal S., Forster M.J., and Lal H. 1994. Oxidative damage, mitochondrial oxidant generation and antioxidant defenses during aging and in response to food restriction in the mouse. *Mech. Ageing Dev.* **74:** 121–133.

Song D.D., Shults C.W., Sisk A., Rockenstein E., and Masliah E. 2004. Enhanced substantia nigra mitochondrial pathology in human alpha-synuclein transgenic mice after treatment with MPTP. *Exp. Neurol.* **186:** 158–172.

Starikovskaya E.B., Sukernik R.I., Schurr T.G., Kogelnik A.M., and Wallace D.C. 1998. Mitochondrial DNA diversity in Chukchi and Siberian Eskimos: Implications for genetic history of ancient Beringia and peopling of the New World. *Am. J. Hum. Genet.* **63:** 1473–1491.

Technikova-Dobrova Z., Sardanelli A.M., Speranza F., Scacco S., Signorile A., Lorusso V., and Papa S. 2001. Cyclic adenosine monophosphate-dependent phosphorylation of mammalian mitochondrial proteins: Enzyme and substrate characterization and functional role. *Biochemistry* **40:** 13941–13947.

Tong J., Schriner S.E., McCleary D., Day B.J., and Wallace D.C. 2007. Life extension through neurofibromin mitochondrial regulation and antioxidant therapy for neurofibromatosis-1 in *Drosophila melanogaster*. *Nat. Genet.* **39:** 476–485.

Torroni A., Lott M.T., Cabell M.F., Chen Y., Lavergne L., and Wallace D.C. 1994. MtDNA and the origin of Caucasians. Identification of ancient Caucasian-specific haplogroups, one of which is prone to a recurrent somatic duplication in the D-loop region. *Am. J. Hum. Genet.* **55:** 760–776.

Torroni A., Petrozzi M., D'Urbano L., Sellitto D., Zeviani M., Carrara F., Carducci C., Leuzzi V., Carelli V., Barboni P., De Negri A., and Scozzari R. 1997. Haplotype and phylogenetic analyses suggest that one European-specific mtDNA background plays a role in the expression of Leber hereditary optic neuropathy by increasing the penetrance of the primary mutations 11778 and 14484. *Am. J. Hum. Genet.* **60:** 1107–1121.

Trifunovic A., Hansson A., Wredenberg A., Rovio A.T., Dufour E., Khvorostov I., Spelbrink J.N., Wibom R., Jacobs H.T., and Larsson N.G. 2005. Somatic mtDNA mutations cause aging phenotypes without affecting reactive oxygen species production. *Proc. Natl. Acad. Sci.* **102:** 17993–17998.

Trifunovic A., Wredenberg A., Falkenberg M., Spelbrink J.N., Rovio A.T., Bruder C.E., Bohlooly-Y M., Gidlof S., Oldfors A., Wibom R., Tornall J., Jacobs H.T., and Larsson N.G. 2004. Premature ageing in mice expressing defective mitochondrial DNA polymerase. *Nature* **429:** 417–423.

Valenzano D.R., Terzibasi E., Genade T., Cattaneo A., Domenici L., and Cellerino A. 2006. Resveratrol prolongs lifespan and retards the onset of age-related markers in a short-lived vertebrate. *Curr. Biol.* **16:** 296–300.

van der Walt J.M., Dementieva Y.A., Martin E.R., Scott W.K., Nicodemus K.K., Kroner C.C., Welsh-Bohmer K.A., Saunders A.M., Roses A.D., Small G.W., Schmechel D.E., Murali Doraiswamy P., Gilbert J.R., Haines J.L., Vance J.M., and Pericak-Vance M.A. 2004. Analysis of European mitochondrial haplogroups with Alzheimer disease risk. *Neurosci. Lett.* **365:** 28–32.

van der Walt J.M., Nicodemus K.K., Martin E.R., Scott W.K., Nance M.A., Watts R.L., Hubble J.P., Haines J.L., Koller W.C., Lyons K., Pahwa R., Stern M.B., Colcher A., Hiner B.C., Jankovic J., Ondo W.G., Allen F.H., Goetz C.G., Small G.W., Mastaglia F., Stajich J.M., McLaurin A.C., Middleton L.T., Scott B.L., Schmechel D.E., Pericak-Vance M.A., and Vance J.M. 2003. Mitochondrial polymorphisms significantly reduce the risk of Parkinson disease. *Am. J. Hum. Genet.* **72:** 804–811.

Vaziri H., Dessain S.K., Ng Eaton E., Imai S.I., Frye R.A., Pandita T.K., Guarente L., and Weinberg R.A. 2001. hSIR2(SIRT1) functions as an NAD-dependent p53 deacetylase. *Cell* **107:** 149–159.

Vermulst M., Bielas J.H., Kujoth G.C., Ladiges W.C., Rabinovitch P.S., Prolla T.A., and Loeb L.A. 2007. Mitochondrial point mutations do not limit the natural lifespan of mice. *Nat. Genet.* **39:** 540–543.

Wallace D.C. 1983. Structure and evolution of organelle DNAs. In *Endocytobiology. II. Intracellular space as oligogenetic ecosystem* (ed. H. Schenk and W. Schwemmler), pp. 87–100. deGruyter, New York.

———. 1992a. Diseases of the mitochondrial DNA. *Annu. Rev. Biochem.* **61:** 1175–1212.

———. 1992b. Mitochondrial genetics: A paradigm for aging and degenerative diseases? *Science* **256:** 628–632.

———. 1999. Mitochondrial diseases in man and mouse. *Science* **283:** 1482–1488.

———. 2001. Mouse models for mitochondrial disease. *Am. J. Med. Genet.* **106:** 71–93.

———. 2005a. Mitochondria and cancer: Warburg addressed. *Cold Spring Harbor Symp. Quant. Biol.* **70:** 363–374.

———. 2005b. A mitochondrial paradigm of metabolic and degenerative diseases, aging, and cancer: A dawn for evolutionary medicine. *Annu. Rev. Genet.* **39:** 359–407.

———. 2007. Why do we have a maternally inherited mitochondrial DNA? Insights from evolutionary medicine. *Annu. Rev. Biochem.* **76:** 781–821.

Wallace D.C. and Lott M.T. 2002. Mitochondrial genes in degenerative diseases, cancer and aging. In *Emery and Rimoin's principles and practice of medical genetics* (ed. D.L. Rimoin et al.), pp. 299–409. Churchill Livingstone, London.

Wallace D.C., Brown M.D., and Lott M.T. 1999. Mitochondrial DNA variation in human evolution and disease. *Gene* **238:** 211–230.

Wallace D.C., Lott M.T., and Procaccio V. 2007. Mitochondrial genes in degenerative diseases, cancer and aging. In *Emery and Rimoin's principles and practice of medical genetics*, 5th edition (ed. D.L. Rimoin et al.), pp. 194–298. Churchill Livingstone, Philadelphia, Pennsylvania.

Wallace D.C., Lott M.T., Brown M.D., and Kerstann K. 2001. Mitochondria and neuro-ophthalmological diseases. In *The metabolic and molecular basis of inherited disease* (ed. C.R. Scriver et al.), pp. 2425–2512. McGraw-Hill, New York.

Wallace D.C., Brown M.D., Schurr T.G., Chen E., Chen Y.-S., Starikovskaya Y.B., and Sukernik R.I. 2000. Global mitochondrial DNA variation and the origin of Native Americans. In *The origin of humankind: Conference Proceedings of the International Symposium*, Venice, 14–15 May, 1998 (ed. M. Aloisi et al.), pp. 9–11. IOS Press, Washington, D.C.

Wallace D.C., Singh G., Lott M.T., Hodge J.A., Schurr T.G., Lezza A.M., Elsas L.J., II, and Nikoskelainen E.K. 1988a. Mitochondrial DNA mutation associated with Leber's hereditary optic neuropathy. *Science* **242:** 1427–1430.

Wallace D.C., Zheng X.X., Lott M.T., Shoffner J.M., Hodge J.A., Kelley R.I., Epstein C.M., and Hopkins L.C. 1988b. Familial mitochondrial encephalomyopathy (MERRF): Genetic, pathophysiological, and biochemical characterization of a mitochondrial DNA disease. *Cell* **55:** 601–610.

Wang Q., Ito M., Adams K., Li B.U., Klopstock T., Maslim A., Higashimoto T., Herzog J., and Boles R.G. 2004. Mitochondrial DNA control region sequence variation in migraine headache and cyclic vomiting syndrome. *Am. J. Med. Genet. A* **131:** 50–58.

Wang Y., Michikawa Y., Mallidis C., Bai Y., Woodhouse L., Yarasheski K.E., Miller C.A., Askanas V., Engel W.K., Bhasin S., and Attardi G. 2001. Muscle-specific mutations accumulate with aging in critical human mtDNA control sites for replication. *Proc. Natl. Acad. Sci.* **98:** 4022–4027.

Wood J.G., Rogina B., Lavu S., Howitz K., Helfand S.L., Tatar M., and Sinclair D. 2004. Sirtuin activators mimic caloric restriction and delay ageing in metazoans. *Nature* **430:** 686–689.

Wu Z., Puigserver P., Andersson U., Zhang C., Adelmant G., Mootha V., Troy A., Cinti S., Lowell B., Scarpulla R.C., and Spiegelman B.M. 1999. Mechanisms controlling mitochondrial biogenesis and respiration through the thermogenic coactivator PGC-1. *Cell* **98:** 115–124.

Yant L.J., Ran Q., Rao L., Van Remmen H., Shibatani T., Belter J.G., Motta L., Richardson A., and Prolla T.A. 2003. The selenoprotein GPX4 is essential for mouse development and protects from radiation and oxidative damage insults. *Free Radic. Biol. Med.* **34:** 496–502.

Zhang J., Asin-Cayuela J., Fish J., Michikawa Y., Bonafe M., Olivieri F., Passarino G., De Benedictis G., Franceschi C., and Attardi G. 2003. Strikingly higher frequency in centenarians and twins of mtDNA mutation causing remodeling of replication origin in leukocytes. *Proc. Natl. Acad. Sci.* **100:** 1116–1121.

2

Sirtuins: A Universal Link between NAD, Metabolism, and Aging

Shin-ichiro Imai
Department of Molecular Biology and Pharmacology
Washington University School of Medicine
St. Louis, Missouri 63110

Leonard P. Guarente
Department of Biology
Massachusetts Institute of Technology
Cambridge, Massachusetts 02139

> *Nature is an endless combination and repetition of a very few laws. She hums the old well-known air through innumerable variations.*
> Ralph Waldo Emerson
> Essay I *History,* Essays: First Series, 1841

FOR THE PAST SEVERAL YEARS, Sir2 family proteins, now called "sirtuins," have been emerging as an evolutionarily conserved, critical regulator for aging and longevity in diverse model organisms, providing a novel paradigm to the field of aging research. The *SIR2* gene was originally identified by Klar et al. (1979) as one of the genes that regulate the a and α mating types of budding yeast, *Saccharomyces cerevisiae.* Subsequent studies have demonstrated that Sir2 has a critical role in the regulation of transcriptional silencing at mating-type loci, telomeres, and ribosomal DNA (rDNA) repeats (Guarente 1999). At mating-type loci and telomeres, the Sir complex that includes Sir2 and the other two Sir proteins, Sir3 and Sir4, forms polymerized, closed chromatin structure, namely, heterochromatin, and silences reporter genes inserted into these genomic loci (Rine and Herskowitz 1987; Gottschling et al. 1990). At rDNA repeats, Sir2 is included in another complex named RENT (*regulator of nucleolar silencing and telophase exit*), along with Net1, Cdc14, and other proteins, and is involved in silencing transcription of pol II reporter genes inserted into rDNA repeats (Bryk et al. 1997; Smith and Boeke 1997; Straight et al. 1999). Sir2-mediated rDNA silencing is also important for suppression of homologous recombination within this highly repetitive rDNA region (Gottlieb and Esposito 1989). Silencing requires specific lysines in the amino-terminal tails of histones H3 and H4 (Thompson et al. 1994; Hecht et al. 1995; Braunstein et al. 1996; Hoppe et al. 2002). In particular, lysine 16 of H4 is essential for silencing of all three loci in the yeast genome. These and other lysine residues of H3 and H4 tails are acetylated in active chromatin but deacetylated in silenced chromatin. In 1993, Braunstein et al. (1993) found that overexpression of Sir2 promoted global deacetylation of histones in yeast, leading to a hypothesis that the biochemical function of Sir2 might be a histone deacetylase. However, early attempts to identify such enzymatic activity were unsuccessful.

The first important clue for the enzymatic activity of Sir2 came from the study of the *Salmonella typhimurium* Sir2-like protein, CobB (Tsang and Escalante-Semerena 1998). CobB can substitute for the function of CobT, a protein that transfers phosphoribose from nicotinic acid mononucleotide to dimethylbenzimidazole in the cobalamin synthesis pathway (Trzebiatowski and Escalante-Semerena 1997). These results suggested that Sir2 might be able to catalyze a related pyridine nucleotide transfer reaction. Indeed, Frye (1999) showed that Sir2 proteins from *Escherichia coli* and humans were able to transfer ^{32}P from [^{32}P]NAD to bovine serum albumin. Subsequent to this study, Tanny et al. (1999) proved that the moiety transferred was, indeed, ADP-ribose, using histones as substrates, leading them to the proposal that the ADP-ribosyltransferase activity of Sir2 was essential for silencing in vivo. Meanwhile, Imai and Guarente noticed that peptides of the amino-terminal tails of histone H3 or H4 could accept ^{32}P from [^{32}P]NAD, but only if the peptides were acetylated (Imai et al. 2000b). A real surprise came when analyzing the Sir2-modified product by mass spectrometry. The relative molecular weight of the product was actually smaller than that of the acetylated substrate by 42 daltons, clearly indicating that the major enzymatic activity of Sir2 was deacetylase and not ADP-ribosyltransferase (Imai et al. 2000a). Furthermore, Sir2 could specifically deacetylate lysine 16 of H4 in an NAD-dependent manner, strongly suggesting that the NAD-dependent deacetylase activity has a vital role in silencing in vivo.

NADH, NADP, and NADPH could not substitute for NAD in this reaction. The NAD-dependent deacetylase activity was also highly conserved between yeast and mouse Sir2. The absolute requirement of NAD for the Sir2 deacetylase activity immediately suggested that Sir2 and closely related homologs might function as sensors of the cellular energy status represented by NAD. Landry et al. (2000b) showed that Sir2 and Hst2, a yeast Sir2 homolog, could catalyze both NAD–nicotinamide exchange reaction and NAD-dependent deacetylation. Smith et al. (2000) also confirmed that Sir2 proteins from yeast, bacteria, and humans exhibited NAD-dependent histone deacetylase activity. The discovery of this evolutionarily conserved NAD-dependent deacetylase activity of Sir2 proteins may have identified one of those "very few laws," providing a novel molecular framework to link between NAD, metabolism, and aging through different organisms.

CATALYTIC MECHANISM OF SIR2 PROTEINS AND THE REGULATION OF THEIR ACTIVITY

The unusual feature of the NAD-dependent deacetylation of Sir2 proteins, now classified as class III histone deacetylases (HDACs), has drawn immediate attention of biochemists and structural biologists to their catalytic mechanism. Tanner et al. (2000) and Tanny et al. (Tanny and Moazed 2001) independently found that the deacetylation reaction was tightly coupled with the cleavage of NAD into nicotinamide and ADP-ribose and the formation of a previously unidentified compound, O-acetyl-ADP-ribose. The crystal structures of an *Archaeoglobus fulgidus* Sir2 homolog, Sir2-Af1, provided the first clue for the detailed catalytic mechanism of this unprecedented reaction (Min et al. 2001). Several other studies also determined crystal structures of yeast, human, and archaeobacterial Sir2 homologs and proposed the structure-based catalytic mechanism of Sir2 proteins (Finnin et al. 2001; Avalos et al. 2002; Zhao et al. 2004). In this model, NAD is initially bound in a pocket structure that can be divided into three regions (sites A, B, and C). At the C site, the carbonyl oxygen of acetyl-lysine initiates a nucleophilic attack on the $C1'$ of the nicotinamide ribose of NAD, releasing nicotinamide from NAD and forming O-alkylamidate intermediate (see Fig. 1) (Avalos et al. 2005). An evolutionarily conserved histidine of Sir2 proteins mediates the internal attack of the $2'$ OH group on this O-alkylamidate intermediate, resulting in deacetylated lysine and $2'$-O-acetyl-ADP-ribose (Sauve et al. 2001; Jackson and Denu 2002). $2'$-O-acetyl-ADP-ribose subsequently equilibrates with $3'$-O-acetyl-ADP-ribose by an intramolecular transesterification reaction.

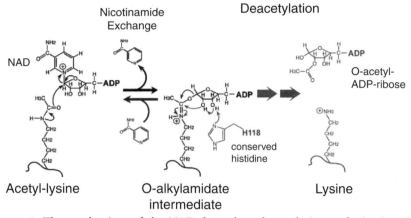

Figure 1. The mechanism of the NAD-dependent deacetylation and nicotinamide exchange reactions catalyzed by Sir2 proteins. For details, see text. (Modified, with permission, from Elsevier, from Avalos et al. 2005.)

If nicotinamide concentration is high, nicotinamide reacts with the O-alkylamidate intermediate and promotes the nicotinamide exchange reaction at the expense of the deacetylation reaction (Fig. 1).

Why do Sir2 proteins require such a complex reaction for deacety-lation? Liou et al. (2005) have recently provided a clue to answer this question. Surprisingly, O-acetyl-ADP-ribose, produced by the Sir2-mediated histone deacetylation reaction, promoted the recruitment of multiple copies of Sir3 into the Sir2/Sir4 complex and induced a dramatic globular-to-cylindrical structural rearrangement in the Sir complex. Therefore, NAD-dependent deacetylation by Sir2 has a critical role in orchestrating both processes of histone deacetylation and silent chromatin assembly in yeast. Any role of O-acetyl-ADP-ribose in higher organisms has not yet been demonstrated, but it is possible that, along with the deacetylation of substrates by Sir2 proteins, this novel reaction product might function as a second messenger to promote the formation of new protein complexes including Sir2 proteins.

Another interesting reaction product is nicotinamide. Nicotinamide can act as a noncompetitive inhibitor of Sir2 proteins in vitro (Landry et al. 2000a; Bitterman et al. 2002). Addition of nicotinamide to culture media at final concentrations of 5–10 mM also inhibits the activity of Sir2 proteins in yeast and mammalian cells (Luo et al. 2001; Bitterman et al. 2002), and depletion of nicotinamide by Pnc1, a yeast nicotinami-dase that converts nicotinamide to nicotinic acid, activates Sir2 activity in vivo in yeast (Anderson et al. 2003). Isonicotinamide antagonizes this inhibitory action of nicotinamide and activates Sir2 activity in vitro and in vivo (Sauve et al. 2005). In yeast, the calculated concentration of endogenous nicotinamide ranges from 10 to 150 μM, which is in a similar range from 50 to 120 μM of IC_{50} values for Sir2 inhibition (Landry et al. 2000a; Bitterman et al. 2002). On the basis of these findings, it has been proposed that nicotinamide functions as an endogenous inhibitor to regulate the activity of Sir2 proteins.

On the other hand, Lin et al. (2004) showed that NADH acts as a competitive inhibitor of Sir2, leading to the proposal that the NAD/NADH ratio is a critical parameter for the regulation of Sir2 activity. In this scenario, the regulation of NAD biosynthesis is also critical, and Revollo et al. (2004) proved that Nampt (nicotinamide phosphoribosyltransferase), the rate-limiting enzyme in the NAD biosynthesis pathway from nicotinamide in vertebrates, regulates mammalian Sir2 activity through NAD biosynthesis. Interestingly, in mammals, nicotinamide concentrations in plasma range from 0.3 μM (human) to 5 μM (mouse) (Bernofsky 1980; Jacobson et al. 1995; Catz et al. 2005), which is consistent with the reported K_M values of Nampt but much lower than the IC_{50} values for Sir2 inhibition (Rongvaux et al. 2002; Revollo et al. 2004). Therefore, the NAD/NADH ratio or NAD itself has a major role in the regulation of the NAD-dependent deacetylase activity of Sir2 proteins in mammals, whereas nicotinamide might be more critical for the activity of Sir2 proteins in lower eukaryotes that tend to have higher endogenous concentrations of nicotinamide.

SIRTUINS—UNIVERSAL REGULATORS FOR AGING AND LONGEVITY

Sir2 proteins are evolutionarily highly conserved in the organisms from bacteria to human (Brachmann et al. 1995; Frye 2000). These Sir2 proteins in different organisms, called "sirtuins," share highly conserved enzymatic core domains (Fig. 2). For the past several years, Sir2 and its closest orthologs have emerged as critical regulators for aging and longevity in experimental model organisms, such as yeast, worms, and flies (Blander and Guarente 2004; Longo and Kennedy 2006). They also mediate anti-aging and life-span-extending effects of caloric restriction, a dietary regimen low in calories without malnutrition that delays aging and extends life span in a wide variety of organisms (Bordone and Guarente 2005). Mammals have seven sirtuins, Sirt1–7, and increasing lines of evidence strongly indicate that mammalian sirtuins might also have a significant role in the regulation of aging and longevity through a broad range of physiological functions (Bordone et al. 2006; Haigis and

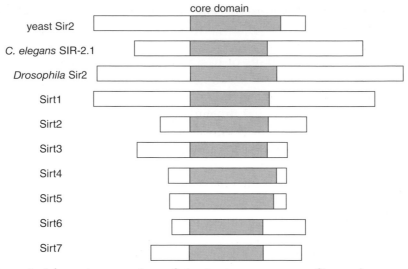

Figure 2. Schematic comparison of sirtuins in yeast, worms, flies, and mammals. Shaded boxes depict enzymatically catalytic core domains that are highly conserved through evolution.

Guarente 2006; Moynihan and Imai 2006). Therefore, through their NAD-dependent enzymatic activities, sirtuins function as a universal link between NAD, metabolism, and aging.

Yeast Sirtuins

The budding yeast *S. cerevisiae* divides asymmetrically to give rise to a larger mother cell and a smaller daughter cell. In this organism, life span is measured by the number of times a mother cell produces daughter cells before it senesces (Mortimer and Johnston 1959). As yeast mother cells age, they become enlarged and wrinkled and have a decreased rate of division (Sinclair et al. 1998). Kennedy et al. (1995) showed that the life-extending *sir4-42* mutation caused the accumulation of the normally telomeric Sir2/3/4 complex at the rDNA repeats (Kennedy et al. 1997). Sinclair and Guarente (1997) then demonstrated that yeast aging was caused by the toxic accumulation of extrachromosomal rDNA circles (ERCs), which arose from homologous recombination between the tandem repeats of rDNA. The large excess of ERCs may titrate essential replication and/or transcription factors, thereby impairing the normal processes of replication and transcription and eventually terminating the life of yeast mother cells. Importantly, an extra copy of the *SIR2* gene decreases ERC formation and extends life span of yeast mother cells, whereas deletion of the *SIR2* gene increases frequency of rDNA recombination and shortens their life span, demonstrating that Sir2 is a uniquely important anti-aging gene in yeast (Kaeberlein et al. 1999). Furthermore, the NAD-dependent deacetylase activity of Sir2 is critical for the maintenance of normal life span in yeast mother cells (Imai et al. 2000a).

In budding yeast, caloric restriction (CR) is achieved either by limiting their glucose availability or by mutating a component in the glucose-sensing PKA signaling pathway, such as the GTP–GDP exchange factor Cdc25 (Lin et al. 2000). Yeast mother cells grown on media containing 0.5% glucose (moderate CR) live significantly longer than those grown on media containing 2% glucose, and yeast mother cells lacking *CDC25* also display a significant extension of life span. The life span extension induced by 0.5% glucose or *cdc25* mutation requires *SIR2*, suggesting that Sir2 is an essential mediator for the effect of moderate CR in yeast. *NPT1*, the gene encoding an enzyme that initiates NAD biosynthesis from nicotinic acid in yeast, is also required for life span extension mediated by CR (Lin et al. 2000). Thus, these findings suggest that CR extends life span by increasing Sir2 activity, which thereby reduces rDNA recombination rates and subsequent accumulation of toxic rDNA circles in yeast.

Moderate CR appears to activate Sir2 activity through at least two independent pathways in yeast (Fig. 3). In one pathway, CR shifts metabolism toward respiration and increases the NAD/NADH ratio by reducing NADH levels (Lin et al. 2002). Deletion of *CYT1*, the gene encoding cytochrome *c*1, abolishes the effect of CR, whereas overexpression of *NDE1* or *NDE2*, genes encoding mitochondrial NADH dehydrogenases, increases the NAD/NADH ratio and life span (Lin et al. 2002, 2004). In the other pathway, CR increases levels of the yeast nicotinamidase Pnc1, which converts the endogenous Sir2 inhibitor nicotinamide into nicotinic acid (Anderson et al. 2003). Deletion of *PNC1* abolishes the ability of CR to increase life span, whereas overexpression of *PNC1* extends life span in a Sir2-dependent manner.

In a different yeast strain, CR induced by 0.5% glucose can extend life span in a Sir2-independent manner (Kaeberlein et al. 2004). In this case, another yeast sirtuin, Hst2, may be responsible for the Sir2-independent life span extension by CR through the suppression of rDNA recombination (see Fig. 3) (Lamming et al. 2005). Thus, there might be functional redundancy among yeast sirtuins to mediate the effect of CR. Interestingly, a more severe regimen of CR (0.05% glucose) extends life span of yeast mother cells by what might be a different mechanism, which

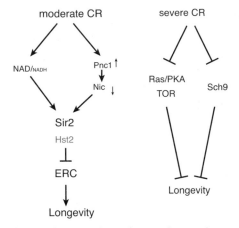

Figure 3. Sir2-dependent and Sir2-independent pathways that mediate the life-span-extending effect of caloric restriction (CR) in yeast. (*Left panel*) Moderate CR increases the NAD/NADH ratio by shifting metabolism toward respiration and decreases nicotinamide through the induction of the yeast nicotinamidase, Pnc1, leading to the activation of Sir2, the suppression of ERC accumulation, and the extension of life span. In certain yeast strains, another yeast sirtuin, Hst2, mediates the life span extension by CR. (*Right panel*) Severe CR decreases the activities of PKA, TOR, and the Akt ortholog Sch9 and extends life span in a Sir2-independent manner.

is independent of both Sir2 and mitochondrial respiration. This SIR2-independent pathway may depend on the nutrient-responsive kinases, Tor and PKA, and the Akt ortholog Sch9 (see Fig. 3) (Kaeberlein et al. 2005a).

A second way to assess yeast survival is by measuring the kinetics of cell death in stationary-phase culture. In this assay, Sir2 did not affect survival in wild-type strains, but in certain long-surviving mutants, deleting *SIR2* actually extended survival time (Fabrizio et al. 2005). Thus, it is possible that under certain extreme conditions, the role of Sir2 in survival may be inverted from its anti-aging function in replicative aging.

Caenorhabditis elegans Sirtuins

There are four sirtuin genes in the *C. elegans* genome. *sir-2.1* has 31% identity to the conserved core domain of yeast Sir2, whereas *sir-2.2, sir-2.3,* and *sir-2.4* have between 10% and 20% identity with the core domain. Of these four genes, *sir-2.1* was demonstrated to have a major role in the regulation of *C. elegans* longevity (Tissenbaum and Guarente 2001). Stable duplication strains containing an extra copy of *sir-2.1*, as well as independent transgenic lines expressing extra copies of *sir-2.1*, all have up to a 50% increase in life span. Similarly, *sir2.1* null mutants have slightly shortened life spans, as well as heightened sensitivity to various stresses (Wang and Tissenbaum 2006).

Genetic analyses using transgenic *sir-2.1* overexpressors initially placed *sir-2.1* in the highly conserved insulin/IGF-1 signaling pathway (IIS), since the extended life span conferred by *sir-2.1* overexpression requires the presence of the forkhead transcription factor *daf-16* and does not further extend the life span of the *daf-2* insulin-like receptor mutant (Tissenbaum and Guarente 2001). However, more refined studies showed that *sir-2.1* actually functions in a stress response pathway parallel to IIS and impinges directly upon DAF-16 (Fig. 4) (Berdichevsky et al. 2006;

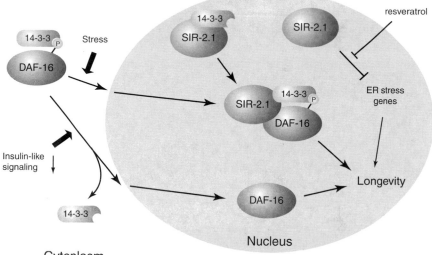

Figure 4. Model for the regulation of life span mediated by SIR-2.1 in *C. elegans*. SIR-2.1 forms a complex with 14-3-3 scaffolding proteins in the nucleus, whereas the forkhead transcription factor DAF-16 is phosphorylated and bound to 14-3-3 proteins in the cytoplasm. Upon heat stress, the DAF-16–14-3-3 complex translocates into the nucleus, and DAF-16 binds to SIR-2.1 in a 14-3-3-dependent manner, resulting in the transcriptional activation of DAF-16 target genes and the extension of life span. Under low insulin-like signaling conditions, DAF-16 becomes unphosphorylated and dissociated from 14-3-3 and translocates into the nucleus. This nuclear DAF-16 does not require SIR-2.1 and 14-3-3 for transcriptional activation and life span extension. Under certain unknown conditions, SIR-2.1 might also have a pro-aging effect through the repression of a class of genes termed *abu* involved in ER stress. Resveratrol extends life span by lifting the SIR-2.1-mediated repression of *abu* genes.

Wang and Tissenbaum 2006). Heat stress, but not reduced insulin signaling, promotes a physical interaction between SIR-2.1 and DAF-16 that depends on the class of 14-3-3 scaffolding proteins (Berdichevsky et al. 2006). In the absence of stress, SIR-2.1 is bound to 14-3-3 proteins in the nucleus, and DAF-16 is bound to 14-3-3 in the cytoplasm (Berdichevsky et al. 2006; Y. Wang et al. 2006). Heat likely triggers the nuclear entry of DAF-16/14-3-3 promoted by JNK kinase (Oh et al. 2005), and the union of SIR-2.1 and DAF-16 then activates DAF-16 target genes that promote stress resistance.

Independently of its effect on DAF-16, *sir-2.1* regulates a class of genes termed *abu* involved in endoplasmic reticulum (ER) stress (Fig. 4). Intriguingly, *sir-2.1* represses these genes, which appears to result in a shortening of life span, since enforced expression of *abu-11* extends life span. This repression of *abu* genes is partially lifted by the plant polyphenol resveratrol, discussed below, which explains the life-extending property of this compound in worms. Thus, like the example of stationary-phase growth described above, *sir-2.1* appears to have a pro-aging function on ER stress genes, which is trumped by its anti-aging effect on DAF-16. We suspect that *sir-2.1* mediates an ER response to a particular kind of stressor, which is not known at this time.

CR and resultant life span extension can be obtained in worms by diluting their bacterial food source (Klass 1977), culturing the worms axenically in the absence of bacteria (Houthoofd et al. 2002; Kaeberlein et al. 2006), or reducing their food intake via genetic manipulations. *eat-2* mutants have defects in pharyngeal pumping which result in a reduced rate of food intake and an extended life span (Lakowski and Hekimi 1998). This life span extension is suppressed in the absence of *sir-2.1* (Wang and Tissenbaum 2006). However, other groups have observed that CR imposed by food dilution can also extend the life span of *sir-2.1* mutant worms (Bishop and Guarente 2007). It will be of great interest to examine whether other *C. elegans* sirtuins have a role in CR in the absence of *sir-2.1*, possibly explaining the differences found thus far. In summary, *C. elegans* SIR-2.1 clearly has a critical role in the regulation of life span and stress resistance, but further investigation will be necessary to clarify the importance of *C. elegans* sirtuins in CR.

Drosophila Sirtuins

There are five sirtuins in *Drosophila melanogaster*, with dSir2 being the closest homolog of the yeast Sir2 (Frye 2000). dSir2 catalyzes the NAD-dependent deacetylation of lysine residues in core histone tails and represses transcription from the hyperacetylated histone-DNA complex in vitro (Parsons et al. 2003). dSir2 is also required for the epigenetic silencing by Polycomb proteins and physically associates with an E(Z) histone methyltransferase complex (Furuyama et al. 2004). Interestingly, whereas yeast Sir2 is localized exclusively to the nucleus, dSir2 exhibits nuclear and cytoplasmic shuttling during embryogenesis prior to its exclusive nuclear localization in the adult fly (Rosenberg and Parkhurst 2002). Similar to the results obtained in yeast and worms, dSir2 also regulates life span. Ubiquitous overexpression of dSir2 in *Drosophila* using tubulin-GAL4 drivers increases life span up to 57%, and tissue-specific overexpression of dSir2 using pan-neuronal ELAV-GAL4 drivers extends life span by up to 52% (Rogina and Helfand 2004). On the other hand, dSir2 null mutant flies show a significantly shorter life span compared to wild-type flies (Astrom et al. 2003). Therefore, the longevity-regulating function of Sir2 proteins, along with the NAD-dependent deacetylase activity, is highly conserved in yeast, worms, and flies.

CR can be achieved in flies by diluting the yeast and glucose in their food source (Clancy et al. 2002). In addition to having an increased life span, calorically restricted flies also show a twofold increase in dSir2 mRNA levels (Rogina et al. 2002). CR-induced life span extension was suppressed in flies that lack dSir2, and CR cannot further extend the life span of dSir2-overexpressing flies, suggesting that dSir2 and CR act through the same pathway to regulate life span (Rogina and Helfand 2004). Finally, dSir2 has also been shown to act downstream from the class I histone deacetylase Rpd3, and CR represses Rpd3 expression, resulting in an increase in dSir2 expression and thereby life span extension in flies (Rogina et al. 2002; Rogina and Helfand 2004). Taken together, the fact that Sir2 proteins regulate life span in response to CR in evolutionarily distant organisms suggests the existence of a universal mechanism that attunes the pace of aging to diet.

MAMMALIAN SIRTUINS—EMERGING ROLES IN METABOLISM, STRESS RESISTANCE, AND AGING

The mammalian sirtuin family comprises seven members, named Sirt1 through Sirt7. Sirt1 has the greatest degree of sequence identity in the highly conserved core domain to yeast Sir2 (see Fig. 2) (Frye 1999). Although the majority of mammalian sirtuin research has focused on the function of Sirt1, recent studies have revealed novel roles for several of the other six sirtuin members (Bordone et al. 2006; Haigis and Guarente 2006; Moynihan and Imai 2006). In this section, therefore, we first review physiological roles of Sirt1 in the regulation of metabolism and stress resistance and then summarize the functions of other mammalian sirtuins.

Sirt1

Although it is not yet proven whether Sirt1 regulates aging and longevity in mammals, increasing lines of evidence clearly indicate that Sirt1 has a critical role in the regulation of metabolism in response to nutrient availability and cell survival in response to stress. Through both regulatory pathways, Sirt1 may affect the pace of aging and mediate the effect of CR in mammals.

Sirt1 in Pancreatic β Cells

Pancreatic β cells have a central role in maintaining glucose homeostasis in mammals by secreting insulin in response to fluctuations in blood glucose levels. Two recent studies have demonstrated that Sirt1 positively regulates glucose-stimulated insulin secretion in pancreatic β cells (Moynihan et al. 2005; Bordone et al. 2006). Moynihan et al. have demonstrated that an increased dosage of Sirt1 in pancreatic β cells enhances insulin secretion in response to glucose and improves glucose tolerance in β-cell-specific Sirt1-overexpressing (BESTO) transgenic mice (Moynihan et al. 2005). Bordone et al. (2006) have also demonstrated that Sirt1-deficient mice and islets show blunted insulin secretion in response to glucose. In pancreatic β cells, Sirt1 regulates expression of genes involved in β-cell physiology and glucose homeostasis (Moynihan et al. 2005). Among them, the uncoupling protein 2 (*Ucp2*) gene, which encodes a mitochondrial inner-membrane protein that uncouples respiration from ATP production, is repressed by Sirt1 (see Fig. 5) (Moynihan et al. 2005; Bordone et al. 2006). The repression of *Ucp2* by Sirt1 increases the proton gradient across the mitochondrial membrane and thereby enhances ATP production. Indeed, BESTO islets show higher ATP levels in response to glucose, whereas Sirt1-knockdown β cells show defects in glucose-stimulated ATP production. This Sirt1-mediated repression of *Ucp2* is alleviated by fasting, which might be explained by a decrease in NAD levels in fasted pancreas (Bordone et al. 2006).

Sirt1 also appears to have an important role in protecting pancreatic β cells against metabolic stresses. Sirt1 protects pancreatic β cells against

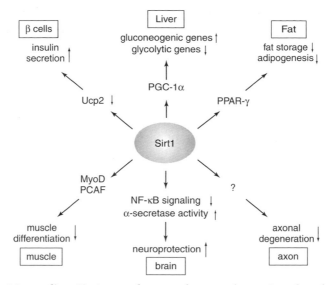

Figure 5. Mammalian Sirt1 as a key regulator orchestrating the physiological response to nutrient availability in different organs and tissues. Sirt1 regulates the expression or the activities of the indicated effectors and mediates diverse physiological responses in organ/tissue-dependent manners.

oxidative stress by deacetylating the forkhead transcription factor Foxo1 and enhancing its transcriptional activity to induce expression of two important β-cell transcription factors, NeuroD and MafA (Kitamura et al. 2005). BESTO islets also show resistance to high levels of free fatty acids (K.A. Moynihan and S. Imai, unpubl.). Although the physiological relevance of these effects should be further investigated, Sirt1 may promote resistance of pancreatic β cells against metabolic stresses under certain pathophysiological conditions, such as type 2 diabetes.

Sirt1 in Liver

The liver has a critical role in maintaining blood glucose levels during fasting by promoting glycogenolysis and gluconeogenesis and repressing glycolysis. Gene expression profiling of calorically restricted livers reveals that CR also promotes gluconeogenesis and inhibits glycolysis (Yamaza et al. 2002). Several reports have shown that nutrient availability influences the expression of Sirt1 in the liver. During fasting and CR, Sirt1 mRNA and protein levels are up-regulated in the liver (Cohen et al. 2004; Nemoto et al. 2004; Al-Regaiey et al. 2005; Rodgers et al. 2005). Both pyruvate and NAD levels are also elevated in the liver after fasting (Rodgers et al. 2005). An increase in pyruvate leads to higher levels of Sirt1 protein, whereas increased NAD levels likely contribute to the enhancement of Sirt1 activity in the fasted state. Upon induction of Sirt1 during fasting, Sirt1 interacts with and deacetylates PGC-1α (peroxisome proliferator-activated receptor-γ coactivator 1α), a key transcriptional regulator of glucose production, in an NAD-dependent manner (Fig. 5) (Rodgers et al. 2005). Sirt1 is required for both the induction of gluconeogenic genes and the repression of glycolytic genes by PGC-1α in cultured hepatocytes. Sirt1 is also required for the PGC-1α-induced increase in glucose production in cultured hepatocytes (Rodgers et al. 2005). These findings suggest that Sirt1 has an important role in maintaining glucose homeostasis in response to changes in nutrient availability by regulating the PGC-1α-mediated gluconeogenic/glycolytic pathways in liver.

Sirt1 in Fat

Lipolysis and mobilization of free fatty acids from adipocytes permits their use by metabolic tissues as an alternative fuel source during fasting. Several reports have shown that Sirt1 mRNA and protein levels increase during fasting and CR in white adipose tissue (WAT) (Cohen et al. 2004; Nemoto et al. 2004). Additionally, in response to fasting, Sirt1 is recruited to the promoters of genes, such as aP2, that are controlled by PPAR-γ (peroxisome proliferator-activated receptor-γ), a nuclear hormone receptor that promotes adipogenesis and fat storage (Picard et al. 2004). Sirt1 interacts with and represses the activity of PPAR-γ through interaction with its corepressors NCoR (nuclear receptor corepressor) and SMRT (silencing mediator for retinoid and thyroid hormone receptor), resulting in the down-regulation of genes driving adipogenesis and fat storage (Fig. 5). Consistent with these findings, fasting-induced fatty acid mobilization is compromised in Sirt1 heterozygous mice (Picard et al. 2004).

Adipocytes also function as an endocrine tissue that secretes a variety of adipokines, including leptin, tumor necrosis factor-α (TNF-α), and adiponectin. It has recently been reported that Sirt1 increases adiponectin transcription in adipocytes by activating Foxo1 and promoting the interaction between Foxo1 and the CCAAT/enhancer-binding protein α (C/EBPα) (Qiao and Shao 2006). Adiponectin levels are also reported to increase in CR, implying that Sirt1 might mediate this effect in response to CR. Interestingly, both Foxo1 and Sirt1 protein levels are reduced in epididymal fat from db/db and high-fat diet-induced obese mice (Qiao and Shao 2006), suggesting a possible involvement of Sirt1 in pathogenesis of type 2 diabetes.

Sirt1 in Muscle

Skeletal muscle is one of the main tissues responsible for the regulation of glucose uptake and insulin sensitivity. Sirt1 has been found to regulate skeletal muscle gene expression and differentiation in response to the NAD/NADH ratio (Fulco et al. 2003). Overexpression of Sirt1 inhibits muscle-specific gene transcription and retards differentiation, whereas reduction of Sirt1 levels results in increased muscle differentiation. Sirt1 forms a complex with and deacetylates both the transcription factor MyoD and the acetyltransferase PCAF, and its NAD-dependent deacetylase activity is required for inhibition of muscle differentiation (Fig. 5) (Fulco et al. 2003). CR increases Sirt1 protein levels, suggesting higher activity of Sirt1, and an important role of Sirt1 in muscle function is suggested by recent findings with resveratrol, a putative Sirt1 activator (see below).

Sirt1 in Brain

CR has been shown to have beneficial effects in animal models of neurodegenerative diseases, including Alzheimer's, Parkinson's, and Huntington diseases (Mattson et al. 2002). Numerous studies have demonstrated that CR promotes the resistance of neurons against a variety of toxicities and enhances their survival. Sirt1 has also been reported to show neuroprotective effects under certain pathological conditions. Two recent studies have demonstrated that Sirt1 mediates neuroprotection against amyloid-β (Aβ) neurotoxicity (J. Chen et al. 2005; Qin et al. 2006). J. Chen et al. (2005) have shown that Sirt1 suppresses NF-κB signaling induced by Aβ in microglia and promotes strong neuroprotective effects (Fig. 5). Qin et al. (2006) have also shown that increasing the dosage or the activity of Sirt1 promotes α-secretase activity and reduces Aβ generation in cultured neurons. Interestingly, CR increases Sirt1 protein levels and NAD content in brain, which coincides with a significant decrease in Aβ contents and Aβ-induced neuropathology in transgenic model mice (Qin et al. 2006). By using C. elegans as a model system, the Sirt1 ortholog, sir-2.1, has been shown to protect against neuronal dysfunction due to polyglutamine toxicity caused by the Huntington disease-associated protein huntingtin, suggesting that Sirt1 might also mediate neuroprotection against Huntington disease (Parker et al. 2005).

Sirt1 also appears to protect axons against injury and toxic insults (Araki et al. 2004). Wallerian degeneration slow (Wlds) mice, which display delayed axonal degeneration, overexpress a fusion protein composed of the ubiquitin assembly protein Ufd2a and the NAD-biosynthetic enzyme Nmnat1. The phenotype of axonal protection following both mechanical and toxic insults has been ascribed to increased Nmnat1 enzymatic activity, and addition of exogenous NAD prior to axonal insult has been found to mimic the effects of Nmnat1 overexpression (Araki et al. 2004; Wang et al. 2005). Araki et al. (2004) have demonstrated that Sirt1 is required for this delayed axonal degeneration (Fig. 5). The Sirt1 dependency has recently been called into question, however, by Wang et al. (2005) because there was no significant difference in axonal degeneration between wild-type and Sirt1 knockout mice. It is interesting to note that the time course of axonal degeneration in these two studies is different, suggesting that there may be two different NAD-dependent mechanisms. Further investigation will be necessary to resolve these discrepancies.

Sirt1 and Stress Resistance

Recent evidence has also shown that Sirt1 regulates several pathways of stress resistance and cell survival (Fig. 6). Following cellular stress or DNA damage, acetylation of the tumor suppressor p53 by the CBP/p300 acetyltransferases leads to the increased transcriptional activity of p53, ultimately causing either cell cycle arrest or apoptosis. Sirt1 physically interacts with and deacetylates p53, thereby inhibiting p53-mediated apoptosis and promoting cell survival in response to oxidative stress and DNA damage (Luo et al. 2001; Vaziri et al. 2001). Overexpression of wild-type Sirt1 reverses the irradiation-induced acetylation of p53 and reduces levels of the p53 target p21, whereas both the Sir2 inhibitor nicotinamide and a catalytically defective Sirt1 inhibit p53 deacetylation and activate p53-dependent apoptosis (Luo et al. 2001; Vaziri et al. 2001). Additionally, it has been reported that Sirt1-deficient mice have highly acetylated p53 and increased levels of irradiation-induced apoptosis in the thymocytes (Cheng et al. 2003). Whereas Sirt1 promotes cell survival through the inhibition of p53-mediated apoptosis, Sirt1 may increase cancer risk in the absence of HIC1 (hypermethylated in cancer 1), a tumor suppressor that cooperates with p53 to suppress age-dependent development of cancer in mice (W. Y. Chen et al. 2005). The loss of HIC1 promotes tumorigenesis by up-regulating Sirt1 expression and thereby attenuating p53 function in normal or cancer cells. Therefore, physiological responses mediated by Sirt1 could be a double-edged sword, modulated by a variety of genetic factors.

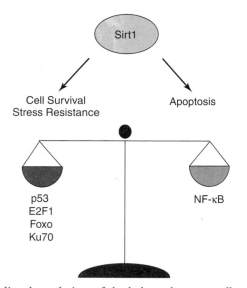

Figure 6. Sirt1-mediated regulation of the balance between cell survival/stress resistance and apoptosis in response to environmental and nutritional input. Sirt1 promotes cell survival and stress resistance through the regulation of p53, E2F1, Foxo, and Ku70. Under certain conditions, Sirt1 may also promote apoptosis. For example, Sirt1 sensitizes cells to TNF-α-induced apoptosis by inhibiting NF-κB function.

E2F1, an E2F family transcription factor that regulates cell cycle and apoptosis, has an important role in the apoptotic response to DNA damage through p53-dependent or p53-independent pathways (Stanelle and Pützer 2006). Whereas E2F1 induces Sirt1 expression at the transcriptional level after DNA damage, Sirt1 inhibits E2F1-mediated apoptosis by physically interacting with and deacetylating E2F1, forming a negative feedback loop (see Fig. 6) (C. Wang et al. 2006).

Sirt1 also promotes cell survival and stress resistance by interacting with and deacetylating Foxo family members (see Fig. 6) (Brunet et al. 2004; Daitoku et al. 2004; Motta et al. 2004; van der Horst et al. 2004), which have a myriad of functions, including DNA-damage repair, cell cycle arrest, and apoptosis (Greer and Brunet 2005). Sirt1 interacts with and deacetylates various Foxo family members, including Foxo1, 3, and 4, in response to oxidative stress. Sirt1-mediated Foxo deacetylation is reported to increase Foxo activity for DNA repair and stress-resistant genes and to decrease it for proapoptotic genes. For example, the activation of Foxo targets *GADD45*, *p27KIP1*, and *MnSod* and the repression of Foxo targets *Fas Ligand* and *Bim* promote cell cycle arrest and DNA repair and inhibit apoptosis (Brunet et al. 2004; Daitoku et al. 2004; Motta et al. 2004; van der Horst et al. 2004).

Another mechanism by which Sirt1 shifts the balance from apoptosis to cell survival is through the interaction with and deacetylation of the DNA repair factor Ku70 (Fig. 6). This interaction results in sequestration of the proapoptotic Bax in the cytoplasmic Bax–Ku70 complex, thereby preventing Bax-mediated, stress-induced apoptosis (Cohen et al. 2004).

Sirt1 can, however, also promote apoptosis under certain conditions (Fig. 6). Sirt1 physically interacts with, deacetylates, and inhibits the function of NF-κB, a prosurvival transcription factor (Yeung et al. 2004). Enhancing Sirt1 activity represses NF-κB-mediated transcription of *cIAP-2*, an anti-apoptotic gene whose product inhibits TNF-α-induced caspase activation, and thereby sensitizes cells to TNF-α-induced apoptosis (Yeung et al. 2004). Thus, depending on the apoptotic stimuli, Sirt1 is able to either promote or prevent cell survival in a context-dependent manner.

Sirt1 and CR

The studies mentioned above strongly indicate that Sirt1 regulates both metabolic and stress responses, which are also induced by CR (see Weindruch et al., this volume), suggesting that Sirt1 may mediate at least some of the effects of this dietary regimen. Two recent studies have provided genetic evidence that links Sirt1 to CR in mice (D. Chen et al. 2005; Nisoli et al. 2005). Nisoli et al. (2005) have demonstrated that CR enhances mitochondrial biogenesis by up-regulating the expression of the endothelial nitric oxide synthase (eNOS). This CR-induced enhancement of mitochondrial biogenesis is blunted in eNOS-deficient mice. Moreover, NO was shown to up-regulate the Sirt1 promoter in cells, and induction of Sirt1 is also blunted in eNOS-deficient mice. Therefore, one can propose a pathway in which CR induces eNOS and NO, thereby up-regulating Sirt1 and mitochondrial biogenesis (Fig. 7). The ability of Sirt1 to deacetylate PGC-1α and activate it may explain the effect on mitochondria (see discussion of resveratrol, below). D. Chen et al. (2005) have examined physical movement during CR in wild-type and Sirt1-deficient mice. Whereas wild-type mice show a significant increase in physical activity (McCarter et al. 1997), Sirt1-deficient mice do not show such an increase even when other metabolic parameters respond to CR. This is not due to a reduced capacity for movement because Sirt1-deficient mice perform as well as or better than wild-type mice in rotarod or treadmill challenges. Interestingly, it has been reported that CR significantly

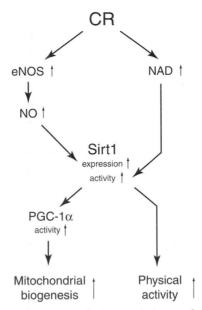

Figure 7. Genetic linkage between caloric restriction and Sirt1 in mammals. CR enhances mitochondrial biogenesis by up-regulating the expression of the endothelial nitric oxide synthase (eNOS), promoting the production of NO and increasing Sirt1 expression. These CR-induced events are blunted in eNOS-deficient mice. The activation of PGC-1α by Sirt1 likely mediates the effect of CR on mitochondria. CR also enhances physical activity by increasing NAD content and up-regulating Sirt1 protein levels in the brain. Sirt1-deficient mice do not show the CR-induced enhancement of physical activity.

increases Sirt1 protein levels and NAD content in the brain (Qin et al. 2006). Therefore, CR might enhance Sirt1 activity in the brain, resulting in an increase in physical activity (Fig. 7).

Sirt2

Sirt2 is a cytoplasmic NAD-dependent protein deacetylase present in the heart, brain, testis, and skeletal muscle (Afshar and Murnane 1999). Functional studies have demonstrated a role for Sirt2 in the regulation of mitosis and cell cycle progression (Dryden et al. 2003). Sirt2 becomes hyperphosphorylated at the G_2/M transition, and the amount of Sirt2 increases during the M phase of the cell cycle (Fig. 8). Overexpression of wild-type Sirt2 prolongs the M phase and delays reentry into the cell cycle. After dephosphorylation of Sirt2 by the protein phosphatase

CDC14B, Sirt2 is ubiquitinated and targeted for degradation by the 26S proteasome (Dryden et al. 2003). Sirt2 has recently also been shown to be responsible for the global decrease in deacetylation of lysine 16 on histone H4 during the G_2/M transition during the cell cycle (Vaquero et al. 2006).

In addition to its role in the regulation of cell cycle progression, Sirt2 has also been found to colocalize with microtubules and to be an NAD-dependent tubulin deacetylase (see Fig. 8) (North et al. 2003). Knockdown of Sirt2 leads to hyperacetylation of tubulin, and Sirt2 colocalizes and interacts with another tubulin deacetylase, HDAC6. Currently, the physiological consequences of Sirt2-mediated tubulin deacetylation are not yet known. Interestingly, Sirt2 levels have been shown to be dramatically reduced in human glioma cell lines, and overexpression of Sirt2 in these cell lines results in a disruption of the microtubule network and a decrease in the number of stable clones (Hiratsuka et al. 2003). Thus, Sirt2 might

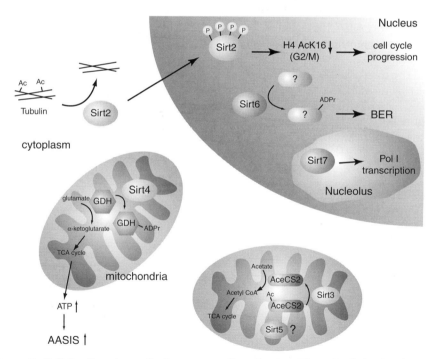

Figure 8. Cellular functions of other mammalian sirtuins. The subcellular localization and the function of Sirt2-7 are summarized. See details in text. (ADPr) ADP-ribose; (AceCS2) acetyl-CoA synthetase 2; (GDH) glutamate dehydrogenase; (AASIS) amino acid-stimulated insulin secretion; (BER) base excision repair; (Pol I), RNA polymerase I.

act as a tumor suppressor gene in human gliomas through regulation of the microtubule network.

Sirt3

Sirt3 is a mitochondrial NAD-dependent protein deacetylase that is highly expressed in metabolically active tissues, such as muscle, liver, kidney, and heart (Onyango et al. 2002; Schwer et al. 2002). The full-length Sirt3 protein is enzymatically inactive, and the amino-terminal portion of the inactive form of Sirt3 is proteolytically cleaved by the matrix processing peptidase (MMP), resulting in the active form of Sirt3 in the mitochondrial matrix (Schwer et al. 2002).

Sirt3 may regulate mitochondrial function and thermogenesis in brown adipocytes (Shi et al. 2005). CR increases Sirt3 expression levels in brown adipocyte tissue (BAT), whereas obesity decreases Sirt3 levels in BAT. Overexpression of Sirt3 in HIB-1B brown adipocyte cells stimulates CREB phosphorylation and increases PGC-1α and UCP1 expression, and both the ADP-ribosyltransferase and the deacetylase activities of Sirt3 are required for these changes. Functional consequences of Sirt3 overexpression in HIB-1B cells include decreased membrane potential, reduced reactive oxygen species production, and increased cellular respiration (Shi et al. 2005).

Two recent studies have identified the mitochondrial acetyl-CoA synthetase 2 (AceCS2) as a substrate for Sirt3 (Fig. 8) (Hallows et al. 2006; Schwer et al. 2006). Along with the cytoplasmic AceCS1, AceCSs produce acetyl-CoA from acetate, CoA, and ATP. Acetyl-CoA generated by AceCS1 in the cytoplasm is used in fatty acid synthesis, whereas acetyl-CoA generated by AceCS2 goes into the TCA cycle, resulting in an increase in CO_2 output in mitochondria (Hallows et al. 2006). Both AceCSs are controlled by reversible acetylation, and AceCS1 and AceCS2 are deacetylated and activated by Sirt1 and Sirt3, respectively (Hallows et al. 2006; Schwer et al. 2006). Interestingly, this regulatory mechanism of AceCS activity is evolutionarily conserved because the *Salmonella enterica* Sir2-like protein CobB also deacetylates and activates AceCS in this organism (Starai et al. 2002). These findings suggest that Sirt3 may have an important role in promoting mitochondrial respiration by utilizing acetate as an alternative energy source, particularly under nutrient-deprived conditions.

Analysis of Sirt3 alleles in selected human populations raises the possibility that Sirt3 is involved in the regulation of aging and longevity. Two independent polymorphisms in the Sirt3 gene have been associated with longevity in human population (Rose et al. 2003; Bellizzi et al. 2005), and it will be important to extend these initial findings.

Sirt4

Sirt4 is a mitochondrial sirtuin that possesses robust mono-ADP-ribosyltransferase activity (Haigis et al. 2006). Sirt4 is expressed ubiquitously, with highest levels in the kidney, heart, brain, and liver. Sirt4 is also expressed specifically in pancreatic β cells. Haigis et al. (2006) have demonstrated that Sirt4 ADP-ribosylates and inhibits glutamate dehydrogenase (GDH), which converts glutamate into α-ketoglutarate in mitochondria (see Fig. 8). It has been known that GDH has a significant role in the regulation of amino-acid-stimulated insulin secretion (AASIS) by gating the flow of amino-acids into central metabolism in pancreatic β cells. Sirt4-deficient mice show reduced ADP-ribosylation of GDH and increased activity. Consistent with the alteration in GDH activity, Sirt4-deficient mice and islets have increased AASIS. Interestingly, CR increases GDH activity in pancreatic β cells and liver, and islets isolated from CR mice show a striking increase in AASIS compared to those from ad libitum-fed mice. Taken together, these findings suggest that Sirt4 activity is down-regulated by CR, resulting in a switch to AASIS in β cells under CR (Fig. 8). A similar mechanism applies to liver (Haigis et al. 2006), where GDH activation promotes amino acid use for gluconeogenesis during CR.

It is remarkable that both Sirt1 and Sirt4 regulate insulin secretion in response to different nutritional inputs, namely, glucose and amino acids, respectively, but in opposing directions. Calorically restricted mice have reduced basal insulin levels over most of the day but show postprandial hypersecretion of insulin compared to ad libitum-fed mice (Richardson et al. 2004). Down-regulation of Sirt4, which might be caused by a possible decrease in mitochondrial NAD levels in β cells under CR, may have an important role in making β cells more sensitive to a surge of nutritional inputs. One interesting idea is that NAD biosynthesis might be shifted from mitochondria to nucleus/cytoplasm, resulting in enhanced Sirt1 activity in the nucleus and reduced Sirt4 activity in mitochondria. Further analysis of energy dynamics during CR, particularly the regulation of NAD biosynthesis, will provide more insight into any functional cooperativity between Sirt1 and Sirt4 in response to CR.

Sirt5

Very little is known about the biological function of Sirt5, which is also localized to the mitochondria and is expressed in a wide variety of tissues, with highest expression in the thymus, heart, and lymphoblasts (Michishita et al. 2005; Mahlknecht et al. 2006). Sirt5 has weak NAD-dependent deacetylase activity and does not show detectable ADP-ribosyltransferase activity (North et al. 2003; Haigis et al. 2006).

Sirt6

Sirt6 is a nuclear, chromatin-associated sirtuin and is expressed ubiquitously, with highest levels in muscle, brain, and thymus (Liszt et al. 2005; Michishita et al. 2005; Mostoslavsky et al. 2006). Although Sirt6 lacks any NAD-dependent deacetylase activity, it possesses robust mono-ADP-ribosyltransferase activity (see Fig. 8) (North et al. 2003; Liszt et al. 2005). Recently, Mostoslavsky et al. (2006) have generated Sirt6-deficient mice and analyzed their phenotypes. Sirt6-deficient mice show acute, severe degenerative phenotypes, including lymphopenia, loss of subcutaneous fat, lordokyphosis, and severe metabolic defects. SIRT6-deficient mouse embryonic fibroblasts (MEFs) are hypersensitive to irradiation-induced DNA damage and possess numerous chromosomal abnormalities, thus revealing a role for SIRT6 in DNA-damage repair and maintenance of genomic stability. These phenotypes appear to be due to a specific defect in base excision repair (BER). Although substrates for Sirt6 are still unknown, it is possible that Sirt6 ADP-ribosylates a component in BER or chromatin and facilitates repair (Fig. 8). Interestingly, Sirt6-deficient mice have reduced insulin-like growth factor-1 (IGF-1) levels and suffer from severe hypoglycemia (Mostoslavsky et al. 2006). It will be important to determine whether the DNA repair and metabolic roles of Sirt6 are independent functions or are mechanistically linked.

Sirt7

Sirt7 lacks any enzymatic activity in vitro (North et al. 2003). Sirt7 is a nucleolar sirtuin that is expressed in metabolically active tissues, such as liver, spleen, and testes, but not muscle, heart, or brain. Sirt7 is associated with transcriptionally active rRNA genes and interacts with both histones and RNA polymerase I (pol I) (see Fig. 8) (Ford et al. 2006). Functional studies of overexpressing and knocking down Sirt7 demonstrate that this sirtuin is a positive regulator of pol I transcription. It will be important to determine whether Sirt7 links diet to ribosome synthesis and cell growth in mitotically active tissues in vivo.

PHARMACOLOGICAL INTERVENTIONS TARGETING MAMMALIAN SIRTUINS

Because genetic and cellular studies suggest that Sirt1 regulates physiological responses to CR, there has been an increasing enthusiasm to find chemical activators for Sirt1 as CR mimetics. Sinclair and his colleagues have reported the first study of this kind (Howitz et al. 2003) by identifying a class of plant polyphenolic compounds, including quercetin, piceatannol, and resveratrol, as activators for yeast Sir2 and mammalian Sirt1. Resveratrol, found largely in red grapes and wine, is associated with numerous health benefits, including cardioprotection, neuroprotection, and cancer suppression (Jang et al. 1997; Middleton et al. 2000; Baur and Sinclair 2006). Resveratrol has been shown to increase yeast life span in a Sir2-dependent manner and does not further extend the life span of calorically restricted yeast (Howitz et al. 2003). Similar results have since been obtained in resveratrol-treated worms and flies (Wood et al. 2004). On the basis of these results, resveratrol has also been linked to roles of Sirt1 in fat mobilization, axonal protection, polyglutamine cytotoxicity, and inhibition of NF-κB (Araki et al. 2004; Picard et al. 2004; Yeung et al. 2004; Parker et al. 2005). However, the specificity of resveratrol for Sirt1 activity has been called into question, because Sirt1 activation by resveratrol in vitro is obtained only in the presence of a specific fluorescent acetyl substrate (Borra et al. 2005; Kaeberlein et al. 2005b). Furthermore, Kaeberlein et al. (2005b) have reported that resveratrol has no effect on life span in three different yeast strain backgrounds. Viswanathan et al. (2005) have shown that resveratrol extends *C. elegans* life span by inhibiting, instead of enhancing, *sir-2.1*-mediated repression of ER stress genes. Finally, resveratrol also appears to inhibit insulin secretion in isolated rat pancreatic islets, which is opposite to the action of Sirt1 in pancreatic β cells (Szkudelski 2006).

Strikingly, two recent studies have reported remarkable physiological actions of resveratrol in mice possibly through the activation of Sirt1 (Baur et al. 2006; Lagouge et al. 2006). Sinclair, de Cabo, and their colleagues have demonstrated that resveratrol (22 mg/kg/day) shifts the physiology of high-fat diet-fed mice toward that of regular diet-fed mice and increases their life expectancy, although it is not clear whether the effects of resveratrol are primarily through Sirt1 (Baur et al. 2006). Auwerx and his colleagues have also demonstrated that resveratrol (400 mg/kg/day) significantly increases the aerobic capacity and the endurance of high-fat diet-fed or control mice in a running exercise by promoting oxidative phosphorylation and mitochondrial biogenesis in skeletal muscle (Lagouge et al. 2006). These effects appear to be explained by a resveratrol-mediated decrease in PGC-1α acetylation and an increase in its transcriptional activity, and these effects were not observed in Sirt1-deficient MEFs. These studies therefore suggest that Sirt1 activators may partly mimic the beneficial effects of CR in mammals.

Another means by which to enhance sirtuin activity is to increase NAD biosynthesis. NAD biosynthesis is mediated by either the de novo pathway from tryptophan or pathways using nicotinic acid and nicotinamide, two different forms of vitamin B_3. In yeast, increasing the dosage of Npt1, the nicotinic acid phosphoribosyltransferase necessary for NAD biosynthesis from nicotinic acid, or Pnc1, the nicotinamidase that converts nicotinic acid to nicotinamide, enhances the silencing activity of Sir2 and extends life span (Anderson et al. 2002, 2003). In vertebrates, NAD biosynthesis predominantly uses nicotinamide rather than nicotinic acid as a precursor for NAD. Increasing the dosage of the nicotinamide phosphoribosyltransferase Nampt, the rate-limiting component in mammalian NAD biosynthesis, enhances Sirt1 activity in mammalian cells (Revollo et al. 2004). Therefore, the manipulation of NAD biosynthesis pathways might provide another pharmacological strategy to convey beneficial effects mediated by Sirt1 and perhaps other sirtuins.

CONCLUSION

For the past several years, sirtuin biology has evolved quite rapidly into one of the major fields in aging research. It is now clear that sirtuins unequivocally link NAD, an important energy currency in living organisms, to the regulation of metabolism and stress resistance in response to nutritional and environmental stimuli. In lower model organisms, this function of sirtuins is critical in controlling longevity and mediating the effects of moderate CR. In mammals, increasing lines of evidence indicate that mammalian sirtuins regulate diverse biological processes that are critical in aging and are also regulated by CR. Although further investigation will be necessary to test whether sirtuins are also involved in longevity in mammals, these studies already provide a novel molecular framework to develop therapeutic/preventive interventions for age-associated diseases, such as obesity, diabetes, and neurodegenerative diseases.

In the next few years, research will focus on detailed analyses of all seven mammalian sirtuins and their possible connections to aging, longevity, and CR. There are several important questions: Do Sirt1 and other sirtuins have critical roles in the regulation of aging and longevity in mammals? If so, in which organ do they have major roles? How do Sirt1 and other sirtuins orchestrate physiological responses to CR? Are they also involved in the pathophysiology of age-associated diseases? Can we develop therapeutic interventions that target sirtuins? It may not be so long before we know whether sirtuin activators can act as real CR mimetics in mammals. Thus, the discovery of the NAD-dependent deacetylase activity of Sir2 proteins has provided a paradigm that links NAD, metabolism, aging, and disease.

ACKNOWLEDGMENTS

We apologize to those whose work is not cited due to space limitations. We thank Kathryn Moynihan and other members of the Imai lab and the Guarente lab for critical discussions and suggestions. S.I. is supported by grants from the National Institute on Aging, Ellison Medical Foundation, American Diabetes Association, Juvenile Diabetes Research Foundation, and the Washington University Clinical Nutrition Research Unit. L.G. is supported by grants from the National Institutes of Health and the Glenn Foundation.

REFERENCES

Afshar G. and Murnane J.P. 1999. Characterization of a human gene with sequence homology to *Saccharomyces cerevisiae SIR2*. *Gene* **234:** 161–168.

Al-Regaiey K.A., Masternak M.M., Bonkowski M., Sun L., and Bartke A. 2005. Long-lived growth hormone receptor knockout mice: Interaction of reduced insulin-like growth factor i/insulin signaling and caloric restriction. *Endocrinology* **146:** 851–860.

Anderson R.M., Bitterman K.J., Wood J.G., Medvedik O., and Sinclair D.A. 2003. Nicotinamide and PNC1 govern lifespan extension by calorie restriction in *Saccharomyces cerevisiae. Nature* **423:** 181–185.

Anderson R., Bitterman K., Wood J., Medvedik O., Cohen H., Lin S., Manchester J., Gordon J., and Sinclair D. 2002. Manipulation of a nuclear NAD+ salvage pathway delays aging without altering steady-state NAD+ levels. *J. Biol. Chem.* **277:** 18881–18890.

Araki T., Sasaki Y., and Milbrandt J. 2004. Increased nuclear NAD biosynthesis and SIRT1 activation prevent axonal degeneration. *Science* **305:** 1010–1013.

Astrom S.U., Cline T.W., and Rine J. 2003. The *Drosophila melanogaster sir2$^+$* gene is nonessential and has only minor effects on position-effect variegation. *Genetics* **163:** 931–937.

Avalos J.L., Bever K.M., and Wolberger C. 2005. Mechanism of sirtuin inhibition by nicotinamide: Altering the NAD(+) cosubstrate specificity of a Sir2 enzyme. *Mol. Cell* **17:** 855–868.

Avalos J.L., Celic I., Muhammad S., Cosgrove M.S., Boeke J.D., and Wolberger C. 2002. Structure of a Sir2 enzyme bound to an acetylated p53 peptide. *Mol. Cell* **10:** 523–535.

Baur J.A. and Sinclair D.A. 2006. Therapeutic potential of resveratrol: The in vivo evidence. *Nat. Rev. Drug Discov.* **5:** 493–506.

Baur J.A., Pearson K.J., Price N.L., Jamieson H.A., Lerin C., Kalra A., Prabhu V.V., Allard J.S., Lopez-Lluch G., Lewis K., et al. 2006. Resveratrol improves health and survival of mice on a high-calorie diet. *Nature* **444:** 337–342.

Bellizzi D., Rose G., Cavalcante P., Covello G., Dato S., De Rango F., Greco V., Maggiolini M., Feraco E., Mari V., et al. 2005. A novel VNTR enhancer within the SIRT3 gene, a human homologue of SIR2, is associated with survival at oldest ages. *Genomics* **85:** 258–263.

Berdichevsky A., Viswanathan M., Horvitz H.R., and Guarente L. 2006. C. elegans SIR-2.1 interacts with 14-3-3 proteins to activate DAF-16 and extend life span. *Cell* **125:** 1165–1177.

Bernofsky C. 1980. Physiology aspects of pyridine nucleotide regulation in mammals. *Mol. Cell. Biochem.* **33:** 135–143.

Bishop N.A. and Guarente L. 2007. Two neurons mediate diet-restriction-induced longevity in C. elegans. *Nature* **447:** 545–549.

Bitterman K.J., Anderson R.M., Cohen H.Y., Latorre-Esteves M., and Sinclair D.A. 2002. Inhibition of silencing and accelerated aging by nicotinamide, a putative negative regulator of yeast sir2 and human SIRT1. *J. Biol. Chem.* **277:** 45099–45107.

Blander G. and Guarente L. 2004. The Sir2 family of protein deacetylases. *Annu. Rev. Biochem.* **73:** 417–435.

Bordone L. and Guarente L. 2005. Calorie restriction, SIRT1 and metabolism: Understanding longevity. *Nat. Rev. Mol. Cell Biol.* **6:** 298–305.

Bordone L., Motta M.C., Picard F., Robinson A., Jhala U.S., Apfeld J., McDonagh T., Lemieux M., McBurney M., Szilvasi A., et al. 2006. Sirt1 regulates insulin secretion by repressing UCP2 in pancreatic beta cells. *PLoS Biol.* **4:** e31.

Borra M.T., Smith B.C., and Denu J.M. 2005. Mechanism of human SIRT1 activation by resveratrol. *J. Biol. Chem.* **280:** 17187–17195.

Brachmann C.B., Sherman J.M., Devine S.E., Cameron E.E., Pillus L., and Boeke J.D. 1995. The *SIR2* gene family, conserved from bacteria to humans, functions in silencing, cell cycle progression, and chromosome stability. *Genes Dev.* **9:** 2888–2902.

Braunstein M., Rose A.B., Holmes S.G., Allis C.D., and Broach J.R. 1993. Transcriptional silencing in yeast is associated with reduced nucleosome acetylation. *Genes Dev.* **7:** 592–604.

Braunstein M., Sobel R.E., Allis C.D., Turner B.M., and Broach J.R. 1996. Efficient transcriptional silencing in *Saccharomyces cerevisiae* requires a heterochromatin histone acetylation pattern. *Mol. Cell. Biol.* **16:** 4349–4356.

Brunet A., Sweeney L.B., Sturgill J.F., Chua K.F., Greer P.L., Lin Y., Tran H., Ross S.E., Mostoslavsky R., Cohen H.Y., et al. 2004. Stress-dependent regulation of FOXO transcription factors by the SIRT1 deacetylase. *Science* **303:** 2011–2015.

Bryk M., Banerjee M., Murphy M., Knudsen K.E., Garfinkel D.J., and Curcio M.J. 1997. Transcriptional silencing of Ty1 elements in the *RDN1* locus of yeast. *Genes Dev.* **11:** 255–269.

Catz P., Shinn W., Kapetanovic I.M., Kim H., Kim M., Jacobson E.L., Jacobson M.K., and Green C.E. 2005. Simultaneous determination of myristyl nicotinate, nicotinic acid, and nicotinamide in rabbit plasma by liquid chromatography-tandem mass spectrometry using methyl ethyl ketone as a deproteinization solvent. *J. Chromatogr. B Analyt. Technol. Biomed. Life Sci.* **829:** 123–135.

Chen D., Steele A.D., Lindquist S., and Guarente L. 2005. Increase in activity during calorie restriction requires Sirt1. *Science* **310:** 1641.

Chen J., Zhou Y., Mueller-Steiner S., Chen L.F., Kwon H., Yi S., Mucke L., and Gan L. 2005. SIRT1 protects against microglia-dependent amyloid-beta toxicity through inhibiting NF-kappaB signaling. *J. Biol. Chem.* **280:** 40364–40374.

Chen W.Y., Wang D.H., Yen R.C., Luo J., Gu W., and Baylin S.B. 2005. Tumor suppressor HIC1 directly regulates SIRT1 to modulate p53-dependent DNA-damage responses. *Cell* **123:** 437–448.

Cheng H.L., Mostoslavsky R., Saito S., Manis J.P., Gu Y., Patel P., Bronson R., Appella E., Alt F. W., and Chua K.F. 2003. Developmental defects and p53 hyperacetylation in Sir2 homolog (SIRT1)-deficient mice. *Proc. Natl. Acad. Sci.* **100:** 10794–10799.

Clancy D.J., Gems D., Hafen E., Leevers S.J., and Partridge L. 2002. Dietary restriction in long-lived dwarf flies. *Science* **296:** 319.

Cohen H.Y., Miller C., Bitterman K.J., Wall N.R., Hekking B., Kessler B., Howitz K.T., Gorospe M., De Cabo R., and Sinclair D.A. 2004. Calorie restriction promotes mammalian cell survival by inducing the SIRT1 deacetylase. *Science* **305:** 390–392.

Daitoku H., Hatta M., Matsuzaki H., Aratani S., Ohshima T., Miyagishi M., Nakajima T., and Fukamizu A. 2004. Silent information regulator 2 potentiates Foxo1-mediated transcription through its deacetylase activity. *Proc. Natl. Acad. Sci.* **101:** 10042–10047.

Dryden S.C., Nahhas F.A., Nowak J.E., Goustin A.S., and Tainsky M.A. 2003. Role for human SIRT2 NAD-dependent deacetylase activity in control of mitotic exit in the cell cycle. *Mol. Cell. Biol.* **23:** 3173–3185.

Fabrizio P., Gattazzo C., Battistella L., Wei M., Cheng C., McGrew K., and Longo V.D. 2005. Sir2 blocks extreme life-span extension. *Cell* **123:** 655–667.

Finnin M.S., Donigian J.R., and Pavletich N.P. 2001. Structure of the histone deacetylase SIRT2. *Nat. Struct. Biol.* **8:** 621–625.

Ford E., Voit R., Liszt G., Magin C., Grummt I., and Guarente L. 2006. Mammalian Sir2 homolog SIRT7 is an activator of RNA polymerase I transcription. *Genes Dev.* **20:** 1075–1080.

Frye R.A. 1999. Characterization of five human cDNAs with homology to yeast SIR2 gene: Sir2-like proteins (Sirtuins) metabolize NAD and may have protein ADP-ribosyltransferase activity. *Biochem. Biophys. Res. Commun.* **260:** 273–279.

———. 2000. Phylogenetic classification of prokaryotic and eukaryotic Sir2-like proteins. *Biochem. Biophys. Res. Commun.* **273:** 793–798.

Fulco M., Schiltz R.L., Iezzi S., King M.T., Zhao P., Kashiwaya Y., Hoffman E., Veech R.L., and Sartorelli V. 2003. Sir2 regulates skeletal muscle differentiation as a potential sensor of the redox state. *Mol. Cell* **12:** 51–62.

Furuyama T., Banerjee R., Breen T.R., and Harte P.J. 2004. SIR2 is required for polycomb silencing and is associated with an E(Z) histone methyltransferase complex. *Curr. Biol.* **14:** 1812–1821.

Gottlieb S. and Esposito R.E. 1989. A new role for a yeast transcriptional silencer gene, *SIR2,* in regulation of recombination in ribosomal DNA. *Cell* **56:** 771–776.

Gottschling D.E., Aparicio O.M., Billington B.L., and Zakian V.A. 1990. Position effect at *S. cerevisiae* telomeres: Reversible repression of Pol II transcription. *Cell* **63:** 751–762.

Greer E.L. and Brunet A. 2005. FOXO transcription factors at the interface between longevity and tumor suppression. *Oncogene* **24:** 7410–7425.

Guarente L. 1999. Diverse and dynamic functions of the Sir silencing complex. *Nat. Genet.* **23:** 281–285.

Haigis M.C. and Guarente L.P. 2006. Mammalian sirtuins—Emerging roles in physiology, aging, and calorie restriction. *Genes Dev.* **20:** 2913–2921.

Haigis M.C., Mostoslavsky R., Haigis K.M., Fahie K., Christodoulou D.C., Murphy A.J., Valenzuela D.M., Yancopoulos G.D., Karow M., Blander G., et al. 2006. SIRT4 inhibits glutamate dehydrogenase and opposes the effects of calorie restriction in pancreatic beta cells. *Cell* **126:** 941–954.

Hallows W.C., Lee S., and Denu J.M. 2006. Sirtuins deacetylate and activate mammalian acetyl-CoA synthetases. *Proc. Natl. Acad. Sci.* **103:** 10230–10235.

Hecht A., Laroche T., Strahl-Bolsinger S., Gasser S.M., and Grunstein M. 1995. Histone H3 and H4 N-termini interact with SIR3 and SIR4 proteins: A molecular model for the formation of heterochromatin in yeast. *Cell* **80:** 583–592.

Hiratsuka M., Inoue T., Toda T., Kimura N., Shirayoshi Y., Kamitani H., Watanabe T., Ohama E., Tahimic C.G., Kurimasa A., and Oshimura M. 2003. Proteomics-based identification of differentially expressed genes in human gliomas: Down-regulation of SIRT2 gene. *Biochem. Biophys. Res. Commun.* **309:** 558–566.

Hoppe G.J., Tanny J.C., Rudner A.D., Gerber S.A., Danaie S., Gygi S.P., and Moazed D. 2002. Steps in assembly of silent chromatin in yeast: Sir3-independent binding of a Sir2/Sir4 complex to silencers and role for Sir2-dependent deacetylation. *Mol. Cell. Biol.* **22:** 4167–4180.

Houthoofd K., Braeckman B.P., Lenaerts I., Brys K., De Vreese A., Van Eygen S., and Vanfleteren J.R. 2002. Axenic growth up-regulates mass-specific metabolic rate, stress resistance, and extends life span in *Caenorhabditis elegans*. *Exp. Gerontol.* **37:** 1371–1378.

Howitz K.T., Bitterman K.J., Cohen H.Y., Lamming D.W., Lavu S., Wood J.G., Zipkin R.E., Chung P., Kisielewski A., Zhang L.L., et al. 2003. Small molecule activators of sirtuins extend *Saccharomyces cerevisiae* lifespan. *Nature* **425:** 191–196.

Imai S., Armstrong C.M., Kaeberlein M., and Guarente L. 2000a. Transcriptional silencing and longevity protein Sir2 is an NAD-dependent histone deacetylase. *Nature* **403:** 795–800.

Imai S., Johnson F.B., Marciniak R.A., McVey M., Park P.U., and Guarente L. 2000b. Sir2: An NAD-dependent histone deacetylase that connects chromatin silencing, metabolism, and aging. *Cold Spring Harbor Symp. Quant. Biol.* **65:** 297–302.

Jackson M.D. and Denu J.M. 2002. Structural identification of 2'- and 3'-O-acetyl-ADP-ribose as novel metabolites derived from the Sir2 family of β-NAD$^+$-dependent histone/protein deacetylases. *J. Biol. Chem.* **277:** 18535–18544.

Jacobson E.L., Dame A.J., Pyrek J.S., and Jacobson M.K. 1995. Evaluating the role of niacin in human carcinogenesis. *Biochimie* **77:** 394–398.

Jang M., Cai L., Udeani G.O., Slowing K.V., Thomas C.F., Beecher C.W., Fong H.H., Farnsworth N.R., Kinghorn A.D., Mehta R.G., et al. 1997. Cancer chemopreventive activity of resveratrol, a natural product derived from grapes. *Science* **275:** 218–220.

Kaeberlein M., McVey M., and Guarente L. 1999. The *SIR2/3/4* complex and *SIR2* alone promote longevity in *Saccharomyces cerevisiae* by two different mechanisms. *Genes Dev.* **13:** 2570–2580.

Kaeberlein M., Kirkland K.T., Fields S., and Kennedy B.K. 2004. Sir2-independent life span extension by calorie restriction in yeast. *PLoS Biol.* **2:** E296.

Kaeberlein T.L., Smith E.D., Tsuchiya M., Welton K.L., Thomas J.H., Fields S., Kennedy B.K., and Kaeberlein M. 2006. Lifespan extension in *Caenorhabditis elegans* by complete removal of food. *Aging Cell* **5:** 487–494.

Kaeberlein M., Powers R.W., III, Steffen K.K., Westman E.A., Hu D., Dang N., Kerr E.O., Kirkland K.T., Fields S., and Kennedy B.K. 2005a. Regulation of yeast replicative life span by TOR and Sch9 in response to nutrients. *Science* **310:** 1193–1196.

Kaeberlein M., McDonagh T., Heltweg B., Hixon J., Westman E.A., Caldwell S., Napper A., Curtis R., DiStefano P.S., Fields S., et al. 2005b. Substrate-specific activation of sirtuins by resveratrol. *J. Biol. Chem.* **280:** 17038–17045.

Kennedy B.K., Austriaco N.R., Zhang J., and Guarente L. 1995. Mutation in the silencing gene *SIR4* can delay aging in *S. cerevisiae. Cell* **80:** 485–496.

Kennedy B.K., Gotta M., Sinclair D.A., Mills K., McNabb D.S., Murthy M., Pak S.M., Laroche T., Gasser S.M., and Guarente L. 1997. Redistribution of silencing proteins from telomeres to the nucleolus is associated with extension of life span in *S. cerevisiae. Cell* **89:** 381–391.

Kitamura Y.I., Kitamura T., Kruse J.P., Raum J.C., Stein R., Gu W., and Accili D. 2005. FoxO1 protects against pancreatic beta cell failure through NeuroD and MafA induction. *Cell Metab.* **2:** 153–163.

Klar A.J.S., Fogel S., and MacLeod K. 1979. *MAR1*—A regulator of the *HMa* and *HMα* loci in *Saccharomyces cerevisiae. Genetics* **93:** 37–50.

Klass M.R. 1977. Aging in the nematode *Caenorhabditis elegans:* Major biological and environmental factors influencing life span. *Mech. Ageing Dev.* **6:** 413–429.

Lagouge M., Argmann C., Gerhart-Hines Z., Meziane H., Lerin C., Daussin F., Messadeq N., Milne J., Lambert P., Elliott P., et al. 2006. Resveratrol improves mitochondrial function and protects against metabolic disease by activating SIRT1 and PGC-1alpha. *Cell* **127:** 1109–1122.

Lakowski B. and Hekimi S. 1998. The genetics of caloric restriction in *Caenorhabditis elegans. Proc. Natl. Acad. Sci.* **95:** 13091–13096.

Lamming D.W., Latorre-Esteves M., Medvedik O., Wong S.N., Tsang F.A., Wang C., Lin S.J., and Sinclair D.A. 2005. HST2 mediates SIR2-independent life-span extension by calorie restriction. *Science* **309:** 1861–1864.

Landry J., Slama J.T., and Sternglanz R. 2000a. Role of NAD$^+$ in the deacetylase activity of the SIR2-like proteins. *Biochem. Biophys. Res. Commun.* **278:** 685–690.

Landry J., Sutton A., Tafrov S.T., Heller R.C., Stebbins J., Pillus L., and Sternglanz R. 2000b. The silencing protein SIR2 and its homologs are NAD-dependent protein deacetylases. *Proc. Natl. Acad. Sci.* **97:** 5807–5811.

Lin S.-J., Defossez P.-A., and Guarente L. 2000. Life span extension by calorie restriction in *S. cerevisiae* requires NAD and *SIR2. Science* **289:** 2126–2128.

Lin S.-J., Ford E., Haigis M., Liszt G., and Guarente L. 2004. Calorie restriction extends yeast life span by lowering the level of NADH. *Genes Dev.* **18:** 12–16.

Lin S.-J., Kaeberlein M., Andalis A.A., Sturtz L.A., Defossez P.-A., Culotta V.C., Fink G.R., and Guarente L. 2002. Calorie restriction extends *Saccharomyces cerevisiae* lifespan by increasing respiration. *Nature* **418:** 344–348.

Liou G.-G., Tanny J.C., Kruger R.G., Walz T., and Moazed D. 2005. Assembly of the SIR complex and its regulation by O-acetyl-ADP-ribose, a product of NAD-dependent histone deacetylation. *Cell* **121:** 515–527.

Liszt G., Ford E., Kurtev M., and Guarente L. 2005. Mouse Sir2 homolog SIRT6 is a nuclear ADP-ribosyltransferase. *J. Biol. Chem.* **280:** 21313–21320.

Longo V.D. and Kennedy B.K. 2006. Sirtuins in aging and age-related disease. *Cell* **126:** 257–268.

Luo J., Nikolaev A.Y., Imai S., Chen D., Su F., Shiloh A., Guarente L., and Gu W. 2001. Negative control of p53 by Sir2α promotes cell survival under stress. *Cell* **107:** 137–148.

Mahlknecht U., Ho A.D., Letzel S., and Voelter-Mahlknecht S. 2006. Assignment of the NAD-dependent deacetylase sirtuin 5 gene (SIRT5) to human chromosome band 6p23 by in situ hybridization. *Cytogenet. Genome Res.* **112:** 208–212.

Mattson M.P., Chan S.L., and Duan W. 2002. Modification of brain aging and neurodegenerative disorders by genes, diet, and behavior. *Physiol. Rev.* **82:** 637–672.

McCarter R.J., Shimokawa I., Ikeno Y., Higami Y., Hubbard G.B., Yu B.P., and McMahan C.A. 1997. Physical activity as a factor in the action of dietary restriction on aging: Effects in Fischer 344 rats. *Aging* **9:** 73–79.

Michishita E., Park J.Y., Burneskis J.M., Barrett J.C., and Horikawa I. 2005. Evolutionarily conserved and nonconserved cellular localizations and functions of human SIRT proteins. *Mol. Biol. Cell* **16:** 4623–4635.

Middleton E., Jr., Kandaswami C., and Theoharides T.C. 2000. The effects of plant flavonoids on mammalian cells: Implications for inflammation, heart disease, and cancer. *Pharmacol. Rev.* **52:** 673–751.

Min J., Landry J., Sternglanz R., and Xu R.M. 2001. Crystal structure of a SIR2 homolog-NAD complex. *Cell* **105:** 269–279.

Mortimer R.K. and Johnston J.R. 1959. Life span of individual yeast cells. *Nature* **183:** 1751–1752.

Mostoslavsky R., Chua K.F., Lombard D.B., Pang W.W., Fischer M.R., Gellon L., Liu P., Mostoslavsky G., Franco S., Murphy M.M., et al. 2006. Genomic instability and aging-like phenotype in the absence of mammalian SIRT6. *Cell* **124:** 315–329.

Motta M.C., Divecha N., Lemieux M., Kamel C., Chen D., Gu W., Bultsma Y., McBurney M., and Guarente L. 2004. Mammalian SIRT1 represses forkhead transcription factors. *Cell* **116:** 551–563.

Moynihan K.A. and Imai S. 2006. Sirt1 as a key regulator orchestrating the response to caloric restriction. *Drug Discov. Today: Dis. Mech.* **3:** 11–17.

Moynihan K.A., Grimm A.A., Plueger M.M., Bernal-Mizrachi E., Ford E., Cras-Meneur C., Permutt M.A., and Imai S. 2005. Increased dosage of mammalian Sir2 in pancreatic β cells enhances glucose-stimulated insulin secretion in mice. *Cell Metab.* **2:** 105–117.

Nemoto S., Fergusson M.M., and Finkel T. 2004. Nutrient availability regulates SIRT1 through a forkhead-dependent pathway. *Science* **306:** 2105–2108.

Nisoli E., Tonello C., Cardile A., Cozzi V., Bracale R., Tedesco L., Falcone S., Valerio A., Cantoni O., Clementi E., et al. 2005. Calorie restriction promotes mitochondrial biogenesis by inducing the expression of eNOS. *Science* **310:** 314–317.

North B.J., Marshall B.L., Borra M.T., Denu J.M., and Verdin E. 2003. The human Sir2 ortholog, SIRT2, is an NAD+-dependent tubulin deacetylase. *Mol. Cell* **11:** 437–444.

Oh S.W., Mukhopadhyay A., Svrzikapa N., Jiang F., Davis R.J., and Tissenbaum H.A. 2005. JNK regulates lifespan in *Caenorhabditis elegans* by modulating nuclear translocation of forkhead transcription factor/DAF-16. *Proc. Natl. Acad. Sci.* **102:** 4494–4499.

Onyango P., Celic I., McCaffery J.M., Boeke J.D., and Feinberg A.P. 2002. SIRT3, a human SIR2 homologue, is an NAD-dependent deacetylase localized to mitochondria. *Proc. Natl. Acad. Sci.* **99:** 13653–13658.

Parker J.A., Arango M., Abderrahmane S., Lambert E., Tourette C., Catoire H., and Neri C. 2005. Resveratrol rescues mutant polyglutamine cytotoxicity in nematode and mammalian neurons. *Nat. Genet.* **37:** 349–350.

Parsons X.H., Garcia S.N., Pillus L., and Kadonaga J.T. 2003. Histone deacetylation by Sir2 generates a transcriptionally repressed nucleoprotein complex. *Proc. Natl. Acad. Sci.* **100:** 1609–1614.

Picard F., Kurtev M., Chung N., Topark-Ngarm A., Senawong T., Oliveira R.M., Leid M., McBurney M.W., and Guarente L. 2004. Sirt1 promotes fat mobilization in white adipocytes by repressing PPAR-γ. *Nature* **429:** 771–776.

Qiao L. and Shao J. 2006. SIRT1 regulates adiponectin gene expression through Foxo1-C/enhancer-binding protein alpha transcriptional complex. *J. Biol. Chem.* **281:** 39915–39924.

Qin W., Yang T., Ho L., Zhao Z., Wang J., Chen L., Zhao W., Thiyagarajan M., MacGrogan D., Rodgers J.T., et al. 2006. Neuronal SIRT1 activation as a novel mechanism underlying the prevention of Alzheimer disease amyloid neuropathology by calorie restriction. *J. Biol. Chem.* **281:** 21745–21754.

Revollo J.R., Grimm A.A., and Imai S. 2004. The NAD biosynthesis pathway mediated by nicotinamide phosphoribosyltransferase regulates Sir2 activity in mammalian cells. *J. Biol. Chem.* **279:** 50754–50763.

Richardson A., Liu F., Adamo M.L., Van Remmen H., and Nelson J.F. 2004. The role of insulin and insulin-like growth factor-I in mammalian ageing. *Best Pract. Res. Clin. Endocrinol. Metab.* **18:** 393–406.

Rine J. and Herskowitz I. 1987. Four genes responsible for a position effect on expression from HML and HMR in *Saccharomyces cerevisiae*. *Genetics* **116:** 9–22.

Rodgers J.T., Lerin C., Haas W., Gygi S.P., Spiegelman B.M., and Puigserver P. 2005. Nutrient control of glucose homeostasis through a complex of PGC-1α and SIRT1. *Nature* **434:** 113–118.

Rogina B. and Helfand S.L. 2004. Sir2 mediates longevity in the fly through a pathway related to calorie restriction. *Proc. Natl. Acad. Sci.* **101:** 15998–16003.

Rogina B., Helfand S.L., and Frankel S. 2002. Longevity regulation by *Drosophila* Rpd3 deacetylase and caloric restriction. *Science* **298:** 1745.

Rongvaux A., Shea R.J., Mulks M.H., Gigot D., Urbain J., Leo O., and Andris F. 2002. Pre-B-cell colony-enhancing factor, whose expression is up-regulated in activated lymphocytes, is a nicotinamide phosphoribosyltransferase, a cytosolic enzyme involved in NAD biosynthesis. *Eur. J. Immunol.* **32:** 3225–3234.

Rose G., Dato S., Altomare K., Bellizzi D., Garasto S., Greco V., Passarino G., Feraco E., Mari V., Barbi C., et al. 2003. Variability of the SIRT3 gene, human silent information regulator Sir2 homologue, and survivorship in the elderly. *Exp. Gerontol.* **38:** 1065–1070.

Rosenberg M.I. and Parkhurst S.M. 2002. *Drosophila* Sir2 is required for heterochromatic silencing and by euchromatic Hairy/E(Spl) bHLH repressors in segmentation and sex determination. *Cell* **109:** 447–458.

Sauve A.A., Moir R.D., Schramm V.L., and Willis I.M. 2005. Chemical activation of Sir2-dependent silencing by relief of nicotinamide inhibition. *Mol. Cell* **17:** 595–601.

Sauve A.A., Celic I., Avalos J., Deng H., Boeke J.D., and Schramm V.L. 2001. Chemistry of gene silencing: The mechanism of NAD+-dependent deacetylation reactions. *Biochemistry* **40:** 15456–15463.

Schwer B., Bunkenborg J., Verdin R.O., Andersen J.S., and Verdin E. 2006. Reversible lysine acetylation controls the activity of the mitochondrial enzyme acetyl-CoA synthetase 2. *Proc. Natl. Acad. Sci.* **103:** 10224–10229.

Schwer B., North B.J., Frye R.A., Ott M., and Verdin E. 2002. The human silent information regulator (Sir)2 homologue hSIRT3 is a mitochondrial nicotinamide adenine dinucleotide-dependent deacetylase. *J. Cell Biol.* **158:** 647–657.

Shi T., Wang F., Stieren E., and Tong Q. 2005. SIRT3, a mitochondrial sirtuin deacetylase, regulates mitochondrial function and thermogenesis in brown adipocytes. *J. Biol. Chem.* **280:** 13560–13567.

Sinclair D.A. and Guarente L. 1997. Extrachromosomal rDNA circles—A cause of aging in yeast. *Cell* **91:** 1033–1042.

Sinclair D., Mills K., and Guarente L. 1998. Aging in *Saccharomyces cerevisiae. Annu. Rev. Microbiol.* **52:** 533–560.

Smith J.S. and Boeke J.D. 1997. An unusual form of transcriptional silencing in yeast ribosomal DNA. *Genes Dev.* **11:** 241–254.

Smith J.S., Brachmann C.B., Celic I., Kenna M.A., Muhammad S., Starai V.J., Avalos J.L., Escalante-Semerena J.C., Grubmeyer C., Wolberger C., and Boeke J.D. 2000. A phylogenetically conserved NAD$^+$-dependent protein deacetylase activity in the Sir2 protein family. *Proc. Natl. Acad. Sci.* **97:** 6658–6663.

Stanelle J. and Pützer B.M. 2006. E2F1-induced apoptosis: Turning killers into therapeutics. *Trends Mol. Med.* **12:** 177–185.

Starai V.J., Celic I., Cole R.N., Boeke J.D., and Escalante-Semerena J.C. 2002. Sir2-dependent activation of acetyl-CoA synthetase by deacetylation of active lysine. *Science* **298:** 2390–2392.

Straight A.F., Shou W., Dowd G.J., Turck C.W., Deshaies R.J., Johnson A.D., and Moazed D. 1999. Net1, a Sir2-associated nucleolar protein required for rDNA silencing and nucleolar integrity. *Cell* **97:** 245–256.

Szkudelski T. 2006. Resveratrol inhibits insulin secretion from rat pancreatic islets. *Eur. J. Pharmacol.* **552:** 176–181.

Tanner K.G., Landry J., Sternglanz R., and Denu J.M. 2000. Silent information regulator 2 family of NAD- dependent histone/protein deacetylases generates a unique product, 1-O-acetyl-ADP-ribose. *Proc. Natl. Acad. Sci.* **97:** 14178–14182.

Tanny J.C. and Moazed D. 2001. Coupling of histone deacetylation to NAD breakdown by the yeast silencing protein Sir2: Evidence for acetyl transfer from substrate to an NAD breakdown product. *Proc. Natl. Acad. Sci.* **98:** 415–420.

Tanny J.C., Dowd G.J., Huang J., Hilz H., and Moazed D. 1999. An enzymatic activity in the yeast Sir2 protein that is essential for gene silencing. *Cell* **99:** 735–745.

Thompson J.S., Ling X., and Grunstein M. 1994. Histone H3 amino terminus is required for telomeric and silent mating locus repression in yeast. *Nature* **369:** 245–247.

Tissenbaum H.A. and Guarente L. 2001. Increased dosage of a *sir-2* gene extends lifespan in *Caenorhabditis elegans. Nature* **410:** 227–230.

Trzebiatowski J.R. and Escalante-Semerena J.C. 1997. Purification and characterization of CobT, the nicotinate-mononucleotide:5,6-dimethylbenzimidazole phosphoribosyltransferase enzyme from *Salmonella typhimurium* LT2. *J. Biol. Chem.* **272:** 17662–17667.

Tsang A.W. and Escalante-Semerena J.C. 1998. CobB, a new member of the SIR2 family of eucaryotic regulatory proteins, is required to compensate for the lack of nicotinate mononucleotide:5,6-dimethylbenzimidazole phosphoribosyltransferase activity in cobT mutants during cobalamin biosynthesis in *Salmonella typhimurium* LT2. *J. Biol. Chem.* **273:** 31788–31794.

van der Horst A., Tertoolen L.G., de Vries-Smits L.M., Frye R.A., Medema R.H., and Burgering B.M. 2004. FOXO4 is acetylated upon peroxide stress and deacetylated by the longevity protein hSir2(SIRT1). *J. Biol. Chem.* **279:** 28873–28879.

Vaquero A., Scher M.B., Lee D.H., Sutton A., Cheng H.L., Alt F.W., Serrano L., Sternglanz R., and Reinberg D. 2006. SirT2 is a histone deacetylase with preference for histone H4 Lys 16 during mitosis. *Genes Dev.* **20:** 1256–1261.

Vaziri H., Dessain S.K., Eaton E.N., Imai S., Frye R.A., Pandita T.K., Guarente L., and Weinberg R.A. 2001. hSIR2[SIRT1] functions as an NAD-dependent p53 deacetylase. *Cell* **107:** 149–159.

Viswanathan M., Kim S.K., Berdichevsky A., and Guarente L. 2005. A role for SIR-2.1 regulation of ER stress response genes in determining *C. elegans* life span. *Dev. Cell* **9:** 605–615.

Wang C., Chen L., Hou X., Li Z., Kabra N., Ma Y., Nemoto S., Finkel T., Gu W., Cress W.D., and Chen J. 2006. Interactions between E2F1 and SirT1 regulate apoptotic response to DNA damage. *Nat. Cell Biol.* **8:** 1025–1031.

Wang J., Zhai Q., Chen Y., Lin E., Gu W., McBurney M.W., and He Z. 2005. A local mechanism mediates NAD-dependent protection of axon degeneration. *J. Cell Biol.* **170:** 349–355.

Wang Y. and Tissenbaum H.A. 2006. Overlapping and distinct functions for a *Caenorhabditis elegans* SIR2 and DAF-16/FOXO. *Mech. Ageing Dev.* **127:** 48–56.

Wang Y., Oh S.W., Deplancke B., Luo J., Walhout A.J., and Tissenbaum H.A. 2006. *C. elegans* 14-3-3 proteins regulate life span and interact with SIR-2.1 and DAF-16/FOXO. *Mech. Ageing Dev.* **127:** 741–747.

Wood J.G., Rogina B., Lavu S., Howitz K., Helfand S.L., Tatar M., and Sinclair D. 2004. Sirtuin activators mimic caloric restriction and delay ageing in metazoans. *Nature* **430:** 686–689.

Yamaza H., Chiba T., Higami Y., and Shimokawa I. 2002. Lifespan extension by caloric restriction: An aspect of energy metabolism. *Microsc. Res. Tech.* **59:** 325–330.

Yeung F., Hoberg J.E., Ramsey C.S., Keller M.D., Jones D.R., Frye R.A., and Mayo M.W. 2004. Modulation of NF-κB-dependent transcription and cell survival by the SIRT1 deacetylase. *EMBO J.* **23:** 2369–2380.

Zhao K., Harshaw R., Chai X., and Marmorstein R. 2004. Structural basis for nicotinamide cleavage and ADP-ribose transfer by NAD(+)-dependent Sir2 histone/protein deacetylases. *Proc. Natl. Acad. Sci.* **101:** 8563–8568.

3

Calorie Restriction in Lower Organisms

Stephen L. Helfand, Johannes H. Bauer, and Jason G. Wood
Department of Molecular Biology, Cell Biology,
and Biochemistry, Division of Biology and Medicine
Brown University, Providence, Rhode Island 02912

CALORIE RESTRICTION OR DIETARY RESTRICTION is the only known intervention to extend life span in a variety of species including yeast, nematodes, flies, spiders, and mammals (Klass 1977; Weindruch and Walford 1988; Austad 1989; Chippindale et al. 1993; Chapman and Partridge 1996; Lin et al. 2000; Masoro 2005; Tatar 2007). The life span extension seen in each of these species has been reported to be as much as 50%. The fact that such an enormously diverse range of species responds to a reduction in caloric intake by extending their healthy life span highlights the importance of understanding this universal biological phenomenon. Therefore, studying model organisms seems likely to uncover the common underlying mechanistic etiologies for calorie restriction as well as to provide insights into the normal process of aging. This chapter summarizes what has been learned about the molecular genetic mechanisms of caloric restriction (CR) or dietary restriction (DR) and discusses issues specifically related to CR and DR in two of the most frequently used model systems in aging research, the round worm nematode, *Caenorhabditis elegans*, and the fruit fly, *Drosophila melanogaster*.

CALORIC RESTRICTION OR DIETARY RESTRICTION?

The term CR is used to indicate a decrease in total calorie intake unrelated to the specific source of the calories (fat, protein, carbohydrate), whereas DR refers to a decrease in specific nutritional components that need not be accompanied by a decrease in calories. In rats and mice, it has been suggested that it is the total amount of calories, or CR, that is important in inducing life span extension, although there remains some

Molecular Biology of Aging ©2008 Cold Spring Harbor Laboratory Press 978-087969824-9

controversy on this matter. In nematodes and flies, the issue of whether a reduction in calories or reduction in specific nutritional components is important is not as well defined, in part due to the major confounding physiological changes on relevant systems, such as reproduction, induced by the alterations in specific nutrients, e.g., reduction of the yeast in food.

GENETIC AND MOLECULAR MECHANISMS UNDERLYING CR/DR LIFE SPAN EXTENSION IN NEMATODES AND FLIES

One of the primary reasons for employing model systems such as nematodes and flies is to make use of their powerful molecular genetic methodologies to confirm the role of candidate genes and identify new genes causally involved in CR/DR life span extension. *C. elegans* has been a very useful model organism for exploring the genetics of aging, and in fact, many of the canonical genetic pathways known to affect life span in higher organisms were first studied in nematodes. However, due to the lack of standardized CR protocols and other problems mentioned below, relatively few genetic studies of CR have taken place in nematodes.

Insulin Signaling

One of the most well-characterized genetic pathways affecting longevity in the nematode is the insulin/insulin-like growth factor-1 (IGF-1) signaling (IIS) pathway (Kenyon 2005). Loss-of-function mutations in members of this pathway, notably *daf-2* (the insulin-like receptor) and *age-1* (a phosphoinositol-3 kinase [PI3K]), reduce signaling through this pathway and cause dramatic increases in life span. IIS in the nematode converges on the forkhead transcription factor DAF-16, causing its phosphorylation and retention in the cytoplasm. When insulin signaling is reduced or eliminated, DAF-16 becomes localized to the nucleus, where it activates transcription of various genes involved in stress and infection resistance, somatic maintenance, and promotion of longevity. The longevity phenotypes of *daf-2* and *age-1* are suppressed by *daf-16* mutations.

All available evidence indicates that, at least in nematodes, CR does not act through the insulin pathway. CR/DR—when performed by bacterial dilution, culture on axenic medium, or the genetic CR mimic *eat-2*—was still able to extend life span in a *daf-16* mutant background in a number of studies (Lakowski and Hekimi 1998; Houthoofd et al. 2003; Kaeberlein et al. 2006; Lee et al. 2006). Furthermore, long-lived *daf-2* mutants experienced a further increase in life span when calorie-restricted by these same methods, again indicating that the two interventions lie in separate pathways.

Indy

Among the first genes speculated to be involved in CR/DR in the fly was the *Indy* (I'm not dead yet) gene. Reduction of *Indy* gene activity leads to life span extension (Rogina et al. 2000). INDY is a transporter of Krebs cycle intermediates primarily localized in the plasma membrane of the fly's fat body cells, oenocytes, and selected portions of the midgut (Knauf et al. 2002). Due to its role as a transporter of Krebs cycle intermediates and expression in tissues associated with uptake, utilization, and storage of nutrients, it was speculated that *Indy* might trigger a change in metabolism similar to CR (Rogina et al. 2000). A homolog of *Indy* in *C. elegans, nac-2*, when reduced using RNA interference (RNAi), also results in significant life span extension (Fei et al. 2004). The NAC-2 protein is found in the intestine of the nematode and affects fat storage, and the induced life span extension has been postulated to be associated with CR/DR.

chico

Although mammalian studies have implicated changes in IIS in CR/DR life span extension, this association is still under intense investigation in both the nematode and fly. The strongest evidence for an association between the life-span-extending effects of IIS and CR/DR in flies is from life span studies of the *chico* mutant on different calorie conditions (Clancy et al. 2002). The *chico* gene is the *Drosophila* ortholog of the insulin-receptor substrate (IRS) and is not found in *C. elegans*. When flies are raised on a range of food with different calorie contents, *chico* flies live longer than control flies on higher-calorie food. However, the chico life span is shorter than control flies on the lower-calorie levels that give maximum life span for the control strain. When median life span is plotted against food concentration, the resulting curve for the *chico* flies is shifted toward higher-calorie food compared to the control flies. These experiments suggest that *chico* and CR may share a similar life-span-extension mechanism. *chico* is so far the only gene of the IIS pathway that has been directly linked to life span extension by CR/DR.

TOR

The IIS and the TOR (target of rapamycin) pathways are both major regulators of cell size and growth. Although they function mostly in parallel, extensive cross-talk exists between these pathways. Activation of IIS leads to activation of protein kinase B/Akt, a component of the TOR and IIS pathways. Akt can phosphorylate and inactivate FoxO transcription factors, downstream effectors of insulin signaling. In addition, Akt can also phosphorylate and inhibit *Tsc2*. *Tsc2* and *Tsc1* act together to inhibit TOR signaling. Interestingly, overexpression of *dTsc1, dTsc2,* or dominant-negative (DN) versions of *dTor* have all been shown to extend *Drosophila* life span (Kapahi et al. 2004). Importantly, when raised on food with decreasing yeast content, flies overexpressing *dTsc* do not show additive effects on life span on this lower-calorie food, suggesting that CR may in part employ the *dTor* signaling pathway to extend life span. Interestingly, it is sufficient to inhibit TOR signaling specifically in the fat body of the fly to achieve life span extension. The fat body is the major metabolic organ with important roles in fat storage, metabolic control, and regulation of immunity, suggesting coordination of these biological processes with longevity regulation. In nematodes, loss-of-function alleles or RNAi of both TOR (*let-363*) and Raptor (*daf-15*) homologs are able to extend life span (Vellai et al. 2003; Jia et al. 2004), as is overexpression of the AMP-activated protein kinase (AMPK) *aak-2*, which inhibits TOR function (Apfeld et al. 2004). However, direct evidence linking this pathway to CR in nematodes is still lacking.

Rpd3

In yeast, a reduction in the Rpd3 histone deacetylase is associated with life span extension (Kim et al. 1999). In flies, CR/DR is associated with a decrease in the levels of *rpd3* and an increase in the levels of the histone deacetylase dSir2 (Rogina et al. 2002). Reduction of *rpd3* gene dosage extends the flies' life span, and no further life span extension is observed when *rpd3* mutant long-lived flies are additionally subjected to CR (Rogina et al. 2002). This suggests that the life span extension associated with a decrease in Rpd3 activity may be related to the CR/DR life-span-extending pathway.

Sir2

An increase in Sir2 expression is associated with life span extension in yeast (Kaeberlein et al. 1999; Lin et al. 2000), *C. elegans* (Tissenbaum and Guarente 2001), and *Drosophila* (Rogina and Helfand 2004). A linkage to CR/DR in flies was suggested when it was shown that dSir2 mRNA is increased in long-lived *rpd3* mutant flies and in CR/DR long-lived flies (Pletcher et al. 2002; Rogina et al. 2002). It has subsequently been shown that the life-span-extending effect of CR/DR is blocked in flies mutant for dSir2 and the life-span-extending effect of dSir2 overexpression is not additive with CR/DR (Rogina and Helfand 2004). These data strongly suggest that dSir2 mediates aspects of CR/DR-dependent life span extension in *Drosophila*.

In *C. elegans*, the role of Sir2 in mediating CR/DR is somewhat unclear. There are four Sir2 homologs in nematodes, and *sir-2.1*, the closest to Sir2, is the only one able to extend life span when present in a single extra copy (Tissenbaum and Guarente 2001). There have been conflicting reports on the interaction of *sir-2.1* with CR/DR, with one study showing that a *sir-2.1* mutant background blocks life span extension by certain *eat-2* alleles (Wang and Tissenbaum 2006), and several others showing that CR still extends life span in this background (Kaeberlein et al. 2006; Lee et al. 2006; Hansen et al. 2007).

p53

A downstream target of Sir2 deacetylase activity in mammals, and presumably also in *Drosophila*, is the tumor suppressor p53. This deacetylation event effectively inhibits the transcriptional activity of p53. Although a decrease in p53 activity would at first glance appear to be problematic, as it would predispose to the generation of cancer, too much active p53 is also disadvantageous. Overexpression of an activated form of p53 in mice leads to a shortened life span and symptoms associated with rapid aging (Tyner et al. 2002; Maier et al. 2004). It was suggested on the basis of these findings that a decrease in p53 activity might lead to a life span extension if the development of cancer could be avoided. Expression of two different dominant-negative versions of *Drosophila* p53 (Dmp53) in the nervous system of the fly does indeed lead to life span extension with no obvious physiological trade-offs (Bauer et al. 2005). Furthermore, the life span of flies on a CR/DR regimen cannot be further extended through expression of dominant-negative Dmp53 (DN-Dmp53), suggesting, as with dSir2, that Dmp53 may mediate aspects of the CR/DR pathway. Finally, it has been shown that the beneficial effects of dSir2 or DN-Dmp53 expression on the fly life span are observed even when their expression is restricted to the adult nervous system. Expression of DN-Dmp53 in other tissues may actually be detrimental to longevity. The finding that dSir2- and Dmp53-induced life span extension is effected through the nervous system and linked to CR/DR life span extension suggests that the nervous system may be a major determinant of CR/DR-mediated longevity.

clk-1

Another class of mutants that can extend the life span of nematodes are the clock genes. The best characterized of these, *clk-1*, encodes a mitochondrial protein necessary for coenzyme Q synthesis. *clk-1* mutants seem to have an overall decreased metabolic rate, with delays in development and behaviors such as pharyngeal pumping, as well as increased life span (Lakowski and Hekimi 1996). CR, as modeled by *eat-2* mutants, was unable to further extend the life span of *clk-1* mutants, suggesting that these two interventions might lie in a common pathway (Lakowski and Hekimi 1998). Clock genes have not yet been tested in flies.

Resveratrol and CR

One of the reasons for constructing a genetic/biochemical pathway for CR/DR life span extension is to allow for the rational development of therapeutic interventions. So far, the best example of this is resveratrol. On the basis of the importance of Sir2 in life span extension of yeast and nematodes, a screen for small molecules that can activate Sir2 in vitro was performed (Howitz et al. 2003). One of the compounds identified, resveratrol, has been shown to extend life span in yeast (Howitz et al. 2003), nematodes, and flies (Wood et al. 2004). In flies and nematodes, Sir2 mutants block the life-span-extending effect of resveratrol, indicating that the life-span-extending effect of resveratrol is Sir2 dependent. Finally, in flies, the life span extension of CR/DR is not further extended by resveratrol, suggesting that resveratrol may be activating part of the CR/DR life-span-extending pathway (Wood et al. 2004).

GENES, TISSUES, CR/DR, AND LIFE SPAN

Several elements of the IIS and TOR pathways, including PTEN, FoxO, and dTsc2, extend fly life span through direct effects on the fat body (Hwangbo et al. 2004; Kapahi et al. 2004). Thus, manipulation of the nutrient-sensing and/or the IIS pathway appears to mediate CR/DR life span extension through effects on the fat body. The CR/DR life-span-extending effects of dSir2 and Dmp53, however, are mediated through changes in the nervous system. This implies that *Drosophila* may have at least two pathways for CR/DR-dependent life span extension: one that functions through metabolic adjustment of the fat body and the other that functions through an effect on the health or physiology of the nervous system. Do these represent two discrete systems independently affecting CR/DR longevity or is there some cross-talk or signaling between the nervous system and the fat body?

Interestingly, in the fly, the nervous system and some elements of the endocrine system are tightly linked. Unlike mammals, where glucose levels are sensed and insulin secreted by the β cells of the pancreas, insulin in the fly is secreted by cells in the brain. The *Drosophila* brain contains two sets of seven median neurosecretory cells, called insulin-producing cells (IPCs), which secrete three different types of insulin (*Drosophila* insulin-like peptides 2, 3, and 5) in response to nutritional changes (Ikeya et al. 2002; Rulifson et al. 2002). Thus, in the fly, the brain can sense nutritional changes and secrete insulin. The presence of insulin-secreting cells in the brain suggests a possible pathway whereby the brain can sense reduced dietary conditions and respond with increased activity of dSir2. The increase in dSir2 activity, possibly through changes in Dmp53 activity, could result in decreased secretion of insulin and suppression of IIS in the fat body, coupled with inhibition of TOR signaling. Additional elements may also be important in this system. For example, although it has not yet been linked to CR/DR, overexpression of JNK in IPCs reduces insulin mRNA, leads to decreased insulin signaling in the fat body, and extends life span (Wang et al. 2005). Such a model is not exclusive and is likely to represent only part of the life-span-extending effects of CR/DR. For example, nutritional changes inducing CR/DR may be sensed in one tissue and the cascade of effectors may be initiated by other tissues such as the gut/fat body (e.g., *Indy*) or nutrient-sensing systems such as the olfactory system (Libert et al. 2007).

Similar tissue-specific effects are seen in the nematode, where a neuroendocrine pathway seems to function to coordinate insulin signaling. Sensory neurons in the brain are important for life span regulation, and they respond to feeding by releasing insulin (DAF-28) into the body cavity (Apfeld and Kenyon 1998; Li et al. 2003). Recent results demonstrate that the sensory ASI neurons are required for life span extension by DR in a *daf-16*-independent fashion. These neuroendocrine cells sense environmental nutrient conditions and presumably respond by secretion of hormones, which might include ILPs (Bishop and Guarente 2007). However, the intestine seems to be the most important organ for sensing and responding to insulin, as intestine-specific expression of DAF-16, the downstream target of the insulin-signaling pathway, is sufficient to restore normal function and life span to a *daf-16* mutant (Libina et al. 2003). This pathway functions to regulate and coordinate insulin signaling in a cell-nonautonomous manner. It appears that in the nematode, as in the fly, sensory cells in the gut and the brain perceive nutrient availability and respond with altered hormonal secretion that ultimately may lead to extended life spans. The

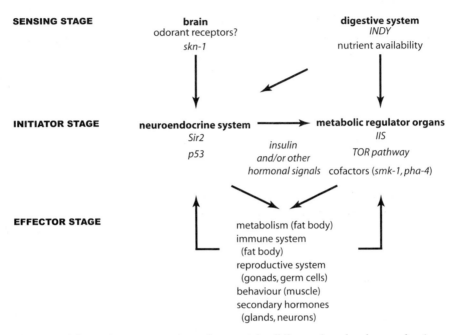

Figure 1. Schematic representation of potential cellular and molecular mechanisms mediating the CR/DR response. In this speculative scheme, the organism continually monitors environmental and nutrient conditions through sensor systems in the brain and the digestive system (sensing stage). This input is relayed to organs responsible for setting the general metabolic state of the organism by comparing and matching general metabolism to perceived environmental conditions (initiator stage). This includes the neuroendocrine system and some of the major metabolic regulatory organs themselves (fat body in flies and intestine in nematodes). Signaling to all of the major metabolic tissues (effector stage) sets a new metabolic state for the organism. Reciprocal signaling from these effector tissues back to the initiator stage tissues ensures a good match to the available nutrient state of the organism. Genes and signaling pathways implicated in mediating the CR/DR response are in italics.

exact nature of these hormonal signals and whether they affect insulin-signaling or related pathways remain to be determined (Fig. 1).

PRACTICAL ASPECTS OF CR AND DR IN FLIES AND NEMATODES

How Are Calories Restricted in a Fly or Nematode?

What do we mean when we use the term calorie restriction or dietary restriction with nematodes and flies? In mammalian studies, the amount of daily intake can be readily controlled and monitored. Individually housed mice, rats, or nonhuman primates can be given a designated ration of food of a specific constituency, and precise information concerning intake and output can be obtained. However, the demographic-sized studies making model organisms such as *C. elegans* or *Drosophila* valuable for studying aging preclude the housing of individuals in most studies. Medflies appear to be one notable exception as studies have been performed in which they have been kept in individual housing (Carey et al. 2005).

The logistical challenge to measuring what each individual fly or nematode consumes is a problem since *C. elegans* and *Drosophila* are typically housed with direct and continuous access to their food. Thus, the ability to restrict food intake and continuously monitor intake and output throughout a life span is impractical. To decrease calorie or dietary intake, studies on flies and some studies on *C. elegans* have resorted to diluting total calories or specific components per food volume (Pletcher et al. 2005; Houthoofd and Vanfleteren 2006).

For nematodes, normally grown on an agar surface with live *Escherichia coli* bacteria, the issue of diluting the amount of *E. coli* and maintaining the volume of bacteria in the face of continued *E. coli* proliferation is problematic. Several studies have used bacterial dilution of the *E. coli* food source, by manipulating either the peptone concentration in the agar medium or the density of the bacterial culture in a liquid medium, in an attempt to limit nutrient availability to nematodes and induce CR/DR (Klass 1977; Houthoofd et al. 2003). Steps such as using *E. coli* killed by antibiotics, heat treatment, or UV irradiation in an effort to prevent bacterial proliferation have also been employed, although it is unclear how nutritionally replete such bacteria remain after these treatments. Complicating these attempts to use bacterial dilution as a method of inducing CR/DR is the fact that the *E. coli* used under lab culture conditions is toxic to nematodes (Garigan et al. 2002). It is unclear how much of the life span extension seen under these methods is attributable to reduction of *E. coli* toxicity and how much represents CR/DR. A number of studies have tried to get around this problem by culturing nematodes with synthetic medium lacking bacteria (axenic medium), leading to large life span extensions (Vanfleteren and Braeckman 1999; Houthoofd et al. 2002, 2003). It has also been shown that leaving adult nematodes on agar plates devoid of *E. coli* can sustain life for a considerable period of time (weeks) (Kaeberlein et al. 2006; Lee et al. 2006). Whether this is a state of calorie restriction or "starvation" is not clear. An additional method of restricting intake in *C. elegans* employs the use of mutations in several of the *eat* genes that affect the pumping capacity of the pharyngeal organ, an approach that is thought to reduce bacterial food intake directly (Lakowski and Hekimi 1998). There are several possible concerns with the assumption that since pumping is decreased in the mutant, it will remain so to the same extent 24 hours a day for the length of the life span. It is not clear whether normal *C. elegans* has periods of decreased pharyngeal pumping during the day or over a period of a few days. If so, then the mutant, if it maintains itself at a constant rate of pumping, could "make up" the difference. In addition, although the *eat* mutation appears to primarily affect pharyngeal pumping, it may also affect other aspects of the animals' musculature or physiology and alter fitness and life span in that manner.

For flies, a number of different strategies have been utilized most of which amount to a dilution of the total or specific constituents that make up "fly food," which may include cornmeal, agar, sucrose, molasses, killed yeast, and, in some cases, live yeast (Chippindale et al. 1993; Chapman and Partridge 1996). Further confusing the issue is the concern over which of the components in food are most important for CR in flies, and this remains the subject of much active research. The difficulty of having food sources that are themselves a complicated and often inconsistent combination of a variety of carbohydrates, lipids, proteins, and small molecules, and which are known to directly and dramatically affect the neuroendocrine and/or reproductive system, is of great concern (e.g., components of yeast are directly linked to female egg laying and fertility in flies). Unlike in mammalian studies where it is thought that well-controlled simple constituents can be used to make up the entire food intake, the use of *E. coli* for nematodes and ill-defined food constituents, such as yeast for flies, has led to difficulties in being able to compare results from different laboratories (Pletcher et al. 2005). The development of a universally used synthetic well-defined nematode or fly CR medium has not yet been achieved.

Does Food Dilution Lead to CR/DR?

The difficulty of measuring food intake in individual flies or nematodes, and the fact that they have free access to their food, raises the question as to whether reducing calorie content in the food should have any effect or if the animals can compensate by eating more. Little has been done to directly measure this in nematodes. The use of mutations that reduce pharyngeal pumping rates should directly decrease intake, although as noted above, studies demonstrating this over a 24-hour period or more have not been done.

It is reasonably clear that flies with continuous access to food are not able to adjust their eating behavior sufficiently to take in enough food to completely compensate for a decrease in the calorie content in their food. Comparison of flies on low-calorie life-span-extending diets show profound physiological changes consistent with the interpretation that they are obtaining fewer calories in their diet. In addition to the extension in life span seen, under lower-calorie conditions, there is a decrease in weight gain, reduction in egg laying, alteration in body composition favoring an increase in glycogen storage, and an increase in spontaneous physical activity especially in males, similar to the increase in physical activity seen with CR in mice (Bross et al. 2005; Chen et al. 2005). This directly indicates that under the typical range of diets thought to be relevant for calorie reduction, as opposed to starvation, flies, despite having constant access to a food source, do not completely compensate for a reduction in calorie content in the food. They may, however, compensate in a way that makes it difficult to be certain just how much fewer calories they are taking in.

What Do Measurements of Intake Tell Us?

A number of studies have examined whether flies on lower-calorie foods take in fewer calories. This has included measuring the ingestion of a dye over a prescribed period of time, documenting the feeding behavior of individual flies over short periods of time, or quantifying the incorporation of a radiolabeled nucleotide (Bross et al. 2005; Carvalho et al. 2005; Min and Tatar 2006). In general, each of these studies confirms that flies on the lower-calorie life-span-extending diet do obtain fewer calories. The studies using ingestion of a dye and observation of proboscis extension over a short period of time suggest little compensation under "low"-calorie diets, although in the dye uptake experiments, an increase in uptake under very low-calorie diets and possibly under high-calorie diets was seen (Bross et al. 2005; Min and Tatar 2006). The study measuring uptake/incorporation of a radiolabeled nucleotide indicates that there is some compensation in incorporation and that there may be some compensation in food intake. However, this compensation is insufficient to bring the low-calorie-fed animals up to the same intake as the higher-calorie-fed animals (Carvalho et al. 2005). The authors caution that due to the possibility of compensation, it may be incorrect to assume that the differences in intake between different calorie food is precisely the same as the differences in the calorie content of each of the foods. The question of whether there is an increase in feeding in the low-calorie food conditions is not completely resolved, but it is clear that over time, flies on the low-calorie life-span-extending food conditions do ingest fewer calories than flies on higher-calorie food sources.

Reproductive Trade-offs and Life Span Extension from CR/DR

The life span extension in response to CR/DR is greater in female flies than in male flies. To date, all studies examining CR/DR in nematodes have been performed on hermaphrodites, and the response of male nematodes to CR/DR is unknown. Although CR/DR does cause a reduction in fecundity and brood size in hermaphrodites, it is not clear whether this has a role in the life span extension seen. In general, mutations in *C. elegans* that lower fertility are not associated with an increase in longevity. In flies, it is not precisely understood what accounts for the sexual differences in response to CR/DR, but it raises the question of how a decrease in calories might impinge upon reproductive physiology and the life span of female flies. It is well known that a reduction in calorie content, especially yeast, significantly decreases egg laying (Chapman and Partridge 1996). It is also known that a decrease in mating and perhaps egg laying can be associated with an increase in female life span. Thus, it has been suggested that some of the differences in the life span extension of females and males in response to CR/DR is the result of this selective effect on female reproduction (Le Bourg and Minois 2005). Recent experiments, however, suggest that a decrease in egg production may not be a major component of the life-span-extending effect of CR/DR in females. Flies with genetic mutations that prevent the production and laying of mature eggs continue to respond to CR/DR by extending life span (Mair et al. 2004). Although a reduction in calories does induce a decrease in egg production, the decrease in egg production may not be the causative element in CR-dependent life span extension. Analysis of genes in flies that extend life span through CR/DR-dependent mechanisms, such as mutations that decrease Rpd3 and manipulations that increase dSir2 levels and decrease Dmp53, do not have a reduction in female fertility, supporting the suggestion that the life-span-extending effects of CR/DR can be dissociated from the effects on female fertility (Rogina et al. 2002; Rogina and Helfand 2004; Bauer et al. 2005).

In mammalian studies, the recognition of a major confounding element of female reproduction during CR/DR has resulted primarily in the use of males in many studies. The use of male flies to examine some of these issues has been suggested, and although this does eliminate the concern of egg production as a major element in energy allocation, it is not yet known what the cost of a reduced nutrient source might be for male reproductive physiology, and how this might impact life span. Further studies on resource allocations will need to be done to sort this out. A reduction of calories, particularly yeast, can dramatically alter the reproductive physiology of *Drosophila*, as well as a variety of other physiological changes. One of the challenges in the near future is to identify which of the many physiological changes induced by a reduction in calories (CR/DR) is directly linked to life span extension. Getting to the core of these issues is of critical importance for understanding the mechanisms by which CR extends life span in flies as well as in other organisms.

Calorie Restriction versus Dietary Changes without a Reduction in Calories?

A study altering diet by varying sugar and yeast independently has suggested that it is the reduction of a specific nutrient, yeast, and not overall calories that may be important in the life-span-extending effect of CR/DR in flies (Mair et al. 2005). Examination of life span while independently varying the amount of sugar and yeast in food containing the same total amount of calories (low yeast/high sugar food vs. high yeast/low sugar food) led to the conclusion that it is the amount of yeast and not the total calories that is the major determinant of life span at least for female flies. The conclusions of this study have been challenged by another report that sought to measure the amount of calories taken in by the flies using combustion calorimetry of both the female flies and their eggs (Min et al. 2007). In this study, it was shown that the total calories taken in by female flies are affected by the food source. Although the low yeast/high sugar diet has the same amount of total calories, flies on the high yeast/low sugar diet take in considerably more calories. Plotting the amount of calories taken in versus life span shows a direct relationship of calories per fly and life span.

These studies point out the problematic issues that may arise when trying to alter specific nutrients, especially when some of these nutrients are known to be associated with altering feeding behavior and reproductive physiology. Yeast is a complex mixture of nutrients including proteins, lipids, and small molecules known to alter the feeding, foraging, and reproductive behavior of flies, especially female flies. Unfortunately, both of these studies were performed using only female flies. It is thus difficult to draw clear conclusions on the issue of the role of total calories versus specific nutrients when some of the nutrients being altered are complex and contain compounds with effects beyond their nutritional value. Studying the effects on males might be of some value in dissecting these confounding issues.

Nematodes subsist under laboratory culture conditions on a single food source—*E. coli*. This complicates attempts to determine which specific components of the diet are responsible for the CR effect. However, successful culture of nematodes under axenic conditions, as well as the observation that nematodes can survive on standard NGM agar medium in the absence of bacteria, suggests that nematodes can take up nutrients from their environment even if *E. coli* are not present. To date, no studies have examined the effects on life span of varying specific components of the diet.

Genetic Background and Life Span Extension from CR

It has recently become apparent that different inbred strains of *D. melanogaster* do not respond the same to CR/DR. For example, although the standard Canton-S strain shows a robust response to CR, with life span extensions of females up to 40–50%, other strains such as w^{1118} show a rather blunted response to changes in calories with increases usually of less than 10%. A recent comparison of high-calorie food, sugar/yeast (SY) 15%, and low-calorie food, SY 5%, found a 22% increase in mean life span for Canton-S female flies, but only a 4% increase in mean life span for w^{1118} female flies. Male mean life span increased 13% for Canton-S and 0% for w^{1118} (Libert et al. 2007). It should not be surprising that the genetic background, and hence metabolic endowment of a particular inbred strain of flies, would have an important role in response to changes in diet. In addition, a number of the studies on CR/DR in flies use flies that are from genetically heterogeneous outbred stocks. Differences in genetic background are likely to be the basis for some of the discrepancies in results and conclusions in the literature. Despite this causing great difficulty in reaching some general conclusions on CR/DR responses for flies when comparing data from different laboratories, it demonstrates the plasticity of the CR/DR response and can provide insight into what elements of the CR/DR response are most robust. Perhaps more importantly, the variation of response to dietary conditions between different strains should provide a fertile ground for exploring the specific molecular genetic elements involved in the CR/DR response through studies of the molecular genetic differences between these different strains. The standardization of the common wild-type *C. elegans* strain (N2), as well as the ability to easily generate large populations of isogenic individuals from a single hermaphrodite, has made these issues less of a concern in nematode studies.

Does CR/DR Change the Rate of Aging or Affect Immediate Mortality?

The data collected from life span studies can be used to assess the possible underlying causes of mortality. Features of the mortality curve, such as the slope and the y intercept, are commonly used to compare populations with different life spans in an effort to try to assess the cause of mortality. The demonstration of a change in the slope between two populations is thought to reflect a change in the rate of aging, whereas a change in the y intercept reflects an affect on the short-term risk of mortality. Analyses of mortality curves from females of an outbreed stock of *Drosophila* on different food diets showed a change in y intercept or change in short-term risk, rather than a change in slope or rate of aging (Mair et al. 2003). This suggests that calorie restriction in flies could affect the immediate risk of death rather than alter the rate of aging. Furthermore, female flies switched from high- to low-calorie conditions (SY 15% to SY 5%) assumed the mortality rate of female flies on a low-calorie food within 48 hours of the switch and demographically showed little evidence of any residual effect on mortality from the fly's previous calorie condition.

A study using Canton-S, a common wild-type inbred strain, confirmed some of these observations on different calorie foods but showed another interesting demographic effect. Although the effect on the mortality curve for female flies on SY 5% versus SY 15% diets is a change in y intercept, only a slight increase in the diet to SY 20% food now revealed a change in slope but not y intercept (Bross et al. 2005). These data suggest that there may be a caloric or dietary threshold, at least for Canton-S female flies, in which a small change in diet alters the mortality curve from one of a change in immediate short-term risk to a change in the rate of aging. Whether this shift to a change in rate of aging at a higher-calorie condition would occur with the outbred flies is not known. One possibility that has been suggested to explain the differences is that the slightly higher SY 20% diet leads to a general toxicity in female flies. This seems unlikely, since the differences between the supposed SY 15% and SY 20% diet have been shown to be less than 20% difference in yeast and sugar due to the manner in which the food is prepared. Furthermore, the female flies on the SY 20% diet continued to lay large numbers of eggs. If the 20% increase in yeast/sugar led to a general toxicity, then females would be expected to have a reduction in egg laying. The fact that one study was performed using an outbred stock (Dahomey) and the other an inbred genetically homogeneous stock (Canton-S) may contribute to the differences seen, but it also suggests that examining full range of diets may yield important information as to how diet and life span interact in *Drosophila*.

In an attempt to understand the molecular changes associated with environmental interventions that change the slope versus the y intercept of the mortality curve, the rate of accumulation of oxidative damage with age was examined for flies living at different ambient temperatures or on different calorie diets. Using a lipid peroxidation marker, 4-hydroxynonenal (4-HNE), it was found that both higher ambient temperature and higher-calorie foods were associated with a more rapid accumulation of 4-HNE (Zheng et al. 2005). Both ambient temperature and CR/DR showed a change in the rate of accumulation of 4-HNE that scaled with respect to life span. These studies show that when life span is altered using ambient temperature or CR/DR, there was a similar response in the rate of accumulation of some types of oxidative damage, a potential molecular marker for aging. This suggests that these two interventions, apparently different demographically in how they are predicted to affect aging, may share some underlying molecular similarities and perhaps mechanisms. Unfortunately, this study did not include an experiment in which flies were switched from a high-calorie diet to a low-calorie diet to determine if the accumulation of oxidative damage at the higher-calorie condition would be cleared or reset to the rate for lower-calorie diet. However, these studies appear to be inconsistent with the hypothesis that CR/DR primarily changes the short-term immediate risk of mortality. Taken together, these studies suggest caution in trying to use mathematical equations on data derived from the age of death of individuals in a population to predict specific molecular or physiological mechanisms of aging.

Typical *C. elegans* life span experiments use smaller numbers of animals. This makes the statistical demographic analyses of mortality curves seen in *Drosophila* studies much less common, necessitating the reporting of survivorship curves.

We are only now beginning to understand the intricacies of nutrient sensing and metabolic regulation in nematodes and flies. The pleiotropic mechanisms of CR/DR are notoriously difficult to reduce from simple experimental analysis. However, the available data make hypothesis testing possible. Our present molecular understanding of CR/DR is very limited. Only a small number of genes have been linked to CR/DR and only mutations in one gene, *dSir2*, have been shown to abolish CR/DR life span effects. Presumably, many more genes have a role in mediating all the pleiotropic effects of CR/DR. However, the genes so far identified, and the molecular and cellular pathways they are associated with, provide a starting framework with which to further dissect the molecular mechanisms of CR/DR. The field is moving beyond phenomenological questions to identifying and testing mechanisms. Tools such as loss-of-function mutants and tissue-specific knockouts or knockdowns for rigorously testing some of the above-outlined scenarios are available, and studies are in progress to test their effects on CR/DR-dependent longevity extension.

CONCLUSIONS

In all organisms examined, from yeast to mammals, the application of a reduction in the intake of calories while still maintaining adequate nutrition (calorie restriction/dietary restriction) induces a physiological state resulting in an extension of healthy life span. It is thought that the induced life span extension—rather than being due to a decrease in metabolic rate and "slowing" down of general life processes—is the result of specific selective physiological changes. Model organisms such as *C. elegans* and *D. melanogaster* are of great value for much needed mechanistic insight in understanding the CR/DR effect on life span extension. Identifying the molecular, cellular, and physiological changes by which a reduction in caloric intake causes an extension of healthy life span is arguably one of the most important questions in modern aging/gerontologic research and should be an achievable goal. Understanding the mechanisms underlying calorie restriction life span extension promises to provide realistic tools for slowing age-dependent decline as well as ameliorating or halting age-related disorders such as cardiac disease, neurodegeneration, diabetes, and cancer.

ACKNOWLEDGMENTS

We apologize to all of our colleagues whose work we could not mention due to space constraints. This work was supported by grants from the National Institute of Aging (AG16667, AG24353, and AG25277), the Donaghue Foundation, and the Ellison Medical Foundation to S.L.H. S.L.H. is an Ellison Medical Research Foundation senior investigator. This research was conducted while J.H.B. was an Ellison Medical Foundation/AFAR senior postdoctoral fellow.

REFERENCES

Apfeld J. and Kenyon C. 1998. Cell nonautonomy of *C. elegans daf-2* function in the regulation of diapause and life span. *Cell* **95:** 199–210.

Apfeld J., O'Connor G., McDonagh T., DiStefano P.S., and Curtis R. 2004. The AMP-activated protein kinase AAK-2 links energy levels and insulin-like signals to life span in *C. elegans*. *Genes Dev.* **18:** 3004–3009.

Austad S.N. 1989. Life extension by dietary restriction in the bowl and doily spider, *Frontinella pyramitela*. *Exp. Gerontol.* **24:** 83–92.

Bauer J.H., Poon P.C., Glatt-Deeley H., Abrams J.M., and Helfand S.L. 2005. Neuronal expression of p53 dominant-negative proteins in adult *Drosophila melanogaster* extends life span. *Curr. Biol.* **15:** 2063–2068.

Bishop N.A. and Guarente L. 2007. Two neurons mediate diet-restriction-induced longevity in *C. elegans*. *Nature* **447:** 545–549.

Bross T.G., Rogina B., and Helfand S.L. 2005. Behavioral, physical, and demographic changes in *Drosophila* populations through dietary restriction. *Aging Cell* **4:** 309–317.

Carey J.R., Liedo P., Muller H.G., Wang J.L., Zhang Y., and Harshman L. 2005. Stochastic dietary restriction using a Markov-chain feeding protocol elicits complex, life history response in medflies. *Aging Cell* **4:** 31–39.

Carvalho G.B., Kapahi P., and Benzer S. 2005. Compensatory ingestion upon dietary restriction in *Drosophila melanogaster*. *Nat. Methods* **2:** 813–815.

Chapman T. and Partridge L. 1996. Female fitness in *Drosophila melanogaster*: An interaction between the effect of nutrition and of encounter rate with males. *Proc. R. Soc. Lond. B Biol. Sci.* **263:** 755–759.

Chen D., Steele A.D., Linquist S., and Guarente L. 2005. Increase in activity during calorie restriction requires Sirt1. *Science* **310:** 1641.

Chippindale A.K., Leroi A.M., Kim S.B., and Rose M.R. 1993. Phenotypic plasticity and selection in *Drosophila* life-history evolution. I. Nutrition and the cost of reproduction. *J. Evol. Biol.* **6:** 171–193.

Clancy D.J., Gems D., Hafen E., Leevers S.J., and Partridge L. 2002. Dietary restriction in long-lived dwarf flies. *Science* **296:** 319.

Fei Y.J., Liu J.C., Inoue K., Zhuang L., Miyake K., Miyauchi S., and Ganapathy V. 2004. Relevance of NAC-2, an Na^+-coupled citrate transporter, to life span, body size and fat content in *Caenorhabditis elegans*. *Biochem. J.* **379:** 191–198.

Garigan D., Hsu A.L., Fraser A.G., Kamath R.S., Ahringer J., and Kenyon C. 2002. Genetic analysis of tissue aging in *Caenorhabditis elegans*: A role for heat-shock factor and bacterial proliferation. *Genetics* **161:** 1101–1112.

Hansen M., Taubert S., Crawford D., Libina N., Lee S.J., and Kenyon C. 2007. Lifespan extension by conditions that inhibit translation in *Caenorhabditis elegans*. *Aging Cell* **6:** 95–110.

Houthoofd K. and Vanfleteren J.R. 2006. The longevity effect of dietary restriction in *Caenorhabditis elegans*. *Exp. Gerontol.* **41:** 1026–1031.

Houthoofd K., Braeckman B.P., Johnson T.E., and Vanfleteren J.R. 2003. Life extension via dietary restriction is independent of the Ins/IGF-1 signalling pathway in *Caenorhabditis elegans*. *Exp. Gerontol.* **38:** 947–954.

Houthoofd K., Braeckman B.P., Lenaerts I., Brys K., De Vreese A., Van Eygen S., and Vanfleteren J.R. 2002. Axenic growth up-regulates mass-specific metabolic rate, stress resistance, and extends life span in *Caenorhabditis elegans*. *Exp. Gerontol.* **37:** 1371–1378.

Howitz K.T., Bitterman K.J., Cohen H.Y., Lamming D.W., Lavu S., Wood J.G., Zipkin R.E., Chung P., Kisielewski A., Zhang L.L., Scherer B., and Sinclair D.A. 2003. Small molecule activators of sirtuins extend *Saccharomyces cerevisiae* lifespan. *Nature* **425:** 191–196.

Hwangbo D.S., Gershman B., Tu M.P., Palmer M., and Tatar M. 2004. *Drosophila* dFOXO controls lifespan and regulates insulin signalling in brain and fat body. *Nature* **429:** 562–566.

Ikeya T., Galic M., Belawat P., Nairz K., and Hafen E. 2002. Nutrient-dependent expression of insulin-like peptides from neuroendocrine cells in the CNS contributes to growth regulation in *Drosophila*. *Curr. Biol.* **12:** 1293–1300.

Jia K., Chen D., and Riddle D.L. 2004. The TOR pathway interacts with the insulin signaling pathway to regulate *C. elegans* larval development, metabolism and life span. *Development* **131:** 3897–3906.

Kaeberlein M., McVey M., and Guarente L. 1999. The SIR2/3/4 complex and SIR2 alone promote longevity in *Saccharomyces cerevisiae* by two different mechanisms. *Genes Dev.* **13:** 2570–2580.

Kaeberlein T.L., Smith E.D., Tsuchiya M., Welton K.L., Thomas J.H., Fields S., Kennedy B.K., and Kaeberlein M. 2006. Lifespan extension in *Caenorhabditis elegans* by complete removal of food. *Aging Cell* **5:** 487–494.

Kapahi P., Zid B.M., Harper T., Koslover D., Sapin V., and Benzer S. 2004. Regulation of lifespan in *Drosophila* by modulation of genes in the TOR signaling pathway. *Curr. Biol.* **14:** 885–890.

Kenyon C. 2005. The plasticity of aging: Insights from long-lived mutants. *Cell* **120:** 449–460.

Kim S., Benguria A., Lai C.Y., and Jazwinski S.M. 1999. Modulation of life-span by histone deacetylase genes in *Saccharomyces cerevisiae*. *Mol. Biol. Cell* **10:** 3125–3136.

Klass M.R. 1977. Aging in the nematode *Caenorhabditis elegans*: Major biological and environmental factors influencing life span. *Mech. Ageing Dev.* **6:** 413–429.

Knauf F., Rogina B., Jiang Z., Aronson P.S., and Helfand S.L. 2002. Functional characterization and immunolocalization of the transporter encoded by the life-extending gene Indy. *Proc. Natl. Acad. Sci.* **99:** 14315–14319.

Lakowski B. and Hekimi S. 1996. Determination of life-span in *Caenorhabditis elegans* by four clock genes. *Science* **272:** 1010–1013.

———. 1998. The genetics of caloric restriction in *Caenorhabditis elegans*. *Proc. Natl. Acad. Sci.* **95:** 13091–13096.

Le Bourg E. and Minois N. 2005. Does dietary restriction really increase longevity in *Drosophila melanogaster*? *Ageing Res. Rev.* **4:** 409–421.

Lee G.D., Wilson M.A., Zhu M., Wolkow C.A., de Cabo R., Ingram D.K., and Zou S. 2006. Dietary deprivation extends lifespan in *Caenorhabditis elegans*. *Aging Cell* **5:** 515–524.

Li W., Kennedy S.G., and Ruvkun G. 2003. daf-28 encodes a *C. elegans* insulin superfamily member that is regulated by environmental cues and acts in the DAF-2 signaling pathway. *Genes Dev.* **17:** 844–858.

Libert S., Zwiener J., Chu X., Vanvoorhies W., Roman G., and Pletcher S.D. 2007. Regulation of *Drosophila* life span by olfaction and food-derived odors. *Science* **315:** 1133–1137.

Libina N., Berman J.R., and Kenyon C. 2003. Tissue-specific activities of *C. elegans* DAF-16 in the regulation of lifespan. *Cell* **115:** 489–502.

Lin S.J., Defossez P.A., and Guarente L. 2000. Requirement of NAD and SIR2 for life-span extension by calorie restriction in *Saccharomyces cerevisiae*. *Science* **289:** 2126–2128.

Maier B., Gluba W., Bernier B., Turner T., Mohammad K., Guise T., Sutherland A., Thorner M., and Scrable H. 2004. Modulation of mammalian life span by the short isoform of p53. *Genes Dev.* **18:** 306–319.

Mair W., Piper M.D., and Partridge L. 2005. Calories do not explain extension of life span by dietary restriction in *Drosophila*. *PLoS Biol.* **3:** e223.

Mair W., Goymer P., Pletcher S.D., and Partridge L. 2003. Demography of dietary restriction and death in *Drosophila*. *Science* **301:** 1731–1733.

Mair W., Sgro C.M., Johnson A.P., Chapman T., and Partridge L. 2004. Lifespan extension by dietary restriction in female *Drosophila melanogaster* is not caused by a reduction in vitellogenesis or ovarian activity. *Exp. Gerontol.* **39:** 1011–1019.

Masoro E.J. 2005. Overview of caloric restriction and ageing. *Mech. Ageing Dev.* **126:** 913–922.

Min K.J. and Tatar M. 2006. *Drosophila* diet restriction in practice: Do flies consume fewer nutrients? *Mech. Ageing Dev.* **127:** 93–96.

Min K.J., Flatt T., Kulaots I., and Tatar M. 2007. Counting calories in *Drosophila* diet restriction. *Exp. Gerontol.* **42:** 247–251.

———. 2007. Comment on Min K.J., Flatt T., Kulaots I., and Tatar M. (2006) "Counting calories in *Drosophila* dietary restriction" in *Exp. Gerontol.* **42:** 247–251. *Exp. Gerontol.* **42:** 253–255.

Pletcher S.D., Libert S., and Skorupa D. 2005. Flies and their golden apples: The effect of dietary restriction on *Drosophila* aging and age-dependent gene expression. *Ageing Res. Rev.* **4:** 451–480.

Pletcher S.D., Macdonald S.J., Marguerie R., Certa U., Stearns S.C., Goldstein D.B., and Partridge L. 2002. Genome-wide transcript profiles in aging and calorically restricted *Drosophila melanogaster*. *Curr. Biol.* **12:** 712–723.

Rogina B. and Helfand S.L. 2004. Sir2 mediates longevity in the fly through a pathway related to calorie restriction. *Proc. Natl. Acad. Sci.* **101:** 15998–16003.

Rogina B., Helfand S.L., and Frankel S. 2002. Longevity regulation by *Drosophila* Rpd3 deacetylase and caloric restriction. *Science* **298:** 1745.

Rogina B., Reenan R.A., Nilsen S.P., and Helfand S.L. 2000. Extended life-span conferred by cotransporter gene mutations in *Drosophila*. *Science* **290:** 2137–2140.

Rulifson E.J., Kim S.K., and Nusse R. 2002. Ablation of insulin-producing neurons in flies: Growth and diabetic phenotypes. *Science* **296:** 1118–1120.

Tatar M. 2007. Diet restriction in *Drosophila melanogaster*. Design and analysis. *Interdiscip. Top. Gerontol.* **35:** 115–136.

Tissenbaum H.A. and Guarente L. 2001. Increased dosage of a *sir-2* gene extends lifespan in *Caenorhabditis elegans*. *Nature* **410:** 227–230.

Tyner S.D., Venkatachalam S., Choi J., Jones S., Ghebranious N., Igelmann H., Lu X., Soron G., Cooper B., Brayton C., Hee Park S., Thompson T., Karsenty G., Bradley A., and Donehower L.A. 2002. p53 mutant mice that display early ageing-associated phenotypes. *Nature* **415:** 45–53.

Vanfleteren J.R. and Braeckman B.P. 1999. Mechanisms of life span determination in *Caenorhabditis elegans*. *Neurobiol. Aging* **20:** 487–502.

Vellai T., Takacs-Vellai K., Zhang Y., Kovacs A.L., Orosz L., and Muller F. 2003. Genetics: Influence of TOR kinase on lifespan in *C. elegans*. *Nature* **426:** 620.

Wang M.C., Bohmann D., and Jasper H. 2005. JNK extends life span and limits growth by antagonizing cellular and organism-wide responses to insulin signaling. *Cell* **121:** 115–125.

Wang Y. and Tissenbaum H.A. 2006. Overlapping and distinct functions for a *Caenorhabditis elegans* SIR2 and DAF-16/FOXO. *Mech. Aging Dev.* **127:** 48–56.

Weindruch R.H. and Walford R.L. 1988. *The retardation of aging and disease by dietary restriction.* C.C. Thomas, Springfield, Illinois.

Wood J.G., Rogina B., Lavu S., Howitz K., Helfand S.L., Tatar M., and Sinclair D. 2004. Sirtuin activators mimic caloric restriction and delay ageing in metazoans. *Nature* **430:** 686–689.

Zheng J., Mutcherson R., II, and Helfand S.L. 2005. Calorie restriction delays lipid oxidative damage in *Drosophila melanogaster. Aging Cell* **4:** 209–216.

4

Evolutionary Theory in Aging Research

Steven N. Austad
Department of Cellular and Structural Biology
Barshop Institute for Longevity and Aging Studies
University of Texas Health Science Center
San Antonio, Texas 78245

Thomas B.L. Kirkwood
Henry Wellcome Laboratory for Biogerontology Research
Institute for Ageing and Health
University of Newcastle
Newcastle upon Tyne NE4 6BE
United Kingdom

EVOLUTIONARILY PLAUSIBLE SCENARIOS OF HOW SENESCENCE, although detrimental to the individual soma, could evolve via natural selection on individuals were first developed more than 50 years ago (Medawar 1952). The premise behind what we will call *classical* evolutionary senescence theory is that, due to the inescapability of death from extrinsic hazards such as predators, environmental degradation, and infectious disease, the cumulative probability of surviving to older and older ages grows ever smaller even in the absence of senescence. Medawar (1952) pointed out that following from this self-evident premise, if traits affecting evolutionary fitness—reproduction and survival—were expressed in an age-specific manner, the power of natural selection to affect the evolutionary fate of these traits would gradually wane with age. Put slightly differently, a lethal genetic allele expressed only in centenarians has no evolutionary fitness disadvantage, both because it is so unlikely to be expressed at all (only about one person in 10,000 survives to 100 years in modern populations) and because reproduction has long since ceased by age 100. By the same logic, an allele improving survival in centenarians confers no evolutionary advantage either. In contrast, similar alleles expressed at age 20 would have an enormous impact on fitness.

Molecular Biology of Aging ©2008 Cold Spring Harbor Laboratory Press 978-087969824-9

From the preceding scenario, senescence could evolve by one or both of the following hypothetical genetic mechanisms: (1) Deleterious alleles that affect survival or reproduction only very late in life, when selection is weak, could accumulate in the genome over evolutionary time by mutation pressure checked only weakly by mutation-selection balance (Medawar 1952) and (2) alleles with antagonistically pleiotropic effects, such that they enhance fitness early in life when selection is strong but depress it late in life when selection is weak, can be actively favored by natural selection because the early salubrious effects will evolutionarily outweigh the later deleterious effects, even when the early effects are small compared with the later effects (Williams 1957).

In the decades following the formulation of this classical theory, quantitative modeling confirmed the intuitive logic of the previous verbal approaches. Hamilton (1966) focused on how age-specific alterations in mortality and fertility schedules would be expected to affect Darwinian fitness, whereas Charlesworth (1980) formulated explicit genetic models to address the same issues. Kirkwood (1977, 1981) also developed a cellular analog of antagonistic pleiotropy, the disposable soma theory, in which aging was seen to be due to the accumulation of random molecular damage modulated by energetic trade-offs among maintenance, growth, and reproduction. In both disposable soma and antagonistic pleiotropy theories, genetic trade-offs between early- and late-life fitness are expected.

An abundance of supportive empirical evidence accumulated in the decades following the development of these theories. A testable prediction of classical theory is that delaying the age-related decay of natural selection's impact, by either delaying reproduction or decreasing environmental hazards, will lead to retarded aging and longer life. Although this prediction has been verified repeatedly, most notably by laboratory evolution in fruit flies (Rose 1989; Zwaan et al. 1995; Sgrò and Partridge 1999; Stearns et al. 2000) and also in a "natural experiment" comparing opossums in a low-hazard island population with a high-hazard mainland population (Austad 1993), there are subtleties of population biology associated with it that should be appreciated (see Neoclassical Theory section below).

In virtually all of these experiments, increased longevity has been associated with a decrement in early-life fitness components, although which component is affected varies. Reduced fecundity is the most common trade-off, although reduced larval viability, increased development time, and decreased body size have also been observed (for review, see Kirkwood and Austad 2000). The identification of specific antagonistically pleiotropic genes has been elegantly revealed by direct competition between *Caenorhabditis elegans* genotypes differing only in whether they carry wild-type versus long-lived alleles at the relevant loci. In one case, the long-lived *daf2* allele was rapidly outcompeted by wild type under all study conditions (Jenkins et al. 2004); in another case, the long-lived *age1* allele had equal fitness with wild type when food was superabundant but went rapidly extinct under conditions of alternating food abundance and shortage (Walker et al. 2000).

MISUNDERSTANDINGS OF EVOLUTION THEORY

Although the logical and empirical underpinnings of the classical evolutionary theory of senescence are extremely strong, there has continued to be a tendency to seek explanations of aging in terms of some kind of adaptive genetic program that specifically limits the individual's life span. This has led to recurring misunderstandings about the genetic basis of aging and longevity.

The attractions of the program concept are easily understood. First, aging is phylogenetically a very widely distributed trait, and in species where senescence occurs, it affects every individual that lives long enough to experience its adverse impacts on fertility and vitality. To many, it therefore seems to make sense that aging exists "for a purpose." Second, there are clear genetic effects on longevity and this leads naturally to supposing that the relevant genes specify some kind of "aging clock." Third, in a postgenome era, when new evidence of genetic causality is being uncovered in many realms of biology, the default assumption that aging is *caused by* gene action preexists in the minds of most of those who come afresh to considering why aging occurs, although this argument is undercut by recent observations that regulation of gene expression deteriorates with age (Bahar et al. 2006). Finally, and despite the evidence that the details of the aging process are intrinsically variable from one individual to another, there is sufficient broad reproducibility about the manifestations of senescence that it just "looks as if it has to be programmed."

The reasons why aging should be programmed are frequently taken for granted, but, if pressed, the commonest suggestions offered are that possession of a fixed limit to life span (1) is beneficial, or even necessary, to prevent the species from overcrowding its environment (Wynne-Edwards 1962) or (2) promotes long-term evolutionary fitness by securing the necessary turnover of generations that allows novel adaptations to be selected (Libertini 1988).

For either of these suggestions to work, it is a necessary prerequisite that intrinsic aging should make a sufficient contribution to natural mortality that the hypothesized selection process is feasible. If an individual dies before senescent effects are apparent, it makes no difference whether or not that individual is endowed with genes that program aging. Such a program can only be fashioned by selection acting to realize the hypothesized benefits of a program for aging *in those individuals who survive to an age when the program takes effect.* It is therefore a problem for program theories of aging that although senescence has been *detected* in many natural populations of species that show evident aging in a captive setting, relatively few individuals survive long enough to be affected by it (Finch 1990; Brunet-Rossinni and Austad 2006). The exception occurs in semelparous species, such as Pacific salmon, that have evolved a life history plan in which there is only a single bout of reproduction. In such species, death of the parent usually occurs soon after reproduction. This is generally the consequence of directing all available resources to maximizing reproductive success, without regard to the subsequent survival of the adult. An important source of misunderstanding of the evolutionary theory of aging has been to regard postreproductive death in semelparous species as an instance of programmed aging, when in fact its evolutionary explanation may be very different (Kirkwood 1985). Even if aging in Pacific salmon is programmed in a certain sense, this is clearly a phenomenon different from the virtually ubiquitous aging of organisms that reproduce multiple times (Austad 2004).

The fact that in iteroparous species—those capable of repeated reproduction—senescence is a small contributor to mortality raises the bar over which any plausible program theory must leap. However, to lower the bar by discovering a case where there is significant age-associated mortality does little to help. This is because the evolutionary mechanism on which program theories depend is that of "group selection," and stronger evidence of an effect of aging makes building the case for group selection even harder.

For an individual, senescence is disadvantageous because fertility and vitality are diminished. Therefore, if there is a program for aging, disruption of this program by mutating the responsible genes is likely to produce a "selfish" advantage. The program theory relies on putting the hypothesized benefit to the population (species or group) higher than the disadvantage to the individual. Herein lies a long recognized drawback (Maynard Smith 1976). For group selection to act successfully against opposing selection at the level of the individual, it is required that the species be distributed among isolated groups and that the introduction of a nonaging mutant into a group leads to the group's extinction. Although theoretical modeling indicates that it is possible to create a scenario in which evolution of programmed aging can occur in a suitably spatially structured environment (Travis 2004), the assumptions necessary to support such examples also indicate the severity of the problems facing any general explanation of aging in terms of a genetic program.

The idea that programmed aging is necessary for, or helps, evolution to occur by securing generational turnover faces problems of an equally difficult nature. First, the argument that the long-term advantage of generating evolutionary novelty outweighs the short-term disadvantage of limiting life span depends on assuming a significant rate of change in the environment. Second, selection on the germ-line mutation and/or recombination rate would seem to be a more direct way to achieve the same result. Third, for iteroparous species, which spread their reproduction across time, the critical factor in determining the rate of turnover of generations is not life span per se, so much as the age at which individuals become reproductively mature. Although there is force to the argument that species with long development times may be limited in their adaptability, the fact that such species also tend to have long life spans does not establish that longevity itself poses a disadvantage. This would be true if a long life span necessitated slow maturation, but the causality is much more likely to be the other way around.

PERCEIVED EMPIRICAL CHALLENGES

Single Gene Mutations with Large Life-extending Effects

The classic evolutionary theory of aging predicts that there is likely to be a large number of genes determining longevity. This follows because in the case of the Medawar/Williams scenarios, the theory predicts the existence of whole classes of alleles of which there might be many specific instances, whereas the disposable soma concept applies generally to a wide array of mechanisms for somatic maintenance and repair. Thus, the discovery of genes with large life-extending effects, initially in C. elegans but subsequently in other species, appears to challenge this prediction. In most cases, the genes with major effects on longevity have been found to regulate central aspects of metabolism, particularly with respect to organismal energetics, such as insulin signaling (Partridge and Gems 2002). Furthermore, the relevant pathways are often involved in the response to environmental modulation such as crowding or variation in nutrient abundance. With hindsight, the discovery of these gene effects should perhaps not have been surprising. The optimal allocation of metabolic resources between competing activities such as maintenance, growth, reproduction, and storage lies at the heart of the physiological evolution of life histories, in particular, the trade-off between investment in maintenance and other activities is the mainstay of the disposable soma theory. Organisms that are subject to varying or unpredictable environments are likely to have evolved a regulatory gene hierarchy that can detect change and adjust metabolism accordingly to a different optimum. This appears to be the case, for example, in rodent calorie restriction and in the genes regulating induction of the long-lived stress-resistant dauer larva in *C. elegans.* The insulin-like growth factor-1 (IGF-1) receptor homolog *daf-2* regulates the transcription factor-encoding *daf-16*, which in turn regulates several hundred genes involved in somatic maintenance functions (Murphy et al. 2003). Single gene mutations with large life-extending effects thus constitute a class of high-level regulatory elements that appear to have evolved to modulate the kinds of genes predicted by the classic evolutionary theory of aging.

Mortality Plateaus

The classic evolutionary theory suggests that death rates should increase progressively with age. It is possible that, in some organisms, factors such as a continuing increase in body size, resulting in turn in declining extrinsic mortality and increasing fecundity, might temporarily stave off or even reverse the age-related decline in the force of natural selection (Hamilton 1966; Charlesworth 1980). Indeed, such an explanation might underlie the considerable longevity of some species of fish, first commented upon in relation to their capacity for continued growth by Bidder (1932). However, it is biologically implausible that growth can be sustained at the rate necessary to prevent an eventual collapse in the force of natural selection due to cumulative mortality, regardless of which specific genetic factors regulate the overall rate of aging.

Observations that mortality rates in fruit flies and medflies reached a plateau or even declined at the oldest ages (Curtsinger et al. 1992) and that the rate of mortality increase in humans decelerates after about age 90 (Vaupel 1997; Thatcher 1999) might therefore appear to challenge this expectation (Charlesworth and Partridge 1997; Pletcher and Curtsinger 1998). Although theoretical models have sought to explain mortality plateaus as being actually predicted by the evolutionary theory (Mueller and Rose 1996), these models have their own problems (Kirkwood 1999).

One life history trait that may contribute to the late-life deceleration of mortality is the cessation of reproduction, which may be hazardous in its own right (Sgrò and Partridge 1999). Several nonevolutionary factors may explain the existence of mortality plateaus as well. First, in genetically heterogeneous populations, some slowing of the mortality rate is expected at old age, even if for each individual within the population the risk of dying continues to increase. The reason is simply that frailer individuals die first, so eventually the survivors comprise only the most robust subset. Such heterogeneity explained mortality leveling in *C. elegans*, since it was observed that an isogenic population showed less evidence of a plateau than a mixed population (Brooks et al. 1994). However, this was not true for the fruit fly where similar mortality leveling was seen in genetically homogeneous and heterogeneous stocks (Curtsinger et al. 1992). Second, even in genetically homogeneous populations, there is marked stochastic variation in the development of the senescent phenotype. Such stochastic events may produce a variance in frailty, which will then result in mortality leveling as outlined above. Third, there may be alterations in behavior of very old animals that can result in a reduction in mortality risk. For example, old flies tend to crawl instead of flying, and very old people may modify their lifestyle as well as receiving extra care. Finally, there may simply be biomechanical constraints that tend to prevent an ever-increasing mortality rate. For example, Gavrilov and Gavrilova (2001) have argued that the principles of reliability engineering may be important for understanding mortality leveling at old age.

Menopause

The occurrence of menopause—the universal cessation of human female fertility at approximately age 50—presents an intriguing evolutionary puzzle, which is sometimes used to support the idea of programmed aging. Why should a woman cease reproducing at a much earlier age relative to her biological life span potential than occurs in other mammals? There are two credible possibilities. First, it may be that humans do not differ from other mammals as much as it might seem. The earlier and more abrupt decline in reproduction seen in women relative to men is observed in a strikingly similar fashion in rats and mice living under the protected conditions of the laboratory (Austad 1994). Moreover, the pattern of the age-specific increase in the probability of undergoing menopause mirrors almost exactly the age-specific increase in the probability of death that aging causes, and this pattern is very different for a clearly adaptive trait such as puberty (Bronikowski and Promislow 2005). Thus, it is possible that for the vast majority of human evolutionary history, women for the most part only survived about as long as their egg supply (the depletion of which is the proximate cause of menopause), and only began experiencing menopause routinely as humans achieved longer and longer lives over the past several millennia.

An alternative, clearly more intriguing, explanation for the evolution of the menopause may be found in the unique combination of circumstances that define the human life history (Kirkwood 1997). The pressure to evolve increased life spans was probably driven by the increase in human brain size, leading to advanced intelligence, tool use, and social living, all of which will have reduced the level of extrinsic mortality and favored increased investments in somatic maintenance. Increased neonatal brain size, however, makes giving birth riskier. The result appears to have been a compromise whereby, in comparison with other mammals, the human infant is born unusually altricial (i.e., requiring extended postnatal development before gaining independence from the mother) while still possessing an unusually large head. This has led to the suggestion that menopause protects older mothers from the risks of late child-bearing, when senescence may make pregnancy and child-bearing less safe, and favors the survival of the mother to raise her existing children to independence. An alternative is that postreproductive females may gain more by contributing to the reproductive success of their offspring, through helping to care for and provision their offspring, than they would gain from attempting further reproduction of their own. The latter "grandmother" hypothesis has attracted powerful empirical support from anthropological studies (Hawkes et al. 1998; Lahdenpera et al. 2004), with evidence that maternal grandmothers can improve nutritional status and survival of children. Recent theoretical modeling indicates, however, that neither of the two hypotheses outlined above—the "maternal mortality" and "grandmother" hypotheses—is in fact adequate on its own. Only when both are taken together in a combined model can they show that menopause does indeed confer an evolutionary advantage (Shanley and Kirkwood 2001). This is important because it may explain why menopause is essentially unique to our species, in which this combination of factors has occurred. In essence, it is this combination, representing a convergence of biological and cultural evolution, that conferred sufficient biological value on older women that menopause evolved as an adaptation to reflect this value in evolving human social groups.

Testing whether menopause has an adaptive basis or merely reflects the absence of selection to increase the functional lifetime of the ovary in line with increasing human longevity is clearly an interesting challenge for evolutionary life history research.

Aging in Unicellular Organisms

The original formulations of the classic evolutionary theory, particularly antagonistic pleiotropy and disposable soma, sought to explain the evolution of somatic mortality, as opposed to the immortality of the germ line. The implication is that unicellular organisms, which lack a soma (in the commonly understood sense), should be immortal. However, during the 1980s, the budding yeast *Saccharomyces cerevisiae* became firmly established as an experimental model for research on aging. In this case, one might stretch a point and suggest that the "mortal" mother cell can, in a sense, be seen as soma, whereas the smaller bud that becomes the daughter can be seen as the germ line (Lai et al. 2002). The same argument might even include the bacterium *Caulobacter crescentus* which divides asymmetrically and also exhibits a form of aging (Ackermann et al. 2003). The case became harder, however, with reports of aging in fission yeast *Schizosaccharomyces pombe* (Barker and Walmsley 1999) and *Escherichia coli* (Stewart et al. 2005).

A resolution to this seeming challenge comes from looking more closely at the molecular and cellular basis of aging (Ackermann et al. 2007). The origin of the aging process in multicellular animals arose from the division of labor between the germ line and soma (Weissman 1889). As soon as the germ-line/soma distinction evolved, only the germ cells carried the responsibility for forming individuals of the next generation, freeing somatic cells to become specialists, such as neurons, muscle cells, or cells in the lens of the eye. This came at a price, however, because the soma then became disposable. What the recent work has shown is that a division of labor exists even in *E. coli*, highlighting the fundamental importance of reproductive *asymmetry* in creating a context for aging to evolve (Kirkwood 1981; Partridge and Barton 1993). The germ-line/soma distinction is a *sufficient* instance of such asymmetry, but it is not a *necessary* one. Although *E. coli* appears to divide symmetrically, in molecular terms, it does not in fact do so. One daughter cell receives the old cell pole and the other cell receives a new pole. The difference is apparently enough to cause a decline in fitness of the daughter that receives the old pole and thus does not benefit from the complete renewal of its molecular structures.

Apoptotic Pathways in Single-cell Organisms

Yeast cells have been described to undergo a process closely resembling the apoptosis of cells in higher eukaryotes, prompting speculation about whether this might represent a form of programmed aging and death (Büttner et al. 2006). The idea that a single-celled organism might undergo suicide begs the question of "Who benefits"? Suggested benefits for yeast apoptosis are that it induces cell death when mating is unsuccessful, thereby eliminating infertile cells, or that in situations where microorganisms cluster together to survive nutrient depletion, it serves to destroy infected or damaged cells, which might otherwise consume dwindling nutrients or spread infection. In these respects, the hypothesized role of apoptosis is rather similar to that in multicellular organisms where (apart from the tissue morphogenetic role of cell death during development) the primary function of apoptosis is to delete cells that are damaged or surplus to requirement, and where it works because cells share essentially the same genome. Although the possibility of such an adaptive role for apoptosis in single-cell organisms is certainly conceivable, particularly in populations of very close genetic kin, we are not aware that the idea has yet been developed in a sufficiently quantitative manner that would substantiate the hypothesized fitness advantage, nor does the idea obviate any of the difficulties outlined earlier in sustaining the theory of programmed aging in multicellular animals.

NEOCLASSICAL THEORY

Two standard predictions from the classical senescence theory have more recently provoked renewed interest of theorists. The first prediction is that for any species with age-structured populations, the evolution of senescence is inescapable, or as stated by Hamilton (1966), "for organisms that reproduce repeatedly, senescence is to be expected as an inevitable consequence of the workings of natural selection." We call this the *senescence inevitability* prediction. A possible empirical challenge to this notion is that limited information on some very long-lived iteroparous species, such as the Rougheye Rockfish (*Sebastes aleutianus*) that can live two centuries (Cailliet et al. 2001), suggests that they may not senesce appreciably (de Bruin et al. 2004), a phenomenon termed negligible senescence (Finch 1990). The second prediction (the *extrinsic hazards* prediction) specifies that low levels of extrinsic hazards, other factors being equal, should lead to the evolution of slow aging and high levels to the evolution of fast aging (Williams 1957). Although this prediction has been supported in both experimental laboratory studies on fruit flies (Stearns et al. 2000) and field observations (Austad 1993) and by comparative biology, in which animals particularly resistant to external hazards because they possess spines or armor, have exceptional abilities to escape danger such as the ability to fly, or live in protected ecological niches, have been noted to live exceptionally long lives and deteriorate physically relatively slowly (Kirkwood and Austad 2000). However, a recent, very detailed, study of guppies derived from two high- and two low-predation stream environments found largely the opposite result (Reznick et al. 2004). Although they reached maturity earlier and reproduced more copiously as standard evolutionary life history theory would predict (Stearns 2004), guppies from the high-predation streams lived and reproduced longer than those from the low-predation streams. Somewhat puzzlingly, given these demographic findings, guppies from the high-predation streams deteriorated more rapidly in a measure of physical performance, maximum acceleration during an escape response, than did the shorter-lived fishes from the low-predation streams.

The standard predictions of classical evolutionary senescence theory rest, of course, on the assumptions of the quantitative models that make the predictions. All mathematical models have implicit simplifying assumptions, and when the assumptions are not met, the models' predictions may no longer hold. For instance, a short by no means exhaustive list of the assumptions implicit in Hamilton's (1966) original model include (1) that the organism modeled reproduces asexually, (2) that population growth is density-independent, and (3) that the Malthusian parameter, r, is a suitable measure of a genotype's evolutionary success. There are numerous well-known conditions such as a rapidly declining population or very strong selection under which assumption 3 is known to be invalid. However, Hamilton's major contribution, and later Charlesworth's (1980), was to note how selection would be altered particularly with respect to the evolution of aging with age-specific life history alterations as opposed to ignoring age-specificity, rather than to make evolutionary predictions that were valid under all conditions (Rose 1991).

Abrams (1993) modified Hamilton's original approach to incorporate a variety of ecological scenarios to ask how general the extrinsic hazards prediction was, and he noted that if the assumption of density independence was violated, the standard prediction does not hold. Moreover, further analyses by the same author pointed out how the exact nature of density dependence—specifically, the extent to which all age classes are affected equally or not—also affects the evolutionary predictions of the model. Given the appropriate conditions, a decrease in extrinsic mortality would be predicted to decrease, increase, or not change. Of course, Hamilton's analysis was never intended to be exhaustive, covering every conceivable population. Rather, it was a first quantitative venture into how natural selection might mold senescence given certain age-specific fitness effects.

The question of the relationship between extrinsic mortality risk and rate of aging has recently been reexamined in an exchange between P.D. Williams et al. (2006) and Caswell (2007), which reiterates the points made in Abrams' (1993) treatment. As Caswell points out, if the extrinsic mortality rate is increased uniformly across all adult age ranges, all that happens is that the population's intrinsic rate of natural increase is depressed by an equal amount, without altering how the force of natural selection changes with age. To change the latter, it is necessary that some additional factor, such as might be mediated, for example, through population density effects, be brought into play. An example can be found in the model used by Kirkwood and Rose (1991) to explore how variation in extrinsic mortality might affect the optimum investment in somatic maintenance and repair. Simply varying extrinsic mortality has no effect on this optimum; however, when the optimization was made subject to the constraint that at the optimum, the intrinsic rate of natural increase in the population was zero, i.e., the population was kept at equilibrium, the result revealed a clear trade-off. The developments in the neoclassical theory, which is being driven both by new experimental data and by new insights into the dynamics of selection within age-structured populations, are helping to illuminate the important interplay between population genetics and population dynamics. In coming to grips with these issues, both population genetic models and optimality models have roles to play.

The senescence inevitability prediction has been a particular focus of theorists recently. An optimality model constructed from standard evolutionary life history variables, including an assumption of trade-offs among these variables plus a concept of vitality, which seeks to measure the fitness consequences of declining fertility and increasing mortality with age, determined that nonaging organisms could evolve under certain conditions (Sozou and Seymour 2004). Under the assumptions of this model, nonaging organisms are most likely to evolve when a small degree of physical deterioration soon after maturity leads to a large and rapid decrease in reproductive fitness in an environment with low extrinsic hazards and little population growth. This would seem to be an unlikely combination of conditions under most plausible ecological scenarios.

Perhaps a more powerful rebuttal to the senescence inevitability prediction uses the same general approach as Hamilton's original formulation (Baudisch 2005). The author points out that the inescapable decline in the force of natural selection with age depends entirely on the manner in which Hamilton parameterized survival. Specifically, Hamilton parameterized the force of natural selection with respect to mortality as the change in fitness, r, associated with the logarithm of an age-specific change in survival. Specifically,

$$\frac{dr}{d \ln p_a} = \frac{\sum_{x=a+1}^{\infty} e^{-rx} l_x m_x}{\sum_{x=0}^{\infty} x\, e^{-rx} l_x m_x}$$

where x is age, l_x is survival to age x, m_x is fecundity at age x, and p_a is the age-specific probability of survival from age a to a + 1, such that $l_a = p_0, p_1, p_2 \ldots p_{a-1}$. Note that by this definition $p_a = 1 - q_a$, where q_a is the age-specific mortality rate and also $p_a = e^{-\mu}$, where μ_a is the mean of the instantaneous death rate, μ, between a and a + 1. Thus, Hamilton chose a reasonable method of parameterizing the force of natural selection, but instead of using $dr/d \ln p_a$, he could have used dr/dp_a, dr/dq_a, $dr/d \ln q_a$, $dr/d \ln \mu_a$ or several other formulations (Baudisch 2005). Interestingly, Baudisch shows that all of these alternative formulations can lead to a force of natural selection that decreases, increases, or remains the same with age depending on the pattern of mortality and fertility at subsequent ages. Specifically, if survival and/or fertility rise sufficiently at later ages, the force of natural selection could increase with age such that genes that enhance somatic maintenance late in life could be evolutionarily advantaged, and conversely. She also notes that Hamilton's formulation implies that the effects of genetic mutations on mortality and fertility are additive—a reasonable assumption—but not the only plausible assumption. Some formulations capture the possibility that the effects of such mutations are proportional on mortality and fertility. Which, if either, of these is actually the case is an empirical question.

Both of the papers discussed above have shown convincingly that senescence is not a mathematically inevitable consequence of growing older. However, whether negligible or even antisenescence could evolve depends ultimately on realistic combinations of ecological conditions and on how genes actually *do* impact age-specific mortality and fertility rates, rather than how they *could* impact them. Thus, although it is theoretically possible that negligible senescence exists somewhere in the animal world, it has not yet been convincingly described for any species. Notably, individuals of all species for which we have reasonable documentation of demography under protected conditions and know something about physiological changes with age do indeed exhibit senescence, even such species such as the bacterium *E. coli* which were previously thought not to age (Stewart et al. 2005). If negligible senescence does exist, it must be exceedingly rare. Long observation (and great patience) would be needed to determine if it truly exists.

SYSTEMS/NETWORK MODELING: WHAT CAN IT CONTRIBUTE?

An important prediction of the evolutionary theory of aging, particularly the disposable soma theory, is that multiple kinds of damage contribute to the lifelong accumulation of molecular and cellular defects that cause senescence. This is both challenging and enabling because although experiments show that the various hypothesized varieties of individual lesions do accumulate during aging, for *none* of them is there compelling evidence that any one of these varieties is by itself sufficient to account for normal age-related frailty, disability, and disease. This has led to recent initiatives to develop "network" theories of aging in which the contributions of the various mechanisms are considered together, thereby allowing for interaction and synergism between different processes (Kirkwood et al. 2003).

A further attraction of the systems approach is that although the various mechanisms comprising the network are likely to operate to some degree in all cell types and in all species, there may be important differences concerning which mechanisms are more important. All cells share a basic vulnerability to damage affecting key macromolecules such as DNA and proteins, particularly when this damage arises from generic sources such as endogenous oxidative stress caused by ROS (reactive oxygen species). However, cells in actively proliferating tissues are more vulnerable than postmitotic cells to suffer somatic mutations and telomere erosion because of the repeated requirement for DNA replication. However, in these tissues, damaged cells are easily replaced so *observed* damage may be lower. Conversely, postmitotic cells are more vulnerable to accumulation of aberrant proteins and metabolic wastes through failure of turnover processes, since in dividing cells, any such accumulation will be diluted by the synthesis of new cellular constituents during mitosis. Thus, although the network of mechanisms underlying cellular aging may share common components across all cell types, the relative importance of these components may differ. Furthermore, if dividing cells reduce their rate of division in later life as a consequence of intrinsic molecular aging, they may then undergo a corresponding shift in the balance between different mechanisms of molecular damage accumulation. Similar considerations apply to differences between species that may differ in the extent of cell renewal or exposure to specific types of molecular damage.

REFERENCES

Abrams P.A. 1993. Does increased mortality favor the evolution of more rapid senescence? *Evolution* **47:** 877–887.

Ackermann M., Stearns S.C., and Jenal U. 2003. Senescence in a bacterium with asymmetric division. *Science* **300:** 1920.

Ackermann M., Chao L., Bergstrom C.T., and Doebeli M. 2007. On the evolutionary origin of aging. *Aging Cell* **6:** 235–244.

Austad S.N. 1993. Retarded senescence in an insular population of Virginia opossums. *J. Zool.* **229:** 695–708.

———. 1994. Menopause: An evolutionary perspective. *Exp. Gerontol.* **29:** 255–263.

———. 2004. Is aging programed? *Aging Cell* **3:** 249–251.

Bahar R., Hartmann C.H., Rodriguez K.A., Denny A.D., Busuttil R.A., Dolle M.E., Calder R.B., Chisholm G.B., Pollock B.H., Klein C.A., and Vijg J. 2006. Increased cell-to-cell variation in gene expression in ageing mouse heart. *Nature* **441:** 1011–1014.

Barker M.G. and Walmsley R.M. 1999. Replicative ageing in the fission yeast *Schizosaccharomyces pombe. Yeast* **15:** 1511–1518.

Baudisch A. 2005. Hamilton's indicators of the force of selection. *Proc. Natl. Acad. Sci.* **102:** 8263–8268.

Bidder G.P. 1932. Senescence. *Br. Med. J.* **2:** 583–585.

Bronikowski A.M. and Promislow D.E. 2005. Testing evolutionary theories of aging in wild populations. *Trends Ecol. Evol.* **20:** 271–273.

Brooks A., Lithgow G.J., and Johnson T.E. 1994. Mortality rates in a genetically heterogeneous population of *Caenorhabditis elegans. Science* **263:** 668–671.

Brunet-Rossinni A.K. and Austad S.N. 2006. Senescence in wild populations of mammals and birds. In *Handbook of the biology of aging* (ed. E.J. Masoro and S.N. Austad), pp. 243–266. Academic Press, San Diego, California.

Büttner S., Eisenberg T., Herker E., Carmona-Gutierrez D., Kroemer G., and Madeo F. 2006. Why yeast cells can undergo apoptosis: Death in times of peace, love and war. *J. Cell Biol.* **175:** 521–525.

Cailliet G.M., Andrews A.H., Burton E.J., Watters D.L., Kline D.E., and Ferry-Graham L.A. 2001. Age determination and validation studies of marine fishes: Do deep-dwellers live longer? *Exp. Gerontol.* **36:** 739–764.

Caswell H. 2007. Extrinsic mortality and the evolution of senescence. *Trends Ecol. Evol.* **22:** 173–174.

Charlesworth B. 1980. *Evolution in age-structured populations.* Cambridge University Press, Cambridge, United Kingdom.

Charlesworth B. and Partridge L. 1997. Ageing: Levelling of the grim reaper. *Curr. Biol.* **7:** R440–R442.

Curtsinger J.W., Fukui H.H., Townsend D.R., and Vaupel J.W. 1992. Demography of genotypes: Failure of the limited life-span paradigm in *Drosophila melanogaster*. *Science* **258:** 461–463.

de Bruin J.P., Gosden R.G., Finch C.E., and Leaman B.M. 2004. Ovarian aging in two species of long-lived rockfish, *Sebastes aleutianus* and *S. alutus*. *Biol. Reprod.* **71:** 1036–1042.

Finch C.E. 1990. *Longevity, senescence, and the genome.* University of Chicago Press, Chicago, Illinois.

Gavrilov L.A. and Gavrilova N.S. 2001. The reliability theory of aging and longevity. *J. Theor. Biol.* **213:** 527–545.

Hamilton W.D. 1966. The moulding of senescence by natural selection. *J. Theor. Biol.* **12:** 12–45.

Hawkes K., O'Connell J.F., Jones N.G., Alvarez H., and Charnov E.L. 1998. Grandmothering, menopause, and the evolution of human life histories. *Proc. Natl. Acad. Sci.* **95:** 1336–1339.

Jenkins N.L., McColl G., and Lithgow G.J. 2004. Fitness cost of extended lifespan in *Caenorhabditis elegans. Proc. Biol. Sci.* **271:** 2523–2526.

Kirkwood T.B.L. 1977. Evolution of ageing. *Nature* **270:** 301–304.

———. 1981. Repair and its evolution: Survival versus reproduction. In *Physiological ecology: An evolutionary approach to resource use* (ed. C.R. Townsend and P. Calow), pp. 165–189. Blackwell Scientific, London, United Kingdom.

———. 1985. Comparative and evolutionary aspects of longevity. In *Handbook of the biology of aging* (ed. C.E. Finch and E.L. Schneider), pp. 27–44. Van Nostrand Reinhold, New York, New York.

———. 1997. The origins of human ageing. *Philos. Trans. R. Soc. Lond. B Biol. Sci.* **352:** 1765–1772.

———. 1999. Evolution, molecular biology and mortality plateaus. In *Molecular biology of ageing: Alfred Benzon Symposium* (ed. V.A. Bohr et al.), vol. 44, pp. 383–390. Munksgaard, Copenhagen, Denmark.

Kirkwood T.B.L. and Austad S.N. 2000. Why do we age? *Nature* **408:** 233–238.

Kirkwood T.B.L. and Rose M.R. 1991. Evolution of senescence: Late survival sacrificed for reproduction. *Philos. Trans. R. Soc. Lond. B Biol. Sci.* **322:** 15–24.

Kirkwood T.B.L., Boys R.J., Gillspie C.S., Proctor C.J., Shanley D.P., and Wilkinson D.J. 2003. Towards an e-biology of ageing: Integrating theory and data. *Nat. Rev. Mol. Cell Biol.* **4:** 243–249.

Lahdenpera M., Lummaa V., Helle S., Tremblay M., and Russell A.F. 2004. Fitness benefits of prolonged post-reproductive lifespan in women. *Nature* **428:** 178–181.

Lai C.Y., Jaruga E., Borghouts C., and Jazwinski S.M. 2002. A mutation in the *ATP2* gene abrogates the age asymmetry between mother and daughter cells of the yeast *Saccharomyces cerevisiae. Genetics* **162:** 73–87.

Libertini G. 1988. An adaptive theory of the increasing mortality with increasing chronological age in populations in the wild. *J. Theor. Biol.* **132:** 145–162.

Maynard Smith J. 1976. Group selection. *Q. Rev. Biol.* **51:** 277–283.

Medawar P.B. 1952. *An unsolved problem in biology.* H.K. Lewis, London, United Kingdom.

Mueller L.D. and Rose M.R. 1996. Evolutionary theory predicts late-life mortality plateaus. *Proc. Natl. Acad. Sci.* **93:** 15249–15253.

Murphy C.T., McCarroll S.A., Bargmann C.I., Fraser A., Kamath R.S., Ahringer J., Li H., and Kenyon C. 2003. Genes that act downstream of DAF-16 to influence the lifespan of *Caenorhabditis elegans. Nature* **424:** 277–283.

Partridge L. and Barton N. 1993. Optimality, mutation and the evolution of ageing. *Nature* **362:** 305–311.

Partridge L. and Gems D. 2002. Mechanisms of ageing: Public or private? *Nat. Rev. Genet.* **3:** 165–175.

Pletcher S.D. and Curtsinger J.W. 1998. Mortality plateaus and the evolution of senescence: Why are the old-age mortality rates so low? *Evolution* **52:** 454–464.

Reznick D.N., Bryant M.J., Roff D., Ghalambor C.K., and Ghalambor D.E. 2004. Effect of extrinsic mortality on the evolution of senescence in guppies. *Nature* **431:** 1095–1099.

Rose M.R. 1989. Genetics of increased lifespan in *Drosophila*. *Bioessays* **11:** 132–135.

———. 1991. *Evolutionary biology of aging.* Oxford University Press, Oxford, United Kingdom.

Sgrò C.M. and Partridge L. 1999. A delayed wave of death from reproduction in *Drosophila*. *Science* **286:** 2521–2524.

Shanley D.P. and Kirkwood T.B.L. 2001. Evolution of the human menopause. *Bioessays* **23:** 282–287.

Sozou P.D. and Seymour R.M. 2004. To age or not to age. *Proc. Biol. Sci.* **271:** 457–463.

Stearns S.C. 2004. *The evolution of life histories.* Oxford University Press, Oxford, United Kingdom.

Stearns S.C., Ackermann M., Doebeli M., and Kaiser M. 2000. Experimental evolution of aging, growth, and reproduction in fruitflies. *Proc. Natl. Acad. Sci.* **97:** 3309–3313.

Stewart E.J., Madden R., Paul G., and Taddei F. 2005. Aging and death in an organism that reproduces by morphologically symmetric division. *PLoS. Biol.* **3:** e45.

Thatcher A.R. 1999. The long-term pattern of adult mortality and the highest attained age. *J.R. Stat. Soc. Ser. A Stat. Soc.* **162:** 5–43.

Travis J.M. 2004. The evolution of programmed death in a spatially structured population. *J. Gerontol. A Biol. Sci. Med. Sci.* **59:** 301–305.

Vaupel J.W. 1997. The remarkable improvements in survival at older ages. *Philos. Trans. R. Soc. Lond. B Biol. Sci.* **352:** 1799–1804.

Walker D.W., McColl G., Jenkins N.L., Harris J., and Lithgow G.J. 2000. Evolution of lifespan in *C. elegans*. *Nature* **405:** 296–297.

Weissman A. 1889. *Essays upon heredity and kindred biological problems.* Clarendon Press, Oxford, United Kingdom.

Williams G.C. 1957. Pleiotropy, natural selection, and the evolution of senescence. *Evolution* **11:** 398–411.

Williams P.D., Day T., Fletcher Q., and Rowe L. 2006. The shaping of senescence in the wild. *Trends Ecol. Evol.* **21:** 458–463.

Wynne-Edwards V.C. 1962. *Animal dispersion in relation to social behaviour.* Oliver and Boyd, Edinburgh, Scotland, United Kingdom.

Zwaan B.J., Bijlsma R., and Hoffmann R.S. 1995. Direct selection of lifespan in *Drosophila melanogaster*. *Evolution* **49:** 649–659.

5

An Overview of the Biology of Aging: A Human Perspective

George M. Martin
Departments of Pathology and Genome Sciences
University of Washington
Seattle, Washington 98195

Caleb E. Finch
Department of Biological Sciences and Davis School of Gerontology
University of Southern California
Los Angeles, California 90089

THE POWER OF GENETIC ANALYSIS IN MODEL ORGANISMS, notably yeast, worms, flies, and mice, has yielded enormous understanding of the genetic modulations of life span, as clearly demonstrated in the pages of this volume. We now have at hand the first demonstrable "public" mechanism for life span modulation—variations in the function of the insulin/insulin-like growth factor-1 (IGF-1) signaling pathway, a signaling transduction network that evolved in nature to enhance reproductive fitness in the context of metabolic trade-offs (Partridge and Gems 2002). These lab creations, however long-lived they may be, would be unlikely to survive as long in the real environments wherein the wild types had evolved (Walker et al. 2000).

Recognizing the divide between the lab and the Darwinian world should not deter us from further digging in the gold mine of model systems and looking for yet other gero-rich veins. We anticipate that other insect species with multifarious diapauses and life history alternates in nature (Brown and Hodek 1983; Finch 1990) might be "tweaked" by investigators to learn their secrets of slowing and accelerating biological time through epigenetically determined alternative developmental phenotypes. Enticing prospects are offered by the social insects, with many examples of life history alternates, best known at present in the 100-fold difference in life spans of worker bees and queen bees. This is becoming an especially attractive area of research, given the growing genome database of social insects (Birney et al. 2006) and the emergence of sociogenomics (Robinson et al. 2005).

Our tidy yeast and invertebrate models will continue to be wonderful for identifying genes of interest to aging and for testing human disease genes. Research on these model systems has not been without its problems, however. First and foremost, we know very little about the details of aging at the cell and physiological levels in the short-lived fly and worm. Do they die of definable diseases, like lab mice? Moreover, normal aging adult flies and worms fail to exhibit at least two key features of human aging. First, they do not exhibit the alterations in proliferative homeostasis that characterize many common late-life human disorders (e.g., osteoarthritis, atherosclerosis, benign prostatic hyperplasia, and benign and malignant neoplasia). (We can be hopeful, however, that basic principles will emerge from studies, in invertebrates, of aberrant germ-line proliferations [Pinkston et al. 2006] and abnormal somatic cell proliferations during development [see, e.g., Beaucher et al. 2007].) Second, although there are fascinating chemical communications between invertebrates capable of changing behavior, humans have unique multigenerational social interactions that are far more complex than those found in the compressed generations of fly and worm. For example, leukocyte telomeres were found to be shorter in identical twins with lower socioeconomic status (Cherkas et al. 2006) and in mothers of children with disabilities, interpreted as a response to stress (Epel et al. 2004). One might also ask, How "wild" is your "wild type?" Although the artificial selection of conventional wild-type strains for fast growth and high fecundity has permitted great progress in genetic analysis, this has come with a price. Wild-caught mice are remarkable for their slower maturation, smaller size, and longer life spans as compared to lab rodents, as shown in important studies by Austad and Miller (Miller et al. 2002). Each wild-type strain likely has some idiosyncratic baggage of special genetic vulnerabilities. Thus, an observed increase in life span following a genetic manipulation might be merely addressing a private modulation of aging that requires replication in additional strains for proper interpretation. Spencer and Promislow (2002) have nicely summarized these issues.

Given the above discussion, and in the interest of suggesting priorities for future research, we point to the most relevant animal for studies of aging, *Homo sapiens*. We believe it timely and urgent to move biogerontological research into the biology of human aging. Yes, yes! We must accept the inconveniently long human life span and also defer to human subject guidelines in not trying to direct mating behavior. Let us consider the advantages. First of all, no animal on earth has revealed more about its individual life history of health and disease and has more diverse and interesting populations. We include the domains of demography, sociology, psychology and psychiatry, economics, physiology, reproductive biology, anatomy, biochemistry, microbiology, pathology, pharmacology, toxicology, among other sources of testable hypotheses in the human sciences. What was missing until recently was the formal genetics of humans. That has now changed. A vast store of information is rapidly accumulating on the human genome. To cite just one advance, we now know that, in addition to millions of single-nucleotide base-pair polymorphisms, there are also many variations in gene copy number, likely the result of repeated unequal crossing-over between flanking repeat sequences (Locke et al. 2006; Sharp et al. 2006). Powerful new statistical and molecular technologies are emerging for whole-genome association and linkage studies. Bioengineering is producing increasingly sensitive and comparatively noninvasive methods for functional studies. Unfortunately, given recent budgetary constraints, at least in the United States, the large grants needed for such extensive, long-term longitudinal studies will be difficult to fund. More on this later.

It is always a good idea to provide some basic definitions in a chapter that puts forth an opinionated overview of opportunities for research in the biology of aging. Everyone understands aging of living organisms to mean a degenerative change over time. Why then use the word senescence? This distinction is useful because some aging changes begin even before birth and only much later lead to the accelerating mortality risks that are the domain of organismic senescence. For example, the number of ovarian oocytes is determined during development and declines irreversibly after birth; even before puberty, half the eggs are gone. Again, fetal arteries have microscopic aggregates of oxidized lipids and activated macrophages that some consider to be prodromal atherosclerotic plaques. Inflammatory processes are already smoldering in arteries at the time of birth (Yamashita et al. 2006). Arterial elastin also begins to age soon after birth: Because of extremely slow turnover (Keeley et al. 2002), elastin accumulates oxidative damage and cross-links that contribute to the inexorable increase of systolic pressure during aging. Thus, biogerontologists, like our colleagues in social gerontology, should be concerned with all stages of the life course (Fig. 1).

Some postmaturational alterations in structure and function are engaged in order to compensate for deleterious changes. One might use

The Six Stages of the Life Course

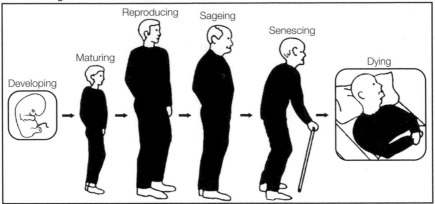

Figure 1. Gene actions at all six stages of the human life cycle have the potential to modulate life span, health span, and duration of life. A period of "sageing" is envisioned as one that increasingly implements compensatory mechanisms for physiological declines; when these fail, senescence ensues.

the term "sageing" to refer to such compensations (Martin 1997). The mechanisms of sageing can be presumed to have evolved for part-time use in young organisms in order to maintain homeostasis during unusual physiological challenges. At any age after 30, each of us will recall giving up something from the prior decade. Sageing is likely to be engaged increasingly with advancing age, when a plethora of physiological declines gradually and insidiously emerge. The ages and stages when sageing fades into frailty and senescence can provide reasonable operational definitions of the times of onset of various senescent phenotypes (Fig. 1).

What phenotypes should we study in *Homo sapiens*? Familial aggregates of exceptional longevity are certainly of interest and are already being examined via linkage analysis, with some preliminary results (Geesaman et al. 2003). The same is true of association studies, some of which have incorporated, as controls, the progeny of centenarians and their ethnically and age-matched spouses or neighbors (Barzilai et al. 2001). As we discuss below, however, there is concern that "Lady Luck" is the major contributor to the achievement of exceptionally long (or short) life spans within a species. Moreover, allelic variations at numerous loci are likely to each contribute only comparatively small effects. Given the exceptional vulnerability of many contemporary human populations to cardiovascular pathology, however, it is perhaps not surprising that relevant loci are beginning to provide positive signals. Besides the well-known advantages of the *ApoE3* allele in favoring later-age arterial and brain health, several other lipoprotein system gene variants are candidates in centenarians (Atzmon et al. 2006). In contrast to the lively interest in genetic research with nonogenarians, centenarians, and even super-centenarians, there has been neglect of the genetic analysis of changes in specific cellular and physiological functions during middle age. This would surely be a disappointment to Nathan Shock, who devoted his career to the groundwork for such studies in the Baltimore Longitudinal Study of Aging.

There are numerous advantages of a research focus on middle-aged subjects. First of all, in contrast to centenarians, enthusiastic and compliant subjects are much more available. Second, there is ample evidence for measurable declines in brain, myocardium, and reproductive functions by the 40s in individuals without any overt pathology. This is indeed predicted by the venerable evolutionary biological theory of aging. In essence, senescence is considered as an array of phenotypes resulting from evolutionary neglect. These adverse phenotypes arise haphazardly as the force of natural selection fades after the median life span, which was about 40 years until very recently (Hamilton 1966; Charlesworth 1994). Third, multiple generations are often available, providing the potential to determine the phase relationships of alleles. Fourth, there is the potential for such methods as sib-pair analysis. Fifth, in contrast to centenarians, assays for physiological functions are not complicated by the effects of multiple comorbidities. Alas, this rosy scenario is deficient because we have yet to develop batteries of highly sensitive, comparatively noninvasive assays for a wide range of physiological functions (Martin 2002). We should be turning to our colleagues in bioengineering for progress in such methods, as our friends in departments of physiology seem to be fully preoccupied with cellular and molecular biology these days.

Returning to our esteemed Nathan Shock, we remember his blunt but jovial admonition to the emerging group of molecular gerontologists of three decades ago: "*Why can't you smartaleck molecular biologists tell me why a danged mouse only lives two years and a darned dog only fifteen?*" We are still mute. No one doubts that species differences in life span are genetically seated. In contrast to the mechanisms underlying *intraspecific* variations in life span (discussed below), the constitutional genome trumps the influence of both stochastic and environmental influences. Our most robust environmental manipulation of life span in lab mice, partial dietary restriction, can boost their longevity by a couple of years or so, but the evolutionary remolding of the basic mammalian genome has created humans living up to 30 times as long (Fig. 2) and bowhead whales living perhaps 60 times as long (George et al. 1999). No

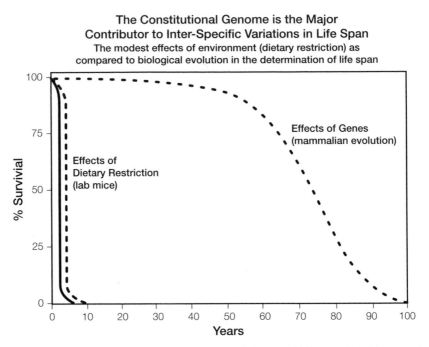

**The Constitutional Genome is the Major
Contributor to Inter-Specific Variations in Life Span**
The modest effects of environment (dietary restriction) as
compared to biological evolution in the determination of life span

Figure 2. The impact of an environmental modulation of life span (in this case, the effects of caloric restriction on the life span of a laboratory mouse) is not very impressive when examined in the context of what evolution can achieve (in this case, the emergence of a long-lived mammal known as *Homo sapiens*).

single trait or group of traits, however, can yet explain much about species differences. At the gross level, the schedule of aging scales approximately with life span. A canonical suite of mammalian aging changes includes reproductive senescence, loss of arterial elasticity, and bone thinning, features that are universal after midlife in mice, dogs, monkeys, and humans. Few expect a single molecular or cell chronometer to govern the rate of aging, and most anticipate a large number of genetic changes, each with small effect in each evolutionary line.

The evolution of the *APOE* locus may be a good model for the study of small, single-gene effects. The ancestral human allele is *APOE4*, the "bad" gene that is a risk factor for vascular events and for accelerated Alzheimer's disease. The *APOE3* allele spread in human populations about 200,000 years ago (Fullerton et al. 2000) and may have been advantageous to health in middle-age caregivers and mentors who have unique roles in human societies (Finch and Sapolsky 1999). The persistence of the life-shortening *APOE4* allele is hypothesized to be protective in lipophilic microbial infections (Martin 1999). Recent evidence, in fact, shows that the ApoE protein mediates the presentation of lipophilic antigens (van den et al. 2005) and that *APOE*4 carriers have less severe damage during hepatitis C virus infections (Wozniak et al. 2002; Fabris et al. 2005).

Gerontologists have puzzled over the observations of marked variations in the life spans of genetically defined organisms despite attempts to control the environment in which they age (Finch and Kirkwood 2000). From worms to human twins, about 30% of the variance in life span is heritable (Finch and Tanzi 1997). The experiments of Vanfleteren et al. (1998) with *Caenorhabditis elegans* are particularly instructive. The usual remarkable variance in the life spans of these genetically identical organisms was observed despite the improved environmental controls provided by growth in suspension culture and the use of axenic media. This extreme variance even applies to long-lived *C. elegans* mutants, some of which do not live as long as members of some members of parental wild-type organisms (Fig. 3) (Kirkwood and Finch 2002). Recent studies of gene expression in young and old human identical twins, including molecular assessments of the regulation of gene expression, have demonstrated discordant epigenetic drifts among the old twin pairs, which we can presume are influenced by both stochastic and environmental influences (Fraga et al. 2005). The best current evidence for a key role of stochastic events comes from experiments with *C. elegans* carried out in the laboratory of Tom Johnson (Rea et al. 2005). Recognizing the observations that resistance to thermal stress (and other stresses) was a typical characteristic of long-lived mutant worms, the Johnson group introduced a transgene bearing the promoter for a small heat shock protein in tandem with a reporter (green fluorescent protein). This construct assesses the degree to which an experimental heat shock induced expression of this promoter (which might serve as a surrogate for the extent of induction of other heat shock loci). The stimulus for gene induction was given during the first day of the worm's adult life. Having at their disposal a worm sorter, the Johnson group successfully separated subsets of living worms with varying degrees of induction of the transgene. The prediction would be that worms with the greatest degree of expression had the longest life spans. This was indeed observed; but the most valuable result of these experiments was that, at a first approximation, these phenotypes were not heritable. Populations of progeny worms derived from either low expressors or strong expressors generated essentially the same distributions of life spans. It therefore seems that, within a species, stochastic events are the predominant contribu-

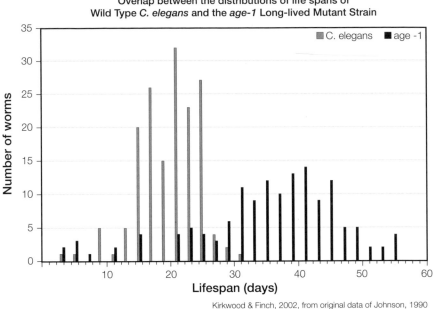

Figure 3. Some individuals within a population of parental wild-type *C. elegans* live longer than some individuals within a long-lived mutant derivative strain. (Reprinted, with permission, from Kirkwood and Finch 2002 [© Nature Publishing Group])

tors to individual differences in longevity. Colloquially, one can say that, for intraspecific differences in life span, Lady Luck trumps the influences of nature and nurture. The molecular basis for being lucky or unlucky remains to be established. It might represent random epigenetic shifts in gene expression or random transcriptional and translational "noise." Perhaps the tissues of aging organisms become genetic mosaics because of accumulating epigenetic drifts in gene expression as well as bona fide somatic mutations. Evidence for the latter is quite compelling and involves both nuclear (Martin et al. 1996) and mitochondrial DNA (Wallace 2005). Transcriptional noise does appear to increase in aging tissues; it could result, in part, from both somatic mutational and epigenetic mechanisms (Bahar et al. 2006).

What are the priorities for future research? Despite our emphasis in this brief overview on the importance of increasing research on aging in our own species, we should certainly bring to fruition the current efforts on the genetic basis for unusual longevities in amenable model organisms. This should certainly include extensions of the *C. elegans* transgene heat shock experiment mentioned above. Transgenes are intrinsically unstable, and therefore, the results should be confirmed with reporters of endogenous loci. Common pathways that are emerging from research in yeast, worms, and flies should be explored via transgenic experiments in mice, as is now being done, for example, for the sirtuins (Sinclair and Guarente 2006). A comprehensive program of this type is in fact in place at the University of Washington, thanks to the support of the Ellison Medical Foundation. A nice feature of that program is the availability of these mouse lines to labs around the world, many of which will be able to implement special expertise in exploring a range of relevant phenotypes. Some special opportunities in the field of mouse genetics continue to be neglected, however. One that particularly comes to mind is the opportunity to rapidly select for variants, produced spontaneously or via mutagenesis, in cultures of pluripotent mouse embryonic stem cells. One such experiment led to the comparatively rapid isolation of an antimutator line that was highly resistant to oxidative stress, a sort of "acceleration of evolution in a Petri dish" (Ogburn et al. 1994). Such cells have the potential for passage through the germ line via chimeric mice (Martin 2005). The relevant loci could then be positionally cloned.

Comparative biogerontological research that embraces molecular and cellular methods has barely begun. An ambitious program in that direction is being initiated at San Antonio. We hope that these studies will include molecular evolutionary investigations of the subset of regulatory species of RNA that are associated with the emergence of long-lived mammalian species. Tools for evolutionary biological research on the massive amount of nonprotein-coding "dark matter" of the genome have recently emerged (Pollard et al. 2006; Ponting and Lunter 2006).

As regards priorities for research in our own species, a number of areas deserve special attention. Women outlive men in nearly all populations; e.g., by 6 years in the United States and Sweden. This is particularly striking among centenarians, groups that are dominated by women. This simple fact trumps all other public genotypic influences on the human life span, yet we do not know why. The sex difference in the distribution of visceral fat (M>F) is among many other adult disease risk factors determined by the sex chromosomes. However, the female survival advantage is clearly present at birth and throughout childhood. Our rodent models may not help much here, because the sex differences in life span are much smaller (Rollo 2002). Although the hermaphroditic *C. elegans* does segregate small numbers of males, there are no pure female controls.

Some of us have been emphasizing the important role of the inflammatory process in aging and late-life disease for many years, and we are finally getting a hearing (Finch and Crimmins 2004; McGeer and McGeer 2004; McGeer et al. 2006; Finch 2007). Inflammation is recognized as a fundamental process in arterial aging and atherosclerosis (Ross 1999; Finch and Kirkwood 2000; Najjar et al. 2005; Finch 2007). Senile plaques in Alzheimer's disease have almost identical inventories of inflammatory markers to those in atheromas (Finch 2005). Tobacco smoke is well known to increase systemic oxidative stress and to accelerate arterial disease. The reduction of inflammation and infection in the 18th and 19th centuries is hypothesized to be a major factor in the remarkable increases of life span, even before antibiotics (Crimmins and Finch 2006; Finch 2007). Other gerontogens are found in the AGEs (advanced glycation end-products) of cooked food that stimulate systemic inflammatory responses (Uribarri et al. 2005). The spread of environmental inflammogens from air pollution is a major cause for concern that may reverse the 20-year global gains in life span in the 20th century.

A neglected avenue of genetic research in *H. sapiens* takes advantage of the special merits of studying middle-aged populations, as noted above. We envisage longitudinal studies of a wide array of highly sensitive and specific assays of physiological function being carried out on at least two distinct and relatively homogeneous ethnic population groups (e.g., Icelanders, Finns, Japanese, and Ashkenazi Jews). The goal would be to identify an elite subset of individuals with negligible declines in one or more of these physiological functions, followed by a search for extreme discordance or concordance among sib pairs (Risch and Zhang 1995, 1996; Zhang and Risch 1996). Alas, the timing for such an expensive long-term study by U.S. scientists could not be worse, given the limitations of available National Institutes of Health funds, particularly for new initiatives, and the enormous national debt, a multigenerational legacy that can only get worse as a result of increasing demands from the military budget. We may well be losing a generation of U.S. biomedical researchers (Weinberg 2006). Sadly, all this comes at a time when biology is flowering.

This discussion, of course, raises the broader question of the long-range goals of the entire enterprise of biogerontological research. We see no sins in the extension of healthy life span and enhanced life spans. There is no intrinsic contradiction between these goals and the goal of enhanced societal productivity and economic welfare. The general public will certainly want technology transfer for increased health spans and life spans on a comparatively fast track. Let us hope that resveratrol and its congeners come to our rescue (Baur and Sinclair 2006)! Meanwhile, the National Institute on Aging-supported Longevity Intervention Program is open to a broader array of potential interventions (http://www.nia.nih.gov/ResearchInformation/ScientificResources/ITPapp.htm). What are sadly lacking in such studies, however, are validated surrogate biomarkers of longevity. Our community has been aware of this deficiency for many decades, but we have made comparatively little progress. Our more immediate task, however, is to do everything we can to identify and ameliorate the special vulnerabilities of individuals to specific aspects of aging processes. A deeper understanding of protective mechanisms, a goal of the research programs we have focused on in this brief review, offers a robust route to a rational program of prevention and treatment of these vulnerable individuals. All of us probably are suffering from some special "private" mechanisms of aging, and thus, all of us have the potential to benefit from such research.

ACKNOWLEDGMENTS

The authors thank Dr. Linda Partridge for helpful suggestions in the development of this manuscript. C.E.F. is grateful for support from the National Institute on Aging and the Elleson Foundation for Medical Research. G.M.M. is grateful for support from National Institutes of Health grants P30 AG01751, PO1 AG01751, U01 AG 07198, and P50 AG05136-17.

REFERENCES

Atzmon G., Rincon M., Schechter C.B., Shuldiner A.R., Lipton R.B., Bergman A., and Barzilai N. 2006. Lipoprotein genotype and conserved pathway for exceptional longevity in humans. *PLoS. Biol.* **4:** e113.

Bahar R., Hartmann C.H., Rodriguez K.A., Denny A.D., Busuttil R.A., Dolle M.E., Calder R.B., Chisholm G.B., Pollock B.H., Klein C.A., and Vijg J. 2006. Increased cell-to-cell variation in gene expression in ageing mouse heart. *Nature* **441:** 1011–1014.

Barzilai N., Gabriely I., Gabriely M., Iankowitz N., and Sorkin J.D. 2001. Offspring of centenarians have a favorable lipid profile. *J. Am. Geriatr. Soc.* **49:** 76–79.

Baur J.A. and Sinclair D.A. 2006. Therapeutic potential of resveratrol: The in vivo evidence. *Nat. Rev. Drug Discov.* **5:** 493–506.

Beaucher M., Goodliffe J., Hersperger E., Trunova S., Frydman H., and Shearn A. 2007. *Drosophila* brain tumor metastases express both neuronal and glial cell type markers. *Dev. Biol.* **301:** 287–297.

Birney E., Andrews D., Caccamo M., Chen Y., Clarke L., Coates G., Cox T., Cunningham F., Curwen V., Cutts T., et al. 2006. Ensembl 2006. *Nucleic Acids Res.* **34:** D556–D561.

Brown V.K. and Hodek I., Eds. 1983. *Diapause and life cycle strategies in insects.* Junk, The Hague.

Charlesworth B. 1994. *Evolution in age-structured populations,* 2nd edition. Cambridge University Press, Cambridge, United Kingdom.

Cherkas L.F., Aviv A., Valdes A.M., Hunkin J.L., Gardner J.P., Surdulescu G.L., Kimura M., and Spector T.D. 2006. The effects of social status on biological aging as measured by white-blood-cell telomere length. *Aging Cell* **5:** 361–365.

Crimmins E.M. and Finch C.E. 2006. Infection, inflammation, height, and longevity. *Proc. Natl. Acad. Sci.* **103:** 498–503.

Epel E.S., Blackburn E.H., Lin J., Dhabhar F.S., Adler N.E., Morrow J.D., and Cawthon R.M. 2004. Accelerated telomere shortening in response to life stress. *Proc. Natl. Acad. Sci.* **101:** 17312–17315.

Fabris C., Toniutto P., Bitetto D., Minisini R., Smirne C., Caldato M., and Pirisi M. 2005. Low fibrosis progression of recurrent hepatitis C in apolipoprotein E epsilon4 carriers: Relationship with the blood lipid profile. *Liver Int.* **25:** 1128–1135.

Finch C.E. 1990. *Longevity, senescence, and the genome.* University of Chicago Press, Chicago, Illinois.

———. 2005. Developmental origins of aging in brain and blood vessels: An overview. *Neurobiol. Aging* **26:** 281–291.

———. 2007. *The biology of human longevity: Inflammation, nutrition, and aging in the evolution of lifespans.* Academic Press, New York.

Finch C.E. and Crimmins E.M. 2004. Inflammatory exposure and historical changes in human life-spans. *Science* **305:** 1736–1739.

Finch C.E. and Kirkwood T.B.L. 2000. *Chance, development, and aging.* Oxford University Press, New York.

Finch C.E. and Sapolsky R.M. 1999. The evolution of Alzheimer disease, the reproductive schedule, and apoE isoforms. *Neurobiol. Aging* **20:** 407–428.

Finch C.E. and Tanzi R.E. 1997. Genetics of aging. *Science* **278:** 407–411.

Fraga M.F., Ballestar E., Paz M.F., Ropero S., Setien F., Ballestar M.L., Heine-Suner D., Cigudosa J.C., Urioste M., Benitez J., et al. 2005. Epigenetic differences arise during the lifetime of monozygotic twins. *Proc. Natl. Acad. Sci.* **102:** 10604–10609.

Fullerton S.M., Clark A.G., Weiss K.M., Nickerson D.A., Taylor S.L., Stengard J.H., Salomaa V., Vartiainen E., Perola M., Boerwinkle E., and Sing C.F. 2000. Apolipoprotein E variation at the sequence haplotype level: Implications for the origin and maintenance of a major human polymorphism. *Am. J. Hum. Genet.* **67:** 881–900.

Geesaman B.J., Benson E., Brewster S.J., Kunkel L.M., Blanche H., Thomas G., Perls T.T., Daly M.J., and Puca A.A. 2003. Haplotype-based identification of a microsomal transfer protein marker associated with the human lifespan. *Proc. Natl. Acad. Sci.* **100:** 14115–14120.

George J.C., Bada J., Zeh J., Scott L., Brown S.E., O'Hara T., and Suydam R. 1999. Age and growth estimates of bowhead whales (*Balaena mysticetus*) via aspartic acid racemization. *Can. J. Zool.* **77:** 571–580.

Hamilton W.D. 1966. The moulding of senescence by natural selection. *J. Theor. Biol.* **12:** 12–45.

Keeley F.W., Bellingham C.M., and Woodhouse K.A. 2002. Elastin as a self-organizing biomaterial: Use of recombinantly expressed human elastin polypeptides as a model for investigations of structure and self-assembly of elastin. *Philos. Trans. R. Soc. Lond. B Biol. Sci.* **357:** 185–189.

Kirkwood T.B. and Finch C.E. 2002. Ageing: The old worm turns more slowly. *Nature* **419**: 794–795.

Locke D.P., Sharp A.J., McCarroll S.A., McGrath S.D., Newman T.L., Cheng Z., Schwartz S., Albertson D.G., Pinkel D., Altshuler D.M., and Eichler E.E. 2006. Linkage disequilibrium and heritability of copy-number polymorphisms within duplicated regions of the human genome. *Am. J. Hum. Genet.* **79**: 275–290.

Martin G.M. 1997. Genetics and the pathobiology of ageing. *Philos. Trans. R. Soc. Lond. B Biol. Sci.* **352**: 1773–1780.

———. 1999. APOE alleles and lipophylic pathogens. *Neurobiol. Aging* **20**: 441–443.

———. 2002. Help wanted: Physiologists for research on aging. *Sci. Aging Knowledge. Environ.* **2002**: vp2.

———. 2005. Genetic engineering of mice to test the oxidative damage theory of aging. *Ann. N.Y. Acad. Sci.* **1055**: 26–34.

Martin G.M., Ogburn C.E., Colgin L.M., Gown A.M., Edland S.D., and Monnat R.J., Jr. 1996. Somatic mutations are frequent and increase with age in human kidney epithelial cells. *Hum. Mol. Genet.* **5**: 215–221.

McGeer P.L. and McGeer E.G. 2004. Inflammation and the degenerative diseases of aging. *Ann. N.Y. Acad. Sci.* **1035**: 104–116.

McGeer P.L., Rogers J., and McGeer E.G. 2006. Inflammation, anti-inflammatory agents and Alzheimer disease: The last 12 years. *J. Alzheimers. Dis.* **9**: 271–276.

Miller R.A., Harper J.M., Dysko R.C., Durkee S.J., and Austad S.N. 2002. Longer life spans and delayed maturation in wild-derived mice. *Exp. Biol. Med.* **227**: 500–508.

Najjar S.S., Scuteri A., and Lakatta E.G. 2005. Arterial aging: Is it an immutable cardiovascular risk factor? *Hypertension* **46**: 454–462.

Ogburn C.E., Turker M.S., Kavanagh T.J., Disteche C.M., Smith A.C., Fukuchi K., and Martin G.M. 1994. Oxygen-resistant multipotent embryonic carcinoma cell lines exhibit antimutator phenotypes. *Somat. Cell Mol. Genet.* **20**: 361–370.

Partridge L. and Gems D. 2002. Mechanisms of ageing: Public or private? *Nat. Rev. Genet.* **3**: 165–175.

Pinkston J.M., Garigan D., Hansen M., and Kenyon C. 2006. Mutations that increase the life span of *C. elegans* inhibit tumor growth. *Science* **313**: 971–975.

Pollard K.S., Salama S.R., Lambert N., Lambot M.A., Coppens S., Pedersen J.S., Katzman S., King B., Onodera C., Siepel A., et al. 2006. An RNA gene expressed during cortical development evolved rapidly in humans. *Nature* **443**: 167–172.

Ponting C.P. and Lunter G. 2006. Evolutionary biology: Human brain gene wins genome race. *Nature* **443**: 149–150.

Rea S.L., Wu D., Cypser J.R., Vaupel J.W., and Johnson T.E. 2005. A stress-sensitive reporter predicts longevity in isogenic populations of *Caenorhabditis elegans*. *Nat. Genet.* **37**: 894–898.

Risch N. and Zhang H. 1995. Extreme discordant sib pairs for mapping quantitative trait loci in humans. *Science* **268**: 1584–1589.

———. 1996. Mapping quantitative trait loci with extreme discordant sib pairs: Sampling considerations. *Am. J. Hum. Genet.* **58**: 836–843.

Robinson G.E., Grozinger C.M., and Whitfield C.W. 2005. Sociogenomics: Social life in molecular terms. *Nat. Rev. Genet.* **6**: 257–270.

Rollo C.D. 2002. Growth negatively impacts the life span of mammals. *Evol. Dev.* **4**: 55–61.

Ross R. 1999. Atherosclerosis—An inflammatory disease. *N. Engl. J. Med.* **340**: 115–126.

Sharp A.J., Hansen S., Selzer R.R., Cheng Z., Regan R., Hurst J.A., Stewart H., Price S.M., Blair E., Hennekam R.C., et al. 2006. Discovery of previously unidentified genomic disorders from the duplication architecture of the human genome. *Nat. Genet.* **38:** 1038–1042.

Sinclair D.A. and Guarente L. 2006. Unlocking the secrets of longevity genes. *Sci. Am.* **294:** 48–51, 54–57.

Spencer C.C. and Promislow D.E. 2002. Genes, culture, and aging flies—What the lab can and cannot tell us about natural genetic variation for senescence. *Sci. Aging Knowledge. Environ.* **2002:** e6.

Uribarri J., Cai W., Sandu O., Peppa M., Goldberg T., and Vlassara H. 2005. Diet-derived advanced glycation end products are major contributors to the body's AGE pool and induce inflammation in healthy subjects. *Ann. N.Y. Acad. Sci.* **1043:** 461–466.

van den E.P., Garg S., Leon L., Brigl M., Leadbetter E.A., Gumperz J.E., Dascher C.C., Cheng T.Y., Sacks F.M., Illarionov P.A., et al. 2005. Apolipoprotein-mediated pathways of lipid antigen presentation. *Nature* **437:** 906–910.

Vanfleteren J.R., De Vreese V.A., and Braeckman B.P. 1998. Two-parameter logistic and Weibull equations provide better fits to survival data from isogenic populations of *Caenorhabditis elegans* in axenic culture than does the Gompertz model. *J. Gerontol. A Biol. Sci. Med. Sci.* **53:** B393–B403.

Walker D.W., McColl G., Jenkins N.L., Harris J., and Lithgow G.J. 2000. Evolution of life-span in *C. elegans*. *Nature* **405:** 296–297.

Wallace D.C. 2005. A mitochondrial paradigm of metabolic and degenerative diseases, aging, and cancer: A dawn for evolutionary medicine. *Annu. Rev. Genet.* **39:** 359–407.

Weinberg R.A. 2006. A lost generation. *Cell* **126:** 9–10.

Wozniak M.A., Itzhaki R.F., Faragher E.B., James M.W., Ryder S.D., and Irving W.L. 2002. Apolipoprotein E-epsilon 4 protects against severe liver disease caused by hepatitis C virus. *Hepatology* **36:** 456–463.

Yamashita T., Freigang S., Eberle C., Pattison J., Gupta S., Napoli C., and Palinski W. 2006. Maternal immunization programs postnatal immune responses and reduces athero-sclerosis in offspring. *Circ. Res.* **99:** e51–e64.

Zhang H. and Risch N. 1996. Mapping quantitative-trait loci in humans by use of extreme concordant sib pairs: Selected sampling by parental phenotypes. *Am. J. Hum. Genet.* **59:** 951–957.

6

p53, Cancer, and Longevity

Lawrence A. Donehower

Departments of Molecular Virology & Microbiology and
Molecular & Cellular Biology
Baylor College of Medicine
Houston, Texas 77030

Arnold J. Levine

Cancer Institute of New Jersey
University of Medicine and Dentistry of New Jersey
New Brunswick, New Jersey 08903 and
The Institute for Advanced Study
Princeton, New Jersey 08540

CANCER IS A PATHOLOGY DIRECTLY LINKED to the aging process (Campisi 2000; Balducci and Beghe 2001). Cancer incidence in the human population increases in an almost geometric fashion with age. Cancer is also a genetic disease, as it arises due in large part to an accumulation of mutations in critical genes that maintain normal cellular division, growth, and homeostasis (Vogelstein and Kinzler 2004). One group of genes that have evolved to minimize the frequency of these mutations and their effects are the tumor suppressor genes. The hundreds of tumor suppressor genes in our genome prevent early cancers from arising by maintaining genomic stability and by eliminating genomically unstable cells before they can become cancer cells. By preventing these early cancers, they are also longevity assurance genes. Vogelstein and Kinzler (1997) have categorized tumor suppressor genes as either "caretakers" or "gatekeepers." The caretakers are a first line of defense against cancer in that they prevent the genome from acquiring new, potentially oncogenic mutations. The gatekeepers are a second line of defense in that they eliminate or arrest those cells that do acquire oncogenic mutations, preventing the emergence of a cancer. The primary mechanisms by which the gatekeepers suppress cancers are apoptosis (programmed cell death) and senescence (irreversible cell cycle arrest) (Campisi 2003). Rodent cancer models in particular have been instrumental in showing that nascent tumors are arrested or eliminated by both apoptosis and senescence (Fig. 1) (Schmitt 2003; Lowe et al. 2004). Mutation or loss of these gatekeeper tumor suppressors almost invariably results in an acceleration of tumor development in model systems (Ghebranious and Donehower 1998).

The title "king of the tumor suppressors" should probably go to p53. p53 is a prototypical gatekeeper tumor suppressor that is a potent inducer of apoptosis, cell cycle arrest, and senescence in response to a wide array of cellular stresses (Fig. 1) (Levine 1997). In particular, DNA damage and aberrantly activated oncogenes activate p53 to induce apoptosis or senescence in order to prevent the emergence of a nascent cancer cell (Vousden and Lu 2002). The importance of p53 in preventing cancer is underlined by the fact that it is the most frequently mutated gene observed in human cancers. Approximately 50% of all human sporadically arising cancers incur loss or mutation in the p53 gene, and it has been estimated that at least 80% of all human cancers have dysfunctional p53 signaling (Levine 1997; Lozano and Elledge 2000). Thus, disruption of normal p53 signaling is likely to be a critical prerequisite for the development or progression of cancer in humans.

The critical role of p53 in preventing cancer has been firmly established. Its status as a longevity assurance gene is also self-evident. Recently, however, evidence has begun to emerge suggesting that p53 may not simply affect life span by preventing or delaying cancers. Paradoxically, p53 may suppress longevity in some circumstances. The evidence is fragmentary and limited, but studies on fly and mouse models, as well as molecular epidemiological studies in humans, suggest a more complicated picture of the role of p53 in aging and longevity. The goal of this chapter is to explore some of the functions of p53 and how these functions relate to potential p53 effects on aging/longevity issues as well as cancer.

THE p53 PATHWAY

The p53 protein and its signal transduction pathway respond to a wide variety of both intrinsic and extrinsic stresses that have the ability to alter the homeostatic mechanisms which preserve genomic integrity and permit the faithful reproduction of the cell (Levine 1997). These stress

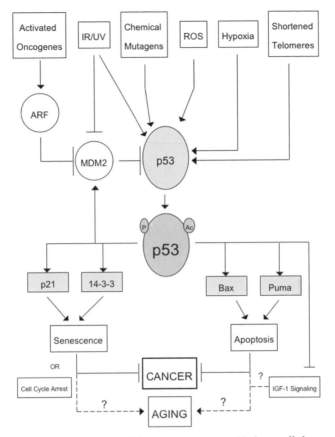

Figure 1. The role of p53 in the cellular stress response. Various cellular stresses (*top*) induce DNA damage that activates p53. Many stresses, such as ultraviolet (UV), ionizing radiation (IR), or reactive oxygen species (ROS), inhibit the E3 ubiquitin ligase Mdm2 that promotes p53 proteolytic degradation, and this leads to stabilization and activation of p53. Stress-activated p53 undergoes posttranslational modification (which includes phosphorylation and acetylation) and translocation to the nucleus. Activated p53 can transcriptionally activate a number of genes, including proapoptotic genes (e.g., Bax and Puma) and genes that induce cell cycle arrest (e.g., p21 and 14-3-3). In some cases, cell cycle arrest induced by p53 is transient, and in other cases, it is permanent, leading to a state of senescence. In addition, p53 up-regulates Mdm2, which can down-regulate p53 through proteolytic degradation, allowing the cell to return to the unstressed state if it does not undergo apoptosis or senescence. p53-induced senescence or apoptosis will prevent the emergence of a damaged or mutated cell that could ultimately become cancerous. A potential side effect of apoptosis and senescence induction could be promotion of aging through depletion of stem and progenitor cell number or self-renewal potential (senescence). In addition to senescence/apoptosis effects, p53 is known to inhibit the IGF-1-signaling pathway, and this could contribute to cancer suppression and alter aging processes (see Fig. 2).

signals include DNA damage, hypoxia, the monitoring of the ribonucle-oside triphosphate pools and ribosomal biogenesis, heat and cold shock, mitotic spindle poisons, proper protein folding in the cell, and even sensing the mutational activation of several diverse oncogenes. The outcomes of these stresses may be p53-regulated cell cycle arrest and repair, senescence or apoptosis, eliminating the damaged clones of cells, and preventing mistakes in cell cycle duplication (Vogelstein et al. 2000). Under normal conditions, the p53 protein is kept at low concentrations and low activity levels by its major negative regulator, Mdm2, an E3 ubiquitin ligase, which confers a very short half-life (6–40 minutes) on the p53 protein. After DNA damage, Mdm2 and p53 are commonly modified by phosphorylation, acetylation, methylation, ubiquitination, neddylation, or sumolation, and the Mdm2 protein polyubiquitinates itself and is degraded in the proteosome (Levine et al. 2006a). The half-life of the p53 protein increases and p53 levels rise, at which time it is activated as a transcription factor, initiating a program of gene expression resulting in cell cycle arrest or cell death (Fig. 1) (Levine et al. 2006a).

Cell cycle arrest by p53 is mediated in part by the synthesis of p21, which inhibits cyclin D1–CDK-4 and mediates G_1 arrest. The up-regulation of 14-3-3-sigma by p53 retains the CDC25C phosphatase in the cell cytoplasm so that it can no longer activate cyclin B for entry from G_2 to M phase of the cell cycle. The presence of mitotic spindle inhibitors, in the absence of a wild-type p53, permits another round of S phase without a cytokinesis resulting in tetraploid cells, a process that is blocked by the wild-type p53 protein (Vogelstein et al. 2000). In addition, the p53 protein regulates the transcription of some DNA-repair functions, so that cell cycle arrest can be reversed and a viable cell can resume a faithful duplication process. Alternatively, an activated p53 protein will up-regulate a variety of proapoptotic proteins such as bax, puma, noxa, and apaf-1, which lead to the release of cytochrome c from mitochondria, caspase activation, and programmed cell death. Little is known about the genes that mediate a p53-regulated senescence, but this process eliminates cells from duplicating mutations that occur at high frequency when cells divide under stressful conditions. All three of these outcomes retain a high level of cell cycle fidelity, which fails in the absence of p53, often resulting in cancers at a very high frequency. Mice without a p53 gene all develop cancers within several months, and heterozygous p53 mutant mice will develop sarcomas and other cancers over a period of a year or more (Donehower et al. 1992; Jacks et al. 1994). Humans who are heterozygous for the p53 gene, and have Li-Fraumeni syndrome, display a 50% cancer incidence by the age of 30 (Malkin et al. 1990).

Polymorphisms in the human Mdm2 gene, which raise the levels of this protein in cells, result in an earlier age of onset of these cancers in p53 heterozygous individuals (Bond et al. 2004) and even in people with wild-type p53 alleles (Bond et al. 2004, 2006a,b).

The p53 signal transduction pathway communicates with a wide variety of cellular processes and molecular signal transduction pathways (Harris and Levine 2005). There are a number of negative feedback loops acting on the p53 pathway that result in the regulation of significant portions of other cell cycle replicative signal transduction pathways. p53 protein levels are autoregulated in the cell (Harris and Levine 2005; Levine et al. 2006a) because p53 transcribes the Mdm2, PIRH-2, and COP-1 genes, which in turn synthesize ubiquitin ligases that can ubiquitinate the p53 protein and mediate its degradation (Lev Bar-Or et al. 2000; Lahav et al. 2004).

A variety of oncogenes (β-catenin, Ras, myc, E2F-1 regulated by retinoblastoma) promote the synthesis of the p19^{ARF} protein, which in turn binds to Mdm2 and reduces its activity as a ubiquitin ligase, raising p53 levels in the cell (Damalas et al. 2001; Zindy et al. 2003; Harris and Levine 2005). Finally, the induction of p53 in a cell by stress frequently initiates the transcription of the truncated forms of p63 and p73, two transcription factors that are close relatives of p53. These so-called delta-N truncated forms of p63 and p73 act by binding to DNA sequences that p53 also can bind to the p53 DNA responsive elements or REs, but they function as negative regulators of p53-mediated transcription because the delta-N forms lack the transcriptional activator domains of these proteins (Grob et al. 2001; Kartasheva et al. 2002; Benard et al. 2003; Murray-Zmijewski et al. 2006). This wide variety of communication circuits helps to positively or negatively regulate p53 and to communicate with many other signal transduction pathways in a cell.

Cells undergoing a p53 response also communicate to other cells in the body. The activation of p53 initiates the transcription of the TSAP-6 gene, which produces a protein that enhances the rate of production of cellular exosomes (Yu et al. 2006). These are small vesicles that are synthesized in the endosomal organelle, engulf cytoplasmic components, and bud out into the media or systemic fluids of the body. These exosomes can mediate enhanced cellular or humoral immunity to proteins presented in the context of class 1 or 2 antigens on the surface of these endosome vesicles (Fevrier and Raposo 2004). In this way, stressed cells communicate with the immune system. The exosomes can also fuse with adjacent cells imparting information (molecules) from the dying cell to its neighbors. The p53 response also regulates the synthesis of a number of secreted proteins that can alter the extracellular matrix (inhibitors of plasminogen activators—maspin, PAI) or influence angiogenesis in the damaged cellular region (thrombospondin) (Levine et al. 2006a). In these ways, p53 responses to stress are communicated at both the molecular and cellular levels as well as systemically.

COMMUNICATION AND INTERACTIONS BETWEEN THE p53 PATHWAY AND THE IGF-1–mTor PATHWAYS

Studies carried out during the past 5–10 years with yeast, worms, flies, and mice have demonstrated that genetic alterations in the insulin-like growth factor-1 (IGF-1) pathway can enhance the longevity of these organisms (Guarente and Kenyon 2000; Guarente and Picard 2005; Kenyon 2005). That signal transduction pathway is in turn negatively regulated, in a p53-dependent fashion, by stress signals that activate p53, which transcriptionally increases the rate of synthesis of genes in the IGF-1–mTor (mammalian target of rapamycin) pathways (Buckbinder et al. 1995; Stambolic et al. 2001; Feng et al. 2005; Jones et al. 2005; Levine et al. 2006b; Z. Feng; A. Levine; both pers. comm.). These p53-regulated gene products each down-modulate the IGF-1 and mTor pathways in response to stresses that would lower the fidelity of the cell division cycle. These regulatory interactions and epistatic communication between the p53 pathway genes and the IGF-1 and mTor pathway genes suggest a role for the p53 protein in contributing to the longevity of an organism.

Figure 2 presents a diagram of the interrelationships and functions of the gene products in the IGF-1 pathway, the mTor pathway, and the p53 pathway (Levine et al. 2006b). In times of abundant nutrient (glucose and amino acids) and mitogen (IGF-1) signaling for division, the IGF-1 protein engages its receptor at the membrane of selected cells, and this results in activation of the receptor and the binding to it of an adapter molecule, commonly the Shc protein. These events attract the phosphoinositol-3 kinase (PI3K) to this membrane complex, which produces phosphoinositol triphosphate (PIP-3) from phosphoinositol diphosphate (PIP-2). PIP-3 activates a variety of lipid kinases (PDK-1 and mTor–rictor), which in turn phosphorylate and activate the AKT kinases (AKT-1,2,3, depending on the cell types). The AKT-1 kinase is translocated into the nucleus where it phosphorylates several FOXO transcription factors, which then leave the nucleus. This changes the pattern of cellular transcription from cell cycle arrest (p27 synthesis) to preparing for cell cycle division (the synthesis of antioxidants, heat shock proteins, etc.) (Brunet et al. 1999;

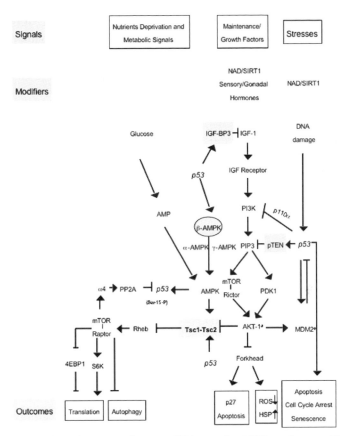

Figure 2. Interactions between the p53–IGF-1 and mTOR pathways. The signals, modifiers, and outcomes for three signal transduction pathways are presented. The major genes and gene products in the p53 pathway initiated by stress signals, the IGF-1 pathway initiated by mitogen signals, and the mTOR pathway initiated by nutrient signals are presented. The nutrient and mitogen signals are each required for cellular growth and division. The mTOR and IGF-1 pathways, however, will be shut down or inhibited in the presence of stress signals that trigger a p53 response. The purpose of blocking the cell growth and division response after stress is to avoid mistakes in cell division that can lead to cancers. Four p53-regulated genes, pTEN, TSC-2, β-AMPkinase and IGF-BP-3 are transcriptionally activated by p53 after a stress signal and block various parts of the mTOR and IGF-1 pathways. The IGF-BP3 protein is secreted from cells and acts to block the IGF-1 protein from engaging with its receptor at the cell surface. pTEN is a lipid phosphatase that degrades PIP-3, which is required to activate mTOR rictor, mTOR raptor, PDK-1, and AKT-1. TSC-2, in a complex with TSC-1, is a GTPase that inactivates the G-protein RHEB that is required for mTOR activity. The β subunit of the AMP kinase activates TSC-2, which in turn inactivates mTOR. In this way, both the mTOR and IGF-1 pathways are negatively regulated by p53.

Blume-Jensen and Hunter 2001; Sarbassov et al. 2005; Levine et al. 2006b). At the same time, AKT-1 phosphorylates Mdm2, which increases its activity and lowers the p53 levels in the cell (Zhou et al. 2001). AKT-1 also phosphorylates and inactivates Bad, a proapoptotic protein that influences cytochrome c release from mitochondria (Datta et al. 1997; del Peso et al. 1997). Third, AKT-1 kinase phosphorylates the TSC-2 protein, which is in a complex with TSC-1, and this complex forms a GTPase that inactivates RHEB, a G protein that positively regulates the mTor–raptor protein kinase (Gao and Pan 2001; Inoki et al. 2002, 2003b). Phosphorylation of TSC-2 by AKT-1 inactivates the TSC-1, -2 complex, activating RHEB, which in turn positively regulates mTor–raptor kinase (Gao and Pan 2001; Inoki et al. 2003b). This mTor–raptor kinase inactivates, via phosphorylation, the 4EBP-1-binding protein and activates, via phosphorylation, the S6 kinase activity. These two proteins regulate the efficiency of translation of a number of mRNAs that influence cell growth and division. The 4EBP-1 protein binds to and inactivates the e-IF4E translational initiation factor, which permits the translation of CAP-dependent mRNAs. Inhibition of the translation of these mRNAs (by 4EBP-1 binding to its initiation factor) slows the rate of cell growth (Hay and Sonenberg 2004). The S6 kinase phosphorylates the ribosomal S6 protein and is thought to enhance the translation of proteins involved in ribosome biogenesis, mitochondrial biogenesis, and oxygen consumption (Thomas 2002; Hannan et al. 2003).

In the absence of nutrients, such as glucose or amino acids, AMP levels rise in the cell (no oxidative phosphorylation to make ATP at high levels), which activates the AMP kinase. The AMP kinase phosphorylates the TSC-2 protein, but unlike the AKT-1 phosphorylation that inhibits this GTPase, the AMP kinase activates the GTPase, which in turn inhibits RHEB and shuts down the mTor–raptor activity (Thomas 2002; Inoki et al. 2003a). Now the 4EBP-1 protein slows translation of CAP-mRNAs, the S6 kinase is not activated (slowing ribosomal biogenesis, mitochondrial biogenesis, and oxygen consumption), and the cell stops growth. The inhibition of mTor–raptor activity results in the activation of autophagy (Kamada et al. 2004; Lum et al. 2005). During the starvation of nutrients, double-membrane vesicles are formed in the endosome compartment of the cell which engulf mitochondria, cytoplasmic components, ribosomes, etc. These autophagic vesicles move to the endosome, and the components are degraded, permitting the recycling of lipids, nucleotides, amino acids, and sugars in the cell. Thus, the mTor kinase pathway senses the nutrient levels in the environment and responds with anabolic or catabolic pathways. Together the mTor and

IGF-1 pathways coordinately regulate a commitment to growth and cell division, via a nutrient-sensing and mitogen-sensing signal for division.

When this commitment to growth and cell division is interrupted by a stress signal, such as DNA damage, the process must be halted so as to prevent mistakes which could produce clones of cells that could result in cancers. The p53 protein responds to the stress signal and shuts down the IGF-1 and mTor signaling pathways by enhancing the synthesis of four gene products; IGF-BP3, PTEN, TSC-2, and the β-1 and -2 subunits of the AMP kinase (Z. Feng; A. Levine; both pers. comm.). The IGF-BP3 protein is secreted into the serum and binds to IGF-1, preventing it from binding to and activating the IGF-1 receptor. The PTEN protein is a phosphatase converting PIP-3 to PIP-2, which no longer activates the AKT-1 kinase. Interestingly, this forms a positive feedback loop for p53 production. PTEN lowers the AKT-1 kinase activity, which lowers the Mdm2 activity, which enhances p53 activity. High levels of p53 also negatively regulate the transcription of the p110-α subunit of the PI3K, shutting down this pathway. p53 activation also results in an enhanced transcription of the TSC-2 gene, and elevated levels of TSC-2 reduce the mTor–raptor activity. The AMP kinase is composed of three protein subunits: α, which is the AMP-binding and regulatory subunit; γ, which is the kinase catalytic subunit; and β, which brings together the regulatory and catalytic subunit and whose level is rate limiting for the AMP kinase activity (Hardie 2005). p53 increases the rate of transcription of the β AMP kinase subunit, increasing the activity of this kinase, which in turn shuts down mTor–raptor activity. Thus, all four p53-regulated genes act to shut down the IGF-1 and mTor pathways after a stress signal (see Fig. 1). Because mutations in the IGF-1 pathway can affect the longevity of a population of organisms that harbor these mutations, it is likely that mutations or polymorphisms in the p53 pathway might well have an impact upon the longevity of a population with such mutations or polymorphisms (Levine et al. 2006b).

p53, APOPTOSIS, SENESCENCE, AND AGING

In vitro and in vivo experiments have repeatedly demonstrated that the anti-oncogenic functions of p53 are mainly due to its ability to induce apoptosis or cell cycle arrest (Campisi 2005; Dimri 2005). Induction of apoptosis can be relatively rapid and extensive following DNA damage. However, Evan and colleagues have recently shown that this rapid p53-induced apoptosis may be relatively unimportant in protection against tumors. In a genetically engineered mouse in which p53 could be turned on and off by drug administration, it was shown that p53 activity during treatment with ionizing radiation (a potent DNA damage inducer) was less important in preventing subsequent tumors than the presence of p53 in the postradiation phase beginning 8 days after treatment (Christophorou et al. 2006). This suggests that the radiation-induced p53-mediated apoptosis itself was not antitumorigenic. Rather, the ability of p53 to prevent tumorigenesis may reside in its apoptotic activities on survivor cells that become oncogenically activated. Consistent with this, the p53 protection is absolutely dependent on p19^{ARF}, a tumor suppressor that activates p53 and is not induced by DNA damage, but by oncogenic disruption of the cell cycle (Christophorou et al. 2006).

When p53 does not induce apoptosis, it may cause a cell cycle arrest. Cell cycle arrest can be transient or permanent, depending on the nature of the p53 stimulus, the magnitude of the stimulus, the cell type, the cell status (transformed or nontransformed), and the cellular environment. p53-mediated cellular senescence can be induced by DNA-damaging agents, oxidative stress, shortened telomeres, alterations in chromatin conformation, and activated oncogenes (Fig. 1) (Campisi 2005). Each inducer may affect p53-induced senescence by a distinct pathway. For example, shortened telomeres may activate an ATM/ATR–Chk1/Ch2 kinase cascade that phosphorylates and enhances p53 function, whereas activated oncogenes may induce p53 by up-regulation of p19^{ARF}, which in turn sequesters Mdm2 and prevents it from mediating p53 degradation (Fig. 1) (Dimri 2005; von Zglinicki et al. 2005).

In vivo evidence of p53-induced senescence is also emerging from studies of animal models. For example, Schmitt et al. (2002) showed that primary murine lymphomas blocked from undergoing apoptosis respond to chemotherapy by engaging a p53-controlled senescence program. However, tumors missing p53 or its upstream regulator p19^{ARF} were poorly responsive to chemotherapy. Other groups have shown that transgenic mice harboring an activated oncogene induce early-stage tumors with high levels of senescent cells that are dependent on p53 (Braig et al. 2005; Chen et al. 2005; Collado et al. 2005). Removal of p53 results in loss of senescent cells and a more aggressive tumor progression.

That p53-induced senescence is a major fail-safe mechanism for cancer prevention is becoming clear. But does p53-induced senescence occur in normal cells of an aging organism, and if so, does it have any pathophysiological relevance to the aging process? Recent studies have allowed the detection of aging-specific molecular markers in aged human and mouse tissues (Dimri et al. 1995; Krishnamurthy et al. 2004).

Development of the senescence-associated β-galactosidase assay by the Campisi laboratory provided a mechanism for detection of senescent cells in vivo (Dimri et al. 1995). Moreover, overexpression of the p16^{INK4A} cyclin-dependent kinase inhibitor was found by Sharpless and colleagues to correlate well with senescence (Krishnamurthy et al. 2004). Both of these markers were found in much higher abundance in aged mouse and human tissues compared to their younger counterparts (Dimri et al. 1995; Krishnamurthy et al. 2004; C. Gatza and L.A. Donehower, unpubl.). Sedivy and colleagues have shown that senescent skin fibroblasts increase exponentially with age in baboons and that this increase is accompanied by activation of the ATM kinase (a p53 activator) and heterochromatinized nuclei (Herbig et al. 2006).

Given that senescent cells do accumulate with age, how might they promote aging phenotypes? Apart from a generalized reduction in tissue functionality, Campisi and colleagues have demonstrated senescent cells may actively promote aging through altered secretory processes (Krtolica et al. 2001; Campisi 2005). Senescent human fibroblasts secrete epithelial growth factors, inflammatory cytokines, and a number of matrix metal-loproteinases (Krtolica et al. 2001). It was demonstrated that senescent fibroblasts could disrupt normal epithelial cell functions and could pro-mote the development of malignant epithelial lesions in vivo (Parrinello et al. 2005). Campisi has hypothesized that "senescent cells can, at least in principle, contribute to age-related changes in tissue structure and function, as well as the development of age-related pathology, including late-life cancer" (Campisi 2005). However, direct evidence that senescent cells affect the aging process in vivo is still lacking.

Another mechanism by which senescence could affect aging was outlined by three recent papers. As noted by Sharpless and colleagues, protein levels of the senescence effector and marker p16^{INKA} increase with age in many tissues (Krishnamurthy et al. 2004). It was also shown that stem and progenitor cells of various lineages (hematopoietic stem cells, pancreatic islet cells, and subventricular zone neuronal progenitors) display decreased proliferation and functionality with age and that functional decline correlated with p16^{INK4A} levels (Janzen et al. 2006; Krishnamurthy et al. 2006; Molofsky et al. 2006). p16^{INK4A}-deficient stem and progenitor cells showed increased proliferative capacity compared to their wild-type counterparts. Thus, the decline of regenerative capacities with age may be due in part to reductions in stem and progenitor cell functions due to accumulation and activation of tumor suppressors such as p16^{INK4A}. Consistent with this interpretation, our laboratory has examined the tumor suppressor p53 for its role in affecting long-term hematopoietic stem cell function with age (Dumble et al. 2007). We found that alteration of p53 dosage had a pronounced effect on stem cell proliferation and regenerative capacity, but primarily in stem cells from aged animals. Reduction in p53 enhanced aged stem cell function, whereas increases in p53 activity over normal resulted in reduced stem cell function. Thus, tumor suppressors such as p16^{INK4A} and p53, which are both potent inducers of senescence, could influence the aging process through effects on regenerative processes mediated by stem and progenitor cells.

p53, AGING, AND LONGEVITY IN INVERTEBRATE AND VERTEBRATE MODEL SYSTEMS

p53 is evolutionarily conserved in that its homologs have been identified in *Drosophila melanogaster* and *C. elegans* as well as in vertebrates (Jin et al. 2000; Ollmann et al. 2000; Derry et al. 2001). Unlike vertebrates, with three p53 family members, p53, p63, and p73, flies, worms, and mollusks have only a single identified homolog (Table 1) (Yang et al. 2002). p53 in *C. elegans* and *Drosophila* appears to be phylogenetically closer to p63, suggesting that p53 may have evolved later in vertebrates (Yang et al. 2002). This is consistent with the fact that flies and worms are somatically postmitotic, short-lived, and not susceptible to tumors, in contrast to vertebrates with renewable tissues, longer life spans, and cancer susceptibility. Thus, one theory is that p53 may have evolved in response to the need for protection of renewable stem and progenitor cell compartments from propagation of defective or damaged cells that could produce a cancer. In vertebrates, p53 is largely dispensable for embryonic development, whereas deficiencies in p63 and p73 result in major developmental aberrations (Moll and Slade 2004). Despite these differences

Table 1. Role of p53 family members in cancer and aging

Organism	p53 Family members	Tumor suppressor	Implicated in aging	Cancer-independent longevity effects
Yeast	none	n.a.	n.a.	n.a.
Worms	Cep-1	no	no	↓p53 longevity↓
Flies	Dmp53	no	yes	↓p53 longevity↑↓
Mice	p53, p63, p73	p53	yes, p53, p63	↑p53 longevity↓
				↓p63 longevity↓
Humans	p53, p63, p73	p53	yes, p53	↓p53 longevity↑

n.a. indicates not applicable.

in overall function, all three family members share the ability to *trans*-activate p53 target genes and induce apoptosis in response to DNA damage.

The *C. elegans* p53 homolog, *Cep-1,* appears to be important for inducing apoptosis and normal meiotic segregation in germ cells subjected to genotoxic stress, but *Cep-1* effects are minimal in stressed adult somatic cells (Derry et al. 2001). There is little evidence that *Cep-1*-deficient mutants have altered longevity, but induction of *Cep-1* overexpression in adult worms resulted in rapid lethality that was caspase-independent (Derry et al. 2001). Recently, Kenyon and colleagues have shown that, even in *C. elegans, Cep-1* can act as a tumor suppressor of sorts (Pinkston et al. 2006). The *gld-1* mutation results in aberrantly dividing germ cells that break out of the gonad and fill the body, killing the animal early. These overproliferating gonadal cells have been likened to early-stage tumors in mammals. The shortened longevity of *gld-1* mutants can be dramatically increased by *daf-2* mutations, which cause increased longevity in *C. elegans* through reduced IGF-1 signaling. *daf-2* mutants appear to extend longevity by induction of apoptosis in the aberrantly dividing germ cells. Interestingly, *Cep-1* was required for the *daf-2* mutant-induced apoptosis, indicating that *Cep-1* has a proapoptotic tumor suppressor effect in this particular aberrantly dividing cell type (Pinkston et al. 2006).

In contrast to worms, *Drosophila* p53 (Dmp53) levels have been manipulated in a way that can result in dramatically altered longevities. Flies lacking functional Dmp53 display reduced fertility and a shorter life span (Lee et al. 2003; Bauer et al. 2005). Irradiated larvae without functional Dmp53 exhibit radiosensitivity and lethality, indicating a Dmp53-induced apoptotic effect. Recently, Helfand and colleagues have shown that mutant dominant-negative versions of p53 expressed in the neuronal compartment of *Drosophila* resulted in significant extensions of life span. These p53-deficient flies had normal physiological activity and reproductive fecundity but were resistant to oxidative stress. When calorie-restricted, the neuronally expressed dominant-negative p53 flies showed no life span advantage over calorie-restricted normal flies, indicating that the p53 effects were in a similar pathway to effects mediated by calorie restriction (Bauer et al. 2005). Interestingly, these longevity effects could only be produced when dominant-negative p53 was expressed in the neurons of the fly; expression in the muscle and fat body actually resulted in modest decreases in life span. The mechanisms of this longevity effect mediated by reduction of p53 dosage in neurons are unclear, but they are likely to be due in part to endocrine effects.

The most useful experimental model for investigating the effects of p53 in cancer and longevity has been the laboratory mouse, particularly genetically engineered versions of the mouse. The first p53 mutant mice to be generated globally expressed dominant-negative forms of p53, and these resulted in reduced survival due to early tumor incidence (Lavigueur et al. 1989). Null alleles of p53 engineered into the mouse germ line also resulted in early tumors (and truncated survival), formally proving that p53 is not only a tumor suppressor, but also a longevity assurance gene (Donehower et al. 1992; Harvey et al. 1993; Jacks et al. 1994; Purdie et al. 1994). Subsequent point mutant variants of p53 "knocked in" to the mouse have also been shown to increase tumor incidence and reduce longevity (Liu et al. 2000, 2004; Olive et al. 2004).

The first clue that p53 could influence longevity in a cancer-independent fashion was provided by reports describing telomerase-deficient (Terc$^{-/-}$) mice. Terc$^{-/-}$ mice that had been crossed for several generations had greatly shortened telomeres which resulted in a number of accelerated-aging phenotypes and reduced longevity (Rudolph et al. 1999). Fibroblasts and testicular cells derived from these mice contained significantly elevated levels of activated p53. Crossing of these mice into a p53-deficient background resulted in a diminution of the aging phenotypes associated with telomerase deficiency (Chin et al. 1999).

Other knockout mice deficient in genes that ensure genomic integrity have exhibited pathologies resembling accelerated aging and have had shortened life spans. These include mice deficient in *Brca1*, *Atm*, and *Ku80* (Vogel et al. 1999; Cao et al. 2003; Wong et al. 2003). These mice often exhibit enhanced p53 signaling or have their aging and reduced longevity phenotypes attenuated by introduction of a p53-deficient background. Such observations prompt the hypothesis that the increased DNA damage in these genetically deficient mice is eliciting an augmented p53 response that may be partially responsible for the resulting accelerated-aging phenotypes.

Another example of this phenomenon was provided by a mouse model of human progeria, a syndrome that is accompanied by a number of early pathologies resembling accelerated aging, including cardiovascular defects and death by the teenage years (Martin 2005). This mouse model was deficient in *Zmpste24*, a gene responsible for processing of lamin A (similar to the defect in human progeria patients), and displayed a very short life span and phenocopied some of the human progeroid syndrome symptoms (Pendas et al. 2002). It was recently reported that the *Zmpste24*-deficient mice had a dramatic up-regulation of numerous p53 target genes, indicating an enhanced p53 response (Varela et al.

2005). In addition, the accelerated-aging phenotypes and reduced life span of the *Zmpste24* null mice were partially rescued by the absence of p53 (Varela et al. 2005).

Activation of the p53 response in some of these genetically engineered mice with accelerated aging suggests that increased p53 activity might be responsible for some of the altered aging and longevity phenotypes. The obvious question is whether direct overexpression of p53 in the mouse produces early aging and shortened life span. It is clear that massive overexpression of p53 is incompatible with normal mouse development, because mice missing Mdm2, an E3 ubiquitin ligase that facilitates p53 degradation, display early embryonic lethality due to accumulation of excessive p53 (Jones et al. 1995; Montes de Oca Luna et al. 1995; de Rozieres et al. 2000). Serrano and colleagues have addressed the overexpression issue in a more subtle way by engineering mice with only a single extra copy of p53 (Garcia-Cao et al. 2002). These "super p53" mice exhibited an augmented p53 response to DNA damage and were also resistant to spontaneous and carcinogen-induced tumors. Importantly, the super p53 mice had an overall longevity similar to that of normal mice, suggesting that there is a window of p53 activity above normal that does not affect development or longevity, yet provides additional protection against cancer (Garcia-Cao et al. 2002). A second mouse, generated by Perry and collaborators, was hypomorphic for Mdm2 activity (Mendrysa et al. 2006). Since Mdm2 facilitates p53 degradation, it was not surprising that the Mdm2 hypomorphs had an enhanced p53 response to DNA damage and were tumor-resistant. Longevity appeared to be normal in these mice.

In contrast to the above reports, two hypermorphic p53 mice have been reported that, like the super p53 mice and Mdm2 hypomorphs, exhibit cancer resistance, but unlike them, display phenotypes of accelerated aging and shortened longevity. The first, the $p53^{+/m}$ mouse, generated by our laboratory, contains a deletion in the first six exons of p53 that results in the expression of a truncated form of p53 that stabilizes and activates wild-type p53 (Tyner et al. 2002). The result appears to be hypermorphic activity of the wild-type p53 present in these mice. The $p53^{+/m}$ mice are resistant to spontaneous cancers with a 6% lifetime incidence versus 45% for $p53^{+/+}$ mice of similar genetic background. However, these $p53^{+/m}$ have a 23% shortened longevity compared to wild-type p53 mice and exhibit a number of accelerated-aging phenotypes, including organ atrophy, osteoporosis, skin atrophy, and reduced tolerance to stresses (Tyner et al. 2002). A second p53 hypermorphic model has been reported by Scrable and colleagues (Maier et al. 2004).

The p44 transgenic mouse expresses a larger naturally occurring truncated isoform of p53 and exhibits a dramatically shortened life span and a number of early-aging phenotypes. The mechanism of these early-aging phenotypes may be in part due to hyperactivation of the IGF-1 signaling axis (Maier et al. 2004).

An important question arises regarding the differences in phenotypes of the four hypermorphic p53 models. All of the models exhibit cancer resistance, but the super p53 mice and Mdm2 hypomorphs have a normal longevity, whereas the p53$^{+/m}$ and p44 models display truncated life spans and accelerated aging. We and Serrano have hypothesized that the super p53 mice maintain a normal, intact regulation of p53, so that increases in p53 activity only appear in times of stress (Klatt and Serrano 2003; Dumble et al. 2004). In contrast, the p53$^{+/m}$ mice may have somewhat abnormal p53 regulation in that there may be low-level chronic activation of p53 that results in the early-aging phenotypes. Thus, the p53$^{+/m}$ mice may phenocopy some of the other accelerated-aging models, such as the Terc$^{-/-}$, Ku80$^{-/-}$, Brca1$^{-/-}$, and Zmpste24$^{-/-}$ mice that are likely to elicit continuous low-level p53 activation due to shortened telomeres, unresolved DNA damage, or cellular stresses. Clearly, further experiments will be required to understand the mechanisms by which p53 influences longevity and aging phenotypes.

As indicated above, in vertebrates, p53 has two other family members, p63 and p73, that share many functions with p53 (transcriptional activation, induction of apoptosis) (Table 1). However, neither p63 nor p73 appears to be a classic tumor suppressor. Both p63 and p73 null mice show extensive developmental abnormalities, and p63 and p73 heterozygotes are not notably prone to early cancers (Perez-Losada et al. 2005; Keyes et al. 2006). However, p63$^{+/-}$ mice have been shown by Mills and colleagues to exhibit decreased longevity and accelerated-aging phenotypes (Keyes et al. 2005). Moreover, p63 deficiency was shown to be a potent inducer of cellular senescence (Keyes et al. 2005). The mechanisms by which this p53 family member affects senescence and aging remain unclear, but one possibility is that some isoforms of p63 may interact with p53 and increase its activity.

POLYMORPHISMS IN THE p53 GENE CAN AFFECT THE CANCER FREQUENCIES AND LONGEVITY OF HUMAN POPULATIONS

The human p53 gene contains 393 codons, and the protein is frequently divided into several domains. Amino acids 1–50 comprise two transcriptional *trans*-activational domains that interact with the RNA polymerase II complex which mediates transcription. Amino acids 50 to about 100 contain a proline-rich domain that may regulate apoptosis. The amino acids between 120 and 280 residues comprise a DNA-binding domain, followed by a protein tetramerization domain, and at the carboxyl terminus, there is a basic set of amino acids that appear to have a regulatory function for p53 activity (Levine et al. 2006a). There are at least nine splice variants of this gene that eliminate or substitute several of these domains (Bourdon et al. 2005; Murray-Zmijewski et al. 2006). At codon 72, in the proline-rich domain, there is a proline-to-arginine substitution or polymorphism whose distribution in populations varies with racial groups and geography (Beckman et al. 1994). At the equator and in African populations, almost everyone is homozygous for the proline allele (Beckman et al. 1994). In Caucasians and Asians, and as one moves north or south from the equator, the frequency of the arginine allele increases (Beckman et al. 1994). This gradient of geographical distribution and the fact that the arginine allele appears later in the evolution of humans does suggest a possible selection pressure for the arginine allele as Caucasians or Asians evolve (or a strong founder effect). Several types of experiments, largely carried out in cell culture, have demonstrated that the proline form of p53 has a lower apoptotic efficiency than the arginine form (Bonafe et al. 2002; Dumont et al. 2003).

Recently, a second polymorphism in the p53 pathway has been described in the Mdm2 gene (Bond et al. 2004). Here, the polymorphism is located in the first intron and the G allele produces higher levels of Mdm2 mRNA and protein, whereas the T allele has two- to fourfold lower Mdm2 levels. High Mdm2 levels mean low p53 activity, and individuals with the G allele have earlier ages of onset of developing cancers and higher frequencies of cancer in a population (Bond et al. 2004, 2006a,b). For example, the odds ratio for developing lung cancer in a population in China was 1.83 for individuals with the G/G alleles (compared to T/T) and 1.47 for those with the Pro/Pro alleles (compared to Arg/Arg). The G/G and Pro/Pro combined population had an odds ratio of 4.56, demonstrating the synergistic and epistatic relationship of these two genes that regulate each other. The population of individuals who were G/G, Pro/Pro and were also smokers (a mutagenic p53 stress signal) had an odds ratio of 10.4 (Zhang et al. 2006), again demonstrating the combination of genetic and environmental variables that act on the p53 pathway in vivo and in humans. Similar results have been found with esophageal cancers and smoking (Hong et al. 2005). Thus, these polymorphisms have a role in response to environmental mutagens and development of diseases such as cancers.

van Heemst et al. (2005) recently carried out a study of the impact of the Pro/Pro and Arg/Arg polymorphisms in the p53 gene on the frequency of developing cancers and on the longevity of this population. Employing a formal meta-analysis of published results from the literature, they demonstrated that individuals with the Tp53 codon 72 Pro/Pro genotype had an increased cancer risk compared to Arg/Arg carriers ($p < 0.05$). In a prospective study of individuals age 85 years and older (carried out over 10 years with 1226 people), they showed that carriers of the Pro/Pro genotype have a 41% increased survival ($p = 0.032$) despite a 2.54-fold increased proportional mortality from cancer ($p = 0.007$). These data suggest that an active p53 (Arg/Arg) may protect against cancer but at a cost in longevity. An efficient p53-mediated response to stress by clones of cells or by stem cells may well protect the organism from developing cancer but could at the same time reduce the stem cell population, resulting in reduced longevity. A similar study carried out with polymorphisms in the p53 gene (codon 72) and the CCR5 gene with 131 long-lived individuals from Novosibirsk and Tyumen came to similar conclusions (Smetannikova et al. 2004). These types of studies are at an early stage in testing the hypothesis that p53 has an important role in longevity of an organism or a population. They require confirmation and extension to larger populations studied with detailed clinical information. However, all of these studies provide a similar hypothesis: An active or strong p53 response to stress protects very well against cancer but at a cost of reduced longevity. A weaker p53 response to stress, although it predisposes the organism to develop cancers, provides the survivors with longer longevity in a population.

FINAL THOUGHTS AND FUTURE DIRECTIONS

In this chapter, we have described some of the functions of p53, particularly those that could potentially influence the aging process. The role of p53 in assuring longevity through prevention of cancer is unquestioned, but whether it affects aging and longevity independently of its cancer prevention activities is still an open question. We have presented some provocative early studies which suggest that, in fact, p53 may promote aging processes in late life. The mechanisms by which p53 could influence the aging process remain unclear. One hypothesis is that tumor suppressors such as p53 have evolved to protect long-lived organisms which maintain renewable tissues (such as stem cells and progenitor cells) into adulthood. Adult tissue stem and progenitor cells are necessary to maintain organ homeostasis throughout the life span of the organism. As cells are lost, they must be replaced in a manner which minimizes genetic changes that are associated with DNA replication and mitosis. Thus, these cells are particularly vulnerable because mutations in them can be propagated to numerous progeny and, unlike differentiated cells, have active cell division programs that can be readily subverted and deregulated to form cancer stem cells. The importance of cancer stem cells and their derivation from normal stem cells is now just beginning to be recognized (Reya et al. 2001; Pardal et al. 2003). p53 is likely to have a major role in this stem cell anticancer function. Indeed, reductions in p53 dosage have been shown to enhance stem cell division as well as other functional attributes (Palacios et al. 1996; Shounan et al. 1996; Hirabayashi et al. 2002; TeKippe et al. 2003; Dumble et al. 2007).

The ability of p53 to potentially suppress the emergence of cancer stem cells has implications for a potential role in aging. It has been hypothesized that aging phenotypes are in part a result of functional depletion of the stem and progenitor cell compartments (Van Zant and Liang 2003; Chen 2004; Pelicci 2004; Sharpless and DePinho 2004; Campisi 2005). The recent studies showing increases of the senescence effector $p16^{INK4A}$ in aged stem and progenitor cells are supportive of this idea (Janzen et al. 2006; Krishnamurthy et al. 2006; Molofsky et al. 2006). The antiproliferative responses of p53 may also have a role in the decline of stem cell functionality with age (Dumble et al. 2007). The enhanced p53 activity observed in some mouse accelerated-aging models could affect the aging process through more robust suppression of stem cell activities as the organism ages. The ability of stem cells to maintain organ homeostasis declines more rapidly, and the result may be the organ atrophy and reduced overall functionality observed in these prematurely aging animals.

The last 10 years have been exciting for our understanding of the mechanisms of aging. The identification of mutations that can extend (or shorten) longevity in multiple model organisms has provided an important new genetic foundation for what had previously been a largely descriptive science. Yet, our understanding of the genetics of aging and longevity is still in a relatively primitive state. The tumor suppressors, particularly p53 and $p16^{INK4A}$, represent strong candidates for genes that regulate the aging process beyond their cancer-suppressive functions. Further experiments will be required to establish p53 as a gene that contributes to the regulation of basic aging processes. Beyond this, the identification of the mechanisms by which p53 regulates aging could provide key insights into this most complex and interesting biological phenomenon.

ACKNOWLEDGMENTS

This work was supported by grants from the National Institute on Aging and the Ellison Medical Foundation to L.A.D.

REFERENCES

Balducci L. and Beghe C. 2001. Cancer and age in the USA. *Crit. Rev. Oncol. Hematol.* **37:** 137–145.

Bauer J.H., Poon P.C., Glatt-Deeley H., Abrams J.M., and Helfand S.L. 2005. Neuronal expression of p53 dominant-negative proteins in adult *Drosophila melanogaster* extends life span. *Curr. Biol.* **15:** 2063–2068.

Beckman G., Birgander R., Sjalander A., Saha N., Holmberg P.A., Kivela A., and Beckman L. 1994. Is p53 polymorphism maintained by natural selection? *Hum. Hered.* **44:** 266–270.

Benard J., Douc-Rasy S., and Ahomadegbe J.C. 2003. TP53 family members and human cancers. *Hum. Mutat.* **21:** 182–191.

Blume-Jensen P. and Hunter T. 2001. Oncogenic kinase signalling. *Nature* **411:** 355–365.

Bonafe M., Salvioli S., Barbi C., Mishto M., Trapassi C., Gemelli C., Storci G., Olivieri F., Monti D., and Franceschi C. 2002. p53 codon 72 genotype affects apoptosis by cytosine arabinoside in blood leukocytes. *Biochem. Biophys. Res. Commun.* **299:** 539–541.

Bond G.L., Menin C., Bertorelle R., Alhorpuro P., Aaltonen L.A., and Levine A.J. 2006a. MDM2 SNP309 accelerates colorectal tumour formation in women. *J. Med. Genet.* **43:** 950–952.

Bond G.L., Hirshfield K.M., Kirchhoff T., Alexe G., Bond E.E., Robins H., Bartel F., Taubert H., Wuerl P., Hait W., et al. 2006b. MDM2 SNP309 accelerates tumor formation in a gender-specific and hormone-dependent manner. *Cancer Res.* **66:** 5104–5110.

Bond G.L., Hu W., Bond E.E., Robins H., Lutzker S.G., Arva N.C., Bargonetti J., Bartel F., Taubert H., Wuerl P., et al. 2004. A single nucleotide polymorphism in the MDM2 promoter attenuates the p53 tumor suppressor pathway and accelerates tumor formation in humans. *Cell* **119:** 591–602.

Bourdon J.C., Fernandes K., Murray-Zmijewski F., Liu G., Diot A., Xirodimas D.P., Saville M.K., and Lane D.P. 2005. p53 isoforms can regulate p53 transcriptional activity. *Genes Dev.* **19:** 2122–2137.

Braig M., Lee S., Loddenkemper C., Rudolph C., Peters A.H., Schlegelberger B., Stein H., Dorken B., Jenuwein T., and Schmitt C.A. 2005. Oncogene-induced senescence as an initial barrier in lymphoma development. *Nature* **436:** 660–665.

Brunet A., Bonni A., Zigmond M.J., Lin M.Z., Juo P., Hu L.S., Anderson M.J., Arden K.C., Blenis J., and Greenberg M.E. 1999. Akt promotes cell survival by phosphorylating and inhibiting a Forkhead transcription factor. *Cell* **96:** 857–868.

Buckbinder L., Talbott R., Velasco-Miguel S., Takenaka I., Faha B., Seizinger B.R., and Kley N. 1995. Induction of the growth inhibitor IGF-binding protein 3 by p53. *Nature* **377:** 646–649.

Campisi J. 2000. Cancer, aging and cellular senescence. *In Vivo* **14:** 183–188.

———. 2003. Cancer and ageing: Rival demons? *Nat. Rev. Cancer* **3:** 339–349.

———. 2005. Senescent cells, tumor suppression, and organismal aging: Good citizens, bad neighbors. *Cell* **120:** 513–522.

Cao L., Li W., Kim S., Brodie S.G., and Deng C.X. 2003. Senescence, aging, and malignant transformation mediated by p53 in mice lacking the Brca1 full-length isoform. *Genes Dev.* **17:** 201–213.

Chen J. 2004. Senescence and functional failure in hematopoietic stem cells. *Exp. Hematol.* **32:** 1025–1032.

Chen Z., Trotman L.C., Shaffer D., Lin H.K., Dotan Z.A., Niki M., Koutcher J.A., Scher H.I., Ludwig T., Gerald W., et al. 2005. Crucial role of p53-dependent cellular senescence in suppression of Pten-deficient tumorigenesis. *Nature* **436:** 725–730.

Chin L., Artandi S.E., Shen Q., Tam A., Lee S.L., Gottlieb G.J., Greider C.W., and DePinho R.A. 1999. p53 deficiency rescues the adverse effects of telomere loss and cooperates with telomere dysfunction to accelerate carcinogenesis. *Cell* **97:** 527–538.

Christophorou M.A., Ringshausen I., Finch A.J., Swigart L.B., and Evan G.I. 2006. The pathological response to DNA damage does not contribute to p53-mediated tumour suppression. *Nature* **443:** 214–217.

Collado M., Gil J., Efeyan A., Guerra C., Schuhmacher A.J., Barradas M., Benguria A., Zaballos A., Flores J.M., Barbacid M., et al. 2005. Tumour biology: Senescence in premalignant tumours. *Nature* **436:** 642.

Damalas A., Kahan S., Shtutman M., Ben-Ze'ev A., and Oren M. 2001. Deregulated beta-catenin induces a p53- and ARF-dependent growth arrest and cooperates with Ras in transformation. *EMBO J.* **20:** 4912–4922.

Datta S.R., Dudek H., Tao X., Masters S., Fu H., Gotoh Y., and Greenberg M.E. 1997. Akt phosphorylation of BAD couples survival signals to the cell-intrinsic death machinery. *Cell* **91:** 231–241.

de Rozieres S., Maya R., Oren M., and Lozano G. 2000. The loss of mdm2 induces p53-mediated apoptosis. *Oncogene* **19:** 1691–1697.

del Peso L., Gonzalez-Garcia M., Page C., Herrera R., and Nunez G. 1997. Interleukin-3-induced phosphorylation of BAD through the protein kinase Akt. *Science* **278:** 687–689.

Derry W.B., Putzke A.P., and Rothman J.H. 2001. *Caenorhabditis elegans* p53: Role in apoptosis, meiosis, and stress resistance. *Science* **294:** 591–595.

Dimri G.P. 2005. What has senescence got to do with cancer? *Cancer Cell* **7:** 505–512.

Dimri G.P., Lee X., Basile G., Acosta M., Scott G., Roskelley C., Medrano E.E., Linskens M., Rubelj I., and Pereira-Smith O., et al. 1995. A biomarker that identifies senescent human cells in culture and in aging skin in vivo. *Proc. Natl. Acad. Sci.* **92:** 9363–9367.

Donehower L.A., Harvey M., Slagle B.L., McArthur M.J., Montgomery C.A., Jr., Butel J.S., and Bradley A. 1992. Mice deficient for p53 are developmentally normal but susceptible to spontaneous tumours. *Nature* **356:** 215–221.

Dumble M., Gatza C., Tyner S., Venkatachalam S., and Donehower L.A. 2004. Insights into aging obtained from p53 mutant mouse models. *Ann. N.Y. Acad. Sci.* **1019:** 171–177.

Dumble M., Moore L., Chambers S.M., Geiger H., Van Zant G., Goodell M.A., and Donehower L.A. 2007. The impact of altered p53 dosage on hematopoietic stem cell dynamics during aging. *Blood* **109:** 1736–1742.

Dumont P., Leu J.I., Della Pietra A.C., III, George D.L., and Murphy M. 2003. The codon 72 polymorphic variants of p53 have markedly different apoptotic potential. *Nat. Genet.* **33:** 357–365.

Feng Z., Zhang H., Levine A.J., and Jin S. 2005. The coordinate regulation of the p53 and mTOR pathways in cells. *Proc. Natl. Acad. Sci.* **102:** 8204–8209.

Fevrier B. and Raposo G. 2004. Exosomes: Endosomal-derived vesicles shipping extracellular messages. *Curr. Opin. Cell Biol.* **16:** 415–421.

Gao X. and Pan D. 2001. TSC1 and TSC2 tumor suppressors antagonize insulin signaling in cell growth. *Genes Dev.* **15:** 1383–1392.

Garcia-Cao I., Garcia-Cao M., Martin-Caballero J., Criado L.M., Klatt P., Flores J.M., Weill J.C., Blasco M.A., and Serrano M. 2002. "Super p53" mice exhibit enhanced DNA damage response, are tumor resistant and age normally. *EMBO J.* **21:** 6225–6235.

Ghebranious N. and Donehower L.A. 1998. Mouse models in tumor suppression. *Oncogene* **17:** 3385–3400.

Grob T.J., Novak U., Maisse C., Barcaroli D., Luthi A.U., Pirnia F., Hugli B., Graber H.U., De Laurenzi V., Fey M.F., et al. 2001. Human delta Np73 regulates a dominant negative feedback loop for TAp73 and p53. *Cell Death Differ.* **8:** 1213–1223.

Guarente L. and Kenyon C. 2000. Genetic pathways that regulate ageing in model organisms. *Nature* **408:** 255–262.

Guarente L. and Picard F. 2005. Calorie restriction—The SIR2 connection. *Cell* **120:** 473–482.

Hannan K.M., Brandenburger Y., Jenkins A., Sharkey K., Cavanaugh A., Rothblum L., Moss T., Poortinga G., McArthur G.A., Pearson R.B., and Hannan R.D. 2003. mTOR-dependent regulation of ribosomal gene transcription requires S6K1 and is mediated by phosphorylation of the carboxy-terminal activation domain of the nucleolar transcription factor UBF. *Mol. Cell. Biol.* **23:** 8862–8877.

Hardie D.G. 2005. New roles for the LKB1 → AMPK pathway. *Curr. Opin. Cell Biol.* **17:** 167–173.

Harris S.L. and Levine A.J. 2005. The p53 pathway: Positive and negative feedback loops. *Oncogene* **24:** 2899–2908.

Harvey M., McArthur M.J., Montgomery C.A., Jr., Butel J.S., Bradley A., and Donehower L.A. 1993. Spontaneous and carcinogen-induced tumorigenesis in p53-deficient mice. *Nat. Genet.* **5:** 225–229.

Hay N. and Sonenberg N. 2004. Upstream and downstream of mTOR. *Genes Dev.* **18:** 1926–1945.

Herbig U., Ferreira M., Condel L., Carey D., and Sedivy J.M. 2006. Cellular senescence in aging primates. *Science* **311:** 1257.

Hirabayashi Y., Matsuda M., Aizawa S., Kodama Y., Kanno J., and Inoue T. 2002. Serial transplantation of p53-deficient hemopoietic progenitor cells to assess their infinite growth potential. *Exp. Biol. Med.* **227:** 474–479.

Hong Y., Miao X., Zhang X., Ding F., Luo A., Guo Y., Tan W., Liu Z., and Lin D. 2005. The role of P53 and MDM2 polymorphisms in the risk of esophageal squamous cell carcinoma. *Cancer Res.* **65:** 9582–9587.

Inoki K., Zhu T., and Guan K.L. 2003a. TSC2 mediates cellular energy response to control cell growth and survival. *Cell* **115:** 577–590.

Inoki K., Li Y., Xu T., and Guan K.L. 2003b. Rheb GTPase is a direct target of TSC2 GAP activity and regulates mTOR signaling. *Genes Dev.* **17:** 1829–1834.

Inoki K., Li Y., Zhu T., Wu J., and Guan K.L. 2002. TSC2 is phosphorylated and inhibited by Akt and suppresses mTOR signalling. *Nat. Cell Biol.* **4:** 648–657.

Jacks T., Remington L., Williams B.O., Schmitt E.M., Halachmi S., Bronson R.T., and Weinberg R.A. 1994. Tumor spectrum analysis in p53-mutant mice. *Curr. Biol.* **4:** 1–7.

Janzen V., Forkert R., Fleming H.E., Saito Y., Waring M.T., Dombkowski D.M., Cheng T., Depinho R.A., Sharpless N.E., and Scadden D.T. 2006. Stem-cell ageing modified by the cyclin-dependent kinase inhibitor p16(INK4a). *Nature* **443:** 421–426.

Jin S., Martinek S., Joo W.S., Wortman J.R., Mirkovic N., Sali A., Yandell M.D., Pavletich N.P., Young M.W., and Levine A.J. 2000. Identification and characterization of a p53 homologue in *Drosophila melanogaster. Proc. Natl. Acad. Sci.* **97:** 7301–7306.

Jones R.G., Plas D.R., Kubek S., Buzzai M., Mu J., Xu Y., Birnbaum M.J., and Thompson C.B. 2005. AMP-activated protein kinase induces a p53-dependent metabolic checkpoint. *Mol. Cell* **18:** 283–293.

Jones S.N., Roe A.E., Donehower L.A., and Bradley A. 1995. Rescue of embryonic lethality in Mdm2-deficient mice by absence of p53. *Nature* **378:** 206–208.

Kamada Y., Sekito T., and Ohsumi Y. 2004. Autophagy in yeast: A TOR-mediated response to nutrient starvation. *Curr. Top. Microbiol. Immunol.* **279:** 73–84.

Kartasheva N.N., Contente A., Lenz-Stoppler C., Roth J., and Dobbelstein M. 2002. p53 induces the expression of its antagonist p73 Delta N, establishing an autoregulatory feedback loop. *Oncogene* **21:** 4715–4727.

Kenyon C. 2005. The plasticity of aging: Insights from long-lived mutants. *Cell* **120:** 449–460.

Keyes W.M., Wu Y., Vogel H., Guo X., Lowe S.W., and Mills A.A. 2005. p63 deficiency activates a program of cellular senescence and leads to accelerated aging. *Genes Dev.* **19:** 1986–1999.

Keyes W.M., Vogel H., Koster M.I., Guo X., Qi Y., Petherbridge K.M., Roop D.R., Bradley A., and Mills A.A. 2006. p63 heterozygous mutant mice are not prone to spontaneous or chemically induced tumors. *Proc. Natl. Acad. Sci.* **103:** 8435–8440.

Kinzler K.W. and Vogelstein B. 1997. Cancer-susceptibility genes. Gatekeepers and caretakers. *Nature* **386:** 761, 763.

Klatt P. and Serrano M. 2003. Engineering cancer resistance in mice. *Carcinogenesis* **24:** 817–826.

Krishnamurthy J., Ramsey M.R., Ligon K.L., Torrice C., Koh A., Bonner-Weir S., and Sharpless N.E. 2006. p16(INK4a) induces an age-dependent decline in islet regenerative potential. *Nature* **443:** 453–457.

Krishnamurthy J., Torrice C., Ramsey M.R., Kovalev G.I., Al-Regaiey K., Su L., and Sharpless N.E. 2004. Ink4a/Arf expression is a biomarker of aging. *J. Clin. Invest.* **114:** 1299–1307.

Krtolica A., Parrinello S., Lockett S., Desprez P.Y., and Campisi J. 2001. Senescent fibroblasts promote epithelial cell growth and tumorigenesis: A link between cancer and aging. *Proc. Natl. Acad. Sci.* **98:** 12072–12077.

Lahav G., Rosenfeld N., Sigal A., Geva-Zatorsky N., Levine A.J., Elowitz M.B., and Alon U. 2004. Dynamics of the p53-Mdm2 feedback loop in individual cells. *Nat. Genet.* **36:** 147–150.

Lavigueur A., Maltby V., Mock D., Rossant J., Pawson T., and Bernstein A. 1989. High incidence of lung, bone, and lymphoid tumors in transgenic mice overexpressing mutant alleles of the p53 oncogene. *Mol. Cell. Biol.* **9:** 3982–3991.

Lee J.H., Lee E., Park J., Kim E., Kim J., and Chung J. 2003. In vivo p53 function is indispensable for DNA damage-induced apoptotic signaling in *Drosophila. FEBS Lett.* **550:** 5–10.

Lev Bar-Or R., Maya R., Segel L.A., Alon U., Levine A.J., and Oren M. 2000. Generation of oscillations by the p53-Mdm2 feedback loop: A theoretical and experimental study. *Proc. Natl. Acad. Sci.* **97:** 11250–11255.

Levine A.J. 1997. p53, the cellular gatekeeper for growth and division. *Cell* **88:** 323–331.

Levine A.J., Hu W., and Feng Z. 2006a. The P53 pathway: What questions remain to be explored? *Cell Death Differ.* **13:** 1027–1036.

Levine A.J., Feng Z., Mak T.W., You H., and Jin S. 2006b. Coordination and communication between the p53 and IGF-1-AKT-TOR signal transduction pathways. *Genes Dev.* **20:** 267–275.

Liu G., McDonnell T.J., Montes de Oca Luna R., Kapoor M., Mims B., El-Naggar A.K., and Lozano G. 2000. High metastatic potential in mice inheriting a targeted p53 missense mutation. *Proc. Natl. Acad. Sci.* **97:** 4174–4179.

Liu G., Parant J.M., Lang G., Chau P., Chavez-Reyes A., El-Naggar A.K., Multani A., Chang S., and Lozano G. 2004. Chromosome stability, in the absence of apoptosis, is critical for suppression of tumorigenesis in Trp53 mutant mice. *Nat. Genet.* **36:** 63–68.

Lowe S.W., Cepero E., and Evan G. 2004. Intrinsic tumour suppression. *Nature* **432:** 307–315.

Lozano G. and Elledge S.J. 2000. p53 sends nucleotides to repair DNA. *Nature* **404:** 24–25.

Lum J.J., Bauer D.E., Kong M., Harris M.H., Li C., Lindsten T., and Thompson C.B. 2005. Growth factor regulation of autophagy and cell survival in the absence of apoptosis. *Cell* **120:** 237–248.

Maier B., Gluba W., Bernier B., Turner T., Mohammad K., Guise T., Sutherland A., Thorner M., and Scrable H. 2004. Modulation of mammalian life span by the short isoform of p53. *Genes Dev.* **18:** 306–319.

Malkin D., Li F.P., Strong L.C., Fraumeni J.F., Jr., Nelson C.E., Kim D.H., Kassel J., Gryka M.A., Bischoff F.Z., and Tainsky M.A., et al. 1990. Germ line p53 mutations in a familial syndrome of breast cancer, sarcomas, and other neoplasms. *Science* **250:** 1233–1238.

Martin G.M. 2005. Genetic modulation of senescent phenotypes in *Homo sapiens*. *Cell* **120:** 523–532.

Mendrysa S.M., O'Leary K.A., McElwee M.K., Michalowski J., Eisenman R.N., Powell D.A., and Perry M.E. 2006. Tumor suppression and normal aging in mice with constitutively high p53 activity. *Genes Dev.* **20:** 16–21.

Moll U.M. and Slade N. 2004. p63 and p73: Roles in development and tumor formation. *Mol. Cancer Res.* **2:** 371–386.

Molofsky A.V., Slutsky S.G., Joseph N.M., He S., Pardal R., Krishnamurthy J., Sharpless N.E., and Morrison S.J. 2006. Increasing p16(INK4a) expression decreases forebrain progenitors and neurogenesis during ageing. *Nature* **443:** 448–452.

Montes de Oca Luna R., Wagner D.S., and Lozano G. 1995. Rescue of early embryonic lethality in mdm2-deficient mice by deletion of p53. *Nature* **378:** 203–206.

Murray-Zmijewski F., Lane D.P., and Bourdon J.C. 2006. p53/p63/p73 isoforms: An orchestra of isoforms to harmonise cell differentiation and response to stress. *Cell Death Differ.* **13:** 962–972.

Olive K.P., Tuveson D.A., Ruhe Z.C., Yin B., Willis N.A., Bronson R.T., Crowley D., and Jacks T. 2004. Mutant p53 gain of function in two mouse models of Li-Fraumeni syndrome. *Cell* **119:** 847–860.

Ollmann M., Young L.M., Di Como C.J., Karim F., Belvin M., Robertson S., Whittaker K., Demsky M., Fisher W.W., Buchman A., et al. 2000. *Drosophila* p53 is a structural and functional homolog of the tumor suppressor p53. *Cell* **101:** 91–101.

Palacios R., Bucana C., and Xie X. 1996. Long-term culture of lymphohematopoietic stem cells. *Proc. Natl. Acad. Sci.* **93:** 5247–5252.

Pardal R., Clarke M.F., and Morrison S.J. 2003. Applying the principles of stem-cell biology to cancer. *Nat. Rev. Cancer* **3:** 895–902.

Parrinello S., Coppe J.P., Krtolica A., and Campisi J. 2005. Stromal-epithelial interactions in aging and cancer: Senescent fibroblasts alter epithelial cell differentiation. *J. Cell Sci.* **118:** 485–496.

Pelicci P.G. 2004. Do tumor-suppressive mechanisms contribute to organism aging by inducing stem cell senescence? *J. Clin. Invest.* **113:** 4–7.

Pendas A.M., Zhou Z., Cadinanos J., Freije J.M., Wang J., Hultenby K., Astudillo A., Wernerson A., Rodriguez F., Tryggvason K., and Lopez-Otin C. 2002. Defective prelamin A processing and muscular and adipocyte alterations in Zmpste24 metalloproteinase-deficient mice. *Nat. Genet.* **31:** 94–99.

Perez-Losada J., Wu D., DelRosario R., Balmain A., and Mao J.H. 2005. p63 and p73 do not contribute to p53-mediated lymphoma suppressor activity in vivo. *Oncogene* **24:** 5521–5524.

Pinkston J.M., Garigan D., Hansen M., and Kenyon C. 2006. Mutations that increase the life span of *C. elegans* inhibit tumor growth. *Science* **313:** 971–975.

Purdie C.A., Harrison D.J., Peter A., Dobbie L., White S., Howie S.E., Salter D.M., Bird C.C., Wyllie A.H., and Hooper, M.L., et al. 1994. Tumour incidence, spectrum and ploidy in mice with a large deletion in the p53 gene. *Oncogene* **9:** 603–609.

Reya T., Morrison S.J., Clarke M.F., and Weissman I.L. 2001. Stem cells, cancer, and cancer stem cells. *Nature* **414:** 105–111.

Rudolph K.L., Chang S., Lee H.W., Blasco M., Gottlieb G.J., Greider C., and DePinho R.A. 1999. Longevity, stress response, and cancer in aging telomerase-deficient mice. *Cell* **96:** 701–712.

Sarbassov D.D., Guertin D.A., Ali S.M., and Sabatini D.M. 2005. Phosphorylation and regulation of Akt/PKB by the rictor-mTOR complex. *Science* **307:** 1098–1101.

Schmitt C.A. 2003. Senescence, apoptosis and therapy—Cutting the lifelines of cancer. *Nat. Rev. Cancer* **3:** 286–295.

Schmitt C.A., Fridman J.S., Yang M., Lee S., Baranov E., Hoffman R.M., and Lowe S.W. 2002. A senescence program controlled by p53 and p16INK4a contributes to the outcome of cancer therapy. *Cell* **109:** 335–346.

Sharpless N.E. and DePinho R.A. 2004. Telomeres, stem cells, senescence, and cancer. *J. Clin. Invest.* **113:** 160–168.

Shounan Y., Dolnikov A., MacKenzie K.L., Miller M., Chan Y.Y., and Symonds G. 1996. Retroviral transduction of hematopoietic progenitor cells with mutant p53 promotes survival and proliferation, modifies differentiation potential and inhibits apoptosis. *Leukemia* **10:** 1619–1628.

Smetannikova M.A., Beliavskaia V.A., Smetannikova N.A., Savkin I.V., Denisova D.V., Ustinov S.N., Maksimov V.N., Shabalin A.V., Bolotnova T.V., and Voevoda M.I. 2004. Functional polymorphism of p53 and CCR5 genes in the long-lived of the Siberian region (translation). *Vestn. Ross. Akad. Med. Nauk.* **2004:** 25–28.

Stambolic V., MacPherson D., Sas D., Lin Y., Snow B., Jang Y., Benchimol S., and Mak T.W. 2001. Regulation of PTEN transcription by p53. *Mol. Cell* **8:** 317–325.

TeKippe M., Harrison D.E., and Chen J. 2003. Expansion of hematopoietic stem cell phenotype and activity in Trp53-null mice. *Exp. Hematol.* **31:** 521–527.

Thomas G. 2002. The S6 kinase signaling pathway in the control of development and growth. *Biol. Res.* **35:** 305–313.

Tyner S.D., Venkatachalam S., Choi J., Jones S., Ghebranious N., Igelmann H., Lu X., Soron G., Cooper B., Brayton C., et al. 2002. p53 mutant mice that display early ageing-associated phenotypes. *Nature* **415:** 45–53.

van Heemst D., Mooijaart S.P., Beekman M., Schreuder J., de Craen A.J., Brandt B.W., Slagboom P.E., and Westendorp R.G. 2005. Variation in the human TP53 gene affects old age survival and cancer mortality. *Exp. Gerontol.* **40:** 11–15.

Van Zant G. and Liang Y. 2003. The role of stem cells in aging. *Exp. Hematol.* **31:** 659–672.

Varela I., Cadinanos J., Pendas A.M., Gutierrez-Fernandez A., Folgueras A.R., Sanchez L.M., Zhou Z., Rodriguez F.J., Stewart C.L., Vega J.A., et al. 2005. Accelerated ageing in mice deficient in Zmpste24 protease is linked to p53 signalling activation. *Nature* **437:** 564–568.

Vogel H., Lim D.S., Karsenty G., Finegold M., and Hasty P. 1999. Deletion of Ku86 causes early onset of senescence in mice. *Proc. Natl. Acad. Sci.* **96:** 10770–10775.

Vogelstein B. and Kinzler K.W. 2004. Cancer genes and the pathways they control. *Nat. Med.* **10:** 789–799.

Vogelstein B., Lane D., and Levine A.J. 2000. Surfing the p53 network. *Nature* **408:** 307–310.

von Zglinicki T., Saretzki G., Ladhoff J., d'Adda di Fagagna F., and Jackson S.P. 2005. Human cell senescence as a DNA damage response. *Mech. Ageing Dev.* **126:** 111–117.

Vousden K.H. and Lu X. 2002. Live or let die: The cell's response to p53. *Nat. Rev. Cancer* **2:** 594–604.

Wong K.K., Maser R.S., Bachoo R.M., Menon J., Carrasco D.R., Gu Y., Alt F.W., and DePinho R.A. 2003. Telomere dysfunction and Atm deficiency compromises organ homeostasis and accelerates ageing. *Nature* **421:** 643–648.

Yang A., Kaghad M., Caput D., and McKeon F. 2002. On the shoulders of giants: p63, p73 and the rise of p53. *Trends Genet.* **18:** 90–95.

Yu X., Harris S.L., and Levine A.J. 2006. The regulation of exosome secretion: A novel function of the p53 protein. *Cancer Res.* **66:** 4795–4801.

Zhang X., Miao X., Guo Y., Tan W., Zhou Y., Sun T., Wang Y., and Lin D. 2006. Genetic polymorphisms in cell cycle regulatory genes MDM2 and TP53 are associated with susceptibility to lung cancer. *Hum. Mutat.* **27:** 110–117.

Zhou B.P., Liao Y., Xia W., Zou Y., Spohn B., and Hung M.C. 2001. HER-2/neu induces p53 ubiquitination via Akt-mediated MDM2 phosphorylation. *Nat. Cell Biol.* **3:** 973–982.

Zindy F., Williams R.T., Baudino T.A., Rehg J.E., Skapek S.X., Cleveland J.L., Roussel M.F., and Sherr C.J. 2003. Arf tumor suppressor promoter monitors latent oncogenic signals in vivo. *Proc. Natl. Acad. Sci.* **100:** 15930–15935.

7

Aging Processes in *Caenorhabditis elegans*

Heidi A. Tissenbaum
Program in Gene Function and Expression
Program in Molecular Medicine
University of Massachusetts Medical School
Worcester, Massachusetts 01605

Thomas E. Johnson
Institute for Behavioral Genetics and
Department of Integrative Physiology
University of Colorado at Boulder
Boulder, Colorado 80309

DETERMINING THE UNDERLYING MOLECULAR MECHANISMS of aging has been the focus of many research studies during the past three decades. At the center of this research are findings derived from the nematode, *Caenorhabditis elegans*. From the original identification of a single mutation that can lengthen life span, through the finding that the insulin/insulin-like growth factor-1 (IGF-1) signaling (IIS) pathway specifies processes affecting life span, *C. elegans* has provided the basis and initial observations for much of what we currently understand about molecular mechanisms of aging.

C. ELEGANS AS A SYSTEM FOR ANALYSIS OF AGING

Studies on *C. elegans* initially began in 1974 with the genetic map initiated by Sydney Brenner (Brenner 1974). Since then, many seminal findings on biological function have been first shown in *C. elegans,* including the initial dissection of programmed cell death (Ellis et al. 1991), the systematic cloning of the genome (Coulson et al. 1986), and the deciphering of the entire DNA sequence (*C. elegans* Consortium 1998).

C. elegans represents a relative newcomer among genetic systems used in aging studies, with few publications prior to 1982 (Epstein et al. 1974). Despite the slow start, *C. elegans* has emerged as a system of choice for aging research, in part because this system allows the use of extensions in life span in detecting genetic mutations. Thus far, such genetic dissection of the *C. elegans* aging process has identified about 200 or more genes that exhibit life span as a result of hypomorphic (reduced function) mutations. Additionally, during the past 25 years or so, two major breakthroughs in aging research derived from *C. elegans* studies, which help to define worms as a powerful system for aging research.

The initial studies, which focused on polygenetic control of aging (Johnson and Wood 1982; Johnson 1987b) and the identification of a single gene that modulated life span, were carried out in *C. elegans*. Klass (1983), using a brute-force approach, identified mutants that had altered life span. Subsequently, all of the mutants were found to be in a single genetic locus, named *age-1*, and were mapped and characterized (Friedman and Johnson 1988a,b; Johnson 1990; Morris et al. 1996). The fundamental importance of this finding is that it suggested for the first time that aging is modulated by genes and that reduction of function could extend life, not just shorten it.

A second breakthrough was the identification of another mutant that alters life span (Age mutant), *daf-2*, which similarly to *age-1*, lengthened the life of adult worms (Kenyon et al. 1993); subsequent molecular cloning and identification of *daf-2* showed it to be an insulin/IGF-1 receptor (Kimura et al. 1997). Taken together, the *C. elegans* system has provided several major advances in the identification of molecular mechanisms of aging.

IS AGING PROGRAMMED?

This topic has led to considerable controversy not only in *C. elegans*, but also in many other studies of aging. Indeed, whole books have been devoted to this subject (Johnson 1987b; Russell 1987). Here, we do not use the word "program" but choose instead to use the words "specify" or "modulate" to describe genes that alter one or more aging-related properties. There is only space for a brief summary here to disuss the idea of regulation of the aging process; a more complete analysis can be found in Lithgow (2006). We suggest that the primary argument stems from the fact that, unlike development, which is clearly of benefit to the organism, aging has not been selected for. Aging arises as a result of lack of selection against late-life deleterious events because such events do not decrease evolutionary "fitness." These are subtle arguments and have frequently been ignored, but they can be proven to be true mathematically and experimentally. The work described in this chapter gains new richness and interpretation when this perspective is used because we can see that aging is a by-product of the normal events specifying nematode physiology rather than a defined "program." For example, in *C. elegans*, it is possible to extend life span up to eightfold (Arantes-Oliveira et al. 2003; Houthoofd et al. 2005). Therefore, since there is no genetic program causing aging, the side effects of normal physiology that limit life can be fairly readily bypassed; the reader is referred to a few recent discussions of these points (Antebi 2005; Lithgow 2006).

INSULIN/IGF-LIKE SIGNALING PATHWAY

Components of the IIS Pathway

The IIS pathway is evolutionarily conserved and found in species ranging from yeast, worms, and flies to humans. At the center of the pathway is the receptor tyrosine kinase DAF-2 that signals to regulate the nuclear/cytoplasmic distribution of the forkhead transcription factor, DAF-16. Numerous studies of this pathway have led to the identification of several regulatory inputs and outputs. This pathway influences several physiological processes, including stress resistance, development, cell cycle, and metabolism (Kenyon 2005).

The regulation of IIS and development is an intriguing connection. In *C. elegans*, in response to increased levels of a secreted pheromone as well as food deprivation (i.e., unfavorable growth conditions) during early larval stage (L1), worms can enter an alternative developmental mode and form dauer larvae (Riddle and Albert 1997). Dauer larvae are morphologically and physiologically distinct from wild-type worms: They are very thin and do not move or feed (Riddle and Albert 1997). Many of the genes that function in the IIS pathway were first isolated based on their effect on *da*uer formation (*daf*). Interestingly, dauers are a nonaging stage of development (Klass and Hirsh 1976; Riddle and Albert 1997). However, mutations that affect dauer may or may not alter adult life span. Although both the IIS pathway and the transforming growth factor-β (TGF-β) signaling pathway modulate the dauer development decision, only genes in the IIS pathway affect life span (Kenyon et al. 1993; Duhon et al. 1996). Therefore, if there is a molecular connection between dauer formation and life span, it is not simply turning off genes that modulate life span in a dauer larva.

The IIS pathway itself is usually shown to begin with the receptor DAF-2, since exact molecular details of the inputs into the pathway are limited (Fig. 1). Upstream of *daf-2*, however, several genes have been identified which are involved in neurological signaling that impinges on the DAF-2 receptor, including *unc-64* (mammalian syntaxin), *unc-13* (mammalian mUNC-13), *unc-18* (mammalian mUNC-18) involved in

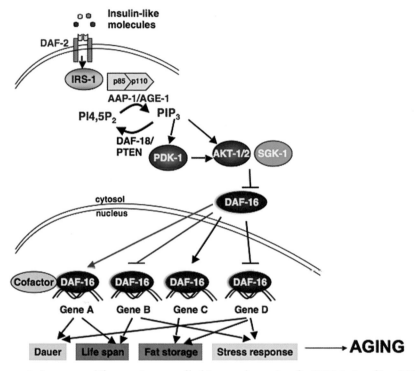

Figure 1. In worms, life span is controlled in part by an insulin/IGF-1 signaling (IIS) pathway. The pathway begins with the *C. elegans* IIS receptor encoded by the *daf-2* gene, is homologous to both mammalian insulin and insulin-like growth factor-1 (IGF-1) receptors (Kimura et al. 1997), and the ligand is not known. Similar to mammalian signaling, downstream from DAF-2, a phosphoinositol-3 kinase (PI3K) pathway negatively regulates DAF-16, the forkhead/HNF3-related transcription factor (FOXO). The conserved PI3K signaling components, AKT-1, AKT-2, SGK-1, and PDK-1 kinases, primarily regulate the translocation and activation of DAF-16. AKT-1/2 and SGK-1 phosphorylate and sequester DAF-16 in the cytosol, preventing it from entering the nucleus and binding to nuclear target genes. Once in the nucleus, DAF-16 *trans*-activates/represses numerous target genes that function in life span regulation, stress response, development, and metabolism. A negative regulator PTEN, encoded by the gene *daf-18*, inhibits the signaling pathway by functioning as a lipid phosphatase. See text for references and further details.

synaptic vesicle fusion during synaptic transmission, and *unc-31*, a CAPS homolog involved in calcium-dependent peptide secretion (Ogawa et al. 1998; Saifee et al. 1998; Ailion et al. 1999; Gems and Riddle 2000; Munoz and Riddle 2003). *unc-31*, *unc-64*, *unc-18*, and *unc-13* are presumed to function in neuronal signaling, leading to release of the insulin-like ligand (Ann et al. 1997; Ailion et al. 1999; Munoz and Riddle 2003) that binds to DAF-2.

Although 38 insulins/insulin-like peptides have been identified (Pierce et al. 2001; Li et al. 2003), no biochemical data have been found to discern which of the peptides bind to DAF-2. Genetic (dauer formation and life span) and expression pattern analyses, however, suggest the possibility that there are at least four potential ligands encoded by the genes *ins-1*, *ins-18*, *daf-28*, and *ins-7* (Kawano et al. 2000; Pierce et al. 2001; Li et al. 2003; Murphy et al. 2003). Because the genetics and phenotypes are well established for *daf-2*, some key predictions can be made for a *daf-2* ligand. Therefore, analysis of the large number of insulins has revealed that some may function as agonists, others as antagonists, of the DAF-2 receptor.

For all of the insulins, the phenotypes tested are life span and dauer formation. *ins-1* is predicted to be an antagonist, since increased dosage of *ins-1* has been shown to promote dauer formation and increase life span (Pierce et al. 2001). However, *ins-18* may function as an agonist, since inhibition of *ins-18* results in an approximately 30–40% increase in mean life span (Kawano et al. 2000). Another possible DAF-2 agonist is encoded by *daf-28*, an insulin-like protein (Li et al. 2003) where dauer-inducing, environmental conditions caused a dramatic decrease in *daf-28*::GFP expression. Additionally, mutation in *daf-28* had a modest increase in life span, although this may be related to the redundancy and complex expression pattern of the *ins* genes in *C. elegans* (Pierce et al. 2001; Li et al. 2003). Finally, microarray analysis comparing *daf-2* mutants to *daf-16; daf-2* double mutants revealed that another putative DAF-2 ligand was encoded by the gene *ins-7* (McElwee et al. 2003; Murphy et al. 2003). However, since *ins-7* was isolated as a DAF-16 target gene, this suggests that *ins-7* functions in a feed-forward mechanism whereby the IIS pathway activates INS-7 and then INS-7 acts on a DAF-2 receptor in a neighboring cell (Murphy et al. 2003). Therefore, to date, analysis has suggested that several of the *ins* genes could encode a DAF-2 ligand, but the biochemical nature of the interaction is still unclear.

Insulin genes have also been identified by other assays which include gene expression analysis changes from development and up to 19 days post-fertilization by microarray analysis (Lund et al. 2002), microarray analysis for genes in the TGF-β signaling pathway important for dauer formation

(Liu et al. 2004) and a genome-wide RNA interference (RNAi) screen for genes required for synapse structure and function (Sieburth et al. 2005).

The only biochemical analysis of any *C. elegans* insulins examined the structure and receptor-binding properties of *ins-6*. This study found that *ins-6* shows structural similiarity to human insulin and can bind and activate the human insulin receptor (Hua et al. 2003). This is somewhat surprising given that experiments examining RNAi inactivation of *ins-6* have revealed no specific phenotype and given that its expression is in the amphid neurons (Pierce et al. 2001), which would suggest an important role in life span regulation.

Li et al. (2003) also examined the redundancy among the insulins by testing the ability of high-copy *ins* transgenes to suppress a *daf-28* mutant. The two *ins* genes most similar to *daf-28, ins-4* and *ins-6*, could functionally complement *daf-28* at high doses, whereas transgenes of *ins* genes that were less similar (*ins-7, ins-9, ins-17,* or a combination of *ins-21-23*) could not (Li et al. 2003). Therefore, there is the possibility of redundancy among the *C. elegans* insulin gene family.

Downstream from *daf-2*, a well-studied PI3K signaling pathway functions to ultimately regulate a forkhead transcription factor (FOXO), DAF-16, through a series of phosphorylation events in a manner similar to that of mammalian DAF-16 homologs (Biggs et al. 1999; Brunet et al. 1999; Kops et al. 1999; Henderson and Johnson 2001; Lee et al. 2001; Lin et al. 2001). The *C. elegans* IIS pathway includes an insulin receptor substrate (IRS-1), a phosphatidylinositol-3-kinase (AGE-1 and AAP-1), a phosphoinosotide-dependent kinase (PDK-1), a serum glucocorticoid kinase (SGK-1), two protein kinase B homologs (also known as akt; AKT-1/2), and a forkhead transcription factor (DAF-16). The pathway also includes a phosphatase (DAF-18 in worms and homologous to the human tumor suppressor PTEN) that acts to counter the activity of AGE-1 (Fig. 1).

By analogy to studies in mammals, it is presumed that the PI3K signaling cascade regulates levels of PIP-2 through recruitment of the AKT-1, AKT-2, and PDK-1 kinases to the plasma membrane where PDK-1 activates AKT and by phosphorylation (Paradis and Ruvkun 1998). AKT and SGK-1 then phosphorylate and sequester DAF-16 in the cytosol (Lin et al. 2001; Hertweck et al. 2004; Oh et al. 2005). In contrast, in the absence of ligands, DAF-16 is not phosphorylated and can then translocate into the nucleus (Henderson and Johnson 2001; Lin et al. 2001). Upon entering the nucleus, DAF-16 binds to and *trans*-activates/represses numerous target genes involved in life span regulation as well as other regulatory processes including stress response, dauer formation, and fat storage (Fig. 1) (Lee et al. 2003a; McElwee et al. 2003; Murphy et al. 2003; Oh et al. 2006).

Consistent with this signaling pathway, reduction-of-function mutations in genes upstream of the IIS pathway (*unc-31, unc-64, unc-18,* or *unc-13*) or in the IIS pathway (*daf-2, age-1, aap-1, pdk-1, sgk-1,* or *akt-1/2*) that down-regulate DAF-16 result in life span extension since DAF-16 is not phosphorylated and can freely enter the nucleus to turn on/off downstream target genes (Dorman et al. 1995; Larsen et al. 1995; Paradis et al. 1999; Wolkow et al. 2002; Hertweck et al. 2004). In contrast, a null mutation of *daf-16* slightly shortens life span and completely suppresses life span extension caused by mutations in the upstream components of the IIS since there is no DAF-16 to enter the nucleus to turn off/on its numerous target genes (Lin et al. 1997; Ogg et al. 1997; Lee et al. 2003a; McElwee et al. 2003; Murphy et al. 2003; Oh et al. 2006).

Sensory Input into the IIS

C. elegans, a soil-dwelling nematode, uses ciliated sensory neurons located in the head and tail to sense its external environment including aversive stimuli and attractants, bacteria (food), and chemical messengers (e.g., pheromones). In the head of the worm, the major sensory neurons are contained within paired structures called amphids. Within the amphid, the ciliated endings of 12 sensory neurons, a sheath cell, and a socket cell together form a pore to the exterior. In the tail, the sensory organ is the phasmid and has a minor function but a similar structure to the amphids (Hedgecock et al. 1985).

Genetic and molecular analysis has identified many genes that affect the function and development of sensory neurons, including classes of genes that affect the animal's ability to detect a variety of chemical stimuli (*che*) and changes in osmolarity (*osm*). Mutations in these genes give rise to animals that are impaired in their ability to detect their environment. These animals frequently demonstrate a dauer phenotype at high temperature, as well as life span extension at many temperatures (Vowels and Thomas 1992; Apfeld and Kenyon 1999). In nematodes, it is possible to directly test the role of amphids in life span by ablating the structures with a laser. Such sensory-deprived worms live approximately 33% longer than nonablated animals (Apfeld and Kenyon 1999). Therefore, lack of sensory input affects both dauer formation and life span, similar to mutations in the IIS pathway. Consistent with this finding, all of the increases in dauer formation and longevity in the *che* and *osm* mutants were found to be suppressed by mutations in *daf-16,* the downstream target of the IIS pathway (Vowels and Thomas 1992; Apfeld and Kenyon 1999; Alcedo and Kenyon 2004). The dependence of *daf-16* for life span extension suggests that sensory cues are transmitted through insulin-like signaling to influence life span. The importance of the sensory neurons in life span regulation by the IIS signaling pathway is also revealed by studies showing that *daf-2* function in neurons is critical for life span regulation (Apfeld and Kenyon 1998; Wolkow et al. 2000). The importance of sensory perception in regulation of life span of higher eukaryotes is still undetermined (Kenyon 2005).

IIS and Germ-line Signaling

The reproductive system can also modulate the life span of *C. elegans,* in part by interacting with the IIS. The *C. elegans* gonad primarily exists as self-fertilizing structured by a gonadal primordium consisting of four cells (Z1, Z2, Z3, Z4) (Hubbard and Greenstein 2000). The cell lineages derived from Z1 and Z4 give rise to the somatic gonad that includes the sheath, spermatheca, and uterus in hermaphrodites and the testis, seminal vesicles, and vas deferens in males. The cell lineages of Z2 and Z3 give rise to the germ cells (Hubbard and Greenstein 2000). Using a laser to destroy distinct germ cells revealed that ablation of the germ-line precursor cells (Z2, Z3) caused an increase in life span of up to 60% (Hsin and Kenyon 1999). Importantly, this was not simply due to sterility, since removal of the entire reproductive system (germ line plus somatic gonad-ablation of Z1, Z2, Z3, Z4) has no effect on life span. Further studies show that this active signal from the reproductive tissue is dependent on *daf-16* such that germ-line ablation in a *daf-16* mutant background has no effect on life span (Hsin and Kenyon 1999). However, the connection of germ-line signaling to the IIS pathway is more complicated. Ablation of the germ line in a long-lived *daf-2* mutant background further extends life span, regardless of the presence/absence of the somatic gonad. One possibility is that the somatic gonad produces a signal that requires DAF-2, and this acts in parallel to counterbalance the germ-line signal to DAF-16 (Hsin and Kenyon 1999). Therefore, signals from the reproductive signal modulate life span in part by acting through the IIS. These signals have alternately been suggested to be new insulin molecules or physiological processes leading, for instance, to mitochondrial expansion (Rea and Johnson 2003).

SIR2 and the IIS

sir2 (silent information regulator) was originally isolated in *Saccharomyces cerevisiae* as a gene important for gene silencing (Rine and Herskowitz 1987). The emergence of the importance of SIR2 in life span stemmed from findings that providing additional copies of SIR2 extends life span in yeast, worms, and flies, suggesting the possibility of an evolutionarily conserved mechanism of aging (Kaeberlein et al. 1999; Tissenbaum and Guarente 2001; Rogina and Helfand 2004). In *C. elegans*, life span extension resulting from overexpressing *sir-2.1* (*C. elegans* SIR2 ortholog) is completely dependent on a wild-type copy of *daf-16* (Tissenbaum and Guarente 2001) and, together with additional genetic analysis, suggests that *sir-2.1* extends life span by functioning through the IIS. However, additional analysis with a deletion mutation of *sir-2.1* revealed that *sir-2.1* and *daf-16* have overlapping and distinct roles in life span regulation, based on the genetic and expression analyses (Wang and Tissenbaum 2006). Most recently, the connection between SIR-2.1 and DAF-16 has been shown to be mediated by the *C. elegans* 14-3-3 proteins. This work suggests that DAF-16 and SIR-2.1 interact in a complex mediated by the 14-3-3 proteins to modulate life span and stress resistance (Berdichevsky et al. 2006; Wang and Tissenbaum 2006). Interestingly, the connection between SIRT1 (mammalian SIR-2.1) and FOXO (mammalian DAF-16) has been well documented (Brunet et al. 2004; Motta et al. 2004).

A second role for SIR-2.1 emerged from microarray analysis using the plant-derived compound resveratrol (Viswanathan et al. 2005). Studies from several systems suggest that resveratrol is a caloric-restriction mimetic that functions as an activator of SIR2 (Howitz et al. 2003; Wood et al. 2004). Although the potential benefits of resveratrol have recently been extended to studies in mice (Baur et al. 2006), the mechanism of resveratrol function through SIR2 is largely in dispute (Borra et al. 2005; Kaeberlein et al. 2005; Viswanathan et al. 2005). In *C. elegans*, worms treated with resveratrol show a small life span increase (10–15%) dependent on *sir-2.1* and independent of *daf-16* (Wood et al. 2004; Viswanathan et al. 2005). Microarray studies on worms treated with resveratrol revealed a large up-regulation of genes involved in the endoplasmic reticulum (ER) stress response to unfolded proteins. Additional analysis challenged the role of resveratrol as an activator of SIR-2.1 and rather suggested that resveratrol extends life span by inhibiting the SIR-2.1-mediated repression of ER stress genes (Viswanathan et al. 2005).

JNK Signaling and IIS

In mammals, the JNK family, a subgroup of the mitogen-activated protein kinase (MAPK) superfamily, is a part of the signal transduction cascade that is activated by cytokines, including tumor necrosis factor (TNF) and interleukin-1 (IL-1), and by exposure to environmental stresses. Additionally, the JNK family has been implicated in critical biological processes such as cancer, development, apoptosis, and cell survival (Davis 2000). Importantly, in mammalian cell culture, components of the JNK signaling pathway have been shown to interact with the insulin signaling pathway through interaction with either the insulin receptor substrate-1 (IRS-1) (Aguirre et al. 2000) or the AKT protein kinase (Kim et al. 2003). Most recently, in worms and flies, JNK overexpression extends life span and increases stress resistance, and this life span extension is dependent on DAF-16/FOXO (Oh et al. 2005; Wang et al. 2005). Further studies in *C. elegans* find that JNK physically interacts with and phosphorylates DAF-16 at sites different from the AKT phosphorylation sites and this results in enhanced nuclear translocation of DAF-16 (Oh et al. 2005). The identification of the JNK pathway connecting to DAF-16 exemplifies the notion that DAF-16 also receives DAF-2-independent signals.

Regulation of DAF-16

DAF-16 is a member of the forkhead transcription factor (FOXO) family that includes AFX (FOXO4), FKHR (FOXO1), and FKHR-L1 (FOXO3a), with greatest homology with FOXO3a. This family of transcription factors binds DNA as a monomer at the consensus binding site (TTG/ATTTAC) (Furuyama et al. 2000). Thus far, many of the inferred molecular characteristics of DAF-16 are deduced from its similarity with other FOXOs, in part due to the difficulty of obtaining good worm extracts for biochemistry because of the thick cuticle of the worm, and in part due to the lack of a robust cell culture system, although one has been reported (Christensen et al. 2002). FOXOs are regulated by phosphorylation on multiple threonine and serine residues (Birkenkamp and Coffer 2003; Van Der Heide et al. 2004) and continuously shuttle between the nucleus and the cytosol with the help of nuclear transporter proteins (Van Der Heide et al. 2004; Vogt et al. 2005). Ultimately, all signals regulate this shuttling of the FOXO from cytoplasm to nucleus and back. When activity of the upstream kinase AKT/PKB is low, the rate of import exceeds that of export to leave FOXO proteins predominantly in the nucleus. Upon activation of AKT/PKB, AKT/PKB phosphorylates FOXO, and the phosphorylated proteins have an increased binding affinity to the 14-3-3 proteins. This results in the release of the FOXO protein from the DNA and its relocalization to the cytosol (Brunet et al. 1999; Cahill et al. 2001). Following translocation to the cytosol, the bound 14-3-3 prevents reentry of FOXO into the nucleus by masking the nuclear localization signal (NLS) (Obsilova et al. 2005) and inhibiting importin binding until FOXO proteins are modified (e.g., dephosphorylation) (Burgering and Kops 2002).

Recently, it has been shown that under conditions of hydrogen peroxide stress, FOXO3A becomes phosphorylated at a conserved site in the forkhead domain by the MST1 kinase (Lehtinen et al. 2006). This results in disruption of its interaction with the 14-3-3 proteins and promotes nuclear entry (Lehtinen et al. 2006). Results suggest that a similar mechanism occurs in C. elegans, based on conservation of the phosphorylation site and genetic data (Lehtinen et al. 2006).

The importance of AKT phosphorylation in regulation of DAF-16 is still somewhat controversial. Due to the limited biochemistry in C. elegans, as well as the lack of cell culture, similar to Drosophila S2 cells, the assays rely primarily on genetic observations that may be more subjective. In one study, absence of AKT phosphorylation is sufficient to cause dauer arrest (Lee et al. 2001). Therefore, this predicts that DAF-16 is the major target of AKT. In contrast, Lin et al. (2001) find that although absence of AKT phosphorylation facilitates DAF-16 entry into the nucleus, it was not sufficient to induce dauer formation or life span extension. Therefore, this suggests a role for unidentified proteins that activate DAF-16 (Lin et al. 2001). Consistent with the latter studies, it has recently been shown that DAF-16 can also be modified by JNK phosphorylation (at sites different from AKT phosphorylation) and that this interaction leads to enhanced nuclear localization following stress (Oh et al. 2005). In addition, more recent studies (Hertweck et al. 2004) identified the C. elegans serum-glucocorticoid kinase SGK-1, which functions at the level of AKT-1/2. Hertweck et al. (2004) found that these three kinases AKT-1/AKT-2/SGK-1 form a protein complex that functions to transduce the PI3K signal and control the localization and activation of DAF-16 by direct phosphorylation.

Modulating DAF-16 transcriptional activity is a major function of IIS in C. elegans, and the key output is regulating nuclear localization of DAF-16. Several different environmental stresses cause nuclear localization of DAF-16 (Henderson and Johnson 2001). Additionally, increasing the dosage of daf-16 results in increased stress resistance and life span while at the same time slowing growth and reproduction (Henderson and Johnson 2001). Taken together, DAF-16 functions to promote longevity in a dosage-dependent manner and is regulated by its entry into the nucleus.

Recently, a potential cofactor for DAF-16 has been identified and is suggested to determine an underlying mechanism for DAF-16 specificity. Genetic, molecular, and physiological analyses of SMK-1 (suppressor of MEK null), initially identified in a genetic suppressor screen of MEK1 mutants in Dictyostelium discoideum, find that SMK-1 is required for DAF-16-dependent regulation of life span (Wolff et al. 2006). Interestingly,

SMK-1 does not affect dauer formation or regulation of life span by the reproductive tissues, two other functions of DAF-16 (Wolff et al. 2006). Transcription and physiological studies show that *smk-1* is required for oxidative stress, innate immunity, and UV stress and is not necessary for the thermal stress function of DAF-16 (Wolff et al. 2006). Therefore, SMK-1 possesses all of the requirements for an IIS-mediated longevity cofactor of DAF-16. Future direct biochemical data are needed to confirm this function, but the idea that DAF-16 may bind to distinct cofactors for different functions (Berdichevsky and Guarente 2006; Wolff et al. 2006) is attractive.

Targets of DAF-16

What is the functional consequence of signaling through the IIS? FOXO is known to signal to many different targets involved in the oxidative stress response (*MnSOD*), DNA repair (*GADD45*), cell cycle arrest (*p27KIP1*), and apoptosis (*BIM* and *Fas ligand*) (Aballay and Ausubel 2002; Birkenkamp and Coffer 2003; Van Der Heide et al. 2004). For *C. elegans,* targets of DAF-16 have been the recent focus of much research. Genetic screens and analysis of individual genes initially suggested a handful of DAF-16 targets. These "targets" were identified using many different approaches and include superoxide dismutase (*sod-3*) (Honda and Honda 1999), tyrosine kinase (*old-1*) (Murakami and Johnson 2001), metallothionein (*mtl-1*) (Barsyte et al. 2001), SCP-like extracellular protein (*scl-1*) (Ookuma et al. 2003), raptor (*daf-15*) (Jia et al. 2004), and small heat shock proteins (Hsu et al. 2003). Targets are typically defined by mutation or overexpression phenotypes that mimic mutations in the IIS pathway or where expression depends on intact DAF-16. Systematic and genome-wide analyses have been employed to identify putative downstream targets of DAF-16. Using cDNA microarrays, a number of targets were identified whose expression level is dependent on DAF-16 (McElwee et al. 2003; Murphy et al. 2003). Using bioinformatics, Lee et al. (2003a) identified 17 targets that were subsequently assayed for their involvement in DAF-16-dependent phenotypes. Many of these "target" genes from the microarray and/or the bioinformatics studies (Lee et al. 2003a; McElwee et al. 2003; Murphy et al. 2003) could be linked to an influence on life span, since they included antioxidant genes (such as superoxide dismutase, metallothionein, catalase, and glutathione-*S*-transferase), metabolic genes (such as apolipoprotein genes, glyoxylate-cycle genes, genes involved in amino acid turnover), small heat shock protein genes, and antibacterial genes, and were shown to directly specify small effects on longevity. These results generally agree with the concept that an increase in cellular defense results in extended life span (Garigan et al. 2002; Gems and McElwee 2005). In agreement with this idea, mammalian FOXO proteins activate stress-response genes when insulin or IGF-1 levels are reduced (Kops et al. 2002; Nemoto and Finkel 2002; Tran et al. 2002).

All these analyses suggested a large number of probable targets but did not determine whether any of these "targets" are indeed direct in vivo targets of DAF-16 or genes affected by downstream pathways. Most recently, a chromatin immunoprecipitation (ChIP)-based cloning strategy in *C. elegans* identified more than 100 putative direct target genes. ChIP fundamentally relies on the physical interaction of a transcription factor and its target promoter. The ChIP studies showed, for the first time, that DAF-16 directly binds to previously known and numerous novel target promoters in *C. elegans* (Oh et al. 2006). In addition, the large number of direct target genes identified suggests that there is a complex regulation downstream from DAF-16.

MUTATIONS IN MITOCHONDRIAL COMPONENTS

There are numerous other Age genes not in the IIS pathway. The largest class comprises those encoding mitochondrial genes that as a class have been referred to as "Mit." It seems almost heretical that reducing electron transport chain (ETC) activity might extend life span but, in *C. elegans*, it does. The Mit class of long-lived mutants are generally reduced-in-function alterations in mitochondrial proteins and typically exhibit a 20–40% increase in mean adult life span, although several have been reported to live much longer (for review, see Henderson et al. 2005). Almost all of the Mit mutants tested act independently of the insulin-like *daf-2* signaling pathway.

Clock Mutants

The Clock (Clk) set of mutants were the first to be found and have a modest Age phenotype (Lakowski and Hekimi 1996); *clk-1* is the best-characterized Mit mutant; it has a defective demethoxyubiquinone (DMQ) monooxygenase, preventing synthesis of 5 hydroxyubiquinone, the penultimate intermediate of ubiquinone (Q) (Jonassen et al. 2001; Stenmark et al. 2001), and consequently accumulates significant but insufficient quantities of DMQ_9 (Miyadera et al. 2002). The Age character in *clk-1* may result from reduced reactive oxidant species (ROS) produced in *clk-1* mitochondria (Miyadera et al. 2002).

Morgan and colleagues (Kayser et al. 2004) extensively characterized mitochondria from *clk-1* and N2 (wild type) and found that complex I in *clk-1* mitochondria operated at about 30% activity using endogenous quinone carriers only (DMQ_9, Q_8, and rhodoquinone). These findings contradicted earlier studies (Felkai et al. 1999; Miyadera et al. 2001; Jonassen et al. 2003), reporting no differences in complex I (NADH oxidoreductase) in *clk-1*. In a set of parallel studies, Kayser et al. (2004) also showed that *clk-1* nematodes display a reduced level of oxidized mitochondrial proteins relative to wild type, suggesting reduced oxidative damage in *clk-1*, consistent with the free-radical hypothesis of aging (Harman 1956). Moreover, the longevity effects of the *clk-1* mutation have been replicated in mice (Liu et al. 2005).

Screens Using RNA Inhibition

Genomic RNAi libraries (Kamath et al. 2003) made it feasible to screen the entire genome for longevity-enhancing, loss-of-function mutations. Both the Kenyon and Ruvkun labs (Dillin et al. 2002; Lee et al. 2003b) found that reducing the function of many mitochondrial genes by RNAi, paradoxically, extended worm life span. Rather disappointing was the lack of overlap between the sets of mutants, but longevity after RNAi is very dependent on exact conditions (S.L. Rea et al., pers. comm.), and the screens performed by the two labs were somewhat different. The Ruvkun group found that 1.8% of the 5690 genes they screened extended life span by 5–30%; 15% encoded mitochondrial proteins; almost all affected ETC components with little preference for any one complex. Most caused a reduction in adult size, which has become a hallmark of the Mit phenotype (S.L. Rea et al., pers. comm.). Many had altered mitochondrial morphology and exhibited no obvious relationship between resistance to hydrogen peroxide or paraquat and life extension (Lee et al. 2003b). The life-span-enhancing effects of the RNAi clones were only observed if the RNAi was fed to animals during the larval period (Dillin et al. 2002), suggesting that specific, mitochondrial dysfunction signals must be sensed some time during development for animals to adapt with an increased life span; perhaps this is a metabolic signal necessary for mitochondrial proliferation associated with maturation and reproduction.

isp-1, lrs-2, and *frh-1*

Several Mit mutants do not exhibit maternal-effect rescue but do display slowed development and rhythmicity. For example, *isp-1(qm50)* is a missense point mutant in the Rieske iron–sulfur protein subunit of complex III and has about 80% increase in mean adult life span at both 20°C and 25°C (Feng et al. 2001). Oxygen consumption is reduced by 60% in L1 larvae (Feng et al. 2001), and this may result in reduced reactive oxygen species (ROS). *lrs-2(mg312)* encodes a truncated mitochondrial tRNA synthetase and has a 200% increase in life span at 20°C but only a 30% increase at 25°C (Lee et al. 2003b). Ventura et al. (2005) found that reducing frataxin (*frh-1*) expression extended life span and resistance to some stressors, making *frh-1* the latest member of the Mit class of long-lived worm mutants. Frataxin is required for proper iron storage, and in humans, defective expression of this mitochondrial protein causes Friedreich's ataxia, characterized by progressive ataxia and reduced life expectancy (Puccio and Koenig 2002). This underscores the possible lack of correspondence between human and worm phenotypes.

Hypotheses for Longevity Extension of the Mit Mutants

One simple explanation for the extended longevity of the Mit mutants is that these animals may generate fewer reactive species. However, biophysical considerations suggest instead that mitochondria of some Mit mutants might become overt ROS generators. Perhaps in this instance, such signals might initiate life-long, or life-lasting, protective responses that are the equivalent of hormesis, where both life span extension and stress resistance can be induced by prior sublethal exposures to exogenous stressors (Lithgow 2001; Rattan 2001; Van Voorhies 2001; Cypser et al. 2006).

Nevertheless, all Mit mutants still need to generate ATP and to get rid of their reducing equivalents. *C. elegans* is capable of employing alternate pathways for generating ATP while simultaneously maintaining redox balance (Foll et al. 1999). Rea and Johnson (2003) proposed that the Mit mutants could be long-lived because they use alternate mechanisms for ATP generation and redox balance, thus lowering ROS production, but this has not been tested.

All of the Mit mutants are characterized by reduced fertility and/or fecundity, if not overt sterility (Wong et al. 1995; Dillin et al. 2002; Lee et al. 2003b; Shibata et al. 2003). It has been suggested (Rea and Johnson 2003) that *clk-1*, and other Mit mutants, may lack the mitochondrial DNA amplification that accompanies germ-line expansion in normal fourth-stage larvae and young adults (Tsang and Lemire 2002). This may also underlie the longevity effects of germ-line ablation (Hsin and Kenyon 1999) and may be the trigger for their increased longevity.

NONGENETIC WAYS TO EXTEND LIFE

Caloric Restriction

Caloric restriction (CR) extends life span. Three methods have been used: reduced bacterial concentration (Klass 1977), growth in axenic media (De Cuyper and Vanfleteren 1982; Houthoofd et al. 2002), and mutants having reduced feeding rates (Eat mutants) (Lakowski and Hekimi 1998). A threefold increase in life span is seen under axenic growth (Houthoofd et al. 2003), whereas, at most, a twofold increase is seen using the other methods. Under CR conditions, worms develop and reproduce slowly and exhibit increased stress resistance (Houthoofd et al. 2002). The increase in life span that occurs by raising animals in axenic media is largely independent of the IIS pathway. The IIS independence of axenic media and Eat mutants is surprising, since in mammals, CR is known to reduce insulin and IGF levels, and it seemed logical to assume that the benefits of mammalian CR stem from reduced IIS. Moreover, DAF-16 is activated after food restriction (Henderson and Johnson 2001). In addition, two groups have recently shown that worms exhibit a significant increase in life span when maintained in the absence of a bacterial food source. This life span extension in response to prolonged starvation is independent of both the IIS and SIR-2.1 (Kaeberlein et al. 2006; Lee et al. 2006).

Hormesis

Hormesis is a phenomenon in which organisms exposed to a stressor develop increased resistance to subsequent exposure to that same stressor. In some cases, this results in life extension as well (Cypser et al. 2006). Hormesis has been observed in response to a broad variety of physical agents and environmental stressors (Minois 2000). Exposing worms to 35°C for 2 hours significantly increases thermotolerance 12 hours later and increases mean life span approximately 23%, (Cypser and Johnson 2002). This treatment up-regulates numerous stress response genes, including the small heat shock genes such as *hsp-16.2* (Link et al. 1999). Cypser and Johnson (2002) found that *daf-12, daf-16,* and *daf-18* were required for heat-induced increased life span, whereas those same genes were only weakly required for the subsequent thermotolerance. This suggests a significant role for a subset of dauer genes in long-term adaptation to stress. Interestingly, the DAF-16 protein is nuclear-localized in response to a variety of stressors and may generally function to prepare the animal for adverse conditions (Henderson and Johnson 2001).

Drug Interventions That Extend Life

Many investigators use *C. elegans* to screen for compounds that lead to life extension (Sampayo et al. 2003). Both wortmanin and LY294002 (inhibitors of the AGE-1 protein PI3K) produced small increases in life (Babar et al. 1999). The most dramatic (>100%) extension of both mean and maximum life span used synthetic compounds with both superoxide dismutase (SOD) and catalase activity (SCMs), thus reducing the levels of intracellular free radicals (Melov et al. 2000). Controversy exists as to replicability and generalness of the effects of these SCMs, and it may be that these compounds have limited usefulness (Keaney and Gems 2003; Sampayo et al. 2003). Evason et al. (2005) showed that ethosuximide and other anticonvulsants approved for human use also extend the life of the worm by as much as 50%. These extensions were partially dependent on *daf-16* but seemed independent of other longevity pathways and mutants. Resveratrol and other sirtuin (histone deacetylase, HDAC) activators have been shown to result in life extension in *C. elegans* as well (Wood et al. 2004), as have polyphenolic compounds obtained from natural sources such as blueberries (Wilson et al. 2006).

OTHER DISCOVERIES

Stress Response

Inhibition of the IIS pathway also leads to large increases in resistance to a variety of stressors. The first aging mutant, *age-1 (hx546)*, shows resistance to a remarkable number of stressors, including hydrogen peroxide (Larsen 1993), paraquat (Vanfleteren 1993), ultraviolet light (Murakami and Johnson 1996), heat (Lithgow et al. 1995), the potent bacterial pathogen *Pseudomonas aeruginosa* (Mahajan-Miklos et al. 1999), MPTP, a compound used to produce a mouse model of Parkinson's disease (Johnson et al. 2002), and to unfolded proteins including Aβ, the peptide responsible for Alzheimer's disease (Cohen et al. 2006). Such strong increases in stress resistance have been reported in other components of the IIS pathway as well. Although correlation does not mean causality, most Age mutants of *C. elegans*, and indeed many other species, also demonstrate increased stress resistance (Finkel and Holbrook 2000). Three groups have used increased stress resistance as a surrogate for increased longevity to select for mutants that increase stress resistance and life span in *C. elegans* (Sampayo et al. 2000; Munoz 2003; de Castro et al. 2004), and one group has used budding yeast (Kennedy et al. 1995).

Experiments using transgenes that promote the increased expression of stress response genes have also been used successfully in *C. elegans*. First, worms bearing a transgene containing the *old-1* tyrosine kinase gene led to both life extension and increased stress resistance in a DAF-16-dependent manner (Murakami and Johnson 1996). Later studies showed that another transgene carrying a muscle-specific promoter and driving HSP70F (also known as *mot-1*) expression produced a 43% increase in life expectancy (Yokoyama et al. 2002). Walker and Lithgow found a slight increase using *hsp-16*, a small heat shock gene (Walker et al. 2003), and Hsu et al. (2003) increased heat shock factor (*hsf-1*), a key regulator of stress response genes, and increased wild-type life span approximately 40%. Both *hsp-16* and *hsf-1* were dependent on a functional *daf-16* gene, suggesting that the transcription factor DAF-16 may work in coordination with *hsf-1* to control the expression of small heat shock genes and other stress-response genes (Hsu et al. 2003; Morley and Morimoto 2003).

Increasing internal stressors or inhibiting responses to stress can shorten life. The *mev-1* mutant, defective in the large cytochrome *c* subunit of mitochondrial complex II, shortens life dramatically (Ishii et al. 1998), and RNAi of *hsf-1* also accelerates tissue aging, shortens life, and is epistatic to *daf-2* (Garigan et al. 2002; Morley and Morimoto 2003). Inhibition of one of four small heat shock genes, *hsp-16.1*, *hsp-16.49*, *hsp-12.6*, or *sip-1*, also shortens life of both wild type and *daf-2(e1370)* animals (Hsu et al. 2003), suggesting a key role for stress-response genes working together with other signaling pathways to influence aging.

Quantitative Trait Loci for Aging

The first long-lived strains to be produced in *C. elegans* were recombinant inbred (RI) strains (Johnson and Wood 1982; Johnson 1987a), not mutants or transgenics. Several labs have since utilized these RI strains to detect and map genes (quantitative trait loci or QTLs) specifying life span, fertility, and other life-history traits. Shook et al. (1996; Shook and Johnson 1999) found four major QTLs; two showed genotype-by-environment interactions, and genetic epistasis and pleiotropy were also detected. Ayyadavara et al. (2001, 2003) have tuned this approach to even higher levels and may be converging on the genes underlying individual QTLs for life span in *C. elegans* (Ayyadevara et al. 2001, 2003). In general, these QTLs specify only one trait and show little antagonistic pleiotropy or trade-offs between different traits.

Biomarkers and Stochastic Effects in Aging

Several studies have revealed that the behavioral declines and tissue degeneration seen in old worms can be slowed by alterations that increase life span. For example, Age mutations lead to both increased mobility and fewer morphological signs of degeneration over time (Hosono et al. 1980; Johnson 1987a; Duhon and Johnson 1995; Garigan et al. 2002; Herndon et al. 2002). The decrease in physical markers of aging in long-lived mutants is consistent with the hypothesis that these manipulations affect the fundamental aging process itself. Another instance of Age mutants being more robust can be seen from the observation that *age-1* mutants had delayed age-related decline of isothermal tracking capability, resulting in a 210% extension (Murakami et al. 2005). There have been suggestions that extension of life also extends reproduction, and although there are certainly instances in which this is true, in general, there seems to be little evidence that both fertility and longevity are regulated concordantly, and fertility in individual worms is not a predictor of individual longevity (Brooks and Johnson 1991; Duhon and Johnson 1995; Shook et al. 1996; Shook and Johnson 1999; Chen et al. 2001).

There is a large variation in age at which worms die, even though populations are typically isogenic and are maintained in a homogeneous environment. Rates of decline in behavioral, physiological, and morphological signs of aging are also quite variable among individuals. This heterogeneity within identical populations reveals the stochastic nature of aging (Kirkwood and Austad 2000). Herndon et al. (2002) pursued this chance variation by classifying aging worms into one of three classes: A, the highly mobile; B, immobile unless prodded; and C, immobile even when prodded (Herndon et al. 2002). These behavioral markers were very good predictors of life span, and C proved to be a better marker for life expectancy than chronological age. Tissue degeneration, particularly in the muscle, became more evident with increasing age, as revealed by myosin::GFP fusions and electron microscopy. *C. elegans* showed signs of increasing sarcopenia with age; sarcomeres became disorganized and contained fewer myosin thick filaments, and cells were frequently smaller and highly invaginated. Interestingly, old neurons showed little change in gross morphology, suggesting that different tissues "age" at different rates (Herndon et al. 2002). In the pharynx, the progression of sarcopenia has been ascribed to contraction-related injury as a factor affecting the initiation and during aging (Chow et al. 2006). In another effort to identify markers that change with age, Lund et al. (2002) analyzed gene expression using whole-genome microarrays to find changes during chronological aging. Using worms at six ages, they applied a rigorous statistical model with multiple replicates and found that only 164 genes showed statistically significant changes in transcript levels with chronological age, less than 1% of the genes on the array. This should be compared with the number of transcripts that change during development, where one or two orders of magnitude more genes show significant changes (Reinke 2002). Expression of heat shock proteins decreased as a class; no changes were seen in genes that respond to oxidative stress. The largest changes were increased expression of certain transposases in older worms. Ibanez-Ventoso et al. (2006) examined the expression of 114 identified microRNAs (miRNAs) during the adult life span, finding that 34 changed expression during adulthood more than twofold, most declining. *lin-4 let-7* miRNAs, as well as the muscle miRNA *miR-1,* showed variation in expression with chronological age. They suggested that miRNAs are potential modulators of age-related decline and that there may be a general reduction of message-specific translational inhibition during aging. Reducing the activity of *lin-4* shortened life span and accelerated tissue aging, whereas overexpression of *lin-4* or reducing *lin-14* activity extended life span (Boehm and Slack 2005). Life span extension was dependent on the DAF-16 and HSF-1 transcription factors. Interestingly, changes in mitochondrial stability had been seen previously (Melov et al. 1994), and these alterations appeared to be reduced in the *age-1* mutant (Melov et al. 1995).

Golden and Melov (2004) compared gene expression in *daf-2* and wild type across the life span and went one step further by performing these analyses in individual worms. The study uncovered a huge amount of variation in gene expression among individual worms as well as many changes between the mutant and wild type. This theme of stochastic gene action and ability to predict subsequent life has been extended in studies (S.L. Rea et al., pers. comm.) in which HSP-16-2 was tagged with green fluorescent protein (GFP) and used as a marker in nematodes showing diverse levels of GFP expression that were separated with a worm sorter. They found that subsequent longevity could be accurately predicted by such a separation, with bright worms being associated with a threefold increase in overall longevity and a somewhat larger difference in life expectancy. This study is the first in the aging literature to accurately predict subsequent longevity based on an assessment at early age.

SUMMARY

Many manipulations (both genetic and environmental) increase life span in *C. elegans.* For all its complexity, however, there does appear to be a common theme. Many alterations that increase life span may be considered to "fool" the animal into signaling that either resources are scarce or damaging agents are present, when in fact they are not. This may lead the animal to invest resources into somatic maintenance and thereby minimize damage to cellular constituents. We have proposed that the increased life spans conferred by alterations in both the TOR and insulin-like pathways function by inappropriately activating food-deprivation pathways (Henderson et al. 2006). Similarly, several gerontogenes are involved in the control of energy metabolism, e.g., the AMP-activated protein kinase (AMPK), where overexpression of *aak-2* increases life span (Curtis et al. 2006). *aak-2* interacts with several pathways known to specify aging in worms; both the IIS and the *sir-2.1* deacetylase were dependent on *aak-2* for their life span extension.

Such a scenario would be disadvantageous in the wild, where reproduction is more important than a longer life span and resources cannot be wasted. (The inability of *age-1* mutants to survive in a competitive experiment with wild-type animals under changing environmental conditions supports this notion [Walker et al. 2000; Jenkins et al. 2004].) With this framework in mind, interventions that increase life span in *C. elegans* can be put into three broad categories. First are nonstressful alterations to the animal that lead to the activation of stress-response pathways under conditions that do not require them. These include manipulation of sensory and signaling pathways. Second are alterations that reduce the availability of resources to a point that repair pathways are activated but are not damaging. One simple example of this is caloric restriction. Third are nonlethal, stressful interventions that stimulate stress genes to a point where the beneficial effects of the stress response outweigh the harmful effects of the stressor—such interventions can be broadly defined as hormesis.

At the end, we thus see that there is no aging program per se. Life is measured by the extent to which resources are directed from reproduction to maintenance; aging is simply a by-product of a much larger program. We call that program Life!

ACKNOWLEDGMENTS

H.A.T. is a William Randolph Hearst Young Investigator and is supported by a Career Award in the Biomedical Sciences from the Burroughs Wellcome Fund, an endowment from the William Randolph Hearst Foundation, and a grant from the National Institute of Aging (R01AG025891). The work of T.E.J. was supported by grants from the National Institutes of Health.

REFERENCES

Aballay A. and Ausubel F.M. 2002. *Caenorhabditis elegans* as a host for the study of host-pathogen interactions. *Curr. Opin. Microbiol.* **5:** 97–101.

Aguirre V., Uchida T., Yenush L., Davis R., and White M.F. 2000. The c-Jun NH_2-terminal kinase promotes insulin resistance during association with insulin receptor substrate-1 and phosphorylation of Ser^{307}. *J. Biol. Chem.* **275:** 9047–9054.

Ailion M., Inoue T., Weaver C.I., Holdcraft R.W., and Thomas J.H. 1999. Neurosecretory control of aging in *Caenorhabditis elegans. Proc. Natl. Acad. Sci.* **96:** 7394–7397.

Alcedo J. and Kenyon C. 2004. Regulation of *C. elegans* longevity by specific gustatory and olfactory neurons. *Neuron* **41:** 45–55.

Ann K., Kowalchyk J.A., Loyet K.M., and Martin T.F.J. 1997. Novel Ca^{2+}-binding protein (CAPS) related to UNC-31 required for Ca^{2+}-activated exocytosis. *J. Biol. Chem.* **272:** 19637–19640.

Antebi A. 2005. Physiology. The tick-tock of aging? *Science* **310:** 1911–1913.

Apfeld J. and Kenyon C. 1998. Cell nonautonomy of *C. elegans daf-2* function in the regulation of diapause and life span. *Cell* **95:** 199–210.

———. 1999. Regulation of lifespan by sensory perception in *Caenorhabditis elegans. Nature* **402:** 804–809.

Arantes-Oliveira N., Berman J.R., and Kenyon C. 2003. Healthy animals with extreme longevity. *Science* **302:** 611.

Ayyadevara S., Ayyadevara R., Hou S., Thaden J.J., and Shmookler Reis R.J. 2001. Genetic mapping of quantitative trait loci governing longevity of *Caenorhabditis elegans* in recombinant-inbred progeny of a Bergerac-BO x RC301 interstrain cross. *Genetics* **157:** 655–666.

Ayyadevara S., Ayyadevara R., Vertino A., Galecki A., Thaden J.J., and Shmookler Reis R.J. 2003. Genetic loci modulating fitness and life span in *Caenorhabditis elegans:* Categorical trait interval mapping in CL2a x Bergerac-BO recombinant-inbred worms. *Genetics* **163:** 557–570.

Babar P., Adamson C., Walker G.A., Walker D.W., and Lithgow G.J. 1999. P13-kinase inhibition induces dauer formation, thermotolerance and longevity in *C. elegans. Neurobiol. Aging* **20:** 513–519.

Barsyte D., Lovejoy D.A., and Lithgow G.J. 2001. Longevity and heavy metal resistance in daf-2 and age-1 long-lived mutants of *Caenorhabditis elegans. FASEB J.* **15:** 627–634.

Baur J.A., Pearson K.J., Price N.L., Jamieson H.A., Lerin C., Kalra A., Prabhu V.V., Allard J.S., Lopez-Lluch G., Lewis K., et al. 2006. Resveratrol improves health and survival of mice on a high-calorie diet. *Nature* **444:** 337–342.

Berdichevsky A. and Guarente L. 2006. A stress response pathway involving sirtuins, forkheads and 14-3-3 proteins. *Cell Cycle* **5:** 2588–2591.

Berdichevsky A., Viswanathan M., Horvitz H.R., and Guarente L. 2006. *C. elegans* SIR-2.1 interacts with 14-3-3 proteins to activate DAF-16 and extend life span. *Cell* **125:** 1165–1177.

Biggs W.H., III, Meisenhelder J., Hunter T., Cavenee W.K., and Arden K.C. 1999. Protein kinase B/Akt-mediated phosphorylation promotes nuclear exclusion of the winged helix transcription factor FKHR1. *Proc. Natl. Acad. Sci.* **96:** 7421–7426.

Birkenkamp K.U. and Coffer P.J. 2003. Regulation of cell survival and proliferation by the FOXO (Forkhead box, class O) subfamily of Forkhead transcription factors. *Biochem. Soc. Trans.* **31:** 292–297.

Boehm M. and Slack F. 2005. A developmental timing microRNA and its target regulate life span in *C. elegans*. *Science* **310:** 1954–1957.

Borra M.T., Smith B.C., and Denu J.M. 2005. Mechanism of human SIRT1 activation by resveratrol. *J. Biol. Chem.* **280:** 17187–17195.

Brenner S. 1974. The genetics of *Caenorhabditis elegans*. *Genetics* **77:** 71–94.

Brooks A. and Johnson T.E. 1991. Genetic specification of life span and self-fertility in recombinant-inbred strains of *Caenorhabditis elegans*. *Heredity* **67:** 19–28.

Brunet A., Bonni A., Zigmond M.J., Lin M.Z., Juo P., Hu L.S., Anderson M.J., Arden K.C., Blenis J., and Greenberg M.E. 1999. Akt promotes cell survival by phosphorylating and inhibiting a Forkhead transcription factor. *Cell* **96:** 857–868.

Brunet A., Sweeney L.B., Sturgill J.F., Chua K.F., Greer P.L., Lin Y., Tran H., Ross S.E., Mostoslavsky R., Cohen H.Y., et al. 2004. Stress-dependent regulation of FOXO transcription factors by the SIRT1 deacetylase. *Science* **303:** 2011–2015.

Burgering B.M. and Kops G.J. 2002. Cell cycle and death control: Long live Forkheads. *Trends Biochem. Sci.* **27:** 352–360.

Cahill C.M., Tzivion G., Nasrin N., Ogg S., Dore J., Ruvkun G., and Alexander-Bridges M. 2001. Phosphatidylinositol 3-kinase signaling inhibits DAF-16 DNA binding and function via 14-3-3-dependent and 14-3-3-independent pathways. *J. Biol. Chem.* **276:** 13402–13410.

C. elegans Sequencing Consortium. 1998. Genome sequence of the nematode *C. elegans:* A platform for investigating biology. *Science* **282:** 2012–2018.

Chen J., Carey J.R., and Ferris H. 2001. Comparative demography of isogenic populations of *Caenorhabditis elegans*. *Exp. Gerontol.* **36:** 431–440.

Chow D.K., Glenn C.F., Johnston J.L., Goldberg I.G., and Wolkow C.A. 2006. Sarcopenia in the *Caenorhabditis elegans* pharynx correlates with muscle contraction rate over lifespan. *Exp. Gerontol.* **41:** 252–260.

Christensen M., Estevez A., Yin X., Fox R., Morrison R., McDonnell M., Gleason C., Miller D.M., III, and Strange K. 2002. A primary culture system for functional analysis of *C. elegans* neurons and muscle cells. *Neuron* **33:** 503–514.

Cohen E., Bieschke J., Perciavalle R.M., Kelly J.W., and Dillin A. 2006. Opposing activities protect against age-onset proteotoxicity. *Science* **313:** 1604–1610.

Coulson A., Sulston J., Brenner S., and Karn J. 1986. Toward a physical map of the genome of the nematode *Caenorhabditis elegans*. *Proc. Natl. Acad. Sci.* **83:** 7821–7825.

Curtis R., O'Connor G., and DiStefano P.S. 2006. Aging networks in *Caenorhabditis elegans:* AMP-activated protein kinase (aak-2) links multiple aging and metabolism pathways. *Aging Cell* **5:** 119–126.

Cypser J.R. and Johnson T.E. 2002. Multiple stressors in *Caenorhabditis elegans* induce stress hormesis and extended longevity. *J. Gerontol. A Biol. Sci. Med. Sci.* **57:** B109–B114.

Cypser J.R., Tedesco P., and Johnson T.E. 2006. Hormesis and aging in *Caenorhabditis elegans*. *Exp. Gerontol.* **41:** 935–939.

Davis R.J. 2000. Signal transduction by the JNK group of MAP kinases. *Cell* **103:** 239–252.

de Castro E., Hegi de Castro S., and Johnson T.E. 2004. Isolation of long-lived mutants in *Caenorhabditis elegans* using selection for resistance to juglone. *Free Radic. Biol. Med.* **37:** 139–145.

De Cuyper C. and Vanfleteren J.R. 1982. Oxygen consumption during development and aging of the nematode *Caenorhabditis elegans*. *Comp. Biochem. Physiol.* **73A:** 283–289.

Dillin A., Hsu A.L., Arantes-Oliveira N., Lehrer-Graiwer J., Hsin H., Fraser A.G., Kamath R.S., Ahringer J., and Kenyon C. 2002. Rates of behavior and aging specified by mitochondrial function during development. *Science* **298:** 2398–2401.

Dorman J.B., Albinder B., Shroyer T., and Kenyon C. 1995. The *age-1* and *daf-2* genes function in a common pathway to control the lifespan of *Caenorhabditis elegans*. *Genetics* **141:** 1399–1406.

Duhon S.A. and Johnson T.E. 1995. Movement as an index of vitality: Comparing wild type and the age-1 mutant of *Caenorhabditis elegans*. *J. Gerontol. A Biol. Sci. Med. Sci.* **50:** B254–B261.

Duhon S.A., Murakami S., and Johnson T.E. 1996. Direct isolation of longevity mutants in the nematode *Caenorhabditis elegans*. *Dev. Genet.* **18:** 144–153.

Ellis R.E., Yuan J.Y., and Horvitz H.R. 1991. Mechanisms and functions of cell death. *Annu. Rev. Cell Biol.* **7:** 663–698.

Epstein H.F., Waterston R.H., and Brenner S. 1974. A mutant affecting the heavy chain of myosin in *Caenorhabditis elegans*. *J. Mol. Biol.* **90:** 291–300.

Evason K., Huang C., Yamben I., Covey D.F., and Kornfeld K. 2005. Anticonvulsant medications extend worm life-span. *Science* **307:** 258–262.

Felkai S., Ewbank J.J., Lemieux J., Labbe J.C., Brown G.G., and Hekimi S. 1999. CLK-1 controls respiration, behavior and aging in the nematode *Caenorhabditis elegans*. *EMBO J.* **18:** 1783–1792.

Feng J., Bussiere F., and Hekimi S. 2001. Mitochondrial electron transport is a key determinant of life span in *Caenorhabditis elegans*. *Dev. Cell* **1:** 633–644.

Finkel T. and Holbrook N.J. 2000. Oxidants, oxidative stress and the biology of ageing. *Nature* **408:** 239–247.

Foll R.L., Pleyers A., Lewandovski G.J., Wermter C., Hegemann V., and Paul R.J. 1999. Anaerobiosis in the nematode *Caenorhabditis elegans*. *Comp. Biochem. Physiol. B Biochem. Mol. Biol.* **124:** 269–280.

Friedman D.B. and Johnson T.E. 1988a. A mutation in the *age-1* gene in *Caenorhabditis elegans* lengthens life and reduces hermaphrodite fertility. *Genetics* **118:** 75–86.

———. 1988b. Three mutants that extend both mean and maximum life span of the nematode, *Caenorhabditis elegans,* define the age-1 gene. *J. Gerontol.* **43:** B102–B109.

Furuyama T., Nakazawa T., Nakano I., and Mori N. 2000. Identification of the differential distribution patterns of mRNAs and consensus binding sequences for mouse DAF-16 homologues. *Biochem. J.* **349:** 629–634.

Garigan D., Hsu A.L., Fraser A.G., Kamath R.S., Ahringer J., and Kenyon C. 2002. Genetic analysis of tissue aging in *Caenorhabditis elegans:* A role for heat-shock factor and bacterial proliferation. *Genetics* **161:** 1101–1112.

Gems D. and McElwee J.J. 2005. Broad spectrum detoxification: The major longevity assurance process regulated by insulin/IGF-1 signaling? *Mech. Ageing Dev.* **126:** 381–387.

Gems D. and Riddle D.L. 2000. Genetic, behavioral and environmental determinants of male longevity in *Caenorhabditis elegans*. *Genetics* **154:** 1597–1610.

Golden T.R. and Melov S. 2004. Microarray analysis of gene expression with age in individual nematodes. *Aging Cell* **3:** 111–124.

Harman D. 1956. Aging: A theory based on free radical and radiation chemistry. *J. Gerontol.* **11:** 298–300.

Hedgecock E.M., Culotti J.G., Thomson J.N., and Perkins L.A. 1985. Axonal guidance mutants of *Caenorhabditis elegans* identified by filling sensory neurons with fluorescein dyes. *Dev. Biol.* **111:** 158–170.

Henderson S.T. and Johnson T.E. 2001. daf-16 integrates developmental and environmental inputs to mediate aging in the nematode *Caenorhabditis elegans*. *Curr. Biol.* **11:** 1975–1980.

Henderson S.T., Bonafe M., and Johnson T.E. 2006. daf-16 protects the nematode *Caenorhabditis elegans* during food deprivation. *J. Gerontol. A Biol. Sci. Med. Sci.* **61:** 444–460.

Henderson S.T., Rea S., and Johnson T.E. 2005. Dissecting the processes of aging using the nematode *Caenorhabditis elegans*. In *Handbook of the biology of aging*, 6th edition (ed. E.J. Masoro and S.N. Austad), pp. 360–399. Academic Press.

Herndon L.A., Schmeissner P.J., Dudaronek J.M., Brown P.A., Listner K.M., Sakano Y., Paupard M.C., Hall D.H., and Driscoll M. 2002. Stochastic and genetic factors influence tissue-specific decline in ageing *C. elegans*. *Nature* **419:** 808–814.

Hertweck M., Gobel C., and Baumeister R. 2004. *C. elegans* SGK-1 is the critical component in the Akt/PKB kinase complex to control stress response and life span. *Dev. Cell.* **6:** 577–588.

Honda Y. and Honda S. 1999. The *daf-2* gene network for longevity regulates oxidative stress resistance and Mn-superoxide dismutase gene expression in *Caenorhabditis elegans*. *FASEB J.* **13:** 1385–1393.

Hosono R., Sato Y., Aizawa S.I., and Mitsui Y. 1980. Age-dependent changes in mobility and separation of the nematode *Caenorhabditis elegans*. *Exp. Gerontol.* **15:** 285–289.

Houthoofd K., Braeckman B.P., Johnson T.E., and Vanfleteren J.R. 2003. Life extension via dietary restriction is independent of the Ins/IGF-1 signalling pathway in *Caenorhabditis elegans*. *Exp. Gerontol.* **38:** 947–954.

Houthoofd K., Braeckman B.P., Lenaerts I., Brys K., De Vreese A., Van Eygen S., and Vanfleteren J.R. 2002. Ageing is reversed, and metabolism is reset to young levels in recovering dauer larvae of *C. elegans*. *Exp. Gerontol.* **37:** 1015–1021.

Houthoofd K., Braeckman B.P., Lenaerts I., Brys K., Matthijssens F., De Vreese A., Van Eygen S., and Vanfleteren J.R. 2005. DAF-2 pathway mutations and food restriction in aging *Caenorhabditis elegans* differentially affect metabolism. *Neurobiol. Aging* **26:** 689–696.

Howitz K.T., Bitterman K.J., Cohen H.Y., Lamming D.W., Lavu S., Wood J.G., Zipkin R.E., Chung P., Kisielewski A., Zhang L.L., et al. 2003. Small molecule activators of sirtuins extend *Saccharomyces cerevisiae* lifespan. *Nature* **425:** 191–196.

Hsin H. and Kenyon C. 1999. Signals from the reproductive system regulate the lifespan of *C. elegans*. *Nature* **399:** 362–366.

Hsu A.L., Murphy C.T., and Kenyon C. 2003. Regulation of aging and age-related disease by DAF-16 and heat-shock factor. *Science* **300:** 1142–1145.

Hua Q.X., Nakagawa S.H., Wilken J., Ramos R.R., Jia W., Bass J., and Weiss M.A. 2003. A divergent INS protein in *Caenorhabditis elegans* structurally resembles human insulin and activates the human insulin receptor. *Genes Dev.* **17:** 826–831.

Hubbard E.J. and Greenstein D. 2000. The *Caenorhabditis elegans* gonad: A test tube for cell and developmental biology. *Dev. Dyn.* **218:** 2–22.

Ibanez-Ventoso C., Yang M., Guo S., Robins H., Padgett R.W., and Driscoll M. 2006. Modulated microRNA expression during adult lifespan in *Caenorhabditis elegans*. *Aging Cell* **5:** 235–246.

Ishii N., Fujii M., Hartman P.S., Tsuda M., Yasuda K., Senoo-Matsuda N., Yanase S., Ayusawa D., and Suzuki K. 1998. A mutation in succinate dehydrogenase cytochrome b causes oxidative stress and ageing in nematodes. *Nature* **394:** 694–697.

Jenkins N.L., McColl G., and Lithgow G.J. 2004. Fitness cost of extended lifespan in *Caenorhabditis elegans*. *Proc. Biol. Sci.* **271:** 2523–2526.

Jia K., Chen D., and Riddle D.L. 2004. The TOR pathway interacts with the insulin signaling pathway to regulate *C. elegans* larval development, metabolism and life span. *Development* **131:** 3897–3906.

Johnson T.E. 1987a. Aging can be genetically dissected into component processes using long-lived lines of *Caenorhabditis elegans. Proc. Natl. Acad. Sci.* **84:** 3777–3781.

———. 1987b. Developmentally programmed aging: Future directions. In *Molecular biological theories of aging* (ed. H.R. Warner et al.), pp. 63–76. Raven Press, New York.

———. 1990. Increased life-span of age-1 mutants in *Caenorhabditis elegans* and lower Gompertz rate of aging. *Science* **249:** 908–912.

Johnson T.E. and Wood W.B. 1982. Genetic analysis of life-span in *Caenorhabditis elegans. Proc. Natl. Acad. Sci.* **79:** 6603–6607.

Johnson T.E., Henderson S., Murakami S., de Castro E., de Castro S.H., Cypser J., Rikke B., Tedesco P., and Link C. 2002. Longevity genes in the nematode *Caenorhabditis elegans* also mediate increased resistance to stress and prevent disease. *J. Inherit. Metab. Dis.* **25:** 197–206.

Jonassen T., Larsen P.L., and Clarke C.F. 2001. A dietary source of coenzyme Q is essential for growth of long-lived *Caenorhabditis elegans clk-1* mutants. *Proc. Natl. Acad. Sci.* **98:** 421–426.

Jonassen T., Davis D.E., Larsen P.L., and Clarke C.F. 2003. Reproductive fitness and quinone content of *Caenorhabditis elegans* clk-1 mutants fed coenzyme Q isoforms of varying length. *J. Biol. Chem.* **278:** 51735–51742.

Kaeberlein M., McVey M., and Guarente L. 1999. The SIR2/3/4 complex and SIR2 alone promote longevity in *Saccharomyces cerevisiae* by two different mechanisms. *Genes Dev.* **13:** 2570–2580.

Kaeberlein M., McDonagh T., Heltweg B., Hixon J., Westman E.A., Caldwell S.D., Napper A., Curtis R., Distefano P.S., Fields S., et al. 2005. Substrate-specific activation of sirtuins by resveratrol. *J. Biol. Chem.* **280:** 17038–17045.

Kaeberlein T.L., Smith E.D., Tsuchiya M., Welton K.L., Thomas J.H., Fields S., Kennedy B.K., and Kaeberlein M. 2006. Lifespan extension in *Caenorhabditis elegans* by complete removal of food. *Aging Cell* **5:** 487–494.

Kamath R.S., Fraser A.G., Dong Y., Poulin G., Durbin R., Gotta M., Kanapin A., Le Bot N., Moreno S., Sohrmann M., et al. 2003. Systematic functional analysis of the *Caenorhabditis elegans* genome using RNAi. *Nature* **421:** 231–237.

Kawano T., Ito Y., Ishiguro M., Takuwa K., Nakajima T., and Kimura Y. 2000. Molecular cloning and characterization of a new insulin/IGF-like peptide of the nematode *Caenorhabditis elegans. Biochem. Biophys. Res. Commun.* **273:** 431–436.

Kayser E.B., Sedensky M.M., Morgan P.G., and Hoppel C.L. 2004. Mitochondrial oxidative phosphorylation is defective in the long-lived mutant clk-1. *J. Biol. Chem.* **279:** 54479–54486.

Keaney M. and Gems D. 2003. No increase in lifespan in *Caenorhabditis elegans* upon treatment with the superoxide dismutase mimetic EUK-8. *Free Radic. Biol. Med.* **34:** 277–282.

Kennedy B.K., Austriaco N.R., Jr., Zhang J., and Guarente L. 1995. Mutation in the silencing gene SIR4 can delay aging in *S. cerevisiae. Cell* **80:** 485–496.

Kenyon C. 2005. The plasticity of aging: Insights from long-lived mutants. *Cell* **120:** 449–460.

Kenyon C., Chang J., Gensch E., Rudner A., and Tabtiang R. 1993. A *C. elegans* mutant that lives twice as long as wild type. *Nature* **366:** 461–464.

Kim A.H., Sasaki T., and Chao M.V. 2003. JNK-interacting protein 1 promotes Akt1 activation. *J. Biol. Chem.* **278:** 29830–29836.

Kimura K.D., Tissenbaum H.A., Liu Y., and Ruvkun G. 1997. daf-2, an insulin receptor-like gene that regulates longevity and diapause in *Caenorhabditis elegans. Science* **277:** 942–946.

Kirkwood T.B. and Austad S.N. 2000. Why do we age? *Nature* **408:** 233–238.

Klass M.R. 1977. Aging in the nematode *Caenorhabditis elegans:* Major biological and environmental factors influencing life span. *Mech. Ageing Dev.* **6:** 413–429.

———. 1983. A method for the isolation of longevity mutants in the nematode *Caenorhabditis elegans* and initial results. *Mech. Ageing Dev.* **22:** 279–286.

Klass M. and Hirsh D. 1976. Non-ageing developmental variant of *Caenorhabditis elegans. Nature* **260:** 523–525.

Kops G.J., de Ruiter N.D., De Vries-Smits A.M., Powell D.R., Bos J.L., and Burgering B.M. 1999. Direct control of the forkhead transcription factor AFX by protein kinase B. *Nature* **398:** 630–634.

Kops G.J., Dansen T.B., Polderman P.E., Saarloos I., Wirtz K.W., Coffer P.J., Huang T.T., Bos J.L., Medema R.H., and Burgering B.M. 2002. Forkhead transcription factor FOXO3a protects quiescent cells from oxidative stress. *Nature* **419:** 316–321.

Lakowski B. and Hekimi S. 1996. Determination of life-span in *Caenorhabditis elegans* by four clock genes. *Science* **272:** 1010–1013.

———. 1998. The genetics of caloric restriction in *Caenorhabditis elegans. Proc. Natl. Acad. Sci.* **95:** 13091–13096.

Larsen P.L. 1993. Aging and resistance to oxidative stress in *Caenorhabditis elegans. Proc. Natl. Acad. Sci.* **90:** 8905–8909.

Larsen P.L., Albert P.S., and Riddle D.L. 1995. Genes that regulate both development and longevity in *Caenorhabditis elegans. Genetics* **139:** 1567–1583.

Lee R.Y., Hench J., and Ruvkun G. 2001. Regulation of *C. elegans* DAF-16 and its human ortholog FKHRL1 by the daf-2 insulin-like signaling pathway. *Curr. Biol.* **11:** 1950–1957.

Lee S.S., Kennedy S., Tolonen A.C., and Ruvkun G. 2003a. DAF-16 target genes that control *C. elegans* life-span and metabolism. *Science* **300:** 644–647.

Lee S.S., Lee R.Y., Fraser A.G., Kamath R.S., Ahringer J., and Ruvkun G. 2003b. A systematic RNAi screen identifies a critical role for mitochondria in *C. elegans* longevity. *Nat. Genet.* **33:** 40–48.

Lee G.D., Wilson M.A., Zhu M., Wolkow C.A., de Cabo R., Ingram D.K., and Zou S. 2006. Dietary deprivation extends lifespan in *Caenorhabditis elegans. Aging Cell* **5:** 515–524.

Lehtinen M.K., Yuan Z., Boag P.R., Yang Y., Villen J., Becker E.B., DiBacco S., de la Iglesia N., Gygi S., Blackwell T.K., and Bonni A. 2006. A conserved MST-FOXO signaling pathway mediates oxidative-stress responses and extends life span. *Cell* **125:** 987–1001.

Li W., Kennedy S.G., and Ruvkun G. 2003. daf-28 encodes a *C. elegans* insulin superfamily member that is regulated by environmental cues and acts in the DAF-2 signaling pathway. *Genes Dev.* **17:** 844–858.

Lin K., Dorman J.B., Rodan A., and Kenyon C. 1997. *daf-16:* An HNF-3/forkhead family member that can function to double the life-span of *Caenorhabditis elegans. Science* **278:** 1319–1322.

Lin K., Hsin H., Libina N., and Kenyon C. 2001. Regulation of the *Caenorhabditis elegans* longevity protein DAF-16 by insulin/IGF-1 and germline signaling. *Nat. Genet.* **28:** 139–145.

Link C.D., Cypser J.R., Johnson C.J., and Johnson T.E. 1999. Direct observation of stress response in *Caenorhabditis elegans* using a reporter transgene. *Cell Stress Chaperones* **4:** 235–242.

Lithgow G.J. 2001. Hormesis—A new hope for ageing studies or a poor second to genetics? *Hum. Exp. Toxicol.* **20:** 301–303 (discussion 319–320).

———. 2006. Why aging isn't regulated: A lamentation on the use of language in aging literature. *Exp. Gerontol.* **41:** 890–893.

Lithgow G.J., White T.M., Melov S., and Johnson T.E. 1995. Thermotolerance and extended life-span conferred by single-gene mutations and induced by thermal stress. *Proc. Natl. Acad. Sci.* **92:** 7540–7544.

Liu, T., Zimmerman K.K., and Patterson G.I. 2004. Regulation of signaling genes by TGFβ during entry into dauer diapause in *C. elegans*. *BMC Dev. Biol.* **4:** 11.

Liu X., Jiang N., Hughes B., Bigras E., Shoubridge E., and Hekimi S. 2005. Evolutionary conservation of the clk-1-dependent mechanism of longevity: Loss of mclk1 increases cellular fitness and lifespan in mice. *Genes Dev.* **19:** 2424–2434.

Lund J., Tedesco P., Duke K., Wang J., Kim S.K., and Johnson T.E. 2002. Transcriptional profile of aging in *C. elegans*. *Curr. Biol.* **12:** 1566–1573.

Mahajan-Miklos S., Tan M.W., Rahme L.G., and Ausubel F.M. 1999. Molecular mechanisms of bacterial virulence elucidated using a *Pseudomonas aeruginosa-Caenorhabditis elegans* pathogenesis model. *Cell* **96:** 47–56.

McElwee J., Bubb K., and Thomas J.H. 2003. Transcriptional outputs of the *Caenorhabditis elegans* forkhead protein DAF-16. *Aging Cell* **2:** 111–121.

Melov S., Hertz G.Z., Stormo G.D., and Johnson T.E. 1994. Detection of deletions in the mitochondrial genome of *Caenorhabditis elegans*. *Nucleic Acids Res.* **22:** 1075–1078.

Melov S., Lithgow G.J., Fischer D.R., Tedesco P.M., and Johnson T.E. 1995. Increased frequency of deletions in the mitochondrial genome with age of *Caenorhabditis elegans*. *Nucleic Acids Res.* **23:** 1419–1425.

Melov S., Ravenscroft J., Malik S., Gill M.S., Walker D.W., Clayton P.E., Wallace D.C., Malfroy B., Doctrow S.R., and Lithgow G.J. 2000. Extension of life-span with superoxide dismutase/catalase mimetics. *Science* **289:** 1567–1569.

Minois N. 2000. Longevity and aging: Beneficial effects of exposure to mild stress. *Biogerontology* **1:** 15–29.

Miyadera H., Kano K., Miyoshi H., Ishii N., Hekimi S., and Kita K. 2002. Quinones in long-lived clk-1 mutants of *Caenorhabditis elegans*. *FEBS Lett.* **512:** 33–37.

Miyadera H., Amino H., Hiraishi A., Taka H., Murayama K., Miyoshi H., Sakamoto K., Ishii N., Hekimi S., and Kita K. 2001. Altered quinone biosynthesis in the long-lived *clk-1* mutants of *Caenorhabditis elegans*. *J. Biol. Chem.* **276:** 7713–7716.

Morley J.F. and Morimoto R.I. 2004. Regulation of longevity in *Caenorhabditis elegans* by heat shock factor and molecular chaperones. *Mol. Biol. Cell* **15:** 657–664.

Morris J.Z., Tissenbaum H.A., and Ruvkun G. 1996. A phosphatidylinositol-3-OH kinase family member regulating longevity and diapause in *Caenorhabditis elegans*. *Nature* **382:** 536–539.

Motta M.C., Divecha N., Lemieux M., Kamel C., Chen D., Gu W., Bultsma Y., McBurney M., and Guarente L. 2004. Mammalian SIRT1 represses forkhead transcription factors. *Cell* **116:** 551–563.

Munoz M.J. 2003. Longevity and heat stress regulation in *Caenorhabditis elegans*. *Mech. Ageing Dev.* **124:** 43–48.

Munoz M.J. and Riddle D.L. 2003. Positive selection of *Caenorhabditis elegans* mutants with increased stress resistance and longevity. *Genetics* **163:** 171–180.

Murakami H., Bessinger K., Hellmann J., and Murakami S. 2005. Aging-dependent and -independent modulation of associative learning behavior by insulin/insulin-like growth factor-1 signal in *Caenorhabditis elegans*. *J. Neurosci.* **25:** 10894–10904.

Murakami S. and Johnson T.E. 1996. A genetic pathway conferring life extension and resistance to UV stress in *Caenorhabditis elegans*. *Genetics* **143:** 1207–1218.

————. 2001. The OLD-1 positive regulator of longevity and stress resistance is under DAF-16 regulation in *Caenorhabditis elegans*. *Curr. Biol.* **11:** 1517–1523.

Murphy C.T., McCarroll S.A., Bargmann C.I., Fraser A., Kamath R.S., Ahringer J., Li H., and Kenyon C. 2003. Genes that act downstream of DAF-16 to influence the lifespan of *Caenorhabditis elegans*. *Nature* **424:** 277–283.

Nemoto S. and Finkel T. 2002. Redox regulation of forkhead proteins through a p66shc-dependent signaling pathway. *Science* **295:** 2450–2452.

Obsilova V., Vecer J., Herman P., Pabianova A., Sulc M., Teisinger J., Boura E., and Obsil T. 2005. 14-3-3 Protein interacts with nuclear localization sequence of forkhead transcription factor FoxO4. *Biochemistry* **44:** 11608–11617.

Ogawa H., Harada S., Sassa T., Yamamoto H., and Hosono R. 1998. Functional properties of the *unc-64* gene encoding a *Caenorhabditis elegans* syntaxin. *J. Biol. Chem.* **273:** 2192–2198.

Ogg S., Paradis S., Gottlieb S., Patterson G.I., Lee L., Tissenbaum H.A., and Ruvkun G. 1997. The Fork head transcription factor DAF-16 transduces insulin-like metabolic and longevity signals in *C. elegans*. *Nature* **389:** 994–999.

Oh S.W., Mukhopadhyay A., Dixit B.L., Raha T., Green M.R., and Tissenbaum H.A. 2006. Identification of direct DAF-16 targets controlling longevity, metabolism and diapause by chromatin immunoprecipitation. *Nat. Genet.* **38:** 251–257.

Oh S.W., Mukhopadhyay A., Svrzikapa N., Jiang F., Davis R.J., and Tissenbaum H.A. 2005. JNK regulates lifespan in *Caenorhabditis elegans* by modulating nuclear translocation of forkhead transcription factor/DAF-16. *Proc. Natl. Acad. Sci.* **102:** 4494–4499.

Ookuma S., Fukuda M., and Nishida E. 2003. Identification of a DAF-16 transcriptional target gene, scl-1, that regulates longevity and stress resistance in *Caenorhabditis elegans*. *Curr. Biol.* **13:** 427–431.

Paradis S. and Ruvkun G. 1998. *Caenorhabditis elegans* Akt/PKB transduces insulin receptor-like signals from AGE-1 PI3 kinase to the DAF-16 transcription factor. *Genes Dev.* **12:** 2488–2498.

Paradis S., Ailion M., Toker A., Thomas J.H., and Ruvkun G. 1999. A PDK1 homolog is necessary and sufficient to transduce AGE-1 PI3 kinase signals that regulate diapause in *Caenorhabditis elegans*. *Genes Dev.* **13:** 1438–1452.

Pierce S.B., Costa M., Wisotzkey R., Devadhar S., Homburger S.A., Buchman A.R., Ferguson K.C., Heller J., Platt D.M., Pasquinelli A.A., et al. 2001. Regulation of DAF-2 receptor signaling by human insulin and *ins-1*, a member of the unusually large and diverse *C. elegans* insulin gene family. *Genes Dev.* **15:** 672–686.

Puccio H. and Koenig M. 2002. Friedreich ataxia: A paradigm for mitochondrial diseases. *Curr. Opin. Genet. Dev.* **12:** 272–277.

Rattan S.I. 2001. Hormesis in biogerontology. *Crit. Rev. Toxicol.* **31:** 663–664.

Rea S. and Johnson T.E. 2003. A metabolic model for life span determination in *Caenorhabditis elegans*. *Dev. Cell* **5:** 197–203.

Rea S.L., Wu D., Cypser J.R., Vaupel J.W., and Johnson T.E. 2005. A stress-sensitive reporter predicts longevity in isogenic populations of *Caenorhabditis elegans*. *Nat. Genet.* **37:** 894–898.

Reinke V. 2002. Functional exploration of the *C. elegans* genome using DNA microarrays. *Nat. Genet.* (suppl.) **32:** 541–546.

Riddle D.L. and Albert P.S. 1997. Genetic and environmental regulation of dauer larva development. In C elegans *II* (ed. D.L. Riddle et al.), pp. 739–768. Cold Spring Harbor Laboratory Press, Cold Spring Harbor, New York.

Rine J. and Herskowitz I. 1987. Four genes responsible for a position effect on expression from HML and HMR in *Saccharomyces cerevisiae*. *Genetics* **116:** 9–22.

Rogina B. and Helfand S.L. 2004. Sir2 mediates longevity in the fly through a pathway related to calorie restriction. *Proc. Natl. Acad. Sci.* **101:** 15998–16003.

Russell R.L. 1987. Evidence for and against the theory of developmentally programmed aging. In *Modern theories of aging* (ed. H.R. Warner et al.), pp. 35–61. Raven Press, New York.

Saifee O., Wei L., and Nonet M.L. 1998. The *Caenorhabditis elegans unc-64* locus encodes a syntaxin that interacts genetically with synaptobrevin. *Mol. Biol. Cell* **9:** 1235–1252.

Sampayo J.N., Jenkins N.L., and Lithgow G.J. 2000. Using stress resistance to isolate novel longevity mutations in *Caenorhabditis elegans*. *Ann. N.Y. Acad. Sci.* **908:** 324–326.

Sampayo J.N., Olsen A., and Lithgow G.J. 2003. Oxidative stress in *Caenorhabditis elegans:* Protective effects of superoxide dismutase/catalase mimetics. *Aging Cell* **2:** 319–326.

Shibata Y., Branicky R., Landaverde I.O., and Hekimi S. 2003. Redox regulation of germline and vulval development in *Caenorhabditis elegans*. *Science* **302:** 1779–1782.

Shook D.R. and Johnson T.E. 1999. Quantitative trait loci affecting survival and fertility-related traits in *Caenorhabditis elegans* show genotype-environment interactions, pleiotropy and epistasis. *Genetics* **153:** 1233–1243.

Shook D.R., Brooks A., and Johnson T.E. 1996. Mapping quantitative trait loci affecting life history traits in the nematode *Caenorhabditis elegans*. *Genetics* **142:** 801–817.

Sieburth D., Ch'ng Q., Dybbs M., Tavazoi M., Kennedy S., Wang D., Dupuy D., Rual J.F., Hill D.E., Vidal M., Ruvkun G., and Kaplan J.M. 2005. Systematic analysis of genes required for synapse structure and function. *Nature* **436:** 510–517.

Stenmark P., Grunler J., Mattsson J., Sindelar P.J., Nordlund P., and Berthold D.A. 2001. A new member of the family of di-iron carboxylate proteins. Coq7 (clk-1), a membrane-bound hydroxylase involved in ubiquinone biosynthesis. *J. Biol. Chem.* **276:** 33297–33300.

Tissenbaum H.A. and Guarente L. 2001. Increased dosage of a sir-2 gene extends lifespan in *Caenorhabditis elegans*. *Nature* **410:** 227–230.

Tran H., Brunet A., Grenier J.M., Datta S.R., Fornace A.J., Jr., DiStefano P.S., Chiang L.W., and Greenberg M.E. 2002. DNA repair pathway stimulated by the forkhead transcription factor FOXO3a through the Gadd45 protein. *Science* **296:** 530–534.

Tsang W.Y. and Lemire B.D. 2002. Mitochondrial genome content is regulated during nematode development. *Biochem. Biophys. Res. Commun.* **291:** 8–16.

Van Der Heide L.P., Hoekman M.F., and Smidt M.P. 2004. The ins and outs of FoxO shuttling: Mechanisms of FoxO translocation and transcriptional regulation. *Biochem. J.* **380:** 297–309.

Van Voorhies W.A. 2001. Metabolism and lifespan. *Exp. Gerontol.* **36:** 55–64.

Vanfleteren J.R. 1993. Oxidative stress and aging in *Caenorhabditis elegans*. *Biochem. J.* **292:** 605–608.

Ventura N., Rea S., Henderson S.T., Condo I., Johnson T.E., and Testi R. 2005. Reduced expression of frataxin extends the lifespan of *Caenorhabditis elegans*. *Aging Cell* **4:** 109–112.

Viswanathan M., Kim S.K., Berdichevsky A., and Guarente L. 2005. A role for SIR-2.1 regulation of ER stress response genes in determining *C. elegans* life span. *Dev. Cell* **9:** 605–615.

Vogt P.K., Jiang H., and Aoki M. 2005. Triple layer control: Phosphorylation, acetylation and ubiquitination of FOXO proteins. *Cell Cycle* **4:** 908–913.

Vowels J.J. and Thomas J.H. 1992. Genetic analysis of chemosensory control of dauer formation in *Caenorhabditis elegans*. *Genetics* **130:** 105–123.

Walker D.W., McColl G., Jenkins N.L., Harris J., and Lithgow G.J. 2000. Evolution of lifespan in *C. elegans*. *Nature* **405:** 296–297.

Walker G.A., Thompson F.J., Brawley A., Scanlon T., and Devaney E. 2003. Heat shock factor functions at the convergence of the stress response and developmental pathways in *Caenorhabditis elegans*. *FASEB J.* **17:** 1960–1962.

Wang M.C., Bohmann D., and Jasper H. 2005. JNK extends life span and limits growth by antagonizing cellular and organism-wide responses to insulin signaling. *Cell* **121:** 115–125.

Wang Y. and Tissenbaum H.A. 2006. Overlapping and distinct functions for a *Caenorhabditis elegans* SIR2 and DAF-16/FOXO. *Mech. Ageing Dev.* **127:** 48–56.

Wilson M.A., Shukitt-Hale B., Kalt W., Ingram D.K., Joseph J.A., and Wolkow C.A. 2006. Blueberry polyphenols increase lifespan and thermotolerance in *Caenorhabditis elegans*. *Aging Cell* **5:** 59–68.

Wolff S., Ma H., Burch D., Maciel G.A., Hunter T., and Dillin A. 2006. SMK-1, an essential regulator of DAF-16-mediated longevity. *Cell* **124:** 1039–1053.

Wolkow C.A., Kimura K.D., Lee M.S., and Ruvkun G. 2000. Regulation of *C. elegans* lifespan by insulinlike signaling in the nervous system. *Science* **290:** 147–150.

Wolkow C.A., Munoz M.J., Riddle D.L., and Ruvkun G. 2002. Insulin receptor substrate and p55 orthologous adaptor proteins function in the *Caenorhabditis elegans* daf-2/insulin-like signaling pathway. *J. Biol. Chem.* **277:** 49591–49597.

Wong A., Boutis P., and Hekimi S. 1995. Mutations in the clk-1 gene of *Caenorhabditis elegans* affect developmental and behavioral timing. *Genetics* **139:** 1247–1259.

Wood J.G., Rogina B., Lavu S., Howitz K., Helfand S.L., Tatar M., and Sinclair D. 2004. Sirtuin activators mimic caloric restriction and delay ageing in metazoans. *Nature* **430:** 686–698.

Yokoyama K., Fukumoto K., Murakami T., Harada S., Hosono R., Wadhwa R., Mitsui Y., and Ohkuma S. 2002. Extended longevity of *Caenorhabditis elegans* by knocking in extra copies of hsp70F, a homolog of mot-2 (mortalin)/mthsp70/Grp75. *FEBS Lett.* **516:** 53–57.

8

Cellular Senescence: A Link between Tumor Suppression and Organismal Aging?

John M. Sedivy, Ursula M. Munoz-Najar, and Jessie C. Jeyapalan
Department of Molecular Biology, Cell Biology
and Biochemistry, and Center for Genomics
and Proteomics, Brown University
Providence, Rhode Island 02903

Judith Campisi
Life Sciences Division, Lawrence Berkeley National
Laboratory, Berkeley, California 94720
Buck Institute for Age Research
Novato, California 94945

THE AGING OF ORGANISMS OCCURS AT VIRTUALLY every level of complexity—from molecules to tissues to organ systems. Between these extremes are the basic units of life: individual cells. Among multicellular organisms, how do cells age? The deterioration of life processes in postmitotic cells—chronological aging—is explored elsewhere in this book. Here, we consider the aging of cells that retain the capacity for proliferation in adult organisms.

Normal somatic cells of higher metazoans, with the exception of germ cells and some stem cells, have a limited proliferative capacity (also referred to as replicative life span). This phenomenon was first formally described by Hayflick and Moorhead (1961), who observed that human fibroblasts, upon explant into cell culture, displayed an initial phase of rapid proliferation followed by a period of declining replicative potential. Eventually, all cells in the culture ceased dividing, but they remained in a viable and stable state. This postmitotic growth arrest was termed replicative senescence (Hayflick 1965) and, later, cellular aging.

Molecular Biology of Aging ©2008 Cold Spring Harbor Laboratory Press 978-087969824-9

The discovery of replicative senescence led to two important hypotheses. The first one proposed that cellular senescence recapitulates aspects of organismal aging and contributes to aging phenotypes in vivo (Hayflick 1985). Although there is mounting evidence to support this idea, it still rests largely on circumstantial evidence. The second hypothesis invoked cellular senescence as a mechanism that suppresses the development of cancer (Sager 1991). There is now substantial evidence to support this hypothesis (Campisi 2005; Hemann and Narita 2007). This chapter focuses on the links among cellular senescence, carcinogenesis, and organismal aging. Understanding the molecular basis of cellular aging may provide important insights into organismal aging and suggest novel strategies for intervening in the aging process.

PHENOTYPES OF SENESCENT CELLS

Senescent cells are postmitotic, i.e., in the absence of experimental manipulation, they fail to undergo proliferation (used here interchangeably with growth) in response to physiological mitogens. Cellular senescence has been studied most extensively in culture, although there is now strong evidence that this process also occurs in vivo. Most senescent cells adopt a characteristic enlarged morphology and show striking changes in gene expression, protein processing, chromatin organization, and metabolism (Cristofalo et al. 2004).

In culture, the fraction of senescent cells increases with passage (Kill et al. 1994), and although the culture typically reaches senescence asynchronously, individual cells execute cell cycle withdrawal rapidly (Herbig et al. 2003, 2004). Senescent cells generally arrest growth with a G_1 DNA content (Pignolo et al. 1998). In addition, many senescent cells are more resistant to apoptotic cell death than their nonsenescent counterparts (Marcotte et al. 2004; Hampel et al. 2005).

The enlarged senescent morphology includes an increase in the size of the nucleus and nucleoli. Senescent cells also exhibit an increase in the number and/or size of lysosomes, vacuoles, and mitochondria (Cristofalo et al. 2004), as well as alterations in the composition of the cytoskeleton that can modify their migration properties (Nishio and Inoue 2005). Additionally, as discussed below, senescent fibroblasts secrete factors that might promote phenotypes associated with aging or age-related disease (Campisi 2005). These features are the result of widespread changes in gene and protein expression (Gonos et al. 1998; Shelton et al. 1999; Schwarze et al. 2002; Semov et al. 2002; Chang et al. 2003; Zhang et al. 2003; Yoon et al. 2004; Pascal et al. 2005; Xie et al. 2005; Cong et al. 2006;

Trougakos et al. 2006; Zdanov et al. 2006). A recent large-scale study identified more than 600 genes that are differentially expressed in senescent fibroblasts (Zhang et al. 2003). Interestingly, gene expression patterns in senescent mammary epithelial cells were substantially different from those in senescent fibroblasts.

Much work remains to be done in this area. The complexity of the expression differences between nonsenescent and senescent cells appears to be dauntingly large, the responses seem to vary between different cell types, and only a small fraction of the changes have been functionally linked with specific senescent phenotypes. Nonetheless, these studies are consistent with observations that the senescence growth arrest is associated with large-scale changes in chromatin organization (Narita et al. 2003; Zhang et al. 2005, 2007; Funayama et al. 2006).

BIOMARKERS OF CELLULAR SENESCENCE

One limitation to understanding the role of cellular senescence in intact organisms is the lack of specific biomarkers that can distinguish senescent cells from quiescent or terminally differentiated cells. The first, and most widely used marker is the senescence-associated β-galactosidase (SA β-gal) (Dimri et al. 1995). This marker allows histochemical staining of senescent cells in freshly frozen tissue, but it is also expressed by nonsenescent cells under certain stressful conditions (such as prolonged confluence or growth factor deprivation), by some differentiated cells, and in certain extracellular structures such as hair follicles (Dimri et al. 1995; Yegorov et al. 1998; Severino et al. 2000; Cristofalo 2005; Yang and Hu 2005). SA β-gal is the product of the *GLB1* gene that encodes lysosomal β-gal (Lee et al. 2006), and the staining may reflect in part the increase in lysosome biogenesis that occurs in many senescent cells (Cristofalo et al. 2004).

More recently, additional senescence markers have been identified and shown to be useful in monitoring the presence of senescent cells both in culture and in vivo. These markers fall into three broad categories: components of signal transduction pathways known to be involved in the establishment and maintenance of senescent states, DNA-damage markers coincident with telomeres, and markers of focal heterochromatin. As discussed in more detail below, none of these markers are common to all senescence states, and most will give positive signals in situations that do not involve senescence. Nevertheless, when used in combination, senescence markers have provided important evidence for the presence of senescent cells in vivo, as well as insights into the mechanisms that generated them.

It is now clear that the loss of cell division capacity observed by Hayflick and colleagues is due in large measure to the shortening and eventual malfunction of telomeres (Harley et al. 1990; Bodnar et al. 1998). Telomeres are stretches of repetitive DNA and associated proteins at the terminal ends of chromosomes. The telomeric structure protects chromosome ends from being recognized as DNA double-strand breaks (DSBs) and thus prevents their degradation or fusion by DNA-repair machineries. Telomeres shorten with each cell division because DNA polymerases cannot fully replicate 3′ termini, a phenomenon referred to as the end-replication problem (Olovnikov 1973). When telomeres become critically short and incapable of end protection, most cells undergo senescence (Shay and Wright 2005).

Telomere-induced senescence has been shown to share many components of the DNA-damage response triggered by DSBs (d'Adda di Fagagna et al. 2004; von Zglinicki et al. 2005). A well-established DNA-damage response marker is the presence of nuclear foci containing γ-H2AX (phosphorylated form of the histone variant H2AX), which localize to sites of DSB. These foci also contain a multitude of proteins involved in the recognition and repair of DSB, as well as signaling proteins (such as the phosphorylated forms of the ATM, CHK1, and CHK2 kinases) that mediate the DNA-damage response checkpoint. Senescent cells contain such foci, which frequently localize to telomeres (d'Adda di Fagagna et al. 2003) and are referred to as telomere dysfunction-induced foci (TIF). TIF are operationally characterized as colocalizations, detectable by immunofluorescence, of γ-H2AX (or other DNA-damage response-associated proteins) with a telomere-binding protein such as TRF1 (Takai et al. 2003) or with telomeric sequences that can be visualized using fluorescence in situ hybridization (FISH) (Herbig et al. 2004; Sedelnikova et al. 2004). TIF are a robust indicator of telomere-initiated senescence in cell culture, although DNA-damage response proteins also localize to telomeres during late S phase in proliferating cells (Verdun and Karlseder 2006). TIF can also be detected in tissue sections (Herbig et al. 2006; Jeyapalan et al. 2007).

Some senescent cells accumulate a distinctive type of facultative heterochromatin designated senescence-associated heterochromatin foci (SAHF). SAHF were first identified as punctate nuclear structures by their propensity to bind dyes that stain DNA (Narita et al. 2003). SAHF are also enriched for heterochromatin-associated proteins such as HP1α, β, and γ, lysine 9-methylated histone H3, and the histone H2A variant macroH2A (Narita et al. 2003; Braig et al. 2005; Zhang et al. 2005). SAHF formation is mediated by several chromatin-modifying proteins, including the histone methyltransferase Suv39h1 and heterochromatin assembly factors such as HIRA and ASF1a (Ye et al. 2007; Zhang et al. 2007). The detection of SAHF is a useful new addition to the portfolio of senescence biomarkers. However, two caveats need to be considered. First, all cells contain non-SAHF heterochromatin, such as pericentromeric chromatin, which is also detected by the above assays. Second, not all senescent human cells contain cytologically detectable SAHF, and the levels of SAHF can vary significantly between different types of senescent cells (Bartkova et al. 2006; Herbig et al. 2006; Jeyapalan et al. 2007). Recent results show that each SAHF contains the condensed chromatin from a single chromosome (Funayama et al. 2006; Zhang et al. 2007).

TRIGGERS OF CELLULAR SENESCENCE

A large number of causes can lead to cellular senescence. As noted above, the first described was repeated cell division resulting in progressive telomere shortening (Hayflick 1965; Harley et al. 1990; Bodnar et al. 1998). Subsequently, it became apparent that many types of stress, including ionizing and UV irradiation, reactive oxygen species, nutrient imbalances, and even suboptimal culture conditions can induce some normal cells to undergo senescence (Ben-Porath and Weinberg 2004). Cells can also senesce in response to changes in chromatin organization, such as those caused by pharmacological agents or the expression of proteins that modify DNA or histones (Neumeister et al. 2002; Bandyopadhyay and Medrano 2003; Narita and Lowe 2004). Finally, activation of some (but not all) oncogenes is an important trigger of senescence in normal cells (Serrano et al. 1997; Collado and Serrano 2006).

Despite the plethora of stimuli that can induce senescence, in the great majority of cases, either or both of two central signaling pathways leading to the activation of the p53 and retinoblastoma (pRB) tumor suppressor proteins appear to be responsible for initiating and maintaining the senescence state. The two key effectors that distinguish these senescence-regulating pathways are the cyclin-dependent kinase (CDK) inhibitors p21 and p16, encoded by the *CDKN1A* and *CDKN2A* genes, respectively (Campisi 2005; Herbig and Sedivy 2006). CDKs, their activators (cyclins), and their inhibitors are the intrinsic regulators of cell cycle progression (Sherr and Roberts 1999). Either *CDKN1A* or *CDKN2A*, or both, are transcriptionally up-regulated in response to virtually all stimuli that induce cellular senescence. The p21 protein binds to several cyclin-CDK complexes and inhibits their activity, whereas p16 is specific for CDK4 and CDK6 and antagonizes their binding to D-type cyclins

(Vidal and Koff 2000). The inhibition of CDKs blocks cell cycle progression in G_1 by maintaining pRB in its active (hypophosphorylated) form, which then in some cases can initiate the formation of SAHF (Narita and Lowe 2004; Zhang et al. 2007).

The p53-p21 and p16-pRB pathways are distinguished by their upstream signaling components. p21 is transcriptionally activated by p53, and this pathway is the primary mediator of both telomere-dependent senescence and senescence caused by genotoxic stress (d'Adda di Fagagna et al. 2004). The p16-pRB pathway mediates many forms of nongenotoxic stress-induced senescence, such as that caused by chromatin perturbations (Ohtani et al. 2004). The up-regulation of p16 is complex and not well understood (Gil and Peters 2006). Oncogene-induced senescence can activate both the p16-pRB and p21-p53 pathways. Oncogenes such as RAS induce p16 by stimulating the activity of Ets transcription factors (Ohtani et al. 2001). These oncogenes also cause premature firing and termination of replication origins, thereby triggering a DNA-damage response and activation of the p21-p53 pathway (Bartkova et al. 2006; Di Micco et al. 2006; Mallette et al. 2007). Despite exceptions such as RAS-induced senescence, there are many examples where one pathway can be activated independently of the other, or with very different kinetics. However, since both pathways entail up-regulated expression of CDK inhibitors, pRB is a common downstream component, and in some cases, the pathways have been found to reinforce each other (Ben-Porath and Weinberg 2004).

TELOMERE-DEPENDENT REPLICATIVE SENESCENCE

As noted above, telomeres form essential protective structures at chromosome ends (de Lange 2002), but the end-replication problem causes progressive telomere shortening with each cell division (Olovnikov 1973). Telomeres are maintained by a specialized RNA-templated polymerase, the telomerase reverse transcriptase (Greider and Blackburn 1985; Cech 2004). The telomerase holoenzyme is composed of a catalytic subunit and a small template RNA. In long-lived or large species (Seluanov et al. 2007), telomerase activity is found only in germ cells and some stem cells and is transiently expressed in some proliferative cells of renewable tissues, such as keratinocytes, intestinal crypt cells, and T cells (Forsyth et al. 2002; Boukamp 2005; Effros 2007). Telomerase may be induced at very low levels during the S phase of many cell types, but this activity is insufficient to prevent telomere shortening (Masutomi et al. 2003). The virtual absence of telomerase activity in the great majority of somatic tissues is often due to the transcriptional down-regulation of the gene encoding the telomerase catalytic subunit. Strikingly, telomerase is expressed by a majority of cancer cells (Kim et al. 1994), which must acquire an ability to ignore or bypass senescence-inducing signals to become fully malignant (Hanahan and Weinberg 2000).

In the absence of telomerase, telomeres shorten by 50–200 bp with each round of replication (Harley et al. 1990; Sfeir et al. 2005). Telomere attrition can be significantly accelerated by reactive oxygen species (von Zglinicki 2002), and rapid changes in length can arise through processes such as homologous recombination (Lansdorp 2005; Britt-Compton et al. 2006). Average telomere lengths in humans range from 15–20 kb at birth to 6–8 kb in senescent fibroblast cultures, but they can be significantly shorter in vivo in highly replicative cell compartments, such as some lymphoid cell subpopulations, or in cancer cells (Rufer et al. 1998). The role of telomere shortening in triggering replicative senescence was demonstrated by the ability of ectopic telomerase expression to immortalize normal somatic cells (Bodnar et al. 1998). Recently, telomerase has been implicated in cellular functions other than telomere maintenance, such as global DNA-damage responses (Masutomi et al. 2005). Such telomerase activities are incompletely understood but may contribute to cancer cell survival (Dong et al. 2005; Konnikova et al. 2005). Despite the widespread expression of telomerase in cancer cells, in a minority of cases, telomeres are maintained by recombination in the absence of telomerase activity (Reddel et al. 2001), indicating that telomere integrity, rather than the presence of telomerase per se, is the key factor for avoiding telomere-dependent senescence (Steinert et al. 2000).

Telomeres end with 3′ overhangs estimated to be 35–105 nucleotides long, with leading-strand overhangs being shorter than lagging-strand overhangs (Chai et al. 2006). The overhangs are folded back and hybridize to internal sequences; the resultant duplex circles are referred to as t-loops (Griffith et al. 1999). Telomeric DNA sequences are bound by telomere-specific proteins TRF1, TRF2, and POT1, which together with other components (TIN2, TPP1, and Rap1) form a complex designated Shelterin, whose function is to prevent the chromosome ends from being recognized as broken or damaged DNA (Ferreira et al. 2004; de Lange 2005). Shelterin has an active role in t-loop formation, and interference with its function, for example, by expressing a dominant-defective form of TRF2 or knocking out POT1, leads to loss of end protection and engagement of the DNA-damage response (Takai et al. 2003; Wu et al. 2006). During repeated cell divisions, telomeres are thought to shorten to such an extent that they are no longer able to form t-loops. The critical telomere length threshold has not been pinpointed (Baird 2005; Britt-Compton et al. 2006); however, it appears that the shortest telomeres trigger senescence (Hemann et al. 2001), and very few dysfunctional telomeres (perhaps as few as one) may be able to trigger a DNA-damage response (Herbig et al. 2006; Jeyapalan et al. 2007).

The importance of the p53-p21 pathway in replicative senescence was first suggested by experiments showing that tumor-virus-encoded oncoproteins that interfere with p53 function extend the replicative life span of human cells, essentially by allowing them to ignore signals from short dysfunctional telomeres (Shay and Wright 1991; Sedivy 1998). A direct role for p53 was suggested by experiments showing that a dominant-defective p53 allele likewise extends proliferative capacity (Bond et al. 1994). p21 was first implicated in the control of senescence when its cDNA was identified in a screen for genes expressed by senescent human fibroblasts (Noda et al. 1994). Subsequently, a genetic knockout of p21 in human fibroblasts was shown to completely bypass the senescence response to short dysfunctional telomeres, causing cells to enter a state of severe genomic instability termed crisis (Brown et al. 1997; Wei and Sedivy 1999).

Although neither the precise structure of dysfunctional telomeres nor the exact mechanism by which they trigger a DNA-damage response are known (Francia et al. 2007), a key early signaling event is recruitment of the ATM kinase to the dysfunctional telomere and its activation by autophosphorylation (d'Adda di Fagagna et al. 2003; Takai et al. 2003; Herbig et al. 2004). Inhibition of ATM activity or reduced ATM expression interrupts the signal, allowing cells with dysfunctional telomeres to resume proliferation (d'Adda di Fagagna et al. 2003). The downstream mediator kinases CHK1 and CHK2 are likewise recruited to telomeres and activated by phosphorylation. CHK2 has a major role in the signaling from dysfunctional telomeres because its ablation interferes with the telomere-initiated senescence response (Gire et al. 2004). The activation of CHK1 and/or CHK2 leads to the phosphorylation and activation of p53, which in turn transcriptionally up-regulates p21. Very similar events—the sequential activation of kinases such as ATM, CHK1, and CHK2, culminating in p53 activation and p21 expression—are elicited when cells experience nontelomeric DNA DSB (d'Adda di Fagagna et al. 2004).

The senescence growth arrest that results from dysfunctional telomeres and other forms of DNA damage likely requires continuous signaling for its maintenance. Inactivation of p53 or DNA-damage response proteins that signal to p53 can reverse the growth arrest, providing the cells do not express high levels of p16 (Gire and Wynford-Thomas 1998; Beausejour et al. 2003; Di Micco et al. 2006). However, senescent cells do not necessarily express high levels of p53 (Atadja et al. 1995; Vaziri et al. 1997), in contrast to cells that experience acute DNA damage. Rather, senescent cells typically contain levels of p53 comparable to those in undamaged cells. In addition, they contain low levels of p53 phosphorylated on serine 15 (d'Adda di Fagagna et al. 2003; Herbig et al. 2004), which is known to promote *trans*-activation of its target genes (Appella and Anderson 2000).

As human fibroblast cultures approach replicative senescence, the up-regulation of p21 and p16, monitored by immunoblotting or northern hybridization, occurs gradually as replicative capacity declines (Stein et al. 1999; Wei et al. 2001). In contrast, when monitored at the single-cell level, the up-regulation is much more rapid, typically within the period of a single cell cycle (Herbig et al. 2003, 2004). These observations explain earlier data showing that cultures approaching replicative senescence contain increasing numbers of nondividing cells (Smith and Whitney 1980; Cristofalo et al. 2004).

p16 probably does not have a direct role in telomere-initiated senescence. Some human fibroblast strains do not up-regulate p16 during replicative senescence (Beausejour et al. 2003). Others, however, up-regulate both p21 and p16 during replicative senescence; in these cases, p21 increases before p16 (Stein et al. 1999), and by single-cell analyses, these increases were found to be independent events (Herbig et al. 2004). In other words, some human fibroblast strains senesce as mosaics, with some cells activating the p53-p21 pathway in response to dysfunctional telomeres, whereas others activate the p16-pRB pathway in response to unknown signals without signs of dysfunctional telomeres (d'Adda di Fagagna et al. 2003; Itahana et al. 2003; Herbig et al. 2004). Moreover, acute telomere upcapping elicited by the expression of a dominant-defective TRF2 protein up-regulates both p21 and p16, but whereas p21 responds within hours, the up-regulation of p16 is seen only after several days (Smogorzewska and de Lange 2002; Jacobs and de Lange 2004). Interestingly, ablation of ATM or p53 in cells that have only up-regulated p21 allows cell cycle reentry, whereas the arrest of cells that have up-regulated p16 cannot be reversed (Beausejour et al. 2003; Herbig et al. 2004). It thus appears likely that although telomere dysfunction rapidly up-regulates the p21-p53 pathway through a DNA-damage response, p16 is up-regulated by a secondary slower mechanism, or indirectly in response to physiological changes set in motion as cells enter the p53-dependent senescent state.

TELOMERE-INDEPENDENT SENESCENCE

A senescence response can be induced by many genotoxic stresses that cause genome-wide DNA damage. For example, exposing cells to hydrogen peroxide (H_2O_2) or a variety of other oxidants (Chen and Ames 1994; Toussaint et al. 2000) results in the rapid engagement of the p53-p21 pathway and cell cycle arrest (von Zglinicki et al. 2005). It has been suggested that telomeres, being G-rich, may be especially vulnerable to reactive oxygen species-induced DNA lesions such as 8-oxo-deoxyguanine (Chen et al. 1998; Oikawa and Kawanishi 1999; von Zglinicki 2001; Szekely et al. 2005). Nontelomeric DNA damage, at least at moderate levels, may be more easily repaired than damaged telomeres, which can undergo rapid catastrophic shortening by a replication-independent, recombination-mediated mechanism (Martens et al. 2000; Bailey et al. 2004; Lansdorp 2005). Whatever the case, the senescence response to general genotoxic stress occurs in the presence of telomerase expression, as well as in cancer cell lines with an intact p53 pathway (de Magalhaes et al. 2002; Gorbunova et al. 2002; Roninson 2003). In contrast, p16 can be induced by stresses unrelated to direct genotoxic stress (Gorbunova et al. 2002; Forsyth et al. 2003; Itahana et al. 2003; Parrinello et al. 2003). For example, chromatin perturbation caused by inhibition of histone deacetylase activity induces senescence by a p16-dependent mechanism (Munro et al. 2004), and many epithelial cells senescence owing to p16 induction without detectable telomere shortening (Brenner et al. 1998; Huschtscha et al. 1998; Rheinwald et al. 2002).

Recent data show that oncogene-induced senescence is an important in vivo cancer defense mechanism in both humans and mice (Woo and Poon 2004; Braig et al. 2005; Chen et al. 2005; Collado et al. 2005; Michaloglou et al. 2005). Full-blown malignancy, then, depends on the acquisition of mutations, most notably in p53 and/or p16, that circumvent the senescence growth arrest. Although recently validated in vivo, oncogene-induced senescence was first described in cell culture, where the mutational activation or overexpression of a variety of pro-mitogenic signaling molecules causes normal cells to undergo senescence (Serrano et al. 1997; Lin et al. 1998; Zhu et al. 1998; Dimri et al. 2000; Bischof et al. 2002; Michaloglou et al. 2005; Bartkova et al. 2006; Mallette et al. 2007). The strength of the senescence response appears to depend on the cell type, the signaling effector, and the level of its activation or overexpression (Hemann and Narita 2007). For example, oncogenic RAS induces senescence in established human fibroblast strains but not in freshly isolated cell populations (Benanti and Galloway 2004). RAS-induced senescence has been reported to require both the p53-p21 and p16-Rb pathways (Serrano et al. 1997; Wei et al. 2003), the p16-Rb pathway alone (Brookes et al. 2002), or the p53-p21 pathway alone (Voorhoeve and Agami 2003). RAS induced the formation of SAHF, whereas E2F1 did not, although both caused a senescence-like growth arrest (Mallette et al. 2007).

In evaluating these disparate data, it is important to consider the magnitude of the signaling imbalance, the cellular context (cell or tissue type) in which it occurs, and cross-talk between the p53-p21 and p16-RB pathways. Hyperproliferative signaling was recently shown to trigger a DNA-damage response, as well as an increase in reactive oxygen species production (Bartkova et al. 2006; Di Micco et al. 2006; Takahashi et al. 2006). The DNA-damage response is likely caused by unscheduled DNA synthesis leading to replication fork collapse and the formation of DNA DSB (Di Micco et al. 2006). The increase in reactive oxygen species may derive from a positive feedback loop in which ROS activates PKC δ and PKC δ signaling sustains production of reactive oxygen species (Lee et al. 1999; Takahashi et al. 2006). Reactive oxygen species might reinforce the DNA-damage response by causing oxidative DNA lesions or might independently signal p16 up-regulation. Engagement of the p53-p21 pathway coupled with p16 induction would be expected to reinforce maintenance of pRB in its active (hypophosphorylated) state and consequent SAHF formation. Context is crucial in this case because the presence of hypophosphorylated pRB in quiescent cells does not lead to SAHF formation. Finally, in human mammary epithelial cells but not fibroblasts, high levels of p16 reduce p53 and p21 protein levels, whereas down-regulation of p16 expression stabilizes p53, resulting in up-regulated p21 expression (Zhang et al. 2006). The mechanism responsible for this reciprocal regulation may derive from the fact that the E2F transcription factor, which is repressed by active pRB (Dyson 1998), stimulates the expression ARF (Bates et al. 1998), which in turn inhibits the major p53-degradation mediator H/Mdm2 (Zhang et al. 1998).

The transcriptional regulation of p16 is complex and not well understood (Gil and Peters 2006; Kim and Sharpless 2006). As noted above, several cell types, such as keratinocytes, melanocytes, or breast epithelial cells, senesce in cell culture in a telomere-independent manner due to the up-regulation of p16. Since in some cases, this senescence response can be prevented or delayed by culture conditions, such as culture on feeder layers (Ramirez et al. 2001; Wright and Shay 2002), it has been termed "culture shock" (Sherr and DePinho 2000). An important unanswered question is: What are the equivalent conditions that up-regulate p16 in vivo? p16 expression increases with age in normal tissues of mice, non-human primates, and humans (Zindy et al. 1997; Herbig et al. 2006;

Ressler et al. 2006), and this rise can be delayed by caloric restriction (Krishnamurthy et al. 2004). Recent findings show that p16 expression increases with age in a variety of stem or progenitor cells (Janzen et al. 2006; Krishnamurthy et al. 2006; Molofsky et al. 2006), which, as discussed below, may have widespread consequences for tissue homeostasis and renewal.

Among the known regulators of p16 are Bmi-1 (polycomb group finger 4, PCGF4), CBX7 (chromobox 7), and Mel-18 (polycomb group finger 2, PCGF2), members of the polycomb group (PcG) of transcriptional repressors (Jacobs et al. 1999; Gil et al. 2004; Bernard et al. 2005; Guo et al. 2007). *PcG* genes were discovered in *Drosophila* and control the transcription of homeotic genes by establishing heritable repressive chromatin states. These chromatin states maintain transcriptional patterns into adulthood, even in the absence of the regulators that initially established the patterning. Bmi-1 expression declines in senescent human fibroblasts and, when overexpressed, extends their replicative life span by repressing p16 expression (Itahana et al. 2003). Bmi-1 and Mel-18 have recently been demonstrated to be transcriptional targets of c-Myc (Guney et al. 2006; Guo et al. 2007). The senescence-associated up-regulation of p16 is rapid and has been likened to a switch between p16-OFF and p16-ON states (Guney and Sedivy 2006). The connection between c-Myc and the p16 switch suggests that a threshold of Bmi-1 is necessary for the maintenance of p16 repression by PcG.

EVOLUTIONARY SIGNIFICANCE OF CELLULAR SENESCENCE

As noted earlier, Hayflick's initial findings spawned two hypotheses regarding the significance of cellular senescence: It suppresses the development of cancer and also contributes to aging phenotypes. How can cellular senescence, or any fundamental process, be both beneficial (tumor suppressive) and detrimental (pro-aging)?

Cancer poses a major challenge to the longevity of organisms with renewable tissues such as mammals. Regeneration and repair are essential for the health of such organisms; however, cell proliferation is also a prime requirement for tumorigenesis (Hanahan and Weinberg 2000). Moreover, proliferating cells are more prone than nondividing cells to acquiring mutations (Busuttil et al. 2006), which are a major driving force behind cancer initiation and progression (Bishop 1995). The risk cancer poses to the longevity of organisms with renewable tissues was mitigated by the evolution of tumor suppressor mechanisms such as cellular senescence.

Complex organisms evolved in an environment replete with extrinsic hazards such as predation, starvation, and infection. Life spans were limited mainly by death due to external catastrophes, and tumor suppressor mechanisms thus only needed to be effective during the period of peak reproduction: a few decades for humans and several months for mice. If tumor suppressive mechanisms had deleterious effects after peak reproductive age, there would be little selective pressure to eliminate them. Indeed, some tumor suppressor mechanisms can be both beneficial and deleterious, depending on the age of the organism (Fig. 1) (Campisi 2003). This idea—that a biological process can be beneficial

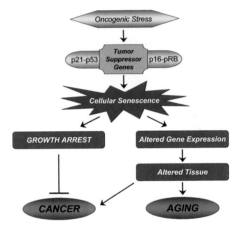

Figure 1. Trade-off between tumor suppression and aging. Many potentially oncogenic stimuli (Oncogenic Stress) activate tumor suppressor genes, the most important of which are those encoding the p53 and pRB proteins. These proteins govern major tumor suppressor pathways, often referred to as the p21-p53 and p16-pRB pathways. Both of these pathways participate in establishing and maintaining cellular senescence. The hallmarks of cellular senescence are an essentially irreversible arrest of cell proliferation and widespread alterations in gene expression. The growth arrest ensures that oncogenically stressed cells cannot divide and hence cannot progress to a cancerous stage. The altered gene expression, however, includes many secreted factors that are proposed to alter the tissue microenvironment. These factors can, in principle, cause the degeneration of normal tissue structures and therefore a loss of tissue function—a characteristic of aging tissues. In addition, these factors might create a proinflammatory and procarcinogenic tissue milieu that can synergize with somatic mutations to promote cancer progression. Young tissues typically harbor very few senescent cells, but these cells tend to accumulate in many tissues at older ages. Thus, although cellular senescence undoubtedly protects organisms from cancer early in life, certain aging phenotypes and age-related diseases, including late-life cancer, may derive from the accumulation of senescent cells later in life.

early in life but counterproductive in old age—comprises an important evolutionary theory of aging termed antagonistic pleiotropy (Williams 1957; Kirkwood and Austad 2000).

CELLULAR SENESCENCE AND TUMOR SUPPRESSION

There is now substantial evidence from cell culture and animal models as well as human studies that the senescence response suppresses the development of malignant tumors. First, virtually all the senescence-inducing stimuli that have been identified are also potentially oncogenic under some circumstances. Second, virtually all cancers during their progression acquire mutations in either the p53-p21 or p16-pRB pathways (or both)—the same major pathways that establish and maintain the senescence growth arrest. As a consequence, most cancer cells have a greatly extended replicative capacity and can achieve replicative immortality with the up-regulation of telomerase (Hanahan and Weinberg 2000). Moreover, mutations in the p53-p21 or p16-pRB pathways seriously compromise the ability of many cancer cells to senesce in response to mitogenic oncogenes and other stresses (Ohtani et al. 2004; Chen et al. 2005; Christophorou et al. 2006).

The tumor suppressive activities of the p53-p21 and p16-pRB pathways are not limited to their ability to induce cellular senescence. These pathways also regulate apoptosis, terminal differentiation, DNA repair, and other processes. The importance of senescence has, however, been underscored by recent research pointing to numerous instances in which loss of the senescence response appears to be a crucial step for the subsequent development of cancer. For example, several strains of mice have been genetically engineered to contain cells that fail to senesce in response to appropriate stimuli, and these mice invariably are cancer-prone (Donehower et al. 1992; Braig et al. 2005; Chen et al. 2005). Likewise, cells from human Li-Fraumeni syndrome patients, which carry mutations in p53 or CHK2 (Lee et al. 2001; Iwakuma et al. 2005), spontaneously overcome senescence at a high frequency (Shay et al. 1995), and these patients are also cancer-prone (Iwakuma et al. 2005). A glimpse at properly working senescence responses was provided by mouse models in which strong mitogenic signaling caused by the overexpression of activated oncogenes, or loss of the PTEN tumor suppressor, resulted in the development of benign lesions composed of senescent cells. Likewise, benign melanocytic naevi in human skin contain cells that express oncogenic BRAF and are senescent. These findings indicate that oncogene-induced senescence occurs in vivo and suggest that the role of the senescence response is to prevent oncogene-expressing cells from pro-gressing to full-blown malignancy (Braig et al. 2005; Chen et al. 2005; Collado et al. 2005; Lazzerini Denchi et al. 2005; Michaloglou et al. 2005). In support of this idea, the malignant tumors that grew out of the benign lesions had inactivated and bypassed the senescence response.

Cellular senescence suppresses tumorigenesis most likely because the development of cancer requires cell proliferation (Hanahan and Weinberg 2000). Thus, any mechanism that stringently opposes the growth of potential cancer cells will a priori antagonize the development of cancer. However, the failure to senesce is not, by itself, sufficient for malignancy. For example, expression of telomerase prevents telomere-initiated senescence but does not confer malignant properties on cells (Morales et al. 1999). Likewise, inactivation of the p53-p21 and/or p16-pRB pathways extends the replicative life span of human cells but does not transform them per se. Rather, such cells eventually suffer telomere dysfunction and genomic instability, and only rare replicatively immortal cells arise from these cell populations (Shay et al. 1993). Even then, the immortal cells may not be tumorigenic because full malignant transformation may also require the activation of mitogenic oncogenes (such as RAS) or loss of additional tumor suppressors that dampen mito-genic signals (such as PTEN) (Braig et al. 2005; Chen et al. 2005; Michaloglou et al. 2005). Thus, the genetic or epigenetic events that avert senescence are a necessary but insufficient step for malignant tumori-genesis (Hanahan and Weinberg 2000).

Failure to senesce is a common feature of tumor cells, but reversal of senescence—the resumption of proliferation by an already senescent cell—has so far been documented only in experimentally manipulated cell cultures (Gire and Wynford-Thomas 1998; Beausejour et al. 2003; Di Micco et al. 2006). Whether this ever occurs in vivo is not known. In culture, the inability to induce p16 allows reversal of senescence upon loss of p53 function (Beausejour et al. 2003). Because normal human tis-sues apparently harbor a few cells with promoter methylated and silenced p16 (Holst et al. 2003), were those cells to senesce in response to a sig-nal such as telomere dysfunction and subsequently lose p53 function, they could in principle resume proliferation. Likewise, senescent p16-negative cells in dysplastic naevi (Michaloglou et al. 2005) could in prin-ciple acquire p53 mutations and resume proliferation. Because nondi-viding cells can acquire mutations (Busuttil et al. 2006), these scenarios are not impossible; however, it is not yet known whether they happen in vivo. Whatever the case, it is now clear that senescence poses a formi-dable, but not insurmountable, barrier to cancer progression.

CELLULAR SENESCENCE AND AGING

The recent availability of a variety of senescence-associated markers has allowed the identification of senescent cells in vivo. Where are putative senescent cells found in organisms? In rodents, primates, and humans, such cells have been found in many renewable tissues, including the vasculature, the hematopoietic system, many epithelial organs, and the stroma surrounding epithelial organs (Dimri et al. 1995; Krtolica and Campisi 2002; Braig et al. 2005; Campisi 2005; Chen et al. 2005; Collado et al. 2005; Lazzerini Denchi et al. 2005; Michaloglou et al. 2005; Herbig et al. 2006; Janzen et al. 2006; Krishnamurthy et al. 2006; Molofsky et al. 2006; Jeyapalan et al. 2007). Notably, the incidence of cells with one or more senescence marker generally increases with age. How abundant are senescent cells in aged organisms? Estimates vary widely depending on the study, species, and tissue, ranging from less than 1% to more than 15%. It is also difficult to know the cause of the senescence response from many of these studies. The recent demonstration that cells with TIF accumulate with age (Herbig et al. 2006) suggests that at least in some tissues, telomere dysfunction can cause a senescence response in vivo. In a preliminary study, the TIF marker revealed significant differences between cell types (high in dermal fibroblasts and low in skeletal myocytes), suggesting that the prevalence of senescent cells varies among tissues (Jeyapalan et al. 2007).

Cells expressing senescence markers are also found at sites of chronic age-related pathology, such as osteoarthritis and atherosclerosis (Chang and Harley 1995; Vasile et al. 2001; Price et al. 2002; Matthews et al. 2006). Thus, senescent cells are associated with aging and age-related diseases in vivo. In addition, they are associated with the early stages of hyperplastic age-associated lesions, consistent with a role in protecting organisms from age-related cancer. As noted earlier, senescent cells are found in benign dysplastic or preneoplastic lesions (Braig et al. 2005; Chen et al. 2005; Collado et al. 2005; Michaloglou et al. 2005) and also in benign prostatic hyperplasia (Castro et al. 2003). They are also found in normal and tumor tissues following DNA-damaging chemotherapy (Schmitt et al. 2002; te Poele et al. 2002; Roninson 2003; Roberson et al. 2005), supporting a role for cellular senescence in suppressing cancer. Senescent cells are therefore found at the appropriate times and places for their proposed roles in both tumor suppression and aging.

An important unanswered question is why do senescent cells accumulate in vivo? In culture, senescent cells are very long-lived and many cell types are also resistant to apoptosis. However, it has been difficult to determine the stability of senescent cells in vivo. Are they also long-lived in vivo or are they somehow removed? If they are cleared, do they accumulate with increasing age and age-related pathology because they are produced at higher rates or because clearance mechanisms are compromised? Recent findings suggest senescent cells, at least those induced by acute p53 activation in murine tumor models, can be cleared by host mechanisms (Ventura et al. 2007; Xue et al. 2007), one of which is the innate immune system (Xue et al. 2007). Nothing is known about how senescent cells are recognized by the immune system or whether there are additional mechanisms to remove them. Likewise, it is not known whether the senescent cells found in vivo have escaped clearance mechanisms, are in the process of being eliminated, or both.

The findings that senescent cells accumulate with age and are found at sites of age-related pathology indicate that they are present at the right time and place, but this does not show that they are actively driving aging phenotypes. How might they do so? First, because senescent cells are incapable of proliferation, they might reduce the capacity for tissue repair and regeneration in renewable tissues. Recent findings support this possibility and suggest that p16-dependent senescence may be responsible for age-associated decrements in neurogenesis, hematopoiesis, and pancreatic function (Janzen et al. 2006; Krishnamurthy et al. 2006; Molofsky et al. 2006). These studies showed that p16 expression increases with age in stem or progenitor cells of the mouse brain, bone marrow, and pancreas, with the rise being apparent even in middle-aged animals. Moreover, it was responsible for suppressing stem cell proliferation because the age-related decline in tissue regeneration was substantially retarded in p16 knockout mice. As expected of mice that lack an important tumor suppressor, their life span was prematurely curtailed by cancer.

The age-dependent accumulation of p16-positive stem and progenitor cells is consistent with the idea that stem cell senescence might at least partly explain the age-related decline in brain and bone marrow function and the development of type II diabetes. However, it is not yet known whether these p16-positive cells are in fact senescent. Likewise, it is not known what stress (if any) is responsible for their increase with aging or whether these cells had initiated neoplastic transformation. Whatever the case, these findings raise the possibility that the tumor-suppressive activity of p16 might be inextricably linked to pro-aging effects, as expected if p16-mediated tumor suppression was antagonistically pleiotropic. A similar trade-off between tumor suppression and aging has already been documented in mice expressing constitutively hyperactive forms of p53: These animals are remarkably tumor-free but show multiple signs of accelerated aging (Tyner et al. 2002; Maier et al. 2004). In these cases, at least some of the accelerated aging may be due to increased sensitivity to senescence-inducing stimuli (Maier et al. 2004).

A second mechanism by which senescent cells might contribute to aging stems from their altered gene expression, specifically, the up-regulation of genes encoding extracellular-matrix-degrading enzymes, inflammatory cytokines, and growth factors (Campisi 2005). These factors can have striking effects on the behavior of neighboring cells as well as distal cells. Factors secreted by senescent cells have been shown to be capable of disrupting normal tissue structures and functions in cell culture models (e.g., the differentiation of mammary epithelial cells or keratinocytes) (Funk et al. 2000; Parrinello et al. 2005). Moreover, senescent-cell-derived soluble factors have been shown to stimulate the growth and angiogenic activity of nearby premalignant cells, both in culture and in vivo (Krtolica et al. 2001; Dilley et al. 2003; Martens et al. 2003; Bavik et al. 2006; Coppe et al. 2006). Thus, senescent cells, which themselves cannot form tumors, may promote the progression of nearby premalignant cells and fuel the development of cancer in aging organisms. These findings also support the idea that the senescence response is antagonistically pleiotropic, balancing the benefits of suppressing cancer in young organisms against the promotion of aging phenotypes later in life.

REMAINING QUESTIONS

Despite its transition from a cell culture curiosity to a potential regulator of cancer and aging, cellular senescence remains enigmatic and continues to pose many questions. Precisely how do the p53 and pRB pathways establish and maintain the senescent growth arrest? Both pathways can trigger reversible cell cycle arrest as well as promote apoptosis, so how are their activities controlled by senescence-inducing signals to impose the essentially permanent arrest and unique senescence-associated phenotypes? The signals that induce p16, both in culture and in vivo, are especially obscure at present. Likewise, very little is known about the mechanisms that cause the apparently deleterious senescent secretory phenotype. How, and why, does this phenotype develop?

Of particular interest are the mechanisms that determine the balance among tumor suppression, tissue regeneration, and aging. As suggested by mouse models, it would be undesirable to reverse the senescence growth arrest, which would allow damaged, stressed, or oncogene-expressing cells to proliferate and consequently accelerate the risk of cancer. But would it be possible to attenuate the pro-aging aspects of cellular senescence, for example, the secretory phenotype, without compromising the senescent growth arrest? The generation of mice carrying additional copies of properly regulated p53 (Garcia-Cao et al. 2002) or the Ink4 locus (p16 and p19^{ARF}) (Matheu et al. 2004), or reduced activity of the negative p53 regulator Mdm2 (Mendrysa et al. 2006), suggests that this might, at least in principle, be possible. These mice develop very little cancer without accelerated aging. However, their life spans were not significantly extended, even though cancer is the major cause of death in mice. It remains to be investigated whether the development of other age-related pathologies was exacerbated. Moreover, the balance between cancer, tissue regeneration, and aging may well differ between short- and long-lived species such as mice and humans. Nevertheless, recent studies raise the prospect that it may be possible, through specific interventions, to ameliorate some of the antagonistically pleiotropic effects of cellular senescence.

REFERENCES

Appella E. and Anderson C.W. 2000. Signaling to p53: Breaking the posttranslational modification code. *Pathol. Biol.* **48:** 227–245.

Atadja P., Wong H., Garkavtsev I., Veillette C., and Riabowol K. 1995. Increased activity of p53 in senescing fibroblasts. *Proc. Natl. Acad. Sci.* **92:** 8348–8352.

Bailey S.M., Brenneman M.A., and Goodwin E.H. 2004. Frequent recombination in telomeric DNA may extend the proliferative life of telomerase-negative cells. *Nucleic Acids Res.* **32:** 3743–3751.

Baird D.M. 2005. New developments in telomere length analysis. *Exp. Gerontol.* **40:** 363–368.

Bandyopadhyay D. and Medrano E.E. 2003. The emerging role of epigenetics in cellular and organismal aging. *Exp. Gerontol.* **38:** 1299–1307.

Bartkova J., Rezaei N., Liontos M., Karakaidos P., Kletsas D., Issaeva N., Vassiliou L.V., Kolettas E., Niforou K., Zoumpourlis V.C., Takaoka M., Nakagawa H., Tort F., Fugger K., Johansson F., Sehested M., Andersen C.L., Dyrskjot L., Orntoft T., Lukas J., Kittas C., Helleday T., Halazonetis T.D., Bartek J., and Gorgoulis V.G. 2006. Oncogene-induced senescence is part of the tumorigenesis barrier imposed by DNA damage checkpoints. *Nature* **444:** 633–637.

Bates S., Phillips A.C., Clark P.A., Stott F., Peters G., Ludwig R.L., and Vousden K.H. 1998. p14ARF links the tumor suppressors RB and p53. *Nature* **395:** 125–129.

Bavik C., Coleman I., Dean J.P., Knudsen B., Plymate S., and Nelson P.S. 2006. The gene expression program of prostate fibroblast senescence modulates neoplastic epithelial cell proliferation through paracrine mechanisms. *Cancer Res.* **66:** 794–802.

Beausejour C.M., Krtolica A., Galimi F., Narita M., Lowe S.W., Yaswen P., and Campisi J. 2003. Reversal of human cellular senescence: Roles of the p53 and p16 pathways. *EMBO J.* **22:** 4212–4222.

Ben-Porath I. and Weinberg R.A. 2004. When cells get stressed: An integrative view of cellular senescence. *J. Clin. Invest.* **113:** 8–13.

Benanti J. and Galloway D. 2004. Normal human fibroblasts are resistant to RAS-induced senescence. *Mol. Cell. Biol.* **24:** 2842–2852.

Bernard D., Martinez-Leal J.F., Rizzo S., Martinez D., Hudson D., Visakorpi T., Peters G., Carnero A., Beach D., and Gil J. 2005. CBX7 controls the growth of normal and tumor-derived prostate cells by repressing the Ink4a/Arf locus. *Oncogene* **24:** 5543–5551.

Bischof O., Kirsh O., Pearson M., Itahana K., Pelicci P.G., and Dejean A. 2002. Deconstructing PML-induced premature senescence. *EMBO J.* **21:** 3358–3369.

Bishop J.M. 1995. Cancer: The rise of the genetic paradigm. *Genes Dev.* **9:** 1309–1315.

Bodnar A.G., Ouellette M., Frolkis M., Holt S.E., Chiu C.P., Morin G.B., Harley C.B., Shay J.W., Lichtsteiner S., and Wright W.E. 1998. Extension of life-span by introduction of telomerase into normal human cells. *Science* **279:** 349–352.

Bond J.A., Wyllie F.S., and Wynford-Thomas D. 1994. Escape from senescence in human diploid fibroblasts induced directly by mutant p53. *Oncogene* **9:** 1885–1889.

Boukamp P. 2005. Skin aging: A role for telomerase and telomere dynamics? *Curr. Mol. Med.* **5:** 171–177.

Braig M., Lee S., Loddenkemper C., Rudolph C., Peters A.H., Schlegelberger B., Stein H., Dorken B., Jenuwein T., and Schmitt C.A. 2005. Oncogene-induced senescence as an initial barrier in lymphoma development. *Nature* **436:** 660–665.

Brenner A.J., Stampfer M.R., and Aldaz C.M. 1998. Increased p16 expression with first senescence arrest in human mammary epithelial cells and extended growth capacity with p16 inactivation. *Oncogene* **17:** 199–205.

Britt-Compton B., Rowson J., Locke M., Mackenzie I., Kipling D., and Baird D.M. 2006. Structural stability and chromosome-specific telomere length is governed by *cis*-acting determinants in humans. *Hum. Mol. Genet.* **15:** 725–733.

Brookes S., Rowe J., Ruas M., Llanos S., Clark P.A., Lomax M., James M.C., Vatcheva R., Bates S., Vousden K.H., Parry D., Gruis N., Smit N., Bergman W., and Peters G. 2002. INK4a-deficient human diploid fibroblasts are resistant to RAS-induced senescence. *EMBO J.* **21:** 2936–2945.

Brown J.P., Wei W., and Sedivy J.M. 1997. Bypass of senescence after disruption of p21CIP1/WAF1 gene in normal diploid human fibroblasts. *Science* **277:** 831–834.

Busuttil R.A., Rubio M., Dolle M.E., Campisi J., and Vijg J. 2006. Mutant frequencies and spectra depend on growth state and passage number in cells cultured from transgenic lacZ-plasmid reporter mice. *DNA Repair* **5:** 52–60.

Campisi J. 2003. Cancer and ageing: Rival demons? *Nat. Rev. Cancer* **3:** 339–349.

———. 2005. Senescent cells, tumor suppression, and organismal aging: Good citizens, bad neighbors. *Cell* **120:** 1–10.

Castro P., Giri D., Lamb D., and Ittmann M. 2003. Cellular senescence in the pathogenesis of benign prostatic hyperplasia. *Prostate* **55:** 30–38.

Cech T.R. 2004. Beginning to understand the end of the chromosome. *Cell* **116:** 273–279.

Chai W., Du Q., Shay J.W., and Wright W.E. 2006. Human telomeres have different overhang sizes at leading versus lagging strands. *Mol. Cell* **21:** 427–435.

Chang B.D., Swift M.E., Shen M., Fang J., Broude E.V., and Roninson I.B. 2003. Molecular determinants of terminal growth arrest induced in tumor cells by a chemotherapeutic agent. *Proc. Natl. Acad. Sci.* **99:** 389–394.

Chang E. and Harley C.B. 1995. Telomere length and replicative aging in human vascular tissues. *Proc. Natl. Acad. Sci.* **92:** 11190–11194.

Chen Q.M. and Ames B.M. 1994. Senescence-like growth arrest induced by hydrogen peroxide in human diploid fibroblast F65 cells. *Proc. Natl. Acad. Sci.* **91:** 4130–4134.

Chen Q.M., Bartholomew J.C., Campisi J., Acosta M., Reagan J.D., and Ames B.M. 1998. Molecular analysis of H_2O_2-induced senescence-like growth arrest in normal human fibroblasts: p53 and Rb control G1 arrest but not cell replication. *Biochem. J.* **332:** 43–50.

Chen Z., Trotman L.C., Shaffer D., Lin H.K., Dotan Z.A., Niki M., Koutcher J.A., Scher H.I., Ludwig T., Gerald W., Cordon-Cardo C., and Pandolfi P.P. 2005. Crucial role of p53-dependent cellular senescence in suppression of Pten-deficient tumorigenesis. *Nature* **436:** 725–730.

Christophorou M.A., Ringshausen I., Finch A.J., Swigart L.B., and Evan G.I. 2006. The pathological response to DNA damage does not contribute to p53-mediated tumour suppression. *Nature* **443:** 214–217.

Collado M. and Serrano M. 2006. The power and the promise of oncogene-induced senescence markers. *Nat. Rev. Cancer* **6:** 472–476.

Collado M., Gil J., Efeyan A., Guerra C., Schuhmacher A.J., Barradas M., Benguria A., Zaballos A., Flores J.M., Barbacid M., Beach D., and Serrano M. 2005. Tumour biology: Senescence in premalignant tumours. *Nature* **436:** 642.

Cong Y.S., Fan E., and Wang E. 2006. Simultaneous proteomic profiling of four different growth states of human fibroblasts, using amine-reactive isobaric tagging reagents and tandem mass spectrometry. *Mech. Ageing Dev.* **127:** 332–343.

Coppe, J.P., Kauser, K., Campisi, J., and Beausejour, C.M. 2006. Secretion of vascular endothelial growth factor by primary human fibroblasts at senescence. *J. Biol. Chem.* **281:** 29568–29574.

Cristofalo V.J. 2005. SA β Gal staining: Biomarker or delusion. *Exp. Gerontol.* **40:** 836–838.

Cristofalo V.J., Lorenzini A., Allen R.G., Torres C., and Tresini M. 2004. Replicative senescence: A critical review. *Mech. Ageing Dev.* **125:** 827–848.

d'Adda di Fagagna F., Teo S.H., and Jackson S.P. 2004. Functional links between telomeres and proteins of the DNA-damage response. *Genes Dev.* **18:** 1781–1799.

d'Adda di Fagagna F., Reaper P.M., Clay-Farrace L., Fiegler H., Carr P., Von Zglinicki T., Saretzki G., Carter N.P., and Jackson S.P. 2003. A DNA damage checkpoint response in telomere-initiated senescence. *Nature* **426:** 194–198.

de Lange T. 2002. Protection of mammalian telomeres. *Oncogene* **21:** 532–540.

———. 2005. Shelterin: The protein complex that shapes and safeguards human telomeres. *Genes Dev.* **19:** 2100–2110.

de Magalhaes J.P., Chainiaux F., Remade J., and Toussaint O. 2002. Stress-induced premature senescence in BJ and hTERT-BJ1 human foreskin fibroblasts. *FEBS Lett.* **523:** 157–162.

Dilley T.K., Bowden G.T., and Chen Q.M. 2003. Novel mechanisms of sublethal oxidant toxicity: Induction of premature senescence in human fibroblasts confers tumor promoter activity. *Exp. Cell Res.* **290:** 38–48.

Di Micco R., Fumagalli M., Cicalese A., Piccinin S., Gasparini P., Luise C., Schurra C., Garre' M., Nuciforo P.G., Bensimon A., Maestro R., Pelicci P.G., and d'Adda di Fagagna F. 2006. Oncogene-induced senescence is a DNA damage response triggered by DNA hyper-replication. *Nature* **444:** 638–642.

Dimri G.P., Itahana K., Acosta M., and Campisi J. 2000. Regulation of a senescence checkpoint response by the E2F1 transcription factor and p14/ARF tumor suppressor. *Mol. Cell. Biol.* **20:** 273–285.

Dimri G.P., Lee X., Basile G., Acosta M., Scott G., Roskelley C., Medrano E.E., Linskens M., Rubelj I., Pereira-Smith O., Peacocke M., and Campisi J. 1995. A biomarker that identifies senescent human cells in culture and in aging skin in vivo. *Proc. Natl. Acad. Sci.* **92:** 9363–9367.

Donehower L.A., Harvey M., Slagke B.L., McArthur M.J., Montgomery C.A., Butel J.S., and Bradley A. 1992. Mice deficient for p53 are developmentally normal but susceptible to spontaneous tumors. *Nature* **356:** 215–221.

Dong C.K., Masutomi K., and Hahn W.C. 2005. Telomerase: Regulation, function and transformation. *Crit. Rev. Oncol. Hematol.* **54:** 85–93.

Dyson N. 1998. The regulation of E2F by pRB-family proteins. *Genes Dev.* **12:** 2245–2262.

Effros R.B. 2007. Telomerase induction in T cells: A cure for aging and disease? *Exp. Gerontol.* **42:** 416–420.

Ferreira M.G., Miller K.M., and Cooper J.P. 2004. Indecent exposure: When telomeres become uncapped. *Mol. Cell* **13:** 7–18.

Forsyth N.R., Wright W.E., and Shay J.W. 2002. Telomerase and differentiation in multicellular organisms: Turn it off, turn it on, and turn it off again. *Differentiation* **69:** 188–197.

Forsyth N.R., Evans A.P., Shay J.W., and Wright W.E. 2003. Developmental differences in the immortalization of lung fibroblasts by telomerase. *Aging Cell* **2:** 235–243.

Francia S., Weiss R., and d'Adda di Fagagna F. 2007. Need telomere maintenance? Call 911. *Cell Div.* **2:** 3.

Funayama R., Saito M., Tanobe H., and Ishikawa F. 2006. Loss of linker histone H1 in cellular senescence. *J. Cell Biol.* **175:** 869–880.

Funk W.D., Wang C.K., Shelton D.N., Harley C.B., Pagon G.D., and Hoeffler W.K. 2000. Telomerase expression restores dermal integrity to in vitro aged fibroblasts in a reconstituted skin model. *Exp. Cell Res.* **258:** 270–278.

Garcia-Cao I., Garcia-Cao M., Martin-Caballero J., Criado L.M., Klatt P., Flores J.M., Weill J.C., Blasco M.A., and Serrano M. 2002. "Super p53" mice exhibit enhanced DNA damage response, are tumor resistant and age normally. *EMBO J.* **21:** 6225–6235.

Gil J. and Peters G. 2006. Regulation of the INK4b-ARF-INK4a tumour suppressor locus: All for one or one for all. *Nat. Rev. Mol. Cell Biol.* **7:** 667–677.

Gil J., Bernard D., Martinez D., and Beach D. 2004. Polycomb CBX7 has a unifying role in cellular lifespan. *Nat. Cell Biol.* **6:** 67–72.

Gire V. and Wynford-Thomas D. 1998. Reinitiation of DNA synthesis and cell division in senescent human fibroblasts by microinjection of anti-p53 antibodies. *Mol. Cell. Biol.* **18:** 1611–1621.

Gire V., Roux P., Wynford-Thomas D., Brondello J.M., and Dulic V. 2004. DNA damage checkpoint kinase Chk2 triggers replicative senescence. *EMBO J.* **23:** 2554–2563.

Gonos E.S., Derventzi A., Kveiborg M., Agiostratidou G., Kassem M., Clark B.F., Jat P.S., and Rattan S.I. 1998. Cloning and identification of genes that associate with mammalian replicative senescence. *Exp. Cell Res.* **240:** 66–74.

Gorbunova V., Seluanov A., and Pereira-Smith O.M. 2002. Expression of human telomerase (hTERT) does not prevent stress-induced senescence in normal human fibroblasts but protects the cells from stress-induced apoptosis and necrosis. *J. Biol. Chem.* **277:** 38540–38549.

Greider C.W. and Blackburn E.H. 1985. Identification of a specific telomere terminal transferase activity in *Tetrahymena* extracts. *Cell* **43:** 405–413.

Griffith J.D., Comeau L., Rosenfield S., Stansel R.M., Bianchi A., Moss H., and de Lange T. 1999. Mammalian telomeres end in a large duplex loop. *Cell* **97:** 503–514.

Guney I. and Sedivy J.M. 2006. Cellular senescence, epigenetic switches and c-Myc. *Cell Cycle* **5:** 2319–2323.

Guney I., Wu S., and Sedivy J.M. 2006. Reduced c-Myc signaling triggers telomere-independent senescence by regulating Bmi-1 and p16(INK4a). *Proc. Natl. Acad. Sci.* **103:** 3645–3650.

Guo W.J., Datta S., Band V., and Dimri G.P. 2007. Mel-18, a polycomb group protein, regulates cell proliferation and senescence via transcriptional repression of Bmi-1 and c-Myc oncoproteins. *Mol. Biol. Cell* **18:** 536–546.

Hampel B., Wagner M., Teis D., Zwerschke W., Huber L.A., and Jansen-Durr P. 2005. Apoptosis resistance of senescent human fibroblasts is correlated with the absence of nuclear IGFBP-3. *Aging Cell* **4:** 325–330.

Hanahan D. and Weinberg R.A. 2000. The hallmarks of cancer. *Cell* **100:** 57–70.

Harley C.B., Futcher A.B., and Greider C.W. 1990. Telomeres shorten during ageing of human fibroblasts. *Nature* **345:** 458–460.

Hayflick L. 1965. The limited in vitro lifetime of human diploid cell strains. *Exp. Cell Res.* **37:** 614–636.

———. 1985. Theories of biological aging. *Exp. Gerontol.* **20:** 145–159.

Hayflick L. and Moorhead P.S. 1961. The serial cultivation of human diploid cell strains. *Exp. Cell Res.* **25:** 585–621.

Hemann M.T. and Narita M. 2007. Oncogenes and senescence: Breaking down in the fast lane. *Genes Dev.* **21:** 1–5.

Hemann M.T., Strong M.A., Hao L.Y., and Greider C.W. 2001. The shortest telomere, not average telomere length, is critical for cell viability and chromosome stability. *Cell* **107:** 67–77.

Herbig U. and Sedivy J.M. 2006. Regulation of growth arrest in senescence: Telomere damage is not the end of the story. *Mech. Ageing Dev.* **127:** 16–24.

Herbig U., Ferreira M., Condel L., Carey D., and Sedivy J.M. 2006. Cellular senescence in aging primates. *Science* **311:** 1257.

Herbig U., Jobling W.A., Chen B.P., Chen D.J., and Sedivy J.M. 2004. Telomere shortening triggers senescence of human cells through a pathway involving ATM, p53, and p21(CIP1), but not p16(INK4a). *Mol. Cell* **14:** 501–513.

Herbig U., Wei W., Dutriaux A., Jobling W.A., and Sedivy J.M. 2003. Real-time imaging of transcriptional activation in live cells reveals rapid up-regulation of the cyclin-dependent kinase inhibitor gene *CDKN1A* in replicative cellular senescence. *Aging Cell* **2:** 295–304.

Holst C.R., Nuovo G.J., Esteller M., Chew K., Baylin S.B., Herman J.G., and Tlsty T.D. 2003. Methylation of p16(INK4a) promoters occurs in vivo in histologically normal human mammary epithelia. *Cancer Res.* **63:** 1596–1601.

Huschtscha L.I., Noble J.R., Neumann A.A., Moy E.L., Barry P., Melki J.R., Clark S.J., and Reddel R.R. 1998. Loss of p16INK4 expression by methylation is associated with lifespan extension of human mammary epithelial cells. *Cancer Res.* **58:** 3508–3512.

Itahana K., Zou Y., Itahana Y., Martinez J.L., Beausejour C., Jacobs J.J., van Lohuizen M., Band V., Campisi J., and Dimri G.P. 2003. Control of the replicative life span of human fibroblasts by p16 and the polycomb protein Bmi-1. *Mol. Cell. Biol.* **23:** 389–401.

Iwakuma T., Lozano G., and Flores E.R. 2005. Li-Fraumeni syndrome: A p53 family affair. *Cell Cycle* **4:** 865–867.

Jacobs J.J. and de Lange T. 2004. Significant role for p16INK4a in p53-independent telomere-directed senescence. *Curr. Biol.* **14:** 2302–2308.

Jacobs J.J., Kieboom K., Marino S., DePinho R.A., and van Lohuizen M. 1999. The oncogene and Polycomb-group gene *bmi-1* regulates cell proliferation and senescence through the *ink4a* locus. *Nature* **397:** 164–168.

Janzen V., Forkert R., Fleming H.E., Saito Y., Waring M.T., Dombkowski D.M., Cheng T., DePinho R.A., Sharpless N.E., and Scadden D.T. 2006. Stem-cell ageing modified by the cyclin-dependent kinase inhibitor p16INK4a. *Nature* **443:** 421–426.

Jeyapalan J.C., Ferreira M., Sedivy J.M., and Herbig U. 2007. Accumulation of senescent cells in mitotic tissue of aging primates. *Mech. Ageing Dev.* **128:** 36–44.

Kill I.R., Faragher R.G., Lawrence K., and Shall S. 1994. The expression of proliferation-dependent antigens during the lifespan of normal and progeroid human fibroblasts in culture. *J. Cell Sci.* **107:** 571–579.

Kim N.W., Piatyszek M.A., Prowse K.R., Harley C.B., West M.D., Ho P.L., Coviello G.M., Wright W.E., Weinrich S.L., and Shay J.W. 1994. Specific association of human telomerase activity with immortal cells and cancer. *Science* **266:** 2011–2015.

Kim W.Y. and Sharpless N.E. 2006. The regulation of INK4/ARF in cancer and aging. *Cell* **127:** 265–275.

Kirkwood T.B. and Austad S.N. 2000. Why do we age? *Nature* **408:** 233–238.

Konnikova L., Simeone M.C., Kruger M.M., Kotecki M., and Cochran B.H. 2005. Signal transducer and activator of transcription 3 (STAT3) regulates human telomerase reverse transcriptase (hTERT) expression in human cancer and primary cells. *Cancer Res.* **65:** 6516–6520.

Krishnamurthy J., Torrice C., Ramsey M.R., Kovalev G.I., Al-Regaiey K., Su L., and Sharpless N.E. 2004. Ink4a/Arf expression is a biomarker of aging. *J. Clin. Invest.* **114:** 1299–1307.

Krishnamurthy J., Ramsey M.R., Ligon K.L., Torrice C., Koh A., Bonner-Weir S., and Sharpless N.E. 2006. p16INK4a induces an age-dependent decline in islet regenerative potential. *Nature* **443:** 453–457.

Krtolica A. and Campisi J. 2002. Cancer and aging: A model for the cancer promoting effects of the aging stroma. *Int. J. Biochem. Cell Biol.* **34:** 1401–1414.

Krtolica A., Parrinello S., Lockett S., Desprez P., and Campisi J. 2001. Senescent fibroblasts promote epithelial cell growth and tumorigenesis: A link between cancer and aging. *Proc. Natl. Acad. Sci.* **98:** 12072–12077.

Lansdorp P.M. 2005. Major cutbacks at chromosome ends. *Trends Biochem Sci.* **30:** 388–395.

Lazzerini Denchi E., Attwooll C., Pasini D., and Helin K. 2005. Deregulated E2F activity induces hyperplasia and senescence-like features in the mouse pituitary gland. *Mol. Cell. Biol.* **25:** 2660–2672.

Lee A.C., Fenster B.E., Ito H., Takeda K., Bae N.S., Hirai T., Yu Z.X., Ferrans V.J., Howard B.H., and Finkel T. 1999. Ras proteins induce senescence by altering the intracellular levels of reactive oxygen species. *J. Biol. Chem.* **274:** 7936–7940.

Lee B.Y., Han J.A., Im J.S., Morrone A., Johung K., Goodwin E.C., Kleijer W.J., DiMaio D., and Hwang E.S. 2006. Senescence-associated β-galactosidase is lysosomal β-galactosidase. *Aging Cell* **5:** 187–195.

Lee S.B., Kim S.H., Bell D.W., Wahrer D.C., Schiripo T.A., Jorczak M.M., Sgroi D.C., Garber J.E., Li F.P., Nichols K.E., Varley J.M., Godwin A.K., Shannon K.M., Harlow E., and Haber D.A. 2001. Destabilization of CHK2 by a missense mutation associated with Li-Fraumeni syndrome. *Cancer Res.* **61:** 8062–8067.

Lin A.W., Barradas M., Stone J.C., van Aelst L., Serrano M., and Lowe S.W. 1998. Premature senescence involving p53 and p16 is activated in response to constitutive MEK/MAPK mitogenic signaling. *Genes Dev.* **12:** 3008–3019.

Maier B., Gluba W., Bernier B., Turner T., Mohammad K., Guise T., Sutherland A., Thorner M., and Scrable H. 2004. Modulation of mammalian life span by the short isoform of p53. *Genes Dev.* **18:** 306–319.

Mallette F.A., Gaumont-Leclerc M.F., and Ferbeyre G. 2007. The DNA damage signaling pathway is a critical mediator of oncogene-induced senescence. *Genes Dev.* **21:** 43–48.

Marcotte R., Lacelle C., and Wang E. 2004. Senescent fibroblasts resist apoptosis by down-regulating caspase-3. *Mech. Ageing Dev.* **125:** 777–783.

Martens J.W., Sieuwerts A.M., Vries J.B., Bosma P.T., Swiggers S.J., Klijn J.G., and Foekens J.A. 2003. Aging of stromal-derived human breast fibroblasts might contribute to breast cancer progression. *Thromb. Haemostasis* **89:** 393–404.

Martens U.M., Chavez E.A., Poon S.S., Schmoor C., and Lansdorp P.M. 2000. Accumulation of short telomeres in human fibroblasts prior to replicative senescence. *Exp. Cell Res.* **256:** 291–299.

Masutomi K., Possemato R., Wong J.M., Currier J.L., Tothova Z., Manola J.B., Ganesan S., Lansdorp P.M., Collins K., and Hahn W.C. 2005. The telomerase reverse transcriptase regulates chromatin state and DNA damage responses. *Proc. Natl. Acad. Sci.* **102:** 8222–8227.

Masutomi K., Yu E.Y., Khurts S., Ben-Porath I., Currier J.L., Metz G.B., Brooks M.W., Kaneko S., Murakami S., DeCaprio J.A., Weinberg R.A., Stewart S.A., and Hahn W.C. 2003. Telomerase maintains telomere structure in normal human cells. *Cell* **114:** 241–253.

Matheu A., Pantoja C., Efeyan A., Criado L.M., Martin-Caballero J., Flores J.M., Klatt P., and Serrano M. 2004. Increased gene dosage of Ink4a/Arf results in cancer resistance and normal aging. *Genes Dev.* **18:** 2736–2746.

Matthews C., Gorenne I., Scott S., Figg N., Kirkpatrick P., Ritchie A., Goddard M., and Bennett M. 2006. Vascular smooth muscle cells undergo telomere-based senescence in human atherosclerosis: Effects of telomerase and oxidative stress. *Circ. Res.* **99:** 156–164.

Mendrysa S.M., O'Leary K.A., McElwee M.K., Michalowski J., Eisenman R.N., Powell D.A., and Perry M.E. 2006. Tumor suppression and normal aging in mice with constitutively high p53 activity. *Genes Dev.* **20:** 16–21.

Michaloglou C., Vredeveld L.C., Soengas M.S., Denoyelle C., Kuilman T., van der Horst C.M., Majoor D.M., Shay J.W., Mooi W.J., and Peeper D.S. 2005. BRAFE600-associated senescence-like cell cycle arrest of human naevi. *Nature* **436:** 720–724.

Molofsky A.V., Slutsky S.G., Joseph N.M., He S., Pardal R., Krishnamurthy J., Sharpless N.E., and Morrison S.J. 2006. Increasing p16INK4a expression decreases forebrain progenitors and neurogenesis during ageing. *Nature* **443:** 448–452.

Morales C.P., Holt S.E., Ouellette M., Kaur K.J., Yan Y., Wilson K.S., White M.A., Wright W.E., and Shay J.W. 1999. Absence of cancer-associated changes in human fibroblasts immortalized with telomerase. *Nat. Genet.* **21:** 115–118.

Munro J., Barr N.I., Ireland H., Morrison V., and Parkinson E.K. 2004. Histone deacetylase inhibitors induce a senescence-like state in human cells by a p16-dependent mechanism that is independent of a mitotic clock. *Exp. Cell Res.* **295:** 525–538.

Narita M. and Lowe S.W. 2004. Executing cell senescence. *Cell Cycle* **3:** 244–246.

Narita M., Nunez S., Heard E., Narita M., Lin A.W., Hearn S.A., Spector D.L., Hannon G.J., and Lowe S.W. 2003. Rb-mediated heterochromatin formation and silencing of E2F target genes during cellular senescence. *Cell* **113:** 703–716.

Neumeister P., Albanese C., Balent B., Greally J., and Pestell R.G. 2002. Senescence and epigenetic dysregulation in cancer. *Int. J. Biochem. Cell Biol.* **34:** 1475–1490.

Nishio K. and Inoue A. 2005. Senescence-associated alterations of cytoskeleton: Extraordinary production of vimentin that anchors cytoplasmic p53 in senescent human fibroblasts. *Histochem. Cell Biol.* **123:** 263–273.

Noda A., Ning Y., Venable S.F., Pereira-Smith O.M., and Smith J.R. 1994. Cloning of senescent cell-derived inhibitors of DNA synthesis using an expression screen. *Exp. Cell Res.* **211:** 90–98.

Ohtani N., Yamakoshi K., Takahashi A., and Hara E. 2004. The p16INK4a-RB pathway: Molecular link between cellular senescence and tumor suppression. *J. Med. Invest.* **51:** 146–153.

Ohtani N., Zebedee Z., Huot T.J., Stinson J.A., Sugimoto M., Ohashi Y., Sharrocks A.D., Peters G., and Hara E. 2001. Opposing effects of Ets and Id proteins on p16INK4a expression during cellular senescence. *Nature* **409:** 1067–1070.

Oikawa S. and Kawanishi S. 1999. Site-specific DNA damage at GGG sequence by oxidative stress may accelerate telomere shortening. *FEBS Lett.* **453:** 365–368.

Olovnikov A.M. 1973. A theory of marginotomy. The incomplete copying of template margin in enzymic synthesis of polynucleotides and biological significance of the phenomenon. *J. Theor. Biol.* **41:** 181–190.

Parrinello S., Coppe J.P., Krtolica A., and Campisi J. 2005. Stromal-epithelial interactions in aging and cancer: Senescent fibroblasts alter epithelial cell differentiation. *J. Cell Sci.* **118:** 485–496.

Parrinello S., Samper E., Krtolica A., Goldstein J., Melov S., and Campisi J. 2003. Oxygen sensitivity severely limits the replicative lifespan of murine fibroblasts. *Nat. Cell Biol.* **5:** 741–747.

Pascal T., Debacq-Chainiaux F., Chretien A., Bastin C., Dabee A.F., Bertholet V., Remacle J., and Toussaint O. 2005. Comparison of replicative senescence and stress-induced premature senescence combining differential display and low-density DNA arrays. *FEBS Lett.* **579:** 3651–3659.

Pignolo R.J., Martin B.G., Horton J.H., Kalbach A.N., and Cristofalo V.J. 1998. The pathway of cell senescence: WI-38 cells arrest in late G1 and are unable to traverse the cell cycle from a true G0 state. *Exp. Gerontol.* **33:** 67–80.

Price J.S., Waters J.G., Darrah C., Pennington C., Edwards D.R., Donell S.T., and Clark I.M. 2002. The role of chondrocyte senescence in osteoarthritis. *Aging Cell* **1:** 57–65.

Ramirez R.D., Morales C.P., Herbert B.S., Rohde J.M., Passons C., Shay J.W., and Wright W.E. 2001. Putative telomere-independent mechanisms of replicative aging reflect inadequate growth conditions. *Genes Dev.* **15:** 398–403.

Reddel R.R., Bryan T.M., Colgin L.M., Perrem K.T., and Yeager T.R. 2001. Alternative lengthening of telomeres in human cells. *Radiat. Res.* **155:** 194–200.

Ressler S., Bartkova J., Niederegger H., Bartek J., Scharffetter-Kochanek K., Jansen-Durr P., and Wlaschek M. 2006. p16INK4A is a robust in vivo biomarker of cellular aging in human skin. *Aging Cell* **5:** 379–389.

Rheinwald J.G., Hahn W.C., Ramsey M.R., Wu J.Y., Guo Z., Tsao H., De Luca M., Catricala C., and O'Toole K.M. 2002. A two-stage, p16(INK4A)- and p53-dependent keratinocyte senescence mechanism that limits replicative potential independent of telomere status. *Mol. Cell. Biol.* **22:** 5157–5172.

Roberson R.S., Kussick S.J., Vallieres E., Chen S.Y., and Wu D.Y. 2005. Escape from therapy-induced accelerated cellular senescence in p53-null lung cancer cells and in human lung cancers. *Cancer Res.* **65:** 2795–2803.

Roninson I.B. 2003. Tumor cell senescence in cancer treatment. *Cancer Res.* **63:** 2705–2715.

Rufer N., Dragowska W., Thornbury G., Roosnek E., and Lansdorp P.M. 1998. Telomere length dynamics in human lymphocyte subpopulations measured by flow cytometry. *Nat. Biotechnol.* **16:** 743–747.

Sager R. 1991. Senescence as a mode of tumor suppression. *Environ. Health Perspect.* **93:** 59–62.

Schmitt C.A., Fridman J.S., Yang M., Lee S., Baranov E., Hoffman R.M., and Lowe S.W. 2002. A senescence program controlled by p53 and p16INK4a contributes to the outcome of cancer therapy. *Cell* **109:** 335–346.

Schwarze S.R., DePrimo S.E., Grabert L.M., Fu V.X., Brooks J.D., and Jarrard D.F. 2002. Novel pathways associated with by passing cellular senescence in human prostate epithelial cells. *J. Biol. Chem.* **277:** 14877–14883.

Sedelnikova O.A., Horikawa I., Zimonjic D.B., Popescu N.C., Bonner W.M., and Barrett J.C. 2004. Senescing human cells and ageing mice accumulate DNA lesions with unrepairable double-strand breaks. *Nat. Cell Biol.* **6:** 168–170.

Sedivy J.M. 1998. Can ends justify the means?: Telomeres and the mechanisms of replicative senescence and immortalization in mammalian cells. *Proc. Natl. Acad. Sci.* **95:** 9078–9081.

Seluanov A., Chen Z., Hine C., Sasahara T.H., Ribeiro A.A., Catania K.C., Presgraves D.C., and Gorbunova V. 2007. Telomerase activity coevolves with body mass not lifespan. *Aging Cell* **6:** 45–52.

Semov A., Marcotte R., Semova N., Ye X., and Wang E. 2002. Microarray analysis of E-box binding-related gene expression in young and replicatively senescent human fibroblasts. *Anal. Biochem.* **302:** 38–51.

Serrano M., Lin A.W., McCurrach M.E., Beach D., and Lowe S.W. 1997. Oncogenic ras provokes premature cell senescence associated with accumulation of p53 and p16INK4a. *Cell* **88:** 593–602.

Severino J., Allen R.G., Balin S., Balin A., and Cristofalo V.J. 2000. Is beta-galactosidase staining a marker of senescence in vitro and in vivo? *Exp. Cell Res.* **257:** 162–171.

Sfeir A.J., Chai W., Shay J.W., and Wright W.E. 2005. Telomere end processing: The terminal nucleotides of human chromosomes. *Mol. Cell* **18:** 131–138.

Shay J.W. and Wright W.E. 1991. Defining the molecular mechanisms of human cell immortalization. *Biochim. Biophys. Acta* **1071:** 1–7.

———. 2005. Senescence and immortalization: Role of telomeres and telomerase. *Carcinogenesis* **26:** 867–874.

Shay J.W., Tomlinson G., Piatyszek M.A., and Gollahon L.S. 1995. Spontaneous in vitro immortalization of breast epithelial cells from a patient with Li-Fraumeni syndrome. *Mol. Cell Biol.* **15:** 425–432.

Shay J.W., Van Der Haegen B.A., Ying Y., and Wright W.E. 1993. The frequency of immortalization of human fibroblasts and mammary epithelial cells transfected with SV40 large T-antigen. *Exp. Cell Res.* **209:** 45–52.

Shelton D.N., Chang E., Whittier P.S., Choi D., and Funk W.D. 1999. Microarray analysis of replicative senescence. *Curr. Biol.* **9:** 939–945.

Sherr C.J. and DePinho R.A. 2000. Cellular senescence: Mitotic clock or culture shock? *Cell* **102:** 407–410.

Sherr C.J. and Roberts J.M. 1999. CDK inhibitors: Positive and negative regulators of G1-phase progression. *Genes Dev.* **13:** 1501–1512.

Smith J.R. and Whitney R.G. 1980. Intraclonal variation in proliferative potential of human diploid fibroblasts: Stochastic mechanism for cellular aging. *Science* **207:** 82–84.

Smogorzewska A. and de Lange T. 2002. Different telomere damage signaling pathways in human and mouse cells. *EMBO J.* **21:** 4338–4348.

Stein G.H., Drullinger L.F., Soulard A., and Dulic V. 1999. Differential roles for cyclin-dependent kinase inhibitors p21 and p16 in the mechanisms of senescence and differentiation in human fibroblasts. *Mol. Cell. Biol.* **19:** 2109–2117.

Steinert S., Shay J.W., and Wright W.E. 2000. Transient expression of human telomerase extends the life span of normal human fibroblasts. *Biochem. Biophys. Res. Commun.* **273:** 1095–1098.

Szekely A.M., Bleichert F., Numann A., Van Komen S., Manasanch E., Ben Nasr A., Canaan A., and Weissman S.M. 2005. Werner protein protects nonproliferating cells from oxidative DNA damage. *Mol. Cell. Biol.* **25:** 10492–10506.

Takahashi A., Ohtani N., Yamakoshi K., Iida S., Tahara H., Nakayama K., Nakayama K.I., Ide T., Saya H., and Hara E. 2006. Mitogenic signalling and the p16INK4a-Rb pathway cooperate to enforce irreversible cellular senescence. *Nat. Cell Biol.* **8:** 1291–1297.

Takai H., Smogorzewska A., and de Lange T. 2003. DNA damage foci at dysfunctional telomeres. *Curr. Biol.* **13:** 1549–1556.

te Poele R.H., Okorokov A.L., Jardine L., Cummings J., and Joel S.P. 2002. DNA damage is able to induce senescence in tumor cells in vitro and in vivo. *Cancer Res.* **62:** 1876–1883.

Toussaint O., Dumont P., Dierick J.F., Pascal T., Frippiat C., Chainiaux F., Sluse F., Eliaers F., and Remacle J. 2000. Stress-induced premature senescence. Essence of life, evolution, stress, and aging. *Ann. N.Y. Acad. Sci.* **908:** 85–98.

Trougakos I.P., Saridaki A., Panayotou G., and Gonos E.S. 2006. Identification of differentially expressed proteins in senescent human embryonic fibroblasts. *Mech. Ageing Dev.* **127:** 88–92.

Tyner S.D., Venkatachalam S., Choi J., Jones S., Ghebranious N., Ingelmann H., Lu X., Soron G., Cooper B., Brayton C., Park S.H., Thompson T., Karsenty G., Bradley A., and Donehower L.A. 2002. p53 mutant mice that display early aging-associated phenotypes. *Nature* **415:** 45–53.

Vasile E., Tomita Y., Brown L.F., Kocher O., and Dvorak H.F. 2001. Differential expression of thymosin beta-10 by early passage and senescent vascular endothelium is modulated by VPF/VEGF: Evidence for senescent endothelial cells in vivo at sites of atherosclerosis. *FASEB J.* **15:** 458–466.

Vaziri H., West M.D., Allsopp R.C., Davison T.S., Wu Y.S., Arrowsmith C.H., Poirier G.G., and Benchimol S. 1997. ATM-dependent telomere loss in aging human diploid fibroblasts and DNA damage lead to the post-translational activation of p53 protein involving poly(ADP-ribose) polymerase. *EMBO J.* **16:** 6018–6033.

Ventura A., Kirsch D.G., McLaughlin M.E., Tuveson D.A., Grimm J., Lintault L., Newman J., Reczek E.E., Weissleder R., and Jacks T. 2007. Restoration of p53 function leads to tumour regression in vivo. *Nature* **445:** 661–665.

Verdun R.E. and Karlseder J. 2006. The DNA damage machinery and homologous recombination pathway act consecutively to protect human telomeres. *Cell* **127:** 709–720.

Vidal A. and Koff A. 2000. Cell-cycle inhibitors: Three families united by a common cause. *Gene* **247:** 1–15.

von Zglinicki T. 2001. Telomeres and replicative senescence: Is it only length that counts? *Cancer Lett.* **168:** 111–116.

———. 2002. Oxidative stress shortens telomeres. *Trends Biochem. Sci.* **27:** 339–344.

von Zglinicki T., Saretzki G., Ladhoff J., d'Adda di Fagagna F., and Jackson S.P. 2005. Human cell senescence as a DNA damage response. *Mech. Ageing Dev.* **126:** 111–117.

Voorhoeve P.M. and Agami R. 2003. The tumor-suppressive functions of the human INK4A locus. *Cancer Cell* **4:** 311–319.

Wei W. and Sedivy J.M. 1999. Differentiation between senescence (M1) and crisis (M2) in human fibroblast cultures. *Exp. Cell Res.* **253:** 519–522.

Wei W., Hemmer R.M., and Sedivy J.M. 2001. Role of p14(ARF) in replicative and induced senescence of human fibroblasts. *Mol. Cell. Biol.* **21:** 6748–6757.

Wei W., Jobling W.A., Chen W., Hahn W.C., and Sedivy J.M. 2003. Abolition of cyclin-dependent kinase inhibitor p16Ink4a and p21Cip1/Waf1 functions permits Ras-induced anchorage-independent growth in telomerase-immortalized human fibroblasts. *Mol. Cell. Biol.* **23:** 2859–2870.

Williams G.C. 1957. Pleiotropy, natural selection, and the evolution of senescence. *Evolution* **11:** 398–411.

Woo R.A. and Poon R.Y. 2004. Activated oncogenes promote and cooperate with chromosomal instability for neoplastic transformation. *Genes Dev.* **18:** 1317–1330.

Wright W.E. and Shay J.W. 2002. Historical claims and current interpretations of replicative aging. *Nat. Biotechnol.* **20:** 682–688.

Wu L., Multani A.S., He H., Cosme-Blanco W., Deng Y., Deng J.M., Bachilo O., Pathak S., Tahara H., Bailey S.M., Deng Y., Behringer R.R., and Chang S. 2006. Pot1 deficiency promotes recombination at telomeres, chromosomal instability, and malignant transformation. *Cell* **126:** 49–62.

Xie L., Tsaprailis G., and Chen Q.M. 2005. Proteomic identification of insulin-like growth factor-binding protein-6 induced by sublethal H_2O_2 stress from human diploid fibroblasts. *Mol. Cell. Proteomics* **4:** 1273–1283.

Xue W., Zender L., Miething C., Dickins R.A., Hernando E., Krizhanovsky V., Cordon-Cardo C., and Lowe S.W. 2007. Senescence and tumour clearance is triggered by p53 restoration in murine liver carcinomas. *Nature* **445:** 656–650.

Yang N.C. and Hu M.L. 2005. The limitations and validities of senescence associated-beta-galactosidase activity as an aging marker for human foreskin fibroblast Hs68 cells. *Exp. Gerontol.* **40:** 813–819.

Ye X., Zerlanko B., Zhang R., Somaiah N., Lipinski M., Salomoni P., and Adams P.D. 2007. Definition of pRB- and p53-dependent and independent steps in HIRA/ASF1a-mediated formation of senescence-associated heterochromatin foci (SAHF). *Mol. Cell. Biol.* **27:** 2452–2465.

Yegorov Y.E., Akimov S.S., Hass R., Zelenin A.V., and Prudovsky I.A. 1998. Endogenous beta-galactosidase activity in continuously nonproliferating cells. *Exp. Cell Res.* **243:** 207–211.

Yoon I.K., Kim H.K., Kim Y.K., Song I.H., Kim W., Kim S., Baek S.H., Kim J.H., and Kim J.R. 2004. Exploration of replicative senescence-associated genes in human dermal fibroblasts by cDNA microarray technology. *Exp. Gerontol.* **39:** 1369–1378.

Zdanov S., Debacq-Chainiaux F., Remacle J., and Toussaint O. 2006. Identification of p38MAPK-dependent genes with changed transcript abundance in H_2O_2-induced premature senescence of IMR-90 hTERT human fibroblasts. *FEBS Lett.* **580:** 6455–6463.

Zhang H., Pan K.H., and Cohen S.N. 2003. Senescence-specific gene expression fingerprints reveal cell-type-dependent physical clustering of up-regulated chromosomal loci. *Proc. Natl. Acad. Sci.* **100:** 3251–3256.

Zhang J., Pickering C.R., Holst C.R., Gauthier M.L., and Tlsty T.D. 2006. p16INK4a modulates p53 in primary human mammary epithelial cells. *Cancer Res.* **66:** 10325–10331.

Zhang R., Chen W., and Adams P.D. 2007. Molecular dissection of formation of senescent associated heterochromatin foci. *Mol. Cell. Biol.* **27:** 2343–2358.

Zhang R., Poustovoitov M.V., Ye X., Santos H.A., Chen W., Daganzo S.M., Erzberger J.P., Serebriiskii I.G., Canutescu A.A., Dunbrack R.L., Pehrson J.R., Berger J.M., Kaufman P.D., and Adams P.D. 2005. Formation of MacroH2A-containing senescence-associated heterochromatin foci and senescence driven by ASF1a and HIRA. *Dev. Cell* **8:** 19–30.

Zhang Y., Xiong Y., and Yarbrough W.G. 1998. ARF promotes MDM2 degradation and stabilizes p53: ARF-INK4a locus deletion impairs both the Rb and p53 tumor suppressor pathways. *Cell* **92:** 725–734.

Zhu J., Woods D., McMahon M., and Bishop J.M. 1998. Senescence of human fibroblasts induced by oncogenic Raf. *Genes Dev.* **12:** 2997–3007.

Zindy F., Quelle D.E., Roussel M.F., and Sherr C.J. 1997. Expression of the p16INK4a tumor suppressor versus other INK4 family members during mouse development and aging. *Oncogene* **15:** 203–211.

9

Genome-wide Views of Aging Gene Networks

Stuart K. Kim

Departments of Developmental Biology and Genetics
Stanford University Medical Center
Stanford, California 94305-5329

AGING IS A COMPLEX PROCESS INVOLVING THE ADDITIVE EFFECTS of many genetic pathways (Kirkwood and Austad 2000). To embrace the complexity of aging, an attractive approach is to use DNA microarrays to scan the entire genome for genes that change expression as a function of age or under conditions when longevity is extended. The list of age-regulated genes provides clues about genetic pathways and mechanisms that underlie the aging process. In addition to single-gene analysis, the combined transcriptional profile of aging can act as a molecular phenotype of old age. During the last 20 years, there has been a great deal of effort to search for biomarkers of aging, and recent studies have shown that expression profiles of aging derived from DNA microarray experiments may provide this long-desired goal.

A gene expression signature for aging is a quantitative phenotype that gives a high-resolution view of the aging process, much like using transcriptional profiles of cancer to inform about their severity or malignancy. Previously, one could recognize old versus young individuals in a photograph, or old versus young tissue on a microscope slide. Now it is possible to recognize old versus young genetic networks by analyzing expression levels of the entire set of age-regulated genes (Fig. 1). Unlike photographs or micrographs, expression data from DNA microarrays are quantitative, and thus it is possible to compare age-related transcriptional profiles between different tissues, between different conditions that affect

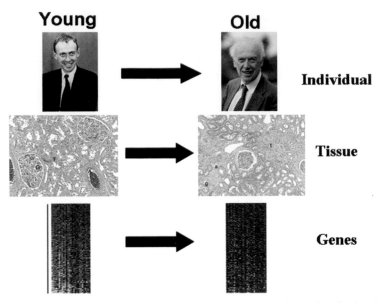

Figure 1. Expression profiles are markers for aging at the molecular level.

longevity, and even between diverse species. Such comparisons are not possible by browsing images of young and old tissues.

Using high-density gene arrays to measure age-related changes in gene expression has been a robust area of research since the development of gene arrays in the late 1990s. Table 1 lists 75 publications that use gene arrays to study aging involving diverse organisms (yeast, worms, flies, mice, rats, monkeys, and humans) and many tissues (including brain, muscle, kidney, liver, blood, eye, prostate, and heart). Each paper presents a set of age-regulated genes that identify biological pathways involved in aging, such as the stress response, the extracellular matrix genes, and the electron transport chain.

In most cases, analysis is constrained to data from only one publication. This is partly because it is difficult to compare DNA microarray data from different publications, as experimental results can be heavily influenced by subtle differences in experimental design and experimental platform. An important informatics challenge for the future will be to develop robust analytical tools that permit the integration of results from diverse experiments. Combining DNA microarray results on aging from multiple labs, different tissues, and different species will reveal which results are reproducible or shared across tissues and which are public or private across species.

Table 1. Listing of DNA microarray studies of aging

Organism	Tissue	References
Yeast		Lau et al. (2003)
C. elegans		Lund et al. (2002); McElwee et al. (2003, 2004); Murphy et al. (2003); McCarroll et al. (2004); Viswanathan et al. (2005); Girardot et al. (2006); Lener et al. (2006)
	individual worms	Golden and Melov (2004)
D. melanogaster		Jin et al. (2001); Pletcher et al. (2002); Seroude et al. (2002); Landis et al. (2004); Kim et al. (2005); Girardot et al. (2006)
Mouse	single heart cells	Bahar et al. (2006)
	liver	Cao et al. (2001); Boylston et al. (2004); Dhahbi et al. (2004, 2005)
	heart	Lee et al. (2002, 2004); Edwards et al. (2003, 2004)
	T cells	Chen et al. (2003); Mo et al. (2003); Han et al. (2006)
	muscle	Lee et al. (1999); Beggs et al. (2004)
	brain	Lee et al. (2000); Prolla (2002); Verbitsky et al. (2004); Cheng et al. (2006)
	macrophages	Chelvarajan et al. (2006)
	olfactory mucosae	Getchell et al. (2004)
	oocytes	Hamatani et al. (2004)
	submandibular gland	Hiratsuka et al. (2002)
Rat	marrow-derived stromal cells	Akavia et al. (2006)
	prostate	Lau et al. (2003); Beggs et al. (2004); Reyes et al. (2005)
	bone	Meyer et al. (2004)
	heart	Dobson et al. (2003)
	muscle	Pattison et al. (2003)
	liver	Thomas et al. (2002)
	brain	Blalock et al. (2003)
	colon	Lee et al. (2001)

(*continued*)

Table 1. (*continued*)

Organism	Tissue	References
	urinary bladder	Lluel et al. (2003)
	kidney	Preisser et al. (2004)
	pituitary	Preisser et al. (2004)
Rhesus monkey	muscle	Kayo et al. (2001)
Human	Hutchison-Guilford progeria syndrome	Cao et al. (2001)
	Werner's syndrome	Ly et al. (2000); Kyng et al. (2003)
	brain	Blalock et al. (2004); Lu et al. (2004); Ricciarelli et al. (2004); Fraser et al. (2005); Liang et al. (2006)
	fibroblasts	Kim et al. (2003)
	kidney	Rodwell et al. (2004); Melk et al. (2005)
	muscle	Welle et al. (2001, 2003, 2004); Giresi et al. (2005); Zahn et al. (2006)
	liver	Thomas et al. (2002)
	eye	Yoshida et al. (2002); Hawse et al. (2004); Segev et al. (2005)
	skin	Lener et al. (2006)
	blood	Visala Rao et al. (2003); Tan et al. (2005)

AGING IN *CAENORHABDITIS ELEGANS*

The nematode *C. elegans* is one of the most attractive and useful organisms for studying aging. *C. elegans* normally has a relatively short life span of 2 weeks, enabling one to readily assess the effects of different mutations or treatments on life span. Elegant genetic screens have identified a large number of mutations that can either extend or shorten life span, including mutations in genes that act in the insulin-like signaling pathway such as *daf-2, age-1,* and *daf-16* (Guarente and Kenyon 2000). Although many genes and genetic pathways have been found that alter life span, relatively little is known about the differences between young and old worms at either the cellular or molecular levels. Nomarski and electron microscopy of old and young worms have been used to examine the change in appearance of organs and cells as a function of age (Garigan et al. 2002; Herndon et al. 2002; Gerstbrein et al. 2005). These studies showed that old worms exhibited degeneration of muscle and intestinal cells but that neural tissue remained relatively unaffected.

To begin to define the aging process at the molecular level, several studies have profiled changes in gene expression in young versus old animals (Lund et al. 2002; McCarroll et al. 2004). Lund et al. (2002) used DNA microarrays to profile expression changes at six time points during aging from young adults to adults near their maximum life span. This study identified 164 age-regulated genes, providing a global view of the molecular changes between young and old animals. Because *C. elegans* has a short life span, and because there are powerful genetic methods to determine gene function, the physiological role in aging for each of these 164 genes can now be dissected. Furthermore, the age-regulated genes can now be tagged with green fluorescent protein (GFP) and used as reporters to indicate the relative age of individual animals while they are still alive.

In addition to the aging process itself, it is interesting to investigate the transcriptional profile associated with dauers (a developmental state induced by starvation that allows worms to live much longer than normal worms) and mutants that affect *C. elegans* life span (such as *daf-2, age-1,* or *daf-16*). A global analysis of gene expression changes in dauers or longevity mutants should provide insight about gene expression programs that promote longevity.

The dauer state is induced by poor growth conditions and allows worms to survive harsh environmental conditions and to live up to 10-fold longer than normal (Wood and Johnson 1994). DNA microarray experiments and serial analysis of gene expression (SAGE) have been used to measure changes in expression in normal worms versus dauers (Jones et al. 2001; Wang and Kim 2003). Both studies revealed an enormous set of genes that are differentially expressed in dauers versus normal worms. Interestingly, the gene expression profile associated with old age shows a strong overlap with the profile associated with dauers (Lund et al. 2002). There is an overall tendency for genes that increase expression in old age to also increase expression in dauers, and vice versa. The common aging transcriptional profile shared between old animals and dauers may represent a common adaptive mechanism(s) specifying increased survival.

To learn how the insulin pathway extends life, several groups have used DNA microarrays to identify transcriptional changes caused by mutations in genes that act in the insulin-like signaling pathway (*daf-2, age-1,* or *daf-16*) (McElwee et al. 2003, 2004). Murphy et al. (2003) identified 70 genes (class 1) that show increased expression in *daf-2* or *age-1* mutants but decreased expression in *daf-16* mutants, indicating that these genes are normally repressed by DAF-2 signaling. They also identified 100 genes (class 2) that show the converse behavior—decreased expression in *daf-2* or *age-1* mutants and increased expression in *daf-16* mutants, indicating that these genes are normally activated by DAF-2 signaling. One of the class-2 genes (*ins-7*) encodes an insulin-like peptide, suggesting that an auto-activation mechanism may control the insulin-like signaling pathway. Slight activation of the DAF-2 receptor would lead to increased expression of the downstream INS-7 peptide, which would lead to further activation of the DAF-2 receptor. RNA interference (RNAi) experiments showed that reduction of the activity of genes that act downstream from DAF-16 (either repressed or activated targets) often affects worm life span. Thus, although the *daf-2* pathway regulates expression of an entire network of genes, individual downstream targets still have important roles in longevity, as reducing the activity of just one target often affects life span.

Comparison of the results from the aging, dauer, and insulin-pathway mutants showed that there was a great deal of overlap between the targets of these pathways (Lund et al. 2002; McElwee et al. 2003, 2004; S.K. Kim, unpubl.). Specifically, genes that increase expression with age tend to increase expression in dauers and in either *daf-2* or *age-1* mutants. This result shows that the mechanisms responsible for long life in insulin-like signaling mutants may be similar to those responsible for long life in dauers. This may not be surprising, as the longevity phenotype is only caused by weak loss-of-function alleles of *daf-2* or *age-1,* whereas strong loss-of-function mutations in these genes cause a dauer constitutive phenotype. Thus, insulin-like signaling mutants may extend longevity by inducing a dauer-like metabolic state in the adult.

AGING IN *DROSOPHILA MELANOGASTER*

Several groups have used DNA microarrays to identify genes that change during aging in *Drosophila* (Jin et al. 2001; Pletcher et al. 2002; Landis et al. 2004; Girardot et al. 2006). Pletcher et al. (2002) used Affymetrix GeneChips to measure gene expression changes in *Drosophila* as a function of aging and caloric restriction. They found 885 probe sets that showed age regulation under normal growth conditions. Next, they examined the effects of caloric restriction on the transcriptional profile for aging, as caloric restriction is known to slow the rate of normal aging in *Drosophila* (Chapman and Partridge 1996). They found that the age-dependent trajectory of gene expression was slowed by caloric restriction in the vast majority of cases. Since the rate of change in expression scales with life span, this result indicates that most of the 885 probe sets report physiological age in addition to chronological age.

Another DNA microarray study provided evidence for the role of oxidative damage during fly aging. If oxidative damage has a major role in aging, then accumulation of oxidatively damaged proteins should induce a stress response in old flies. To test this, Landis et al. (2004) compared the transcriptional profiles of aging with expression profiles following oxygen stress (flies grown in the presence of 100% oxygen). Both aging and oxidative stress caused an increase in expression of purine biosynthesis, heat shock, antioxidant, and innate immune response genes. Thus, old flies showed evidence of oxidative stress supporting the oxidative damage theory of aging. Taken together, these two DNA microarray experiments show that the transcriptional profile of aging can be used as a biomarker for fly aging. The transcriptional profile of aging is accelerated by a treatment that accelerates aging (i.e., oxygen stress) and retarded by a treatment that slows the rate of aging (i.e., caloric restriction).

AGING IN MICE

In mice, DNA microarray experiments have been used to profile age-related changes in gene expression for many tissues (Table 1), including brain (Lee et al. 2000; Prolla 2002; Verbitsky et al. 2004; Cheng et al. 2006), liver (Cao et al. 2001; Boylston et al. 2004; Dhahbi et al. 2004, 2005), and heart (Lee et al. 2002, 2004; Edwards et al. 2003, 2004). For each of these tissues, the transcriptional aging profiles of normally fed mice and calorically restricted mice were compared. In each case, caloric restriction attenuated changes in gene expression of the age-regulated genes. This result indicates that the genes depict physiological aging in addition to chronological aging, because the rate of change in expression scales with life span, and that the entire group of aging genes can be used as a molecular phenotype for aging.

Increased Transcriptional Noise in Aging Mice

Bahar et al. (2006) have recently provided evidence for a global change in the robustness of the transcriptional network as a function of age. As individuals age, there is an increase in levels of oxidative damage to transcription factors and chromatin proteins and there is an accumulation of damage to somatic DNA. These events could impair the ability of gene regulatory networks to function efficiently and thus lead to increased transcriptional variability of downstream target genes in individual cells (transcriptional noise) in old age.

Bahar et al. (2006) used reverse transcriptase–polymerase chain reaction (RT-PCR) to measure expression of a set of randomly chosen genes in single cardiomyocytes in young and old mice. They found that in young mice, cardiomyocytes typically express genes at similar levels, but that in old mice, individual cells have varying levels of gene expression. Small stochastic changes in expression, up or down, in a particular cell could lead to large changes in cell physiology when amplified across a large number of genes. Aging caused by a global increase in transcriptional noise is a new and exciting potential mechanism for cellular senescence.

These authors suggest that increased transcriptional noise found in old mice might be caused by stochastic accumulation of somatic DNA damage. Consistent with this hypothesis, they show that oxidative damage to cells in tissue culture can also lead to increased transcriptional noise. Such stochastic somatic mutations would affect expression of different genes in different cells within an individual as it ages. If so, the variance in gene expression between different cells would increase with age. Under this hypothesis, expression in young animals would be relatively steady in different cells, as young promoters would have low levels of DNA damage. Expression in old animals would show higher levels of variance between cells due to stochastic differences in promoter damage.

In addition to DNA damage, several other mechanisms could also explain the increase in transcriptional noise as a function of age. For example, oxidative damage to proteins such as transcription factors or chromatin components could decrease their ability to properly regulate gene expression in old age. The severity of damage to the transcriptional apparatus may vary between individual cells in old animals, resulting in differing levels of expression between cells and, thus, increased transcriptional noise.

AGING IN HUMANS

Many DNA microarray experiments have analyzed aging in various human tissues. Table 1 contains 24 references that profile gene expression changes in eight human tissues and two human progerias.

Transcriptional Profiles as a Biomarker for Human Aging

Individuals age at different rates, such that some 80 year olds remain relatively fit, whereas others become frail (Fig. 2). A biomarker of aging must be able to depict physiological age (the relative health of an individual)

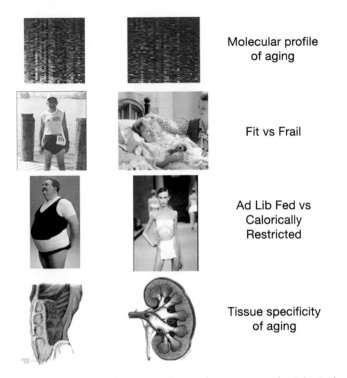

Molecular profile
of aging

Fit vs Frail

Ad Lib Fed vs
Calorically
Restricted

Tissue specificity
of aging

Figure 2. Expression profiles of aging can be used to measure physiological age, compare caloric restriction to normal eating, and investigate common aging patterns in different tissues.

rather than simply mark chronological age (the birth date of an individual). In their analysis of kidney aging, Rodwell et al. (2004) showed that age-regulated markers could predict physiological age. They used Affymetrix GeneChips to measure levels of gene expression in kidney samples from 72 patients between 27 and 92 years of age, and they identified a set of 447 age-regulated genes. Next, they tested whether the overall level of expression of the 447 genes was a signature of physiological age for different individuals. The relative amount of physiological aging for each of the patients used in the study was measured by examining the morphological and histological appearance of the kidney. A remarkable similarity was found between the transcriptional aging profiles and the physiological states of the kidney in different individuals. Some elderly persons displayed an aging signature similar to that of people who were 30–40 years younger, and these individuals also had relatively healthy kidneys. Others displayed an aging signature similar to people who were 10–20 years older, and these patients showed poor kidney health.

A similar experiment was performed by Zahn et al. to analyze aging of human muscle (Rodwell et al. 2004; Zahn et al. 2006). They identified a set of 250 age-regulated genes in skeletal muscle from 81 patients ranging from 16 to 89 years of age. As muscle ages, there is a shift in the relative numbers of type II (slow twitch) to type I (fast twitch) muscle fibers, and the type II/type I muscle fiber ratio can be used as a rough estimate of the physiological age of muscle. Similar to the aging expression profile for the kidney, the overall expression behavior of this set of age-regulated muscle genes correlated with the physiological as well as chronological age of the muscle sample (Zahn et al. 2006). In summary, these results indicate that the overall expression behavior of the age-regulated kidney and muscle genes is a biomarker that depicts the physiological age of human kidney or muscle.

DNA Damage and Human Aging

A long-held view is that accumulation of somatic mutations has a key role in decreasing cellular function in old age. Recent genomics studies by Lu et al. (2004) have provided intriguing new evidence for the role of DNA damage in aging of the human brain. In their study, they used postmortem samples of the frontal pole of 30 individuals ranging in age from 26 to 106 and found about 440 genes that change expression with age. They then measured the rate of aging by examining changes in gene expression at different points during life. Interestingly, they found that most of the age-related changes occurred in people between 43 and 73 years old. Gene expression patterns in the brain were more consistent in the young and the old, but they showed large variability in people who were middle-aged.

Next, Lu et al. asked whether DNA damage may have a role in age-related changes in gene expression. They devised a PCR-based assay to measure levels of damaged DNA in brain samples from old and young patients. They isolated genomic DNA from brain samples and then treated it with formamidopyrimidine-DNA glycosylase (FPG), which is an N-glycosidase and AP-lyase that selectively releases damaged bases from DNA. Damaged DNA would be selectively cleaved by FPG and would show lower levels of amplification in a PCR assay. They examined the promoters of 30 different genes, and they observed that DNA damage targeted the promoters of genes that are repressed with age but did not affect the promoters of genes that are not affected or that increase expression with age.

These findings suggest that accelerated DNA damage may contribute to reduced gene expression in the human brain after age 40. There may be sites in the genome that are exceptionally prone to DNA damage, based on chromatin accessibility and GC content. These sites would be expected to include promoter regions that have high-GC content and are not protected by transcription-coupled repair. Genes that are located near hot spots for DNA damage may show decreased expression levels in old age due to high levels of stochastic DNA damage. Thus, detrimental effects caused by decreased expression with age may be stochastic (the result of unfortunate location of unrelated genes near regions of DNA damage), rather than biologically coherent (changes in the transcriptional regulation of key pathways such as the oxidative stress or DNA-repair pathways).

USING AGING BIOMARKERS TO DISCOVER DRUGS THAT EXTEND LIFE SPAN

A long-held dream is to find drugs or dietary supplements that slow the rate of human aging. Amazingly, this goal may not be just wishful thinking, as several drugs have shown intriguing preliminary results in life span assays. The main problem in the discovery of longevity drugs is not that suitable small molecules are rare, but rather that it is very difficult to know which drug has an interesting effect on life span. The most direct approach to find a longevity-extending drug is to screen small molecules using life span as an assay, which is difficult to do in mice and nearly impossible to do in humans. Rather than using life span as a screen for drugs, an attractive alternative is to use DNA microarrays to find small molecules that cause changes in gene expression (a transcriptional profile) similar to those associated with aging or caloric restriction. The DNA microarray experiments can be performed in days or weeks, compared to life span extension experiments that require years or decades. Thus, transcriptional profiling with DNA microarray experiments could be used as a surrogate for a life span assay and could provide a shortcut to identifying lead anti-aging compounds.

Resveratrol (3, 4', 5-trihydroxystilbene) is a plant-derived polyphenolic compound that has been shown to extend the life span of yeast, worms, flies, and mice (Howitz et al. 2003; Wood et al. 2004; Viswanathan et al. 2005). Resveratrol was initially found through its ability to activate the catalytic activity of mammalian ortholog SIRT1 and yeast Sir2p in vitro (Howitz et al. 2003). However, it is currently unclear whether life span extension caused by resveratrol acts through Sir2 in vivo (Kaeberlein et al. 2005).

Addition of resveratrol to the diet of obese mice lengthens their life span. Increased life span is accompanied by increased insulin sensitivity, reduced insulin-like growth factor activity, and increased AMP-activated kinase activity (Baur et al. 2006). Baur et al. used DNA microarrays to analyze the effects of resveratrol on the livers of 18-month-old mice. They first identified sets of genes that changed expression when mice were fed a high-calorie diet versus a standard diet. Next, they showed that resveratrol counteracted the changes in gene expression caused by the high-calorie diet. Specifically, of 153 gene pathways altered by a high-calorie diet, resveratrol reversed the change in expression for 144. This result strongly indicates that the beneficial effects of resveratrol are similar to those of caloric restriction, involving shifting cellular physiology from the high-calorie state toward the calorically restricted state.

In worms, DNA microarray studies have implicated a different mechanism for life span extension by resveratrol (Viswanathan et al. 2005). DNA microarrays were used to profile gene expression in worms grown on media either with or without resveratrol. The expression signature caused by resveratrol induced a family of genes encoding prion-like glutamine/asparagine-rich proteins involved in endoplasmic reticulum (ER) stress response to unfolded proteins. One of the ER stress response genes (*abu-11*) induced by resveratrol was shown to be a key determinant of *C. elegans* life span. RNAi of *abu-11* abolished resveratrol-mediated extension of life span, and overexpression of *abu-11* extended the life span of transgenic animals. These results suggest that *C. elegans* aging may be caused by the presence of unfolded proteins in the ER. Resveratrol may reduce levels of unfolded proteins and extend life span by inducing expression of ER stress genes such as *abu-11*.

Dhahbi et al. (2005) used DNA microarrays to show that metformin causes changes in gene expression similar to those caused by caloric restriction. Metformin is one of the most widely prescribed drugs for type 2 diabetes. Metformin increases insulin sensitivity in the liver and muscle and decreases glucose production by the liver (Radziuk et al. 2003). Metformin acts by increasing the activity of AMP-activated protein kinase (Shaw et al. 2005) and has been shown to extend the life span of mice (Dilman and Anisimov 1980; Anisimov et al. 2003). Dhahbi et al. analyzed the expression pattern of genes in the livers of mice that had been fed metformin for 8 weeks (Dhahbi et al. 2005; Spindler 2006). They found that metformin induced a transcriptional response that was similar to caloric restriction. On the basis of these results, further work on the effects of metformin on the life spans of mice and humans is crucial.

A popular hypothesis is that accumulation of oxidative damage is a major cause of aging. Lee et al. (2004) used DNA microarrays to study the effects of two dietary antioxidants (α-lipoic acid and Coenzyme Q10) on gene expression in mouse hearts. They analyzed expression in hearts from young and old mice that had been either calorically restricted, fed α-lipoic acid, or fed Coenzyme Q10. They found that dietary supplementation with either α-lipoic acid or Coenzyme Q10 resulted in transcriptional alterations consistent with a state of reduced oxidative stress in the heart. They also found that these two drugs slowed age-related changes in gene expression, similar to caloric restriction but to a lesser extent. These results indicate that α-lipoic acid or Coenzyme Q10 may reduce oxidative stress and may also have a beneficial effect on aging.

COMMON PATHWAYS FOR AGING IN DIVERSE TISSUES

Some aspects of aging only affect specific tissues; examples include progressive weakness of muscle, declining synaptic function in the brain, and decreased filtration rate in the kidney. Other aspects of aging occur in all cells regardless of their tissue type, such as the accumulation of oxidative damage, DNA damage, and telomere shortening. Genome-wide searches for gene expression changes during aging would include both types of expression changes, and it would be interesting to discern which expression changes are tissue-specific and which are common to all tissues. Genes that are age-regulated in all tissues would reveal genes involved in core mechanisms that underlie cellular aging.

Several papers have found similarity in the transcriptional aging profiles of different parts of the same human tissue. Rodwell et al. (2004) found that the transcriptional profiles for aging in the cortex and medulla of the kidney were very similar. Likewise, Fraser et al. (2005) found that age-related gene expression changes were similar among five different regions of the human cerebral cortex. However, they did not find any evidence for similarity between age-related expression changes in the cerebral cortex and in the cerebellum.

Recently, Zahn et al. (2006) have discovered genetic pathways that show common age regulation in kidney, brain, and muscle. They used Affymetrix GeneChips to analyze expression in 81 skeletal muscle samples from patients aged 16 to 86 years and found 250 age-regulated muscle genes. Next, they compared their muscle-aging results to previously published data on kidney and brain aging. Although most of the age-related changes were tissue-specific, they found evidence for common age regulation of five genetic pathways in all three tissues. Specifically, there is an overall increase in expression of the extracellular matrix genes, the cell growth genes, and complement activation genes in all three tissues. Conversely, there is an overall decrease in expression of chloride transport genes and electron transport genes in all three tissues.

COMPARISON OF AGING PATTERNS ACROSS SPECIES

DNA microarray experiments have defined molecular profiles for aging for worms, flies, mice, rats, monkeys, and humans (Table 1). An important question to answer is which parts of the aging transcriptional profile are shared between species (public) and which are specific for only one species (private). On the one hand, many biological processes involved in aging are shared across diverse species. For example, oxidative stress and DNA damage have been postulated to have a role in aging in nearly all animals. Caloric restriction or mutations in genes such as the insulin-like receptor gene (*daf-2*) or *sir2* affect longevity in diverse species from yeast to mice. These common themes for aging suggest that there might also be genes that show common patterns of age regulation in different species.

On the other hand, evolutionary theory postulates that there should be little or no overlap in the particular degeneration pathway of one species with that of distantly related species. Unlike most other biological processes, there is strong reason to believe that the aging process is not evolutionarily conserved per se (Kirkwood and Austad 2000). As a rule, wild animals do not live long enough to grow old. For example, more than 90% of mice die in their first year in the wild, and thus, only a very small fraction of wild mice achieve their maximal life span of 3 years (Kirkwood and Austad 2000). The median human life span in the United States was 47 years in 1900 (Statistics 2006), and relatively few individuals lived to maximal life span during human evolution (Martin 2002). In the wild, life span is limited by disease and predation, not by aging. Events that occur late in life escape the force of natural selection, and the aging process is unlikely to be evolutionarily conserved per se. Thus, there is little reason to believe that genetic and molecular events that occur late in life in model organisms such as the mouse would be conserved in humans.

Early work by McCarroll et al. (2004) characterized gene expression profiles for normal aging in flies and worms. These investigators reported several biological processes that seemed to share a similar overall pattern of change with respect to age in flies and worms. However, the shared similarity between flies and worms is probably not related to aging, as most of the expression changes occur in young adults rather than old animals. Furthermore, the analysis used in the paper was flawed because it greatly overestimated the statistical significance of the results (Melov and Hubbard 2004). Nevertheless, this early paper contained a number of insightful ideas about ways to analyze large-scale gene expression data and to compare emergent properties from transcriptional profiles of aging across different species.

Zahn et al. (2006) compared transcriptional profiles for aging in humans to aging profiles for mice, flies, and worms. They asked whether any of the biological processes that showed age-dependent changes in expression in humans also showed similar age regulation in mice, flies, or worms. Of the five genetic pathways that were age-regulated in humans, one (the electron transport chain genes) showed similar age regulation in the other three species (Zahn et al. 2006; J. Zahn, unpubl.). Compared to humans, mice age 20 to 30 times faster, flies age 400 times faster, and worms age 2000 times faster. Given this large difference in life span, it is interesting that there is a similar twofold decrease in overall expression of the electron transport chain in each species. Expression of the electron transport chain may be a particularly interesting biomarker for aging because it scales with life span in species that are distantly related.

THE ROAD AHEAD

We are just beginning to analyze aging at the systems level. In the future, a key issue to address is which age-regulated genes are merely biomarkers for aging or which may control the rate of aging. Already, initial genetic experiments in worms indicate that a large number of age-regulated genes have a role in specifying normal life span. Another key issue is to determine the upstream regulatory pathways that govern changes in expression of the aging biomarkers. This point will yield key insights regarding mechanisms of aging. Showing that age-related expression of these genes is caused by cell stress would strongly support the oxidative damage theory of aging, whereas showing that the genes are regulated by signaling pathways important for development could provide direct support for the antagonistic pleiotropic theory of aging. Finally, an unsolved challenge is to develop new ways to analyze entire gene networks that can explain how similar genomes (such as mouse and human) give rise to organisms with vastly different life spans. Perhaps systems-level analysis of the human and mouse genetic networks will show how overall robustness of the network sets limits on maximal life span for a species.

ACKNOWLEDGMENTS

I acknowledge members of the Kim lab and Art Owen for comments on the manuscript. S.K.K. was funded by the National Institutes of Health and by the Ellison Medical Scholars program.

REFERENCES

Akavia U.D., Shur I., Rechavi G., and Benayahu D. 2006. Transcriptional profiling of mesenchymal stromal cells from young and old rats in response to Dexamethasone. BMC Genomics 7: 95.

Anisimov V.N., Semenchenko A.V., and Yashin A.I. 2003. Insulin and longevity: Antidiabetic biguanides as geroprotectors. Biogerontology 4: 297–307.

Bahar R., Hartmann C.H., Rodriguez K.A., Denny A.D., Busuttil R.A., Dolle M.E., Calder R.B., Chisholm G.B., Pollock B.H., Klein C.A., and Vijg J. 2006. Increased cell-to-cell variation in gene expression in ageing mouse heart. Nature 441: 1011–1014.

Baur J.A., Pearson K.J., Price N.L., Jamieson H.A., Lerin C., Kalra A., Prabhu V.V., Allard J. S., Lopez-Lluch G., Lewis K., et al. 2006. Resveratrol improves health and survival of mice on a high-calorie diet. Nature 444: 337–342.

Beggs M.L., Nagarajan R., Taylor-Jones J.M., Nolen G., Macnicol M., and Peterson C.A. 2004. Alterations in the TGFbeta signaling pathway in myogenic progenitors with age. Aging Cell 3: 353–361.

Blalock E.M., Geddes J.W., Chen K.C., Porter N.M., Markesbery W.R., and Landfield P.W. 2004. Incipient Alzheimer's disease: Microarray correlation analyses reveal major transcriptional and tumor suppressor responses. Proc. Natl. Acad. Sci. 101: 2173–2178.

Blalock E.M., Chen K.C., Sharrow K., Herman J.P., Porter N.M., Foster T.C., and Landfield P.W. 2003. Gene microarrays in hippocampal aging: Statistical profiling identifies novel processes correlated with cognitive impairment. J. Neurosci. 23: 3807–3819.

Boylston W.H., Gerstner A., DeFord J.H., Madsen M., Flurkey K., Harrison D.E., and Papaconstantinou J. 2004. Altered cholesterologenic and lipogenic transcriptional profile in livers of aging Snell dwarf (Pit1dw/dwJ) mice. Aging Cell 3: 283–296.

Cao S.X., Dhahbi J.M., Mote P.L., and Spindler S.R. 2001. Genomic profiling of short- and long-term caloric restriction effects in the liver of aging mice. Proc. Natl. Acad. Sci. 98: 10630–10635.

Chapman T. and Partridge L. 1996. Female fitness in Drosophila melanogaster: An interaction between the effect of nutrition and of encounter rate with males. Proc. Biol. Sci. 263: 755–759.

Chelvarajan R.L., Liu Y., Popa D., Getchell M.L., Getchell T.V., Stromberg A.J., and Bondada S. 2006. Molecular basis of age-associated cytokine dysregulation in LPS-stimulated macrophages. J. Leukoc. Biol. 79: 1314–1327.

Chen J., Mo R., Lescure P.A., Misek D.E., Hanash S., Rochford R., Hobbs M., and Yung R.L. 2003. Aging is associated with increased T-cell chemokine expression in C57BL/6 mice. *J. Gerontol. A Biol. Sci. Med. Sci.* **58:** 975–983.

Cheng X.R., Zhou W.X., Zhang Y.X., Zhou D.S., Yang R.F., and Chen L.F. 2007. Differential gene expression profiles in the hippocampus of senescence-accelerated mouse. *Neurobiol. Aging* **28:** 497–506.

Dhahbi J.M., Mote P.L., Fahy G.M., and Spindler S.R. 2005. Identification of potential caloric restriction mimetics by microarray profiling. *Physiol. Genomics* **23:** 343–350.

Dhahbi J.M., Kim H.J., Mote P.L., Beaver R.J., and Spindler S.R. 2004. Temporal linkage between the phenotypic and genomic responses to caloric restriction. *Proc. Natl. Acad. Sci.* **101:** 5524–5529.

Dilman V.M. and Anisimov V.N. 1980. Effect of treatment with phenformin, diphenylhydantoin or L-dopa on life span and tumour incidence in C3H/Sn mice. *Gerontology* **26:** 241–246.

Dobson J.G., Jr., Fray J., Leonard J.L., and Pratt R.E. 2003. Molecular mechanisms of reduced beta-adrenergic signaling in the aged heart as revealed by genomic profiling. *Physiol. Genomics* **15:** 142–147.

Edwards M.G., Sarkar D., Klopp R., Morrow J.D., Weindruch R., and Prolla T.A. 2003. Age-related impairment of the transcriptional responses to oxidative stress in the mouse heart. *Physiol. Genomics* **13:** 119–127.

———. 2004. Impairment of the transcriptional responses to oxidative stress in the heart of aged C57BL/6 mice. *Ann. N.Y. Acad. Sci.* **1019:** 85–95.

Fraser H.B., Khaitovich P., Plotkin J.B., Paabo S., and Eisen M.B. 2005. Aging and gene expression in the primate brain. *PLoS Biol.* **3:** e274.

Garigan D., Hsu A.L., Fraser A.G., Kamath R.S., Ahringer J., and Kenyon C. 2002. Genetic analysis of tissue aging in *Caenorhabditis elegans:* A role for heat-shock factor and bacterial proliferation. *Genetics* **161:** 1101–1112.

Gerstbrein B., Stamatas G., Kollias N., and Driscoll M. 2005. In vivo spectrofluorimetry reveals endogenous biomarkers that report healthspan and dietary restriction in *Caenorhabditis elegans. Aging Cell* **4:** 127–137.

Getchell T.V., Peng X., Green C.P., Stromberg A.J., Chen K.C., Mattson M.P., and Getchell M.L. 2004. In silico analysis of gene expression profiles in the olfactory mucosae of aging senescence-accelerated mice. *J. Neurosci. Res.* **77:** 430–452.

Girardot F., Lasbleiz C., Monnier V., and Tricoire H. 2006. Specific age-related signatures in *Drosophila* body parts transcriptome. *BMC Genomics* **7:** 69.

Giresi P.G., Stevenson E.J., Theilhaber J., Koncarevic A., Parkington J., Fielding R.A., and Kandarian S.C. 2005. Identification of a molecular signature of sarcopenia. *Physiol. Genomics* **21:** 253–263.

Golden T.R. and Melov S. 2004. Microarray analysis of gene expression with age in individual nematodes. *Aging Cell* **3:** 111–124.

Guarente L. and Kenyon C. 2000. Genetic pathways that regulate ageing in model organisms. *Nature* **408:** 255–262.

Hamatani T., Falco G., Carter M.G., Akutsu H., Stagg C.A., Sharov A.A., Dudekula D.B., VanBuren V., and Ko M.S. 2004. Age-associated alteration of gene expression patterns in mouse oocytes. *Hum. Mol. Genet.* **13:** 2263–2278.

Han S.N., Adolfsson O., Lee C.K., Prolla T.A., Ordovas J., and Meydani S. N. 2006. Age and vitamin E-induced changes in gene expression profiles of T cells. *J. Immunol.* **177:** 6052–6061.

Hawse J.R., Hejtmancik J.F., Horwitz J., and Kantorow M. 2004. Identification and functional clustering of global gene expression differences between age-related cataract and
clear human lenses and aged human lenses. *Exp. Eye Res.* **79:** 935–940.

Herndon L.A., Schmeissner P.J., Dudaronek J.M., Brown P.A., Listner K.M., Sakano Y.,
Paupard M.C., Hall D.H., and Driscoll M. 2002. Stochastic and genetic factors influence tissue-specific decline in ageing *C. elegans. Nature* **419:** 808–814.

Hiratsuka K., Kamino Y., Nagata T., Takahashi Y., Asai S., Ishikawa K., and Abiko Y. 2002.
Microarray analysis of gene expression changes in aging in mouse submandibular
gland. *J. Dent. Res.* **81:** 679–682.

Howitz K.T., Bitterman K.J., Cohen H.Y., Lamming D.W., Lavu S., Wood J.G., Zipkin R.E.,
Chung P., Kisielewski A., Zhang L.L., et al. 2003. Small molecule activators of sirtuins
extend *Saccharomyces cerevisiae* lifespan. *Nature* **425:** 191–196.

Jin W., Riley R.M., Wolfinger R.D., White K.P., Passador-Gurgel G., and Gibson G. 2001.
The contributions of sex, genotype and age to transcriptional variance in *Drosophila
melanogaster. Nat. Genet.* **29:** 389–395.

Jones S.J., Riddle D.L., Pouzyrev A.T., Velculescu V.E., Hillier L., Eddy S.R., Stricklin S.L.,
Baillie D.L., Waterston R., and Marra M.A. 2001. Changes in gene expression associated with developmental arrest and longevity in *Caenorhabditis elegans. Genome Res.*
11: 1346–1352.

Kaeberlein M., McDonagh T., Heltweg B., Hixon J., Westman E.A., Caldwell S.D., Napper
A., Curtis R., DiStefano P.S., Fields S., et al. 2005. Substrate-specific activation of sirtuins by resveratrol. *J. Biol. Chem.* **280:** 17038–17045.

Kayo T., Allison D.B., Weindruch R., and Prolla T.A. 2001. Influences of aging and caloric
restriction on the transcriptional profile of skeletal muscle from rhesus monkeys. *Proc.
Natl. Acad. Sci.* **98:** 5093–5098.

Kim H., Lee D.K., Choi J.W., Kim J.S., Park S.C., and Youn H.D. 2003. Analysis of the
effect of aging on the response to hypoxia by cDNA microarray. *Mech. Ageing Dev.*
124: 941–949.

Kim S.N., Rhee J.H., Song Y.H., Park D.Y., Hwang M., Lee S.L., Kim J.E., Gim B.S., Yoon
J.H., Kim Y.J., and Kim-Ha J. 2005. Age-dependent changes of gene expression in the
Drosophila head. *Neurobiol. Aging* **26:** 1083–1091.

Kirkwood T.B. and Austad S.N. 2000. Why do we age? *Nature* **408:** 233–238.

Kyng K.J., May A., Kolvraa S., and Bohr V.A. 2003. Gene expression profiling in Werner
syndrome closely resembles that of normal aging. *Proc. Natl. Acad. Sci.* **100:**
12259–12264.

Landis G.N., Abdueva D., Skvortsov D., Yang J., Rabin B.E., Carrick J., Tavare S., and Tower
J. 2004. Similar gene expression patterns characterize aging and oxidative stress in
Drosophila melanogaster. Proc. Natl. Acad. Sci. **101:** 7663–7668.

Lau K.M., Tam N.N., Thompson C., Cheng R.Y., Leung Y.K., and Ho S.M. 2003. Age-
associated changes in histology and gene-expression profile in the rat ventral prostate.
Lab. Invest. **83:** 743–757.

Lee C.K., Weindruch R., and Prolla T.A. 2000. Gene-expression profile of the ageing brain
in mice. *Nat. Genet.* **25:** 294–297.

Lee C.K., Klopp R.G., Weindruch R., and Prolla T.A. 1999. Gene expression profile of
aging and its retardation by caloric restriction. *Science* **285:** 1390–1393.

Lee C.K., Allison D.B., Brand J., Weindruch R., and Prolla T.A. 2002. Transcriptional profiles associated with aging and middle age-onset caloric restriction in mouse hearts.
Proc. Natl. Acad. Sci. **99:** 14988–14993.

Lee C.K., Pugh T.D., Klopp R.G., Edwards J., Allison D.B., Weindruch R., and Prolla T.A. 2004. The impact of alpha-lipoic acid, coenzyme Q10 and caloric restriction on life span and gene expression patterns in mice. *Free Radic. Biol. Med.* **36:** 1043–1057.

Lee H.M., Greeley G.H., Jr., and Englander E.W. 2001. Age-associated changes in gene expression patterns in the duodenum and colon of rats. *Mech. Ageing Dev.* **122:** 355–371.

Lener T., Moll P.R., Rinnerthaler M., Bauer J., Aberger F., and Richter K. 2006. Expression profiling of aging in the human skin. *Exp. Gerontol.* **41:** 387–397.

Liang W.S., Dunckley T., Beach T.G., Grover A., Mastroeni D., Walker D.G., Caselli R.J., Kukull W.A., McKeel D., Morris J.C., et al. 2006. Gene expression profiles in anatomically and functionally distinct regions of the normal aged human brain. *Physiol. Genomics* **28:** 311–322.

Lluel P., Palea S., Ribiere P., Barras M., Teillet L., and Corman B. 2003. Increased adrenergic contractility and decreased mRNA expression of NOS III in aging rat urinary bladders. *Fundam. Clin. Pharmacol.* **17:** 633–641.

Lu T., Pan Y., Kao S.Y., Li C., Kohane I., Chan J., and Yankner B.A. 2004. Gene regulation and DNA damage in the ageing human brain. *Nature* **429:** 883–891.

Lund J., Tedesco P., Duke K., Wang J., Kim S.K., and Johnson T.E. 2002. Transcriptional profile of aging in *C. elegans*. *Curr. Biol.* **12:** 1566–1573.

Ly D.H., Lockhart D.J., Lerner R.A., and Schultz P.G. 2000. Mitotic misregulation and human aging. *Science* **287:** 2486–2492.

Martin G.M. 2002. Gene action in the aging brain: An evolutionary biological perspective. *Neurobiol. Aging* **23:** 647–654.

McCarroll S.A., Murphy C.T., Zou S., Pletcher S.D., Chin C.S., Jan Y.N., Kenyon C., Bargmann C.I., and Li H. 2004. Comparing genomic expression patterns across species identifies shared transcriptional profile in aging. *Nat. Genet.* **36:** 197–204.

McElwee J., Bubb K., and Thomas J.H. 2003. Transcriptional outputs of the *Caenorhabditis elegans* forkhead protein DAF-16. *Aging Cell* **2:** 111–121.

McElwee J.J., Schuster E., Blanc E., Thomas J.H., and Gems D. 2004. Shared transcriptional signature in *Caenorhabditis elegans* Dauer larvae and long-lived daf-2 mutants implicates detoxification system in longevity assurance. *J. Biol. Chem.* **279:** 44533–44543.

Melk A., Mansfield E.S., Hsieh S.C., Hernandez-Boussard T., Grimm P., Rayner D.C., Halloran P.F., and Sarwal M.M. 2005. Transcriptional analysis of the molecular basis of human kidney aging using cDNA microarray profiling. *Kidney Int.* **68:** 2667–2679.

Melov S. and Hubbard A. 2004. Microarrays as a tool to investigate the biology of aging: A retrospective and a look to the future. *Sci. Aging Knowledge Environ.* **2004:** re7.

Meyer M.H., Etienne W., and Meyer R.A., Jr. 2004. Altered mRNA expression of genes related to nerve cell activity in the fracture callus of older rats: A randomized, controlled, microarray study. *BMC Musculoskelet. Disord.* **5:** 24.

Mo R., Chen J., Han Y., Bueno-Cannizares C., Misek D.E., Lescure P.A., Hanash S., and Yung R.L. 2003. T cell chemokine receptor expression in aging. *J. Immunol.* **170:** 895–904.

Murphy C.T., McCarroll S.A., Bargmann C.I., Fraser A., Kamath R.S., Ahringer J., Li H., and Kenyon C. 2003. Genes that act downstream of DAF-16 to influence the lifespan of *Caenorhabditis elegans*. *Nature* **424:** 277–283.

Pattison J.S., Folk L.C., Madsen R.W., and Booth F.W. 2003. Selected contribution: Identification of differentially expressed genes between young and old rat soleus muscle during recovery from immobilization-induced atrophy. *J. Appl. Physiol.* **95:** 2171–2179.

Pletcher S.D., Macdonald S.J., Marguerie R., Certa U., Stearns S.C., Goldstein D.B., and Partridge L. 2002. Genome-wide transcript profiles in aging and calorically restricted *Drosophila melanogaster. Curr. Biol.* **12:** 712–723.

Preisser L., Houot L., Teillet L., Kortulewski T., Morel A., Tronik-Le Roux D., and Corman B. 2004. Gene expression in aging kidney and pituitary. *Biogerontology* **5:** 39–47.

Prolla T.A. 2002. DNA microarray analysis of the aging brain. *Chem. Senses* **27:** 299–306.

Radziuk J., Bailey C.J., Wiernsperger N.F., and Yudkin J.S. 2003. Metformin and its liver targets in the treatment of type 2 diabetes. *Curr. Drug Targets Immune Endocr. Metab. Disord.* **3:** 151–169.

Reyes I., Reyes N., Iatropoulos M., Mittelman A., and Geliebter J. 2005. Aging-associated changes in gene expression in the ACI rat prostate: Implications for carcinogenesis. *Prostate* **63:** 169–186.

Ricciarelli R., d'Abramo C., Massone S., Marinari U., Pronzato M., and Tabaton M. 2004. Microarray analysis in Alzheimer's disease and normal aging. *IUBMB Life* **56:** 349–354.

Rodwell G.E., Sonu R., Zahn J.M., Lund J., Wilhelmy J., Wang L., Xiao W., Mindrinos M., Crane E., Segal E., et al. 2004. A transcriptional profile of aging in the human kidney. *PLoS Biol.* **2:** e427.

Segev F., Mor O., Segev A., Belkin M., and Assia E.I. 2005. Downregulation of gene expression in the ageing lens: A possible contributory factor in senile cataract. *Eye* **19:** 80–85.

Seroude L., Brummel T., Kapahi P., and Benzer S. 2002. Spatio-temporal analysis of gene expression during aging in *Drosophila melanogaster. Aging Cell* **1:** 47–56.

Shaw R.J., Lamia K.A., Vasquez D., Koo S.H., Bardeesy N., Depinho R.A., Montminy M., and Cantley L.C. 2005. The kinase LKB1 mediates glucose homeostasis in liver and therapeutic effects of metformin. *Science* **310:** 1642–1646.

Spindler S.R. 2006. Use of microarray biomarkers to identify longevity therapeutics. *Aging Cell* **5:** 39–50.

Statistics. 2006. *Health, United States, 2006.* National Center for Health Statistics. Hyattsville, Maryland.

Tan Q., Christensen K., Christiansen L., Frederiksen H., Bathum L., Dahlgaard J., and Kruse T.A. 2005. Genetic dissection of gene expression observed in whole blood samples of elderly Danish twins. *Hum. Genet.* **117:** 267–274.

Thomas R.P., Guigneaux M., Wood T., and Evers B.M. 2002. Age-associated changes in gene expression patterns in the liver. *J. Gastrointest. Surg.* **6:** 445–454.

Verbitsky M., Yonan A.L., Malleret G., Kandel E.R., Gilliam T.C., and Pavlidis P. 2004. Altered hippocampal transcript profile accompanies an age-related spatial memory deficit in mice. *Learn. Mem.* **11:** 253–260.

Visala Rao D., Boyle G.M., Parsons P.G., Watson K., and Jones G.L. 2003. Influence of ageing, heat shock treatment and in vivo total antioxidant status on gene-expression profile and protein synthesis in human peripheral lymphocytes. *Mech. Ageing Dev.* **124:** 55–69.

Viswanathan M., Kim S.K., Berdichevsky A., and Guarente L. 2005. A role for SIR-2.1 regulation of ER stress response genes in determining *C. elegans* life span. *Dev. Cell* **9:** 605–615.

Wang J. and Kim S.K. 2003. Global analysis of dauer gene expression in *Caenorhabditis elegans. Development* **130:** 1621–1634.

Welle S., Brooks A., and Thornton C.A. 2001. Senescence-related changes in gene expression in muscle: Similarities and differences between mice and men. *Physiol. Genomics* **5:** 67–73.

Welle S., Brooks A.I., Delehanty J.M., Needler N., and Thornton C.A. 2003. Gene expression profile of aging in human muscle. *Physiol. Genomics* **14:** 149–159.

Welle S., Brooks A.I., Delehanty J.M., Needler N., Bhatt K., Shah B., and Thornton C.A. 2004. Skeletal muscle gene expression profiles in 20-29 year old and 65-71 year old women. *Exp. Gerontol.* **39:** 369–377.

Wood J.G., Rogina B., Lavu S., Howitz K., Helfand S.L., Tatar M., and Sinclair D. 2004. Sirtuin activators mimic caloric restriction and delay ageing in metazoans. *Nature* **430:** 686–689.

Wood W.B. and Johnson T.E. 1994. Aging. Stopping the clock. *Curr. Biol.* **4:** 151–153.

Yoshida S., Yashar B.M., Hiriyanna S., and Swaroop A. 2002. Microarray analysis of gene expression in the aging human retina. *Invest. Ophthalmol. Vis. Sci.* **43:** 2554–2560.

Zahn J.M., Sonu R., Vogel H., Crane E., Mazan-Mamczarz K., Rabkin R., Davis R.W., Becker K.G., Owen A.B., and Kim S.K. 2006. Transcriptional profiling of aging in human muscle reveals a common aging signature. *PLoS Genet.* **2:** e115.

10

Aging in Mammalian Stem Cells and Other Self-renewing Compartments

Derrick J. Rossi
Stanford University School of Medicine
Stanford Institute for Stem Cell Biology and Regenerative Medicine
Stanford, California 94305-5324

Norman E. Sharpless
Department of Medicine and Genetics
The Lineberger Comprehensive Cancer Center
The University of North Carolina
Chapel Hill, North Carolina 27599-7295

LONG-LIVED METAZOANS MUST REPLACE A VARIETY of lost or consumed cells at a furious pace. For example, an adult human replaces about 1% of their 20 trillion red blood cells every day through de novo synthesis. Similarly staggering rates of cell division are at work to produce new cells in the gut, skin, and bone marrow throughout life. Additionally, certain tissues (e.g., memory lymphocytes and pancreatic β cells) possess a potential for facultative growth in the adult organism; i.e., under certain circumstances (e.g., viral infection and pregnancy), these normally quiescent cells can reenter the cell cycle to increase the mass of a given tissue through regulated proliferation. To offset the high cellular turnover rate in such tissues and avoid the onset of tissue-specific hypoplasia and atrophy, many mammalian tissues contain reservoirs of stem cells capable of generating terminally differentiated effector cell types. The unique cellular property that enables stem cells to maintain such function throughout is their ability to produce large numbers of differentiated cell types while also self-renewing themselves so that their reserves do not become depleted over time. Several lines of evidence—foremost of which is evidence indicating that aged tissues characteristically exhibit a diminished capacity to maintain homeostasis or return to homeostasis after exposure to stress—has implicated stem cell decline in the aging process. In this chapter, we review some of the evidence to support the notion that certain aspects of mammalian aging result from an age-dependent decline in the function of self-renewing stem cells, and we discuss the molecular mechanisms thought to underlie stem cell aging.

PROPERTIES OF SELF-RENEWAL RELEVANT TO AGING

A few characteristics of self-renewing cells are of particular relevance to a discussion of stem cell aging. First, most proliferation in an adult mammal is not self-renewing. In fact, self-renewal appears to be a specific and unusual property of stem cells and other rare cell types that are capable of limited numbers of self-renewing divisions after periods of G_0 dormancy. Strictly speaking, not all self-renewing cells are "stem cells," with the difference being that adult tissue-specific stem cells (e.g., hematopoietic stem cells) are multipotent in that they can give rise to differentiated progeny of several different types. In contrast, there are a few types of self-renewing cells, such as pancreatic β cells and memory T cells, that have a restricted potential for differentiation, generating progeny similar to the parental cell (a class of cell sometimes termed unipotent progenitor). Clearly, the distinction between tissue-specific stem cells and unipotent progenitors is somewhat arbitrary, with the latter representing a special case of the former. For convenience, we use the term stem cell in reference to both types of adult self-renewing cells in this chapter unless otherwise indicated. In contrast, we use the term progenitor to designate cell types derived from stem cells that are relatively undifferentiated but that are nonetheless incapable of long-term self-renewal. Note that neither of these classes is to be confused with totipotent embryonic stem (ES) cells which are not further discussed in this chapter.

Many stem cells are believed to produce differentiated cells through a series of increasingly more committed progenitor intermediates. This hierarchical structuring has been characterized in greatest detail in the hematopoietic system, where long-term hematopoietic stem cells (HSCs) give rise to a number of multipotent progenitor subsets that retain full lineage potential yet have a limited capacity for self-renewal (Bryder et al. 2006). These multipotent progenitors in turn give rise to oligopotent progenitors (Kondo et al. 1997; Akashi et al. 2000), which in turn give rise to more lineage-restricted progenitors from which all of the mature blood cells eventually arise (Fig. 1). Each of these stem and progenitor

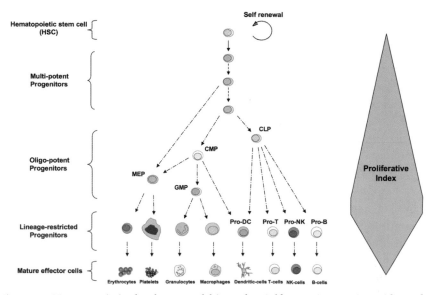

Figure 1. Hematopoietic developmental hierarchy. Self-renewing HSCs reside at the top of the hierarchy, giving rise to a number of multipotent progenitors. Multipotent progenitors give rise to oligopotent progenitors, including the common lymphoid progenitor (CLP), which gives rise to mature B lymphocytes, T lymphocytes, and natural killer (NK) cells. The common myeloid progenitor (CMP) gives rise to granulocyte/macrophage progenitors (GMPs) that differentiate into monocytes/macrophages and granulocytes and megakaryocyte/erythrocyte progenitors (MEPs) that differentiate into megakaryocytes/platelets and erythrocytes. Development from the oligopotent progenitors through to mature blood cells proceeds through a number of intermediate progenitors (not shown). The developmental passage of HSCs through multipotent progenitors, oligopotent progenitors, and lineage-specific progenitors is generally associated with increases in proliferative index, although this trend is not absolute and has not been resolved for all stages of development.

cell subsets can be isolated and purified to near homogeneity by differences in immunophenotype. The hierarchical differentiation scheme of hematopoietic differentiation has several intrinsic advantages. First, it allows an enormous amplification in the numbers of differentiated cells from a single stem cell by combining subsequent steps in differentiation with enormous proliferative potential (Hodgson and Bradley 1984; Passegue et al. 2005). Second, the multistep differentiation scheme also has the advantage of putting limited proliferative demand on the stem cells themselves, such that these cells divide very infrequently (Bradford et al. 1997; Cheshier et al. 1999), with greater than 90% of stem cells in the G_0 phase of the cell cycle (Yamazaki et al. 2006), whereas downstream progenitors cycle much more rapidly (Passegue et al. 2005). Third, the minimal proliferative pressure on stem cells has the benefit of not subjecting them to the potentially mutagenic hazards of DNA replication and cell division and may thus contribute to the integrity and longevity required of these cells. Additionally, because G_0 is a relatively metabolically inactive phase of the cell cycle, it has been suggested that stem cells may be subjected to lower levels of damage-inducing metabolic side products and reactive oxygen species than more metabolically active differentiated cell types (Rossi et al. 2005).

The newly appreciated and prominent role for stem cells in many adult tissues is also germane to this discussion. Although once believed to be restricted to tissues with high turnover rates, stem cells are now known to be involved in controlling homeostasis in many tissues, including those with lower turnover rates such as the brain (Weissman 2000). Although reasonable scientists may disagree over the physiological importance of such cells in the adult organ, their existence in rodents is beyond dispute, and their existence in human tissue is strongly supported. Given this fact, it is reasonable to postulate that at least some characteristics of aging once thought to be solely "degenerative" might in fact reflect a diminution of regenerative capacity by the resident stem cells of the affected tissue. For example, it can be conjectured that degenerative joint disease may not solely result from the loss of synovium and articular cartilage through years of use but may also, in part, reflect a failure to replace such cells by a resident stem cell. Although beyond the scope of this chapter, it is not at all obvious which features of aging reflect solely tissue degeneration versus stem cell functional decline (or some combination thereof), and we believe that elucidating the relative contributions of these two mechanisms of aging on a tissue-by-tissue basis will be a principal focus of future research in mammalian aging. Moreover, because stem cells reside in specialized "niches" where they receive and respond to external cues to proliferate, differentiate, or self-renew (Schofield 1978; Moore and Lemischka 2006), it seems likely that the stem cell microenvironment exerts significant influence on stem cell aging (see below).

Finally, it is important to note that the capacity for self-renewal, although critical for the lifelong function of stem cells, is potentially dangerous at the organismal level (Reya et al. 2001; Campisi 2003; Krishnamurthy et al. 2004). This is because genomic lesions that evade repair in stem cells can be propagated to their self-renewing progeny and in such a way can accumulate during the organism's life span. In particular, mutations that provide a growth or survival advantage in a stem cell compartment provide positive selection for the mutant cell, with malignant transformation being the possible end point of the accumulation of multiple cancer-promoting (or oncogenic) mutations within an individual self-renewing clone. In contrast, mutations in proliferating cells that are fated to terminally differentiate or die are thought to have little or no oncogenic potential unless the mutagenic process establishes the capacity for self-renewal on otherwise non-self-renewing cells. Because stem cells are exposed to a similar array of endogenous and exogenous mutagens as other cells in the body, oncogenic events in the stem cell compartment should occur throughout the mammalian life span. To offset this possibility, stem cells have evolved a number of mechanisms aimed at maintaining genomic integrity beyond that of other proliferating cells. For example, stem cells express high levels of several ABC/MDR transporter genes (Zhou et al. 2001; Rossi et al. 2005) whose products function in cytoprotection by virtue of their ability to efflux xenobiotic genotoxins from the cell. Additionally, at least some stem cells appear to have the ability to retain the replicated strand of DNA during mitosis and to pass the duplicated, and possibly error-containing, strand to more differentiated progeny (Cairns 1975; Potten et al. 2002; Karpowicz et al. 2005; Shinin et al. 2006). It has been conjectured that stem cells may preferentially utilize certain mechanisms of DNA repair that are less likely to be error-prone compared to other proliferating cells (Cairns 2002). Finally, when mutations occur despite these error-prevention mechanisms, evolutionarily perfected tumor-suppressor mechanisms (e.g., apoptosis and senescence) exist that sense malignant growth and censor would-be malignant stem cell clones. This relationship between self-renewal cells and cancer prevention predicts that such tumor suppressor mechanisms may also inadvertently contribute to aging, by contributing to stem cell attrition. We cover experimental support for this notion below (see Molecular Mechanism of Stem Cell Aging).

EVIDENCE THAT STEM CELL FUNCTION DECLINES WITH AGING

In chapters 2, 7, 11, and 17, elegant genetic work from invertebrate systems is detailed which describes the pathways that influence the rate of aging in these model systems. These lower metazoans, however, to date have been less useful in understanding the mechanisms and pathways that regulate tissue regeneration and repair and thereby influence this aspect of mammalian aging. Studies directed at these ends have been largely carried out in rodents with confirmatory analyses in nonrodent mammals, including humans, where possible. In fact, the bulk of such studies have analyzed HSCs in mice, which are the most well-characterized tissue-specific stem cells.

Hematopoietic Stem Cell Aging

Advancing age is accompanied by a number of pathophysiological changes in the hematopoietic system, whose etiology suggests a loss of homeostatic control. The most clinically significant of these are the diminution and decreased competence of the adaptive immune system in the elderly (Linton and Dorshkind 2004), the dramatically increased incidence of myelogenous diseases including myelodysplasia and myeloid leukemias that accompanies aging (Lichtman and Rowe 2004), and the onset of anemia in the elderly (Beghe et al. 2004; Guralnik et al. 2004). Elderly patients are also more likely to suffer adverse side effects and prolonged toxicity from myelosuppressive drugs (e.g., chemotherapy for certain cancers; see, e.g., Brunello et al. 2005; Appelbaum et al. 2006; Lenhoff et al. 2006), and increasing donor age in bone marrow transplantation is a predictor of transplant-related mortality (Buckner et al. 1984; Ash et al. 1991; Kollman et al. 2001; Castro-Malaspina et al. 2002; Yakoub-Agha et al. 2006), suggesting that the reconstituting ability of HSCs from old donors is intrinsically compromised. Finally, the finding of familial aplastic anemia, a profound form of bone marrow failure, in patients with congenital defects in telomerase activity is of particular interest in this regard (see below, Molecular Mechanism of Stem Cell Aging, Senescence/Cell Cycle Inhibitors, as well as Chapter 20). These provocative findings in humans, however, are greatly strengthened by studies in rodents where HSC can be purified to near homogeneity and transplanted, to quantitatively evaluate functional autonomy.

Perhaps counterintuitively, reserves of HSC have been shown to be greatly expanded with age in many (but not all) laboratory strains of mice, through quantitation by immunophenotyping (Morrison et al. 1996; Sudo et al. 2000; Rossi et al. 2005), side population activity (Rossi et al. 2005; Pearce et al. 2006), or cobblestone assays (de Haan et al. 1997; de Haan and Van Zant 1999). Significantly, the age-dependent expansion of HSCs observed in the steady state was shown to be a transplantable, cell-autonomous property of HSC aging, such that even upon transplantation into young recipients, HSCs from aged donors still had a greater propensity to self-renew and give rise to phenocopies of themselves than HSCs from young donors (Rossi et al. 2005; Pearce et al.

2006). A second surprising aspect of HSC biology is that these stem cells can be serially transplanted through successive generations of recipients, as demonstrated in seminal experiments by Harrison and colleagues, who showed that HSCs from C57BL/6 mice could function for more than 8 years and thus far exceed the lifetime of the original donor (Harrison 1979). These experiments thereby established that replicative HSC exhaustion does not occur during periods of normal aging.

This is not to say, however, that HSCs do not exhibit signs of intrinsic aging in the physiological setting. Indeed, numerous studies have documented considerable impact on HSC function with advancing age, including alterations in their mobilization properties (Xing et al. 2006), diminished reconstitution and homing capability (Morrison et al. 1996; Sudo et al. 2000; Chen 2004; Liang et al. 2005; Rossi et al. 2005; Xing et al. 2006), and a loss of lymphoid lineage potential in favor of myeloid lineage commitment with age (Sudo et al. 2000; Kim et al. 2003; Liang et al. 2005; Rossi et al. 2005). Interestingly, this skewing of lineage potential with age was shown to be underwritten by systematic changes in the expression of a cadre of lineage specification genes with age, suggesting that coordinate regulatory control mechanisms become deregulated with age in stem cells (Rossi et al. 2005). Along these lines, it is noteworthy that numerous genes involved in higher-order chromosome dynamics, chromatin remodeling, and epigenetic regulation of gene expression are differentially expressed during HSC aging (Rossi et al. 2005, 2007a).

Moreover, several forms of marrow toxins (chemotherapy, ionizing radiation, etc.) severely limit long-term HSC function and hasten stem cell exhaustion (Gardner et al. 1997; Knudsen et al. 1999; Boccadoro et al. 2002; Meng et al. 2003; Ito et al. 2004, 2006; Wang et al. 2006). Thus, it is possible that mice in the wild would exhibit more pronounced stem cell attrition upon greater exposure to environmental stresses that are not normally encountered by laboratory-housed strains of mice.

Aging in Other Tissue-specific Stem Cells

Analyses of stem cells in nonhematopoietic tissues are frequently complicated by the fact that assays used to evaluate stem cell function, such as transplantation, can be technically difficult, if not impossible, to perform for certain types of stem cells. Moreover, whereas hematopoietic stem and progenitor cells can be enriched to near purity through a combination of cell-surface phenotype and fluorescence-activated cell sorting, strategies for enriching other tissue-specific stem cells are just beginning to be developed. With this said, numerous lines of evidence suggest that stem cells from a variety of tissues suffer functional decline with advancing age.

For example, a decline with age in the number of new neurons produced by neural stem cells (NSCs) in the dentate gyrus has been well documented (Kuhn et al. 1996), whereas the numbers and proliferation of NSCs in the subventricular zone diminish (Maslov et al. 2004; Molofsky et al. 2006), along with their neurogenic potential resulting in an impairment in fine olfactory discrimination (Enwere et al. 2004). In a more blatant example of stem cell functional decline, hair graying in aged mice and humans has been linked to incomplete melanocyte stem cell maintenance (Lang et al. 2005; Nishimura et al. 2005).

Analyses of the insulin-producing β cell of the pancreatic islet illustrate some of the challenges of studying aging in non-HSC tissues that are difficult to transplant and in which the self-renewing cell cannot be significantly enriched. A decline in islet proliferation and regenerative capacity with aging has been well documented in rodent models, and a relative failure of β-cell mass appears to have a role in human type II diabetes (Butler et al. 2003; Yoon et al. 2003). More than 1% of β cells in the islets of young mice are proliferating under steady-state conditions, but this frequency declines by more than tenfold after a year of aging (middle-aged for a mouse) (Fig. 2). β-cell proliferation is known to require cdk4 activity (Rane et al. 1999; Tsutsui et al. 1999), the biochemical target of cdk inhibitors such as $p16^{INK4}$, which significantly accumulates with aging in the islet (Nielsen et al. 1999; Krishnamurthy et al. 2006). The finding that germ-line $p16^{INK4a}$ deficiency largely abrogates the replicative failure of β cells with aging suggests that the age-induced expression of $p16^{INK4a}$ in part has a causal role in this process (Krishnamurthy et al. 2006). The interpretation of this observation, however, establishes a weakness of studies of self-renewal in most nonhematopoietic tissues. The lack of facile transplantation models hinders the determination of which cell or compartment is failing with aging in this setting. For example, in this system, $p16^{INK4a}$ could promote aging in the β cells themselves (inhibiting replication of a unipotent progenitor), in a putative pancreatic stem cell (inhibiting "islet neogenesis"), or in another tissue that influences islet proliferation in a noncell-autonomous manner.

Indeed, noncell-autonomous factors have been shown to affect stem cell aging in other systems. For example, Rando and colleagues have shown that the aging of muscle satellite cells is greatly influenced by the aged microenvironment through the demonstration that satellite cells of

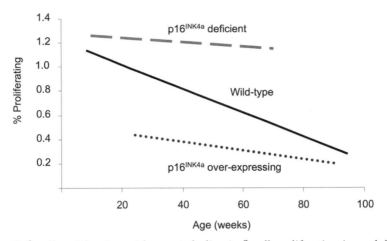

Figure 2. β-cell proliferation with age. A decline in β-cell proliferation is modulated by the expression of the p16^{INK4a} tumor suppressor. Excess activity correlates with reduced proliferation, whereas decreased activity affords a resistance to β-cell aging.

aged mice that were surgically joined to young mice through hetero-chronic parabiosis to establish a conjoined circulatory system were more capable of responding to injury than in the aged setting alone (Conboy et al. 2005). These experiments suggest that the aged milieu, hormonal or otherwise, significantly influences satellite cell aging, although there may also be a cell-autonomous component to satellite cell aging that will be revealed in different experimental settings. By the same token, whereas much of the functional decline that characterizes HSC aging has been shown to be cell-autonomous (Rossi et al. 2005), the influence of the stem cell niche and bone marrow microenvironment on HSC biol-ogy is undeniable, and thus it would be surprising if the aged microen-vironment did not also influence HSC aging to some degree. Although this distinction between cell-autonomous and noncell-autonomous aging is important, it is also worth noting that aging is ultimately intrin-sic to the individual. For example, the finding that serum from aged mice inhibits satellite cell proliferation changes the focus of study from the satellite cell in the muscle to determining the tissue outside of the mus-cle which elaborates, or fails to elaborate, the relevant endocrine factor that regulates satellite cell proliferation. The ability to conditionally tar-get genes in a tissue-specific manner will facilitate the determination of which cells in which compartments are aging intrinsically and which are aging "by proxy."

MOLECULAR MECHANISM OF STEM CELL AGING

The mechanisms whereby old self-renewing cells remember their prior proliferative history have not been clearly elucidated. Characteristic alterations of gene expression have been noted by genome-wide microarray screen performed on whole tissues (see Chapter 9) as well as purified HSCs (Rossi et al. 2005) with aging. These unbiased approaches are powerful and effective at characterizing the expression state of old cells, but they do not directly address the cause of these changes in old cells. Several genetic and epigenetic mechanisms have been implicated to contribute to stem cell functional decline with age.

DNA Damage

The accumulation of somatic damage to cellular macromolecules (proteins, lipids, RNA, DNA) is considered to be a major cause of cellular attrition and aging in most organisms (Kirkwood 1977, 2005). Although all cellular macromolecules are susceptible to damage, the cellular target most intimately linked with aging to date is DNA. This is because, unlike most other cellular polymers such as proteins and RNA, DNA is neither appreciably turned over nor recycled. Moreover, the amount of damage that DNA is subjected to is staggering, with estimates suggesting that every cell in the human body incurs upward of several thousand lesions daily due to spontaneous depurination and hydrolysis alone (Lindahl 1993), although other workers have suggested lower mutational frequencies (Bielas et al. 2006). To offset this damage and maintain genomic integrity, cells have evolved a number of pathways that respond to, and repair, different types of lesions (Hoeijmakers 2001). For example, global nucleotide excision repair (NER) functions to remove a wide class of helix-distorting lesions that can interfere with nucleotide pairing, whereas transcription-coupled nucleotide excision repair (TC-NER) repairs similar lesions in actively transcribed genes. Base excision repair (BER), on the other hand, repairs smaller chemical modifications of bases that can disrupt nucleotide pairing and lead to miscoding, whereas more severe types of damage, such as double-strand breaks and interstrand cross-links, are dealt with by nonhomologous end joining (NHEJ) and homologous recombination pathways. Due to the imperfect nature of DNA-repair systems, however, some damage evades repair and accumulates over time (Hamilton et al. 2001; Stevnsner et al. 2002; Khaidakov et al. 2003).

That insufficiencies in genomic maintenance contribute to aging is supported by studies of human segmental progeroid syndromes such as

Cockayne syndrome (CS), trichothiodystrophy (TTD), Werner syndrome (WS), Bloom syndrome, Rothmund-Thomson syndrome, and ataxia telangiectasia (AT), all of which have been found to result from mutations of genes whose products either directly function in the repair of DNA or function to mediate cellular responses to damage (Martin and Oshima 2000; Bohr 2002). In support of this, mice bearing mutations in mediators of numerous DNA-repair pathways have also been found to develop a variety of phenotypes suggestive of accelerated aging (Rudolph et al. 1999; Vogel et al. 1999; de Boer et al. 2002; Tyner et al. 2002; Cao et al. 2003; Wong et al. 2003; Mostoslavsky et al. 2006). Similarly, mice engineered with a proofreading defect in the mitochondrial DNA (mtDNA) polymerase resulting in a mutator phenotype exhibit phenotypes consistent with accelerated aging (Trifunovic et al. 2004; Kujoth et al. 2005), suggesting that the mitochondrial genome may contribute to aging, although this notion has been recently questioned by Prolla, Loeb, and colleagues (Vermulst et al. 2007). The role of mitochondrial function in aging is further discussed in Chapter 1. It should be noted, however, that it is not clear whether or not human segmental progeria syndromes, or the murine strains that model them accurately, represent normal physiological aging at an accelerated pace or simply reflect novel pathologies that share certain phenotypic characteristics of aged individuals.

In the damage-accrual model of aging, DNA damage accumulates in cells as they age, and when accumulated damage becomes sufficiently disruptive, it can drive cells to (1) malignant transformation, (2) cellular senescence, (3) programmed cell death, or (4) dysfunction. If this aging paradigm is considered within the context of stem cell biology, stem cells unable to repair accumulated damage would be continually eliminated from the functional stem cell pool as cells are driven to dysfunction, senescence, or apoptosis. If this process proceeds to a point where depletion of stem cell reserves surpasses levels of stem cell self-renewal and differentiation, then homeostatic failure—the physiological hallmark of aging—ensues (Fig. 3). Alternatively, if unrepaired DNA damage proves to be sufficiently mutagenic to lead to the malignant transformation of stem cells, then cancer, another frequent characteristic of aging, could manifest (Fig. 3). As mentioned, stem cells have been postulated as targets for malignant transformation because they are long-lived and possess the molecular machinery for self-renewal.

Evidence supporting the notion that genomic instability and DNA damage critically affect stem cell function has primarily been documented in the hematopoietic system. HSCs from mice deficient in

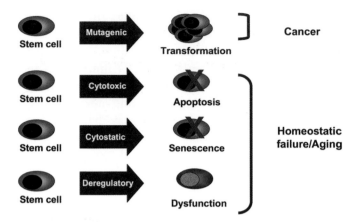

Figure 3. Impact of DNA damage on stem cell function. Model for the outcomes of the accumulation of different types of DNA damage on stem cell biology with age.

FancD1/Brca2 (Navarro et al. 2006) or Msh2 (Reese et al. 2003) have reduced repopulating abilities, whereas Ercc1-deficient mice exhibit multi-lineage cytopenias suggestive of stem/progenitor cell attrition (Prasher et al. 2005). Studies by Ito and colleagues have shown that reactive oxygen species (ROS) limit the functional capacity of HSCs in ATM-deficient mice (Ito et al. 2004), in a p38-mitogen-activated protein kinase (MAPK)-dependent manner (Ito et al. 2006). Finally, the observation that advancing age exacerbates the functional decline of HSC in mice deficient in several genomic maintenance pathways including telomere maintenance, NER, and NHEJ, provides strong genetic evidence that the accrual of DNA damage and genomic instability in stem cells progressively limits their functional capacity with age (Rossi et al. 2007b).

Senescence/Cell Cycle Inhibitors

Senescence (see Chapter 8) is a specialized form of terminal differentiation induced by a variety of stimuli including alterations of telomere length and structure, some forms of DNA damage (e.g., oxidative stress), and activation of certain oncogenes (e.g., see Campisi 2001; Wright and Shay 2002). Senescence requires activation of the retinoblastoma (Rb) and/or p53 protein and expression of their regulators such as p16^{INK4a} and p14^{ARF} (Alcorta et al. 1996; Kamijo et al. 1997; Sage et al. 2003; Stein et al. 1999; see also Chapter 6). These proteins were first identified by their importance in tumor suppression, and the evidence suggesting that senescence prevents cancer is unimpeachable (for review, see Campisi 2005; Kim and Sharpless 2006). Senescence is characterized by a few markers such as senescence-associated-β-galactosidase (SA-β-gal) (Dimri et al. 1995) and senescence-associated heterochromatic foci (SAHF) (Narita et al. 2003). However, the in vivo study of this process has been slowed by the lack of reliable markers that work across tissues and species.

The expression of several markers of senescence increases with aging. Campisi and colleagues showed an increase in SA-β-gal activity in skin with aging (Dimri et al. 1995). Similar increases in SA-β-gal have been seen in aging rodent and human tissues including heart, kidney, pancreatic islet, atherosclerotic plaques, and lung (Melk et al. 2003, 2004; Krishnamurthy et al. 2004; Sone and Kagawa 2005; Urbanek et al. 2005; Matthews et al. 2006; Tsuji et al. 2006). Likewise, Sherr and colleagues reported a marked increase in the expression of p16^{INK4a}, an effector of senescence, in many tissues with aging (Zindy et al. 1997). This finding has been extended by several groups in multiple mammalian species to a large number of aging tissues in health and disease (Nielsen et al. 1999; Krishnamurthy et al. 2004; Melk et al. 2004; Michaloglou et al. 2005; Urbanek et al. 2005; Enomoto et al. 2006; Gray-Schopfer et al. 2006; Herbig et al. 2006; Menzel et al. 2006; Ressler et al. 2006; Sasaki et al. 2006; Tsuji et al. 2006). Given the large change in p16^{INK4a} expression with aging (more that tenfold in many tissues) and the relative ease of quantitating its expression, this finding has led to the proposal that p16^{INK4a} expression could be used as a biomarker of physiological, as opposed to chronological, age (Krishnamurthy et al. 2004).

Although the expression of senescence markers clearly accompanies mammalian aging, some evidence suggests that senescence mechanisms actually contribute to aging. Caloric restriction (CR) potently retards aging in many species including mammals (see Chapters 2, 3, and 15). Correspondingly, CR or other dietary changes retard or even abolish the age-induced increase in the expression of senescence markers such as SA-β-gal and p16^{INK4a} (Krishnamurthy et al. 2004; Sone and Kagawa 2005). Provocatively, CR, like p16^{INK4a} deficiency (Janzen et al. 2006), has been suggested to enhance stem cell function with aging (Chen et al. 2003). Additionally, in humans, a single-nucleotide polymorphism (SNP) at the *INK4/ARF* locus has been associated with a marked resistance to frailty, a reproducible and commonly used clinical indicator of age-related fitness in the elderly (D. Melzer, pers. comm.). Therefore, these correlative data have provided additional evidence for a possible role for senescence-promoting molecules in the aging process.

Experiments employing genetically engineered mice have supported this view. Work using p16^{INK4a}-deficient and overexpressing mice to study self-renewal in HSCs, NSCs, and pancreatic islets has suggested that p16^{INK4a} expression in part contributes to the replicative failure of these tissues with aging (Stepanova and Sorrentino 2005; Janzen et al. 2006; Krishnamurthy et al. 2006; Molofsky et al. 2006). In all three cell types, p16^{INK4a} deficiency partially abrogated an age-induced decline in proliferation and functional decline in each tissue. The effects of p16^{INK4a} loss were consistent across these self-renewing tissues of vastly different biologic properties—in true stem cells (HSCs and NSCs) as well as unipotent progenitors (pancreatic β cells). Therefore, p16^{INK4a} appears to be capable of promoting aging in disparate tissues that are developmentally distinct. In no organ studied, however, did p16^{INK4a} loss completely abrogate the effects of aging, indicating that p16^{INK4a}-independent aging occurs in each of these compartments.

Similar evidence exists to support a role for p53 in mammalian aging. Although it has been difficult to show an increase in p53 activation with aging, modest increases in p53 target transcripts such as p21^{CIP} have been reported in aging tissues (Enomoto et al. 2006; Krishnamurthy et al. 2006; Matthews et al. 2006; Menzel et al. 2006; Tsuji et al. 2006). Moreover, genetically engineered animals possessing augmented p53 function also support a role for this archetypal tumor suppressor in stem cell aging (Tyner et al. 2002; Maier et al. 2004). These results are informed by recent studies showing that p53 expression in HSCs modulates their function with aging, indicating that p53 activation per se compromises the HSC compartment (TeKippe et al. 2003; Dumble et al. 2007). Specifically, mice lacking p53 demonstrate enhanced HSC activity, and mice with increased p53 demonstrate decreased HSC activity. Therefore, p53 appears to exert age-promoting effects in concert with its well-established, beneficial anticancer activity.

It should be noted, however, that these data do not establish that p16^{INK4a} and p53 promote aging through the induction of senescence. For example, the effect of p16^{INK4a} on HSC aging may be unrelated to senescence or other effects on the cell cycle, since purified HSCs from aged mice have an equal capacity to enter cycle and give rise to progeny as HSCs from young mice in rigorous clonal assays (Morrison et al. 1996; Sudo et al. 2000; D.J. Rossi, unpubl.). Thus, although the induction of senescence in some tissues seems likely, it is also possible that these proteins merely decrease the frequency of cell cycle entry in the absence of senescence or, in the case of p53, by inducing apoptosis rather than senescence. The distinction between these possible mechanisms is important for a few reasons, perhaps most importantly that senescence is expected to be irreversible, whereas increased apoptosis or decreased cell cycle entry possibly could be reversed.

The stimuli that activate these senescence-promoting pathways with aging are not clearly elucidated and likely differ among species and tissue type (Fig. 4). Clearly, a wide variety of noxious stimuli induce p16^{INK4a} and p53, including ionizing radiation, reactive oxygen species, telomere dysfunction, and replicative stress (for review, see von Zglinicki et al. 2005; Kim and Sharpless 2006). For p53, the specific signaling cascades that detect DNA damage and activate p53 are reasonably well described: ATM and ATR kinases phosphorylate a host of targets, including CHK2, which in turn activates p53 (for review, see Motoyama and Naka 2004; von Zglinicki et al. 2005). Additionally, ARF is a potent activator of p53, and therefore the increased expression of ARF with aging noted in rodents (Zindy et al. 1997; Krishnamurthy et al. 2004) would be expected to contribute to an age-induced p53 activation as well. Importantly, however, ARF expression does not appear to increase as markedly with aging in humans. The molecular pathways that activate

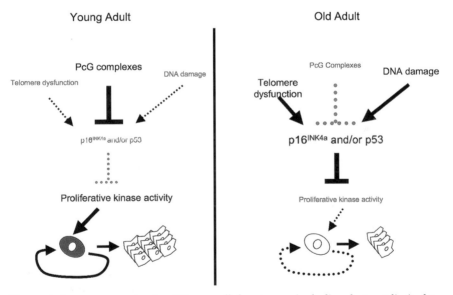

Figure 4. Senescence and aging. Diverse cellular stresses including, but not limited to, telomere dysfunction and other forms of DNA damage increase with age, inducing the senescence-promoting effects of p16^{INK4a} and p53. The PcG complexes appear to repress *Ink4a/Arf* activation and thereby modulate stem cell function, although, to date, no direct proof has established a reduction in PcG activity with aging.

the *Ink4a/Arf* locus in response to such stresses are less well understood, although roles for E2F (DeGregori et al. 1997; Aslanian et al. 2004; Kotake et al. 2007) and MAPK signaling (ERK and p38) (Zhu et al. 1998; Wang et al. 2002; Bulavin et al. 2004; Satyanarayana et al. 2004) have been suggested, the latter possibly through a direct transcriptional effect of Ets transcription factors (Ohtani et al. 2001; Huot et al. 2002). A possible role for polycomb group (PcG) complexes with *Ink4a/Arf* regulation with aging deserves particular consideration.

PcG complexes appear to repress transcription through covalent modifications of histones (for review, see Pasini et al. 2004; Sparmann and van Lohuizen 2006). The relationship of PcG complexes to *INK4a/ARF/INK4b* expression has been particularly provocative because Bmi1 is required for stem cell maintenance of HSCs (Park et al. 2003) and NSCs (Molofsky et al. 2005), and the ability of these complexes to create heritable epigenetic marks might suggest a nongenetic mechanism whereby cells could carry the memory of prior aging-related exposures. At least three PcG proteins (Bmi1, Cbx7, and Mel18) have been reported to repress transcripts encoded by the *Ink4a/Arf/Ink4b* locus ($p16^{Ink4a}$, $p15^{Ink4b}$, and *ARF*) (Jacobs et al. 1999; Itahana et al. 2003; Gil et al. 2004). Loss of *Bmi1* in mice is associated with failure to maintain diverse self-renewing stem cells (e.g., HSCs and NSCs), which can in large part be rescued by *Ink4a/Arf* deficiency (Jacobs et al. 1999; Molofsky et al. 2003, 2005; Park et al. 2003; Bruggeman et al. 2005). In contrast, homeotic transformations of the axial skeleton noted in mice lacking *Bmi1* are not rescued by loss of *Ink4a/Arf* (Jacobs et al. 1999). These results prove that $p16^{INK4a}$ and Arf are the intrinsic mediators of the stem cell exhaustion phenotype in Bmi1-deficient mice. Intriguingly, enforced expression of another PcG protein, Ezh2, in HSCs has been shown to greatly augment their performance in serial transplant (Kamminga et al. 2006), further suggesting that PcG complexes are limiting during stem cell aging, at least under the stressful conditions of serial transplant. Most recently, occupancy of the $p16^{INK4a}$ promoter by Bmi1 and Ezh2 has been shown through chromatin immunoprecipitation, and these associations require Rb-family proteins (Kotake et al. 2007). In contrast to murine cells, in human cells, Bmi1 only appears to regulate the expression of $p16^{INK4a}$, but not that of ARF. Therefore, it is tempting to speculate that this mouse and human difference explains the marked increase with aging in *Arf* expression in mice, but not humans. Thus far, however, it has not been possible to show a consistent relationship between *Bmi1* mRNA expression and aging (N.E. Sharpless, unpubl.), although studies of PcG complex function with aging are under way.

Telomere Dysfunction

Telomeres are nucleoprotein complexes at the chromosome ends that consist of many double-stranded TTAGGG repeats, a 3′ single-strand overhang, and associated telomere-binding proteins (see Chapter 20). DNA polymerase cannot fully replicate the ends of linear DNA duplexes, and in the absence of telomerase, chromosomes shorten slightly with every cell division. Telomerase is a specialized RNA–protein complex that is responsible for the de novo synthesis and maintenance of telomere repeats. The telomerase holoenzyme consists of a telomerase RNA component (TERC) that serves as a template for the addition of repeats, and a few protein components, including telomerase reverse transcriptase (TERT). In the absence of telomerase activity, progressive telomere shortening appears to occur with each cell division and eventually triggers an alteration in telomere structure that is sensed as a DNA double-strand break. Some debate exists as to which tissues in adult mammals express sufficient telomerase activity to prevent telomere attrition. Mice appear to express the enzymatic activity more promiscuously, but in humans, the activity appears predominantly restricted to certain germ cell compartments and some somatic early stem/progenitor compartments. In such somatic tissues, TERT expression is coupled to proliferation and may only serve to slow rather than prevent telomere attrition, despite the presence of the activity as determined by biochemical assays. It is also worth noting that additional activities of TERT have also been suggested in stem cell compartments. For example, Artandi, Blasco, and colleagues have shown a telomerase-independent effect of TERT on proliferation of stem cells of the cutaneous hair follicle (Flores et al. 2005; Sarin et al. 2005). In accord with these findings, several telomere-length-independent effects of TERT have been noted in murine cancer models (Gonzalez-Suarez et al. 2001; Artandi et al. 2002; Chang et al. 2003). At present, the biochemical bases of these telomere-independent effects remain unclear.

Whereas telomere dysfunction is clearly a potent tumor suppressor mechanism, the role of telomere-mediated checkpoints in stem cell aging has not been clearly defined and may differ by species. Landmark studies by DePinho, Greider, and colleagues have demonstrated that mice harbor long telomeres that do not undergo appreciable shortening during the normal murine life span, and therefore, telomere dysfunction is unlikely to be a major cause of stem cell exhaustion in this species (Blasco et al. 1997; Lee et al. 1998). In accord with this notion, HSCs with enforced TERT do not demonstrate enhanced function in assays of serial proliferation (Allsopp et al. 2003b). In contrast, HSCs from mice in which telomerase has been inactivated have demonstrated that telomere length ultimately limits HSC replicative capacity during serial transplantation (Allsopp et al. 2003a). Similarly, the activation of senescence and apoptosis in many replicating and self-renewing compartments, such as the intestinal crypt and the testes, reveals that telomere dysfunction clearly compromises stem cell replicative function in a telomerase-deficient setting (Blasco et al. 1997; Lee et al. 1998; Rudolph et al. 1999). However, since murine telomeres are extremely long, telomere shortening does not appear to be a major barrier to stem cell function with physiological aging in mice.

Given these results in model systems, the crucial question is whether sufficient telomere attrition occurs in human stem cells, which possess shorter telomeres and more regulated telomerase activity, during normal aging to activate an intrinsic DNA-damage response with attendant compromise of stem cell replicative function. The data here are mixed: Humans who harbor short telomeres because of congenital deficiencies of the telomerase complex develop an age-related failure of bone marrow function and other self-renewing compartments (Mitchell et al. 1999; Vulliamy et al. 2001; Yamaguchi et al. 2005) (see also Chapter 20). Moreover, telomere shortening has been shown to precede the development of overt cirrhosis in patients with chronic hepatitis of various etiologies (Kitada et al. 1995; Urabe et al. 1996; Miura et al. 1997; Wiemann et al. 2002). Finally, some studies have demonstrated a relationship between telomere length in peripheral blood lymphocytes (PBLs) and the onset of certain diseases associated with aging. Such studies in nonneoplastic diseases have shown that PBL telomere lengths can provide predictive information on the risk of developing atherosclerosis (Samani et al. 2001; Obana et al. 2003) and overall mortality (Cawthon et al. 2003). On the other hand, although telomere length shortens with human aging in many tissues, telomere length does not strictly correlate with age (Frenck et al. 1998). Likewise, in congenital telomere deficiency syndromes, many individuals of such kindreds who harbor the defective telomerase component do not demonstrate any overt phenotype. In fact, TERT and TERC deficiency in humans shows strong anticipation; i.e., the phenotypic effects of the mutation can be subtle or absent in many carriers but are more severe in subsequent generations, exactly analogous to telomerase-deficient mice. We believe the data favor the possibility of both telomere-independent and telomere-dependent aging of stem cells in humans.

Telomere dysfunction appears to activate the classic tumor suppressor mechanisms of senescence and apoptosis (Fig. 4). As mentioned previously, telomere dysfunction is sensed in a biochemically similar manner to a double-strand break, which potently induces p53, and telomere dysfunction in cultured human and mouse cells has been shown to induce p53-mediated senescence and other checkpoint responses such as apoptosis. Although more controversial, telomere dysfunction also appears to induce $p16^{INK4a}$ (Jacobs and de Lange 2004), albeit with greatly delayed kinetics compared to p53. For example, $p16^{INK4a}$ expression in human fibroblasts undergoing telomere-shortening-induced senescence occurs 7–14 days later than p53 activation, presumably in response to the same DNA-damage-like stimulus. On the other hand, the placement of human epithelial cells and murine cells in culture has been shown to rapidly induce $p16^{INK4a}$ in the absence of telomere shortening (Foster et al. 1998; Huschtscha et al. 1998; Kiyono et al. 1998; Zindy et al. 1998). Therefore, both p53 and $p16^{INK4a}$ appear to be induced by telomere-dependent and -independent stimuli.

In mice, many of the phenotypic consequences of telomere dysfunction in stem cell compartments can be rescued by p53 deficiency (Chin et al. 1999). In particular, apoptosis in the intestinal crypt and testis is nearly completely abrogated by p53 deficiency in mice with dysfunctional telomeres. Senescence, however, may also have a role in mediating the effects of telomere dysfunction, as suggested in a recent study by Rudolph and colleagues (Choudhury et al. 2007). The authors showed that deficiency of $p21^{CIP}$, a p53 transcriptional target that potently inhibits the cell cycle, rescued many of the hypoproliferative phenotypes in the intestine and bone marrow of mice with telomere dysfunction. This benefit in the stem cell compartments remarkably correlated with an enhanced life span in mice with telomere dysfunction. These data suggest the intriguing possibility that p53 promotes aging through $p21^{CIP}$ activation in the context of telomere dysfunction by impairing stem cell proliferation and function.

CONCLUSIONS

Demographic studies indicate that the percentage of elderly adults is at a historical high and will continue to climb for at least the next few decades in most developed regions of the world. Concomitant increases in the prevalence of age-related degenerative and malignant conditions and associated morbidities will thus place a heavy burden on future health care resources. The need to develop therapeutic strategies aimed at treating pathophysiological conditions in the elderly is therefore both medically and economically relevant. An increasing body of evidence suggests that at least some aspects of mammalian aging may be underwritten by a progressive stem cell functional decline. Several possible molecular mechanisms—senescence, apoptosis, or other forms of dysfunction—have been suggested whereby stem cell function could be compromised with aging. In particular, aging or cancer appears to represent either the success or failure, respectively, of potent tumor suppressor mechanisms. These new understandings of the age-induced compromise of stem cell function offer hope to identify a means to retard or even reverse stem cell aging.

REFERENCES

Akashi K., Traver D., Miyamoto T., and Weissman I.L. 2000. A clonogenic common myeloid progenitor that gives rise to all myeloid lineages. *Nature* **404:** 193–197.

Alcorta D.A., Xiong Y., Phelps D., Hannon G., Beach D., and Barrett J.C. 1996. Involvement of the cyclin-dependent kinase inhibitor p16 (INK4a) in replicative senescence of normal human fibroblasts. *Proc. Natl. Acad. Sci.* **93:** 13742–13747.

Allsopp R.C., Morin G.B., DePinho R., Harley C.B., and Weissman I.L. 2003a. Telomerase is required to slow telomere shortening and extend replicative lifespan of HSCs during serial transplantation. *Blood* **102:** 517–520.

Allsopp R.C., Morin G.B., Horner J.W., DePinho R., Harley C.B., and Weissman I.L. 2003b. Effect of TERT over-expression on the long-term transplantation capacity of hematopoietic stem cells. *Nat. Med.* **9:** 369–371.

Appelbaum F.R., Gundacker H., Head D.R., Slovak M.L., Willman C.L., Godwin J.E., Anderson J.E., and Petersdorf S.H. 2006. Age and acute myeloid leukemia. *Blood* **107:** 3481–3485.

Artandi S.E., Alson S., Tietze M.K., Sharpless N.E., Ye S., Greenberg R.A., Castrillon D.H., Horner J.W., Weiler S.R., Carrasco R.D., and DePinho R.A. 2002. Constitutive telomerase expression promotes mammary carcinomas in aging mice. *Proc. Natl. Acad. Sci.* **99:** 8191–8196.

Ash R.C., Horowitz M.M., Gale R.P., van Bekkum D.W., Casper J.T., Gordon-Smith E.C., Henslee P.J., Kolb H.J., Lowenberg B., and Masaoka T., et al. 1991. Bone marrow transplantation from related donors other than HLA-identical siblings: Effect of T cell depletion. *Bone Marrow Transplant.* **7:** 443–452.

Aslanian A., Iaquinta P.J., Verona R., and Lees J.A. 2004. Repression of the Arf tumor suppressor by E2F3 is required for normal cell cycle kinetics. *Genes Dev.* **18:** 1413–1422.

Beghe C., Wilson A., and Ershler W.B. 2004. Prevalence and outcomes of anemia in geriatrics: A systematic review of the literature. *Am. J. Med.* (suppl. 7A) **116:** 3S–10S.

Bielas J.H., Loeb K.R., Rubin B.P., True L.D., and Loeb L.A. 2006. Human cancers express a mutator phenotype. *Proc. Natl. Acad. Sci.* **103:** 18238–18242.

Blasco M.A., Lee H.W., Hande M.P., Samper E., Lansdorp P.M., DePinho R.A., and Greider C.W. 1997. Telomere shortening and tumor formation by mouse cells lacking telomerase RNA. *Cell* **91:** 25–34.

Boccadoro M., Palumbo A., Bringhen S., Merletti F., Ciccone G., Richiardi L., Rus C., Bertola A., Giaccone L., Omede P., and Musto P. 2002. Oral melphalan at diagnosis hampers adequate collection of peripheral blood progenitor cells in multiple myeloma. *Haematologica* **87**: 846–850.

Bohr V.A. 2002. Human premature aging syndromes and genomic instability. *Mech. Ageing Dev.* **123**: 987–993.

Bradford G.B., Williams B., Rossi R., and Bertoncello I. 1997. Quiescence, cycling, and turnover in the primitive hematopoietic stem cell compartment. *Exp. Hematol.* **25**: 445–453.

Bruggeman S.W., Valk-Lingbeek M.E., van der Stoop P.P., Jacobs J.J., Kieboom K., Tanger E., Hulsman D., Leung C., Arsenijevic Y., Marino S., and van Lohuizen M. 2005. Ink4a and Arf differentially affect cell proliferation and neural stem cell self-renewal in Bmi1-deficient mice. *Genes Dev.* **19**: 1438–1443.

Brunello A., Basso U., Pogliani C., Jirillo A., Ghiotto C., Koussis H., Lumachi F., Iacobone M., Vamvakas L., and Monfardini S. 2005. Adjuvant chemotherapy for elderly patients (> or =70 years) with early high-risk breast cancer: A retrospective analysis of 260 patients. *Ann. Oncol.* **16**: 1276–1282.

Bryder D., Rossi D.J., and Weissman I.L. 2006. Hematopoietic stem cells: The paradigmatic tissue-specific stem cell. *Am. J. Pathol.* **169**: 338–346.

Buckner C.D., Clift R.A., Sanders J.E., Stewart P., Bensinger W.I., Doney K.C., Sullivan K.M., Witherspoon R.P., Deeg H.J., and Appelbaum F.R. 1984. Marrow harvesting from normal donors. *Blood* **64**: 630–634.

Bulavin D.V., Phillips C., Nannenga B., Timofeev O., Donehower L.A., Anderson C.W., Appella E., and Fornace A.J., Jr. 2004. Inactivation of the Wip1 phosphatase inhibits mammary tumorigenesis through p38 MAPK-mediated activation of the p16(Ink4a)-p19(Arf) pathway. *Nat. Genet.* **36**: 343–350.

Butler A.E., Janson J., Bonner-Weir S., Ritzel R., Rizza R.A., and Butler P.C. 2003. Beta-cell deficit and increased beta-cell apoptosis in humans with type 2 diabetes. *Diabetes* **52**: 102–110.

Cairns J. 1975. Mutation selection and the natural history of cancer. *Nature* **255**: 197–200.

———. 2002. Somatic stem cells and the kinetics of mutagenesis and carcinogenesis. *Proc. Natl. Acad. Sci.* **99**: 10567–10570.

Campisi J. 2001. Cellular senescence as a tumor-suppressor mechanism. *Trends Cell Biol.* **11**: S27–S31.

———. 2003. Cellular senescence and apoptosis: How cellular responses might influence aging phenotypes. *Exp. Gerontol.* **38**: 5–11.

———. 2005. Suppressing cancer: The importance of being senescent. *Science* **309**: 886–887.

Cao L., Li W., Kim S., Brodie S.G., and Deng C.X. 2003. Senescence, aging, and malignant transformation mediated by p53 in mice lacking the Brca1 full-length isoform. *Genes Dev.* **17**: 201–213.

Castro-Malaspina H., Harris R.E., Gajewski J., Ramsay N., Collins R., Dharan B., King R., and Deeg H.J. 2002. Unrelated donor marrow transplantation for myelodysplastic syndromes: Outcome analysis in 510 transplants facilitated by the National Marrow Donor Program. *Blood* **99**: 1943–1951.

Cawthon R.M., Smith K.R., O'Brien E., Sivatchenko A., and Kerber R.A. 2003. Association between telomere length in blood and mortality in people aged 60 years or older. *Lancet* **361**: 393–395.

Chang S., Khoo C.M., Naylor M.L., Maser R.S., and DePinho R.A. 2003. Telomere-based crisis: Functional differences between telomerase activation and ALT in tumor progression. *Genes Dev.* **17**: 88–100.

Chen J. 2004. Senescence and functional failure in hematopoietic stem cells. *Exp. Hematol.* **32:** 1025–1032.

Chen J., Astle C.M., and Harrison D.E. 2003. Hematopoietic senescence is postponed and hematopoietic stem cell function is enhanced by dietary restriction. *Exp. Hematol.* **31:** 1097–1103.

Cheshier S.H., Morrison S.J., Liao X., and Weissman I.L. 1999. In vivo proliferation and cell cycle kinetics of long-term self-renewing hematopoietic stem cells. *Proc. Natl. Acad. Sci.* **96:** 3120–3125.

Chin L., Artandi S.E., Shen Q., Tam A., Lee S.L., Gottlieb G.J., Greider C.W., and DePinho R.A. 1999. p53 deficiency rescues the adverse effects of telomere loss and cooperates with telomere dysfunction to accelerate carcinogenesis. *Cell* **97:** 527–538.

Choudhury A.R., Ju Z., Djojosubroto M.W., Schienke A., Lechel A., Schaetzlein S., Jiang H., Stepczynska A., Wang C., Buer J., et al. 2007. Cdkn1a deletion improves stem cell function and lifespan of mice with dysfunctional telomeres without accelerating cancer formation. *Nat. Genet.* **39:** 99–105.

Conboy I.M., Conboy M.J., Wagers A.J., Girma E.R., Weissman I.L., and Rando T.A. 2005. Rejuvenation of aged progenitor cells by exposure to a young systemic environment. *Nature* **433:** 760–764.

de Boer J., Andressoo J.O., de Wit J., Huijmans J., Beems R.B., van Steeg H., Weeda G., van der Horst G.T., van Leeuwen W., Themmen A.P., Meradji M., and Hoeijmakers J.H. 2002. Premature aging in mice deficient in DNA repair and transcription. *Science* **296:** 1276–1279.

DeGregori J., Leone G., Miron A., Jakoi L., and Nevins J.R. 1997. Distinct roles for E2F proteins in cell growth control and apoptosis. *Proc. Natl. Acad. Sci.* **94:** 7245–7250.

de Haan G. and Van Zant G. 1999. Dynamic changes in mouse hematopoietic stem cell numbers during aging. *Blood* **93:** 3294–3301.

de Haan G., Nijhof W., and Van Zant G. 1997. Mouse strain-dependent changes in frequency and proliferation of hematopoietic stem cells during aging: Correlation between lifespan and cycling activity. *Blood* **89:** 1543–1550.

Dimri G.P., Lee X., Basile G., Acosta M., Scott G., Roskelley C., Medrano E.E., Linskens M., Rubelj I., and Pereira-Smith O., et al. 1995. A biomarker that identifies senescent human cells in culture and in aging skin in vivo. *Proc. Natl. Acad. Sci.* **92:** 9363–9367.

Dumble M., Moore L., Chambers S.M., Geiger H., Van Zant G., Goodell M.A., and Donehower L.A. 2007. The impact of altered p53 dosage on hematopoietic stem cell dynamics during aging. *Blood* **109:** 1736–1742.

Enomoto K., Mimura T., Harris D.L., and Joyce N.C. 2006. Age differences in cyclin-dependent kinase inhibitor expression and rb hyperphosphorylation in human corneal endothelial cells. *Invest. Ophthalmol. Vis. Sci.* **47:** 4330–4340.

Enwere E., Shingo T., Gregg C., Fujikawa H., Ohta S., and Weiss S. 2004. Aging results in reduced epidermal growth factor receptor signaling, diminished olfactory neurogenesis, and deficits in fine olfactory discrimination. *J. Neurosci.* **24:** 8354–8365.

Flores I., Cayuela M.L., and Blasco M.A. 2005. Effects of telomerase and telomere length on epidermal stem cell behavior. *Science* **309:** 1253–1256.

Foster S.A., Wong D.J., Barrett M.T., and Galloway D.A. 1998. Inactivation of p16 in human mammary epithelial cells by CpG island methylation. *Mol. Cell. Biol.* **18:** 1793–1801.

Frenck R.W., Jr., Blackburn E.H., and Shannon K.M. 1998. The rate of telomere sequence loss in human leukocytes varies with age. *Proc. Natl. Acad. Sci.* **95:** 5607–5610.

Gardner R.V., Astle C.M., and Harrison D.E. 1997. Hematopoietic precursor cell exhaustion is a cause of proliferative defect in primitive hematopoietic stem cells (PHSC) after chemotherapy. *Exp. Hematol.* **25:** 495–501.

Gil J., Bernard D., Martinez D., and Beach D. 2004. Polycomb CBX7 has a unifying role in cellular lifespan. *Nat. Cell Biol.* **6:** 67–72.

Gonzalez-Suarez E., Samper E., Ramirez A., Flores J.M., Martin-Caballero J., Jorcano J.L., and Blasco M.A. 2001. Increased epidermal tumors and increased skin wound healing in transgenic mice overexpressing the catalytic subunit of telomerase, mTERT, in basal keratinocytes. *EMBO J.* **20:** 2619–2630.

Gray-Schopfer V.C., Cheong S.C., Chong H., Chow J., Moss T., Abdel-Malek Z.A., Marais R., Wynford-Thomas D., and Bennett D.C. 2006. Cellular senescence in naevi and immortalisation in melanoma: A role for p16? *Br. J. Cancer* **95:** 496–505.

Guralnik J.M., Eisenstaedt R.S., Ferrucci L., Klein H.G., and Woodman R.C. 2004. Prevalence of anemia in persons 65 years and older in the United States: Evidence for a high rate of unexplained anemia. *Blood* **104:** 2263–2268.

Hamilton M.L., Van Remmen H., Drake J.A., Yang H., Guo Z.M., Kewitt K., Walter C.A., and Richardson A. 2001. Does oxidative damage to DNA increase with age? *Proc. Natl. Acad. Sci.* **98:** 10469–10474.

Harrison D.E. 1979. Mouse erythropoietic stem cell lines function normally 100 months: Loss related to number of transplantations. *Mech. Ageing Dev.* **9:** 427–433.

Herbig U., Ferreira M., Condel L., Carey D., and Sedivy J.M. 2006. Cellular senescence in aging primates. *Science* **311:** 1257.

Hodgson G.S. and Bradley T.R. 1984. In vivo kinetic status of hematopoietic stem and progenitor cells as inferred from labeling with bromodeoxyuridine. *Exp. Hematol.* **12:** 683–687.

Hoeijmakers J.H. 2001. Genome maintenance mechanisms for preventing cancer. *Nature* **411:** 366–374.

Huot T.J., Rowe J., Harland M., Drayton S., Brookes S., Gooptu C., Purkis P., Fried M., Bataille V., Hara E., Newton-Bishop J., and Peters G. 2002. Biallelic mutations in p16(INK4a) confer resistance to Ras- and Ets-induced senescence in human diploid fibroblasts. *Mol. Cell. Biol.* **22:** 8135–8143.

Huschtscha L.I., Noble J.R., Neumann A.A., Moy E.L., Barry P., Melki J.R., Clark S.J., and Reddel R.R. 1998. Loss of p16INK4 expression by methylation is associated with life-span extension of human mammary epithelial cells. *Cancer Res.* **58:** 3508–3512.

Itahana K., Zou Y., Itahana Y., Martinez J.L., Beausejour C., Jacobs J.J., Van Lohuizen M., Band V., Campisi J., and Dimri G.P. 2003. Control of the replicative life span of human fibroblasts by p16 and the polycomb protein Bmi-1. *Mol. Cell. Biol.* **23:** 389–401.

Ito K., Hirao A., Arai F., Takubo K., Matsuoka S., Miyamoto K., Ohmura M., Naka K., Hosokawa K., Ikeda Y., and Suda T. 2006. Reactive oxygen species act through p38 MAPK to limit the lifespan of hematopoietic stem cells. *Nat. Med.* **12:** 446–451.

Ito K., Hirao A., Arai F., Matsuoka S., Takubo K., Hamaguchi I., Nomiyama K., Hosokawa K., Sakurada K., Nakagata N., et al. 2004. Regulation of oxidative stress by ATM is required for self-renewal of haematopoietic stem cells. *Nature* **431:** 997–1002.

Jacobs J.J. and de Lange T. 2004. Significant role for p16INK4a in p53-independent telomere-directed senescence. *Curr. Biol.* **14:** 2302–2308.

Jacobs J.J., Kieboom K., Marino S., DePinho R.A., and van Lohuizen M. 1999. The onco-gene and Polycomb-group gene bmi-1 regulates cell proliferation and senescence through the ink4a locus. *Nature* **397:** 164–168.

Janzen V., Forkert R., Fleming H.E., Saito Y., Waring M.T., Dombkowski D.M., Cheng T., DePinho R.A., Sharpless N.E., and Scadden D.T. 2006. Stem-cell ageing modified by the cyclin-dependent kinase inhibitor p16INK4a. *Nature* **443:** 421–426.

Kamijo T., Zindy F., Roussel M.F., Quelle D.E., Downing J.R., Ashmun R.A., Grosveld G., and Sherr C.J. 1997. Tumor suppression at the mouse INK4a locus mediated by the alternative reading frame product p19ARF. *Cell* **91:** 649–659.

Kamminga L.M., Bystrykh L.V., de Boer A., Houwer S., Douma J., Weersing E., Dontje B., and de Haan G. 2006. The Polycomb group gene Ezh2 prevents hematopoietic stem cell exhaustion. *Blood* **107:** 2170–2179.

Karpowicz P., Morshead C., Kam A., Jervis E., Ramunas J., Cheng V., and van der Kooy D. 2005. Support for the immortal strand hypothesis: Neural stem cells partition DNA asymmetrically in vitro. *J. Cell Biol.* **170:** 721–732.

Khaidakov M., Heflich R.H., Manjanatha M.G., Myers M.B., and Aidoo A. 2003. Accumulation of point mutations in mitochondrial DNA of aging mice. *Mutat. Res.* **526:** 1–7.

Kim M., Moon H.B., and Spangrude G.J. 2003. Major age-related changes of mouse hematopoietic stem/progenitor cells. *Ann. N.Y. Acad. Sci.* **996:** 195–208.

Kim W.Y. and Sharpless N.E. 2006. The regulation of INK4/ARF in cancer and aging. *Cell* **127:** 265–275.

Kirkwood T.B. 1977. Evolution of ageing. *Nature* **270:** 301–304.

————. 2005. Understanding the odd science of aging. *Cell* **120:** 437–447.

Kitada T., Seki S., Kawakita N., Kuroki T., and Monna T. 1995. Telomere shortening in chronic liver diseases. *Biochem. Biophys. Res. Commun.* **211:** 33–39.

Kiyono T., Foster S.A., Koop J.I., McDougall J.K., Galloway D.A., and Klingelhutz A.J. 1998. Both Rb/p16INK4a inactivation and telomerase activity are required to immortalize human epithelial cells. *Nature* **396:** 84–88.

Knudsen L.M., Rasmussen T., Jensen L., and Johnsen H.E. 1999. Reduced bone marrow stem cell pool and progenitor mobilisation in multiple myeloma after melphalan treatment. *Med. Oncol.* **16:** 245–254.

Kollman C., Howe C.W., Anasetti C., Antin J.H., Davies S.M., Filipovich A.H., Hegland J., Kamani N., Kernan N.A., King R., et al. 2001. Donor characteristics as risk factors in recipients after transplantation of bone marrow from unrelated donors: The effect of donor age. *Blood* **98:** 2043–2051.

Kondo M., Weissman I.L., and Akashi K. 1997. Identification of clonogenic common lymphoid progenitors in mouse bone marrow. *Cell* **91:** 661–672.

Kotake Y., Cao R., Viatour P., Sage J., Zhang Y., and Xiong Y. 2007. pRB family proteins are required for H3K27 trimethylation and Polycomb repression complexes binding to and silencing p16INK4alpha tumor suppressor gene. *Genes Dev.* **21:** 49–54.

Krishnamurthy J., Ramsey M.R., Ligon K.L., Torrice C., Koh A., Bonner-Weir S., and Sharpless N.E. 2006. p16INK4a induces an age-dependent decline in islet regenerative potential. *Nature* **443:** 453–457.

Krishnamurthy J., Torrice C., Ramsey M.R., Kovalev G.I., Al-Regaiey K., Su L., and Sharpless N.E. 2004. Ink4a/Arf expression is a biomarker of aging. *J. Clin. Invest.* **114:** 1299–1307.

Kuhn H.G., Dickinson-Anson H., and Gage F.H. 1996. Neurogenesis in the dentate gyrus of the adult rat: Age-related decrease of neuronal progenitor proliferation. *J. Neurosci.* **16:** 2027–2033.

Kujoth G.C., Hiona A., Pugh T.D., Someya S., Panzer K., Wohlgemuth S.E., Hofer T., Seo A.Y., Sullivan R., Jobling W.A., et al. 2005. Mitochondrial DNA mutations, oxidative stress, and apoptosis in mammalian aging. *Science* **309:** 481–484.

Lang D., Lu M.M., Huang L., Engleka K.A., Zhang M., Chu E.Y., Lipner S., Skoultchi A., Millar S.E., and Epstein J.A. 2005. Pax3 functions at a nodal point in melanocyte stem cell differentiation. *Nature* **433:** 884–887.

Lee H.W., Blasco M.A., Gottlieb G.J., Horner J.W., II, Greider C.W., and DePinho R.A. 1998. Essential role of mouse telomerase in highly proliferative organs. *Nature* **392:** 569–574.

Lenhoff S., Hjorth M., Westin J., Brinch L., Backstrom B., Carlson K., Christiansen I., Dahl I.M., Gimsing P., Hammerstrom J., et al. 2006. Impact of age on survival after intensive therapy for multiple myeloma: A population-based study by the Nordic Myeloma Study Group. *Br. J. Haematol.* **133:** 389–396.

Liang Y., Van Zant G., and Szilvassy S.J. 2005. Effects of aging on the homing and engraftment of murine hematopoietic stem and progenitor cells. *Blood* **106:** 1479–1487.

Lichtman M.A. and Rowe J.M. 2004. The relationship of patient age to the pathobiology of the clonal myeloid diseases. *Semin. Oncol.* **31:** 185–197.

Lindahl T. 1993. Instability and decay of the primary structure of DNA. *Nature* **362:** 709–715.

Linton P.J. and Dorshkind K. 2004. Age-related changes in lymphocyte development and function. *Nat. Immunol.* **5:** 133–139.

Maier B., Gluba W., Bernier B., Turner T., Mohammad K., Guise T., Sutherland A., Thorner M., and Scrable H. 2004. Modulation of mammalian life span by the short isoform of p53. *Genes Dev.* **18:** 306–319.

Martin G.M. and Oshima J. 2000. Lessons from human progeroid syndromes. *Nature* **408:** 263–266.

Maslov A.Y., Barone T.A., Plunkett R.J., and Pruitt S.C. 2004. Neural stem cell detection, characterization, and age-related changes in the subventricular zone of mice. *J. Neurosci.* **24:** 1726–1733.

Matthews C., Gorenne I., Scott S., Figg N., Kirkpatrick P., Ritchie A., Goddard M., and Bennett M. 2006. Vascular smooth muscle cells undergo telomere-based senescence in human atherosclerosis: Effects of telomerase and oxidative stress. *Circ. Res.* **99:** 156–164.

Melk A., Schmidt B.M., Takeuchi O., Sawitzki B., Rayner D.C., and Halloran P.F. 2004. Expression of p16INK4a and other cell cycle regulator and senescence associated genes in aging human kidney. *Kidney Int.* **65:** 510–520.

Melk A., Kittikowit W., Sandhu I., Halloran K.M., Grimm P., Schmidt B.M., and Halloran P.F. 2003. Cell senescence in rat kidneys in vivo increases with growth and age despite lack of telomere shortening. *Kidney Int.* **63:** 2134–2143.

Meng A., Wang Y., Van Zant G., and Zhou D. 2003. Ionizing radiation and busulfan induce premature senescence in murine bone marrow hematopoietic cells. *Cancer Res.* **63:** 5414–5419.

Menzel O., Migliaccio M., Goldstein D.R., Dahoun S., Delorenzi M., and Rufer N. 2006. Mechanisms regulating the proliferative potential of human CD8+ T lymphocytes overexpressing telomerase. *J. Immunol.* **177:** 3657–3668.

Michaloglou C., Vredeveld L.C., Soengas M.S., Denoyelle C., Kuilman T., van der Horst C.M., Majoor D.M., Shay J.W., Mooi W.J., and Peeper D.S. 2005. BRAFE600-associated senescence-like cell cycle arrest of human naevi. *Nature* **436:** 720–724.

Mitchell J.R., Wood E., and Collins K. 1999. A telomerase component is defective in the human disease dyskeratosis congenita. *Nature* **402:** 551–555.

Miura N., Horikawa I., Nishimoto A., Ohmura H., Ito H., Hirohashi S., Shay J.W., and Oshimura M. 1997. Progressive telomere shortening and telomerase reactivation during hepatocellular carcinogenesis. *Cancer Genet. Cytogenet.* **93:** 56–62.

Molofsky A.V., He S., Bydon M., Morrison S.J., and Pardal R. 2005. Bmi-1 promotes neural stem cell self-renewal and neural development but not mouse growth and survival by repressing the p16Ink4a and p19Arf senescence pathways. *Genes Dev.* **19:** 1432–1437.

Molofsky A.V., Pardal R., Iwashita T., Park I.K., Clarke M.F., and Morrison S.J. 2003. Bmi-1 dependence distinguishes neural stem cell self-renewal from progenitor proliferation. *Nature* **425:** 962–967.

Molofsky A.V., Slutsky S.G., Joseph N.M., He S., Pardal R., Krishnamurthy J., Sharpless N.E., and Morrison S.J. 2006. Increasing p16INK4a expression decreases forebrain progenitors and neurogenesis during ageing. *Nature* **443:** 448–452.

Moore K.A. and Lemischka I.R. 2006. Stem cells and their niches. *Science* **311:** 1880–1885.

Morrison S.J., Wandycz A.M., Akashi K., Globerson A., and Weissman I.L. 1996. The aging of hematopoietic stem cells. *Nat. Med.* **2:** 1011–1016.

Mostoslavsky R., Chua K.F., Lombard D.B., Pang W.W., Fischer M.R., Gellon L., Liu P., Mostoslavsky G., Franco S., Murphy M.M., et al. 2006. Genomic instability and aging-like phenotype in the absence of mammalian SIRT6. *Cell* **124:** 315–329.

Motoyama N. and Naka K. 2004. DNA damage tumor suppressor genes and genomic instability. *Curr. Opin. Genet. Dev.* **14:** 11–16.

Narita M., Nunez S., Heard E., Lin A.W., Hearn S.A., Spector D.L., Hannon G.J., and Lowe S.W. 2003. Rb-mediated heterochromatin formation and silencing of E2F target genes during cellular senescence. *Cell* **113:** 703–716.

Navarro S., Meza N.W., Quintana-Bustamante O., Casado J.A., Jacome A., McAllister K., Puerto S., Surralles J., Segovia J.C., and Bueren J.A. 2006. Hematopoietic dysfunction in a mouse model for Fanconi anemia group D1. *Mol. Ther.* **14:** 525–535.

Nielsen G.P., Stemmer-Rachamimov A.O., Shaw J., Roy J.E., Koh J., and Louis D.N. 1999. Immunohistochemical survey of p16INK4A expression in normal human adult and infant tissues. *Lab. Invest.* **79:** 1137–1143.

Nishimura E.K., Granter S.R., and Fisher D.E. 2005. Mechanisms of hair graying: Incomplete melanocyte stem cell maintenance in the niche. *Science* **307:** 720–724.

Obana N., Takagi S., Kinouchi Y., Tokita Y., Sekikawa A., Takahashi S., Hiwatashi N., Oikawa S., and Shimosegawa T. 2003. Telomere shortening of peripheral blood mononuclear cells in coronary disease patients with metabolic disorders. *Intern. Med.* **42:** 150–153.

Ohtani N., Zebedee Z., Huot T.J., Stinson J.A., Sugimoto M., Ohashi Y., Sharrocks A.D., Peters G., and Hara E. 2001. Opposing effects of Ets and Id proteins on p16INK4a expression during cellular senescence. *Nature* **409:** 1067–1070.

Park I.K., Qian D., Kiel M., Becker M.W., Pihalja M., Weissman I.L., Morrison S.J., and Clarke M.F. 2003. Bmi-1 is required for maintenance of adult self-renewing haematopoietic stem cells. *Nature* **423:** 302–305.

Pasini D., Bracken A.P., and Helin K. 2004. Polycomb group proteins in cell cycle progression and cancer. *Cell Cycle* **3:** 396–400.

Passegue E., Wagers A.J., Giuriato S., Anderson W.C., and Weissman I.L. 2005. Global analysis of proliferation and cell cycle gene expression in the regulation of hematopoietic stem and progenitor cell fates. *J. Exp. Med.* **202:** 1599–1611.

Pearce D.J., Anjos-Afonso F., Ridler C.M., Eddaoudi A., and Bonnet D. 2007. Age-dependent increase in side population distribution within hematopoiesis: Implications for our understanding of the mechanism of aging. *Stem Cells* **25:** 828–350.

Potten C.S., Owen G., and Booth D. 2002. Intestinal stem cells protect their genome by selective segregation of template DNA strands. *J. Cell Sci.* **115:** 2381–2388.

Prasher J.M., Lalai A.S., Heijmans-Antonissen C., Ploemacher R.E., Hoeijmakers J.H., Touw I.P., and Niedernhofer L.J. 2005. Reduced hematopoietic reserves in DNA interstrand crosslink repair-deficient Ercc1-/- mice. *EMBO J.* **24:** 861–871.

Rane S.G., Dubus P., Mettus R.V., Galbreath E.J., Boden G., Reddy E.P., and Barbacid M. 1999. Loss of Cdk4 expression causes insulin-deficient diabetes and Cdk4 activation results in beta-islet cell hyperplasia. *Nat. Genet.* **22:** 44–52.

Reese J.S., Liu L., and Gerson S.L. 2003. Repopulating defect of mismatch repair-deficient hematopoietic stem cells. *Blood* **102:** 1626–1633.

Ressler S., Bartkova J., Niederegger H., Bartek J., Scharffetter-Kochanek K., Jansen-Durr P., and Wlaschek M. 2006. p16INK4A is a robust in vivo biomarker of cellular aging in human skin. *Aging Cell* **5:** 379–389.

Reya T., Morrison S.J., Clarke M.F., and Weissman I.L. 2001. Stem cells, cancer, and cancer stem cells. *Nature* **414:** 105–111.

Rossi D.J., Bryder D., and Weissman I.L. 2007a. Hematopoietic stem cell aging: Mechanism and consequence. *Exp. Gerontol.* **42:** 385–390.

Rossi D.J., Bryder D., Seita J., Nussenzweig A., Hoeijmakers J., and Weissman I.L. 2007b. Deficiencies in DNA damage repair limit the function of haematopoietic stem cells with age. *Nature* **447:** 725–729.

Rossi D.J., Bryder D., Zahn J.M., Ahlenius H., Sonu R., Wagers A.J., and Weissman I.L. 2005. Cell intrinsic alterations underlie hematopoietic stem cell aging. *Proc. Natl. Acad. Sci.* **102:** 9194–9199.

Rudolph K.L., Chang S., Lee H.W., Blasco M., Gottlieb G.J., Greider C., and DePinho R.A. 1999. Longevity, stress response, and cancer in aging telomerase-deficient mice. *Cell* **96:** 701–712.

Sage J., Miller A.L., Perez-Mancera P.A., Wysocki J.M., and Jacks T. 2003. Acute mutation of retinoblastoma gene function is sufficient for cell cycle re-entry. *Nature* **424:** 223–228.

Samani N.J., Boultby R., Butler R., Thompson J.R., and Goodall A.H. 2001. Telomere shortening in atherosclerosis. *Lancet* **358:** 472–473.

Sarin K.Y., Cheung P., Gilison D., Lee E., Tennen R.I., Wang E., Artandi M.K., Oro A.E., and Artandi S.E. 2005. Conditional telomerase induction causes proliferation of hair follicle stem cells. *Nature* **436:** 1048–1052.

Sasaki M., Ikeda H., Sato Y., and Nakanuma Y. 2006. Decreased expression of Bmi1 is closely associated with cellular senescence in small bile ducts in primary biliary cirrhosis. *Am. J. Pathol.* **169:** 831–845.

Satyanarayana A., Greenberg R.A., Schaetzlein S., Buer J., Masutomi K., Hahn W.C., Zimmermann S., Martens U., Manns M.P., and Rudolph K.L. 2004. Mitogen stimulation cooperates with telomere shortening to activate DNA damage responses and senescence signaling. *Mol. Cell. Biol.* **24:** 5459–5474.

Schofield R. 1978. The relationship between the spleen colony-forming cell and the haemopoietic stem cell. *Blood Cells* **4:** 7–25.

Shinin V., Gayraud-Morel B., Gomes D., and Tajbakhsh S. 2006. Asymmetric division and cosegregation of template DNA strands in adult muscle satellite cells. *Nat. Cell Biol.* **8:** 677–687.

Sone H. and Kagawa Y. 2005. Pancreatic beta cell senescence contributes to the pathogenesis of type 2 diabetes in high-fat diet-induced diabetic mice. *Diabetologia* **48:** 58–67.

Sparmann A. and van Lohuizen M. 2006. Polycomb silencers control cell fate, development and cancer. *Nat. Rev. Cancer* **6:** 846–856.

Stein G.H., Drullinger L.F., Soulard A., and Dulic V. 1999. Differential roles for cyclin-dependent kinase inhibitors p21 and p16 in the mechanisms of senescence and differentiation in human fibroblasts. *Mol. Cell. Biol.* **19:** 2109–2117.

Stepanova L. and Sorrentino B.P. 2005. A limited role for p16Ink4a and p19Arf in the loss of hematopoietic stem cells during proliferative stress. *Blood* **106:** 827–832.

Stevnsner T., Thorslund T., de Souza-Pinto N.C., and Bohr V.A. 2002. Mitochondrial repair of 8-oxoguanine and changes with aging. *Exp. Gerontol.* **37:** 1189–1196.

Sudo K., Ema H., Morita Y., and Nakauchi H. 2000. Age-associated characteristics of murine hematopoietic stem cells. *J. Exp. Med.* **192:** 1273–1280.

TeKippe M., Harrison D.E., and Chen J. 2003. Expansion of hematopoietic stem cell phenotype and activity in Trp53-null mice. *Exp. Hematol.* **31:** 521–527.

Trifunovic A., Wredenberg A., Falkenberg M., Spelbrink J.N., Rovio A.T., Bruder C.E., Bohlooly Y.M., Gidlof S., Oldfors A., Wibom R., et al. 2004. Premature ageing in mice expressing defective mitochondrial DNA polymerase. *Nature* **429:** 417–423.

Tsuji T., Aoshiba K., and Nagai A. 2006. Alveolar cell senescence in patients with pulmonary emphysema. *Am. J. Respir. Crit. Care Med.* **174:** 886–893.

Tsutsui T., Hesabi B., Moons D.S., Pandolfi P.P., Hansel K.S., Koff A., and Kiyokawa H. 1999. Targeted disruption of CDK4 delays cell cycle entry with enhanced p27(Kip1) activity. *Mol. Cell. Biol.* **19:** 7011–7019.

Tyner S.D., Venkatachalam S., Choi J., Jones S., Ghebranious N., Igelmann H., Lu X., Soron G., Cooper B., Brayton C., et al. 2002. p53 mutant mice that display early ageing-associated phenotypes. *Nature* **415:** 45–53.

Urabe Y., Nouso K., Higashi T., Nakatsukasa H., Hino N., Ashida K., Kinugasa N., Yoshida K., Uematsu S., and Tsuji T. 1996. Telomere length in human liver diseases. *Liver* **16:** 293–297.

Urbanek K., Torella D., Sheikh F., De Angelis A., Nurzynska D., Silvestri F., Beltrami C.A., Bussani R., Beltrami A.P., Quaini F., et al. 2005. Myocardial regeneration by activation of multipotent cardiac stem cells in ischemic heart failure. *Proc. Natl. Acad. Sci.* **102:** 8692–8697.

Vermulst M., Bielas J.H., Kujoth G.C., Ladiges W.C., Rabinovitch P.S., Prolla T.A., and Loeb L.A. 2007. Mitochondrial point mutations do not limit the natural lifespan of mice. *Nat. Genet.* **39:** 540–543.

Vogel H., Lim D.S., Karsenty G., Finegold M., and Hasty P. 1999. Deletion of Ku86 causes early onset of senescence in mice. *Proc. Natl. Acad. Sci.* **96:** 10770–10775.

von Zglinicki T., Saretzki G., Ladhoff J., d'Adda di Fagagna F., and Jackson S.P. 2005. Human cell senescence as a DNA damage response. *Mech. Ageing Dev.* **126:** 111–117.

Vulliamy T., Marrone A., Goldman F., Dearlove A., Bessler M., Mason P.J., and Dokal I. 2001. The RNA component of telomerase is mutated in autosomal dominant dyskeratosis congenita. *Nature* **413:** 432–435.

Wang W., Chen J.X., Liao R., Deng Q., Zhou J.J., Huang S., and Sun P. 2002. Sequential activation of the MEK-extracellular signal-regulated kinase and MKK3/6-p38 mitogen-activated protein kinase pathways mediates oncogenic ras-induced premature senescence. *Mol. Cell. Biol.* **22:** 3389–3403.

Wang Y., Schulte B.A., Larue A.C., Ogawa M., and Zhou D. 2006. Total body irradiation selectively induces murine hematopoietic stem cell senescence. *Blood* **107:** 358–366.

Weissman I.L. 2000. Stem cells: Units of development, units of regeneration, and units in evolution. *Cell* **100:** 157–168.

Wiemann S.U., Satyanarayana A., Tsahuridu M., Tillmann H.L., Zender L., Klempnauer J., Flemming P., Franco S., Blasco M.A., Manns M.P., and Rudolph K.L. 2002. Hepatocyte telomere shortening and senescence are general markers of human liver cirrhosis. *FASEB J.* **16:** 935–942.

Wong K.K., Maser R.S., Bachoo R.M., Menon J., Carrasco D.R., Gu Y., Alt F.W., and DePinho R.A. 2003. Telomere dysfunction and Atm deficiency compromises organ homeostasis and accelerates ageing. *Nature* **421:** 643–648.

Wright W.E. and Shay J.W. 2002. Historical claims and current interpretations of replicative aging. *Nat. Biotechnol.* **20:** 682–688.

Xing Z., Ryan M.A., Daria D., Nattamai K.J., Van Zant G., Wang L., Zheng Y., and Geiger H. 2006. Increased hematopoietic stem cell mobilization in aged mice. *Blood* **108:** 2190–2197.

Yakoub-Agha I., Mesnil F., Kuentz M., Boiron J.M., Ifrah N., Milpied N., Chehata S., Esperou H., Vernant J.P., Michallet M., et al. 2006. Allogeneic marrow stem-cell transplantation from human leukocyte antigen-identical siblings versus human leukocyte antigen-allelic-matched unrelated donors (10/10) in patients with standard-risk hematologic malignancy: A prospective study from the French Society of Bone Marrow Transplantation and Cell Therapy. *J. Clin. Oncol.* **24:** 5695–5702.

Yamaguchi H., Calado R.T., Ly H., Kajigaya S., Baerlocher G.M., Chanock S.J., Lansdorp P.M., and Young N.S. 2005. Mutations in TERT, the gene for telomerase reverse transcriptase, in aplastic anemia. *N. Engl. J. Med.* **352:** 1413–1424.

Yamazaki S., Iwama A., Takayanagi S., Morita Y., Eto K., Ema H., and Nakauchi H. 2006. Cytokine signals modulated via lipid rafts mimic niche signals and induce hibernation in hematopoietic stem cells. *EMBO J.* **25:** 3515–3523.

Yoon K.H., Ko S.H., Cho J.H., Lee J.M., Ahn Y.B., Song K.H., Yoo S.J., Kang M.I., Cha B.Y., Lee K.W., et al. 2003. Selective beta-cell loss and alpha-cell expansion in patients with type 2 diabetes mellitus in Korea. *J. Clin. Endocrinol. Metab.* **88:** 2300–2308.

Zhou S., Schuetz J.D., Bunting K.D., Colapietro A.M., Sampath J., Morris J.J., Lagutina I., Grosveld G.C., Osawa M., Nakauchi H., and Sorrentino B.P. 2001. The ABC transporter Bcrp1/ABCG2 is expressed in a wide variety of stem cells and is a molecular determinant of the side-population phenotype. *Nat. Med.* **7:** 1028–1034.

Zhu J., Woods D., McMahon M., and Bishop J.M. 1998. Senescence of human fibroblasts induced by oncogenic Raf. *Genes Dev.* **12:** 2997–3007.

Zindy F., Quelle D.E., Roussel M.F., and Sherr C.J. 1997. Expression of the p16INK4a tumor suppressor versus other INK4 family members during mouse development and aging. *Oncogene* **15:** 203–211.

Zindy F., Eischen C.M., Randle D.H., Kamijo T., Cleveland J.L., Sherr C.J., and Roussel M.F. 1998. Myc signaling via the ARF tumor suppressor regulates p53-dependent apoptosis and immortalization. *Genes Dev.* **12:** 2424–2433.

11

Yeast, a Feast: The Fruit Fly *Drosophila* as a Model Organism for Research into Aging

Linda Partridge

Centre for Research on Ageing
Department of Biology
University College London
London WC1E 6BT, United Kingdom

John Tower

Molecular and Computational Biology Program
Department of Biological Sciences
University of Southern California
Los Angeles, California 90089-2910

RESEARCH INTO AGING HAS BEEN GALVANIZED BY THE DISCOVERY of mutations in single genes that extend life span and evolutionary conservation of their effects. An environmental intervention—dietary restriction—also extends life span in evolutionarily diverse animals. These discoveries have opened the way to using laboratory model organisms to understand human aging. Invertebrate species, budding yeast *Saccharomyces cerevisiae*, the nematode worm *Caenorhabditis elegans*, and the fruit fly *Drosophila melanogaster*, have a vital role in this process of discovery. Their ease of culture and handling in the laboratory and short life spans (~3 days in yeast, ~3 weeks in the worm, and ~3 months in *Drosophila*) mean that much more rapid progress can be made than in the mouse, whose life spans are about 3 years. Completion of the genome sequences for the invertebrates and their closely related species, together with the development of many genetic and other resources, also make them powerful experimental systems. Each of these organisms has strengths and weaknesses for research into aging. We highlight here some of the particular strengths of *Drosophila* and uses to which they have been put and could be put in the future. The jaw-dropping genetics applied to a complex tissue structure approaching that of vertebrates leaves *Drosophila* flying at the front edge of aging research.

The time has long passed when a full review of the biology of aging in the fruit fly *D. melanogaster* could be usefully accommodated in a single chapter. *Drosophila* has been an established model organism for work on aging since the field began and has been the engine of a recent explosion of data. Rather, we aim to provide some indications of the peculiar strengths and weaknesses of the fly for work on aging and some pointers to some new areas of enquiry that seem to us interesting for future work. Our account here is therefore selective rather than exhaustive and probably more useful for those interested in some possible future developments than in a comprehensive account of all that has been achieved so far.

A BRIEF HISTORY OF THE USE OF *DROSOPHILA* IN RESEARCH INTO AGING

D. melanogaster made its earliest appearance in research into aging as a cheap and convenient laboratory organism for studying environmental effects on life span, for instance, of nutrition and temperature (Loeb and Northrop 1917; Pearl 1921; Alpatov 1930). The use of *Drosophila* as a laboratory model organism for genetics began in the 1920s in the laboratory of T.H. Morgan (Kohler 1994), leading to both the chromosome theory of heredity and the production of many single-gene mutations with interesting phenotypes that were subjected to further analysis. However, these phenotypes did not include life span or the rate of aging. Until the 1990s, evolutionary thought and quantitative genetic approaches largely dominated work on why and how aging occurs. *Drosophila* also had a critical role here, as an almost ideal organism for testing evolutionary theories of aging.

Aging presents a paradox from an evolutionary standpoint. It is an unconditionally deleterious trait because it limits the capability of organisms to make a genetic contribution to the next generation. It is not inevitable, because some organisms age very slowly and others seem not to age at all, and the rate of aging can differ markedly between related species. These considerations attracted interest from evolutionary biologists starting in the nineteenth century (Weismann 1891) and culminated in the two main theories for the evolution of aging: mutation accumulation (Medawar 1946, 1952) and pleiotropy (Williams 1957). Fundamental to both theories was the role of extrinsic hazard from disease, predation, and accidents, which can lead to a declining force of natural selection with age. Aging can then evolve as a side effect of mutation pressure (the mutation accumulation theory) or of late-life deleterious effects of mutants that are beneficial to the young (the pleiotropy theory). In either case, aging is not selected for. It evolves as a side effect, not because of natural selection in its favor.

Drosophila is an excellent model organism in which to test the relative importance of these two routes to the evolution of aging. Because it is a diploid outbreeding organism, fly populations harbor natural quantitative genetic variation for most traits. A revolution in the field of aging began in the 1980s with the demonstration that genes affecting life span could be readily manipulated in the laboratory by selecting populations of *Drosophila* for late-life reproduction to create longer-lived flies (Rose and Charlesworth 1980; Luckinbill et al. 1984). A great deal of experimental work on this standing genetic variation, and on new mutations that arose in the fly genome, established that both mutation accumulation and pleiotropy seem to have some role in the evolution of aging, with stronger evidence for the importance of pleiotropy (for review, see Partridge and Gems 2002; Hughes and Reynolds 2005). In *Drosophila*, natural genetic variation that results in slow aging is associated with a lower reproductive rate in both sexes (for review, see Hughes and Reynolds 2005). One of the predictions to emerge from the theoretical and empirical work on the evolution of aging was that the trait was expected to be highly polygenic and that its mechanisms were unlikely to be evolutionarily conserved (for review, see Martin 2002). This view implied both that single-gene functions would not be capable of slowing down aging and that discoveries about mechanisms of aging in experimental organisms would be irrelevant to humans.

The conceptual foundations to our understanding of aging were jolted by the discovery of single-gene mutations that increased life span and oxidative stress resistance in the nematode worm *C. elegans* (Klass 1983; Friedman and Johnson 1988; Kenyon et al. 1993; Larsen 1993), and that turned out to encode components of an invertebrate insulin/insulin-like growth factor (IGF) signaling (IIS) pathway (Kimura et al. 1997; Lin et al. 1997; Ogg et al. 1997). Soon the work on quantitative genetics of aging in *Drosophila* began to include studies of the effects of specific genes, with the onset of mutant screens and the discovery of *Methuselah* (Lin et al. 1998), and the manipulation of mortality rate and life span with transgenes such as *hsp70* (Tatar et al. 1997) and *Cu/ZnSOD* (Parkes et al. 1998; Sun and Tower 1999). Shortly afterward, work on the IIS pathway in *Drosophila* established evolutionary conservation of the effects of IIS signaling on life span between *C. elegans* and *Drosophila* (Clancy et al. 2001; Tatar et al. 2001). More recently, many other manipulations of single genes that extend *Drosophila* life span have been reported, and the emphasis of work on aging in the fly has shifted toward the identification and analysis of evolutionarily conserved pathways affecting life span.

STRENGTHS AND WEAKNESSES OF *DROSOPHILA* AS A LABORATORY MODEL ORGANISM FOR RESEARCH INTO AGING

The budding yeast *S. cerevisiae*, *C. elegans*, *Drosophila* and the mouse are now the standard laboratory organisms for research into biology of aging. They all have the power that comes from completed genome sequences in addition to well-characterized biology and experimental reagents accumulated over many years of work. The fact that they are all laboratory model organisms means that they are by no means a random subset of natural biodiversity in aging. They are all easy to culture, in part because they have high reproductive rates and relatively short life spans. It will be important to confirm whether discoveries made with these organisms apply to other species, particularly long-lived animals with low reproductive rates such as humans, and there is undoubtedly much to be learned from the great biodiversity in the biology of aging in nature (Austad 2001; Partridge and Gems 2006).

Work on aging in *Drosophila* can in general be done much more rapidly than in the mouse because of its shorter life span and the greater speed with which genetic manipulations can be conducted. However, the fly also has disadvantages. The most obvious is that strains cannot yet be held frozen, and maintenance of fly stocks is therefore a considerable part of the effort in any fly laboratory. The approximately 3-month fly life span is also a considerable disadvantage compared with the about 3-week life span of *C. elegans,* and it means that experimental work on flies is a lot slower. The problems with genetic background are also much more acute in *Drosophila* than in *C. elegans*. Life span and related traits such as stress resistance, fecundity, and behavior are continuously varying traits, subject to both environmental and quantitative genetic variation. *Drosophila* is an outbreeding diploid organism, like mice and humans. This means that its life span and fecundity can be subject to marked effects from inbreeding depression and heterosis (Vermeulen and Bijlsma 2004; Swindell and Bouzat 2006). Laboratory strains are typically inbred, and different laboratory "wild types" such as Oregon-R and Canton-S differ substantially in their phenotypes (Tower 1996; Spencer et al. 2003) and show marked heterosis for fitness-related traits when crossed with each other. *C. elegans* is much less subject to problems with inbreeding and heterosis, because it is a self-fertilizing hermaphrodite and is therefore habitually inbred, which means that the deleterious recessive alleles responsible for inbreeding depression are rapidly exposed to natural selection and hence tend to be eliminated from the population. It is always essential, in any organism, that work on genetic effects on aging be conducted using strains where the mutant or transgene is placed in a genetic background as near as possible identical to that of the control. For *Drosophila* and other outbreeders, it may also be necessary to build in additional steps to avoid heterosis in crosses, and this is particularly necessary where heterozygous genetic effects are claimed, as is frequently the case where the GAL4/UAS system is used to drive expression of transgenes. Specific controls for genetic background effects must be performed, such as using conditional transgene expression, multiple heteroallelic combinations, and the backcrossing of mutants or transgenes into more than one genetic background.

In addition, *Drosophila* has certain disadvantages relative to *C. elegans* for genetic screens for genes affecting life span and related traits. For loss-of-function screens, double-stranded RNA interference (RNAi) is much more easily conducted in *C. elegans,* where simple feeding will suffice (Montgomery 2004), than it is in *Drosophila* where transgenic methods and/or direct injection of RNA is required (Carthew 2001; Allikian et al. 2002). Mutant strains of *C. elegans* have been created where RNAi has even greater effectiveness, such as in the nervous system (Chapin et al. 2007; Schmitz et al. 2007), and it is hoped that genetic manipulation of *Drosophila* might make it amenable to RNAi via feeding in the future. Despite a bumpy start, RNAi screens in cultured *Drosophila* cells show promise for the genome-wide identification of genes belonging to conserved regulatory pathways, such as RNAi itself, and may ultimately inform aging research (Casacuberta and Pardue 2006; Dorner et al. 2006; Perrimon and Mathey-Prevot 2007). Because of the relative simplicity of RNAi screens in *C. elegans* and gene inactivation in yeast, it is therefore not surprising that the majority of the initial discoveries of single-gene mutations, pathways, and mechanisms that negatively regulate life span have been made in those models. In contrast, like yeast, *Drosophila* can be readily transformed and mutated using transposable elements, such as the P element, and this allows transgenes and endogenous genes to be readily overexpressed. This has allowed research in *Drosophila* to identify genes that positively regulate life span and that have tissue-specific effects (Parkes et al. 1998; Seong et al. 2001; Landis et al. 2003).

An obvious question about any model organism used in aging research is: What does it die of? Although the cause of death in *Drosophila* remains unknown, the fly has pioneered in certain studies of aging-related pathology. This began in the 1970s with the characterization of aging-induced muscle deterioration, mitochondrial abnormalities, and subcellular inclusions by Miquel et al. (1979) and continues with the molecular characterization of such events today (Walker and Benzer 2004; Zheng et al. 2005). Correlating with the deterioration of the aging fly at the tissue and ultrastructural levels are dramatic functional declines in activity, behavior, and reproduction (Leffelaar and Grigliatti 1984; Grotewiel et al. 2005; Waskar et al. 2005; Martin and Grotewiel 2006; Wallenfang et al. 2006). However, given the enormous amount of work involved in characterizing the aging phenotypes of different tissues, it is rather unlikely that this aspect of fly biology will ever be quite as well known as it is in mammals.

Drosophila and the other model systems, such as yeast and *C. elegans*, have outstanding strengths for research into aging. However, the ultimate aim of much of this research is to improve human health. It is therefore important that work with laboratory model organisms focuses on those aspects of the aging process likely to be of relevance to mammals. Aging and life span studies with the mouse and rat are slow and expensive and hence mostly undertaken only when there is clear evidence of evolutionary conservation of mechanisms. *Drosophila* is a key model organism for establishing such evolutionary conservation, for example, the evidence of an evolutionarily conserved effect on aging of the IIS pathway and dietary restriction (DR) (for review, see Tatar et al. 2003; Bartke and Brown-Borg 2004; Kenyon 2005; Piper et al. 2005; Giannakou and Partridge 2007). *Drosophila* has also been key in establishing evolutionary conservation of the effects of other pathways on aging, such as reactive oxygen species (ROS) metabolism (Hekimi and Guarente 2003; Longo 2004; Landis and Tower 2005; Osiewacz and Scheckhuber 2006), and this is a context in which the fly will continue to have a key role. Indeed, further comparative work on different model organisms, including *Drosophila,* is needed to establish whether the conservation of mechanisms found at the level of signaling pathways extends to the specific biochemical and molecular processes that are regulated by these pathways and to the types of molecular damage against which they might protect to enable increased longevity (Pletcher et al. 2007).

Thanks to 100 years of genetics research, we have the unmatched ability to manipulate the genome and control gene expression in *Drosophila* relative to other multicellular organisms. *Drosophila* are readily transformed and mutated using the P transposable element and other transposons (Bellen et al. 2004), and chemical and irradiation-induced mutagenesis is routine. Cell lineages are readily marked in *Drosophila* using heterologous recombination proteins such as yeast FLP and its target sequence FRT (Golic and Lindquist 1989) and more recently Cre/Lox (Rodin and Georgiev 2005). Introduction of the yeast transcription factor GAL4 into flies allows for convenient activation of transgenes via its binding-site upstream activation sequence (UAS) (Fischer et al. 1988). FLP-induced recombination between chromosomes in dividing *Drosophila* cells can be used to make mutations homozygous in a clone of cells, thereby allowing for the rapid screening of genes for developmental phenotypes at various stages throughout development and gametogenesis. With a system called MARCM (mosaic analysis with a repressible cell marker), the mutant cells lose a transgene encoding the transcriptional repressor GAL80, thereby allowing other transgenes activated by GAL4 to be expressed and mark the clone (Lee and Luo 2001). These approaches have allowed detailed study of nervous system cell lineages and connectivity and the genes required for such development (Truman et al. 2004; Slack et al. 2006).

The wealth of genetic markers available for mutant and mosaic animals has been added to by the rainbow of fluorescent proteins that can be expressed in transgenic flies. As first demonstrated in both *C. elegans* (Chalfie et al. 1994) and *Drosophila* (Wang and Hazelrigg 1994), the jellyfish green fluorescent protein (GFP) and its spectral cousins (dsRED, mCherry, yellow, cyan, etc.) allow for convenient assay of gene expression due to their ability to fluoresce without any added reagent except light. These "vital" (i.e., they can be in scored in living cells and animals) fluorescent markers have produced a minirevolution in developmental biology research, for example, in the cell lineage studies, and their application to aging research is just beginning. Longitudinal analysis has been key to our understanding of the physiology of aging in humans (Lakatta 2000) and is starting to be applied in a significant way to flies (Muller et al. 2001; Carey et al. 2006) and *C. elegans* (Golden and Melov 2004; Huang et al. 2004). The ability to longitudinally assay gene expression with transgenic GFP reporters has revealed genes whose expression level is partially predictive of remaining life span, such as *hsp-16.2* in *C. elegans* (Rea et al. 2005) and *Drosomycin* in *Drosophila* (Landis et al. 2004). When a modified P element called the "protein trap" inserts into *Drosophila* genes, it creates protein-GFP fusions at high frequency, resulting in a library of hundreds of reporter strains begging to be assayed in longitudinal experiments (Buszczak et al. 2007). GFP has therefore paved a way to the discovery of the Golden Fleece of aging research: reliable biomarkers of aging that can be measured in young animals and humans (Ingram et al. 2001; Roth et al. 2002).

Exquisitely fine-scale, tissue-specific control of transgene expression is possible in *Drosophila* by virtue of the combinatorial system referred to as "GAL4/UAS," where the enhancers and regulatory sequences of endogenous genes are hijacked to control the expression of transgenes (Brand and Perrimon 1993). In addition, conditional gene expression systems have been adapted for the fly, including those where transgene expression is regulated by simply feeding flies the drugs doxycycline (Bello et al. 1998; Bieschke et al. 1998) or RU486/Mifepristone (Osterwalder et al. 2001; Roman et al. 2001). When the conditional systems are combined with the GAL4/UAS system, it allows for tissue-specific and conditional transgene expression over three orders of magnitude (Stebbins et al. 2001; Ford et al. 2007). When used to drive expression of inverted repeat constructs, these systems also allow for conditional gene inactivation via RNAi (Allikian et al. 2002; Kirby et al. 2002). Because they are particularly useful in aging research in dealing with the genetic background problem and time/age as a variable, the conditional systems are likely to be increasingly applied to *Drosophila* aging studies (see, e.g., Giannakou et al. 2007). In the conditional transgenic systems, control and experimental flies have identical genetic backgrounds, and transgene expression can be turned on or turned off in essentially any tissue at any stage of the life cycle. After lagging for some time behind the mouse, targeted gene disruption/knockins are becoming routine in *Drosophila* (Xie and Golic 2004) and promise to further accelerate the pace of aging research. Introduction of the heterologous PhiC31 recombination system into *Drosophila* allows for large transgenic insertions ($\geq$100 kb) as well as insertion of different constructs into the same chromosomal location, thereby facilitating control of chromosomal position effects on transgene expression (Venken et al. 2006). Chromosomal deficiencies and translocations can be readily created that span desired genomic intervals, by using P-element insertions and transposase (Cooley et al. 1990), male recombination (Carney et al. 2004), and rare-cutting restriction enzymes (Egli et al. 2004).

Drosophila shares many features with mammals that are absent in yeast and *C. elegans*. Unlike *C. elegans*, it has two sexes and some of the same sex differences. For instance, the larger effect of alterations of the IIS pathway on the female than on the male life span in *Drosophila* may also be observed in mammals (Bartke 2005; Piper et al. 2005). In *Drosophila*, as opposed to *C. elegans*, tissues are well-differentiated, and clear homologs exist for virtually every human organ, including heart, fat, and Malpighian tubules, the functional equivalent of the mammalian kidney (Chintapalli et al. 2007). The most obvious mammalian tissue lacking in *Drosophila* is the calcareous endoskeleton. *Drosophila* is often said to be postmitotic as an adult, but, in fact, it has dividing cells both in the gut and in the reproductive system (Decotto and Spradling 2005; Micchelli and Perrimon 2006; Ohlstein and Spradling 2006, 2007). Flies also show marked circadian rhythms (Bae and Edery 2006; Rosato et al. 2006), they sleep (Cirelli 2003; Hendricks and Sehgal 2004; Huber et al. 2004; Swinderen 2005), and they show both cognitive and behavioral decline during aging (Grotewiel et al. 2005; Martin and Grotewiel 2006), all features of mammals.

A large fraction of human genes, including many involved in disease, have conserved orthologs in *Drosophila* (Rubin 2000). Humans often have two or more duplicated forms of genes that are present as a single copy in *Drosophila*, with some degree of functional redundancy between them. Analysis of gene function not only is faster, but can also be conducted with fewer complications from the presence of paralogs in the fly. Somewhat paradoxically, for human genes where *Drosophila* lacks an ortholog, considerable progress in revealing mechanisms has been made by inserting the human gene into the fly genome, because the consequences can be studied without the complication of the presence of the endogenous gene. These genetic features, together with the sexual dimorphism and clear tissue differentiation of *Drosophila*, have made the fly a powerful model organism for studying the etiology of human diseases, including the major aging-related killer diseases: cardiovascular disease, neurodegeneration, and cancer. As well as presenting an excellent set of tools for genetic analysis, an important advantage of using *Drosophila* to understand human disease is the ability to conduct genome-wide screens for mutations in other genes that modulate the phenotypes associated with the disease model. This means that in addition to testing hypotheses, hypothesis-free investigations can be conducted. The fly can also be used to screen for drugs that extend life span or ameliorate disease phenotypes.

An important advance in recent years has been the use of demographic information to understand underlying mechanisms of aging (Finch et al. 1990). To obtain accurate information on mortality rates at specific ages, especially later ages, large starting populations are required (Vaupel 1997; Vaupel et al. 1998; Gendron et al. 2003). Rodents are quite unsuitable for this kind of work because of the enormous expense involved in maintaining them. *Drosophila*, in contrast, is an excellent organism for this type of work (Curtsinger et al. 1992). It is relatively straightforward and inexpensive to rear and monitor thousands of individuals in a single experiment. It is also possible to obtain detailed fecundity records on individuals and to monitor, for instance, behavioral decline and others traits that can be nonlethally sampled. Demographic information has been used to address a number of important issues, such as the way in which mortality rates change with age (Carey et al. 1992; Vaupel 1997; Vaupel et al. 1998; Johnson et al. 2001; Oeppen and Vaupel 2002) and the roles of risk and damage in mediating the effect of interventions that affect aging (Mair et al. 2003; Partridge et al. 2005; Kaeberlein et al. 2006).

The discoveries that mutations in single genes can extend life span and that their effects can show evolutionary conservation have galvanized research into aging. However, it is important that, in the course of illuminating the areas opened up by these findings, we do not put out the lights behind us. Quantitative and evolutionary genetics continue to illuminate new pathways and natural genetic variation for the rate of aging, and they are essential for putting work on single mutants into a proper context. The single-gene mutations that extend life span in the laboratory typically have large effects and are highly pleiotropic, whereas quantitative genetic variants are more subtle and prone to be sex-specific and environmentally labile, features that will repay further study (Shmookler Reis et al. 2006). Laboratory model organisms inhabit a peculiar environment, in which they are physically constrained from taking normal levels of exercise, are often held at abnormally high densities, and are in close proximity to a limitless food supply. To understand how some of the mutations that affect aging under laboratory conditions are likely to affect the animals under more normal ecological circumstances, we need model organisms that can be studied under more natural conditions and where the evolution of natural genetic variants at the same loci can be understood. *Drosophila* is an excellent organism for this kind of quantitative genetic and evolutionary work.

In the remainder of this chapter, we elaborate on some of these strengths of *Drosophila* and give some pointers to promising future lines of investigation.

SOME PROMISING DIRECTIONS FOR RESEARCH INTO AGING IN *DROSOPHILA*

Sex Differences

By far, the most striking genetic determinant of life span in flies and humans is sex (chromosomal gender), in that females tend to live longer (Burger and Promislow 2004; Graves et al. 2006). In an evolutionary context, sex and aging are thought to be inextricably intertwined. However, how this translates to the molecular and genetic regulation of individual life span presents a problem at the forefront of research on aging. Ultimately, any sex-specific differences in life span and aging phenotypes must be under the control of the sex determination mechanism unique to that species, and it remains to be determined if downstream effects on life span will likewise fall into the species-specific ("Private") category, or rather will impact on species-general ("Public") mechanisms such as IIS signaling and genome maintenance (Kenyon 2005; Martin 2005). Sexual differentiation is linked to increased mortality across a variety of species, including *Hydra* (Yoshida et al. 2006), salmon (Morbey and Ydenberg 2003), and bamboo (Isagi et al. 2004). In *Drosophila* and mammals, most life span quantitative trait loci (QTL), transgenes, and life span interventions tend to affect one sex more than the other (Jackson et al. 2002; Leips and Mackay 2002; Burger and Promislow 2004; Magwere et al. 2004; Bartke 2005; Masoro 2005; Tower 2006). This is perhaps not surprising when we consider that the sexes are different genotypes: *Drosophila* and humans are both heterogametic, with females X/X and males X/Y. These different genotypes therefore present unique environments for the expression of genes affecting life span (Nuzhdin et al. 1997; Mackay and Anholt 2006). Studies in rodents have long documented the effect of chromosomal gender on aging-related changes in gene expression, such as the altered responsiveness to androgens in liver and prostate tissue (Chatterjee et al. 1989; Prins et al. 1996), and the preferential induction of the inflammatory response protein T-kininogen in aging males (Acuna-Castillo et al. 2006). Increased life span in Ames dwarf mice is associated with a decrease in sexually dimorphic gene expression in the liver (Amador-Noguez et al. 2005), and one intriguing idea is that it may be sexual differentiation, rather than reproduction per se, that exacts a physiological cost and contributes to aging and increased mortality. The synthesis of steroid hormones requires complex oxidative pathways; perhaps these generate some life-shortening toxins that are disposed of by IIS-regulated Phase II enzymes.

Sex-determination pathways evolve rapidly and show great diversity in mechanisms across species (Graham et al. 2003; Pires-Dasilva 2007). The genes at the top of the hierarchy change rapidly, for example, *Sxl* in *D. melanogaster* (Tominaga et al. 2002), *Tra* in the medfly, and, traditionally, the *SRY* gene in humans (Ottolenghi et al. 2007). However, the pathways converge on conserved components downstream, such as the Doublesex transcription factors (Huang et al. 2005; Le Bras and Van Doren 2006; Veith et al. 2006; Zhang et al. 2006). As research progresses on sex determination mechanisms for germ-line and somatic cells, the fly and human pathways appear to be increasingly similar in overall design, with increasing implication of the *Xist* gene in humans (Casper and Van Doren 2006; Tower 2006). The fact that both sex-determination pathway and gametogenesis genes are rapidly evolving in insects and humans suggests that these genes are under active (sexual) selection, and as such, these genes may be more likely to exhibit pleiotropy and affect aging (Rice 1992; Maklakov et al. 2005; Hunt et al. 2006; Rice et al. 2006). The *Drosophila* sex determination pathway interacts in genetically and molecularly defined ways with conserved tumorigenesis and developmental signaling pathways such as *hedgehog* (Horabin 2005) and *Notch* (Penn and Schedl 2007), suggesting candidate outputs to life span pathways. The effect of the sex determination pathway on aging and life span may extend across generations, as in both *Drosophila* and humans, the mother contributes copious gene products to both male and female offspring (the "maternal effect"), and these gene products can potentially affect that individual's aging and life span. Maternal effects on certain life history traits have been demonstrated in *Drosophila* (Faurby et al. 2005; Thomson and Lasko 2005) and beetle (Fox et al. 2004), and maternal effects are certain to be an active area of future research into *Drosophila* aging. The striking parallel in humans is the effects of maternal age and prenatal conditions on the individual's subsequent health and life span (Hawley 2003; Barker 2004). In humans, naturally occurring mutations inform on the mechanisms of sex determination (Ottolenghi et al. 2007; Wilhelm et al. 2007) and may prove useful in studying aging effects. In *Drosophila,* X-Y interchanges can be readily generated using the *Chlamydomonas reinhardtii* I-ICreI homing endonuclease to cut within the rDNA repeats on each chromosome and initiate reciprocal chromosomal translocations (Maggert and Golic 2005), providing one avenue for study in the fly.

Interactions between the sexes are important determinants of life span in both sexes of *Drosophila.* Male life span is shortened by exposure to females, and this appears to be mainly a consequence of movement, including courtship (Cordts and Partridge 1996), with little cost of mating itself. Modulations of ovarian activity can have a modest effect on life span (Sgrò and Partridge 1999; Barnes et al. 2006). However, the life span of females can be greatly shortened by the presence of males. The main cause is a cost of mating (Fowler and Partridge 1989; Kuijper et al. 2006), which is exacerbated when females are well-fed, lay more eggs, use up their sperm stores more rapidly, and mate more frequently (Chapman and Partridge 1996). This cost of mating is attributable to molecules in the seminal fluid of the male (Chapman et al. 1995), with the sex peptide having a major role (Wigby and Chapman 2005) and at least three other proteins potentially implicated (Mueller et al. 2007). The sex peptide also induces ovulation, especially in matings with virgin females, and it reduces female receptivity to subsequent mating with other males (Chapman et al. 2003; Liu and Kubli 2003; Peng et al. 2005a). The fact that it is also toxic to females suggests that there may be a conflict between the sexes mediated by the sex peptide, an idea for which there is support from experimental, evolutionary studies (Wigby and Chapman 2004; Lew et al. 2006; Long et al. 2006; Rice et al. 2006). These profound effects of the sex peptide on female life history also imply that this molecule acts on some of the mechanisms that control major life history decisions. There has therefore been considerable interest in identifying the receptor(s) for the sex peptide and understanding its mode of action. Despite some progress (Peng et al. 2005a,b; Carvalho et al. 2006; Soller et al. 2006; Barnes et al. 2007), no receptor for the sex peptide has yet been identified, and this is an interesting challenge for future work.

Drosophila Models of Human Aging-related Diseases

Drosophila is increasingly being used for the creation of models for human diseases, including cancer, infectious disease, and neurodegeneration. For this approach to be useful, it is essential that the fly model display phenotypes that are a good reflection of the human disease state. Humans and flies share many genes and cellular mechanisms and, particularly at the level of individual cells and cell-to-cell interactions, the resemblances between them are great (O'Kane 2003). About 77% of the genes implicated in specific human diseases have one or more *Drosophila* homologs (Reiter et al. 2001; and see http://superfly.ucsd.edu/homophila/). Many fly models of human diseases have been created, for instance, by inserting the human gene or mutant versions thereof into the fly, as has been done, for instance, for fly models of Alzheimer's disease (Crowther et al. 2004, 2006). Alternatively, the endogenous fly ortholog can be altered to contain the corresponding mutation to the human disease state. For instance, Seidner et al. (2006) introduced into the fly presenilin gene mutations that, in the human ortholog, are associated with different severities of Alzheimer's disease and found that their activity in the fly was tightly associated with the human age of onset values. Both genes and chemicals are rapidly being tested for the ability to enhance or suppress the mutant phenotypes in the fly (O'Kane 2003; Crowther et al. 2006; Whitworth et al. 2006), thereby leading to candidates for human disease interventions. The ability to do rapid enhancer/suppressor screens and to test drugs for effects means that *Drosophila* will be a rich source of candidates for human interventions over the next several years.

Circadian Rhythms and Detoxification

An emerging area in aging research is the role of circadian rhythms in regulating cellular and organismal detoxification and life span. Genetically engineered mice with disrupted circadian rhythms show decreased life span and increased rates of cancer (Lee 2006). Several of the genes found to be specifically induced during *Drosophila* aging (Pletcher et al. 2002; Landis et al. 2004) have homologs in *C. elegans* that are downstream targets of IIS signaling and are implicated in mediating life span extension (McElwee et al. 2004). These genes have been found to be under circadian control in mammals and flies and include genes encoding Phase II/ xenobiotic response-type proteins such as cytochrome P450s, oxidoreductases, carboxyl esterases, UDP-glucuronosyl transferases, sulfotransferases, and GSTs (glutathione-*S*-transferase) (Gachon et al. 2006; King-Jones et al. 2006; Palanker et al. 2006). Circadian control pathways are being intensively investigated in the fly and they appear to be increasingly similar to those in mammals, including intriguing links to DNA repair and the oxidative stress response (Collins et al. 2006; Gotter 2006; von Schantz et al. 2006). Like humans, the fly receives key inputs from both light and temperature, whereas *C. elegans,* for example, lives in the soil and may not respond well to light. Strikingly, flies have been used to demonstrate how both aging and oxidative stress can disrupt sleep, thereby providing an ideal model for the similar age-related deterioration of sleep observed in humans (Shaw et al. 2002; Koh et al. 2006).

Sarcopenia

One hallmark of aging in humans is sarcopenia, or muscle-wasting (Marzetti and Leeuwenburgh 2006). In mammals and *Drosophila,* the muscle exhibits striking deterioration during aging, marked by abnormal mitochondria, oxidative damage, expression of oxidative stress response and apoptotic marker genes, and apoptotic-like events as suggested by DNA fragmentation (Takahashi et al. 1970; Wheeler et al. 1995; Walker and Benzer 2004; Kujoth et al. 2005; Zheng et al. 2005). Apoptosis (programmed cell death) can be divided into several subtypes in flies and mammals, most of which appear to involve the mitochondria and oxidative stress in one way or another (Cashio et al. 2005; Bredesen et al. 2006;

Cereghetti and Scorrano 2006; McBride et al. 2006). Interestingly, yeast mortality is reported to have similarities to mammalian apoptosis, such as the implication of mitochondria and oxidative stress (Osiewacz and Scheckhuber 2006). The worm also exhibits a preferential deterioration of muscle tissue during aging (Garigan et al. 2002; Herndon et al. 2002; Chow et al. 2006). The extent to which misregulation of apoptotic pathways might contribute to aging phenotypes, and why this should occur preferentially in certain tissues, such as mammalian skeletal muscle and *Drosophila* flight and leg muscle, will be an active area of future research.

Immune Response

The innate immune response is the first and most important defense against pathogens in humans and is highly conserved in *Drosophila*, including the NF-κB signaling pathway. The interaction between the innate immune response and aging is an area of active investigation (Dionne et al. 2006; Libert et al. 2006). One possibility is that there may be obligatory trade-offs between life span and an effective or sustained immune response. For example, forced activation of NF-κB pathway immune signaling in *Drosophila* fat body increased pathogen resistance, but at a cost of reduced life span. Moreover, the expression level of transgenic reporter constructs consisting of antimicrobial peptide gene promoters fused to GFP was partially predictive of remaining life span in young flies (Landis et al. 2004). Much of the cellular immune system is conserved between flies and humans, including circulating hemocyte cells of various differentiated types that coordinate to identify, kill, and engulf invaders (Zettervall et al. 2004). *Drosophila* does not have the recombinagenic antibody system of humans in which exons are shuffled to create diversity in immune recognition surfaces on immunoglobulin proteins. However, striking new research shows that *Drosophila* fat body and hemocyte immune cells generate diversity in the immunoglobulin-related immune recognition protein Dscam by alternative splicing of exon sequences (Watson et al. 2005), making the fly immune response appear increasingly similar to that in humans in overall design. Microbes are reported to have benefits for *Drosophila* life span under certain conditions (Brummel et al. 2004); however, powerful antibiotics such as doxycycline have little effect on *Drosophila* life span under other conditions (Bieschke et al. 1998; Landis et al. 2003). In *C. elegans,* immune pathways are implicated in life span regulation (Garigan et al. 2002; Troemel et al. 2006). It will be important in the future to determine if pathways defined as "immune" do indeed affect life span by interaction with intracellular or extracellular pathogens or rather function through some related mechanism such as autophagy (Singh and Aballay 2006a,b; Wolff et al. 2006).

Dividing Cells and Cancer

The role of stem cells in human aging is of increasing interest, both as a potential cause and as a possible target for interventions in aging-related diseases (see Chapter 10). Stem cells are characterized by their ability to undergo asymmetric divisions in which they give rise to a new stem cell (self-renewal) as well as a daughter cell (transit cell) that will in turn give rise to two or more differentiated cell types. The ability of the stem cell to renew is regulated by the distinct tissue microenvironment in which the stem cells reside, the stem cell niche (Nystul and Spradling 2006). The distinction between stem cell and transit cell is controlled by the preferential segregation of regulatory molecules to one cell. During *Drosophila* embryogenesis, the central nervous system (CNS) develops from a ventral neuroectoderm, similar to the development of human CNS from a dorsal neuroectoderm. Approximately one in five cells of the *Drosophila* neuroectoderm delaminate into the embryo to form an ordered array of neural stem cells. When these neural stem cells divide, asymmetric segregation of differentiation factors to the daughter (transit) cell induces differentiation. These factors include the homeodomain transcription factor Prospero and the tumor suppressor homolog Brain tumor (Brat), which are thought to repress expression of genes required for self-renewal in the daughter cell (Bello et al. 2006; Choksi et al. 2006; Lee et al. 2006).

In the germ-line stem cells of the *Drosophila* gonads, the distinction between stem cell and daughter (transit) cell is accomplished by segregation of existing cell-cell contacts to the stem cell (oriented plane of division), thereby ensuring that the stem cell stays attached to the somatic cells that comprise the niche. In both *Drosophila* and humans, the production of gametes is dependent on the coordinated division and function of germ-line and somatic stem cells in the gonads. Fertility decreases with age in flies and humans and, in humans, this decrease correlates with deterioration in gamete quality in both males and females (Wyrobek et al. 2006). Signals from the gonad appear to affect organismal life span in both worms (Hsin and Kenyon 1999; Arantes-Oliveira et al. 2002) and mice (Cargill et al. 2003), but apparently not in *Drosophila* (Barnes et al. 2006). In *Drosophila* gonads, the continued production of gametes is supported by the continued division of germ-line stem cells that produce the egg and sperm, as well as by somatic stem cells that produce specialized tissues that support, nourish, and signal the development of the gametes. In the ovary, germ-line stem cells divide to produce the oocyte and supporting nurse cells, whereas somatic stem cells divide to produce the surrounding escort cells and follicular epithelium, and all but the oocyte are destroyed by apoptosis prior to fertilization. In this way, the production of the egg chamber in the fly is analogous to the production of oocytes in the human in that it involves the hormonally controlled, cyclical proliferation and destruction of germ-line and somatic tissues. The organization of the testis is highly analogous, where the germ-line stem cells divide to produce syncytia that are surrounded by somatic escort cells (Decotto and Spradling 2005). Specific genes and signaling pathways have been identified that are required for maintenance of both germ-line and somatic stem cells in the *Drosophila* gonads, and many or most are conserved in function in mammalian stem cell maintenance (Nystul and Spradling 2006). These include signaling pathways involved in cell cycle control and human tumorigenesis such as Wnt, Hh, BMP, and JAK/STAT, as well as components of the structures through which the stem cells are connected to somatic cells in the niche, such as cadherin and β-catenin. Research in flies is directed toward investigating whether the aging-related defects in gametogenesis result from the accumulation of mutations in the stem cells themselves, a lack of correct signals from other cells, or a combination of both (Waskar et al. 2005; Wallenfang et al. 2006).

Stem cell populations and somatic tissue regeneration is clearly one way that humans achieve their long life span. Nowhere is this more dramatic than in the gut, where the cellular lining turns over every few days. One hazard inherent to dividing cells is cancer, and human cells appear to have intrinsic division-counting mechanisms to help prevent overgrowth, such as telomere erosion (Campisi 2005). Tumor-suppressor genes such as *p53* may limit cancer growth in the gut and other tissues, yet have the pleiotropic effect of limiting organismal life span (Campisi 2003; Bauer et al. 2005; Radtke and Clevers 2005). Traditionally, it was thought that there were no dividing cells in the adult fly outside the gonads due to the lack of incorporation of labeled nucleotides and absent mitotic figures (Bozuck 1972). However, recent reports characterize dividing stem cells and their descendents in the adult fly midgut (Micchelli and Perrimon 2006; Ohlstein and Spradling 2006, 2007). The new data are based on FLP-recombination lineage marking but, interestingly, although it is stated that "only dividing cells can undergo [FLP/FRT] recombination" (Ohlstein and Spradling 2006), other studies report abundant FLP-based recombination and resultant *lacZ* transgene expression in postmitotic tissues throughout the fly (Sun and Tower 1999). The midgut has long been known to be a site of dramatic alterations and expansions in mitochondrial morphology during aging (Anton-Erxleben et al. 1983). The midgut is also reported to be the most active site of mitochondrial gene expression in the adult (Garesse and Kaguni 2005). Combined with reports that both mitochondria (Spees et al. 2006) and nuclear DNA (Bergsmedh et al. 2001) can move between cells under appropriate conditions, one additional possible explanation for the DNA replication observed in the fly midgut might be mitochondrial proliferation and/or increases in nuclear ploidy. So far, there is no evidence as to the possible function or requirement of the gut stem cells for normal life span, and this is likely to be an interesting area for future research.

A small revolution has taken place recently in cancer research, with the realization that the growth of tumors, like many normal tissues, may be supported by small populations of stem cells, in this case, "cancer stem cells" (Yang and Wechsler-Reya 2007). Cancer is characterized by uncontrolled cell growth plus tissue invasion (metastasis), and the ability both to divide indefinitely and to migrate through other tissues are hallmarks of germ-line stem cells during the development of *Drosophila* and other model organisms (Kunwar et al. 2006), where these processes can be dissected in detail. The fly has several genetic models for both cell proliferation and tissue invasion, with key genes conserved with human tumorigenesis pathways (Naora and Montell 2005; Korenjak and Brehm 2006; Wodarz and Gonzalez 2006; Vidal and Cagan 2006). Flies carry orthologs of human oncogenes and tumor suppressors and have been invaluable for understanding the basic functions of many of these genes, for example, the PTEN tumor suppressor that functions in the conserved IIS pathway and affects both epithelial cell proliferation during development and adult life span (Hwangbo et al. 2004; Hariharan and Bilder 2006; Ford et al. 2007). *Drosophila* can surround invading pathogens and necrotic tissue with melanotic material to create inclusions sometimes referred to as "melanotic tumors" (Minakhina and Steward 2006). There are no reports that adult flies normally die due to misregulated cell growth or tissue invasion, but we do not know what they die of under most circumstances, so it remains a possibility. It might not be prudent to assume that the cell growth observed in the adult midgut is beneficial—perhaps every fly dies with colon cancer.

Neurodegeneration

Drosophila has a complex nervous system that contains about 100,000 neurons. It shows clear differentiation between brain regions involved in different functions such as vision, learning, and memory. Its sensory capabilities, particularly vision and chemosensation, are well-characterized. Many of the basic cellular and molecular features of neuronal development and function are conserved between *Drosophila* and mammals. Fly models of several neurodegenerative disorders including Alzheimer's disease (AD) (Crowther et al. 2004, 2005, 2006; Chee et al. 2006; Kinghorn et al. 2006), Parkinson's disease (PD) (Feany and Bender 2000; Whitworth et al. 2006), and Huntington's disease (HD) (Marsh et al. 2003) have been produced and recapitulate many of the features of human diseases (for reviews, see Feany and Bender 2000; Bilen and Bonini 2005; Marsh and Thompson 2006). This makes them valuable material for pursuing hypotheses about the function and malfunction of disease-associated genes in humans and for conducting screens.

Sometimes, fly models of human neurodegenerative disease have produced unexpected and informative phenotypes. For instance, parkin is an E3 ubiquitin protein ligase, and loss-of-function mutations in the gene encoding it in humans cause early-onset PD. *Drosophila* has a parkin ortholog. Mutants in these genes cause loss of dopaminergic neurons in the fly brain, which is characteristic of PD in humans. However, the mutants in the fly also cause muscle loss and male sterility, neither of which are present in human PD (Greene et al. 2003). These findings could be used to argue that the fly parkin mutants are a poor model for PD. However, mitochondrial defects are a conserved feature of parkin mutants, and the male sterility of parkin mutants in *Drosophila* is attributable to a mitochondrial defect in sperm that could also be important in muscle loss. Similarly, the results of several studies have implicated oxidative stress and damage as an important underlying biochemical process in the etiology of PD (Greene et al. 2005; Menzies et al. 2005; Meulener et al. 2005; Yang et al. 2005, 2006; Carey et al. 2006). The seemingly disparate phenotypes seen in the fly model may therefore have commonalities at the biochemical level that are also important in PD (Tower 2006; Whitworth et al. 2006; Yang et al. 2006).

Several neurodegenerative diseases in humans, including AD and PD, are associated with the formation of insoluble protein aggregates and the death of the neurons. There has been considerable debate about the significance of the association between the aggregates and the neuronal death. One line of thought regards the aggregates as toxic, and part of the causal chain of events in the etiology of the disease, whereas another views them as the product of cellular defense mechanisms against a different toxic moiety, such as monomers or oligomers of the same protein. It is essential to resolve this issue, since development of therapies relies upon identification of steps in the process to disease rather than off-path events such as the side effects of the organism's defense mechanisms. Some familial forms of AD are associated with mutations in the gene encoding Tau, a microtubule-associated protein that becomes hyper-phosphorylated and forms intraneuronal fibrillary tangles in AD. However, overexpression of the mutant form of Tau in *Drosophila* neurons resulted in neurodegeneration, but without the formation of neurofibrillary tangles, implying that at least part of the toxicity is not attributable to the presence of the tangles (Wittmann et al. 2001). Similarly, *Drosophila* models of PD have been used to demonstrate uncoupling of aggregates of α-synulein from neurodegeneration (Auluck and Bonini 2002; Peng et al. 2005b).

Drosophila disease models have also been valuable for hypothesis testing in vivo. For instance, recent work has identified signaling pathways and cellular processes that lead to or ameliorate Tau-induced toxicity (Khurana et al. 2006; Dias-Santagata et al. 2007; Fulga et al. 2007) and that alleviate toxicity of aggregate-prone proteins (Steffan et al. 2004; Berger et al. 2006; Williams et al. 2006). HD and several other neurodegenerative diseases are caused by expansion and aggregation of polyglutamine (poly-Q) encoding repeat triplets, within otherwise unrelated proteins. For HD, the poly-Q expansion occurs within huntingtin, a protein of unknown function. In humans, individuals with the same degree of poly-Q expansion in huntingtin can show very different ages of onset of HD. A *Drosophila* model of HD was used to demonstrate that the presence of normal repeat-length polyglutamine peptides, non-toxic in themselves, accelerates formation of aggregates and neuronal toxicity and may act as an in vivo modifier of HD pathology in humans (Slepko et al. 2006).

One of the main advantages of using *Drosophila* to understand human neurodegeneration, and other diseases, is the ability to conduct genome-wide screens for mutations in other genes that modulate the phenotypes associated with the disease model, either enhancing or suppressing them. Such screens do not require a priori knowledge of the function of the disease gene, and they can be conducted relatively rapidly and can identify potential drug targets. For instance, a screen for enhancers and suppressors of mutant parkin phenotypes identified loss of function of glutathione *S*-transferase-1 (GstS1) as an enhancer. A strength of *Drosophila* is that it is possible to proceed directly to an experimental test of an interaction identified in a screen, and overexpression of GstS1 significantly suppressed loss of dopaminergic neurons in the parkin mutants (Greene et al. 2005).

Drosophila is also a powerful model for in vivo testing of the efficacy of amelioration of disease processes by genetic interventions and drugs that have first been identified by screens in other organisms. For instance, vesicle trafficking between the endoplasmic reticulum and the Golgi was identified as one of the processes first to be disrupted after expression of α-synuclein, and a screen for toxicity modifiers identified proteins that were involved in this same step. In vivo testing in *Drosophila* established that the human ortholog of one of these could protect against dopaminergic neuron loss in a PD model (Cooper et al. 2006). Similarly, a yeast-based screen identified small-molecule inhibitors of poly-Q aggregation, which were then validated in a mammalian cell model of HD (Zhang et al. 2005). In vivo testing in a *Drosophila* model of HD was then used to confirm the efficacy of compounds handed on from mammalian cell screens (Zhang et al. 2005; Desai et al. 2006). Lithium was first demonstrated to protect against toxicity from poly-Q proteins in mammalian cell models and then also to do so in a *Drosophila* model, and further genetic analysis pointed to Wnt signaling as mediating the protective effects (Berger et al. 2005).

Heart Disease

Heart disease in humans is a leading cause of death and of reduced quality of life. Its incidence is strongly age-related. *Drosophila* is the only invertebrate model organism that has a heart, and there is a high degree of evolutionary conservation of mechanisms of development of the organ between flies and mammals (Long et al. 2006; Serluca and Fishman 2006; Spees et al. 2006). Cardiac function in flies can be impaired to a large extent without causing death in *Drosophila* (Wessells and Bodmer 2004, 2007a,b), because oxygen is supplied to the tissue mainly through the tracheal system rather than by the blood. A system for visualizing cardiac function in awake flies using optical coherence tomography has been developed and has been used to produce a fly model of dilated cardiomyopathy (Osiewacz and Scheckhuber 2006). Natural arrhyhmias increase with aging in *Drosophila*. If the fly heart is stressed electrically, it can arrest or show uncoordinated fibrillation, and the incidence of these responses increases with age. Lowered expression of a ATP-sensitive potassium (K_{ATP}) channel *dSUR* is implicated in the functional decline with age (Akasaka et al. 2006), and activation of *dSUR* protects against stress-induced heart failure. These findings recapitulate features of mammalian heart failure, and this fly model is likely to be valuable for discovering interventions that can ameliorate human heart disease (Bier and Bodmer 2004; Ocorr et al. 2007).

Interestingly, mutations in the fly insulin receptor and in target of rapamycin (TOR), which extend life span in wild-type flies, also greatly reduce the incidence of stress-induced heart failures in older flies (Wessells et al. 2004). This finding has important implications. Many diseases including the predominant killers—heart disease, cancer, and neurodegeneration—increase in incidence with age. Indeed, age has been described as the most potent of all carcinogens. The finding that slowing down the normal aging process by mutations also delays the onset of aging-related disease, in this case heart disease, suggests that the normal aging process itself is a risk factor for disease. Slowing down aging by mutations might therefore delay the onset of multiple aging-related diseases, opening up the prospect of a broad-spectrum preventative medicine for the diseases of aging.

The Drosophila Malpighian (Renal) Tubule and Kidney Disease

The fly excretory organ, the Malpighian tubule, is a promising, but as yet largely unexploited, model of human kidney disease (Dow and Davies 2006). The tubules are long thin tubes that are connected to the hindgut. There are extensive similarities between the development of the insect Malpighian tubule and the mammalian kidney. For instance, both derive from two cell populations: ectodermal epithelial buds and the surrounding mesenchymal mesoderm. In Drosophila, the mesenchmal-derived cell population differentiates as a physiologically and morphologically distinctive class of stellate cells, requiring the cell to undergo a mesenchyme-to-epithelial transition, similar to mammalian kidney development (Jung et al. 2005). The normal incorporation of the stellate cells and later physiological activity of the mature tubule require the activity of HiBRIS, the fly ortholog of mammalian NEPHRIN (Denholm et al. 2003). The genetic regulation of the patterned tubular branching that gives rise to the Malpighian tubule has been well-characterized in Drosophila (Hatton-Ellis et al. 2007), with possible implications for similar processes in mammals.

RNA transcript profiling has identified the genes that are specifically expressed in the tubule, compared with the whole fly. This was made possible by a heroic feat of dissection of 30,000 of these tiny organs. Many of the genes expressed in the tubules failed to be detected as expressed in the whole fly. Of the top 200 genes in the list of tubule-specific expression, only 18 had been previously named, a highly significant underrepresentation. Of the known genes, transporters for organic solutes were greatly overrepresented, and the tubule is likely to have an active role in the removal of a broad range of metabolites and xenobiotics (Dow and Davies 2006). The tubule has also been found to have a role in protective immunity and can sense bacterial challenge and mount a killing response independently of the fat body, the classical insect immune tissue. This antibacterial response was associated with up-regulation of expression of nitric oxide synthase, and forced overexpression of NOS (nitric oxide synthase) in the principal cells of the Malpighian tubules increased resistance to bacterial challenge (McGettigan et al. 2005).

The RNA expression profile for the tubules allowed comparison with humans. The tubule-enriched genes were compared with the Homophila database of *Drosophila* genes of known human disease homologs (http://superfly.ucsd.edu/homophila/). Fifty genes associated with human diseases were associated with up-regulation of the *Drosophila* ortholog at least threefold in tubules, far more than expected by chance. Furthermore, several of these 50 genes have human kidney phenotypes. For instance, *rosy* encodes xanthine oxidase, and mutation in either human or fly produces xanthinuria, severe nephrolithiasis, with concomitant distortion of the tubules in flies. In both species, a high-water/low-purine diet ameliorates lethal effects. A small protein, bc10, has been shown to be down-regulated in transition from early-stage to invasive bladder carcinoma. Its homolog is highly abundant and moderately enriched in tubules (Dow and Davies 2006). Human and *Drosophila* renal phenotypes may hence be quite similar over a wide range of properties, making this system a promising one for detailed analysis of cellular processes leading to kidney disease in humans.

Quantitative and Evolutionary Genetics

Many of the single-gene mutations that slow down aging in *Drosophila* are pleiotropic, with effects on other traits such as fecundity, stress resistance, and metabolism. From the point of view of human health, we need to understand the mechanisms by which mutations extend life span and hence the extent to which any or all of these pleiotropic associations are inevitable. We also need to understand the consequences of these traits for organismal functioning in different kinds of environments, for instance, with varying food supply and activity levels. In addition, genetic variants affecting life span in nature that are subject to natural selection may be a special subclass of those that can affect life span, in that they will also be selected to be free of deleterious pleiotropic effects on other traits. Major insights into these issues are likely to come from investigating the properties of natural, quantitative genetic variation for the rate of aging, which may or may not involve variation in the same genes as those identified in laboratory screens. We also need to understand the ways in which this genetic variation becomes subject to natural selection under different circumstances in nature.

Quantitative genetic analysis of life span in *Drosophila* has been conducted using both inbred strains (see, e.g., Nuzhdin et al. 1997) and lines that have been artificially selected for life span (see, e.g., Wilson et al. 2006). The resulting QTL show both sex-specific (Nuzhdin et al. 1997; Pasyukova et al. 2004) and environment-specific (Pasyukova et al. 2004) effects on life span. QTL for life span also often show epistatic interactions with one another (Pasyukova et al. 2004). Several candidate genes for QTL for life span have been identified by quantitative complementation tests (De Luca et al. 2003; Pasyukova et al. 2004), and future work on the mechanistic basis of these effects will be revealing. An interesting recent development has been the use of QTL mapping in conjunction with RNA transcript profiling, which has identified multiple genes and pathways for further analysis (Lai et al. 2007).

Methuselah was first discovered to affect *D. melanogaster* life span through single-gene mutations isolated in the laboratory (Bello et al. 1998). A comparison of the genomes of three *Drosophila* species revealed that the gene has one of the fastest evolutionary rates of all genes in the genome, and the commonest single-nucleotide polymorphism (SNP) haplotype for the gene also shows latitudinal variation (Schmidt et al. 2000). This latitudinal pattern did not extend to the adjacent genes, and this and other evidence strongly suggest that the latitudinal variation is caused by natural selection (Duvernell et al. 2003). However, data have not yet been produced on any latitudinal pattern in life span in nature, and it is not clear what the phenotypic consequences are of this SNP haplotype variation. Understanding the selective mechanisms at work will reveal the way in which this gene affects organismal function in nature and hence the way in which natural selection acts upon it.

Alterations in the IIS pathway can have substantial effects upon aging in *Drosophila* in the laboratory. Some recent work provides intriguing findings about how natural genetic variation for a component of this pathway may have an adaptive role in flies in nature. Fly populations from higher latitudes show a higher incidence of diapause, a type of reproductive quiescence produced in response to low temperature and shortened photoperiod (Schmidt et al. 2005). Variation in diapause, between a low and high latitude population, has been shown, by using deletions and genomic rescue constructs for the gene, to be causally associated with variation in Dp110, the type-1 phosphoinositol-3 kinase (PI3K) of the *Drosophila* IIS pathway, with the nervous system implicated as a key effector tissue (Williams et al. 2006). It will be important to understand any connection between the diapause phenotype and life span and any implications for the ecology of life span.

CONCLUSIONS

O' Pioneer, *Drosophila*

The banding pattern of the giant polytene chromosomes from the *Drosophila* salivary gland provided the first genome map, and to this day, the unique accessibility of the *Drosophila* genome continues to lead genetics research. Of the three major model systems for aging, *Drosophila* combines speed of experimentation with complex tissue structures and behaviors: For example, strikingly unique male and female fighting styles are controlled by alternate splicing of the same *fruitless* gene (Vrontou et al. 2006). Its genetic transparency makes *Drosophila* the key organism for determining evolutionary conservation of aging mechanisms, as well as for the investigation of basic evolutionary processes in aging. *Drosophila*'s dominance of comparative studies seems likely to continue, with the sequencing of 12 related fly species' genomes nearly complete. Such information should be extremely powerful if and when it can be combined with analysis of aging in these species. Our understanding of processes as diverse as immune response, genome maintenance (Wang et al. 2006), and stem cell division (Hatfield et al. 2005) has been revolutionized by the discovery of the ubiquitous role that small RNA species have, and nowhere is this advance more rapid than in *Drosophila*, where an intersection with life span studies seems to be imminent.

Mitochondria, We Ponder 'Ya

Temperature was the first and will likely be one of the next hot areas for research on aging in model organisms as it dramatically alters life span and metabolism in each invertebrate system. Each of the evolutionarily conserved pathways for life span regulation, DR, IIS, and ROS metabolism, converges on central metabolic regulatory molecules such as TOR, AMPK (AMP-activated protein kinase), and PTEN that serve at the nexus of growth and nutritional and metabolic signaling. If a unifying theme is possible, it might be nuclear-mitochondrial communication (McBride et al. 2006). For example, a dramatic correlate of reduced IIS signaling in

C. elegans is the up-regulation of the mitochondrial manganese-superoxide dismutase (*MnSOD*) gene (Honda and Honda 1999), and both reduced IIS (Clancy et al. 2001; Tatar et al. 2001) and overexpression of *SOD* (Parkes et al. 1998; Sun et al. 2002, 2004) are sufficient to extend life span in *Drosophila*. These data suggest a longevity-promoting retrograde signal of H_2O_2, reminiscent of data from yeast (Jazwinski 2005; Osiewacz and Scheckhuber 2006). These ideas are consistent with the identification of IIS signaling and mitochondrial genes as primary life span regulators in the worm (Kenyon 2005; Kimura et al. 2007), as well as autophagy (Schmitz et al. 2007), since that is the central pathway for mitochondrial turnover. Strikingly, the *Drosophila* gene *sun* (*stunted*), identified as the ligand for *Methuselah,* is also the ε subunit of mitochondrial ATP synthase and has maternal effects (Kidd et al. 2005). Deterioration of reciprocal nuclear-mitochondrial signaling due to pleiotropic gene effects, such as perhaps sexual differentiation, might be manifested at the molecular level as a failure to maintain mitochondrial and nuclear genome integrity, which might both produce and be exacerbated by oxidative stress and in turn activate apoptotic pathways. In humans, it is the mitochondrial genes such as *Pink* and *Parkin* that are being implicated in aging-related diseases such as PD, and their effects are being modeled in the fly (Yang et al. 2006). Increasing data implicate the mitochondria and ROS in the regulation of human telomere erosion (Rodier et al. 2005), thereby implicating nuclear-mitochondrial redox signaling in the cellular senescence models. Tying experimental data on DR into this general model is more problematic: The fly in the ointment is that Kabil et al. (2007) found that DR does not elevate *SOD* activity in flies and that longevity associated with *Drosophila* IIS pathway *chico* mutations was not associated with increased *MnSOD* activity. Moreover, although DR caused alterations in mitochondria, there was no detectable alteration in mitochondrial ROS production associated with the increased longevity (Magwere et al. 2006a). One possible solution to this conundrum is that perhaps SOD activity need only be elevated in a relatively small subset of fly tissues or cells to effect increased life span, such as copper-zinc superoxide dismutase (Cu/ZnSOD) in the motoneurons (Parkes et al. 1998), and therefore, the relevant increase might not always be apparent in whole-fly extracts. It is also important to keep in mind that Cu/ZnSOD and MnSOD appear to have partially additive effects (Sun et al. 2004) and may not have the same mechanism or target tissue for causing increased life span. Consistent with the idea of a critical tissue and level for SOD overexpression, there are *Drosophila* transgenic strains where SOD activity is significantly elevated in whole-fly extracts yet life span is not altered (Parkes et al. 1998; Orr et al. 2003) and where excess overexpression of MnSOD in certain tissues is toxic (Ford et al. 2007). Feeding animals antioxidants may be able to ameliorate the toxic effects of ROS in certain vital tissues (Melov et al. 2001; Keaney et al. 2004; Magwere et al. 2006b), yet still fail to affect tissues critical for life span extension. Certainly, the degree to which the life span effects of IIS, DR, and ROS metabolism may or may not be reconciled with hypotheses for species; general aging mechanisms will be an impetus for research on *Drosophila* for some time to come.

REFERENCES

Acuna-Castillo C., Leiva-Salcedo E., Gomez C.R., Perez V., Li M., Torres C., Walter R., Murasko D.M., and Sierra F. 2006. T-kininogen: A biomarker of aging in Fisher 344 rats with possible implications for the immune response. *J. Gerontol. A Biol. Sci. Med. Sci.* **61:** 641–649.

Akasaka T., Klinedinst S., Ocorr K., Bustamante E.L., Kim S.K., and Bodmer R. 2006. The ATP-sensitive potassium (KATP) channel-encoded dSUR gene is required for *Drosophila* heart function and is regulated by tinman. *Proc. Natl. Acad. Sci.* **103:** 11999–12004.

Allikian M.J., Deckert-Cruz D., Rose M.R., Landis G.N., and Tower J. 2002. Doxycycline-induced expression of sense and inverted-repeat constructs modulates phosphogluconate mutase (Pgm) gene expression in adult *Drosophila melanogaster*. *Genome Biol.* **3:** research0021.

Alpatov W.W. 1930. Experimental studies on the duration of life. XII. The influence of different feeding during the larval and imaginal stages on the duration of life of the imago of *Drosophila melanogaster*. *Am. Nat.* **64:** 37–55.

Amador-Noguez D., Zimmerman J., Venable S., and Darlington G. 2005. Gender-specific alterations in gene expression and loss of liver sexual dimorphism in the long-lived Ames dwarf mice. *Biochem. Biophys. Res. Commun.* **332:** 1086–1100.

Anton-Erxleben F., Miquel J., and Philpott D.E. 1983. Fine-structural changes in the midgut of old *Drosophila melanogaster*. *Mech. Aging Dev.* **23:** 265–276.

Arantes-Oliveira N., Apfeld J., Dillin A., and Kenyon C. 2002. Regulation of life-span by germ-line stem cells in *Caenorhabditis elegans*. *Science* **295:** 502–505.

Auluck P.K. and Bonini N.M. 2002. Pharmacological prevention of Parkinson disease in *Drosophila*. *Nat. Med.* **8:** 1185–1186.

Austad S.N. 2001. An experimental paradigm for the study of slowly aging organisms. *Exp. Gerontol.* **36:** 599–605.

Bae K. and Edery I. 2006. Regulating a circadian clock's period, phase and amplitude by phosphorylation: Insights from *Drosophila*. *J. Biochem.* **140:** 609–617.

Barker D.J. 2004. The developmental origins of adult disease. *J. Am. Coll. Nutr.* (suppl. 6) **23:** 588S–595S.

Barnes A.I., Boone J.M., Partridge L., and Chapman T. 2007. A functioning ovary is not required for sex peptide to reduce receptivity to mating in *D. melanogaster*. *J. Insect Physiol.* **53:** 343–348.

Barnes A.I., Boone J.M., Jacobson J., Partridge L., and Chapman T. 2006. No extension of lifespan by ablation of germ line in *Drosophila*. *Proc. Biol. Sci.* **273:** 939–947.

Bartke A. 2005. Minireview: Role of the growth hormone/insulin-like growth factor system in mammalian aging. *Endocrinology* **146:** 3718–3723.

Bartke A. and Brown-Borg H. 2004. Life extension in the dwarf mouse. *Curr. Top. Dev. Biol.* **63:** 189–225.

Bauer J.H., Poon P.C., Glatt-Deeley H., Abrams J.M., and Helfand S.L. 2005. Neuronal expression of p53 dominant-negative proteins in adult *Drosophila melanogaster* extends life span. *Curr. Biol.* **15:** 2063–2068.

Bellen H.J., Levis R.W., Liao G., He Y., Carlson J.W., Tsang G., Evans-Holm M., Hiesinger P.R., Schulze K.L., Rubin G.M., Hoskins R.A., and Spradling A.C. 2004. The BDGP gene disruption project: Single transposon insertions associated with 40% of *Drosophila* genes. *Genetics* **167:** 761–781.

Bello B., Reichert H., and Hirth F. 2006. The brain tumor gene negatively regulates neural progenitor cell proliferation in the larval central brain of *Drosophila*. *Development* **133:** 2639–2648.

Bello B., Resendez-Perez D., and Gehring W.J. 1998. Spatial and temporal targeting of gene expression in *Drosophila* by means of a tetracycline-dependent transactivator system. *Development* **125:** 2193–2202.

Berger Z., Ttofi E.K., Michel C.H., Pasco M.Y., Tenant S., Rubinsztein D.C., and O'Kane C.J. 2005. Lithium rescues toxicity of aggregate-prone proteins in *Drosophila* by perturbing Wnt pathway. *Hum. Mol. Genet.* **14:** 3003–3011.

Berger Z., Ravikumar B., Menzies F.M., Oroz L.G., Underwood B.R., Pangalos M.N., Schmitt I., Wullner U., Evert B.O., O'Kane C.J., and Rubinsztein D.C. 2006. Rapamycin alleviates toxicity of different aggregate-prone proteins. *Hum. Mol. Genet.* **15:** 433–442.

Bergsmedh, A., Szeles A., Henriksson M., Bratt A., Folkman M.J., Spetz A.L., and Holmgren L. 2001. Horizontal transfer of oncogenes by uptake of apoptotic bodies. *Proc. Natl. Acad. Sci.* **98:** 6407–6411.

Bier E. and Bodmer R. 2004. *Drosophila*, an emerging model for cardiac disease. *Gene* **342:** 1–11.

Bieschke E.T., Wheeler J.C., and Tower J. 1998. Doxycycline-induced transgene expression during *Drosophila* development and aging. *Mol. Gen. Genet.* **258:** 571–579.

Bilen J. and Bonini N.M. 2005. *Drosophila* as a model for human neurodegenerative disease. *Annu. Rev. Genet.* **39:** 153–171.

Bozuck A.N. 1972. DNA synthesis in the absence of somatic cell division associated with aging in *Drosophila* subobscura. *Exp. Gerontol.* **7:** 147–156.

Brand A.H. and Perrimon N. 1993. Targeted gene expression as a means of altering cell fates and generating dominant phenotypes. *Development* **118:** 401–415.

Bredesen D.E., Rao R.V., and Mehlen P. 2006. Cell death in the nervous system. *Nature* **443:** 796–802.

Brummel T., Ching A., Seroude L., Simon A.F., and Benzer S. 2004. *Drosophila* lifespan enhancement by exogenous bacteria. *Proc. Natl. Acad. Sci.* **101:** 12974–12979.

Burger J.M. and Promislow D.E. 2004. Sex-specific effects of interventions that extend fly life span. *Sci. Aging Knowledge Environ.* **2004:** pe30.

Buszczak M., Paterno S., Lighthouse D., Bachman J., Plank J., Owen S., Skora A., Nystul T., Ohlstein B., Allen A., Wilhelm J., Murphy T., Levis B., Matunis E., Srivali N., Hoskins R., and Spradling A. 2007. The Carnegie protein trap library: A versatile tool for *Drosophila* developmental studies. *Genetics* **175:** 1505–1531.

Campisi J. 2003. Cancer and ageing: Rival demons? *Nat. Rev. Cancer* **3:** 339–349.

———. 2005. Senescent cells, tumor suppression, and organismal aging: Good citizens, bad neighbors. *Cell* **120:** 513–522.

Carey J.R., Liedo P., Orozco D., and Vaupel J.W. 1992. Slowing of mortality rates at older ages on large medfly cohorts. *Science* **258:** 457–461.

Carey J.R., Papadopoulos N., Kouloussis N., Katsoyannos B., Muller H.G., Wang J.L., and Tseng Y.K. 2006. Age-specific and lifetime behavior patterns in *Drosophila melanogaster* and the Mediterranean fruit fly, *Ceratitis capitata. Exp. Gerontol.* **41:** 93–97.

Cargill S.L., Carey J.R., Muller H.G., and Anderson G. 2003. Age of ovary determines remaining life expectancy in old ovariectomized mice. *Aging Cell* **2:** 185–190.

Carney G.E., Robertson A., Davis M.B., and Bender M. 2004. Creation of EcR isoform-specific mutations in *Drosophila melanogaster* via local P element transposition, imprecise P element excision, and male recombination. *Mol. Genet. Genomics* **271:** 282–290.

Carthew R.W. 2001. Gene silencing by double-stranded RNA. *Curr. Opin. Cell Biol.* **13:** 244–248.

Carvalho G.B., Kapahi P., Anderson D.J., and Benzer S. 2006. Allocrine modulation of feeding behavior by the Sex Peptide of *Drosophila. Curr. Biol.* **16:** 692–696.

Casacuberta E. and Pardue M.L. 2006. RNA interference has a role in regulating *Drosophila* telomeres. *Genome Biol.* **7:** 220.

Cashio P., Lee T.V., and Bergmann A. 2005. Genetic control of programmed cell death in *Drosophila melanogaster. Semin. Cell Dev. Biol.* **16:** 225–235.

Casper A. and Van Doren M. 2006. The control of sexual identity in the *Drosophila* germline. *Development* **133:** 2783–2791.

Cereghetti G.M. and Scorrano L. 2006. The many shapes of mitochondrial death. *Oncogene* **25:** 4717–4724.

Chalfie M., Tu Y., Euskirchen G., Ward W.W., and Prasher D.C. 1994. Green flourescent protein as a marker for gene expression. *Science* **263:** 802–805.

Chapin A., Correa P., Maguire M., and Kohn R. 2007. Synaptic neurotransmission protein UNC-13 affects RNA interference in neurons. *Biochem. Biophys. Res. Commun.* **354:** 1040–1044.

Chapman T. and Partridge L. 1996. Female fitness in *Drosophila melanogaster:* An interaction between the effect of nutrition and of encounter rates with males. *Proc. Roy. Soc. Ser. B* **263:** 755–759.

Chapman T., Liddle L.F., Kalb J.M., Wolfner M.F., and Partridge L. 1995. Cost of mating in *Drosophila melanogaster* females mediated by male accessory gland products. *Nature* **373:** 241–244.

Chapman T., Bangham J., Vinti G., Seifried B., Lung O., Wolfner M.F., Smith H.K., and Partridge L. 2003. The sex peptide of *Drosophila melanogaster:* Female post-mating responses analyzed by using RNA interference. *Proc. Natl. Acad. Sci.* **100:** 9923–9928.

Chatterjee B., Fernandes G., Yu B.P., Song C., Kim J.M., Demyan W., and Roy A.K. 1989. Calorie restriction delays age-dependent loss in androgen responsiveness of the rat liver. *FASEB J.* **3:** 169–173.

Chee F., Mudher A., Newman T.A., Cuttle M., Lovestone S., and Shepherd D. 2006. Overexpression of tau results in defective synaptic transmission in *Drosophila* neuromuscular junctions. *Biochem. Soc. Trans.* **34:** 88–90.

Chintapalli V.R., Wang J., and Dow J.A. 2007. Using FlyAtlas to identify better *Drosophila melanogaster* models of human disease. *Nat. Genet.* **39:** 715–720.

Choksi S.P., Southall T.D., Bossing T., Edoff K., de Wit E., Fischer B.E., van Steensel B., Micklem G., and Brand A.H. 2006. Prospero acts as a binary switch between self-renewal and differentiation in *Drosophila* neural stem cells. *Dev. Cell* **11:** 775–789.

Chow D.K., Glenn C.F., Johnston J.L., Goldberg I.G., and Wolkow C.A. 2006. Sarcopenia in the *Caenorhabditis elegans* pharynx correlates with muscle contraction rate over lifespan. *Exp. Gerontol.* **41:** 252–260.

Cirelli C. 2003. Searching for sleep mutants of *Drosophila melanogaster*. *Bioessays* **25:** 940–949.

Clancy D.J., Gems D., Harshman L.G., Oldham S., Stocker H., Hafen E., Leevers S.J., and Partridge L. 2001. Extension of life-span by loss of CHICO, a *Drosophila* insulin receptor substrate protein. *Science* **292:** 104–106.

Collins B., Mazzoni E.O., Stanewsky R., and Blau J. 2006. *Drosophila* CRYPTOCHROME is a circadian transcriptional repressor. *Curr. Biol.* **16:** 441–449.

Cooley L., Thompson D., and Spradling A.C. 1990. Constructing deletions with defined endpoints in *Drosophila*. *Proc. Natl. Acad. Sci.* **87:** 3170–3173.

Cooper A.A., Gitler A.D., Cashikar A., Haynes C.M., Hill K.J., Bhullar B., Liu K., Strathearn K.E., Liu F., Cao S., Caldwell K.A., Caldwell G.A., Marsischky G., Kolodner R.D., Labaer J., Rochet J.C., Bonini N.M., and Lindquist S. 2006. Alpha-synuclein blocks ER-Golgi traffic and Rab1 rescues neuron loss in Parkinson's models. *Science* **313:** 324–328.

Cordts R. and Partridge L. 1996. Courtship reduces longevity of male *Drosophila melanogaster*. *Anim. Behav.* **52:** 269–278.

Crowther D.C., Kinghorn K.J., Page R., and Lomas D.A. 2004. Therapeutic targets from a *Drosophila* model of Alzheimer's disease. *Curr. Opin. Pharmacol.* **4:** 513–516.

Crowther D C., Page R., Chandraratna D., and Lomas D.A. 2006. A *Drosophila* model of Alzheimer's disease. *Methods Enzymol.* **412:** 234–255.

Crowther D.C., Kinghorn K.J., Miranda E., Page R., Curry J.A., Duthie F.A., Gubb D.C., and Lomas D.A. 2005. Intraneuronal Abeta, non-amyloid aggregates and neurodegeneration in a *Drosophila* model of Alzheimer's disease. *Neuroscience* **132:** 123–135.

Curtsinger J.W., Fukui H.H., Townsend D.R., and Vaupel J.W. 1992. Demography of genotypes: Failure of the limited life-span paradigm in *Drosophila melanogaster*. *Science* **258:** 461–463.

De Luca M., Roshina N.V., Geiger-Thornsberry G.L., Lyman R.F., Pasyukova E.G., and Mackay T.F. 2003. Dopa decarboxylase (Ddc) affects variation in *Drosophila* longevity. *Nat. Genet.* **34:** 429–433.

Decotto E. and Spradling A.C. 2005. The *Drosophila* ovarian and testis stem cell niches: Similar somatic stem cells and signals. *Dev. Cell* **9:** 501–510.

Denholm B., Sudarsan V., Pasalodos-Sanches S., Artero R., Lawrence P., Maddrell S., Baylies M., and Skaer H. 2003. Dual origin of the renal tubules in *Drosophila*: Mesodermal cells integrate and polarize to establish secretory function. *Curr. Biol.* **13:** 1052–1057.

Desai U.A., Pallos J., Ma A.A., Stockwell B.R., Thompson L.M., Marsh J.L., and Diamond M.I. 2006. Biologically active molecules that reduce polyglutamine aggregation and toxicity. *Hum. Mol. Genet.* **15:** 2114–2124.

Dias-Santagata D., Fulga T.A., Duttaroy A., and Feany M.B. 2007. Oxidative stress mediates tau-induced neurodegeneration in *Drosophila*. *J. Clin. Invest.* **117:** 236–245.

Dionne M.S., Pham L.N., Shirasu-Hiza M., and Schneider D.S. 2006. Akt and FOXO dysregulation contribute to infection-induced wasting in *Drosophila*. *Curr. Biol.* **16:** 1977–1985.

Dorner S., Lum S., Kim M., Paro R., Beachy P.A., and Green R. 2006. A genomewide screen for components of the RNAi pathway in *Drosophila* cultured cells. *Proc. Natl. Acad. Sci.* **103:** 11880–11885.

Dow J.A. and Davies S.A. 2006. The Malpighian tubule: Rapid insights from post-genomic biology. *J. Insect Physiol.* **52:** 365–378.

Duvernell D.D., Schmidt P.S., and Eanes W.F. 2003. Clines and adaptive evolution in the *Methuselah* gene region in *Drosophila melanogaster*. *Mol. Ecol.* **12:** 1277–1285.

Egli D., Hafen E., and Schaffner W. 2004. An efficient method to generate chromosomal rearrangements by targeted DNA double-strand breaks in *Drosophila melanogaster*. *Genome Res.* **14:** 1382–1393.

Faurby S., Kjaersgaard A., Pertoldi C., and Loeschcke V. 2005. The effect of maternal and grandmaternal age in benign and high temperature environments. *Exp. Gerontol.* **40:** 988–996.

Feany M. and Bender W. 2000. A *Drosophila* model of Parkinson's disease. *Nature* **404:** 394–398.

Finch C.E., Pike M.C., and Witten M. 1990. Slow mortality rate accelerations during aging in some animals approximate that of humans. *Science* **249:** 902–905.

Fischer J.A., Giniger E., Maniatis T., and Ptashne M. 1988. GAL4 activates transcription in *Drosophila*. *Nature* **332:** 853–856.

Ford D., Hoe N., Landis G.N., Tozer K., Luu A., Bhole D., Badrinath A., and Tower J. 2007. Alteration of *Drosophila* life span using condititonal, tissue-specific expression of transgenes triggered by doxycycline or RU486/Mifepristone. *Exp. Gerontol.* **42:** 483–497.

Fowler K. and Partridge L. 1989. A cost of mating in female fruit flies. *Nature* **338:** 760–761.

Fox C.W., Czesak M.E., and Wallin W.G. 2004. Complex genetic architecture of population differences in adult lifespan of a beetle: Nonadditive inheritance, gender differences, body size and a large maternal effect. *J. Evol. Biol.* **17:** 1007–1017.

Friedman D.B. and Johnson T.E. 1988. A mutation in the *age-1* gene in *Caenorhabditis elegans* lengthens life and reduces hermaphrodite fertility. *Genetics* **118:** 75–86.

Fulga T.A., Elson-Schwab I., Khurana V., Steinhilb M.L., Spires T.L., Hyman B.T., and Feany M.B. 2007. Abnormal bundling and accumulation of F-actin mediates tau-induced neuronal degeneration in vivo. *Nat. Cell Biol.* **9:** 139–148.

Gachon F., Olela F.F., Schaad O., Descombes P., and Schibler U. 2006. The circadian PAR-domain basic leucine zipper transcription factors DBP, TEF, and HLF modulate basal and inducible xenobiotic detoxification. *Cell Metab.* **4:** 25 36.

Garesse R. and Kaguni L.S. 2005. A *Drosophila* model of mitochondrial DNA replication: Proteins, genes and regulation. *IUBMB Life* **57:** 555–561.

Garigan D., Hsu A.L., Fraser A.G., Kamath R.S., Ahringer J., and Kenyon C. 2002. Genetic analysis of tissue aging in *Caenorhabditis elegans:* A role for heat-shock factor and bacterial proliferation. *Genetics* **161:** 1101–1112.

Gendron C.M., Minois N., Fabrizio P., Longo V.D., Pletcher S.D., and Vaupel J.W. 2003. Biodemographic trajectories of age-specific reproliferation from stationary phase in the yeast *Saccharomyces cerevisiae* seem multiphasic. *Mech. Ageing Dev.* **124:** 1059–1063.

Giannakou M.E. and Partridge L. 2007. Role of insulin-like signalling in *Drosophila* lifespan. *Trends Biochem. Sci.* **32:** 180–188.

Giannakou M.E., Goss M., Jacobson J., Vinti G., Leevers S.J., and Partridge L. 2007. Dynamics of the action of dFOXO on adult mortality in *Drosophila*. *Aging Cell* **6:** 429–438.

Golden T.R. and Melov S. 2004. Microarray analysis of gene expression with age in individual nematodes. *Aging Cell* **3:** 111–124.

Golic K.G. and Lindquist S. 1989. The FLP recombinase of yeast catalyzes site-specific recombination in the *Drosophila* genome. *Cell* **59:** 499–509.

Gotter A.L. 2006. A Timeless debate: Resolving TIM's noncircadian roles with possible clock function. *Neuroreport* **17:** 1229–1233.

Graham P., Penn J.K., and Schedl P. 2003. Masters change, slaves remain. *Bioessays* **25:** 1–4.

Graves B.M., Strand M., and Lindsay A.R. 2006. A reassessment of sexual dimorphism in human senescence: Theory, evidence, and causation. *Am. J. Hum. Biol.* **18:** 161–168.

Greene J.C., Whitworth A.J., Andrews L.A., Parker T.J., and Pallanck L.J. 2005. Genetic and genomic studies of *Drosophila* parkin mutants implicate oxidative stress and innate immune responses in pathogenesis. *Hum. Mol. Genet.* **14:** 799–811.

Greene J.C., Whitworth A.J., Kuo I., Andrews L.A., Feany M.B., and Pallanck L.J. 2003. Mitochondrial pathology and apoptotic muscle degeneration in *Drosophila* parkin mutants. *Proc. Natl. Acad. Sci.* **100:** 4078–4083.

Grotewiel M.S., Martin I., Bhandari P., and Cook-Wiens E. 2005. Functional senescence in *Drosophila melanogaster*. *Ageing Res. Rev.* **4:** 372–397.

Hariharan I.K. and Bilder D. 2006. Regulation of imaginal disc growth by tumor-suppressor genes in *Drosophila*. *Annu. Rev. Genet.* **40:** 335–361.

Hatfield S.D., Shcherbata H.R., Fischer K.A., Nakahara K., Carthew R.W., and Ruohola-Baker H. 2005. Stem cell division is regulated by the microRNA pathway. *Nature* **435:** 974–978.

Hatton-Ellis E., Ainsworth C., Sushama Y., Wan S., VijayRaghavan K., and Skaer H. 2007. Genetic regulation of patterned tubular branching in *Drosophila*. *Proc. Natl. Acad. Sci.* **104:** 169–174.

Hawley R.S. 2003. Human meiosis: Model organisms address the maternal age effect. *Curr. Biol.* **13:** R305–R307.

Hekimi S. and Guarente L. 2003. Genetics and the specificity of the aging process. *Science* **299:** 1351–1354.

Hendricks J.C. and Sehgal A. 2004. Why a fly? Using *Drosophila* to understand the genetics of circadian rhythms and sleep. *Sleep* **27:** 334–342.

Herndon L.A., Schmeissner P.J., Dudaronek J.M., Brown P.A., Listner K.M., Sakano Y., Paupard M.C., Hall D.H., and Driscoll M. 2002. Stochastic and genetic factors influence tissue-specific decline in ageing *C. elegans*. *Nature* **419:** 808–814.

Honda Y. and Honda S. 1999. The *daf-2* gene network for longevity regulates oxidative stress resistance and *Mn-superoxide dismutase* gene expression in *Caenorhabditis elegans*. *FASEB J.* **13:** 1385–1393.

Horabin J.I. 2005. Splitting the Hedgehog signal: Sex and patterning in *Drosophila*. *Development* **132:** 4801–4810.

Hsin H. and Kenyon C. 1999. Signals from the reproductive system regulate the lifespan of *C. elegans*. *Nature* **399:** 362–366.

Huang C., Xiong C., and Kornfeld K. 2004. Measurements of age-related changes of physiological processes that predict lifespan of *Caenorhabditis elegans*. *Proc. Natl. Acad. Sci.* **101:** 8084–8089.

Huang X., Hong C.S., O'Donnell M., and Saint-Jeannet J.P. 2005. The doublesex-related gene, *XDmrt*4, is required for neurogenesis in the olfactory system. *Proc. Natl. Acad. Sci.* **102:** 11349–11354.

Huber R., Hill S.L., Holladay C., Biesiadecki M., Tononi G., and Cirelli C. 2004. Sleep homeostasis in *Drosophila melanogaster*. *Sleep* **27:** 628–639.

Hughes K.A. and Reynolds R.M. 2005. Evolutionary and mechanistic theories of aging. *Annu. Rev. Entomol.* **50:** 421–445.

Hunt J., Jennions M.D., Spyrou N., and Brooks R. 2006. Artificial selection on male longevity influences age-dependent reproductive effort in the black field cricket *Teleogryllus commodus*. *Am. Nat.* **168:** E72–E86.

Hwangbo D.S., Gershman B., Tu M.P., Palmer M., and Tatar M. 2004. *Drosophila* dFOXO controls lifespan and regulates insulin signalling in brain and fat body. *Nature* **429:** 562–566.

Ingram D.K., Nakamura E., Smucny D., Roth G.S., and Lane M.A. 2001. Strategy for identifying biomarkers of aging in long-lived species. *Exp. Gerontol.* **36:** 1025–1034.

Isagi Y., Shimada K., Kushima H., Tanaka N., Nagao A., Ishikawa T., OnoDera H., and Watanabe S. 2004. Clonal structure and flowering traits of a bamboo [*Phyllostachys pubescens* (Mazel) Ohwi] stand grown from a simultaneous flowering as revealed by AFLP analysis. *Mol. Ecol.* **13:** 2017–2021.

Jackson A.U., Galecki A.T., Burke D.T., and Miller R.A. 2002. Mouse loci associated with life span exhibit sex-specific and epistatic effects. *J. Gerontol. A Biol. Sci. Med. Sci.* **57:** B9–B15.

Jazwinski S.M. 2005. The retrograde response links metabolism with stress responses, chromatin-dependent gene activation, and genome stability in yeast aging. *Gene* **354:** 22–27.

Johnson T.E., Wu D., Tedesco P., Dames S., and Vaupel J.W. 2001. Age-specific demographic profiles of longevity mutants in *Caenorhabditis elegans* show segmental effects. *J. Gerontol. A Biol. Sci. Med. Sci.* **56:** B331–B339.

Jung A.C., Denholm B., Skaer H., and Affolter M. 2005. Renal tubule development in *Drosophila:* A closer look at the cellular level. *J. Am. Soc. Nephrol.* **16:** 322–328.

Kabil H., Partridge L., and Harshman L.G. 2007. Superoxide dismutase activities in long-lived *Drosophila melanogaster* females: chico (1) genotypes and dietary dilution. *Biogerontology* **8:** 201–208.

Kaeberlein T.L., Smith E.D., Tsuchiya M., Welton K.L., Thomas J.H., Fields S., Kennedy B.K., and Kaeberlein M. 2006. Lifespan extension in *Caenorhabditis elegans* by complete removal of food. *Aging Cell* **5:** 487–494.

Keaney M., Matthijssens F., Sharpe M., Vanfleteren J., and Gems D. 2004. Superoxide dismutase mimetics elevate superoxide dismutase activity in vivo but do not retard aging in the nematode *Caenorhabditis elegans*. *Free Radic. Biol. Med.* **37:** 239–250.

Kenyon C. 2005. The plasticity of aging: Insights from long-lived mutants. *Cell* **120:** 449–460.

Kenyon C., Chang J., Gensch E., Rudner A., and Tabtiang R. 1993. A *C. elegans* mutant that lives twice as long as wild type. *Nature* **366:** 461–464.

Khurana V., Lu Y., Steinhilb M.L., Oldham S., Shulman J.M., and Feany M.B. 2006. TOR-mediated cell-cycle activation causes neurodegeneration in a *Drosophila* tauopathy model. *Curr. Biol.* **16:** 230–241.

Kidd T., Abu-Shumays R., Katzen A., Sisson J.C., Jimenez G., Pinchin S., Sullivan W., and Ish-Horowicz D. 2005. The epsilon-subunit of mitochondrial ATP synthase is required for normal spindle orientation during the *Drosophila* embryonic divisions. *Genetics* **170:** 697–708.

Kimura K., Tanaka N., Nakamura N., Takano S., and Ohkuma S. 2007. Knockdown of mitochondrial heat shock protein 70 promotes progeria-like phenotypes in *Caenorhabditis elegans*. *J. Biol. Chem.* **282:** 5910–5918.

Kimura K.D., Tissenbaum H.A., Liu Y., and Ruvkun G. 1997. *daf-2*, an insulin receptor-like gene that regulates longevity and diapause in *Caenorhabditis elegans*. *Science* **277:** 942–946.

King-Jones K., Horner M.A., Lam G., and Thummel C.S. 2006. The DHR96 nuclear receptor regulates xenobiotic responses in *Drosophila*. *Cell Metab.* **4:** 37–48.

Kinghorn K.J., Crowther D.C., Sharp L.K., Nerelius C., Davis R.L., Chang H.T., Green C., Gubb D.C., Johansson J., and Lomas D.A. 2006. Neuroserpin binds Abeta and is a neuroprotective component of amyloid plaques in Alzheimer's disease. *J. Biol. Chem.* **281:** 29268–29277.

Kirby K., Hu J., Hilliker A.J., and Phillips J.P. 2002. RNA interference-mediated silencing of Sod2 in *Drosophila* leads to early adult-onset mortality and elevated endogenous oxidative stress. *Proc. Natl. Acad. Sci.* **99:** 16162–16167.

Klass M.R. 1983. A method for the isolation of longevity mutants in the nematode *Caenorhabditis elegans* and initial results. *Mech. Ageing Dev.* **22:** 279–286.

Koh K., Evans J.M., Hendricks J.C., and Sehgal A. 2006. A *Drosophila* model for age-associated changes in sleep:wake cycles. *Proc. Natl. Acad. Sci.* **103:** 13843–13847.

Kohler R.E. 1994. *Lords of the fly:* Drosophila *genetics and the experimental life.* University of Chicago Press, Chicago, Illinois.

Korenjak M. and Brehm A. 2006. The retinoblastoma tumour suppressor in model organisms: New insights from flies and worms. *Curr. Mol. Med.* **6:** 705–711.

Kuijper B., Stewart A.D., and Rice W.R. 2006. The cost of mating rises nonlinearly with copulation frequency in a laboratory population of *Drosophila melanogaster*. *J. Evol. Biol.* **19:** 1795–1802.

Kujoth G.C., Hiona A., Pugh T.D., Someya S., Panzer K., Wohlgemuth S.E., Hofer T., Seo A.Y., Sullivan R., Jobling W.A., Morrow J.D., Van Remmen H., Sedivy J.M., Yamasoba T., Tanokura M., Weindruch R., Leeuwenburgh C., and Prolla T.A. 2005. Mitochondrial DNA mutations, oxidative stress, and apoptosis in mammalian aging. *Science* **309:** 481–484.

Kunwar P.S., Siekhaus D.E., and Lehmann R. 2006. In vivo migration: A germ cell perspective. *Annu. Rev. Cell Dev. Biol.* **22:** 237–265.

Lai C.Q., Parnell L.D., Lyman R.F., Ordovas J.M., and Mackay T.F. 2007. Candidate genes affecting *Drosophila* life span identified by integrating microarray gene expression analysis and QTL mapping. *Mech. Ageing Dev.* **128:** 237–249.

Lakatta E.G. 2000. Cardiovascular aging in health. *Clin. Geriatr. Med.* **16:** 419–444.

Landis G.N. and Tower J. 2005. Superoxide dismutase evolution and life span regulation. *Mech. Ageing Dev.* **126:** 365–379.

Landis G.N., Bhole D., and Tower J. 2003. A search for doxycycline-dependent mutations that increase *Drosophila melanogaster* life span identifies the VhaSFD, Sugar baby, filamin, fwd and Cctl genes. *Genome Biol.* **4:** R8.

Landis G.N., Abdueva D., Skvortsov D., Yang J., Rabin B.E., Carrick J., Tavare S., and Tower J. 2004. Similar gene expression patterns characterize aging and oxidative stress in *Drosophila melanogaster*. *Proc. Natl. Acad. Sci.* **101:** 7663–7668.

Larsen P.L. 1993. Aging and resistance to oxidative damage in *Caenorhabditis elegans*. *Proc. Natl. Acad. Sci.* **90:** 8905–8909.

Le Bras S. and Van Doren M. 2006. Development of the male germline stem cell niche in *Drosophila*. *Dev. Biol.* **294:** 92–103.

Lee C.C. 2006. Tumor suppression by the mammalian *Period* genes. *Cancer Causes Control* **17:** 525–530.

Lee C.Y., Wilkinson B.D., Siegrist S.E., Wharton R.P., and Doe C.Q. 2006. Brat is a Miranda cargo protein that promotes neuronal differentiation and inhibits neuroblast self-renewal. *Dev. Cell* **10:** 441–449.

Lee T. and Luo L. 2001. Mosaic analysis with a repressible cell marker (MARCM) for *Drosophila* neural development. *Trends Neurosci.* **24:** 251–254.

Leffelaar D. and Grigliatti T. 1984. Age-dependent behavior loss in adult *Drosophila melanogaster. Dev. Genet.* **4:** 211–227.

Leips J. and Mackay T.F. 2002. The complex genetic architecture of *Drosophila* life span. *Exp. Aging Res.* **28:** 361–390.

Lew T.A., Morrow E.H., and Rice W.R. 2006. Standing genetic variance for female resistance to harm from males and its relationship to intralocus sexual conflict. *Evolution Int. J. Org. Evolution* **60:** 97–105.

Libert S., Chao Y., Chu X., and Pletcher S.D. 2006. Trade-offs between longevity and pathogen resistance in *Drosophila melanogaster* are mediated by NFkappaB signaling. *Aging Cell* **5:** 533–543.

Lin K., Dorman J.B., Rodan A., and Kenyon C. 1997. *daf-16:* An HNF-3/forkhead family member that can function to double the life-span of *Caenorhabditis elegans. Science* **278:** 1319–1322.

Lin Y.-J., Seroude L., and Benzer S. 1998. Extended life-span and stress resistance in the *Drosophila* mutant *methuselah. Science* **282:** 943–946.

Liu H. and Kubli E. 2003. Sex-peptide is the molecular basis of the sperm effect in *Drosophila melanogaster. Proc. Natl. Acad. Sci.* **100:** 9929–9933.

Loeb J. and Northrop J.H. 1917. On the influence of food and temperature upon the duration of life. *J. Biol. Chem.* **32:** 103–121.

Long T.A., Montgomerie R., and Chippindale A.K. 2006. Quantifying the gender load: Can population crosses reveal interlocus sexual conflict? *Philos. Trans. R. Soc. Lond. B Biol. Sci.* **361:** 363–374.

Longo V.D. 2004. Ras: The other pro-aging pathway. *Sci. Aging Knowledge Environ.* **2004:** pe36.

Luckinbill L.S., Arking R., Clare M.J., Cirocco W.C., and Buck S.A. 1984. Selection for delayed senescence in *Drosophila melanogaster. Evolution* **38:** 996–1003.

Mackay T.F. and Anholt R.R. 2006. Of flies and man: *Drosophila* as a model for human complex traits. *Annu. Rev. Genomics Hum. Genet.* **7:** 339–367.

Maggert K.A. and Golic K.G. 2005. Highly efficient sex chromosome interchanges produced by I-CreI expression in *Drosophila. Genetics* **171:** 1103–1114.

Magwere T., Chapman T., and Partridge L. 2004. Sex differences in the effect of dietary restriction on life span and mortality rates in female and male *Drosophila melanogaster. J. Gerontol. A Biol. Sci. Med. Sci.* **59:** 3–9.

Magwere T., Goodall S., Skepper J., Mair W., Brand M.D., and Partridge L. 2006a. The effect of dietary restriction on mitochondrial protein density and flight muscle mitochondrial morphology in *Drosophila. J. Gerontol. A Biol. Sci. Med. Sci.* **61:** 36–47.

Magwere T., West M., Riyahi K., Murphy M.P., Smith R.A., and Partridge L. 2006b. The effects of exogenous antioxidants on lifespan and oxidative stress resistance in *Drosophila melanogaster. Mech. Ageing Dev.* **127:** 356–370.

Mair W., Goymer P., Pletcher S.D., and Partridge L. 2003. Demography of dietary restriction and death in *Drosophila. Science* **301:** 1731–1733.

Maklakov A.A., Kremer N., and Arnqvist G. 2005. Adaptive male effects on female ageing in seed beetles. *Proc. Biol. Sci.* **272:** 2485–2489.

Marsh J.L. and Thompson L.M. 2006. *Drosophila* in the study of neurodegenerative disease. *Neuron* **52:** 169–178.

Marsh J.L., Pallos J., and Thompson L.M. 2003. Fly models of Huntington's disease. *Hum. Mol. Genet.* (spec. no. 2) **12:** R187–R193.

Martin G.M. 2002. Keynote: Mechanisms of senescence—Complificationists versus simplificationists. *Mech. Aging Dev.* **123:** 65–73.

———. 2005. Genetic modulation of senescent phenotypes in *Homo sapiens*. *Cell* **120:** 523–532.

Martin I. and Grotewiel M.S. 2006. Distinct genetic influences on locomotor senescence in *Drosophila* revealed by a series of metrical analyses. *Exp. Gerontol.* **41:** 877–881.

Marzetti E. and Leeuwenburgh C. 2006. Skeletal muscle apoptosis, sarcopenia and frailty at old age. *Exp. Gerontol.* **41:** 1234–1238.

Masoro E.J. 2005. Overview of caloric restriction and ageing. *Mech. Ageing Dev.* **126:** 913–922.

McBride H.M., Neuspiel M., and Wasiak S. 2006. Mitochondria: More than just a powerhouse. *Curr. Biol.* **16:** R551–R560.

McElwee J.J., Schuster E., Blanc E., Thomas J.H., and Gems D. 2004. Shared transcriptional signature in *Caenorhabditis elegans* Dauer larvae and long-lived *daf-2* mutants implicates detoxification system in longevity assurance. *J. Biol. Chem.* **279:** 44533–44543.

McGettigan J., McLennan R.K., Broderick K.E., Kean L., Allan A.K., Cabrero P., Regulski M.R., Pollock V.P., Gould G.W., Davies S.A., and Dow J.A. 2005. Insect renal tubules constitute a cell-autonomous immune system that protects the organism against bacterial infection. *Insect Biochem. Mol. Biol.* **35:** 741–754.

Medawar P.B. 1946. Old age and natural death. *Mod. Q.* **2:** 30–49.

———. 1952. *An unsolved problem of biology.* H.K. Lewis, London, United Kingdom.

Melov S., Doctrow S.R., Schneider J.A., Haberson J., Patel M., Coskun P.E., Huffman K., Wallace D.C., and Malfroy B. 2001. Lifespan extension and rescue of spongiform encephalopathy in superoxide dismutase 2 nullizygous mice treated with superoxide dismutase-catalase mimetics. *J. Neurosci.* **21:** 8348–8353.

Menzies F.M., Yenisetti S.C., and Min K.T. 2005. Roles of *Drosophila* DJ-1 in survival of dopaminergic neurons and oxidative stress. *Curr. Biol.* **15:** 1578–1582.

Meulener M., Whitworth A.J., Armstrong-Gold C.E., Rizzu P., Heutink P., Wes P.D., Pallanck L.J., and Bonini N.M. 2005. *Drosophila* DJ-1 mutants are selectively sensitive to environmental toxins associated with Parkinson's disease. *Curr. Biol.* **15:** 1572–1577.

Micchelli C.A. and Perrimon N. 2006. Evidence that stem cells reside in the adult *Drosophila* midgut epithelium. *Nature* **439:** 475–479.

Minakhina S. and Steward R. 2006. Melanotic mutants in *Drosophila:* Pathways and phenotypes. *Genetics* **174:** 253–263.

Miquel J., Economos A.C., Bensch K.G., Atlan H., and Johnson J.E., Jr. 1979. Review of cell aging in *Drosophila* and mouse. *Age* **2:** 78–88.

Montgomery M.K. 2004. The use of double-stranded RNA to knock down specific gene activity. *Methods Mol. Biol.* **260:** 129–144.

Morbey Y.E. and Ydenberg R.C. 2003. Timing games in the reproductive phenology of female pacific salmon (*Oncorhynchus spp.*). *Am. Nat.* **161:** 284–298.

Mueller J.L., Page J.L., and Wolfner M.F. 2007. An ectopic expression screen reveals the protective and toxic effects of *Drosophila* seminal fluid proteins. *Genetics* **175:** 777–783.

Muller H.G., Carey J.R., Wu D., Liedo P., and Vaupel J.W. 2001. Reproductive potential predicts longevity of female Mediterranean fruitflies. *Proc. Biol. Sci.* **268:** 445–450.

Naora H. and Montell D.J. 2005. Ovarian cancer metastasis: Integrating insights from disparate model organisms. *Nat. Rev. Cancer* **5:** 355–366.

Nuzhdin S.V., Pasyukova E.G., Dilda C.L., Zeng Z.B., and Mackay T.F. 1997. Sex-specific quantitative trait loci affecting longevity in *Drosophila melanogaster*. *Proc. Natl. Acad. Sci.* **94:** 9734–9739.

Nystul T.G. and Spradling A.C. 2006. Breaking out of the mold: Diversity within adult stem cells and their niches. *Curr. Opin. Genet. Dev.* **16:** 463–468.

O'Kane C.J. 2003. Modelling human diseases in *Drosophila* and *Caenorhabditis. Semin. Cell Dev. Biol.* **14:** 3–10.

Ocorr K., Akasaka T., and Bodmer R. 2007. Age-related cardiac disease model of *Drosophila. Mech. Ageing Dev.* **128:** 112–116.

Oeppen J. and Vaupel J.W. 2002. Demography. Broken limits to life expectancy. *Science* **296:** 1029–1031.

Ogg S., Paradis S., Gottlieb S., Patterson G.I., Lee L., Tissenbaum H.A., and Ruvkun G. 1997. The Fork head transcription factor DAF-16 transduces insulin-like metabolic and longevity signals in *C. elegans. Nature* **389:** 994–999.

Ohlstein B. and Spradling A. 2006. The adult *Drosophila* posterior midgut is maintained by pluripotent stem cells. *Nature* **439:** 470–474.

———. 2007. Multipotent *Drosophila* intestinal stem cells specify daughter cell fates by differential notch signaling. *Science* **315:** 988–992.

Orr W.C., Mockett R.J., Benes J.J., and Sohal R.S. 2003. Effects of overexpression of copper-zinc and manganese superoxide dismutases, catalase, and thioredoxin reductase genes on longevity in *Drosophila melanogaster. J. Biol. Chem.* **278:** 26418–26422.

Osiewacz H.D. and Scheckhuber C.Q. 2006. Impact of ROS on ageing of two fungal model systems: *Saccharomyces cerevisiae* and *Podospora anserina. Free Radic. Res.* **40:** 1350–1358.

Osterwalder T., Yoon K.S., White B.H., and Keshishian H. 2001. A conditional tissue-specific transgene expression system using inducible GAL4. *Proc. Natl. Acad. Sci.* **98:** 12596–12601.

Ottolenghi C., Uda M., Crisponi L., Omari S., Cao A., Forabosco A., and Schlessinger D. 2007. Determination and stability of sex. *Bioessays* **29:** 15–25.

Palanker L., Necakov A.S., Samson H.M., Ni R., Hu C., Thummel C.S., and Krause H.M. 2006. Dynamic regulation of *Drosophila* nuclear receptor activity in vivo. *Development* **133:** 3549–3562.

Parkes T.L., Elia A.J., Dickinson D., Hilliker A.J., Phillips J.P., and Boulianne G.L. 1998. Extension of *Drosophila* lifespan by overexpression of human *SOD1* in motorneurons. *Nat. Genet.* **19:** 171–174.

Partridge L. and Gems D. 2002. Mechanisms of ageing: Public or private? *Nat. Rev. Genet.* **3:** 165–175.

———. 2006. Beyond the evolutionary theory of ageing, from functional genomics to evo-gero. *Trends Ecol. Evol.* **21:** 334–340.

Partridge L., Pletcher S.D., and Mair W. 2005. Dietary restriction, mortality trajectories, risk and damage. *Mech. Ageing Dev.* **126:** 35–41.

Pasyukova E.G., Roshina N.V., and Mackay T.F. 2004. Shuttle craft: A candidate quantitative trait gene for *Drosophila* lifespan. *Aging Cell* **3:** 297–307.

Pearl R. 1921. The biology of death. VI. Experimental studies on the duration of life. *The Scientific Monthly* **13:** 144–164.

Peng J., Zipperlen P., and Kubli E. 2005a. *Drosophila* sex-peptide stimulates female innate immune system after mating via the Toll and Imd pathways. *Curr. Biol.* **15:** 1690–1694.

Peng J., Chen S., Busser S., Liu H., Honegger T., and Kubli E. 2005b. Gradual release of sperm bound sex-peptide controls female postmating behavior in *Drosophila*. *Curr. Biol.* **15**: 207–213.

Penn J.K. and Schedl P. 2007. The master switch gene sex-lethal promotes female development by negatively regulating the N-signaling pathway. *Dev. Cell* **12**: 275–286.

Perrimon N. and Mathey-Prevot B. 2007. Applications of high-throughput RNA interference screens to problems in cell and developmental biology. *Genetics* **175**: 7–16.

Piper M.D.W., Selman C., McElwee J.J., and Partridge L. 2005. Models of insulin signalling and longevity. *Drug Discov. Today: Dis. Models* **2**: 249–256.

Pires-Dasilva A. 2007. Evolution of the control of sexual identity in nematodes. *Semin. Cell Dev. Biol.* (in press).

Pletcher S.D., Kabil H., and Partridge L. 2007. Chemical complexity and the genetics of aging. *Annu. Rev. Ecol. Syst.* (in press).

Pletcher S.D., Macdonald S.J., Marguerie R., Certa U., Stearns S.C., Goldstein D.B., and Partridge L. 2002. Genome-wide transcript profiles in aging and calorically restricted *Drosophila melanogaster*. *Curr. Biol.* **12**: 712–723.

Prins G.S., Jung M.H., Vellanoweth R.L., Chatterjee B., and Roy A.K. 1996. Age-dependent expression of the androgen receptor gene in the prostate and its implication in glandular differentiation and hyperplasia. *Dev. Genet.* **18**: 99–106.

Radtke F. and Clevers H. 2005. Self-renewal and cancer of the gut: Two sides of a coin. *Science* **307**: 1904–1909.

Rea S.L., Wu D., Cypser J.R., Vaupel J.W., and Johnson T.E. 2005. A stress-sensitive reporter predicts longevity in isogenic populations of *Caenorhabditis elegans*. *Nat. Genet.* **37**: 894–898.

Reiter L.T., Potocki L., Chien S., Gribskov M., and Bier E. 2001. A systematic analysis of human disease-associated gene sequences in *Drosophila melanogaster*. *Genome Res.* **11**: 1114–1125.

Rice W.R. 1992. Sexually antagonistic genes: Experimental evidence. *Science* **256**: 1436–1439.

Rice W.R., Stewart A.D., Morrow E.H., Linder J.E., Orteiza N., and Byrne P.G. 2006. Assessing sexual conflict in the *Drosophila melanogaster* laboratory model system. *Philos. Trans. R. Soc. Lond. B Biol. Sci.* **361**: 287–299.

Rodier F., Kim S.H., Nijjar T., Yaswen P., and Campisi J. 2005. Cancer and aging: The importance of telomeres in genome maintenance. *Int. J. Biochem. Cell Biol.* **37**: 977–990.

Rodin S. and Georgiev P. 2005. Handling three regulatory elements in one transgene: Combined use of cre-lox, FLP-FRT, and I-SceI recombination systems. *Biotechniques* **39**: 871–876.

Roman G., Endo K., Zong L., and Davis R.L. 2001. P[Switch], a system for spatial and temporal control of gene expression in *Drosophila melanogaster*. *Proc. Natl. Acad. Sci.* **98**: 12602–12607.

Rosato E., Tauber E., and Kyriacou C.P. 2006. Molecular genetics of the fruit-fly circadian clock. *Eur. J. Hum. Genet.* **14**: 729–738.

Rose M. and Charlesworth B. 1980. A test of evolutionary theories of senescence. *Nature* **287**: 141–142.

Roth G.S., Lane M.A., Ingram D.K., Mattison J.A., Elahi D., Tobin J.D., Muller D., and Metter E.J. 2002. Biomarkers of caloric restriction may predict longevity in humans. Science **297**: 811.

Rubin G.M. 2000. Biological annotation of the *Drosophila* genome sequence. *Novartis Found. Symp.* **229**: 79–83.

Schmidt P.S., Duvernell D.D., and Eanes W.F. 2000. Adaptive evolution of a candidate gene for aging in *Drosophila*. *Proc. Natl. Acad. Sci.* **97**: 10861–10865.

Schmidt P.S., Matzkin L., Ippolito M., and Eanes W.F. 2005. Geographic variation in diapause incidence, life-history traits, and climatic adaptation in *Drosophila melanogaster*. *Evolution Int. J. Org. Evolution* **59**: 1721–1732.

Schmitz C., Kinge P., and Hutter H. 2007. Axon guidance genes identified in a large-scale RNAi screen using the RNAi-hypersensitive *Caenorhabditis elegans* strain nre-1(hd20) lin-15b(hd126). *Proc. Natl. Acad. Sci.* **104**: 834–839.

Seidner G.A., Ye Y., Faraday M.M., Alvord W.G., and Fortini M.E. 2006. Modeling clinically heterogeneous presenilin mutations with transgenic *Drosophila*. *Curr. Biol.* **16**: 1026–1033.

Seong K.H., Ogashiwa T., Matsuo T., Fuyama Y., and Aigaki T. 2001. Application of the gene search system to screen for longevity genes in *Drosophila*. *Biogerontology* **2**: 209–217.

Serluca F.C. and Fishman M.C. 2006. Big, bad hearts: From flies to man. *Proc. Natl. Acad. Sci.* **103**: 3947–3948.

Sgrò C.M. and Partridge L. 1999. A delayed wave of death from reproduction in *Drosophila*. *Science* **286**: 2521–2524.

Shaw P.J., Tononi G., Greenspan R.J., and Robinson D.F. 2002. Stress response genes protect against lethal effects of sleep deprivation in *Drosophila*. *Nature* **417**: 287–291.

Shmookler Reis R.J., Kang P., and Ayyadevara S. 2006. Quantitative trait loci define genes and pathways underlying genetic variation in longevity. *Exp. Gerontol.* **41**: 1046–1054.

Singh V. and Aballay A. 2006a. Heat shock and genetic activation of HSF-1 enhance immunity to bacteria. *Cell Cycle* **5**: 2443–2446.

Singh V. and Aballay A. 2006b. Heat-shock transcription factor (HSF)-1 pathway required for *Caenorhabditis elegans* immunity. *Proc. Natl. Acad. Sci.* **103**: 13092–13097.

Slack C., Somers W.G., Sousa-Nunes R., Chia W., and Overton P.M. 2006. A mosaic genetic screen for novel mutations affecting *Drosophila* neuroblast divisions. *BMC Genet.* **7**: 33.

Slepko N., Bhattacharyya A.M., Jackson G.R., Steffan J.S., Marsh J.L., Thompson L.M., and Wetzel R. 2006. Normal-repeat-length polyglutamine peptides accelerate aggregation nucleation and cytotoxicity of expanded polyglutamine proteins. *Proc. Natl. Acad. Sci.* **103**: 14367–14372.

Soller M., Haussmann I.U., Hollmann M., Choffat Y., White K., Kubli E., and Schafer M.A. 2006. Sex-peptide-regulated female sexual behavior requires a subset of ascending ventral nerve cord neurons. *Curr. Biol.* **16**: 1771–1782.

Spees J.L., Olson S.D., Whitney M.J., and Prockop D.J. 2006. Mitochondrial transfer between cells can rescue aerobic respiration. *Proc. Natl. Acad. Sci.* **103**: 1283–1288.

Spencer C.C., Howell C.E., Wright A.R., and Promislow D.E. 2003. Testing an 'aging gene' in long-lived *Drosophila* strains: Increased longevity depends on sex and genetic background. *Aging Cell* **2**: 123–130.

Stebbins M.J., Urlinger S., Byrne G., Bello B., Hillen W., and Yin J.C. 2001. Tetracycline-inducible systems for *Drosophila*. *Proc. Natl. Acad. Sci.* **98**: 10775–10780.

Steffan J.S., Agrawal N., Pallos J., Rockabrand E., Trotman L.C., Slepko N., Illes K., Lukacsovich T., Zhu Y.Z., Cattaneo E., Pandolfi P.P., Thompson L.M., and Marsh J.L. 2004. SUMO modification of Huntingtin and Huntington's disease pathology. *Science* **304**: 100–104.

Sun J. and Tower J. 1999. FLP recombinase-mediated induction of Cu/Zn-superoxide dismutase transgene expression can extend the life span of adult *Drosophila melanogaster* flies. *Mol. Cell. Biol.* **19**: 216–228.

Sun J., Molitor J., and Tower J. 2004. Effects of simultaneous over-expression of Cu/ZnSOD and MnSOD on *Drosophila melanogaster* life span. *Mech. Ageing Dev.* **125:** 341–349.

Sun J., Folk D., Bradley T.J., and Tower J. 2002. Induced overexpression of mitochondrial Mn-superoxide dismutase extends the life span of adult *Drosophila melanogaster.* *Genetics* **161:** 661–672.

Swindell W.R. and Bouzat J.L. 2006. Inbreeding depression and male survivorship in *Drosophila:* Implications for senescence theory. *Genetics* **172:** 317–327.

Swinderen B. 2005. The remote roots of consciousness in fruit-fly selective attention? *Bioessays* **27:** 321–330.

Takahashi A., Philpott D.E., and Miquel J. 1970. Electron microscope studies on aging *Drosophila melanogaster.* 3. Flight muscle. *J. Gerontol.* **25:** 222–228.

Tatar M., Bartke A., and Antebi A. 2003. The endocrine regulation of aging by insulin-like signals. *Science* **299:** 1346–1351.

Tatar M., Khazaeli A.A., and Curtsinger J.W. 1997. Chaperoning extended life. *Nature* **390:** 30.

Tatar M., Kopelman A., Epstein D., Tu M.P., Yin C.M., and Garofalo R.S. 2001. A mutant *Drosophila* insulin receptor homolog that extends life-span and impairs neuro-endocrine function. *Science* **292:** 107–110.

Thomson T. and Lasko P. 2005. Tudor and its domains: Germ cell formation from a Tudor perspective. *Cell Res.* **15:** 281–291.

Tominaga K., Olgun A., Smith J.R., and Pereira-Smith O.M. 2002. Genetics of cellular senescence. *Mech. Ageing Dev.* **123:** 927–936.

Tower J. 1996. Aging mechanisms in fruit files. *Bioessays* **18:** 799–807.

———. 2006. Sex-specific regulation of aging and apoptosis. *Mech. Ageing Dev.* **127:** 705–718.

Troemel E.R., Chu S.W., Reinke V., Lee S.S., Ausubel F.M., and Kim D.H. 2006. p38 MAPK regulates expression of immune response genes and contributes to longevity in *C. elegans. PLoS Genet.* **2:** e183.

Truman J.W., Schuppe H., Shepherd D., and Williams D.W. 2004. Developmental architecture of adult-specific lineages in the ventral CNS of *Drosophila. Development* **131:** 5167–5184.

Vaupel J.W. 1997. Trajectories of mortality at advanced ages. In *Between Zeus and the salmon: The biodemography of longevity* (ed. K.W. Wachter and C.E. Finch), pp. 17–37. National Academy Press, Washington, D.C.

Vaupel J.W., Carey J.R., Christensen K., Johnson T.E., Yashin A.I., Holm N.V., Iachine I.A., Kannisto V., Khazaeli A.A., Liedo P., Longo V.D., Zeng Y., Manton K.G., and Curtsinger J.W. 1998. Biodemographic trajectories of longevity. *Science* **280:** 855–860.

Veith A.M., Klattig J., Dettai A., Schmidt C., Englert C., and Volff J.N. 2006. Male-biased expression of X-chromosomal DM domain-less Dmrt8 genes in the mouse. *Genomics* **88:** 185–195.

Venken K.J., He Y., Hoskins R.A., and Bellen H.J. 2006. P[acman]: A BAC transgenic platform for targeted insertion of large DNA fragments in *D. melanogaster. Science* **314:** 1747–1751.

Vermeulen C.J. and R. Bijlsma R. 2004. Changes in mortality patterns and temperature dependence of lifespan in *Drosophila melanogaster* caused by inbreeding. *Heredity* **92:** 275–281.

Vidal M. and Cagan R.L. 2006. *Drosophila* models for cancer research. *Curr. Opin. Genet. Dev.* **16:** 10–16.

von Schantz M., Jenkins A., and Archer S.N. 2006. Evolutionary history of the vertebrate period genes. *J. Mol. Evol.* **62:** 701–707.

Vrontou E., Nilsen S.P., Demir E., Kravitz E.A., and Dickson B.J. 2006. *fruitless* regulates aggression and dominance in *Drosophila*. *Nat. Neurosci.* **9:** 1469–1471.

Walker D.W. and Benzer S. 2004. Mitochondrial "swirls" induced by oxygen stress and in the *Drosophila* mutant hyperswirl. *Proc. Natl. Acad. Sci.* **101:** 10290–10295.

Wallenfang M.R., Nayak R., and DiNardo S. 2006. Dynamics of the male germline stem cell population during aging of *Drosophila melanogaster*. *Aging Cell* **5:** 297–304.

Wang S. and Hazelrigg T. 1994. Implications for bcd mRNA localization from spatial distribution of exu protein in *Drosophila* oogenesis. *Nature* **369:** 400–403.

Wang X.H., Aliyari R., Li W.X., Li H.W., Kim K., Carthew R., Atkinson P., and Ding S.W. 2006. RNA interference directs innate immunity against viruses in adult *Drosophila*. *Science* **312:** 452–454.

Waskar M., Li Y., and Tower J. 2005. Stem cell aging in the *Drosophila* ovary. *Age* **27:** 201–212.

Watson F.L., Puttmann-Holgado R., Thomas F., Lamar D.L., Hughes M., Kondo M., Rebel V.I., and Schmucker D. 2005. Extensive diversity of Ig-superfamily proteins in the immune system of insects. *Science* **309:** 1874–1878.

Weismann A. 1891. *On heredity*. Clarendon Press, Oxford, United Kingdom.

Wessells R.J. and Bodmer R. 2004. Screening assays for heart function mutants in *Drosophila*. *Biotechniques* **37:** 58–64.

Wessells R.J., Fitzgerald E., Cypser J.R., Tatar M., and Bodmer R. 2004. Insulin regulation of heart function in aging fruit flies. *Nat. Genet.* **36:** 1275–1281.

———. 2007a. Cardiac aging. *Semin. Cell Dev. Biol.* **18:** 111–116.

———. 2007b. Age-related cardiac deterioration: insights from *Drosophila*. *Front. Biosci.* **12:** 39–48.

Wheeler J.C., Bieschke E.T., and Tower J. 1995. Muscle-specific expression of *Drosophila* hsp70 in response to aging and oxidative stress. *Proc. Natl. Acad. Sci.* **92:** 10408–10412.

Whitworth A.J., Wes P.D., and Pallanck L.J. 2006. *Drosophila* models pioneer a new approach to drug discovery for Parkinson's disease. *Drug Discov. Today* **11:** 119–126.

Wigby S. and Chapman T. 2004. Female resistance to male harm evolves in response to manipulation of sexual conflict. *Evolution Int. J. Org. Evolution* **58:** 1028–1037.

———. 2005. Sex peptide causes mating costs in female *Drosophila melanogaster*. *Curr. Biol.* **15:** 316–321.

Wilhelm D., Palmer S., and Koopman P. 2007. Sex determination and gonadal development in mammals. *Physiol. Rev.* **87:** 1–28.

Williams A., Jahreiss L., Sarkar S., Saiki S., Menzies F.M., Ravikumar B., and Rubinsztein D.C. 2006. Aggregate-prone proteins are cleared from the cytosol by autophagy: Therapeutic implications. *Curr. Top. Dev. Biol.* **76:** 89–101.

Williams G.C. 1957. Pleiotropy, natural selection and the evolution of senescence. *Evolution* **11:** 398–411.

Wilson R.H., Morgan T.J., and Mackay T.F. 2006. High-resolution mapping of quantitative trait loci affecting increased life span in *Drosophila melanogaster*. *Genetics* **173:** 1455–1463.

Wittmann C.W., Wszolek M.F., Shulman J.M., Salvaterra P.M., Lewis J., Hutton M., and Feany M.B. 2001. Tauopathy in *Drosophila:* Neurodegeneration without neurofibrillary tangles. *Science* **293:** 711–714.

Wodarz A. and Gonzalez C. 2006. Connecting cancer to the asymmetric division of stem cells. *Cell* **124:** 1121–1123.

Wolff S., Ma H., Burch D., Maciel G.A., Hunter T., and Dillin A. 2006. SMK-1, an essential regulator of DAF-16-mediated longevity. *Cell* **124:** 1039–1053.

Wyrobek A.J., Eskenazi B., Young S., Arnheim N., Tiemann-Boege I., Jabs E.W., Glaser R.L., Pearson F.S., and Evenson D. 2006. Advancing age has differential effects on DNA damage, chromatin integrity, gene mutations, and aneuploidies in sperm. *Proc. Natl. Acad. Sci.* **103:** 9601–9606.

Xie H.B. and Golic K.G. 2004. Gene deletions by ends-in targeting in *Drosophila melanogaster*. *Genetics* **168:** 1477–1489.

Yang Y., Gehrke S., Imai Y., Huang Z., Ouyang Y., Wang J.W., Yang L., Beal M.F., Vogel H., and Lu B. 2006. Mitochondrial pathology and muscle and dopaminergic neuron degeneration caused by inactivation of *Drosophila* Pink1 is rescued by Parkin. *Proc. Natl. Acad. Sci.* **103:** 10793–10798.

Yang Y., Gehrke S., Haque M.E., Imai Y., Kosek J., Yang L., Beal M.F., Nishimura I., Wakamatsu K., Ito S., Takahashi R., and Lu B. 2005. Inactivation of *Drosophila* DJ-1 leads to impairments of oxidative stress response and phosphatidylinositol 3-kinase/ Akt signaling. *Proc. Natl. Acad. Sci.* **102:** 13670–13675.

Yang Z.J. and Wechsler-Reya R.J. 2007. Hit 'em where they live: Targeting the cancer stem cell niche. *Cancer Cell* **11:** 3–5.

Yoshida K., Fujisawa T., Hwang J.S., Ikeo K., and Gojobori. 2006. Degeneration after sexual differentiation in hydra and its relevance to the evolution of aging. *Gene* **385:** 64–70.

Zettervall C.J., Anderl I., Williams M.J., Palmer R., Kurucz E., Ando I., and Hutmark D. 2004. A directed screen for genes involved in *Drosophila* blood cell activation. *Proc. Natl. Acad. Sci.* **101:** 14192–14197.

Zhang W., Li B., Singh R., Narendra U., Zhu L., and Weiss M.A. 2006. Regulation of sexual dimorphism: Mutational and chemogenetic analysis of the doublesex DM domain. *Mol. Cell. Biol.* **26:** 535–547.

Zhang X., Smith D.L., Merlin A.B., Engemann S., Russel D.E., Roark M., Washington S.L., Maxwell M.M., Marsh J.L., Thompson L.M., Wanker E.E., Young A.B., Housman D.E., Bates G.P., Sherman M.Y., and Kazantsev A.G. 2005. A potent small molecule inhibits polyglutamine aggregation in Huntington's disease neurons and suppresses neurodegeneration in vivo. *Proc. Natl. Acad. Sci.* **102:** 892–897.

Zheng J., Edelman S.W., Tharmarajah G., Walker D.W., Pletcher S.D., and Seroude L. 2005. Differential patterns of apoptosis in response to aging in *Drosophila*. *Proc. Natl. Acad. Sci.* **102:** 12083–12088.

12

DNA Repair and Aging

Vilhelm A. Bohr, David M. Wilson III, and Nadja C. de Souza-Pinto
Laboratory of Molecular Gerontology, National Institute on Aging
National Institutes of Health, Baltimore, Maryland 21224

Ingrid van der Pluijm and J.H. Hoeijmakers
Department of Genetics, Center for Biomedical Genetics, Erasmus
University Medical Center, Rotterdam, The Netherlands

ONE MAJOR THEORY OF AGING IS BASED ON THE NOTION that DNA damage accumulates with age. This damage causes genomic instability and cellular dysfunction via its effect on chromosome replication, transcription, and other DNA metabolic transactions. Overall, DNA damage has the potential for several deleterious effects: It may cause cell death, impaired function, or permanent proliferative arrest (senescence), or it may cause mutations. Whereas the latter is associated primarily with cancer, all of these appear to contribute to the aging process. This notion is supported by studies using cells from human segmental progerias or mouse models with defined DNA-repair defects. Besides stochastic accumulation of DNA damage, there is also a genetic component to aging involving cellular defenses against DNA damage and the insulin-like growth factor-1 (IGF-1)-somatotrophic axis that controls metabolism, which is influenced by DNA damage and stress.

DNA DAMAGE

DNA is constantly subjected to a barrage of undesirable chemical modifications introduced by internal sources including metabolic reactive oxygen species (ROS) and by external sources such as ubiquitous environmental physical radiation and numerous chemical agents (Lindahl 1993; Wogan et al. 2004). The most abundant damages generated are DNA base lesions, although sugar and DNA backbone modifications arise as well. In particular, normal cellular oxidative processes and ROS attack can generate a large number of oxidative DNA base damages (see examples in Fig. 1): Currently, more than 70 different oxidative base lesions have been identified (Dizdaroglu et al. 2002; Sander et al. 2005). Exposure to external environmental agents can induce bulky adducts in DNA, as well as photoproducts and cross-links. It now appears that endogenous metabolic processes can also induce helix-distorting lesions in DNA, including interstrand cross-links (Niedernhofer et al. 2003). In addition, X- or γ-irradiation can cause both single-strand break (SSB) and double-strand break (DSB) formation (Ward 1988). Finally, an emerging class of DNA injury is abortive enzyme–DNA intermediates such as protein–DNA or adenosyl monophospate (AMP)–DNA adducts (Connelly and Leach 2004). Indeed, with advancements in detection and identification technologies, the spectrum of DNA lesions has increased significantly in recent years. It has been estimated that the number of DNA lesions produced per cell per day easily exceeds 10^4, thus mandating effective DNA-repair systems for the maintenance of genome integrity. In fact, our genome could not have grown to the enormous size that it has reached now without concomitant advancement of DNA repair.

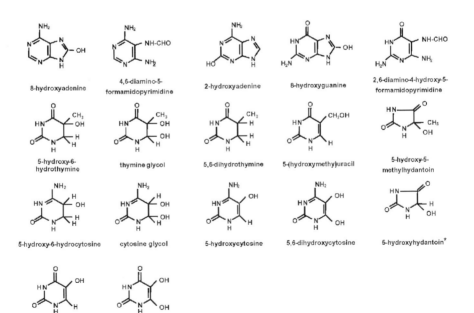

Figure 1. Chemical structures of several oxidative base lesions. Identity of the depicted base damage is indicated below.

DNA-REPAIR PATHWAYS

To cope with the various forms of genetic damage, several DNA-repair pathways have evolved (Fig. 2). Each of these systems has traditionally been viewed to target a specific class of DNA lesions, but increasingly, overlap between pathways has emerged. Base excision repair (BER) is the major system for handling many types of oxidative lesions and mono-functional base modifications such as alkylation of DNA (Wilson and Bohr 2007). BER involves an initial reaction by a DNA glycosylase to remove an inappropriate base, followed by strand incision, termini "cleanup," single-nucleotide or long-patch (LP) repair synthesis, and nick ligation (see details below). For coping with bulky lesions that distort the

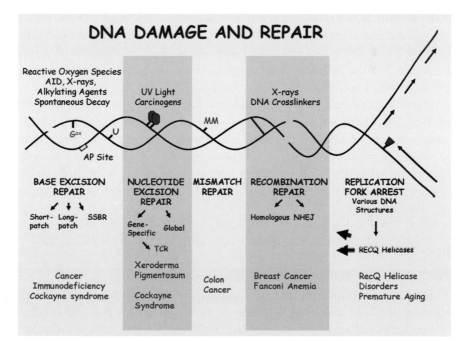

Figure 2. Various forms of DNA damage and specific repair responses. Some general forms of genetic damage are depicted within the double-stranded DNA helical molecule. These include uracil (U), 8–hydroxyguanine (G^{ox}), abasic (AP) sites, single-strand gaps, ultraviolet (UV) light- or carcinogen-induced adducts, mismatched nucleotides (MM), X-ray-induced double-strand breaks, and interstrand cross-links. The causative agent is indicated above, and the major repair pathway is denoted below, as are related subpathways. See text for further details regarding the repair responses. RecQ helicases, as is described briefly in the text, deal with complex DNA structures and promote resolution of stalled replication forks which arise during duplication of difficult sequences or upon DNA polymerase collision with a blocking lesion. Diseases associated with defects in a particular pathway are shown at the bottom.

helical structure of DNA, such as photoproducts (UV dimers) or carcinogen-induced adducts, the nucleotide excision repair (NER) pathway comes into play (Gillet and Scharer 2006). Here, a series of molecular reactions take place that lead to the recognition and removal of the DNA lesion and subsequent insertion of a newly synthesized stretch (a patch) of DNA (see details below). Many of the proteins involved in NER belong to the class of xeroderma pigmentosum (XP) gene products. The subpathways of this system include transcription-coupled repair (TCR) and global genome repair (GGR). The Cockayne syndrome (CS) proteins have specific roles in TCR, which is a form of repair that occurs specifically in actively transcribed genomic regions (Andressoo and Hoeijmakers 2005). Another corrective system is mismatch repair (MMR), whereby replication-associated nucleotide incorporation errors (mismatches) or small hairpin structures resulting from misalignment during replication are recognized and removed, and DNA is resynthesized (Jiricny 2006). In the case of DNA DSBs or interstrand cross-links, lesions that cause severe problems for the cell, the components of recombination repair operate (Wyman and Kanaar 2006). This system also has two major subpathways: nonhomologous end joining (NHEJ), which involves the direct ligation of two DNA DSB ends, often in an error-prone manner, and homologous recombination (HR), which has a major role in the handling of DNA interstrand cross-links and displays generally high fidelity due to its use of the intact sister chromatid for repair when DNA is replicated. We concentrate here primarily on the BER and NER pathways and their relationship to aging.

BER AND AGING

During the process of generating energy in aerobic organisms, ROSs are created as by-products of oxidative phosphorylation at the mitochondrial membrane. The "free radical theory of aging" proposes that the gradual accumulation of oxidative damage to cellular macromolecules gives rise to organism dysfunction and, eventually, death (Harman 1956). Among the primary targets of ROS are nucleic acids, particularly DNA, besides RNA, proteins, lipids, membranes, and organelles. As described herein, data suggest that the accumulation of oxidative DNA damage, which can block transcription and replication and lead to mutagenesis and cell death/senescence over the life span of the organism, drives the aging process and promotes the development of cancer and age-related disease.

As mentioned above, the primary pathway for coping with oxidative DNA damage is BER. This process involves the following five major steps: (1) base damage removal by a DNA glycosylase, (2) incision of the phosphodiester backbone at the resulting abasic (AP) site by an AP endonuclease, (3) DNA termini cleanup by a diesterase or lyase, (4) replacement of the damaged nucleotide by a DNA polymerase, and (5) sealing of the final nick by a DNA ligase (Fig. 3). To date, more than eight different DNA glycosylases have been identified in mammals (including TDG, MBD4, UDG, MPG, MYH1, OGG1, NEIL1, and NTH1; see more below, see Glossary of Abbreviations) (Stivers and Jiang 2003; Huffman et al. 2005). These proteins recognize a specific subset of base modifications, hydrolyzing the N-glycosylic bond to initiate BER. Collectively, the DNA glycosylases have the capacity to remove a wide range of base damage and display considerable overlap in their substrate specificities. This feature likely explains the relatively minor phenotype of single glycosylase knockouts in mice (Parsons and Elder 2003). The major, if not sole, AP endonuclease is APE1 (Wilson and Barsky 2001; Demple and Sung 2005). APE1 exhibits a robust capacity to incise at AP sites in DNA, creating an SSB with a 3'-hydroxyl terminus and a 5'-abasic residue

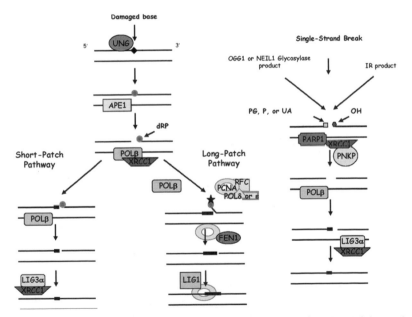

Figure 3. Mammalian BER pathway. The five major steps of BER, and its various subpathways, are described in detail in the text. Definitions for the depicted proteins are also provided in the text. Phosphoglycolate (PG), phosphate (P), and unsaturated aldehyde (UA), which are all possible 3'-blocking termini. Deoxyribose phosphate (dRP), hydroxyl (OH), which are candidate nonconventional 5' termini.

(i.e., the deoxyribose phosphate, dRP). APE1 also possesses the ability to remove certain 3′-blocking damages, such as phosphoglycolates, from SSB gaps and nicks (Suh et al. 1997). The next enzyme of BER is DNA polymerase β (POLβ) (Wilson 1998). POLβ displays the capacity to both fill short gaps in DNA and remove 5′-dRP groups. Subsequently, either a complex of XRCC1/ligase 3α (LIG3) or the protein DNA ligase 1 (LIG1) seals the remaining nick (Tomkinson et al. 2006).

BER can proceed via either a single-nucleotide replacement (i.e., short-patch [SP]-BER) or a multiple nucleotide strand displacement (i.e., LP-BER) reaction (Dogliotti et al. 2001). In the former case, SP-BER most commonly involves POLβ (Fig. 3, left). In the latter case, LP-BER typically engages the replicative, proliferating cell nuclear antigen (PCNA)-dependent polymerases, DNA polymerases δ (POLδ) or ε (POLε), although POLβ can also be involved (Fig. 3, center). Upon strand displacement, the flap endonuclease FEN1 is required to remove the 5′-flap structure. LP-BER is thought primarily to be called upon to deal with 5′-terminal ends that are not effective substrates for the dRP lyase activity of POLβ (Demple and DeMott 2002). LP-BER is also thought to almost exclusively invoke LIG1 for the final step of the response.

It is noteworthy that several enzymes exist to deal with specific chemistries that arise in the context of SSBs (Caldecott 2003). These proteins are more commonly considered components of single-strand-break repair (SSBR), a subpathway of BER (Fig. 3, right). In particular, PNK removes 3′-phosphate groups that arise by free radical attack of DNA or as intermediates of the BER process (Wiederhold et al. 2004); TDP1 excises 3′-topoisomerase I complexes that are covalently trapped by the presence of nearby DNA damage or by exposing cells to the chemotherapeutic agent camptothecin (Pouliot et al. 1999; Connelly and Leach 2004); aprataxin (encoded by *APTX*) resolves abortive 5′-AMP products left behind by failed ligation reactions (Ahel et al. 2006). Notably, links to defective SSBR and neurodegenerative disease have recently emerged. In particular, *TDP1* is mutated in the recessive inherited disorder spinocerebellar ataxia with axonal neuropathy-1 (SCAN1) (Takashima et al. 2002; El Khamisy et al. 2005), and *APTX* is defective in ataxia-oculomotor apraxia type 1 (AOA1) (Date et al. 2001; Moreira et al. 2001). Studies are now beginning to unravel the connections of mammalian BER to various age-associated phenomena in detail.

Presently, there is limited evidence linking defects in the core mammalian BER components (i.e., the DNA glycosylases, AP endonuclease, DNA POLβ, or the DNA ligases) with premature aging. This may stem from the fact that most core BER homozygous knockouts lead to embryonic lethality in mice (Wilson and Thompson 1997). In addition, extensive analysis of life span in mouse mutants not completely deficient for BER capacity is lacking. Nonetheless, experiments in the yeast *Saccharomyces cerevisiae* have found that defined, complex mutations in components of BER result in mutation accumulation and shortened chronological life span (Maclean et al. 2003). Moreover, a recent study by Cabelof et al. (2006), addressing the role of BER in mammalian life expectancy quantitatively, found that POLβ heterozygous (+/−) mice exhibit increased tumor development and an accelerated mortality rate, supporting the role of genomic maintenance in both carcinogenesis and aging. More extensive analysis of other haploinsufficient BER animals is warranted.

Although further studies are required to more exhaustively probe the correlation of deficiencies in the core mammalian BER proteins with aging, defects in several factors that have nonessential, auxiliary functions in BER are firmly established with premature aging characteristics. Most notably, both the Werner syndrome helicase/exonuclease (WRN) and the SWI2/SNF2 DNA-dependent ATPase Cockayne syndrome B (CSB) protein, which operate in various capacities to regulate BER mechanics, are directly linked to human segmental progerias (Licht et al. 2003; Bohr 2005) (see more below). WRN functions in several capacities to regulate BER, for example, in modulating the strand displacement activity of DNA POLβ and LP-BER (Harrigan et al. 2003, 2006). The molecular role of CSB in BER is presently less clear, yet studies indicate a contribution to mediating the repair of oxidative base damage (Tuo et al. 2001, 2003). Thus, evidence indicates connections of two key aging-related proteins to BER, and studies are proceeding to more thoroughly unravel the function(s) of WRN and CSB in the BER response. In addition to BER, as discussed below, these proteins appear to have critical roles in other DNA repair and metabolic processes as well.

Besides WRN and CSB, other potential BER auxiliary proteins influence life expectancy and the aging process. For instance, a recent report indicates that a deficiency in one of the mammalian Sir2 homologs, SIRT6, results in genome instability and leads to aging-associated degenerative phenotypes in mice (Mostoslavsky et al. 2006). Although the precise role of SIRT6 in BER is presently unclear, it was speculated that this protein may modulate POLβ activity by way of deacetylation or by facilitating chromatin accessibility. In addition, correlative data suggest a link between cellular poly(ADP-ribose) formation capacity, which is due mainly to the enzymatic activity of the strand-break response protein PARP-1, and mammalian longevity (Burkle et al. 2005). The notion that the BER-related PARP-1 protein regulates life span is reinforced by evidence of a physical and functional interaction with WRN (von Kobbe et al. 2004). Moreover, there is genetic evidence linking the molecular functions of WRN and PARP-1 (Lebel et al. 2003). As with PARP-1, there is a general trend for BER activities to diminish with age (Lu et al. 2004; Wilson and McNeill 2007), and numerous studies have demonstrated an accumulation of oxidative DNA damage during the life span of an organism (Barja 2004). However, with the current advancement in technologies to quantify the levels of DNA damage, this viewpoint may adjust.

Although the role of BER in aging is still under investigation, the link of defects in this repair response with age-associated disease, such as cancer, has been firmly established. Specifically, studies have identified a genetic linkage between functionally relevant mutations in the DNA-repair glycosylase mutY homolog, MYH, to predisposition for hereditary nonpolyposis colorectal cancer (HNPCC) in humans (Al Tassan et al. 2002). In addition, several commonly observed cancer-associated mutations in *POLβ* have been shown to produce an error-prone "mutator" polymerase, which, when expressed in mouse C127 cells, induces cellular transformation (Sweasy et al. 2005). Several other connections between cancer-associated mutations or population variants and cancer susceptibility/predisposition have been suggested, but data on most of these are conflicting or inconclusive (Hung et al. 2005; Wilson and Bohr 2007). Data from Miller and colleagues indicate that complex mutations in the DNA glycosylases *MYH* and *OGG1* in mice lead to increased cancer susceptibility (Xie et al. 2004). Laboratories are now more exhaustively interrogating the association of impaired BER activity with spontaneous and exposure-dependent cancer risk in both animal models and defined human cohorts. Besides tumorigenesis, genetic mutations in the uracil DNA glycosylase (*UNG*) have been linked to the primary immunodeficiency condition, hyper IgM syndrome (HIGM) (Imai et al. 2003). Furthermore, targeted knockout of the DNA glycosylase, NEIL1, in mice uncovered a role for this protein in the prevention of diseases associated with metabolic syndrome, namely, severe obesity (Vartanian et al. 2006). As the studies of BER in both humans and animal models expand, it is likely that additional, perhaps unanticipated, roles for this repair process in aging and age-associated disease will emerge.

MITOCHONDRIAL DNA REPAIR AND AGING

Mitochondria contain their own genomes, a 16.5-kb circular molecule that encodes the minimal translation machinery (i.e., two ribosomal RNAs and a restricted set of tRNAs) and a specific subset of hydrophobic proteins involved in oxidative phosphorylation. All other proteins required for mitochondrial function (including the DNA-repair proteins described below) are encoded by the nuclear genome and need to be imported from the cytoplasm. Since mitochondria are the primary source of endogenous ROS, the "mitochondrial theory of aging" proposes that the accumulation of oxidative damage and mutations in the mitochondrial genome specifically leads to mitochondrial dysfunction and, consequently, cell loss (Harman 1981). Given its proximity to the inner mitochondrial membrane, where oxidants are formed, mitochondrial DNA is believed to be particularly sensitive to oxidation. Until recently, it was thought that no DNA repair existed in these organelles, since DNA damage was found to accumulate in mitochondria, particularly in association with aging (Ames et al. 1995). The earlier observations that mammalian mitochondria did not repair UV-induced DNA damage contributed to this notion (Clayton et al. 1974). Recent advances, however, have demonstrated that mitochondria do indeed contain DNA damage responses, with BER serving as the major repair mechanism (Bogenhagen et al. 2001; Dianov et al. 2001; LeDoux and Wilson 2001). Table 1 lists our current understanding of the DNA-repair pathways in mitochondria relative to those found in the nucleus, and although the mitochondrial repair machinery is physically separated from the nucleus, there appears to be significant cross-talk between these compartments.

Mitochondrial BER is similar in nature to nuclear BER in terms of the five major molecular steps (see above), although the factors involved are typically unique protein products. For instance, mitochondrial (*UNG1*) and nuclear (*UNG2*) forms of uracil-DNA glycosylase (UDG) are generated by alternative splicing and transcription from different positions in the *UNG* gene (Nilsen et al. 1997). Similarly, nuclear and mitochondrial adenine-DNA-glycosylases are encoded by alternatively spliced forms of the *MYH* gene (Ohtsubo et al. 2000). The oxoguanine

Table 1. Mammalian DNA repair: Nuclear vs. mitochondrial

Pathway	Sub-Pathways	Nuclear	Mito
BER	long patch	Y	N
	short patch	Y	Y
NER	general genome	Y	N
	gene-specific	Y	N
	transcription associated	Y	N
MMR		Y	Y?
Recombination	HR	Y	?
	NHEJ	Y	N

DNA glycosylase (OGG1) is the primary enzyme for the repair of oxidized purines in both the nuclear and mitochondrial DNA (Klungland et al. 1999b; de Souza-Pinto et al. 2001a), and in humans, two distinct OGG1 isoforms are present: α-OGG1, which localizes to the nucleus and mitochondria, and β-OGG1, which localizes only to mitochondria (Takao et al. 1998; Nishioka et al. 1999). Because purified recombinant β-OGG1 does not show significant 8-oxoG activity in vitro, it was proposed that α-OGG1 is the predominant enzyme responsible for 8-oxoG glycosylase activity in mitochondria, whereas the function of β-OGG1 has yet to be identified (Hashiguchi et al. 2004). The human endonuclease III homolog, NTH1, has a putative mitochondrial targeting sequence (MTS) (Takao et al. 1998), and its mitochondrial presence has been established by several groups. NTH1 appears to be the major mitochondrial glycosylase for thymine glycol incision (Karahalil et al. 2003). NEILs are a recently identified group of bifunctional DNA glycosylases—i.e., enzymes that not only remove base damage, but also incise at the resulting AP site via a β- or β,δ-elimination reaction mechanism, which are mammalian orthologs of *Escherichia coli MutM/Nei* protein. Both NEIL1 and NEIL2 have a broad substrate specificity in vitro, including but not limited to 8-oxoG, 2,6-diamino-4-hydroxy-5-formamidopyrimidine (FapyG), 4,6-diamino-5-formamidopyrimidine (FapyA), 5-hydroxycytosine (5-OH-C), and 5-hydroxyuracil (5-OH-U). NEIL1 is found in mouse liver mitochondria (Hu et al. 2005), suggesting that this enzyme is involved in repairing mitochondrial DNA oxidative base damage.

The mitochondrial AP-endonuclease activity appears to be provided by an isoform of the APE1 molecule found in the nucleus that has been posttranslationally modified (Chattopadhyay et al. 2006). Mitochondria have a sole and distinct DNA polymerase, polymerase γ (POLγ), which participates in both DNA repair and replication. POLγ in fact has both polymerase and deoxyribose-phosphate (dRp) lyase activity on a BER intermediate substrate (Longley et al. 1998). The mitochondrial ligase is a unique form of DNA LIG3, created by differential translation initiation (Lakshmipathy and Campbell 2000). Using a uracil-containing oligonucleotide duplex, we have found that this substrate is processed by rat mitochondrial extracts strictly via a one-nucleotide or SP-BER subpathway (Stierum et al. 1999). Recent studies have shown that the level of expression of mitochondrial BER proteins relates to cellular survival and that the specific targeting of BER enzymes to mitochondria protects cells from oxidative stress-induced cell death (Harrison et al. 2005). Moreover, deficiencies in mitochondrial DNA repair may associate with neurodegeneration (Weissman et al. 2007).

Recent reports that mice expressing a proofreading dead DNA POLγ show an accelerated aging phenotype similar to normal aging bolster the "mitochondrial theory of aging" (Trifunovic et al. 2004; Kujoth et al. 2005). In addition, oxidative damage to mitochondrial DNA in the heart and brain is inversely related to maximum life span of mammals, suggesting that accumulation of mitochondrial DNA damage has a causative role in the various disorders that are associated with aging, cancer, and neurodegeneration (Barja and Herrero 2000). We found four times more 8-oxoG in mitochondrial DNA isolated from 23-month-old rat liver mitochondria than in mitochondrial DNA isolated from 6-month-old animals, whereas no significant change was found in nuclear DNA from the same animals (Hudson et al. 1998). In postmitotic tissues, such as heart and brain, or in tissues with high energy demand, such as muscle, loss of mitochondrial function would have the greatest impact on normal physiology. Support for this comes from observations that changes in mitochondrial function with age have been detected in several experimental models and that diseases caused by mutations in the mitochondrial DNA primarily affect muscle, heart, and brain (Wallace 2005; Lesnefsky and Hoppel 2006; Lin and Beal 2006). There are several lines of evidence supporting this theory, but there is also evidence against it. For example, knockout mice for *OGG1*, which accumulate 20-fold more 8-oxoG in their mitochondrial DNA than normal mice, do not have an apparent aging phenotype (Klungland et al. 1999b; de Souza-Pinto et al. 2001b). Furthermore, although mutations and deletions in mitochondrial DNA accumulate with age, the overall frequency of these mutations is relatively low, and the significance is being discussed and investigated (Trifunovic 2006). Because the field of mitochondrial DNA exploration is quite young, further work is needed to clarify the precise deficiencies that arise in DNA repair and metabolism in the organelles with aging.

NER AND AGING

NER is a multistep process that deals mainly with helix-distorting lesions, such as UV-induced 6-4 photoproducts (PPs) and cyclobutanepyrimidine dimers (CPDs). Other substrates for NER include bulky chemical adducts, intrastrand cross-links, and the oxygen free-radical-induced helix-distorting 5', 8-purine cyclodeoxynucleotides (Brooks et al. 2000; Kuraoka et al. 2000). NER is an evolutionarily conserved damage-removal mechanism that bears similarity to the UvrABC system in prokaryotes. The mammalian NER "cut and patch" process involves at least 30 proteins and consists of several distinct but highly interwoven steps: (1) tentative lesion identification, based on local base-pairing interruption; (2) local opening of the double helix; (3) lesion verification and identification of the damage-containing strand; (4) dual incision of the lesion-carrying strand at the borders of the helix opening, leading to excision of the DNA injury as part of a 24–32 oligonucleotide; and (5) gap-filling DNA synthesis followed by DNA ligation (Fig. 4) (de Laat et al. 1999; Wood 1999; Hoeijmakers 2001; Gillet and Scharer 2006).

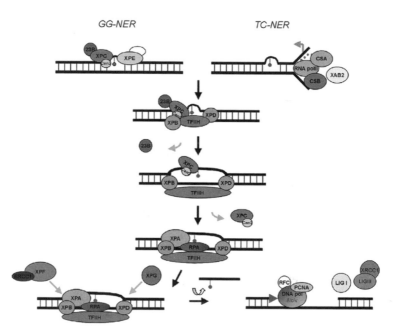

Figure 4. Repair of damaged DNA by nucleotide excision repair (NER). The mechanism consists of two subpathways, global genome (GG-) (*left*) and transcription coupled (TC-) NER (*right*). DNA-damage recognition requires XPC/hHRad23B for recognition of helical distortion anywhere in the genome (GG-NER). TC-NER is initiated when RNA polymerase II progression is blocked by damage in the transcribed strand of DNA. XPC/hHRad23B or stalled RNA polymerase II complex induces the recruitment of basal transcription factor TFIIH followed by XPG to the damaged site. The XPB and XPD helicases, which are part of the TFIIH complex, unwind the DNA around the lesion. The XPA/RPA complex is recruited to the damaged lesion: XPA verifies the DNA damage and RPA serves to stabilize the open intermediate by binding the undamaged strand. Next, XPF/ERCC1 is recruited: XPG and ERCC1/XPF, respectively, cleave at the 3′ and 5′ site of the damaged strand, thereby removing a 24–32-base oligonucleotide containing the damage. The resulting gap is filled in by DNA polymerase δ, ε, and κ, together with Ligase I and III. (Adapted, with permission, from de Boer and Hoeijmakers 2000.)

NER consists of two subpathways: global genome NER (GG-NER) and transcription-coupled NER (TC-NER). GG-NER has to scan the whole genome for lesions, and its rate depends on the type of lesion. For instance, in UV-induced lesions, repair of 6-4PPs, which distort the DNA double helix severely, is fast throughout the genome—typically completed in a few hours. In contrast, GG-NER of CPDs, which mildly distort the double helix, is relatively slow—in human cells taking 24 hours for about 70% removal; in rodent cells, repair is even slower or virtually absent (Mitchell and Nairn 1989). To eliminate blocked transcription because of slowly repaired lesions for too long, mammalian cells have evolved the TC-NER pathway, which selectively removes lesions from the transcribed strand of active genes when they obstruct the transcription machinery. TC-NER is considered a fast and efficient process (Bohr et al. 1985; Mellon et al. 1986, 1987; Hanawalt 2002), although it clearly depends on the intensity of transcription of a specific gene and on the degree of inhibition of the elongating RNA polymerase (RNAP). In agreement, TC-NER was previously shown to remove CPD and 6-4PP lesions with similar rates (van Hoffen et al. 1995).

GG-NER and TC-NER are mechanistically related, except for the initial damage-recognition step (Fig. 4). In GG-NER, the XPC/hHR23B complex and the UV-damaged DNA-binding protein (UV-DDB; composed of the DDB1 and DDB2/XPE subunits) are involved in lesion recognition. In TC-NER, the CSB and CSA proteins are thought to be involved in early stages of damage detection and processing of the arrested transcription machinery to make the lesion accessible for the core NER machinery. After DNA-damage recognition by either GG- or TC-NER-specific components, TFIIH binds the DNA, followed by XPA and RPA. Together, these factors allow subsequent helix unwinding by the TFIIH complex, verification and demarcation of the damaged site, and excision of the damaged strand by the XPG and ERCC1/XPF endonucleases. Finally, the repair reaction is completed by the regular replication machinery consisting of POLδ/ε and POLκ (Ogi and Lehmann 2006) and LIG1/LIG3, which together fill the resulting 24- to 32-base-long single-stranded DNA gap and seal the remaining nick.

Generally, although the GG-NER apparatus will reduce cell death or senescence after DNA damage, its primary importance is preventing mutagenesis. Like replication, GG-NER covers the entire genome and therefore has a major impact on the removal of general lesions that otherwise might trigger mutational events. This theory is consistent with the fact that XPC mutants specifically deficient in GG-NER, but with potent TC-NER, exhibit mainly a strong (skin) cancer predisposition, relatively mild sun sensitivity, and, except for neurodegeneration in older XPC patients, hardly any premature aging. TC-NER, on the other hand, is very important for removing lesions that block transcription in a very small, but vital, compartment of the genome: the transcribed strand of active genes. This is essential for recovery of RNA synthesis after DNA damage, and since gene expression is indispensable for cell viability, this repair process regulates cell survival responses. However, TC-NER and the broader TCR reaction, which may also involve nonhelix-distorting DNA lesions that block transcription, have hardly any impact on mutagenesis (Satoh et al. 1993; Satoh and Lindahl 1994; Dianov et al. 1999), as they deal only with a tiny fraction of the genome. This may explain why CSA and CSB patients, deficient in TCR but with proficient GG-NER, show clear sun sensitivity, a reduced incidence of cancer, and dramatically accelerated aging (see more below).

Deficiencies in NER genes have been associated with three rare autosomal recessive UV-sensitive disorders: XP, CS, and a photosensitive form of the brittle hair disorder trichothiodystrophy (TTD) (Nance and Berry 1992; Bootsma et al. 2001; Lehmann 2003). All NER disorders share photosensitivity but differ markedly in UV-induced skin cancer proneness as well as in the occurrence of additional predominantly premature aging pathologies.

Xeroderma Pigmentosum

XP is characterized by striking sun sensitivity, hyperpigmentation, and high (>1000-fold) UV-induced cancer susceptibility in sun-exposed parts of the skin (Fig. 5) (Bootsma et al. 2001; Cleaver 2005). Complementation analysis has revealed that excision-deficient XP is due to a mutation in one of seven genes: XPA–XPG (for functions in the NER reaction, see Fig. 4 and above). Progressive neurological dysfunction is found in 20% of XP cases (mainly with mutations in XPA, G, and D), due to degeneration of neurons that initially develop normally (Rapin et al. 2000). Neurological symptoms include microcephaly, peripheral neuropathy, sensorineuronal deafness, and loss of reflexes, followed by ataxia and mental retardation or dementia. The age at onset and rate of progression vary greatly between patients. A possible explanation for the onset of neurological abnormalities in XP patients is that unrepaired endogenous NER lesions in neurons trigger apoptosis (Kuraoka et al. 2000). The most severe early-onset neurological subtype of XP (mostly caused by mutations in XPA) is De Sanctis–Cacchione syndrome (XP-DSC). In addition, this disease involves dwarfing, hypogonadism, and a 10–20-fold increase

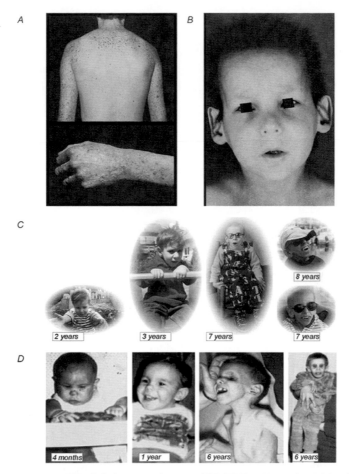

Figure 5. Human NER-deficient syndromes: XP, TTD, CS, and XP/CS. (A) Back of an XP patient. Note the increased freckling on the UV-exposed skin. (B) TTD patient. (C) Patient with CS at different ages. Note the progressive nature of the disease, with close to normal appearance at the age of 2 years and the typical CS appearance at the age of 7 years with deep-set eyes, prominent ears, and profound cachexia. (D) Patient XP20BE with XP/CS at different ages. Note the progressive nature of the disease, with normal appearance at the age of 4 months and 1.5 years, and the typical CS appearance at the age of 6 years with deep-set eyes, prominent ears, and profound cachexia. (B, Adapted from McCuaig et al. [1993]; C, adapted from Share and Care Cockayne Syndrome Network [http://www.cockayne-syndrome.org], with permission of Eric Bixel; D, adapted from Lindenbaum et al. [2001].)

in frequency of several types of internal cancers (DeSanctis 1932). These observations reveal that besides UV-induced DNA damage, endogenously generated NER-relevant lesions are implicated in the degenerative symptoms and the internal cancers associated with the above forms of XP, in which different degrees of defects in GG-NER and TC-NER are combined.

Cockayne Syndrome

Complementation analysis has identified two genes responsible for CS: CSA and CSB. Both of these gene products contribute to TC-NER, but not to GG-NER. Moreover, evidence suggests that CSB has additional functions outside of NER, possibly in global BER (see above) and in mediating general transcription. The clinical features of CS are severe and diverse. In affected patients, growth is retarded in a progressive manner, culminating in cachectic dwarfism and a "bird-like" face resulting from reduced amounts of subcutaneous fat. Furthermore, patients present with dental caries and skeletal abnormalities such as kyphosis. The mean age at death is 12.5 years, with a range from neonatal to 55 years. The most common cause of death is pneumonia as a result of general atrophy and cachexia. However, the most persistent features of CS are the neurological symptoms: delayed psychomotor development, mental retardation, microcephaly, and ataxia (Rapin et al. 2000), with multifocal patchy demyelination in cerebral and cerebellar cortex, dilated ventricles, calcium deposits in basal ganglia and cerebral cortex, and loss of neurons, which is most pronounced in the cerebellum (Itoh et al. 1999). Since head circumference is relatively normal at birth, cessation of brain growth is clearly of postnatal origin. Additionally, most patients have sensorineuronal hearing loss, cataracts, pigmentary degeneration of the retina, and a progressive "salt and pepper"-like fundus. Finally, despite a DNA-repair defect, CS patients have never been reported to develop cancer.

Trichthiodystrophy

The hallmarks of TTD are photosensitivity, ichthyosis, brittle hair and nails, intellectual impairment, decreased fertility, and short stature (referred to by the acronym PIBIDS) (Fig. 5B) (Itin et al. 2001; Lehmann 2001). A significant proportion of these patients have a dramatically shortened life span. In addition, skeletal abnormalities such as osteopenia, axial osteosclerosis, peripheral osteoporosis (McCuaig et al. 1993), and kyphosis (Norwood

1964), as well as retardation of skeletal age, are often observed. Whereas the brittle hair and nails and the ichthyotic skin are unique to TTD patients, other features are strikingly similar to CS, including the absence of cancer predisposition. The abnormalities typical for this disease are generally noted around birth. Up to now, mutations in three genes have been found to cause UV-sensitive TTD, namely, *XPD*, *XPB*, and *TTDA*. A mouse model for TTD has established this rare condition as a premature aging disorder (de Boer et al. 2002), which is consistent with the notion that many of the symptoms are quite similar to CS (Martin 2005).

Mutations in either *XPD* or *XPB* can give rise to three distinct disorders: XP, XP/CS, or TTD (de Boer and Hoeijmakers 2000; Lehmann 2001). The XPB and XPD proteins are involved not only in NER, but—as part of the multisubunit transcription factor IIH (TFIIH) complex—also in transcription. Therefore, the TTD phenotype might be a combination of a repair deficiency together with a transcription deficiency (Bootsma and Hoeijmakers 1993). The transcriptional defect in TTD is likely responsible for the brittle hair and ichthyosis phenotype (Bootsma et al. 2001; Vermeulen et al. 2001), as TTD mutations in *XPD* are found to cause instability of TFIIH. This instability results in depletion of the transcription initiation factor prior to the completion of terminal differentiation of hair and skin, which are consequently left unfinished. Fascinatingly, non-UV-sensitive forms of TTD also exist. As the name suggests, patients display all TTD features except for cutaneous photosensitivity. Recently, the first gene that causes this form of TTD was identified: *C7orf11*. Although no role in UV sensitivity is evident, this gene was found to localize to the nucleus and is therefore presumed to be involved in transcription rather than DNA repair (Nakabayashi ct al. 2005).

The complex involvement of NER genes in these different inherited human disorders that display cancer and/or segmental aging features underscores the importance of this repair pathway in both processes. Whereas the UV sensitivity and cancer predisposition can be well explained by the repair gene defect, this poorly explains many of the non-UV-related features, which have been, so far, puzzling. To this end, the generation of mouse models mimicking these disorders has shed some light on the underlying pathogenic mechanism as well as the cause of the non-UV-related phenotypes.

Mouse Models for NER Disorders

An extensive panel of NER-defective mouse mutants has been generated that, overall, display strong resemblance to their human paradigm NER disorders. Mouse mutants for XP, notably, mutants for XPC (completely deficient in GG-NER) (Sands et al. 1995; Cheo et al. 1997), UV-DDB/XPE (partially deficient in GG-NER) (Itoh et al. 2004), and XPA (complete defect in GG- and TC-NER for distorting damage) (de Vries et al. 1995; Nakane et al. 1995), exhibit virtually all hallmark XP features: UV hypersensitivity, strongly enhanced UV/chemical-induced skin cancer susceptibility, and—as far as investigated—increased spontaneous tumorigenesis (Hollander et al. 2005), elevated spontaneous (Giese et al. 1999) and damage-induced mutagenesis (de Vries et al. 1997), and the predicted cellular parameters typical for the specific NER defect. The accelerated neurological degeneration associated with the human syndromes at older age is, however, not clearly apparent in the animal models, nor are other accelerated aging features.

Mouse mutants for CS (*CSA* and *CSB*) display some of the features of the human syndrome, including photohypersensitivity, growth cessation later in life, neurological features, and enhanced photoreceptor cell loss, but in contrast to CS, they exhibit normal development (van der Horst et al. 1997, 2002). Moreover, the CS symptoms in the *CSA* and *CSB* knockout mice are overall relatively mild compared to the severity of the human condition. Interestingly, upon exposure to UV or the chemical carcinogen dimethylbenz(a)anthracene (DMBA), the mice exhibit only a modestly enhanced skin cancer susceptibility (van der Horst et al. 1997), but otherwise, the TCR defect appears to be protective against cancer development (Brash et al. 2001). At first sight, this finding seems to be in contradiction with a DNA-repair defect. However, as indicated above, TCR primarily serves to rescue transcription when it is blocked by lesions and thereby dictates cellular survival from DNA damage (Ljungman and Lane 2004). Consequently, TCR deficiencies will result in increased cell death and senescence, which in effect protects from cancer, but at the expense of enhancing aging.

Because the XPD helicase as a subunit of the basal transcription initiation factor TFIIH is absolutely essential, only point mutations in this gene are viable, giving rise to XP, XP/CS, or TTD. To generate a TTD mouse model, the R722W point mutation of a human patient was mimicked in the mouse germ line (de Boer et al. 1998). Repair parameters including partial defects in GG-NER and TCR are indistinguishable from the human disease. Phenotypically, the animals also faithfully recapitulate many symptoms present in the patient: *XPD*TTD mice have a reduction of hair-specific cysteine-rich matrix proteins resulting in brittle hair, which is a hallmark feature of the human disorder. Additionally, *XPD*TTD animals develop progeroid symptoms such as growth delay, osteoporosis (trabecular bone loss in the femur), osteosclerosis, kyphosis, gray hair, cachexia, and a reduced fertility and life span (de Boer et al. 2002). Apart from the observed aging pheno-type, *XPD*TTD animals have modestly enhanced skin carcinogenesis after UV exposure, but incidence of spontaneous cancer is reduced (Wijnhoven et al. 2005). A knockin *XPD* mouse model for the rare combined cancer and progeroid disorder XPCS was recently generated (*XPD*-XP-CS mice) by mimicking a patient-derived point mutation in XPD (*XPD*G602D). Strikingly, XPDXPCS mice are more skin-cancer-prone than *XPA*$^{-/-}$ animals. Moreover, XPDXPCS mice also displayed symptoms of segmental progeria, including cachexia and progressive loss of germinal epithelium. These data indicate that cancer or seg-mental progeria can be experimentally modulated by different point mutations in a single DNA-repair-associated gene as was previously dis-cussed for the human *XPD* gene (Andressoo et al. 2006).

XPG$^{-/-}$ null mice suffer from postnatal growth failure and death prior to weaning (Harada et al. 1999). Additionally, the intestine and spleen were found to be relatively small, and the liver displayed remark-ably small hepatocytes. In addition, primary mouse embryonic fibrob-lasts (MEFs) experienced premature senescence, early onset of immortalization, and accumulation of p53. Apart from its role in NER, XPG has been associated with BER of thymine glycol lesions and is thought to be involved in TCR of oxidative DNA damage. A similarly severe phenotype was found in the *XPG*D811STOP/D811STOP mouse, in which the last 360 amino acids of the XPG protein are deleted (Shiomi et al. 2004). In yet other XPG mutants, a point mutation ren-dering the nuclease catalytic site inactive (*XPG*D811A/D811A) (Shiomi et al. 2004), or deletion of exon 15 of the *XPG* gene (*XPG*Δex15/Δex15), resulted in UV sensitivity without any severe phenotype (Wang et al. 2006). Since both the D811A and Δex15 mutations only affected the NER function of XPG, these results suggest that the severe phenotype observed in the *XPG*$^{-/-}$ mouse is at least in part due to a function of XPG outside the context of GG-NER/TC-NER. The phenotype of *XPG* mutant mice can be rationalized with the other mutants and with the human XP-G patients, displaying XP combined with CS on the basis of dramatically accelerated aging.

A defect in the *ERCC1* gene (giving rise to a complete null mutant) results in viable mice with a severe phenotype (McWhir et al. 1993; Weeda et al. 1997). At birth and through the first week of life, *ERCC1*$^{-/-}$ mice are quite similar to wild-type littermates, but then they rapidly develop an accelerated, segmental aging phenotype. Besides cachexia and sarcopenia, $ERCC1^{-/-}$ mice display characteristics of dystonia and progressive ataxia. In addition, ferritin deposition in spleen, progressive renal and liver dysfunction, as well as hepatocellular polyploidization and intranuclear inclusions in the liver were observed, for example, symptoms associated with age-related pathology in mammals (McWhir et al. 1993; Weeda et al. 1997). Additionally, $ERCC1^{-/-}$ mice demonstrate multilineage cytopenia and fatty replacement of bone marrow, similar to old wild-type mice, together with reduced hematopoietic progenitors and stress erythropoiesis (Prasher et al. 2005). $ERCC1^{-/-}$ mutant cells not only are compromised in NER and cross-link repair, but also undergo rapid, premature replicative senescence. In line with this observation, p53 accumulation has been detected in liver, kidney, and brain, and p21 accumulation was observed in $ERCC1^{-/-}$ MEFs (Melton et al. 1998; Chipchase et al. 2003; Niedernhofer et al. 2006). Since $XPA^{-/-}$ mice, which are totally NER-deficient, do not show clear signs of aging, the progeria observed in the $ERCC1^{-/-}$ mice cannot be due solely to their NER defect. Instead, the involvement of ERCC1 in cross-link repair could give rise to the accumulation of interstrand cross-links, which are extremely toxic DNA lesions. This idea is supported by the phenotype of mice carrying an inactivating mutation in the partner of ERCC1, XPF (Tian et al. 2004). Although not as extensively investigated, XPF mutants display principally the same characteristics as $ERCC1^{-/-}$ mice. These parallels extend to the human situation in which a new premature aging syndrome has been identified, caused by a severe mutation in XPF, with striking resemblance to the $ERCC1^{-/-}$ and XPF mutant mice (Niedernhofer et al. 2006), highlighting the relevance of the mouse models for the parallel human disorders and the link between compromised DNA repair and acceleration of aging.

Interestingly, when XPDTTD mice were crossed to $XPA^{-/-}$ animals to examine the effect of a total NER deficiency in the XPDTTD mouse model, this resulted in a dramatically accelerated premature aging phenotype (de Boer et al. 2002), which included retarded growth, cachexia, disturbed gait, spinal kyphosis, and severely shortened life span. Likewise, when TCR-deficient $CSB^{-/-}$ or $CSA^{-/--}$ mice were crossed to either $XPA^{-/-}$ or $XPC^{-/-}$ mice, compromising also GG-NER and leading to increased levels of constitutive DNA damage, this resulted in a very severe phenotype, including growth retardation, neurological abnormalities, and an extremely short life span (Murai et al. 2001; van der Pluijm et al. 2006). In contrast, $XPC^{-/-}/XPA^{-/-}$, $CSB^{-/-}/CSA^{-/-}$, and $XPC^{-/-}/XPD$TTD animals were viable and quite normal.

Taken together, the findings described above indicate that a complete GG-NER deficiency in combination with a TCR defect, as in CSA, CSB, or specific XPD mutations, induces a dramatic premature aging phenotype. Similar, but not in all respects identical, phenotypes have been observed for the above-mentioned $ERCC1^{-/-}$, XPF mutant, and $XPG^{-/-}$mice, consistent with the idea that these animals also display aging features. Moreover, when the previously mentioned $XPG\Delta ex15/\Delta ex15$ mouse was crossed to $XPA^{-/-}$ mice to generate double-mutant mice, this resulted in a similarly severe phenotype (Shiomi et al. 2005). Thus, it appears that the more severe the GG-NER defect in the context of a TCR deficiency, the more severe the premature aging. Because these animals have not been subjected to exogenous DNA-damaging agents, this points toward the contribution of unrepaired endogenous DNA lesions to the observed aging phenotypes. Significantly, the nature of the relevant, causative damage remains unknown.

PROGEROID MOUSE MUTANTS IN OTHER DNA STABILITY SYSTEMS

The connection between compromised genome stability and characteristics of premature aging in the mouse extends beyond NER mutants. In fact, the $ERCC1/XPF$ mice and a human XPF patient disclose the likely contribution of cross-link repair to dramatic progeroid symptoms (Niedernhofer et al. 2006). In addition, $Ku80^{-/-}$ mice, defective in NHEJ, exhibit several aging phenotypes including osteoporosis, premature growth failure, incomplete plate closure, atrophic skin, liver pathology, sepsis, cancer, and a shortened life span. Furthermore, the sensitivity of $Ku80^{-/-}$ animals to ROS is exacerbated as well (Nussenzweig et al. 1996; Zhu et al. 1996; Gu et al. 2000). Specific mutants in several genes implicated in DNA-damage-induced cell cycle checkpoints also display evidence of premature aging.

p53 has a key role in cell cycle control as well as in apoptosis. It has been known for a long time as a tumor suppressor gene, and inactivation in mice ($p53^{-/-}$ animals) logically gives rise to cancer (Jacks et al. 1994). p53 can also promote repair and has recently been connected to aging. A $p53m/+$ mouse that expresses a mutant p53 protein with enhanced p53 activity (Tyner et al. 2002) exhibits dwarfism, osteopenia, generalized organ atrophy, lymphoid atrophy, osteoporosis, atrophic skin, diminished stress tolerance, decreased cancer incidence, and a decreased life span. Thus, modulated forms of p53 can either give rise to cancer or to a decreased life span with reduced cancer incidence, which is reminiscent of some of the NER-deficient disorders.

HUMAN SEGMENTAL PROGERIAS

A number of human diseases are associated with dramatic progression of the aging phenotype at an early stage in life. In some cases, these conditions display striking similarities to the normal aging processes. However, they remain disease conditions and are termed segmental progerias, as they reflect many, but not all, signs and symptoms of normal aging (see, e.g., the NER disorders described above). Nevertheless, molecular studies of these diseases have proven to be remarkably important in advancing our understanding of the aging process. Several segmental progerias are listed in Table 2, and common to all these premature aging diseases is that they harbor a defect in DNA repair. Some of these have been discussed above (e.g., XP, CS, and TTD) and the others are described in more detail here (Fig. 6).

Werner syndrome (WS) presents probably the most classic aging phenotype, where patients at an early age display a multitude of features commonly associated with normal aging, such as short stature, typical faces, premature graying, hair loss, cataracts, atrophy of skin and subcutaneous tissue, and acral sclerosis (Epstein et al. 1966; Yu et al. 1996). Most patients die before the age of 50 years, either from complications of atherosclerotic vascular disease or from malignancy. Notably, the gene expression pattern in WS resembles that seen in the normal aging process (Kyng et al. 2003). Werner protein (WRN) is a helicase and exonuclease, which functions in at least two pathways of DNA repair: BER and recombination (Lee et al. 2005; Ozgenc and Loeb 2005). In addition, WRN has important functions in telomere maintenance (Opresko et al. 2005), i.e., preserving the tips of chromosomes that serve to protect against chromosome attrition and genomic instability.

Bloom syndrome (BS) and Rothmund–Thomson syndrome (RTS) each stem from a deficiency in a protein, BLM and RECQ4, respectively, that has homology with members of the bacterial RecQ helicase family (Ellis et al. 1995; Kitao et al. 1999). The RecQ group of helicases, which also includes WRN, function as "guardians of the genome" in many DNA metabolic processes, including repair, recombination, and the resolution of complex DNA structures (Sharma et al. 2006). In vitro, the different recQ helicases share similar biochemical properties, although they have different interaction partners and operate in different pathways, and the patients have different symptoms. BS is perhaps more accurately considered a premature cancer syndrome, rather than a premature aging disease, since patients at an early age develop a spectrum of cancers normally seen in old individuals (Cheok et al. 2005). Bloom protein (BLM) participates in many DNA

Table 2. Segmental progerias: Premature aging disorders

- Werner syndrome
- Cockayne syndrome
- Xeroderma pigmentosum
- Bloom syndrome (cancer)
- Rothmud–Thomson syndrome
- Hutchinson–Gilford (progeria)
- Trichothiodystrophy

transactions, particularly in recombination repair, functioning as a structure-specific DNA helicase, as well as in Holliday-junction branch migration and the annealing of complementary single-stranded DNA molecules. RTS is an autosomal recessive genodermatosis characterized by a poikilodermatous rash starting in infancy, small stature, skeletal abnormalities, juvenile cataracts, and predisposition to specific cancers (Kellermayer 2006). Cells from RTS patients display severe genomic instability, a common feature among the premature aging diseases and indicative of a DNA-repair deficiency. Interestingly, RTS protein does not exhibit any obvious helicase activity, yet it possesses a DNA-dependent ATPase function and DNA single-strand annealing capacity (Macris et al. 2006). Its precise role in DNA and/or RNA metabolism, however, remains unknown.

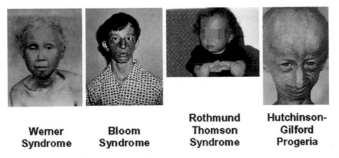

		Rothmund	Hutchinson-
Werner	Bloom	Thomson	Gilford
Syndrome	Syndrome	Syndrome	Progeria

Figure 6. Images of individuals suffering from rare premature aging disorders. Specific disease is indicated below. Shown is a 48-year-old Japanese-American WS patient presenting the characteristic hair loss, graying of the hair, and old physical features. The BS patient displays some of the characteristic symptoms of this disease, including sunlight sensitivity and discolored skin. Note the skin rash (poikiloderma) and skeletal dysplasia common for RTS patients. Seen in the HG patient are the distinctive facial features, characterized by a disproportionately small face in comparison to the head, an undeveloped jaw, abnormally prominent eyes, and a small nose. See text for further details regarding these segmental progerias.

Hutchinson–Gilford (HG) disease is associated with mutations in Lamin A, a component of the nuclear membrane (Eriksson et al. 2003). This disease is characterized by extreme short stature, low body weight, early loss of hair, lipodystrophy, scleroderma, decreased joint mobility, osteolysis, and facial features that resemble aged persons (Hennekam 2006). Defects in Lamin A processing or variant forms of the Lamin A protein have been found to disturb normal DNA-damage responses and enhance chromosomal instability, suggesting at least an indirect role for the protein in DNA repair (Liu et al. 2005; Lans and Hoeijmakers 2006).

SCENARIO FOR THE LINK BETWEEN DNA DAMAGE, REPAIR, AND (PREMATURE) AGING

The above mouse models and human genetic disorders demonstrate that, depending on the type and severity of the DNA repair defect, aging can be promoted and simultaneously cancer can be reduced (as in CS and TTD), cancer can be enhanced with minor acceleration of aging (as in XP), or both can be augmented (as with XP-CS). A common element in these cases is the extent to which TCR is compromised. An additional parameter is the degree to which GG-NER is impaired in combination with a TCR defect; in the absence of GG-NER, even more lesions will arrest transcription. The predicted consequence is enhanced cell death, cellular dysfunction, and senescence, which will lead to exhaustion of the cell-renewal capacity and disturbed tissue homeostasis and thus promote early aging. This interpretation is consistent with the other repair-deficient mutants, which exhibit in part different manifestations of premature aging and are defective in repair of other types of lesions that generally are cytotoxic. It also provides an explanation for why progeroid syndromes are segmental and differ from each other. Each organ and tissue has a different metabolic profile, resulting in different spectra of DNA lesions and, thus, dependence on different DNA repair systems. Defects in different repair systems will affect organs and tissues differentially, explaining much of the heterogeneity within the class of progeroid syndromes. This holds for ERCC1/XPF mutants carrying additional defects in cross-link repair and NHEJ mutants that are deficient in repair of very cytotoxic DSBs. Finally, p53 and several other mutants in DNA-damage-induced cell cycle checkpoints and apoptosis also are compatible with the above scenario.

Recent extensive gene expression profiling has disclosed a striking resemblance of the overall shift in expression in the liver of 15-day-old dramatic progeroid $CSB^{-/-}/XPA^{-/-}$ and $ERCC1^{-/-}$ mice with that of 2.5-year-old wild-type mice. Interestingly, the expression changes involved a suppression of the IGF-1 somatotropic axis that has been extensively linked with life span extension (see elsewhere in this book). Concomitantly, a pronounced shift in metabolism was noted from growth and proliferation to maintenance and repair, involving the carbohydrate, oxidative phosphorylation, fatty acid metabolism, and peroxisome biogenesis. A similar metabolic shift has been observed in long-lived dwarfs and upon caloric restriction (Spindler 2005). Antioxidant defenses were up-regulated in the progeroid mutants, whereas inflammatory responses and protein glycosylation pathways appeared up-regulated in the old wild-type animals but not in the accelerated aging mutants (Niedernhofer et al. 2006; van der Pluijm et al. 2006). The most logical interpretation is that accumulation of DNA damage in the TCR/NER and NER/cross-link repair mutants triggers a "survival" response that bears strong similarities with responses triggered upon natural aging and responses that are constitutively up-regulated in long-lived dwarf mice and in caloric-restricted wild-type animals. These findings highlight the parallels between premature aging, normal aging, and life span extension and underscore the validity of progeroid mouse mutants for the normal process of aging. The notion that such a wide range of repair mutants are associated with such a broad array of premature aging phenotypes in both humans and mice and that some of these mutants resemble natural aging to a striking degree suggests that accumulation of endogenously generated damage to DNA is an important—albeit not the only—cellular factor responsible for aging.

Figure 7 highlights some of the possible consequences of DNA damage, if it persists in the genome. This can lead to genome instability via various steps, indicated in the figure, such as replication and transcription errors and blockage, and to the activation of other, indicated pathways. The suggestion is that accumulated DNA damage may be an important causal factor in the aging process, as documented by experiments in cells and in animal systems.

PERSPECTIVES

There have been tremendous advances in the field of DNA repair in recent years, and the increased mechanistic insight has also proven very useful toward the understanding of the basic mechanisms of aging. This development greatly encourages further study in the emerging field of

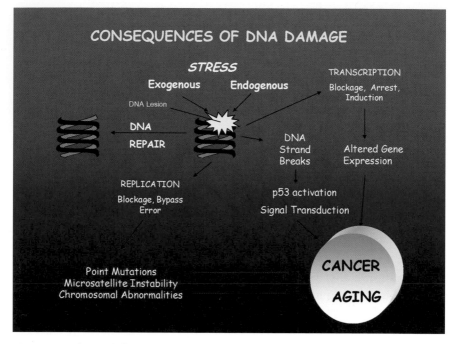

Figure 7. Pathway defects caused by persisting DNA damage in the genome can give rise to the genome instability, which is a hallmark feature of aging and cancer.

DNA metabolism in aging and holds great promise for further mechanistic insight and the development of therapeutic strategies for the intervention against age-associated diseases, which represent a huge problem in current and future health care.

ACKNOWLEDGMENTS

Some of this work was supported by funds from the National Institute on Aging (NIA), National Institutes of Health Intramural Research Program, The Netherlands Organization for Scientific Research (NWO) through the foundation of the Research Institute of Diseases of the Elderly, as well as by grants from SenterNovem IOP-Genomics (IGE03009), National Institutes of Health (IPO1 AG17242-02), National Institute of Environmental Health Sciences (NIEHS) (IUO1 ES011044), the European Community (QRTL-1999-02002; LSHC-CT-2005-512113), Dutch Cancer Society (EUR 99-2004), the NIA Program Project, the NIEHS center, and the Human Frontier Science Program. J.H. is CSO of DNage.

GLOSSARY OF ABBREVIATIONS AND GENE NOMENCLATURE

6-4PPs: 6-4 UV-induced photoproduct

AP: Apurinic/apyrimidinic

APE1: AP-endonuclease 1

BER: base excision repair

BLM: Bloom syndrome mutated protein

CPD: Cyclopyrimidine dimer

CS: Cockayne syndrome

CSA/B: CS complementation groups A/B

DDB1–2: Damaged DNA-binding protein 1–2

dRP: Deoxyribose phosphate

DSB: Double-strand break

ERCC1: Excision repair cross-complementation 1

GGR (GG-NER): Global genome (nucleotide excision) repair

HR: Homologous recombination

IGF-1: Insulin-like growth factor-1

LIG1: DNA ligase 1

LIG3: DNA ligase 3

MBD4: Methyl CpG-binding domain protein 4

MMR: Mismatch repair

MPG: Methylpurine DNA glycosylase

MYH1: Mut Y homolog 1

NEIL1–2: Nei-like homologs 1–2

NER: Nucleotide excision repair

NHEJ: Nonhomologous end joining

NTH1: Endonuclease III homolog 1

OGG1: Oxoguanine DNA glycosylase 1

PARP1: Poly-ADP-ribose polymerase 1

PCNA: Proliferating cell nuclear antigen

PNK: Polynucleotide kinase

POLβ: DNA polymerase β

RECQ4: Human RecQ homolog 4

RPA: Replication protein A

SITR6: Sirt2 deacetylase homolog 6

SSB: Single-strand break

SSBR: Single-strand-break repair

SWI2/SNF2: Transcriptional activator family of proteins with chromatin remodeling activity

TCR (TC-NER): Transcription-coupled (nucleotide excision) repair

TDG: Thymine DNA glycosylase

TDP1: Tyrosyl-DNA phosphodiesterase

TFIIH: Transcription factor II H

TTD: Trichothiodystrophy

UDG: Uracil DNA glycosylase

WRN: Werner syndrome mutated protein

XPA-G: Xeroderma pigmentosum complementation groups A-G

XRCC1: X-ray cross-complementation 1

REFERENCES

Ahel I., Rass U., El-Khamisy S.F., Katyal S., Clements P.M., McKinnon P.J., Caldecott K.W., and West S.C. 2006. The neurodegenerative disease protein aprataxin resolves abortive DNA ligation intermediates. *Nature* **443:** 713–716.

Al Tassan N., Chmiel N.H., Maynard J., Fleming N., Livingston A.L., Williams G.T., Hodges A.K., Davies D.R., David S.S., Sampson J.R., and Cheadle J.P. 2002. Inherited variants of MYH associated with somatic G:C—>T:A mutations in colorectal tumors. *Nat. Genet.* **30:** 227–232.

Ames B.N., Shigenaga M.K., and Hagen T.M. 1995. Mitochondrial decay in aging. *Biochim. Biophys. Acta* **1271:** 165–170.

Andressoo J.O. and Hoeijmakers J.H. 2005. Transcription-coupled repair and premature ageing. *Mutat. Res.* **577:** 179–194.

Andressoo J.O., Mitchell J.R., de Wit J., Hoogstraten D., Volker M., Toussaint W., Speksnijder E., Beems R.B., van Steeg H., and Jans J. 2006. An Xpd mouse model for the combined xeroderma pigmentosum/Cockayne syndrome exhibiting both cancer predisposition and segmental progeria. *Cancer Cell* **10:** 121–132.

Barja G. 2004. Free radicals and aging. *Trends Neurosci.* **27:** 595–600.

Barja G. and Herrero A. 2000. Oxidative damage to mitochondrial DNA is inversely related to maximum life span in the heart and brain of mammals. *FASEB J.* **14:** 312–318.

Bogenhagen D.F., Pinz K.G., and Perez-Jannotti R.M. 2001. Enzymology of mitochondrial base excision repair. *Prog. Nucleic Acid Res. Mol. Biol.* **68:** 257–271.

Bohr V.A. 2005. Deficient DNA repair in the human progeroid disorder, Werner syndrome. *Mutat. Res.* **577:** 252–259.

Bohr V.A., Smith C.A., Okumoto D.S., and Hanawalt P.C. 1985. DNA repair in an active gene: Removal of pyrimidine dimers from the DHFR gene of CHO cells is much more efficient than in the genome overall. *Cell* **40:** 359–369.

Bootsma D. and Hoeijmakers J.H. 1993. DNA repair. Engagement with transcription. *Nature* **363:** 114–115.

Bootsma D., Kraemer K.H., Cleaver J.E., and Hoeijmakers J.H.J. 2001. Nucleotide excision repair syndromes: xeroderma pigmentosum, Cockayne syndrome and trichothiodystrophy. In *The metabolic and molecular basis of inherited disease* (ed. C.R. Scriver et al.), pp. 677–703. McGraw-Hill, New York.

Brash D.E., Wikonkal N.M., Remenyik E., van der Horst G.T., Friedberg E.C., Cheo D.L., van Steeg H., Westerman A., and van Kranen H.J. 2001. The DNA damage signal for

Mdm2 regulation, Trp53 induction, and sunburn cell formation in vivo originates from actively transcribed genes. *J. Invest. Dermatol.* **5:** 1234–1240.

Brooks P.J., Wise D.S., Berry D.A., Kosmoski J.V., Smerdon M.J., Somers R.L., Mackie H., Spoonde A.Y., Ackerman E.J., Coleman K., et al. 2000. The oxidative DNA lesion 8,5'-(S)-cyclo-2'-deoxyadenosine is repaired by the nucleotide excision repair pathway and blocks gene expression in mammalian cells. *J. Biol. Chem.* **275:** 22355–22362.

Burkle A., Diefenbach J., Brabeck C., and Beneke S. 2005. Ageing and PARP. *Pharmacol. Res.* **52:** 93–99.

Cabelof D.C., Ikeno Y., Nyska A., Busuttil R.A., Anyangwe N., Vijg J., Matherly L.H., Tucker J.D., Wilson S.H., Richardson A., and Heydari A.R. 2006. Haploinsufficiency in DNA polymerase beta increases cancer risk with age and alters mortality rate. *Cancer Res.* **66:** 7460–7465.

Caldecott K.W. 2003. XRCC1 and DNA strand break repair. *DNA Repair* **2:** 955–969.

Chattopadhyay R., Wiederhold L., Szczesny B., Boldogh I., Hazra T.K., Izumi T., and Mitra S. 2006. Identification and characterization of mitochondrial abasic (AP)-endonuclease in mammalian cells. *Nucleic Acids Res.* **34:** 2067–2076.

Cheo D.L., Ruven H.J., Meira L.B., Hammer R.E., Burns D.K., Tappe N.J., van Zeeland A.A., Mullenders L.H., and Friedberg E.C. 1997. Characterization of defective nucleotide excision repair in XPC mutant mice. *Mutat. Res.* **374:** 1–9.

Cheok C.F., Bachrati C.Z., Chan K.L., Ralf C., Wu L., and Hickson I.D. 2005. Roles of the Bloom's syndrome helicase in the maintenance of genome stability. *Biochem. Soc. Trans.* **33:** 1456–1459.

Chipchase M.D., O'Neill M., and Melton D.W. 2003. Characterization of premature liver polyploidy in DNA repair (Ercc1)-deficient mice. *Hepatology* **38:** 958–966.

Clayton D.A., Doda J.N., and Friedberg E.C. 1974. The absence of a pyrimidine dimer repair mechanism in mammalian mitochondria. *Proc. Natl. Acad. Sci.* **71:** 2777–2781.

Cleaver J.E. 2005. Cancer in xeroderma pigmentosum and related disorders of DNA repair. *Nat. Rev. Cancer* **5:** 564–573.

Connelly J.C. and Leach D.R. 2004. Repair of DNA covalently linked to protein. *Mol. Cell* **13:** 307–316.

Date H., Onodera O., Tanaka H., Iwabuchi K., Uekawa K., Igarashi S., Koike R., Hiroi T., Yuasa T., Awaya Y., et al. 2001. Early-onset ataxia with ocular motor apraxia and hypoalbuminemia is caused by mutations in a new HIT superfamily gene. *Nat. Genet.* **29:** 184–188.

de Boer J. and Hoeijmakers J.H. 2000. Nucleotide excision repair and human syndromes. *Carcinogenesis* **21:** 453–460.

de Boer J., de Wit J., van Steeg H., Berg R.J., Morreau H., Visser P., Lehmann A.R., Duran M., Hoeijmakers J.H., and Weeda G. 1998. A mouse model for the basal transcription/DNA repair syndrome trichothiodystrophy. *Mol. Cell* **1:** 981–990.

de Boer J., Andressoo J.O., de Wit J., Huijmans J., Beems R.B., van Steeg H., Weeda G., van der Horst G.T.J., van Leeuwen W., Themmen A.P.N., et al. 2002. Premature aging in mice deficient in DNA repair and transcription. *Science* **296:** 1276–1279.

de Laat W.L., Jaspers N.G., and Hoeijmakers J.H. 1999. Molecular mechanism of nucleotide excision repair. *Genes Dev.* **7:** 768–785.

Demple B. and DeMott M.S. 2002. Dynamics and diversions in base excision DNA repair of oxidized abasic lesions. *Oncogene* **21:** 8926–8934.

Demple B. and Sung J.S. 2005. Molecular and biological roles of Ape1 protein in mammalian base excision repair. *DNA Repair* **4:** 1442–1449.

338   V.A. Bohr et al.

DeSanctis C.C.A. 1932. L'ídiozia xerodermica. *Riv. Sper. Frentiatr. Med. Leg. Alienazioni Ment.* **56**: 269–292.

de Souza-Pinto N.C., Hogue B.A., and Bohr V.A. 2001a. DNA repair and aging in mouse liver: 8–oxodG glycosylase activity increase in mitochondrial but not in nuclear extracts. *Free Radic. Biol. Med.* **30**: 916–923.

de Souza-Pinto N.C., Eide L., Hogue B.A., Thybo T., Stevnsner T., Seeberg E., Klungland A., and Bohr V.A. 2001b. Repair of 8–oxodeoxyguanosine lesions in mitochondrial DNA depends on the oxoguanine DNA glycosylase (OGG1) gene and 8–oxoguanine accumulates in the mitochondrial DNA of OGG1–defective mice. *Cancer Res.* **61**: 5378–5381.

de Vries A., Dolle M.E., Broekhof J.L., Muller J.J., Kroese E.D., van Kreijl C.F., Capel P.J., Vijg J., and van Steeg H. 1997. Induction of DNA adducts and mutations in spleen, liver and lung of XPA-deficient/lacZ transgenic mice after oral treatment with benzo[a]pyrene: Correlation with tumour development. *Carcinogenesis* **18**: 2327–2332.

de Vries A., van Oostrom C.T., Hofhuis F.M., Dortant P.M., Berg R.J., de Gruijl F.R., Wester P.W., van Kreijl C.F., Capel P.J., and van Steeg H. 1995. Increased susceptibility to ultraviolet-B and carcinogens of mice lacking the DNA excision repair gene XPA. *Nature* **377**: 169–173.

Dianov G., Bischoff C., Sunesen M., and Bohr V. A. 1999. Repair of 8–oxoguanine in DNA is deficient in Cockayne syndrome group B cells. *Nucleic Acids Res.* **27**: 1365–1368.

Dianov G.L., Souza-Pinto N., Nyaga S.G., Thybo T., Stevnsner T., and Bohr V.A. 2001. Base excision repair in nuclear and mitochondrial DNA. *Prog. Nucleic Acid Res. Mol. Biol.* **68**: 285–297.

Dizdaroglu M., Jaruga P., Birincioglu M., and Rodriguez H. 2002. Free radical-induced damage to DNA: Mechanisms and measurement. *Free Radic. Biol. Med.* **32**: 1102–1115.

Dogliotti E., Fortini P., Pascucci B., and Parlanti E. 2001. The mechanism of switching among multiple BER pathways. *Prog. Nucleic Acid Res. Mol. Biol.* **68**: 3–27.

El Khamisy S.F., Saifi G.M., Weinfeld M., Johansson F., Helleday T., Lupski J.R., and Caldecott K.W. 2005. Defective DNA single-strand break repair in spinocerebellar ataxia with axonal neuropathy-1. *Nature* **434**: 108–113.

Ellis N.A., Groden J., Ye T.Z., Straughen J., Lennon D.J., Ciocci S., Proytcheva M., and German J. 1995. The Bloom's syndrome gene product is homologous to RecQ helicases. *Cell* **83**: 655–666.

Epstein C.J., Martin G.M., Schultz A.L., and Motulsky A.G. 1966. Werner's syndrome a review of its symptomatology, natural history, pathologic features, genetics and relationship to the natural aging process. *Medicine* **45**: 177–221.

Eriksson M., Brown W.T., Gordon L.B., Glynn M.W., Singer J., Scott L., Erdos M.R., Robbins C.M., Moses T.Y., Berglund P., et al. 2003. Recurrent de novo point mutations in lamin A cause Hutchinson-Gilford progeria syndrome. *Nature* **423**: 293–298.

Giese H., Dolle M.E., Hezel A., van Steeg H., and Vijg J. 1999. Accelerated accumulation of somatic mutations in mice deficient in the nucleotide excision repair gene XPA. *Oncogene* **18**: 1257–1260.

Gillet L.C. and Scharer O.D. 2006. Molecular mechanisms of mammalian global genome nucleotide excision repair. *Chem. Rev.* **106**: 253–276.

Gu Y., Sekiguchi J., Gao Y., Dikkes P., Frank K., Ferguson D., Hasty P., Chun J., and Alt F.W. 2000. Defective embryonic neurogenesis in Ku-deficient but not DNA-dependent protein kinase catalytic subunit-deficient mice. *Proc. Natl. Acad. Sci.* **97**: 2668–2673.

Hanawalt P.C. 2002. Subpathways of nucleotide excision repair and their regulation. *Oncogene* 21: 8949–8956.

Harada Y.N., Shiomi N., Koike M., Ikawa M., Okabe M., Hirota S., Kitamura Y., Kitagawa M., Matsunaga T., Nikaido O., and Shiomi T. 1999. Postnatal growth failure, short life span, and early onset of cellular senescence and subsequent immortalization in mice lacking the xeroderma pigmentosum group G gene. *Mol. Cell. Biol.* 19: 2366–2372.

Harman D. 1956. Aging: A theory based on free radical and radiation chemistry. *J. Gerontol.* 11: 298–300.

———— 1981. The aging process. *Proc. Natl. Acad. Sci.* 78: 7124–7128.

Harrigan J.A., Opresko P.L., von Kobbe C., Kedar P.S., Prasad R., Wilson S.H., and Bohr V.A. 2003. The Werner syndrome protein stimulates DNA polymerase beta strand displacement synthesis via its helicase activity. *J. Biol. Chem.* 278: 22686–22695.

Harrigan J.A., Wilson D.M., III, Prasad R., Opresko P.L., Beck G., May A., Wilson S.H., and Bohr V.A. 2006. The Werner syndrome protein operates in base excision repair and cooperates with DNA polymerase beta. *Nucleic Acids Res.* 34: 745–754.

Harrison J.F., Hollensworth S.B., Spitz D.R., Copeland W.C., Wilson G.L., and LeDoux S.P. 2005. Oxidative stress-induced apoptosis in neurons correlates with mitochondrial DNA base excision repair pathway imbalance. *Nucleic Acids Res.* 33: 4660–4671.

Hashiguchi K., Stuart J.A., de Souza-Pinto N.C., and Bohr V.A. 2004. The C-terminal alphaO helix of human Ogg1 is essential for 8–oxoguanine DNA glycosylase activity: The mitochondrial beta-Ogg1 lacks this domain and does not have glycosylase activity. *Nucleic Acids Res.* 32: 5596–5608.

Hennekam R.C. 2006. Hutchinson-Gilford progeria syndrome: Review of the phenotype. *Am. J. Med. Genet. A* 140: 2603–2624.

Hoeijmakers J.H. 2001. Genome maintenance mechanisms for preventing cancer. *Nature* 411: 366–374.

Hollander M.C., Philburn R.T., Patterson A.D., Velasco-Miguel S., Friedberg E.C., Linnoila R.I., and Fornace A.J., Jr. 2005. Deletion of XPC leads to lung tumors in mice and is associated with early events in human lung carcinogenesis. *Proc. Natl. Acad. Sci.* 102: 13200–13205.

Hu J., de Souza-Pinto N.C., Haraguchi K., Hogue B.A., Jaruga P., Greenberg M.M., Dizdaroglu M., and Bohr V.A. 2005. Repair of formamidopyrimidines in DNA involves different glycosylases: Role of the OGG1, NTH1, and NEIL1 enzymes. *J. Biol. Chem.* 280: 40544–40551.

Hudson E.K., Hogue B.A., Souza-Pinto N.C., Croteau D.L., Anson R.M., Bohr V.A., and Hansford R.G. 1998. Age-associated change in mitochondrial DNA damage. *Free Radic. Res.* 29: 573–579.

Huffman J.L., Sundheim O., and Tainer J.A. 2005. DNA base damage recognition and removal: New twists and grooves. *Mutat. Res.* 577: 55–76.

Hung R.J., Hall J., Brennan P., and Boffetta P. 2005. Genetic polymorphisms in the base excision repair pathway and cancer risk: a HuGE review. *Am. J. Epidemiol.* 162: 925–942.

Imai K., Slupphaug G., Lee W.I., Revy P., Nonoyama S., Catalan N., Yel L., Forveille M., Kavli B., Krokan H.E., et al. 2003. Human uracil-DNA glycosylase deficiency associated with profoundly impaired immunoglobulin class-switch recombination. *Nat. Immunol.* 4: 1023–1028.

Itin P.H., Sarasin A., and Pittelkow M.R. 2001. Trichothiodystrophy: Update on the sulfur-deficient brittle hair syndromes. *J. Am. Acad. Dermatol.* 44: 891–920.

Itoh M., Hayashi M., Shioda K., Minagawa M., Isa F., Tamagawa K., Morimatsu Y., and Oda M. 1999. Neurodegeneration in hereditary nucleotide repair disorders. *Brain Dev.* **21:** 326–333.

Itoh T., Cado D., Kamide R., and Linn S. 2004. DDB2 gene disruption leads to skin tumors and resistance to apoptosis after exposure to ultraviolet light but not a chemical carcinogen. *Proc. Natl. Acad. Sci.* **101:** 2052–2057.

Jacks T., Remington L., Williams B.O., Schmitt E.M., Halachmi S., Bronson R.T., and Weinberg R.A. 1994. Tumor spectrum analysis in p53–mutant mice. *Curr. Biol.* **4:** 1–7.

Jiricny J. 2006. The multifaceted mismatch-repair system. *Nat. Rev. Mol. Cell Biol.* **7:** 335–346.

Karahalil B., de Souza-Pinto N.C., Parsons J.L., Elder R.H., and Bohr V.A. 2003. Compromised incision of oxidized pyrimidines in liver mitochondria of mice deficient in NTH1 and OGG1 glycosylases. *J. Biol. Chem.* **278:** 33701–33707.

Kellermayer R. 2006. The versatile RECQL4. *Genet. Med.* **8:** 213–216.

Kitao S., Shimamoto A., Goto M., Miller R.W., Smithson W.A., Lindor N.M., and Furuichi Y. 1999. Mutations in RECQL4 cause a subset of cases of Rothmund-Thomson syndrome. *Nat. Genet.* **22:** 82–84.

Klungland A., Hoss M., Gunz D., Constantinou A., Clarkson S.G., Doetsch P.W., Bolton P.H., Wood R.D., and Lindahl T. 1999a. Base excision repair of oxidative DNA damage activated by XPG protein. *Mol. Cell* **3:** 33–42.

Klungland A., Rosewell I., Hollenbach S., Larsen E., Daly G., Epe B., Seeberg E., Lindahl T., and Barnes D.E. 1999b. Accumulation of premutagenic DNA lesions in mice defective in removal of oxidative base damage. *Proc. Natl. Acad. Sci.* **96:** 13300–13305.

Kujoth G.C., Hiona A., Pugh T.D., Someya S., Panzer K., Wohlgemuth S.E., Hofer T., Seo A.Y., Sullivan R., Jobling W.A., et al. 2005. Mitochondrial DNA mutations, oxidative stress, and apoptosis in mammalian aging. *Science* **309:** 481–484.

Kuraoka I., Bender C., Romieu A., Cadet J., Wood R.D., and Lindahl T. 2000. Removal of oxygen free-radical-induced 5′,8–purine cyclodeoxynucleosides from DNA by the nucleotide excision-repair pathway in human cells. *Proc. Natl. Acad. Sci.* **97:** 3832–3837.

Kyng K.J., May A., Kolvraa S., and Bohr V.A. 2003. Gene expression profiling in Werner syndrome closely resembles that of normal aging. *Proc. Natl. Acad. Sci.* **100:** 12259–12264.

Lakshmipathy U. and Campbell C. 2000. Mitochondrial DNA ligase III function is independent of Xrcc1. *Nucleic Acids Res.* **28:** 3880–3886.

Lans H. and Hoeijmakers J.H. 2006. Cell biology: Ageing nucleus gets out of shape. *Nature* **7080:** 32–34.

Lebel M., Lavoie J., Gaudreault I., Bronsard M., and Drouin R. 2003. Genetic cooperation between the Werner syndrome protein and poly(ADP-ribose) polymerase-1 in preventing chromatid breaks, complex chromosomal rearrangements, and cancer in mice. *Am. J. Pathol.* **162:** 1559–1569.

LeDoux S.P. and Wilson G.L. 2001. Base excision repair of mitochondrial DNA damage in mammalian cells. *Prog. Nucleic Acid Res. Mol. Biol.* **68:** 273–284.

Lee J.W., Harrigan J., Opresko P.L., and Bohr V.A. 2005. Pathways and functions of the Werner syndrome protein. *Mech. Ageing Dev.* **126:** 79–86.

Lehmann A.R. 2001. The xeroderma pigmentosum group D (XPD) gene: One gene, two functions, three diseases. *Genes Dev.* **15:** 15–23.

———— 2003. DNA repair-deficient diseases, xeroderma pigmentosum, Cockayne syndrome and trichothiodystrophy. *Biochimie* **85:** 1101–1111.

Lesnefsky E.J. and Hoppel C.L. 2006. Oxidative phosphorylation and aging. *Ageing Res. Rev.* **5:** 402–433.

Licht C.L., Stevnsner T., and Bohr V.A. 2003. Cockayne syndrome group B cellular and biochemical functions. *Am. J. Hum. Genet.* **73:** 1217–1239.

Lin M.T. and Beal M.F. 2006. Mitochondrial dysfunction and oxidative stress in neurodegenerative diseases. *Nature* **443:** 787–795.

Lindahl T. 1993. Instability and decay of the primary structure of DNA. *Nature* **362:** 709–715.

Lindenbaum Y., Dickson D., Rosenbaum P., Kraemer K., Robbins I., and Rapin I. 2001. Xeroderma pigmentosum/cockayne syndrome complex: First neuropathological study and review of eight other cases. *Eur. J. Paediatr. Neurol.* **5:** 225–242.

Liu B., Wang J., Chan K.M., Tjia W.M., Deng W., Guan X., Huang J.D., Li K.M., Chau P.Y., Chen D.J., et al. 2005. Genomic instability in laminopathy-based premature aging. *Nat. Med.* **11:** 780–785.

Ljungman M. and Lane D.P. 2004. Transcription—Guarding the genome by sensing DNA damage. *Nat. Rev. Cancer* **4:** 727–737.

Longley M.J., Prasad R., Srivastava D.K., Wilson S.H., and Copeland W.C. 1998. Identification of 5'-deoxyribose phosphate lyase activity in human DNA polymerase gamma and its role in mitochondrial base excision repair in vitro. *Proc. Natl. Acad. Sci.* **95:** 12244–12248.

Lu T., Pan Y., Kao S.Y., Li C., Kohane I., Chan J., and Yankner B.A. 2004. Gene regulation and DNA damage in the ageing human brain. *Nature* **429:** 883–891.

Maclean M.J., Aamodt R., Harris N., Alseth I., Seeberg E., Bjoras M., and Piper P.W. 2003. Base excision repair activities required for yeast to attain a full chronological life span. *Aging Cell* **2:** 93–104.

Macris M.A., Krejci L., Bussen W., Shimamoto A., and Sung P. 2006. Biochemical characterization of the RECQ4 protein, mutated in Rothmund-Thomson syndrome. *DNA Repair* **5:** 172–180.

Martin G.M. 2005. Genetic modulation of senescent phenotypes in *Homo sapiens. Cell* **120:** 523–532.

McCuaig C., Marcoux D., Rasmussen J.E., Werner M.M., and Gentner N.E. 1993. Trichothiodystrophy associated with photosensitivity, gonadal failure, and striking osteosclerosis. *J. Am. Acad. Dermatol.* **28:** 820–826.

McWhir J., Selfridge J., Harrison D.J., Squires S., and Melton D.W. 1993. Mice with DNA repair gene (ERCC-1) deficiency have elevated levels of p53, liver nuclear abnormalities and die before weaning. *Nat. Genet.* **5:** 217–224.

Mellon I., Spivak G., and Hanawalt P.C. 1987. Selective removal of transcription-blocking DNA damage from the transcribed strand of the mammalian DHFR gene. *Cell* **51:** 241–249.

Mellon I., Bohr V.A., Smith C.A., and Hanawalt P.C. 1986. Preferential DNA repair of an active gene in human cells. *Proc. Natl. Acad. Sci.* **83:** 8878–8882.

Melton D.W., Ketchen A.M., Nunez F., Bonatti-Abbondandolo S., Abbondandolo A., Squires S., and Johnson R.T. 1998. Cells from ERCC1–deficient mice show increased genome instability and a reduced frequency of S-phase-dependent illegitimate chromosome exchange but a normal frequency of homologous recombination. *J. Cell Sci.* **111:** 395–404.

Mitchell D.L. and Nairn R.S. 1989. The biology of the (6-4) photoproduct. *Photochem. Photobiol.* **49:** 805–819.

Moreira M.C., Barbot C., Tachi N., Kozuka N., Uchida E., Gibson T., Mendonca P., Costa M., Barros J., Yanagisawa T., et al. 2001. The gene mutated in ataxia-ocular apraxia 1 encodes the new HIT/Zn-finger protein aprataxin. *Nat. Genet.* **29:** 189–193.

Mostoslavsky R., Chua K.F., Lombard D.B., Pang W.W., Fischer M.R., Gellon L., Liu P., Mostoslavsky G., Franco S., Murphy M.M., et al. 2006. Genomic instability and aging-like phenotype in the absence of mammalian SIRT6. *Cell* **124:** 315–329.

Murai M., Enokido Y., Inamura N., Yoshino M., Nakatsu Y., van der Horst G.T., Hoeijmakers J.H., Tanaka K., and Hatanaka H. 2001. Early postnatal ataxia and abnormal cerebellar development in mice lacking Xeroderma pigmentosum Group A and Cockayne syndrome Group B DNA repair genes. *Proc. Natl. Acad. Sci.* **98:** 13379–13384.

Nakabayashi K., Amann D., Ren Y., Saarialho-Kere U., Avidan N., Gentles S., MacDonald J.R., Puffenberger E.G., Christiano A.M., Martinez-Mir A., et al. 2005. Identification of C7orf11 (TTDN1) gene mutations and genetic heterogeieity in nonphotosensitive trichothiodystrophy. *Am. J. Hum. Genet.* **76:** 510–516.

Nakane H., Takeuchi S., Yuba S., Saijo M., Nakatsu Y., Murai H., Nakatsuru Y., Ishikawa T., Hirota S., and Kitamura Y.1995. High incidence of ultraviolet-B-or chemical-carcinogen-induced skin tumours in mice lacking the xeroderma pigmentosum group A gene. *Nature* **377:** 165–168.

Nance M.A. and Berry S.A. 1992. Cockayne syndrome: Review of 140 cases. *Am. J. Med. Genet.* **42:** 68–84.

Niedernhofer L.J., Daniels J.S., Rouzer C.A., Greene R.E., and Marnett L.J. 2003. Malondialdehyde, a product of lipid peroxidation, is mutagenic in human cells. *J. Biol. Chem.* **278:** 31426–31433.

Niedernhofer L.J., Garinis G.A., Raams A., Lalai S.A., Robinson A.R., Appeldoorn E., Odijk H., Oostendorp R., Ahmad A., van Leeuwen W., et al. 2006. A novel progeria caused by a DNA repair defect reveals that genotoxic stress supresses the somatotroph axis. *Nature* **444:** 1038–1043.

Nilsen H., Otterlei M., Haug T., Solum K., Nagelhus T.A., Skorpen F., and Krokan H.E. 1997. Nuclear and mitochondrial uracil-DNA glycosylases are generated by alternative splicing and transcription from different positions in the UNG gene. *Nucleic Acids Res.* **25:** 750–755.

Nishioka K., Ohtsubo T., Oda H., Fujiwara T., Kang D., Sugimachi K., and Nakabeppu Y. 1999. Expression and differential intracellular localization of two major forms of human 8–oxoguanine DNA glycosylase encoded by alternatively spliced OGG1 mRNAs. *Mol. Biol. Cell* **10:** 1637–1652.

Norwood W. 1964. The Marinesco-Sjogren syndrome. *J. Pediatr.* **65:** 431–437.

Nussenzweig A., Chen C., da Costa Soares V., Sanchez M., Sokol K., Nussenzweig M.C., and Li G.C. 1996. Requirement for Ku80 in growth and immunoglobulin V(D)J recombination. *Nature* **382:** 551–555.

Ogi T. and Lehmann A.R. 2006. The Y-family DNA polymerase kappa (pol kappa) functions in mammalian nucleotide-excision repair. *Nat. Cell Biol.* **8:** 640–642.

Ohtsubo T., Nishioka K., Imaiso Y., Iwai S., Shimokawa H., Oda H., Fujiwara T., and Nakabeppu Y. 2000. Identification of human MutY homolog (hMYH) as a repair enzyme for 2–hydroxyadenine in DNA and detection of multiple forms of hMYH located in nuclei and mitochondria. *Nucleic Acids Res.* **28:** 1355–1364.

Opresko P.L., Mason P.A., Podell E.R., Lei M., Hickson I.D., Cech T.R., and Bohr V.A. 2005. POT1 stimulates RecQ helicases WRN and BLM to unwind telomeric DNA substrates. *J. Biol. Chem.* **280**: 32069–32080.

Ozgenc A. and Loeb L.A. 2005. Current advances in unraveling the function of the Werner syndrome protein. *Mutat. Res.* **577**: 237–251.

Parsons J.L. and Elder R.H. 2003. DNA N-glycosylase deficient mice: A tale of redundancy. *Mutat. Res.* **531**: 165–175.

Pouliot J.J., Yao K.C., Robertson C.A., and Nash H.A. 1999. Yeast gene for a Tyr-DNA phosphodiesterase that repairs topoisomerase I complexes. *Science* **286**: 552–555.

Prasher J.M., Lalai A.S., Heijmans-Antonissen C., Ploemacher R.E., Hoeijmakers J.H., Touw I.P., and Niedernhofer L.J. 2005. Reduced hematopoietic reserves in DNA interstrand crosslink repair-deficient Ercc1-/- mice. *EMBO J.* **24**: 861–871.

Rapin I., Lindenbaum Y., Dickson D.W., Kraemer K.H., and Robbins J.H. 2000. Cockayne syndrome and xeroderma pigmentosum. *Neurology* **55**: 1442–1449.

Sander M., Cadet J., Casciano D.A., Galloway S.M., Marnett L.J., Novak R.F., Pettit S.D., Preston R.J., Skare J.A., Williams G.M., et al. 2005. Proceedings of a workshop on DNA adducts: Biological significance and applications to risk assessment. *Toxicol. Appl. Pharmacol.* **1**: 1–20.

Sands A.T., Abuin A., Sanchez A., Conti C.J., and Bradley A. 1995. High susceptibility to ultraviolet-induced carcinogenesis in mice lacking XPC. *Nature* **377**: 162–165.

Satoh M.S. and Lindahl T. 1994. Enzymatic repair of oxidative DNA damage. *Cancer Res.* **54**: 1899s-1901s.

Satoh M.S., Jones C.J., Wood R.D., and Lindahl T. 1993. DNA excision-repair defect of xeroderma pigmentosum prevents removal of a class of oxygen free radical-induced base lesions. *Proc. Natl. Acad. Sci.* **90**: 6335–6339.

Sharma S., Doherty K.M., and Brosh R.M., Jr. 2006. Mechanisms of RecQ helicases in pathways of DNA metabolism and maintenance of genomic stability. *Biochem. J.* **398**: 319–337.

Shiomi N., Mori M., Kito S., Harada Y.N., Tanaka K., and Shiomi T. 2005. Severe growth retardation and short life span of double-mutant mice lacking Xpa and exon 15 of Xpg. *DNA Repair* **4**: 351–357.

Shiomi N., Kito S., Oyama M., Matsunaga T., Harada Y.N., Ikawa M., Okabe M., and Shiomi T. 2004. Identification of the XPG region that causes the onset of Cockayne syndrome by using Xpg mutant mice generated by the cDNA-mediated knock-in method. *Mol. Cell. Biol.* **24**: 3712–3719.

Spindler S.R. 2005. Rapid and reversible induction of the longevity, anticancer and genomic effects of caloric restriction. *Mech. Ageing Dev.* **9**: 960–966.

Stierum R.H., Dianov G.L., and Bohr V.A. 1999. Single-nucleotide patch base excision repair of uracil in DNA by mitochondrial protein extracts. *Nucleic Acids Res.* **27**: 3712–3719.

Stivers J.T. and Jiang Y.L. 2003. A mechanistic perspective on the chemistry of DNA repair glycosylases. *Chem. Rev.* **103**: 2729–2759.

Suh D., Wilson D.M., III, and Povirk L.F. 1997. 3′-Phosphodiesterase activity of human apurinic/apyrimidinic endonuclease at DNA double-strand break ends. *Nucleic Acids Res.* **25**: 2495–2500.

Sweasy J.B., Lang T., Starcevic D., Sun K.W., Lai C.C., Dimaio D., and Dalal S. 2005. Expression of DNA polymerase {beta} cancer-associated variants in mouse cells results in cellular transformation. *Proc. Natl. Acad. Sci.* **102**: 14350–14355.

Takao M., Aburatani H., Kobayashi K., and Yasui A. 1998. Mitochondrial targeting of human DNA glycosylases for repair of oxidative DNA damage. *Nucleic Acids Res.* **26:** 2917–2922.

Takashima H., Boerkoel C.F., John J., Saifi G.M., Salih M.A., Armstrong D., Mao Y., Quiocho F.A., Roa B.B., Nakagawa M., et al. 2002. Mutation of TDP1, encoding a topoisomerase I-dependent DNA damage repair enzyme, in spinocerebellar ataxia with axonal neuropathy. *Nat. Genet.* **32:** 267–272.

Tian M., Shinkura R., Shinkura N., and Alt F.W. 2004. Growth retardation, early death, and DNA repair defects in mice deficient for the nucleotide excision repair enzyme XPF. *Mol. Cell. Biol.* **24:** 1200–1205.

Tomkinson A.E., Vijayakumar S., Pascal J.M., and Ellenberger T. 2006. DNA ligases: Structure, reaction mechanism, and function. *Chem. Rev.* **106:** 687–699.

Trifunovic A. 2006. Mitochondrial DNA and ageing. *Biochim. Biophys. Acta* **1757:** 611–617.

Trifunovic A., Wredenberg A., Falkenberg M., Spelbrink J.N., Rovio A.T., Bruder C.E., Bohlooly Y., Gidlof S., Oldfors A., Wibom R., et al. 2004. Premature ageing in mice expressing defective mitochondrial DNA polymerase. *Nature* **429:** 417–423.

Tuo J., Jaruga P., Rodriguez H., Bohr V.A., and Dizdaroglu M. 2003. Primary fibroblasts of Cockayne syndrome patients are defective in cellular repair of 8–hydroxyguanine and 8–hydroxyadenine resulting from oxidative stress. *FASEB J.* **17:** 668–674.

Tuo J., Muftuoglu M., Chen C., Jaruga P., Selzer R.R., Brosh R.M., Jr., Rodriguez H., Dizdaroglu M., and Bohr V.A. 2001. The Cockayne syndrome group B gene product is involved in general genome base excision repair of 8–hydroxyguanine in DNA. *J. Biol. Chem.* **276:** 45772–45779.

Tyner S.D., Venkatachalam S., Choi J., Jones S., Ghebranious N., Igelmann H., Lu X., Soron G., Cooper B., and Brayton C. 2002. p53 Mutant mice that display early ageing-associated phenotypes. *Nature* **415:** 45–53.

van der Horst G.T., Meira L., Gorgels T.G., de Wit J., Velasco-Miguel S., Richardson J.A., Kamp Y., Vreeswijk M.P., Smit B., and Bootsma D. 2002. UVB radiation-induced cancer predisposition in Cockayne syndrome group A (Csa) mutant mice. *DNA Repair* **1:** 143–157.

van der Horst G.T., van Steeg H., Berg R.J., van Gool A.J., de Wit J., Weeda G., Morreau H., Beems R.B., van Kreijl C.F., and de Gruijl F.R.1997. Defective transcription-coupled repair in Cockayne syndrome B mice is associated with skin cancer predisposition. *Cell* **89:** 425–435.

van der Pluijm I., Garinis G.A., Brandt R.M.C., Gorgels T.G.M.F., Wijnhoven S.W., Diderich K.E.M., de Wit J., Mitchell J.R., van Oostrom C., Beems R., et al. 2006. Impaired genome maintenance suppresses the growth hormone–insulin-like growth factor 1 axis in mice with Cockayne syndrome. *PLOS Biol.* **5:** e2.

van Hoffen A., Venema J., Meschini R., van Zeeland A.A., and Mullenders L.H. 1995. Transcription-coupled repair removes both cyclobutane pyrimidine dimers and 6-4 photoproducts with equal efficiency and in a sequential way from transcribed DNA in xeroderma pigmentosum group C fibroblasts. *EMBO J.* **2:** 360–367.

Vartanian V., Lowell B., Minko I.G., Wood T.G., Ceci J.D., George S., Ballinger S.W., Corless C.L., McCullough A.K., and Lloyd R.S. 2006. The metabolic syndrome resulting from a knockout of the NEIL1 DNA glycosylase. *Proc. Natl. Acad. Sci.* **103:** 1864–1869.

Vermeulen W., Rademakers S., Jaspers N.G., Appeldoorn E., Raams A., Klein B., Kleijer W.J., Hansen L.K., and Hoeijmakers J.H. 2001. A temperature-sensitive disorder in basal transcription and DNA repair in humans. *Nat. Genet.* **27:** 299–303.

von Kobbe C., Harrigan J.A., Schreiber V., Stiegler P., Piotrowski J., Dawut L., and Bohr V.A. 2004. Poly(ADP-ribose) polymerase 1 regulates both the exonuclease and helicase activities of the Werner syndrome protein. *Nucleic Acids Res.* **32:** 4003–4014.

Wallace D.C. 2005. A mitochondrial paradigm of metabolic and degenerative diseases, aging, and cancer: A dawn for evolutionary medicine. *Annu. Rev. Genet.* **39:** 359–407.

Wang F., Saito Y., Shiomi T., Yamada S., Ono T., and Ikehata H. 2006. Mutation spectrum in UVB-exposed skin epidermis of a mildly-affected Xpg-deficient mouse. *Environ. Mol. Mutagen.* **47:** 107–116.

Ward J.F. 1988. DNA damage produced by ionizing radiation in mammalian cells: Identities, mechanisms of formation, and reparability. *Prog. Nucleic Acid Res. Mol. Biol.* **35:** 95–125.

Weeda G., Donker I., de Wit J., Morreau H., Janssens R., Vissers C.J., Nigg A., van Steeg H., Bootsma D., and Hoeijmakers J.H. 1997. Disruption of mouse ERCC1 results in a novel repair syndrome with growth failure, nuclear abnormalities and senescence. *Curr. Biol.* **7:** 427–439.

Weissman L., de Souza-Pinto N.C., Stevnsner T., and Bohr V.A. 2007. DNA repair, mitochondria, and neurodegeneration. *Neuroscience* **145:** 1318–1329.

Wiederhold L., Leppard J.B., Kedar P., Karimi-Busheri F., Rasouli-Nia A., Weinfeld M., Tomkinson A.E., Izumi T., Prasad R., Wilson S.H., et al. 2004. AP endonuclease-independent DNA base excision repair in human cells. *Mol. Cell* **15:** 209–220.

Wijnhoven S.W., Beems R.B., Roodbergen M., van den Berg J., Lohman P.H., Diderich K., van der Horst G.T., Vijg J., Hoeijmakers J.H., and van Steeg H. 2005. Accelerated aging pathology in ad libitum fed Xpd(TTD) mice is accompanied by features suggestive of caloric restriction. *DNA Repair* **4:** 1314–1324.

Wilson D.M., III and Barsky D. 2001. The major human abasic endonuclease: Formation, consequences and repair of abasic lesions in DNA. *Mutat. Res.* **485:** 283–307.

Wilson D.M., III and Bohr V.A. 2007. The mechanics of base excision repair, and its relationship to aging and disease. *DNA Repair* **6:** 544–559.

Wilson D.M., III and McNeill D.R. 2007. Base excision repair and the central nervous system. *Neuroscience* **145:** 1187–1200.

Wilson D.M., III and Thompson L.H. 1997. Life without DNA repair. *Proc. Natl. Acad. Sci.* **94:** 12754–12757.

Wilson S.H. 1998. Mammalian base excision repair and DNA polymerase beta. *Mutat. Res.* **407:** 203–215.

Wogan G.N., Hecht S.S., Felton J.S., Conney A.H., and Loeb L.A. 2004. Environmental and chemical carcinogenesis. *Semin. Cancer Biol.* **14:** 473–486.

Wood R.D. 1999. DNA damage recognition during nucleotide excision repair in mammalian cells. *Biochimie* **81:** 39–44.

Wyman C. and Kanaar R. 2006. DNA double-strand break repair: All's well that ends well. *Annu. Rev. Genet.* **40:** 363–383.

Xie Y., Yang H., Cunanan C., Okamoto K., Shibata D., Pan J., Barnes D.E., Lindahl T., McIlhatton M., Fishel R., and Miller J.H. 2004. Deficiencies in mouse Myh and Ogg1 result in tumor predisposition and G to T mutations in codon 12 of the K-ras onco-gene in lung tumors. *Cancer Res.* **64:** 3096–3102.

Yu C.E., Oshima J., Fu Y.H., Wijsman E.M., Hisama F., Alisch R., Matthews S., Nakura J., Miki T., Ouais S., Martin G.M., et al. 1996. Positional cloning of the Werner's syndrome gene. *Science* **272:** 258–262.

Zhu C., Bogue M.A., Lim D.S., Hasty P., and Roth D.B. 1996. Ku86–deficient mice exhibit severe combined immunodeficiency and defective processing of V(D)J recombination intermediates. *Cell* **86:** 379–389.

13

Extended Life Span in Mice with Reduction in the GH/IGF-1 Axis

John J. Kopchick

Edison Biotechnology Institute and Department of Biomedical Sciences
College of Osteopathic Medicine
Ohio University, Athens, Ohio 45701

Andrzej Bartke

Southern Illinois University School of Medicine
Springfield, Illinois 62794-9628

Darlene E. Berryman

School of Human and Consumer Sciences
College of Health and Human Services
Ohio University, Athens, Ohio 45701

IN THE QUEST TO IDENTIFY FACTORS THAT CONTRIBUTE TO AGING, a group of closely related molecules have repeatedly emerged as critical players. Insulin-like growth factor-1 (IGF-1)-like proteins and their downstream intracellular signaling molecules have been shown to be associated with life span in fruit flies and nematodes (Clancy et al. 2001; Tatar et al. 2001). More recently, several mouse models with reduced growth hormone (GH) and/or IGF-1 signaling have also been shown to have extended life spans as compared to control/normal siblings. Evaluation of these mouse models, as well as mice with altered GH signaling that do not show increased life span, has offered some clues as to factors that contribute to aging. For example, it is clear that the role of the GH/IGF-1 axis overlaps partially but is distinct from the effect of calorie restriction, a well-recognized means to delay aging (Bonkowski et al. 2006). Although the mechanisms linking the GH/IGF-1 axis with delayed aging remain to be determined, there are some commonalities among the various strains of long-lived mice, such as reduced IGF-1 signaling, enhanced insulin sensitivity, improved stress resistance, and protection from carcinogenesis, which likely contribute to the improvement in longevity. Future studies comparing these mouse models will undoubtedly provide valuable insight into additional factors that contribute to the extension of life span.

GH/IGF-1 AXIS

Overview

The GH/IGF-1 axis refers to the combined actions of growth hormone (GH) and insulin-like growth factor-1 (IGF-1) (Fig. 1). GH is produced and secreted by somatotrophic cells of the anterior pituitary gland. The regulation of GH synthesis and secretion is predominantly controlled by the balance among the hypothalamic hormones, GH-releasing hormone (GHRH), ghrelin, somatostatin (SS), and serum IGF-1 levels (Lin-Su and Wajnrajch 2002). The actions of GH affect not only growth, as the name implies, but also cellular differentiation and metabolism. This can be seen by the drastic alterations of lipid, protein, carbohydrate, and mineral metabolism as a function of GH action (Davidson 1987). GH exerts its effect by interacting with specific GH receptors (GHRs) on the surface of target tissues, resulting in activation of the Janus kinase (JAK) II and signal transducer and activator of transcription (STAT) 5 pathway, as well as other intracellular signaling systems (Kopchick and Andry 2000). In addition to directly affecting target tissues, GH stimulates the synthesis and release of IGF-1 from many tissues, most notably, the liver. IGF-1, a protein structurally related to insulin, has its own important role in the regulation of cellular and tissue function. The primary action of IGF-1 is mediated by binding primarily to the IGF-I receptor, but effects can also be mediated through the IGF-II receptor and the insulin receptor. The stability of the IGFs and their interaction with their receptors are mediated by specific IGF-binding proteins (IGFBPs), which are found in the circulation. Once IGF-1 binds to the IGF-1 receptor, which possesses tyrosine kinase activity, signaling to the nucleus and mitochondria primarily through the mitogen-activated protein kinase (MAPK) and PI3K/Akt pathways is activated. The main biological effect of IGF-I is to stimulate cell growth and differentiation and to decrease apoptosis, but it also exerts a variety of insulin-like effects in vivo and in vitro (for review, see Cohen 2006). Since GH and IGF-1 have related yet distinct metabolic effects and affect most tissues in the body, the GH/IGF-1 axis can be considered multifaceted and pervasive.

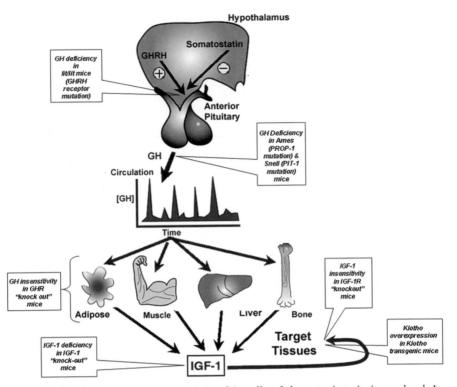

Figure 1. GH is secreted from somatotrophic cells of the anterior pituitary gland. It binds to GH receptors on a variety of cells/tissues including adipose, muscle, liver, and bone, and elicits production of IGF-1. Thus, as GH levels increase, so do those of IGF-1; conversely, as GH levels decrease, IGF-1 levels decrease. Ames and Snell mice are GH deficient due to mutations affecting GH gene transcription. lit/lit mice are also GH-deficient since they possess a mutation in the GHRH receptor; thus, they cannot respond to GHRH, which controls GH synthesis and secretion. All three strains of mice lack GH, are dwarf, and have low levels of IGF-1. Since GH inhibits insulin action, these animals are more insulin-sensitive than wild-type mice; thus, they have relatively low levels of insulin. GHR gene-disrupted or "knockout" mice lack GHR, thus, are GH insensitive, dwarf, and have low levels of IGF-1. They are also insulin-sensitive since they are GH-resistant. IGF-1-disrupted or IGF-1 receptor gene-disrupted mice lack IGF-1 or IGF-1R, respectively. Klotho mice express the klotho protein in the serum, which is thought to bind and repress intracellular signals of insulin and IGF-1 (Kurosu et al. 2005). (Adapted from Kopchick and Andry 2000 © Elsevier.)

GH/IGF-1 Axis and Aging: The Controversy

Several lines of evidence have implicated the GH/IGF-1 axis in aging. GH synthesis and secretion are well documented to decline with healthy aging in all mammalian species studied to date (Sonntag et al. 1980; Crew et al. 1987; Muller et al. 1993). Accordingly, as GH secretion declines, there is a concomitant decrease in IGF-1 levels. This decline, at least in rats, dogs, and humans, appears to be due to reduced release of GH-releasing hormone along with an increased release of somatostatin from the hypothalamus (Sonntag et al. 1985; Cella et al. 1996; Giustina and Veldhuis 1998). The reduction in GH secretion with healthy aging begins shortly after puberty and is correlated with increased percentage of total body and visceral fat, decreased physical fitness, and physiological declines in estrogen and androgen concentrations (Weltman et al. 1994; Veldhuis et al. 1995, 1997). Thus, the natural declines in GH and IGF-1 hint at the possibility that some age-related degenerative processes may be attributed to the actions of the GH/IGF-1 axis.

Although there is general agreement that the GH/IGF-1 axis is less potent with aging, the debate arises over whether this decline could be considered beneficial or detrimental to aging and aging-related problems. In 1990, Rudman (Rudman et al. 1990) reported that administration of GH in elderly men for 6 months could increase lean body mass, decrease adipose tissue mass, increase skin thickness, and improve bone density at least in one specific site. These benefits sparked the anti-aging status of GH, at least for an industry eager to promote youth to an aging public. Similar benefits to body composition, fitness, and bone density with varying GH dosing regimens have been reported by many other groups (Amato et al. 2000; Florakis et al. 2000; Pincelli et al. 2003; Boguszewski et al. 2005). However, it is also apparent that abnormally high levels of GH increase the incidence of morbidity and mortality in both rodent models and humans (Rajasoorya et al. 1994; Orme et al. 1998; Bartke 2003; Sheppard 2005). Although this may not reveal the role of normal levels of GH on disease progression or aging, low-dose GH administration has been reported to increase diabetes and glucose intolerance in healthy adults (Florakis et al. 2000; Blackman et al. 2002) and to increase mortality in patients that are critically ill (Takala et al. 1999). Importantly, a recent meta-analysis of more than 18 distinct studies on healthy elderly treated with GH concluded that although there were positive changes in body composition with GH administration in the elderly, there was also an increase in adverse events such as soft-tissue edema, carpal tunnel syndrome, and gynocomastia in addition to the increased risk of developing diabetes (Liu et al. 2007). Thus, GH, despite having some positive effects on various aging ailments, is unlikely to be the answer to our society's aging woes.

Several mutant invertebrate models have revealed that a reduction in insulin or IGF-1 signaling may be beneficial for increasing longevity. For example, mutations of the insulin/IGF-1-like receptor (DAF-2) in nematodes and of the insulin-IGF receptor (InR) in fruit flies (Clancy et al. 2001; Tatar et al. 2001) decrease the activity of the insulin/IGF-1-like pathway and cause an impressive extension of life span in these species. Since then, several mouse models with a reduction in the activity of the GH/IGF-1 axis have been shown to exhibit increased life span, suggesting that suppression of GH, IGF-1, or their downstream signaling pathways could be beneficial for aging. In this chapter, we discuss the phenotypes of these various mouse mutants with increased longevity, compare the phenotypes of these mice with the effect of calorie restriction (CR), and suggest potential mechanisms for the increased longevity.

INCREASED LONGEVITY IN MICE WITH ALTERED ACTIVITY IN THE GH/IGF-1 AXIS

Studies of several spontaneous and engineered mutant strains of mice with reduced activity of the GH/IGF-1 axis suggest that mechanisms similar to those acting in nematodes and fruit flies also control longevity in mammals. Complete phenotypic and molecular characterization of these mouse models is beyond the scope of this review, but several common traits are compared in Table 1.

Snell and Ames Dwarf Mice

In 1929, Snell first described mice with hereditary dwarfism in which mature mice were one-fourth the size of normal littermates (Snell 1929). These mice, as well as dwarf mice from a spontaneous mutation in the same gene from the Jackson Labs, were later found to be homozygous for a defect in Pit-1 or pituitary-specific transcription factor-1 (Li et al. 1990). Since Pit-1 protein is required for differentiation of hormone-specific cell types in the anterior pituitary, homozygous mutant mice lack GH, prolactin (PRL), and thyroid-stimulating hormone (TSH), producing cells resulting in severe endocrine deficiency of all three hormones. In 1961, researchers at Iowa State University reported a recessive autosomal mutation that caused dwarfism in mice and that was not due to

Table 1. Phenotypic comparison of mutant mice with reduced activity of GH/IGF-1 axis with calorie-restricted mice

Mouse model	Ames dwarf[a] (Prop1[df])	Snell dwarf[a] (Pit1[dw])	Ghrh[b] (lit/lit)	GHR−/−[c]	IGF1R+/−[d]	p66[shc]−/−[e]	Klotho transgenic[f]	Calorie-restricted[g]	GHA transgenic[h]
Primary effect	GH, PRL, TSH deficiency	GH, PRL, TSH deficiency	GH deficiency	GH deficiency/ GH insensitivity	Partial IGF-1 insensitivity	Stress resistance	Inhibit IGF-1 and insulin signaling		Reduced GH signaling
% Increase in mean life span	49%♂ 68%♀	26%♂ 42%♀	23%♂ 25%♀	26–55%♂ 16–38%♀	33%♀	30%	20–31%♂ 19%♀	varies	normal
Plasma levels									
GH	↓	↓	↓↓	↑	n.a.	n.a.	n.a.	↓	↑
IGF-1	↓↓	↓↓	↓↓	↓↓	↑	n.a.	n.a.	↓	↓
Glucose	↓	↓	n.a.	↓	↔	↔	↔	↓	↔
Insulin	↓	↓	n.a.	↓↓	↔	↔	↑	↓	↔
Growth									
Body size	↓↓	↓↓	↓↓	↓↓	↔	↔	↔	↓	↓
% body fat	↑	↑	↑↑	↑	n.a.	n.a.	n.a.	↓	↑

Reproduction

Sexual Maturation	delayed	delayed	n.a.	delayed	normal	delayed	n.a.	n.a.	delayed	n.a.
Fertility	↓↓	↓↓	n.a.	→	↔	→	n.a.	→	→	n.a.

Energy Balance

Food intake	n.a.	n.a.	↔	←	↔	←	↔	↔	→	←
Metabolic rate	→	→	↔	←	↔	←	↔	↔	→	n.a.

(n.a.) No available data. ↑ Refers to elevated levels; ↔ indicates normal levels, and ↓ refers to reduced levels.

[a] As reviewed by Tatar et al. (2003) and by Liang et al. (2003).

[b] Donahue and Beamer (1993); Godfrey et al. (2001); Flurkey et al. (2001); Puche et al. (2002); Kasukawa et al. (2003).

[c] Zhou et al. (1997); Coschigano et al. (2000, 2003); Keene et al. (2002); Berryman et al. (2004, 2006).

[d] Holzenberger et al. (2003); Holzenberger (2004).

[e] Migliaccio et al. (1999).

[f] Kurosu et al. (2005).

[g] As reviewed by Tatar et al. (2003).

[h] Chen et al. (1991a,b); Coschigano et al. (2003 and unpubl.); Berryman et al. (2004).

defects in the Pit-1 gene (Schaible and Gowen 1961). These mice, named the Ames dwarf mice, were determined to have a point mutation in the PROP-1 transcription factor (prophet of Pit-1 or paired-like homeodomain transcription factor in the Prop-1), which is upstream of Pit-1 and is required for Pit-1 activation (Andersen et al. 1995). Like Snell mice, homozygous Ames dwarf mutants have deficiencies in GH, PRL, and TSH.

Because the Ames and Snell dwarf mice share a defect in the same pathway, these mice have similar phenotypes (Table 1). Snell and Ames dwarf mice are characterized by female sterility and severely reduced circulating levels of insulin, IGF-1, glucose, and thyroid hormones (for review, see Bartke et al. 1998). Both models are also remarkably long lived and exhibit a greater than 40% increase in longevity. For Ames dwarf mice, females were reported to have an impressive 68% increase in life span (1206 ± 32 days vs. 718 ± 45 days), whereas males showed a 49% increase in mean life span (1076 ± 56 days vs. 723 ± 54 days) (Brown-Borg et al. 1996). This extension in life span for Ames dwarf mice has been consistently reported, varying from 35% to 69%, depending on diet (Bartke et al. 2001, 2004; Ikeno et al. 2003). For Snell dwarf mice, the extension in life span is less consistent and seems to vary with genetic background, housing, and breeding practices of the mice. Two reports suggest that Snell mice have increased life span varying from a 42% increase with no significant gender difference (1178 ± 235 days vs. 832 ± 158 days for controls) (Flurkey et al. 2001) to a 50% increase in only female mice (Flurkey et al. 2002).

The Lit/Lit Mouse

Another dwarf mouse model with increased life span is the lit/lit mouse with a missense mutation in the extracellular domain of the GHRH receptor (Godfrey et al. 1993). Due to the spontaneous mutation in the GHRH receptor, the lit/lit mice (C57BL/6J) are deficient in GH but have no apparent defect in PRL or TSH as with the Snell or Ames dwarf mice. lit/lit mice are 50–75% smaller than wild-type mice and have reduced serum IGF-1 levels and marked increase in adiposity (Donahue and Beamer 1993; Puche et al. 2002). Flurkey et al. (2001) reported that lit/lit mice in the C57BL/6J background are longer lived than +/– heterozygous controls with a 23% increase in males (1093 ± 186 days vs. 886 ± 148 days) and a 25% increase in females (1070 ± 127 days vs. 857 ± 169 days). Thus, the lit/lit mice provided evidence that defects solely in the GH/IGF-1 axis were sufficient to increase longevity.

GH Receptor/Binding Protein Knockout (GHR$^{-/-}$) Mice

Mice with a disruption of the GH receptor/binding protein gene (GHR$^{-/-}$) were first described by Zhou et al. (1997). Because the GHR is absent, these mice are dwarf and GH-resistant or -insensitive and are characterized by elevated levels of GH yet markedly reduced levels of IGF-1 (Coschigano et al. 2003). Other notable features of these mice include delayed puberty (Keene et al. 2002), decreased fasting insulin and glucose levels (Coschigano et al. 2003; Liu et al. 2004), and diminished pancreatic islet size (Liu et al. 2004). The GHR$^{-/-}$ mice also are protected from both diabetes-induced nephropathy (Bellush et al. 2000) and prostate and mammary carcinogenesis (Wang et al. 2005; Zhang et al. 2006). Although longevity is increased in GHR$^{-/-}$ mice, the increase again varies according to background strain. In 2000, Coschigano et al. (2002) reported that GHR$^{-/-}$ mice maintained on a mixed genetic background had an increase in mean life span of 55% in males to 38% in females. When the GHR$^{-/-}$ mice were backcrossed into the C57BL/6J strain, life span was still significantly increased, but to a lesser degree, with an average increase of 26% and 16% in males and females, respectively (Coschigano et al. 2003). As with the lit/lit mouse, longevity data in GHR$^{-/-}$ mice implicate the GH/IGF and not other pituitary hormones with aging.

Mice with Defects in IGF-1, IGF-1 Receptor, or IGF-1 Signaling

One common feature of the mouse models mentioned thus far with increased longevity is a very low level of circulating IGF-1. So what happens to mice that are lacking in either IGF-1 or its receptor? Unfortunately, mice with a gene disruption in either IGF-1 or the IGF-1 receptor die at birth or shortly after, hampering the ability to study their role on longevity (Liu et al. 1993). However, mice heterozygous for the IGF-1 receptor gene disruption (IGF-IR$^{+/-}$), which have a 50% reduction in receptor levels, do show a marginal increase in life span for females, exhibiting a 33% significant increase (Holzenberger 2004). Male mice did not show a statistically significant increase. Interestingly, the female mice have an increased life span and are not dwarf, unlike the previous models discussed, suggesting that dwarfism is not a requirement for the extended life span in these models. It is important to note that an increase in life span, although observed for female IGF-IR$^{+/-}$ mice, was much less pronounced than that observed in GHR$^{-/-}$ mice, suggesting the intriguing possibility that GH-dependent IGF-1 signaling may be highly relevant for a robust increase in longevity.

Another protein related to IGF-1 receptor function is a specific protein isoform, p66[shc], of the proto-oncogene *shc* locus. Shc proteins mediate growth factor mitogenic actions by activation of the MAPK pathway. Ligand binding and subsequent activation of IGF-1R, as well as other growth factor receptors, results in recruitment of p66[shc] to the receptors, where it is phosphorylated and activated. The p66 isoform has recently been suggested to regulate mitochondrial oxidative capacity, with its absence increasing the reliance on glycolysis as opposed to oxidative metabolism (Nemoto et al. 2006). Mice lacking this IGF-1R substrate, p66[shc] –/–, live 30% longer than their normal littermates and have enhanced resistance to oxidative stress (Migliaccio et al. 1999). Interestingly, p66[shc] –/– mice were also recently shown to be protected against age-related endothelial dysfunction (Francia et al. 2004), angiotensin-II-induced myocardial damage (Graiani et al. 2005), some behavioral aspects of aging (Berry et al. 2007), and diabetic glomerulopathy (Menini et al. 2006), all presumably by reducing oxidative stress.

Klotho Transgenic Mice

The Klotho protein is expressed principally in the distal tubules of the kidney and the choroid plexus in the brain (Kuro-o et al. 1997; Koh et al. 2001), but there is also evidence that the protein functions as humoral factor. Specifically, the Klotho protein appears to bind and repress intracellular signals of insulin and IGF-1 (Kurosu et al. 2005). Interest in this protein as an aging agent initially came from mice homozygous for a mutation in *Klotho* that exhibited accelerated aging and premature death (Kuro-o et al. 1997). More recently, overexpression of *klotho* in mice was shown to extend life span, with mice carrying the transgene outliving wild-type controls by 20–31% in males and by 19% in females (Kurosu et al. 2005). In contrast to previously discussed insulin-sensitive mouse models with increased life span, the *klotho* transgenic mice are hyperinsulinemic and insulin-resistant.

A MOUSE WITH A REPRESSED GH/IGF-1 AXIS BUT WITHOUT INCREASED LONGEVITY

Not all mouse models with reduced GH or IGF-1 levels show improvements in life span. One example is mice that express a transgene for a growth hormone antagonist (GHA). The GHA described by Chen et al. (1991a,b) competes with endogenous GH for GHR binding and results in a marked reduction, but not elimination, of GH-induced intracellular signaling. As a result, GHA transgenic mice are dwarf and have reduced levels of IGF-1. However, unlike many other models discussed, the GHA transgenic mice are reported to have no extension in life span as compared to littermate controls (Coschigano et al. 2003). This model offers an interesting exception to the models previously mentioned. That is, although the mice are dwarf due to a partial inhibition of GH action and concomitant decrease in IGF-1, presumably the decrease in the GH/IGF-1 axis is not sufficient to increase life span. Several key phenotypic differences besides life span were reported between GHA and GHR$^{-/-}$ mice that may offer insight as to the key features relevant to improved life span. Although they have circulating levels of IGF-1 that are significantly lower than control littermates, the reduction is much less pronounced than in other models such as the GHR$^{-/-}$ mice (80% decrease in GHR$^{-/-}$, whereas GHA have only a 20–25% decrease) (Coschigano et al. 2003). Another striking difference between GHR$^{-/-}$ and GHA mice is the weight-gain profile. The weights of GHA transgenic mice eventually approach that of control littermates, whereas GHR$^{-/-}$ remain substantially dwarf even in older mice (Coschigano et al. 2003).

EXCESS IN GH/IGF-1 SIGNALING RESULTS IN REDUCED LONGEVITY

Further evidence that the GH/IGF-1 axis is an important contributor to life span is offered by mice with an excess in GH and IGF-1 signaling. Several labs have developed transgenic mice that express a GH transgene. The excess GH levels in these mice result in a fairly uniform phenotype regardless of genetic background. That is, GH transgenic mice are larger in size and have reduced adiposity and advanced puberty (for review, see Bartke 2003). As would be expected based on the excess in GH signaling, these mice also have several endocrine abnormalities, with IGF-1 and insulin levels being elevated despite normoglycemia. Importantly, these mice are hyperinsulinemic and insulin-resistant (J. Kim et al., in prep.). Thus, in many ways, these mice are phenotypically opposite to the mice just described with repressed GH/IGF-1 signaling. Most importantly, life span of these GH transgenic mice is drastically reduced (~50%) compared to nontransgenic controls (Wolf et al. 1993; Bartke et al. 1998, 2002; Berryman et al. 2004), emphasizing the pivotal role of this axis in aging. Although not proven, it is speculated that the cause of the reduced life span in GH transgenic mice may be hypertension (Bohlooly et al. 2001), lipid abnormalities (Frick et al. 2001), vascular alterations (Bohlooly et al. 2001; Andersson et al. 2006a,b), cardiovascular deficits (Bollano et al. 2000; Fu et al. 2000), kidney damage (Doi et al. 1991; Peten and Striker 1994), and reduced antioxidant defense (Hauck and Bartke 2001).

MECHANISMS OF DELAYED AGING

Distinct from Caloric Restriction?

Caloric restriction (CR) is recognized as one of the most consistent and effective means to delay aging. Interestingly, many phenotypic character-istics of animals subjected to CR are shared by most of the mouse mod-els just described with disruption in GH and/or IGF-1 activity. That is, CR results in reduced body weight, reduced GH and IGF-1 levels, and decreased plasma levels of insulin and glucose, as has been observed in Ames, Snell, lit/lit, and $GHR^{-/-}$ dwarf mice. Other similarities, in most but not all models, include reduced fertility, delayed puberty, and reduced levels of thyroid hormones. Based on these similarities, it is feasible that CR and reductions in the GH/IGF-1 axis may increase life span through similar processes. Thus, one possible reason for the increased life span in the mouse models described could be voluntary food restriction in the mouse lines with altered GH/IGF-1 signaling. However, controlling food intake seems unlikely, as increased life span is noted in $GHR^{-/-}$ mice and Ames dwarf mice, both of which eat more calories than littermate con-trols when food consumption is normalized to body weight (Coschigano et al. 2003), although absorptive capacity in these models has not been assessed. Furthermore, IGF receptor heterozygotes display increased life span, but are not dwarf and have similar calorie consumption as nonmu-tant siblings (Holzenberger 2004).

To directly address whether CR and reduction in the GH/IGF-1 axis are working via distinct or overlapping mechanisms to increase longevity, several dwarf models have been exposed to various regimens of CR. In 2001, the long-lived Ames dwarf mice were subjected to CR (Bartke et al. 2001). In this study, the Ames dwarf mice experienced further extension of average and maximum longevity, suggesting that dwarfism and CR do work through distinct mechanisms. The difference is highlighted by sev-eral documented differences in expression of genes related to insulin sig-naling between normal and Ames dwarf mice subjected to 30% CR (Masternak et al. 2004). Interestingly, results with another dwarf mouse model, the $GHR^{-/-}$ mice, were quite different. A 30% CR did not increase average longevity in either gender of GHR$^{-/-}$ mice and had only a slight impact on maximum longevity in GHR$^{-/-}$ females (Bonkowski et al. 2006). Although this may suggest that GH resistance/insensitivity and CR work via similar mechanisms in this mouse model, gene expression profiling still suggests some distinctions. For example, hepatic mRNA levels of IGF-1 are severely reduced, whereas levels of numerous other genes (AKT1, AKT2, insulin receptor, and insulin receptor substrate-1) are significantly elevated in GHR$^{-/-}$ liver, and CR does not alter expression of these liver genes in comparison to normal mice (Miller et al. 2002b; Al-Regaiey et al. 2005; Masternak et al. 2005). Numerous other examples of differences in gene expression in other tissues have been reported (Miller et al. 2002b; Argentino et al. 2005a,b; Masternak et al. 2006), suggesting that GHR$^{-/-}$ mice are not mere mimetics of CR animals.

Other Potential Mechanisms

Potential mechanisms that link the GH/IGF-1 axis with the control of longevity can be identified by comparing and contrasting the phenotypes of long-lived mutants with CR mice as well as with GHA mice, which experience a relatively mild reduction in IGF-1 levels and no increase in longevity (Table 1). Although commonalities among these mouse lines may offer insight into the mechanisms responsible for increased life span, it is also possible that the combination of different traits unique to each mouse line is most relevant for life span extension. Thus, understanding the mechanism responsible for increased longevity will only be possible once all of these mice have been more fully characterized. Although not all mouse models mentioned have been subjected to thorough characterization, several common traits emerge as potential mechanisms and are worthy of further exploration. These traits include reduced body size, reduced insulin levels along with improved insulin sensitivity, improved oxidative stress resistance, and reduced incidence of cancer.

Reduced Body Size

There is considerable evidence in rodents (Miller et al. 2002a; Rollo 2002), dogs (Galis et al. 2007), and humans (Samaras et al. 2003) that body size is negatively correlated with longevity. Indeed, several mouse models with increased life span (Snell and Ames dwarf, lit/lit mice, GHR$^{-/-}$ mice) are notably smaller than normal siblings. CR also results in smaller animals if diet manipulation is initiated early in life. However, smaller body size does not ensure increased longevity, as demonstrated by GHA mice, and it is not required for increased longevity, as Klotho transgenic mice, p66^{shc}, and IGF-IR$^{+/-}$ female mice display no reduction in body size.

Reduced Insulin Levels along with Improved Insulin Sensitivity

Insulin resistance is well documented to be a risk factor for a number of major diseases, such as cardiovascular disease and type 2 diabetes mellitus. Moreover, plasma levels of insulin are inversely correlated with survival in humans (Roth et al. 2002), whereas centenarians are reported to have excellent sensitivity to insulin (Paolisso et al. 1996). Therefore, it is reasonable to hypothesize that the reduction in insulin release and improved insulin sensitivity could be, in part, responsible for increased longevity. GHR$^{-/-}$ mice, other long-lived dwarf mice, and CR mice have reduced plasma insulin levels and improved insulin sensitivity (Dominici et al. 2002; Liu et al. 2004). Because CR GHR$^{-/-}$ mice do not show a further enhancement in longevity, the extreme insulin sensitivity found in these mice may be at an ultimate level for improved life span.

Klotho transgenic mice, however, exhibit increased insulin resistance despite an improvement in longevity (Kurosu et al. 2005). This outlier is thoroughly discussed in a recent review (Bartke 2006) in which it is suggested that Klotho transgenic and other dwarf mice still share a reduction in the signaling downstream from IGF-1 and insulin receptors, albeit by different mechanisms. Given the established cross-talk between the GH/IGF-1 axis and insulin, reduced insulin signaling remains a viable mechanism for the increase in longevity observed with the animals described here.

Improved Oxidative Stress Resistance

Reactive oxygen species (ROS) and free radicals are involved in a variety of physiological and pathological processes, including degenerative diseases. Accumulation of oxidative damage to DNA, proteins, and membrane lipids has also been implicated in aging and occurs due, in part, to an imbalance between generation rates of ROS and the activity and amount of antioxidant defense systems. There is evidence that several of the mouse models with reduced activity in the GH/IGF-1 axis have an improved resistance to oxidative stressors either by improved antioxidant defense or through reduced production of ROS. The p66^{shc} –/– mice display increased resistance to the oxidative stress induced by paraquat, which generates superoxide anions upon cellular intake, and are characterized by a decreased incidence of aging-associated diseases by reportedly reducing oxidative stress (Francia et al. 2004; Graiani et al. 2005; Menini et al. 2006; Berry et al. 2007). Other models also suggest increased tolerance toward oxidative stress in mouse models with reduced activity of the GH/IGF-1 axis. For example, Ames dwarf mice have reduced lipid, protein, and DNA oxidation in brain in comparison to normal siblings (Brown-Borg et al. 2001; Sanz et al. 2002) and increased antioxidant defense in skeletal muscle (Romanick et al. 2004). Likewise, fibroblasts derived from Snell dwarfs are resistant to a variety of cellular stressors, including UV light, heat, heavy metal exposure, and paraquat (Murakami et al. 2003). Finally, IGF-1R$^{+/-}$ mice live longer than IGF$^{+/+}$ mice when both are challenged with paraquat (Holzenberger et al. 2003). Data for resistance to oxidative stress are less consistent for GHR$^{-/-}$ mice, with these mice showing increase, decrease, or no change in oxidative measures (Hauck et al. 2002; Wolf et al. 2005; Beyea et al. 2006).

Reduced Incidence of Cancer

Available data support that GH/IGF-1 status may influence neoplastic tissue growth. Specifically, IGF-1 exerts potent effects on key stages of cancer development, having been implicated in many types of cancer (for review, see Bustin and Jenkins 2001; Ciampolillo et al. 2005; Larsson et al. 2005). Thus, it might be expected that reduction in the GH/IGF-1 axis would result in a reduction of cancer incidence. Indeed, GHR$^{-/-}$ mice have protection from prostate carcinogenesis (Wang et al. 2005) and estrogen-independent mammary carcinogenesis (Zhang et al. 2006). Likewise, Ames dwarf mice are reported to have delayed occurrence of total neoplastic lesions and reduced incidence of adenocarcinoma in lung (Ikeno et al. 2003). This reduction in cancer incidence is also shared with calorie-restricted rodents (Weindruch et al. 1988). GHA mice also share some reduction in, at least, mammary cancer incidence (Pollak et al. 2001). However, the GHA mice do not have an increase in longevity, indicating that the reduction in cancer incidence may not be the sole factor important in influencing longevity.

SUMMARY

The common feature shared among the long-lived mice described in this review is a decrease in IGF-I activity either through decreasing GH action, by increasing IGF-1 resistance (IGF-1R$^{+/-}$ mice), or by inhibiting IGF-1 signaling (Klotho transgenics). IGF-1 is a potent anabolic hormone that increases cellular metabolism and proliferation, enhances the function of numerous tissues, and participates in glucose homeostasis. The similarities in the insulin/IGF-1 and homologous regulatory systems between invertebrate and mammal models with increased longevity suggest that a fundamental mechanism of aging is likely evolutionarily conserved. The challenge for the future will be to unravel the molecular link or links that tie these models together.

ACKNOWLEDGMENTS

J.J.K. is supported in part by the State of Ohio's Eminent Scholars Program which includes a gift from Milton and Lawrence Goll, by DiAthegen LLC, by U.S. Army MRMC #W81XWH-04-1-0201, by National Institutes of Health grant R01 CA099904, by a World Anti-Doping (WADA) grant, and by a National Institute on Aging (NIA) grant AG19899. D.E.B. is supported by funds from the Diabetes Research Initiative of the Appalachian Rural Health Institute at Ohio University and by funds from the National Institute of Diabetes and Digestive and Kidney Diseases (grant K01-DK064905). A.B. is supported by NIA grants AG 19899 and U19 AG023122, by the Ellison Medical Foundation, and by the Southern Illinois University Geriatrics Medicine and Research Initiative.

REFERENCES

Al-Regaiey K.A., Masternak M.M., Bonkowski M., Sun L., and Bartke A. 2005. Long-lived growth hormone receptor knockout mice: Interaction of reduced insulin-like growth factor i/insulin signaling and caloric restriction. *Endocrinology* **146:** 851–860.
Amato G., Mazziotti G., Di Somma C., Lalli E., De Felice G., Conte M., Rotondi M., Pietrosante M., Lombardi G., Bellastella A., et al. 2000. Recombinant growth hormone (GH) therapy in GH-deficient adults: A long-term controlled study on daily versus thrice weekly injections. *J. Clin. Endocrinol. Metab.* **85:** 3720–3725.
Andersen B., Pearse R.V., II, Jenne K., Sornson M., Lin S.C., Bartke A., and Rosenfeld M.G. 1995. The Ames dwarf gene is required for Pit-1 gene activation. *Dev. Biol.* **172:** 495–503.
Andersson I.J., Ljungberg A., Svensson L., Gan L.M., Oscarsson J., and Bergstrom G. 2006a. Increased atherosclerotic lesion area in apoE deficient mice overexpressing bovine growth hormone. *Atherosclerosis* **188:** 331–340.
Andersson I.J., Johansson M.E., Wickman A., Bohlooly Y.M., Klintland N., Caidahl K., Gustafsson M., Boren J., Gan L.M., and Bergstrom G. 2006b. Endothelial dysfunction in growth hormone transgenic mice. *Clin. Sci.* **110:** 217–225.

Argentino D.P., Dominici F.P., Al-Regaiey K., Bonkowski M.S., Bartke A., and Turyn D. 2005a. Effects of long-term caloric restriction on early steps of the insulin-signaling system in mouse skeletal muscle. *J. Gerontol. A Biol. Sci. Med. Sci.* **60:** 28–34.

Argentino D.P., Dominici F.P., Munoz M.C., Al-Regaiey K., Bartke A., and Turyn D. 2005b. Effects of long-term caloric restriction on glucose homeostasis and on the first steps of the insulin signaling system in skeletal muscle of normal and Ames dwarf (Prop1df/Prop1df) mice. *Exp. Gerontol.* **40:** 27–35.

Bartke A. 2003. Can growth hormone (GH) accelerate aging? Evidence from GH-transgenic mice. *Neuroendocrinology* **78:** 210–216.

———. 2006. Long-lived Klotho mice: New insights into the roles of IGF-1 and insulin in aging. *Trends Endocrinol. Metab.* **17:** 33–35.

Bartke A., Chandrashekar V., Bailey B., Zaczek D., and Turyn D. 2002. Consequences of growth hormone (GH) overexpression and GH resistance. *Neuropeptides* **36:** 201–208.

Bartke A., Brown-Borg H.M., Bode A.M., Carlson J., Hunter W.S., and Bronson R.T. 1998. Does growth hormone prevent or accelerate aging? *Exp. Gerontol.* **33:** 675–687.

Bartke A., Wright J.C., Mattison J.A., Ingram D.K., Miller R.A., and Roth G.S. 2001. Extending the lifespan of long-lived mice. *Nature* **414:** 412.

Bartke A., Peluso M.R., Moretz N., Wright C., Bonkowski M., Winters T.A., Shanahan M.F., Kopchick J.J., and Banz W.J. 2004. Effects of soy-derived diets on plasma and liver lipids, glucose tolerance, and longevity in normal, long-lived and short-lived mice. *Horm. Metab. Res.* **36:** 550–558.

Bellush L.L., Doublier S., Holland A.N., Striker L.J., Striker G.E., and Kopchick J.J. 2000. Protection against diabetes-induced nephropathy in growth hormone receptor/binding protein gene-disrupted mice. *Endocrinology* **141:** 163–168.

Berry A., Capone F., Giorgio M., Pelicci P.G., de Kloet E.R., Alleva E., Minghetti L., and Cirulli F. 2007. Deletion of the life span determinant p66(Shc) prevents age-dependent increases in emotionality and pain sensitivity in mice. *Exp. Gerontol.* **42:** 37–45.

Berryman D.E., List E.O., Coschigano K.T., Behar K., Kim J.K., and Kopchick J.J. 2004. Comparing adiposity profiles in three mouse models with altered GH signaling. *Growth Horm. IGF Res.* **14:** 309–318.

Berryman D.E., List E.O., Kohn D.T., Coschigano K.T., Seeley R.J., and Kopchick J.J. 2006. Effect of growth hormone on susceptibility to diet-induced obesity. *Endocrinology* **147:** 2801–2808.

Beyea J.A., Sawicki G., Olson D.M., List E., Kopchick J.J., and Harvey S. 2006. Growth hormone (GH) receptor knockout mice reveal actions of GH in lung development. *Proteomics* **6:** 341–348.

Blackman M.R., Sorkin J.D., Munzer T., Bellantoni M.F., Busby-Whitehead J., Stevens T.E., Jayme J., O'Connor K.G., Christmas C., Tobin J.D., et al. 2002. Growth hormone and sex steroid administration in healthy aged women and men: A randomized controlled trial. *J. Am. Med. Assoc.* **288:** 2282–2292.

Boguszewski C.L., Meister L.H., Zaninelli D.C., and Radominski R.B. 2005. One year of GH replacement therapy with a fixed low-dose regimen improves body composition, bone mineral density and lipid profile of GH-deficient adults. *Eur. J. Endocrinol.* **152:** 67–75.

Bohlooly Y.M., Carlson L., Olsson B., Gustafsson H., Andersson I.J., Tornell J., and Bergstrom G. 2001. Vascular function and blood pressure in GH transgenic mice. *Endocrinology* **142:** 3317–3323.

Bollano E., Omerovic E., Bohlooly-y M., Kujacic V., Madhu B., Tornell J., Isaksson O., Soussi B., Schulze W., Fu M.L., et al. 2000. Impairment of cardiac function and bioenergetics in adult transgenic mice overexpressing the bovine growth hormone gene. *Endocrinology* **141:** 2229–2235.

Bonkowski M.S., Rocha J.S., Masternak M.M., Al Regaiey K.A., and Bartke A. 2006. Targeted disruption of growth hormone receptor interferes with the beneficial actions of calorie restriction. *Proc. Natl. Acad. Sci.* **103:** 7901–7905.

Brown-Borg H.M., Borg K.E., Meliska C.J., and Bartke A. 1996. Dwarf mice and the ageing process. *Nature* **384:** 33.

Brown-Borg H.M., Johnson W.T., Rakoczy S., and Romanick M. 2001. Mitochondrial oxidant generation and oxidative damage in Ames dwarf and GH transgenic mice. *J. Am. Aging Assoc.* **24:** 85–100.

Bustin S.A. and Jenkins P.J. 2001. The growth hormone-insulin-like growth factor-I axis and colorectal cancer. *Trends Mol. Med.* **7:** 447–454.

Cella S.G., Luceri M., Cattaneo L., Torsello A., and Muller E.E. 1996. Somatostatin withdrawal as generator of pulsatile GH release in the dog: A possible tool to evaluate the endogenous GHRH tone? *Neuroendocrinology* **63:** 481–488.

Chen W.Y., White M.E., Wagner T.E., and Kopchick J.J. 1991a. Functional antagonism between endogenous mouse growth hormone (GH) and a GH analog results in dwarf transgenic mice. *Endocrinology* **129:** 1402–1408.

Chen W.Y., Wight D.C., Mehta B.V., Wagner T.E., and Kopchick J.J. 1991b. Glycine 119 of bovine growth hormone is critical for growth-promoting activity. *Mol. Endocrinol.* **5:** 1845–1852.

Ciampolillo A., De Tullio C., and Giorgino F. 2005. The IGF-I/IGF-I receptor pathway: Implications in the pathophysiology of thyroid cancer. *Curr. Med. Chem.* **12:** 2881–2891.

Clancy D.J., Gems D., Harshman L.G., Oldham S., Stocker H., Hafen E., Leevers S.J., and Partridge L. 2001. Extension of life-span by loss of CHICO, a *Drosophila* insulin receptor substrate protein. *Science* **292:** 104–106.

Cohen P. 2006. Overview of the IGF-I system. *Horm. Res.* (suppl. 1) **65:** 3–8.

Coschigano K.T., Clemmons D., Bellush L.L., and Kopchick J.J. 2000. Assessment of growth parameters and life span of GHR/BP gene-disrupted mice. *Endocrinology* **141:** 2608–2613.

Coschigano K.T., Holland A.N., Riders M.E., List E.O., Flyvbjerg A., and Kopchick J.J. 2003. Deletion, but not antagonism, of the mouse growth hormone receptor results in severely decreased body weights, insulin, and insulin-like growth factor I levels and increased life span. *Endocrinology* **144:** 3799–3810.

Crew M.D., Spindler S.R., Walford R.L., and Koizumi A. 1987. Age-related decrease of growth hormone and prolactin gene expression in the mouse pituitary. *Endocrinology* **121:** 1251–1255.

Davidson M.B. 1987. Effect of growth hormone on carbohydrate and lipid metabolism. *Endocr. Rev.* **8:** 115–131.

Doi T., Striker L.J., Kimata K., Peten E.P., Yamada Y., and Striker G.E. 1991. Glomerulosclerosis in mice transgenic for growth hormone. Increased mesangial extracellular matrix is correlated with kidney mRNA levels. *J. Exp. Med.* **173:** 1287–1290.

Dominici F.P., Hauck S., Argentino D.P., Bartke A., and Turyn D. 2002. Increased insulin sensitivity and upregulation of insulin receptor, insulin receptor substrate (IRS)-1 and IRS-2 in liver of Ames dwarf mice. *J. Endocrinol.* **173:** 81–94.

Donahue L.R. and Beamer W.G. 1993. Growth hormone deficiency in 'little' mice results in aberrant body composition, reduced insulin-like growth factor-I and insulin-like growth factor-binding protein-3 (IGFBP-3), but does not affect IGFBP-2, -1 or -4. *J. Endocrinol.* **136:** 91–104.

Florakis D., Hung V., Kaltsas G., Coyte D., Jenkins P.J., Chew S.L., Grossman A.B., Besser G.M., and Monson J.P. 2000. Sustained reduction in circulating cholesterol in adult hypopituitary patients given low dose titrated growth hormone replacement therapy: A two year study. *Clin. Endocrinol.* **53:** 453–459.

Flurkey K., Papaconstantinou J., and Harrison D.E. 2002. The Snell dwarf mutation Pit1(dw) can increase life span in mice. *Mech. Ageing. Dev.* **123:** 121–130.

Flurkey K., Papaconstantinou J., Miller R.A., and Harrison D.E. 2001. Lifespan extension and delayed immune and collagen aging in mutant mice with defects in growth hormone production. *Proc. Natl. Acad. Sci.* **98:** 6736–6741.

Francia P., delli Gatti C., Bachschmid M., Martin-Padura I., Savoia C., Migliaccio E., Pelicci P.G., Schiavoni M., Luscher T.F., Volpe M., and Cosentino F. 2004. Deletion of p66shc gene protects against age-related endothelial dysfunction. *Circulation* **110:** 2889–2895.

Frick F., Bohlooly Y.M., Linden D., Olsson B., Tornell J., Eden S., and Oscarsson J. 2001. Long-term growth hormone excess induces marked alterations in lipoprotein metabolism in mice. *Am. J. Physiol. Endocrinol. Metab.* **281:** E1230–E1239.

Fu M.L., Tornell J., Schulze W., Hoebeke J., Isaksson O.G., Sandstedt J., and Hjalmarson A. 2000. Myocardial hypertrophy in transgenic mice overexpressing the bovine growth hormone (bGH) gene. *J. Intern. Med.* **247:** 546–552.

Galis F., Van Der Sluijs I., Van Dooren T.J., Metz J.A., and Nussbaumer M. 2007. Do large dogs die young? *J. Exp. Zool. B Mol. Dev. Evol.* **3088:** 119–126.

Giustina A. and Veldhuis J.D. 1998. Pathophysiology of the neuroregulation of growth hormone secretion in experimental animals and the human. *Endocr. Rev.* **19:** 717–797.

Godfrey P., Rahal J.O., Beamer W.G., Copeland N.G., Jenkins N.A., and Mayo K.E. 1993. GHRH receptor of little mice contains a missense mutation in the extracellular domain that disrupts receptor function. *Nat. Genet.* **4:** 227–232.

Graiani G., Lagrasta C., Migliaccio E., Spillmann F., Meloni M., Madeddu P., Quaini F., Padura I.M., Lanfrancone L., Pelicci P., and Emanueli C. 2005. Genetic deletion of the p66Shc adaptor protein protects from angiotensin II-induced myocardial damage. *Hypertension* **46:** 433–440.

Hauck S.J. and Bartke A. 2001. Free radical defenses in the liver and kidney of human growth hormone transgenic mice: Possible mechanisms of early mortality. *J. Gerontol. A Biol. Sci. Med. Sci.* **56:** B153–B162.

Hauck S.J., Aaron J.M., Wright C., Kopchick J.J., and Bartke A. 2002. Antioxidant enzymes, free-radical damage, and response to paraquat in liver and kidney of long-living growth hormone receptor/binding protein gene-disrupted mice. *Horm. Metab. Res.* **34:** 481–486.

Holzenberger M. 2004. The GH/IGF-I axis and longevity. *Eur. J. Endocrinol.* (suppl. 1) **151:** S23–S27.

Holzenberger M., Dupont J., Ducos B., Leneuve P., Geloen A., Even P.C., Cervera P., and Le Bouc Y. 2003. IGF-1 receptor regulates lifespan and resistance to oxidative stress in mice. *Nature* **421:** 182–187.

Ikeno Y., Bronson R.T., Hubbard G.B., Lee S., and Bartke A. 2003. Delayed occurrence of fatal neoplastic diseases in ames dwarf mice: Correlation to extended longevity. *J. Gerontol. A Biol. Sci. Med. Sci.* **58:** 291–296.

Kasukawa Y., Baylink D.J., Guo R., and Mohan S. 2003. Evidence that sensitivity to growth hormone (GH) is growth period and tissue type dependent: Studies in GH-deficient lit/lit mice. *Endocrinology* **144:** 3950–3957.

Keene D.E., Suescun M.O., Bostwick M.G., Chandrashekar V., Bartke A., and Kopchick J.J. 2002. Puberty is delayed in male growth hormone receptor gene-disrupted mice. *J. Androl.* **23:** 661–668.

Koh N., Fujimori T., Nishiguchi S., Tamori A., Shiomi S., Nakatani T., Sugimura K., Kishimoto T., Kinoshita S., Kuroki T., and Nabeshima Y. 2001. Severely reduced production of klotho in human chronic renal failure kidney. *Biochem. Biophys. Res. Commun.* **280:** 1015–1020.

Kopchick J.J. and Andry J.M. 2000. Growth hormone (GH), GH receptor, and signal transduction. *Mol. Genet. Metab.* **71:** 293–314.

Kuro-o M., Matsumura Y., Aizawa H., Kawaguchi H., Suga T., Utsugi T., Ohyama Y., Kurabayashi M., Kaname T., Kume E., et al. 1997. Mutation of the mouse klotho gene leads to a syndrome resembling ageing. *Nature* **390:** 45–51.

Kurosu H., Yamamoto M., Clark J.D., Pastor J.V., Nandi A., Gurnani P., McGuinness O.P., Chikuda H., Yamaguchi M., Kawaguchi H., et al. 2005. Suppression of aging in mice by the hormone Klotho. *Science* **309:** 1829–1833.

Larsson O., Girnita A., and Girnita L. 2005. Role of insulin-like growth factor 1 receptor signalling in cancer. *Br. J. Cancer* **92:** 2097–2101.

Li S., Crenshaw E.B., III, Rawson E.J., Simmons D.M., Swanson L.W., and Rosenfeld M.G. 1990. Dwarf locus mutants lacking three pituitary cell types result from mutations in the POU-domain gene pit-1. *Nature* **347:** 528–533.

Liang H., Masoro E.J., Nelson J.F., Strong R., McMahan C.A., and Richardson A. 2003. Genetic mouse models of extended lifespan. *Exp. Gerontol.* **38:** 1353–1364.

Lin-Su K. and Wajnrajch M.P. 2002. Growth hormone releasing hormone (GHRH) and the GHRH receptor. *Rev. Endocr. Metab. Disord.* **3:** 313–323.

Liu H., Bravata D.M., Olkin I., Nayak S., Roberts B., Garber A.M., and Hoffman A.R. 2007. Systematic review: The safety and efficacy of growth hormone in the healthy elderly. *Ann. Intern. Med.* **146:** 104–115.

Liu J.L., Coschigano K.T., Robertson K., Lipsett M., Guo Y., Kopchick J.J., Kumar U., and Liu Y.L. 2004. Disruption of growth hormone receptor gene causes diminished pancreatic islet size and increased insulin sensitivity in mice. *Am. J. Physiol. Endocrinol. Metab.* **287:** E405–E413.

Liu J.P., Baker J., Perkins A.S., Robertson E.J., and Efstratiadis A. 1993. Mice carrying null mutations of the genes encoding insulin-like growth factor I (Igf-1) and type 1 IGF receptor (Igf1r). *Cell* **75:** 59–72.

Masternak M.M., Al-Regaiey K.A., Del Rosario Lim M.M., Jimenez-Ortega V., Panici J.A., Bonkowski M.S., and Bartke A. 2005. Effects of caloric restriction on insulin pathway gene expression in the skeletal muscle and liver of normal and long-lived GHR-KO mice. *Exp. Gerontol.* **40:** 679–684.

Masternak M.M., Al-Regaiey K., Bonkowski M.S., Panici J., Sun L., Wang J., Przybylski G.K., and Bartke A. 2004. Divergent effects of caloric restriction on gene expression in normal and long-lived mice. *J. Gerontol. A Biol. Sci. Med. Sci.* **59:** 784–788.

Masternak M. M., Al-Regaiey K.A., Del Rosario Lim M.M., Jimenez-Ortega V., Panici J.A., Bonkowski M.S., Kopchick J.J., Wang Z., and Bartke A. 2006. Caloric restriction and growth hormone receptor knockout: Effects on expression of genes involved in insulin action in the heart. *Exp. Gerontol.* **41:** 417–429.

Menini S., Amadio L., Oddi G., Ricci C., Pesce C., Pugliese F., Giorgio M., Migliaccio E., Pelicci P., Iacobini C., and Pugliese C. 2006. Deletion of p66Shc longevity gene protects against experimental diabetic glomerulopathy by preventing diabetes-induced oxidative stress. *Diabetes* **55:** 1642–1650.

Migliaccio E., Giorgio M., Mele S., Pelicci G., Reboldi P., Pandolfi P.P., Lanfrancone L., and Pelicci P.G. 1999. The p66shc adaptor protein controls oxidative stress response and life span in mammals. *Nature* **402:** 309–313.

Miller R.A., Harper J.M., Galecki A., and Burke D.T. 2002a. Big mice die young: Early life body weight predicts longevity in genetically heterogeneous mice. *Aging Cell* **1:** 22–29.

Miller R.A., Chang Y., Galecki A.T., Al-Regaiey K., Kopchick J.J., and Bartke A. 2002b. Gene expression patterns in calorically restricted mice: Partial overlap with long-lived mutant mice. *Mol. Endocrinol.* **16:** 2657–2666.

Muller E.E., Cella S.G., De Gennaro Colonna V., Parenti M., Cocchi D., and Locatelli V. 1993. Aspects of the neuroendocrine control of growth hormone secretion in ageing mammals. *J. Reprod. Fertil. Suppl.* **46:** 99–114.

Murakami S., Salmon A., and Miller R.A. 2003. Multiplex stress resistance in cells from long-lived dwarf mice. *FASEB J.* **17:** 1565–1566.

Nemoto S., Combs C.A., French S., Ahn B.H., Fergusson M.M., Balaban R.S., and Finkel T. 2006. The mammalian longevity-associated gene product p66shc regulates mitochondrial metabolism. *J. Biol. Chem.* **281:** 10555–10560.

Orme S.M., McNally R.J., Cartwright R.A., and Belchetz P.E. 1998. Mortality and cancer incidence in acromegaly: A retrospective cohort study (United Kingdom Acromegaly Study Group). *J. Clin. Endocrinol. Metab.* **83:** 2730–2734.

Paolisso G., Gambardella A., Ammendola S., D'Amore A., Balbi V., Varricchio M., and D'Onofrio F. 1996. Glucose tolerance and insulin action in healty centenarians. *Am. J. Physiol.* **270:** E890–E894.

Peten E.P. and Striker L.J. 1994. Progression of glomerular diseases. *J. Intern. Med.* **236:** 241–249.

Pincelli A.I., Bragato R., Scacchi M., Branzi G., Osculati G., Viarengo R., Leonetti G., and Cavagnini F. 2003. Three weekly injections (TWI) of low-dose growth hormone (GH) restore low normal circulating IGF-I concentrations and reverse cardiac abnormalities associated with adult onset GH deficiency (GHD). *J. Endocrinol. Invest.* **26:** 420–428.

Pollak M., Blouin M.J., Zhang J.C., and Kopchick J.J. 2001. Reduced mammary gland carcinogenesis in transgenic mice expressing a growth hormone antagonist. *Br. J. Cancer* **85:** 428–430.

Puche R.C., Alloatti R., and Chapo G. 2002. Growth and development of male "little" mice assessed with Parks' theory of feeding and growth. *Growth Dev. Aging* **66:** 71–78.

Rajasoorya C., Holdaway I.M., Wrightson P., Scott D.J., and Ibbertson H.K. 1994. Determinants of clinical outcome and survival in acromegaly. *Clin. Endocrinol.* **41:** 95–102.

Rollo C.D. 2002. Growth negatively impacts the life span of mammals. *Evol. Dev.* **4:** 55–61.

Romanick M.A., Rakoczy S.G., and Brown-Borg H.M. 2004. Long-lived Ames dwarf mouse exhibits increased antioxidant defense in skeletal muscle. *Mech. Ageing Dev.* **125:** 269–281.

Roth G.S., Lane M.A., Ingram D.K., Mattison J.A., Elahi D., Tobin J.D., Muller D., and Metter E.J. 2002. Biomarkers of caloric restriction may predict longevity in humans. *Science* **297:** 811.

Rudman D., Feller A.G., Nagraj H.S., Gergans G.A., Lalitha P.Y., Goldberg A.F., Schlenker R.A., Cohn L., Rudman I.W., and Mattson D.E. 1990. Effects of human growth hormone in men over 60 years old. *N. Engl. J. Med.* **323:** 1–6.

Samaras T.T., Elrick H., and Storms L.H. 2003. Is height related to longevity? *Life Sci.* **72:** 1781–1802.

Sanz A., Bartke A., and Barja G. 2002. Long-lived Ames dwarf mice: Oxidative damage to mitochondrial DNA in heart and brain. *J. Am. Aging Assoc.* **25:** 119–122.

Schaible R. and Gowen J.W. 1961. A new dwarf mouse. *Genetics* **46:** 896.

Sheppard M.C. 2005. GH and mortality in acromegaly. *J. Endocrinol. Invest.* **28:** 75–77.

Snell G.D. 1929. Dwarf, a new Mendelian recessive character of the house mouse. *Proc. Natl. Acad. Sci.* **15:** 733–734.

Sonntag W.E., Hylka V.W., and Meites J. 1985. Growth hormone restores protein synthesis in skeletal muscle of old male rats. *J. Gerontol.* **40:** 689–694.

Sonntag W.E., Steger R.W., Forman L.J., and Meites J. 1980. Decreased pulsatile release of growth hormone in old male rats. *Endocrinology* **107:** 1875–1879.

Takala J., Ruokonen E., Webster N.R., Nielsen M.S., Zandstra D.F., Vundelinckx G., and Hinds C.J. 1999. Increased mortality associated with growth hormone treatment in critically ill adults. *N. Engl. J. Med.* **341:** 785–792.

Tatar M., Bartke A., and Antebi A. 2003. The endocrine regulation of aging by insulin-like signals. *Science* **299:** 1346–1351.

Tatar M., Kopelman A., Epstein D., Tu M.P., Yin C.M., and Garofalo R.S. 2001. A mutant *Drosophila* insulin receptor homolog that extends life-span and impairs neuroendocrine function. *Science* **292:** 107–110.

Veldhuis J.D., Iranmanesh A., and Weltman A. 1997. Elements in the pathophysiology of diminished growth hormone (GH) secretion in aging humans. *Endocrine* **7:** 41–48.

Veldhuis J.D., Liem A.Y., South S., Weltman A., Weltman J., Clemmons D.A., Abbott R., Mulligan T., Johnson M.L., and Pincus S., et al. 1995. Differential impact of age, sex steroid hormones, and obesity on basal versus pulsatile growth hormone secretion in men as assessed in an ultrasensitive chemiluminescence assay. *J. Clin. Endocrinol. Metab.* **80:** 3209–3222.

Wang Z., Prins G.S., Coschigano K.T., Kopchick J.J., Green J.E., Ray V.H., Hedayat S., Christov K.T., Unterman T.G., and Swanson S.M. 2005. Disruption of growth hormone signaling retards early stages of prostate carcinogenesis in the C3(1)/T antigen mouse. *Endocrinology* **146:** 5188–5196.

Weindruch R., Naylor P.H., Goldstein A.L., and Walford R.L. 1988. Influences of aging and dietary restriction on serum thymosin alpha 1 levels in mice. *J. Gerontol.* **43:** B40–B42.

Weltman A., Weltman J.Y., Hartman M.L., Abbott R.D., Rogol A.D., Evans W.S., and Veldhuis J.D. 1994. Relationship between age, percentage body fat, fitness, and 24-hour growth hormone release in healthy young adults: Effects of gender. *J. Clin. Endocrinol. Metab.* **78:** 543–548.

Wolf E., Kahnt E., Ehrlein J., Hermanns W., Brem G., and Wanke R. 1993. Effects of long-term elevated serum levels of growth hormone on life expectancy of mice: Lessons from transgenic animal models. *Mech. Ageing Dev.* **68:** 71–87.

Wolf N., Penn P., Pendergrass W., Van Remmen H., Bartke A., Rabinovitch P., and Martin G.M. 2005. Age-related cataract progression in five mouse models for anti-oxidant protection or hormonal influence. *Exp. Eye Res.* **81:** 276–285.

Zhang X., Mehta R.G., Lantvit D.D., Coschigano K.T., Kopchick J.J., Green J.E., Hedayat S., Christov K.T., Ray V.H., Unterman T.G., and Swanson S.M. 2006. Inhibition of estrogen independent mammary carcinogenesis by disruption of growth hormone signaling. *Carcinogenesis* **28:** 143–150.

Zhou Y., Xu B.C., Maheshwari H.G., He L., Reed M., Lozykowski M., Okada S., Cataldo L., Coschigamo K., Wagner T.E., et al. 1997. A mammalian model for Laron syndrome produced by targeted disruption of the mouse growth hormone receptor/binding protein gene (the Laron mouse). *Proc. Natl. Acad. Sci.* **94:** 13215–13220.

14

Alzheimer's Disease: Genetics, Pathogenesis, Models, and Experimental Therapeutics

Philip C. Wong and Donald L. Price
Departments of Pathology, Neurology, Neuroscience,
and the Division of Neuropathology
The Johns Hopkins University School of Medicine
Baltimore, Maryland 21205-2196

Lars Bertram and Ronald E. Tanzi
Department of Neurology
Massachusetts General Hospital
Charlestown, Massachusetts 02129

ALZHEIMER'S DISEASE (AD), MANIFEST AS PROGRESSIVE LOSS of memory and cognitive impairments, affects more than 4 million individuals in the United States (Brookmeyer et al. 1998; Mayeux 2003; Cummings 2004; Wong et al. 2006). The index case, a middle-aged woman with behavioral disturbances and dementia, was described more than 100 years ago (Goedert and Spillantini 2006a; Hardy 2006a; Roberson and Mucke 2006; Small and Gandy 2006; Mandkelkow et al. 2007). Due to the postwar baby boom and increased life expectancy, the elderly are the most rapidly growing segment of our society and the number of persons with AD is predicted to triple over the next several decades. Prevalence, cost of care, impact on individuals and caregivers, and lack of mechanism-based treatments make AD one of the most challenging diseases of this new century (Price et al. 1998; Wong et al. 2002, 2006; Selkoe and Schenk 2003; Citron 2004a,b; Cummings 2004; Walsh and Selkoe 2004). This dementia syndrome results from dysfunction and death of neurons in specific brain regions/circuits, particularly those populations of neurons participating in memory and cognitive functions (Whitehouse et al. 1982; Hyman et al. 1984; Braak and Braak 1991, 1994; West et al. 1994, 2000, 2004; Price et al. 1998). The characteristic neuropathology of AD includes intracellular accumulations of phosphorylated Tau assembled in paired helical filaments (PHFs) within neurofibrillary tangles (NFTs) and abnormal neuritis, as well as extracellular Aβ peptide oligomers that, as aggregates, are at the core of neuritic amyloid plaques and represent sites of synaptic disconnection (Lee et al. 2001; Wong et al. 2002; Walsh and Selkoe 2004).

Age is a major risk factor for putative sporadic AD, and inheritance of mutations in several genes causes autosomal dominant familial AD (fAD). In fAD, mutant genes encoding the amyloid precursor protein (APP) or the presenilins (PS1 and PS2) influence the levels and/or character of Aβ peptides, which are generated via APP cleavages by the activities of β-amyloid cleaving enzyme1 (BACE1) and γ-secretase (the PS, Nct, Pen-2, Aph-1 multiprotein complex). Increases in *APP* gene dosage occur in families with *APP* duplications and in individuals with Down's syndrome (Trisomy 21) which have an extra copy of *APP*. Moreover, alleles of other genes, including ApoE4, are risks for putative sporadic disease (Price et al. 1998; Tanzi and Bertram 2001; Bertram and Tanzi 2005; Hardy 2006b).

Investigators have taken advantage of the knowledge of the disease to design symptomatic therapies for AD: The demonstration of abnormalities of basal forebrain neurons and cholinergic deficits in the cortex and hippocampus lead to the introduction of cholinesterase inhibitors for treatment, and the evidence of involvement in glutamatergic systems in hippocampal and cortical circuitry in AD, coupled with information about glutamate excitotoxicity (mediated by NMDA-R), lead to trials of NMDA-R (*N*-methyl-D-aspartate receptor) antagonists. Both of these strategies have modest symptomatic benefits in some patients.

Building on the identification of the Aβ peptide sequence by Dr. George Glenner and on genetic studies of the identification of *APP* and *PS* genes that encode proteins critical for APP processing, investigators have generated a variety of in vitro and in vivo models relevant to understanding pathogenic mechanisms. Particularly valuable are transgenic and knockout (KO) mice (McGowan et al. 2006) that, respectively, recapitulate some of the anatomical and biochemical pathologies of AD or alter the levels/activities of secretase enzymes critical in pathways leading to disease. For example, mice overexpressing mutant *APP/PS1* exhibit several age-associated abnormalities: elevated brain levels of Aβ42, the presence of synaptotoxic Aβ oligomers, neuritic plaques, and impairments in memory. To gain insights into potential therapeutic targets, genes encoding the proamyloidogenic secretases have been targeted. *BACE1*$^{-/-}$ mice are viable and do not produce Aβ. More significantly, concerning potential therapies, *APPswe;PS1ΔE9;BACE1⁻/⁻* mice do not form Aβ deposits or plaques nor do they show memory deficits. However, although inhibition of BACE1 is an attractive antiamyloidogenic treatment strategy, there are potential confounds associated with profound reductions in BACE1 activity. Similarly, although γ-secretase inhibition/modulation or Aβ immunotherapy—designed to reduce formation of Aβ42 or to enhance clearance, respectively—shows efficacy in mice, these manipulations are associated with adverse events. Nevertheless, studies of these models have greatly enhanced our understanding of amyloid-related disease mechanisms, led to identification of therapeutic targets, allowed testing of novel mechanism-based treatments, and provided awareness and insights regarding potential side effects. With this information, new disease mechanism-based strategies are now being developed to reduce production of Aβ, modify the nature (length) of Aβ peptides to shorter forms/less toxic species, reduce formation of oligomeric species, decrease the impact of toxic peptides, enhance clearance of Aβ, and attenuate aberrant conformations of Tau leading to NFTs.

In this chapter, we discuss the clinical syndromes of mild cognitive impairment (MCI) and AD, diagnostic tests including the great value of imaging and measures of biomarkers, and the neuropathology/biochemistry of the disease. We outline results of genetic approaches to identify genes conferring causation and risk. Subsequently, we describe how investigators have used transgenic and gene-targeted animals to create disease models (i.e., mice expressing mutant transgenes) and to identify therapeutic opportunities (targeting of genes encoding proteins implicated in disease pathways). These investigations of model systems have delineated the efficacies and potential toxicities of these manipulations. As beneficial outcomes and safety issues are defined, some of these therapeutic approaches are entering human trials. In concert with clinical, imaging, and biomarker studies of value for early diagnosis and for assessing outcomes of therapeutic trials, we believe that these new disease-modifying therapies will have major impacts on the health and care of the elderly.

CLINICAL FEATURES AND LABORATORY STUDIES

Many elderly individuals exhibit MCI, characterized by a memory complaint and impairments on formal testing associated with intact general cognition and preserved daily activities. This syndrome, particularly the amnestic form of MCI (aMCI), is regarded as a transitional stage between normal aging and early AD or as an initial manifestation of AD

(Morris et al. 2001; Petersen et al. 2001, 2006; Petersen 2003; Jicha et al. 2006; Markesbery et al. 2006). Subsequently, patients with AD go on to develop progressive difficulties with memory and a variety of cognitive functions (Morris et al. 2001; Morris and Price 2001; Petersen et al. 2001; Cummings 2004; Nestor et al. 2004). In the late stages, these individuals become profoundly demented.

Clinicians rely on histories, physical, neurological, and psychiatric examinations, and neuropsychological tests for diagnosis (Albert et al. 2001; Cummings 2004; Nestor et al. 2004). Studies of biomarkers and brain imaging are very promising for diagnosis and assessing outcomes of anti-amyloid treatments. In cases of AD, the levels of Aβ peptides in cerebrospinal fluid (CSF) are often low and levels of CSF Tau, particularly conformationally altered Tau, may be higher than controls (Sunderland et al. 2003). Imaging studies (Albert et al. 2001; Cummings 2004; Klunk et al. 2004; Nestor et al. 2004) of value include magnetic resonance imaging (MRI), which often discloses atrophy of specific regions of the brain, particularly the hippocampus and entorhinal cortex (Cummings 2004; Nestor et al. 2004), and positron emission tomography (PET) using ^{18}F deoxyglucose (FDG) or single-photon emission computerized tomography (SPECT), which discloses decreased glucose utilization and early reductions in regional blood flow in the parietal and temporal lobes, respectively (Nestor et al. 2004). Several years ago, it was demonstrated that the PET patterns of brain labeling following administration of a brain-penetrant ^{11}C-labeled thioflavin derivative (Pittsburgh compound B [PIB]), which binds to Aβ with high affinity, reflect the Aβ burden in the brain (Klunk et al. 2004). Significantly, on the basis of studies of transgenic models of amyloidosis in the central nervous system (CNS), it has been suggested that efflux of Aβ from brain to plasma may serve as a measure of Aβ brain burden (DeMattos et al. 2002). More recently, inverse relationships have been demonstrated between the amyloid load in brain (as assessed by PET amyloid imaging) and levels of Aβ in CSF (Fagan et al. 2005). These biomarker and imaging studies should allow clinicians to make a more accurate diagnosis of AD in early stages (McKhann et al. 1984; Jicha et al. 2006; Markesbery et al. 2006; Petersen et al. 2006) and, presumably, allow assessment of the efficacies of new antiamyloid therapeutics.

NEUROPATHOLOGY AND BIOCHEMISTRY OF AD

The clinical manifestations of AD stem from abnormalities involving brain regions/neural systems containing populations of neurons essential for memory, learning, and cognitive performance (West et al. 1994; Price and Sisodia 1998; West et al. 2000; West et al. 2004). Damaged systems include the basal forebrain cholinergic system, neuronal circuits in amygdala, hippocampus, nerve cells in entorhinal/limbic cortices, and cells in neocortex (Whitehouse et al. 1982; Coyle et al. 1983; Hyman et al. 1984; Braak and Braak 1991, 1994; Jicha et al. 2006; Markesbery et al. 2006; Petersen et al. 2006). The character, abundance, and distributions of abnormalities (i.e., Aβ burden, diffuse Aβ deposits, neuritic plaques, and tangles) are thought to correlate with clinical state in several cognitively characterized cohorts: controls, individuals with aMCI, and cases of eAD (Markesbery et al. 2006). In this study, no differences were present in the number of diffuse plaques between subject groups. However, in cases of aMCI, tangles were significantly increased in the ventral medial temporal lobe regions as compared to controls. Individuals with aMCI exhibited increased numbers of neuritic plaques in neocortical regions as compared to controls but were fewer in these cases than in persons with eAD who also showed greater numbers of NFTs and neuritic plaques in both frontal lobes and temporal regions. Memory deficits correlated most closely with the abundance of NFTs in CA1 of the hippocampus and in the entorhinal cortex, leading the authors to conclude that tangles, particularly in the medial temporal lobe, are more important than amyloid deposits in the progression from normal to MCI to eAD (Markesbery et al. 2006). Other studies (Jicha et al. 2006; Petersen et al. 2006) suggest that aMCI reflects a transitional state in the evolution of AD. Because the regional distributions of NFTs correlate most closely with the degree of clinical impairments from aged healthy controls to individuals with aMCI to cases of AD, the spread of NFTs beyond the medial temporal lobe is hypothesized to be most closely linked to the development of dementia. Postmortem neuropathological examinations of clinically well-characterized older subjects indicate that the lesions of AD, i.e., plaques and tangles, precede the clinical onset of MCI or AD dementia by many years (Troncoso et al. 1996; Schmitt et al. 2000; Morris and Price 2001).

Cellular abnormalities within these regional neural circuits include the presence within neurons of conformationally altered isoforms of Tau comprising the PHFs in NFTs, neurites, and neuropil threads (Lee et al. 2001; Goedert and Spillantini 2006b; Mandkelkow et al. 2007). A variety of axonal pathologies include terminal clubs and axonal varicosities, which are observed in AD, in aged memory-impaired Rhesus monkeys with Aβ deposits, and in some lines of transgenic mice expressing AD-linked mutant transgenes (Kitt et al. 1984, 1985; Selkoe et al. 1987; Martin et al. 1994; Price and Sisodia 1998; Lazarov et al. 2005b; Stokin et al. 2005). Because the abundant Aβ-containing neuritic plaques represent sites of synaptic disconnection in regions receiving inputs from disease-damaged populations of neurons, the neuritic clubs have been interpreted as disconnected terminals (Martin et al. 1994); the varicosities are presumed to represent focal perturbations of axonal transport. Generic and transmitter-specific synaptic markers are reduced in the target fields of damaged nerve cells (Whitehouse et al. 1982; Coyle et al. 1983; Sze et al. 1997; and see below). Local astroglial and microglial responses are particularly associated with plaques (Akiyama et al. 2000). Thus, the clinical manifestations of aMCI and AD reflect disruptions of synaptic communication in subsets of neural circuits associated with degeneration of axon terminals, followed by degeneration of axons, a process that ultimately leads to the death of neurons (Whitehouse et al. 1982; Coyle et al. 1983; Hyman et al. 1984; Braak and Braak 1991, 1994).

In one hypothetical model proposed to mechanistically link Aβ peptides and phosphorylated Tau PHFs, Aβ42 species, liberated at terminals, oligomerize to form Aβ assemblies or Aβ-derived diffusible ligands (ADDLs) (Lambert et al. 1998; Hartley et al. 1999; McLean et al. 1999; Walsh et al. 1999; Klein et al. 2001; Weninger and Yankner 2001; Wang et al. 2002; Gong et al. 2003; Kawarabayashi et al. 2004; Cleary et al. 2005; Lesne et al. 2006). These toxic Aβ entities impact on pre-/postsynaptic targets, including glutamate receptors, and are associated with synaptic dysfunction and, ultimately, disconnection of terminals from postsynaptic targets (Wong et al. 2002, 2006). Subsequently, a retrograde signal (of unknown nature), which originates at damaged terminals, triggers the activation of kinases (or the inhibition of phosphatases) in cell bodies, leading to the elevation of hyperphosphorylation of Tau which becomes associated with conformational changes in this protein leading to the formation of PHFs and, eventually, NFTs (Lee et al. 2001; Goedert and Spillantini 2006b; Mandkelkow et al. 2007). Disturbances of the cytoskeleton and biology in the neuron alter axonal transport (Price and Sisodia 1998; Lazarov et al. 2005b; Stokin et al. 2005; Wong et al. 2006), which can, in turn, compromise the functions and viability of neurons. Eventually, disconnected nerve cells die (Whitehouse et al. 1982; Hyman et al. 1984; Braak and Braak 1994; Lee et al. 2001; Goedert and Spillantini 2006b), and "tombstone" extracellular tangles are all that remain of neurons destroyed by disease.

GENETICS: FAMILIAL AD AND RISK FACTORS

The genetics of AD is complex and heterogeneous and exhibits an age-related pattern: Rare and highly penetrant early-onset familial AD mutations in the genes encoding APP (*APP*, chromosome 21), PS1 (*PSEN1*, chromosome 14), and PS2 (*PSEN2*, chromosome 1) are transmitted in an autosomal dominant fashion, whereas late-onset AD without obvious familial segregation ("sporadic AD") is thought to reflect the influences of several genetic and nongenetic risk factors (Price et al. 1998; Ghiso and Wisniewski 2004; Bertram and Tanzi 2005; Hardy 2006b). Approximately two thirds of the more than 200 currently known early-onset AD mutations are found in *PSEN1*, where they give rise to the youngest-onset ages and result in the fastest clinical progression of cognitive decline and dementia (for an up-to-date overview, see the "AD & FTD Mutation Database" at http://www.molgen.ua.ac.be/ADMutations/). Despite the large degree of locus and allelic heterogeneity, the majority of the known early-onset AD mutations have been found to share one common biochemical mechanism that results in either increased levels of all Aβ species and/or higher relative amounts of toxic Aβ42, following the modulation of APP cleavage by BACE1 or γ-secretase (Price et al. 1998; Ghiso and Wisniewski 2004). In addition, it was recently shown that certain forms of early-onset familial AD can also be caused by duplications of the *APP* locus itself (Rovelet-Lecrux et al. 2006), lending further support to the "amyloid hypothesis" of AD. This finding is also in line with the long-known observation that patients with Down's syndrome, which is caused by an extra copy of chromosome 21 (including the *APP*), develop AD pathology relatively early in life (Solitare and Lamarche 1966). Accordingly, the only known Down's syndrome case without AD neuropathology was found to have only partial duplication of chromosome 21, which did not encompass the *APP* locus (Prasher et al. 1998).

Together, early-onset autosomal dominant AD represents less than 5% of all AD cases. The vast majority of patients show a much later-onset age, usually in the seventh or eighth decade of life, and often show no obvious signs of familial transmission. However, a recent study on a large sample of monozygotic and dizygotic twins estimated the heritability of late-onset AD to be among the highest of all adult-onset diseases (possibly up to 80%), suggesting that genes have a most prominent role in this form of AD as well (Gatz et al. 2006), probably via the complex interplay of a multitude of different susceptibility alleles. In the quest for deciphering the underlying and largely unknown matrix of AD-risk genes, a vast body of evidence has been accrued in the scientific literature during the past 20 years, represented by well over 1000 studies genetically implicating or excluding certain alleles as risk factors for AD. Since the late 1970s, more than 1200 different genetic variants (or "polymorphisms") in more than 400 different genes have been tested for genetic association with AD (for an up-to-date overview of AD genetic association studies, see the

"AlzGene Database" at http://www.alzgene.org) (Bertram et al. 2007). However, with the exception of one single locus, the apolipoprotein E gene (ApoE) on chromosome 19q13 (Saunders et al. 1993; Strittmatter et al. 1993), none of these putative AD candidates have been unequivocally proven to consistently influence disease risk or onset age in more than a handful of samples. Instead, most proclaimed "novel AD genes" have actually been followed by a large number of conflicting reports, showing either no effects or opposite effects. Even *ApoE-ε4*, despite its unequivocal genetic association with AD, is neither necessary nor sufficient to actually *cause* the disease. Rather, it appears to operate as a genetic risk modifier by decreasing the age of onset in a dose-dependent manner (Blacker et al. 1997; Meyer et al. 1998). The biochemical consequences of *ApoE-ε4* in pathogenesis of AD are not yet fully understood, but they have been proposed to influence Aβ-aggregation/clearance, cerebrovascular events, and/or cholesterol homeostasis (for review, see Mahley et al. 2006).

The possible functional implications for the remaining putative AD susceptibility genes are even less clear and appear to involve a number of different pathophysiological pathways. A recent study systematically meta-analyzing AD genetic association studies (published until December of 2005) has identified a total of 13 non-APOE-related showing statistically significant, but very modest, risk effects based on published genotype distributions in AD cases versus controls (Bertram et al. 2007). The potential biochemical consequences on AD neuropathogenesis among loci most strongly implied by these analyses included changes in Aβ generation (*PSEN1* [see above], *ESR1* [estrogen receptor-α]), Aβ aggregation (*CST3* [cystatin C], *PRNP* [prion protein]), and Aβ degradation (*IDE* [insulin-degrading enzyme], *ACE* [angiotensin-converting enzyme]), as well as the induction of oxidative stress and the increased aggregation of hyperphosphorlylated Tau protein (*TF* [transferrin]) (for review, see Bertram et al. 2007). It is likely that the effects exerted by the putative risk alleles in these genes are further modified by gene-gene and gene-environment interactions, as well as inflammatory reactions. Finally, it can be expected that the results of large-scale and—with respect to gene/protein function—unbiased genetic association studies, which have recently become technically feasible and affordable, will add a number of hitherto unknown molecular mechanisms to the already existing hypotheses. These, together with the results from conventional candidate gene-based approaches and systematic meta-analyses, will hopefully provide an altogether more complete picture of the driving forces behind the genetic predisposition to AD.

APP AND PS FAMILIES OF PROTEINS

APP and APLPs

Members of the *APP* gene family (*APP*, *APLP1*, and *APLP2*), encode type I transmembrane proteins whose functions have not been fully defined (Cao and Sudhof 2001; Wong et al. 2006). APP is abundant in the nervous system, rich in neurons, and transported rapidly anterograde in axons to terminals (Koo et al. 1989; Sisodia et al. 1993; Buxbaum et al. 1998; Lazarov et al. 2005b). At the β and β sites (see below), APP is cleaved by activities of BACE1, producing amino-terminal secreted ectodomain (APPs) and APP carboxy-terminal fragments. Cleavages by the γ-secretase complex generate Aβ peptides and intracellular fragments including an APP intracellular domain (AICD) (Vassar et al. 1999; Cai et al. 2001; Li et al. 2003; Selkoe and Kopan 2003; Citron 2004a,b; Iwatsubo 2004; Laird et al. 2005; Ma et al. 2005). The *APPswe* mutation greatly enhances BACE1 cleavage at the +1 site of Aβ, resulting in substantial elevations in levels of Aβ peptides. APP_{717} mutations promote γ-secretase cleavages to increase secretion of Aβ42, the most toxic Aβ peptide. These *APP* mutations alter the processing of APP and increase the production of Aβ peptides or the amounts of the more toxic Aβ42; other *APP* mutations enhance local fibril formation and vascular amyloidosis (Ghiso and Wisniewski 2004).

PS1 and PS2

These two highly homologous 43–50-kD multipass transmembrane proteins (Sherrington et al. 1995) are involved in regulated intramembranous cleavages of a variety of transmembrane proteins, including APP and Notch 1, signaling molecules necessary for cell-fate decisions (Selkoe and Kopan 2003). PSs are endoproteolytically cleaved by a "presenilinase" to form an amino-terminal approximately 28-kD fragment and a carboxy-terminal approximately 18-kD fragment (Thinakaran et al. 1997), both of which are critical components of the γ-secretase complex (Selkoe and Kopan 2003; Iwatsubo 2004). As mentioned above, nearly 50% of early-onset cases of fAD are linked to more than 100 different mutations in *PS1* (Sherrington et al. 1995; Bertram and Tanzi 2005; Hardy 2006b). A relatively small number of *PS2* mutations also cause autosomal dominant fAD (Price et al. 1998; Bertram and Tanzi 2005). The majority of abnormalities in *PS* genes are missense mutations that enhance γ-secretase activities to increase the levels of Aβ42 peptides.

APP and Secretases

APP

Cleavage by β- and γ-secretases releases the ectodomain of APP (APPs), liberates a cytosolic fragment termed AICD, and generates several species of Aβ peptides (Fig. 1). In the CNS >> PNS (peripheral nervous system), APP and the proamyloidogenic secretases are carried by fast anterograde axonal transport (Buxbaum et al. 1998; Lazarov et al. 2005b); at terminals, Aβ peptides are generated by sequential endoproteolytic cleavages by BACE1 (at the Aβ +1 and +11 sites) to generate APP-β carboxy-terminal fragments (APP-βCTFs) (Cai et al. 2001; Luo et al. 2001) and by the γ-secretase complex (at several sites varying from Aβ 36, 38, 40, 42, 43) to form Aβ species peptides (Li et al. 2003; Citron 2004a,b; Iwatsubo 2004; Ma et al. 2005). The intramembranous cleavages of APP-βCTF by γ-secretase releases an AICD (Cao and Sudhof 2001), which can form a complex with Fe65, a nuclear adapter protein (Cao and Sudhof 2001); Fe65 and AICD or Fe65 alone (in a novel conformation) can gain access to the nucleus to influence gene transcription (Cao and Sudhof 2001), a signaling mechanism analogous to that occurring in the Notch1 (NICD) pathway (Selkoe and Kopan 2003; Iwatsubo 2004; Barrick and Kopan 2006). It has been suggested that AICD signaling may have a role in learning and memory, a hypothesis outlined below (Laird et al. 2005).

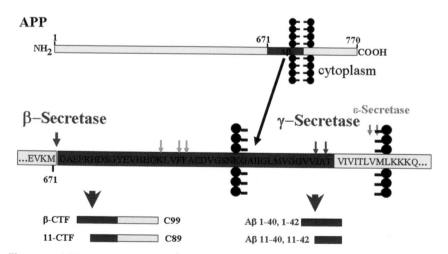

Figure 1. APP, secretases, and Aβ amyloid. An illustration of various Aβ fragments from the cleavage of APP by secretases.

BACE1

This transmembrane aspartyl protease is directly involved in the cleavage of APP at the +11 > +1 sites of Aβ in APP (Vassar et al. 1999; Farzan et al. 2000; Cai et al. 2001; Luo et al. 2001; Laird et al. 2005). In the CNS, BACE1 is localized in a variety of presynaptic terminals (Laird et al. 2005). Brain cells from $BACE1^{-/-}$ mice (Cai et al. 2001; Luo et al. 2001; Laird et al. 2005) do not produce Aβ1-40/42 and Aβ11–40/42, indicating that BACE1 is the neuronal β-secretase (Cai et al. 2001; Luo et al. 2001; Laird et al. 2005). As compared to wild-type APP, APPswe is cleaved approximately 100-fold more efficiently at the +1 site, resulting in a greater increase in BACE1 cleavage products.

γ-Secretase

Essential for the regulated intramembranous proteolysis of a variety of transmembrane proteins, this multiprotein catalytic complex includes PS1 and PS2; nicastrin (Nct), a type I transmembrane glycoprotein; and Aph-1 and Pen-2, two multipass transmembrane proteins (Goutte et al. 2002; Kimberly et al. 2003; Li et al. 2003; Selkoe and Kopan 2003; Iwatsubo 2004; Ma et al. 2005; Serneels et al. 2005). PS contains aspartyl residues that have roles in intramembranous cleavage; substitutions of aspartate residues at D257 in TM 6 and at D385 in TM 7 are reported to reduce secretion of Aβ and cleavage of Notch1 in vitro (Wolfe et al. 1999; Selkoe and Kopan 2003). The functions of the various γ-secretase proteins and their interactions in the complex are not yet fully defined. It has been suggested that the ectodomain of Nct may be important in substrate recognition and binding of amino-terminal stubs (of APP and other transmembrane proteins) generated by sheddases (i.e., BACE1 for APP) (Shah et al. 2005). After substrate docking occurs, γ-secretase cleavage takes place. In one model, Aph-1 and Nct form a precomplex that interacts with PS; subsequently, Pen-2 enters the complex where it is critical for the "presenilinase" cleavage of PS into two fragments. In concert, this complex is responsible for γ-secretase cleavages of APP, Notch, and a variety of other transmembrane proteins (Wolfe et al. 1999; Li et al. 2003; Selkoe and Kopan 2003; Iwatsubo 2004; Serneels et al. 2005).

TACE and BACE2

These proteins are expressed at low levels in neurons of the CNS. In other cells in other organs, APP is cleaved endoproteolytically within the Aβ sequence through alternative nonamyloidogenic pathways: α-secretase or

TACE (tumor necrosis factor-α converting enzyme) cleaves between Aβ residues 16 and 17 (Sisodia et al. 1990); BACE2 cleaves between residues 19 and 20 and 20 and 21 (Farzan et al. 2000). These cleavages, which occur in nonneural tissues, preclude the formation of Aβ peptides and serve to protect these cells/organs from Aβ amyloidosis (Wong et al. 2001).

GENETIC MODELS OF Aβ AMYLOIDOSIS AND AD-LINKED TAUOPATHIES

Aβ Amyloidosis

Investigators have taken advantage of information from genetics to create transgenic models of amyloidosis (McGowan et al. 2006; Savonenko et al. 2006). Mice expressing *APPswe* or *APP₇₁₇* (with or without mutant *PS1*) develop Aβ amyloidosis in the CNS (Mucke et al. 2000; Savonenko et al. 2005, 2006; Lesne et al. 2006). Mutant *APP;PS1* mice have an accelerated disease due to an increased level associated with the presence of diffuse Aβ deposits and neuritic plaques in the hippocampus and cortex. With age, levels of Aβ peptides, particularly Aβ42, increase significantly in the brain (Borchelt et al. 1996, 1997; Jankowsky et al. 2004; Savonenko et al. 2006), and oligomeric species, variously termed ADDLs, Aβ × 56, etc., appear in the CNS (Hartley et al. 1999; Klein et al. 2001; Walsh et al. 2002; Wang et al. 2002; Gong et al. 2003; Kawarabayashi et al. 2004; Cleary et al. 2005; Klyubin et al. 2005; Lesne et al. 2006). Mice carrying mutant transgenes exhibit Aβ deposits surrounded by swollen neurites; glial responses become conspicuous in regions showing neuritic plaques (Savonenko et al. 2006). Depending on the nature of the mouse strain, transgene construct, type of mutation, and level of expression, some lines of mice show evidence of amyloidosis in vessels (Calhoun et al. 1999; Nicolas et al. 2002; Ghiso and Wisniewski 2004; Lazarov et al. 2005a). In forebrain regions, the density of synaptic terminals is reduced (Savonenko et al. 2005), and levels of cholinergic markers (in cortex and hippocampus) and somatostatin (in cortex) are modestly reduced. In some settings, there are deficiencies in synaptic transmission (Chapman et al. 1999; Savonenko et al. 2006). Moreover, some lines of mice show evidence of degeneration of subsets of neurons (Calhoun et al. 1998). Interestingly, *APPswe/ind* mice, whose transgene is regulated by doxycycline (Dox), have high levels of transgene expression and exhibit amyloidosis in the brain. Dox decreases expression (95%) and reduces Aβ production to levels of nontransgenic animals; amyloidosis is reduced, but clearance of amyloid plaques is slow, i.e., mice with mutant *APP* expression suppressed for 6 months still show a significant Aβ burden.

Behavioral studies of these lines of mice, including those generated at Hopkins (Savonenko et al. 2003, 2005, 2006), disclose deficits in spatial reference memory (Morris water maze task) and episodic-like memory (repeated reversal and radial water maze tasks). Although *APPswe/PS1ΔE9* mice develop plaques at 6 months of age, all genotypes are indistinguishable in all cognitive tests from nontransgenic animals. However, 18-month-old *APPswe/PS1ΔE9* mice do not perform all cognitive tasks as well as control mice. Relationships exist between deficits in episodic-like memory tasks and total Aβ loads in the brain (Savonenko et al. 2005, 2006). Collectively, these studies of *APPswe/PS1ΔE9* mice suggest that some form of Aβ (ultimately associated with amyloid deposition) disrupts circuits critical for memory, with episodic-like memory being most sensitive to the toxic effects of Aβ. Behavioral deficits have been linked to the presence of Aβ oligomers (see below) and can be reversed by antibody-mediated reductions of levels of Aβ in the brain (Cleary et al. 2005; Klyubin et al. 2005; Lesne et al. 2006; and see below). Although these transgenic lines do not reproduce the full phenotype of AD, these mice are very useful subjects for research designed to link behavior and Aβ amyloidosis, to delineate disease mechanisms, and to test novel therapies (Savonenko et al. 2006).

As indicated above, a variety of Aβ species, ranging from monomers, oligomers, structural assemblies, and amyloid deposits in neuritic plaques, have been suggested to have important roles in impairing synaptic communication (Lambert et al. 1998; Klein et al. 2001; Walsh et al. 2002; Wang et al. 2002). The pool of insoluble Aβ (or plaques) is believed to exist in equilibrium with peptides in interstitial fluid (Cirrito et al. 2003). Significantly, systemic administration of Aβ antibodies increases levels of Aβ in plasma, and the magnitude of this elevation correlates with amyloid burden in the cortex and hippocampus (DeMattos et al. 2002). In one study, a naturally secreted Aβ peptide was injected into the ventricular system of rats and inhibited long-term potentiation (LTP) in the hippocampus (Klyubin et al. 2005); the adverse activity of this peptide was completely blocked by the injection of a monoclonal Aβ antibody, and active immunization was less effective in rescuing functions (Klyubin et al. 2005). These observations are consistent with the concept that oligomers are the toxic entity and that these species are both necessary and sufficient to perturb learned behavior (Cleary et al. 2005; Klyubin et al. 2005). More recently, studies of TG2576 mice suggested that extracellular accumulations of 56-kD soluble amyloid assemblies (termed Aβ*56), purified from the brains of memory-impaired mice, interfere with memory when delivered to young rats (Lesne et al. 2006).

AD-linked Tauopathies

Early efforts to express mutant *tau* transgenes in mice did not lead to striking clinical phenotypes or pathology (Goedert and Spillantini 2006b). The paucity of Tau abnormalities in various lines of mutant mice with Aβ abnormalities may be related to differences in Tau isoforms expressed in this species (Xu et al. 2002). More recently, mice overexpressing *tau* show clinical signs for the most part attributable to degeneration of motor axons (Lee et al. 2001). When prion or Thy1 promoters are used to drive *tau*$_{P301L}$ (a mutation linked to autosomal dominant frontotemporal dementia with parkinsonism), some brain and spinal cord neurons develop tangles (Gotz et al. 2001). Injection of Aβ42 fibrils into specific brain regions of *tau*$_{P301L}$ mice increases the number of tangles in those neurons projecting to sites of Aβ injection. Mice expressing *APPswe/tau*$_{P301L}$ exhibit enhanced tangle-like pathology in the limbic system and olfactory cortex (Lewis et al. 2001). A triple-transgenic mouse, created by microinjecting *APPswe* and *tau*$_{P301L}$ into single cells derived from monozygous *PS1*$_{M146V}$ knockin mice (Oddo et al. 2003), developed age-related plaques and tangles as well as deficits in LTP, which appear to antedate overt pathology (Oddo et al. 2003). A forebrain-specific *P301L tau* mouse that can be regulated by the tetracycline responsive element (TRE) develops NFTs associated with neuronal loss in the forebrain and behavioral impairments. After suppression of Tau expression by Dox, memory functions recover and neuron numbers stabilize, but NFTs continue to accumulate. These authors conclude that in this model, NFTs are not sufficient to cause cognitive decline or death of neurons. The various lines of mice bearing both mutant *tau* and *APP* (or *APP/PS1*) or mutant *tau* mice injected with Aβ may not be ideal models of fAD because the presence of the *tau* mutation alone is associated with the development of tangles and disease.

TARGETING OF GENES ENCODING AMYLOIDOGENIC SECRETASES

To begin to understand the functions of some of the proteins thought to have roles in AD, investigators have targeted a variety of genes including *BACE1*, *PS1*, *Nct*, and *Aph-1*.

BACE1$^{-/-}$ Mice

These animals mate successfully and exhibit no obvious pathology (Cai et al. 2001; Luo et al. 2001; Laird et al. 2005; Savonenko et al. 2006). *BACE1*$^{-/-}$ neurons do not cleave at the +1 and +11 sites of Aβ, and the production of Aβ peptides is abolished (Cai et al. 2001; Luo et al. 2001; Laird et al. 2005), establishing that BACE1 is the neuronal β-secretase required to generate the amino termini of Aβ. However, $BACE1^{-/-}$ mice show altered performance on some tests of cognition and emotion (Laird et al. 2005; Savonenko et al. 2006) (see below); the former deficits can be rescued by overexpression of APP transgenes. BACE1 null mice manifest alterations in both hippocampal synaptic plasticity and performance on tests of cognition and emotion (Laird et al. 2005); the memory deficits (but not emotional alterations) in $BACE1^{-/-}$ mice are prevented by coexpressing APPswe;PS1ΔE9 transgenes. This observation suggests that APP processing influences cognition/memory and that the other potential substrates of BACE1 may have roles in neural circuits related to emotion. More recently, it has been discovered that genetic deletion of BACE1 causes hypomyelination in the PNS and CNS. These mice have a delay in myelination, reduced thickness of myelin sheaths, and decreased myelin markers in the PNS and CNS. These phenotypes reflect alterations in the biology of neuregulin1 (NRG1), which is known to be a signal by which axons communicate with ensheathing cells to control myelination during development. It is thought that BACE1 cleaves NRG1 and processed NRG1 regulates myelination by phosphorylation of Akt. In $BACE^{-/-}$ mice, full-length NRG1 is increased and cleavage products are decreased; levels of phosphorylated Akt dimininished. These several results establish that BACE1 and APP/NRG1 processing pathways are critical for cognitive, emotional, and synaptic functions and for myelination during development of the PNS and CNS.

$PS1^{-/-}$ Mice

Embryos develop severe abnormalities of the axial skeleton, ribs, and spinal ganglia, a lethal outcome that resembles a partial $Notch1^{-/-}$ phenotype (Shen et al. 1997; Wong et al. 1997). $PS1^{-/}$ cells secrete decreased levels of Aβ (De Strooper et al. 1998; Li et al. 2003) due to the fact that PS1 (along with PS2, Nct, Aph-1 and Pen-2) is a component of the γ-secretase complex that carries out the S3 intramembranous cleavage of Notch1 (De Strooper et al. 1999; Li et al. 2003; Selkoe and Kopan 2003). Without γ-secretase, NICD is not released from the plasma membrane and cannot reach the nucleus to provide a signal to initiate transcriptional processes essential for cell-fate decisions (Selkoe and Kopan 2003; Barrick and Kopan 2006). Significantly, conditional PS1/PS2 targeted mice are impaired in memory and in hippocampal synaptic plasticity (Saura et al. 2004), raising questions as to the roles of loss of PS functions in neurodegeneration and AD (Herget et al. 1998; Delacourte et al. 1999). It is important to note that $PS1^{-/-}$ mice, whose lethal phenotype is rescued through neuronal expression of $PS1$, develop skin cancer; this outcome was interpreted initially to reflect deregulation of the β-catenin pathway (Xia et al. 2001) but may operate through other mechanisms.

$Nct^{-/-}$ Mice

Embryos die early and exhibit several patterning defects (Li et al. 2003), including abnormal segmentation of somites; this phenotype closely resembles that seen in $PS1/PS2^{-/-}$ embryos. Importantly, $Nct^{-/-}$ cells do not secrete Aβ peptides, whereas $Nct^{+/-}$ cells show reductions of approximately 50% (Li et al. 2003). The failure of $Nct^{-/-}$ cells to generate Aβ peptides is accompanied by accumulation of APP carboxy-terminal fragments. Importantly, $Nct^{+/-}$ mice develop tumors of the skin, a phenotype accelerated by reducing $PS1$ and $P53$, both of which manipulations exacerbate the tumor phenotype. The formation of these tumors appear to reflect decreased γ-secretase activities and, potentially, influence on the epidermal growth factor receptor (EGFR), an oncogene implicated in head and neck tumors in humans (Li et al. 2007).

$Aph\text{-}1a^{-/-}$ Mice

$Aph\text{-}1a$, $Aph\text{-}1b$, and $Aph\text{-}1c$ encode four distinct Aph-1 isoforms: Aph-1aL and Aph-1aS (derived from differential splicing of $Aph\text{-}1a$, $Aph\text{-}1b$, and $Aph\text{-}1c$ (Ma et al. 2005; Serneels et al. 2005). $Aph\text{-}1a^{-/-}$ embryos have patterning defects that resemble but are not identical to those of $Notch1$, Nct, or PS null embryos (Ma et al. 2005; Serneels et al. 2005). Moreover, in $Aph\text{-}1a^{-/-}$-derived cells, the levels of Nct, PS fragments, and Pen-2 are decreased, and there is a concomitant reduction in levels of the high-molecular-weight γ-secretase complex and a decrease in secretion of Aβ (Ma et al. 2005). In $Aph\text{-}1a^{-/-}$ cells, other mammalian Aph-1 isoforms can restore the levels of Nct, PS, and Pen-2 (Ma et al. 2005; Serneels et al. 2005).

EXPERIMENTAL TREATMENTS AND THERAPEUTICS

Models relevant to amyloidogenesis provide an opportunity to test the influence of ablations or knockdowns of specific genes, the modulation of cleavage patterns influencing peptide neurotoxicity, and the enhancement of clearance and/or degradation of Aβ (Li et al. 2003; Monsonego and Weiner 2003; Citron 2004a,b; Walsh and Selkoe 2004; Cleary et al.

2005; Klyubin et al. 2005; Laird et al. 2005; Lesne et al. 2006; Savonenko et al. 2006). In this chapter, we comment on selected studies that illustrate experimental strategies directed at specific therapeutic targets that are predicted to provide mechanism-based therapeutic benefits to patients with AD (Savonenko et al. 2006).

Reduction in β-secretase Activity

Deletion of *BACE1* in *APPswe;PS1ΔE9* mice prevents both Aβ deposition and age-associated cognitive abnormalities that occur in this model (Laird et al. 2005; Masliah et al. 2005). Significantly, *BACE1$^{-/-}$*; *APPswe;PS1ΔE9* mice do not develop Aβ deposits or age-associated abnormalities in working memory that occur in the *APPswe;PS1ΔE9* model of Aβ amyloidosis (Borchelt et al. 1996; Laird et al. 2005). Similarly, *BACE1$^{-/-}$* TG2576 mice appear to be rescued from age-dependent memory deficits and physiological abnormalities in other models of amyloidosis (Ohno et al. 2004; Savonenko et al. 2006; Kobayashi et al. 2007). Moreover, Aβ deposits are sensitive to *BACE1* dosage and can be efficiently cleared from regions of the CNS when *BACE1* is silenced at these sites (Laird et al. 2005; Singer et al. 2005). Inhibitors of β-secretase, conjugated to carrier peptides, have been described to reduce activity of inhibitors in vitro and in vivo (following intraperitoneal injection of compounds into TG2576 mice) (Chang et al. 2004). New approaches using conditional expression systems, RNA interference (RNAi) silencing, or manipulations of transcription will allow investigators to examine the pathogenesis of diseases and to assess the degrees of reversibility of the disease processes (Ohno et al. 2004; Laird et al. 2005; Singer et al. 2005). Results of these approaches will provide a better understanding of the mechanisms that lead to disease and aid in the design of new treatments.

Although BACE1 is a very attractive therapeutic target (Citron 2004a,b), several potential problems exist with this approach. First, the BACE1 catalytic site is quite large, and it is uncertain whether it will be possible to achieve adequate brain penetration of a compound of sufficient size that will act in vivo. Second, *BACE1* null mice manifest alterations in both hippocampal synaptic plasticity and performance on tests of cognition and emotion (Laird et al. 2005); the memory deficits (but not emotional alterations) in *BACE1$^{-/-}$* mice are prevented by coexpressing *APPswe;PS1ΔE9* transgenes. This discovery suggests that APP processing influences cognition/memory and that the other potential substrates of *BACE1* may have roles in neural circuits related to emotion. Third, as described above, genetic deletion of *BACE1* causes hypomyelination in the

PNS and CNS, which reflect alterations in the NRG–Akt pathway. Although inhibition of β-secretase activity is an exciting therapeutic opportunity, future studies will be needed to assess possible mechanism-based side effects that may occur with inhibition of *BACE1* (Chang et al. 2004; Laird et al. 2005; Savonenko et al. 2006; Wong et al. 2006).

Inhibition of γ-secretase Activity

Both genetic reduction and pharmacological inhibition of γ-secretase activity decrease production of Aβ peptides in cell-free and cell-based systems and reduce levels of Aβ in mutant mice with Aβ amyloidosis (Li et al. 2007). Thus, γ-secretase activity is a significant target for therapy (Li et al. 2003; Saura et al. 2004; Ma et al. 2005; Wong et al. 2006). However, γ-secretase activity is also essential for processing of Notch and a variety of other transmembrane proteins, which are critical for many properties of cells including lineage specification and cell growth during embryonic development (Shen et al. 1997; Wong et al. 1997; Li et al. 2003; Selkoe and Kopan 2003; Wolfe and Kopan 2004; Wong et al. 2004; Ma et al. 2005). Significantly, one inhibitor of γ-secretase (LY–411, 575) reduces production of Aβ but also has profound effects on T- and B-cell development and on the appearance of intestinal mucosa (proliferation of goblet cells, increased mucin in gut lumen, and crypt necrosis) (Milano et al. 2004; Wong et al. 2004; Barten et al. 2005). Moreover, although *Nct*[+/−] *APPswe;PS1ΔE9* mice show reduced levels of Aβ and amyloid plaques, these mice also develop skin tumors (Li et al. 2007), presumably, in part, to reduction of γ-secretase activity and the role of Notch as a tumor suppressor in skin (Xia et al. 2001; Nicolas et al. 2003). Thus, clinicians carrying out trials of this inhibitor must be alert to several potential adverse events associated with inhibition of this enzyme complex.

γ-Secretase Modulation by NSAID Compounds

Retrospective epidemiological studies suggest that significant exposure to NSAIDs (nonsteroidal anti-inflammatory drugs) reduces risk of AD (Anthony et al. 2000), an outcome initially interpreted as related to suppression of the well-documented inflammatory process occurring in brains of AD cases (Akiyama et al. 2000; Lim et al. 2000; Cummings 2004). However, more recent in vitro studies indicate that a subset of NSAID compounds in this class can modulate secretase cleavages to shorter less-toxic Aβ species without altering the processing of Notch or other transmembrane proteins (Weggen et al. 2001). Moreover, short-term treatment of mutant mice appears to have some benefit in terms of lowering Aβ and plaque pathology (Lim et al. 2000). This strategy is now being evaluated in a Phase-3 clinical trial of Furizin.

CLEARANCE

Removal of Source Aβ

Investigations utilizing lesions of the entorhinal cortex or perforant pathway (Lazarov et al. 2002; Sheng et al. 2002, 2003) to remove APP, the source of Aβ, by lesioning cell bodies or axons/terminals transporting APP to terminals significantly reduce levels of Aβ and amyloid plaques in target fields.

Activation of Aβ-cleaving Proteases

Recently, investigators have attempted to influence levels of Aβ-degrading enzymes to promote degradation and clearance (Hulette et al. 1995; Iwata et al. 2000, 2001, 2004; Vekrellis et al. 2000; Carson and Turner 2002; Farris et al. 2003; Lauritzen and Gold 2003; Marr et al. 2003). Increasing local levels of two metalloproteases, insulin-degrading enzyme (IDE) and neprilysis (NEP), both of which cleave Aβ, reduces levels of the amyloid peptide in the presence of the protease (Iwata et al. 2000, 2004; Vekrellis et al. 2000; Carson and Turner 2002; Farris et al. 2003; Leissring et al. 2003; Marr et al. 2003; Miller et al. 2003). However, the challenge with this strategy includes difficulties in controlling the regulation of these enzymes and the possible off-target effects of these proteases cleaving non-Aβ targets (other proteins important for normal functions).

Aβ Immunotherapy

To date, the most exciting findings regarding clearance of Aβ come from studies using active and passive Aβ immunotherapy (Monsonego and Weiner 2003; Selkoe and Schenk 2003; Federoff and Bowers 2005; Savonenko et al. 2006). In treatment trials in mutant mice, both Aβ immunization (with Freund's adjuvant) and passive transfer of Aβ antibodies reduce levels of Aβ and plaque burden (Schenk et al. 1999; Bard et al. 2000; Morgan et al. 2000; DeMattos et al. 2001, 2002; Dodart et al. 2002; Kotilinek et al. 2002; Monsonego and Weiner 2003; Wilcock et al.

2003, 2004b; Hutton and McGowan 2004; Oddo et al. 2004; Federoff and Bowers 2005; Klyubin et al. 2005). Although the mechanisms of enhanced clearance are not settled (Federoff and Bowers 2005; Wong et al. 2006), investigators have proposed at least two not mutually exclusive hypotheses: (1) A very small amount of Aβ antibody enters the brain, binds to Aβ peptides, promotes the disassembly of fibrils, and, via the Fc antibody domain, encourages activated microglia to enter the affected regions and remove Aβ (Schenk et al. 1999). (2) Serum antibodies serve as "a sink" for the amyloid peptides derived from the brain and entering the circulation, thus changing the equilibrium of Aβ in different compartments and promoting removal of Aβ from the CNS (Morgan et al. 2000; DeMattos et al. 2002; Dodart et al. 2002; Cirrito et al. 2003). Whatever the mechanisms, Aβ immunotherapy in mutant mice is successful in partially clearing Aβ, in attenuating learning and behavioral deficits in several different cohorts of mutant *APP* or *APP/PS1* mice, and in partially reducing Tau abnormalities in the triple-transgenic mice (Morgan et al. 2000; Dodart et al. 2002; Kotilinek et al. 2002; Hutton and McGowan 2004; Oddo et al. 2004; Sigurdsson et al. 2004; Wilcock et al. 2004a,b; Savonenko et al. 2006).

Several problems, however, have been associated with Aβ immunotherapy. In the presence of congophilic angiopathy, brain hemorrhages may be associated with immunotherapy (Pfeifer et al. 2002; Gandy and Walker 2004), perhaps because congophilic angiopathy can weaken vascular walls (Winkler et al. 2001; Herzig et al. 2004), and, potentially, immunotherapeutic removal of some intramural vascular amyloid could contribute to the rupture of damaged vessels and to local bleeding. Some local bleeding has been documented in the brain of treated cases. More significant is the evidence that a subset of patients receiving Aβ vaccination with adjuvant develop meningoencephalitis (see below). The trials in mice are useful for testing efficacy, but they are not necessarily predictive of adverse events in humans.

Anti-Aβ Aggregation Strategies

Another therapeutic strategy aimed at clearing cerebral Aβ targets is the oligomerization and aggregation of the peptide (Tanzi et al. 2004). This approach can also serve to "neutralize" the toxicity of Aβ, since increasing evidence indicates that soluble Aβ oligomers disrupt synaptic activity (Tanzi 2005). One of the most useful clues for therapies aimed at impeding Aβ oligomerization, during the past decade, is the fact that metals, e.g., zinc and copper, are required for the aggregation of Aβ

(Bush and Tanzi 2002). On the basis of this discovery, compounds that can attract metals away from the Aβ peptide (and transfer them to other metal-binding proteins) have been tested both in animal models and in human clinical trials. Notably, the discontinued antibiotic and metal-complexing agent, clioquinol, was shown to significantly reduce Aβ burden and plaque load in TG2576 APP mice (Cherny et al. 2001). This compound was then tested in a small Phase-2 clinical trial for treatment of AD, where it was shown to lower plasma Aβ levels and attenuate cognitive decline in AD patients (Ritchie et al. 2003). Another antiaggregation approach utilizes Aβ-binding drugs, such as tramiprosate (3-amino-1-propanesulfonic acid; Alzhemed), which has been shown to be effective in reducing Aβ polymerization in vitro and in lowering plaque deposition in vivo (Aisen 2005). Thus, in addition to therapies aimed at attenuating Aβ generation, it is highly likely that efficacious anti-Aβ aggregation strategies will also emerge during the next 5 years.

CLINICAL APPROACHES TO ALZHEIMER'S DISEASE

Presently available treatments for AD include cholinesterase inhibitors, (Cummings 2004; Winblad et al. 2006), glutamate antagonists that influence the effects of the transmitter on specific receptors, including NMDA-R, which have roles in excitotoxicity (Reisberg et al. 2003; Cummings 2004), and pharmacological agents useful for behavioral disturbances (Cummings 2004).

Cholinesterase Inhibitors

The "cholinergic hypothesis" of AD is based, in part, on the demonstration that cholinergic markers, including acetylcholinesterase (AChE) and choline acetyltransferase (ChAT), are reduced in the brains of patients (Bowen et al. 1976a,b; Davies and Maloney 1976a,b; Perry et al. 1977). The index case showing loss of basal forebrain cholinergic neurons in AD was a 72-year-old man with a 14-year history of dementia and brain abnormalities consistent with AD (Whitehouse et al. 1981, 1982; Coyle et al. 1983). Cholinergic deficits in the cortex of patients with AD were reported to correlate with the severity of intellectual impairment and with the regional densities of senile plaques (Blessed et al. 1968; E.K. Perry et al. 1978a,b; T.L. Perry et al. 1978; Francis et al. 1985; Francis and Bowen 1985).

For many years, investigators have examined the effects of direct agonists, AChE inhibitors, and precursor loading. Present-day cholinomimetic approaches use anticholinesterases with long half-lives and predilection for central cholinergic synapses in an attempt to maximize the amount of available acetylcholine in the postsynaptic receptors (Cummings 2004; Winblad et al. 2006). FDA-approved compounds include tacrine, donepezil, rivastigmine, and galantamine; donepezil and rivastigmine have been shown to have some efficacy and may slow progressive cognitive deterioration compared to placebo (Cummings 2004).

Postsynaptic cholinergic receptor agonists have also been tried (Bymaster et al. 1994; Nishizaki et al. 2000; Zhao et al. 2001), but many of these drugs, especially muscarinic agonists, have encountered problems because of autonomic, gastrointestinal, and motoric side effects.

NMDA Antagonists

More recently, Memantine, a noncompetitive, low-affinity, open-channel NMDA-R antagonist (Lipton 2004), has been used to treat patients with AD (Reisberg et al. 2003; Cummings 2004). This drug enters the open receptor channel, and because it does not accumulate at the site, it does not interfere with synaptic transmission. The drug is thought to impact on excitotoxicity mediated by excessive glutamate stimulation of NMDA-R (increased Ca^{++} influx through the channel). The benefits on clinical disease have been modest (Cummings 2004).

Aβ Immunotherapy

Individuals receiving vaccinations with preaggregated Aβ and an adjuvant (followed by a booster) develop antibodies that recognize Aβ in the brain and vessels (Hock et al. 2002; Selkoe and Schenk 2003; Schenk et al. 2004; Federoff and Bowers 2005). Unfortunately, although Phase-1 trials with Aβ peptide and adjuvant vaccination were not associated with any adverse events, Phase-2 trials detected complications (meningoencephalitis) in a subset of patients and were suspended (Hock et al. 2003; Monsonego and Weiner 2003; Nicoll et al. 2003; Schenk et al. 2004; Bayer et al. 2005; Masliah et al. 2005). The pathology in the index case, consistent with T-cell meningitis (Nicoll et al. 2003), was interpreted to show some clearance of Aβ deposits, but some regions contained a relatively high density of tangles, neuropil threads, and vascular amyloidosis (Nicoll et al. 2003). Aβ immunoreactivity was sometimes associated with microglia, and T cells were conspicuous in subarachnoid space and around some vessels (Nicoll et al. 2003). In another case, there was a significant reduction in amyloid deposits in the absence of clinical evidence of encephalitis (Masliah et al. 2005). Although the trial was stopped, assessment of cognitive functions in a small subset of patients (30) who received vaccination and booster immunizations disclosed that patients who generated Aβ antibodies (as measured by a new assay) appeared to have a slower decline in several functional measures (Hock et al. 2003). The events occurring in this subset of patients illustrate the challenges of extrapolating outcomes in mutant mice to human trials. Investigators are attempting to make new antigens/adjuvant formulations that do not stimulate T-cell-mediated immunologic attack and are pursuing in parallel passive immunization approaches (Monsonego and Weiner 2003; Selkoe and Schenk 2003; Schenk et al. 2004; Federoff and Bowers 2005; Zamora et al. 2006).

β- and γ-secretase Inhibition and γ-secretase Modulation

The rationale for these approaches are outlined above. At least one drug is in clinical trials: FLURIZAN™ is a selective γ-secretase modulator that reduces levels of the toxic peptide Aβ42 in cultured human cells; in animal models, it reduced the levels of the toxic peptide Aβ42 by approximately 70%. Although it is now in Phase III, it is not yet clear whether this compound will lower Aβ42 levels in human brain. Another γ-secretase modulator, E2012 (Eisai), is currently in Phase I trials.

Anti-Aβ Aggregation Therapies

Currently, there are two major therapeutics in clinical trials aimed at treating AD by targeting the oligomerization and aggregation of Aβ. Furthest along is Alzhemed (Neurochem), which binds Aβ and impairs aggregation. It is currently in Phase III clinical trials. In a Phase II clinical trial of Alzhemed in 58 mild-to-moderate AD patients, the drug was well-tolerated and reduced CSF Aβ42 levels in patients; however, no psychometric score differences in cognitive testing were found between patients and placebo groups during the 3-month double-blind period. The second therapeutic in this category is a compound that strips zinc and copper from Aβ and thereby inhibits oligomerization based on the original finding that metals drive the aggregation of Aβ (Bush et al. 1994). Following the Phase II test of clioquinol (PBT1; Prana Biotechnology, LTD), the drug was discontinued for use in humans due to a di-iodo contaminant. An iodine-free derivative of clioquinol (PBT2;

Prana Biotechnology, LTD) has since been developed and proceeded from Phase I to Phase II trials.

CONCLUSIONS

Investigators have more accurately defined MCI and early AD, developed diagnostic approaches using imaging and biomarkers, and characterized the stages of pathology correlated with these observations with clinical features of MCI and eAD. Genetic studies have provided information regarding the roles of autosomal dominant mutations in *APP* and *PS* genes, and the dose-dependent risks of the ApoE4 alleles. Other loci are being assessed. Parallel studies of AD and of genetically engineered models of Aβ amyloidosis (and the tauopathies) have greatly increased our understanding of pathogenic mechanisms, possible therapeutic targets, and potential mechanism-based treatments designed to benefit patients with AD. Imaging, biomarkers, and genetic studies in concert with investigations of in vitro and in vivo models have brought this field to the threshold of implementing novel treatments based on an understanding of the neurobiology, neuropathology, and biochemistry of this illness. Moreover, a variety of tools, including amyloid imaging and measurements of Aβ flux between compartments, have the potential to assess the outcome efficacies of treatments. It is anticipated that discoveries during the next few years will lead to the design of new mechanism-based therapies that can be tested in animal models, and, eventually, these approaches will be introduced successfully into the clinic for the benefit of patients with this devastating illness.

ACKNOWLEDGMENTS

The authors thank the many colleagues who have worked at JHMI as well as those at other institutions for their contributions to some of the original work cited in this review and for their helpful discussions. Aspects of this work were supported by grants from the U.S. Public Health Service (AGO05146, NS41438, NS45150, NS049088, AG14248, NS1058017) as well as funds from the Metropolitan Life Foundation, Adler Foundation, Alzheimer's Association, CART Foundation, Merck Research Laboratories, and Bristol-Myers Squibb Foundation. Because of space constraints, the citations are limited. Additional references relevant to the research can be found in Wong et al. (2002, 2006); Laird et al. (2005); Savonenko et al. (2006); and Price D.L. (2006) (Alzheimer: 100 years and beyond. In *Research and perspectives in Alzheimer's disease* [ed. Y. Christen]. Springer, New York).

REFERENCES

Aisen P.S. 2005. The development of anti-amyloid therapy for Alzheimer's disease: From secretase modulators to polymerisation inhibitors. *CNS Drugs* **19:** 989–996.

Akiyama H., Barger S.W., Barnum S., Bradt B., Bauer J., Cole G.M., Cooper N.R., Eikelenboom P., Emmerling M., Fiebich B.L., Finch C.E., Frautschy S., Griffin W.S., Hampel H., Hull M., Landreth G., Lue L., Mrak R., Mackenzie I.R., McGeer P.L., O'Banion M.K., Pachter J., Pasinetti G., Plata-Salaman C., and Rogers J., et al. 2000. Inflammation and Alzheimer's disease. *Neurobiol. Aging* **21:** 383–421.

Albert M.S., Moss M.B., Tanzi R., and Jones K. 2001. Preclinical prediction of AD using neuropsychological tests. *J. Int. Neuropsychol. Soc.* **7:** 631–639.

Anthony J.C., Breitner J.C., Zandi P.P., Meyer M.R., Jurasova I., Norton M.C., and Stone S.V. 2000. Reduced prevalence of AD in users of NSAIDs and H2 receptors antagonists: The Cache County study. *Neurology* **54:** 2066–2071.

Bard F., Cannon C., Barbour R., Burke R.L., Games D., Grajeda H., Guido T., Hu K., Huang J., Johnson-Wood K., Khan K., Kholodenko D., Lee M., Lieberburg I., Motter R., Nguyen M., Soriano F., Vasquez N., Weiss K., Welch B., Seubert P., Schenk D., and Yednock T. 2000. Peripherally administered antibodies against amyloid beta-peptide enter the central nervous system and reduce pathology in a mouse model of Alzheimer disease. *Nat. Med.* **6:** 916–919.

Barrick D. and Kopan R. 2006. The Notch transcription activation complex makes its move. *Cell* **124:** 883–885.

Barten D.M., Guss V.L., Corsa J.A., Loo A., Hansel S.B., Zheng M., Munoz B., Srinivasan K., Wang B., Robertson B.J., Polson C.T., Wang J., Roberts S.B., Hendrick J.P., Anderson J.J., Loy J.K., Denton R., Verdoorn T.A., Smith D.W., and Felsenstein K.M. 2005. Dynamics of β-amyloid reductions in brain, cerebrospinal fluid, and plasma of β-amyloid precursor protein transgenic mice treated with a γ-secretase inhibitor. *J. Pharmacol. Exp. Ther.* **312:** 635–643.

Bayer A.J., Bullock R., Jones R.W., Wilkinson D., Paterson K.R., Jenkins L., Millais S.B., and Donoghue S. 2005. Evaluation of the safety and immunogenicity of synthetic Abeta42 (AN1792) in patients with AD. *Neurology* **64:** 94–101.

Bertram L. and Tanzi R.E. 2005. The genetic epidemiology of neurodegenerative disease. *J. Clin. Invest.* **115:** 1449–1457.

Bertram L., McQueen M.B., Mullin K., Blacker D., and Tanzi R.E. 2007. Systematic meta-analyses of Alzheimer disease genetic association studies: The AlzGene database. *Nat. Genet.* **39:** 17–23.

Blacker D., Haines J.L., Rodes L., Terwedow H., Go R.C., Harrell L.E., Perry R.T., Bassett S.S., Chase G., Meyers D., Albert M.S., and Tanzi R. 1997. ApoE-4 and age at onset of Alzheimer's disease: The NIMH genetics initiative. *Neurology* **48:** 139–147.

Blessed G., Tomlinson B.E., and Roth M. 1968. The association between quantitative measures of dementia and of senile change in the cerebral grey matter of elderly subjects. *Br. J. Psychiatry* **114:** 797–811.

Borchelt D.R., Ratovitski T., Van Lare J., Lee M.K., Gonzales V., Jenkins N.A., Copeland N.G., Price D.L., and Sisodia S.S. 1997. Accelerated amyloid deposition in the brains of transgenic mice coexpressing mutant presenilin 1 and amyloid precursor proteins. *Neuron* **19:** 939–945.

Borchelt D.R., Thinakaran G., Eckman C.B., Lee M.K., Davenport F., Ratovitsky T., Prada C.M., Kim G., Seekins S., Yager D., Slunt H.H., Wang R., Seeger M., Levey A.I., Gandy S.E., Copeland N.G., Jenkins N.A., Price D.L., Younkin S.G., and Sisodia S.S. 1996.

Familial Alzheimer's disease-linked presenilin 1 variants elevate Abeta1-42/1-40 ratio in vitro and in vivo. *Neuron* **17:** 1005–1013.

Bowen D.M., Smith C.B., White P., and Davison A.N. 1976a. Neurotransmitter-related enzymes and indices of hypoxia in senile dementia and other abiotrophies. *Brain* **99:** 459–496.

———. 1976b. Senile dementia and related abiotrophies: Biochemical studies on histologically evaluated human postmortem specimens. In *Neurobiology of aging* (ed. R.D. Terry and S. Gershon), pp. 361–389. Raven Press, New York.

Braak H. and Braak E. 1991. Neuropathological staging of Alzheimer-related changes. *Acta Neuropathol.* **82:** 239–259.

———. 1994. Pathology of Alzheimer's disease. In *Neurodegenerative diseases* (ed. D.B. Calne), pp. 585–613. W.B. Saunders, Philadelphia, Pennsylvania.

Brookmeyer R., Gray S., and Kawas C. 1998. Projections of Alzheimer's disease in the United States and the public health impact of delaying disease onset. *Am. J. Public Health* **88:** 1337–1342.

Bush A.I. and Tanzi R.E. 2002. The galvanization of beta-amyloid in Alzheimer's disease. *Proc. Natl. Acad. Sci.* **99:** 7317–7319.

Bush A.I., Pettingell W.H., Multhaup G., Paradis M., Vonsattel J.P., Gusella J.F., Beyreuther K.T., Masters C.L., and Tanzi R.E. 1994. Rapid induction of Alzheimer Aβ amyloid formation by zinc. *Science* **265:** 1464–1467.

Buxbaum J.D., Thinakaran G., Koliatsos V., O'Callahan J., Slunt H.H., Price D.L., and Sisodia S.S. 1998, Alzheimer amyloid protein precursor in the rat hippocampus: Transport and processing through the perforant path. *J. Neurosci.* **18:** 9629–9637.

Bymaster F.P., Wong D.T., Mitch C.H., Ward J.S., Calligaro D.O., Schoepp D.D., Shannon H.E., Sheardown M.J., Olesen P.H., and Suzdak P.D. 1994. Neurochemical effects of the M1 muscarinic agonist xanomeline (LY246708/NNC11-0232). *J. Pharmacol. Exp. Ther.* **269:** 282–289.

Cai H., Wang Y., McCarthy D., Wen H., Borchelt D.R., Price D.L., and Wong P.C. 2001. BACE1 is the major beta-secretase for generation of Abeta peptides by neurons. *Nat. Neurosci.* **4:** 233–234.

Calhoun M.E., Wiederhold K.H., Abramowski D., Phinney A.L., Probst A., Sturchler-Pierrat C., Staufenbiel M., Sommer B., and Jucker M. 1998. Neuron loss in APP transgenic mice. *Nature* **395:** 755–756.

Calhoun M.E., Burgermeister P., Phinney A.L., Stalder M., Tolnay M., Wiederhold K.H., Abramowski D., Sturchler-Pierrat C., Sommer B., Staufenbiel M., and Jucker M. 1999. Neuronal overexpression of mutant amyloid precursor protein results in prominent deposition of cerebrovascular amyloid. *Proc. Natl. Acad. Sci.* **96:** 14088–14093.

Cao X. and Sudhof T.C. 2001. A transcriptionally [correction of transcriptively] active complex of APP with Fe65 and histone acetyltransferase Tip60. *Science* **293:** 115–120.

Carson J.A. and Turner A.J. 2002. Beta-amyloid catabolism: Roles for neprilysin (NEP) and other metallopeptidases? *J. Neurochem.* **81:** 1–8.

Chang W.P., Koelsch G., Wong S., Downs D., Da H., Weerasena V., Gordon B., Devasamudram T., Bilcer G., Ghosh A.K., and Tang J. 2004. In vivo inhibition of Abeta production by memapsin 2 (beta-secretase) inhibitors. *J. Neurochem.* **89:** 1409–1416.

Chapman P.F., White G.L., Jones M.W., Cooper-Blacketer D., Marshall V.J., Irizarry M., Younkin L., Good M.A., Bliss T.V.P., Hyman B.T., Younkin S.G., and Hsiao K.K. 1999. Impaired synaptic plasticity and learning in aged amyloid precursor protein transgenic mice. *Nat. Neurosci.* **2:** 271–276.

Cherny R.A., Atwood C.S., Xilinas M.E., Gray D.N., Jones W.D., McLean C.A., Barnham K.J., Volitakis I., Fraser F.W., Kim Y., Huang X., Goldstein L.E., Moir R.D., Lim J.T., Beyreuther K.T., Zheng H., Tanzi R.E., Masters C.L., and Bush A.I. 2001. Treatment with a copper-zinc chelator markedly and rapidly inhibits β-amyloid accumulation in Alzheimer's disease transgenic mice. *Neuron* **30:** 665–676.

Cirrito J.R., May P.C., O'Dell M.A., Taylor J.W., Parsadanian M., Cramer J.W., Audia J.E., Nissen J.S., Bales K.R., Paul S.M., DeMattos R.B., and Holtzman D.M. 2003. In vivo assessment of brain interstitial fluid with microdialysis reveals plaque-associated changes in amyloid-beta metabolism and half-life. *J. Neurosci.* **23:** 8844–8853.

Citron M. 2004a. Strategies for disease modification in Alzheimer's disease. *Nat. Rev. Neurosci.* **5:** 677–685.

———. 2004b. Beta-secretase inhibition for the treatment of Alzheimer's disease: Promise and challenge. *Trends Pharmacol. Sci.* **25:** 92–97.

Cleary J.P., Walsh D.M., Hofmeister J.J., Shankar G.M., Kuskowski M.A., Selkoe D.J., and Ashe K.H. 2005. Natural oligomers of the amyloid-beta protein specifically disrupt cognitive function. *Nat. Neurosci.* **8:** 79–84.

Coyle J.T., Price D.L., and DeLong M.R. 1983. Alzheimer's disease: A disorder of cortical cholinergic innervation. *Science* **219:** 1184–1190.

Cummings J.L. 2004. Alzheimer's disease. *N. Engl. J. Med.* **351:** 56–67.

Davies P. and Maloney A.J.F. 1976a. Selective loss of central cholinergic neurons in Alzheimer senile dementia. *Nature* **288:** 279–280.

———. 1976b. Selective loss of central cholinergic neurons in Alzheimer's disease. *Lancet* **2:** 1403.

Delacourte A., David J.P., Sergeant N., Buee L., Wattez A., Vermersch P., Ghozali F., Fallet-Bianco C., Pasquier F., Lebert F., Petit H., and Di Menza C. 1999. The biohemical pathway of neurofibrillary degeneration in aging and Alzheimer's disease. *Neurology* **52:** 1158–1165.

DeMattos R.B., Bales K.R., Cummins D.J., Paul S.M., and Holtzman D.M. 2002. Brain to plasma amyloid-beta efflux: A measure of brain amyloid burden in a mouse model of Alzheimer's disease. *Science* **295:** 2264–2267.

DeMattos R.B., Bales K.R., Cummins D.J., Dodart J.C., Paul S.M., and Holtzman D.M. 2001. Peripheral anti-Aβ antibody alters CNS and plasma Aβ clearance and decreases brain Aβ burden in a mouse model of Alzheimer's disease. *Proc. Natl. Acad. Sci.* **98:** 8850–8855.

De Strooper B., Saftig P., Craessaerts K., Vanderstichele H., Guhde G., Annaert W.G., Von Figura K., and Van Leuven F. 1998. Deficiency of presenilin-1 inhibits the normal cleavage of amyloid precursor protein. *Nature* **391:** 387–390.

De Strooper B., Annaert W.G., Cupers P., Saftig P., Craessaerts K., Mumm J.S., Schroeter E.H., Schrijvers V., Wolfe M.S., Ray W.J., Goate A., and Kopan R. 1999. A presenilin-1-dependent gamma-secretase-like protease mediates release of Notch intracellular domain. *Nature* **398:** 518–522.

Dodart J.C., Bales K.R., Gannon K.S., Greene S.J., DeMattos R.B., Mathis C., DeLong C.A., Wu S., Wu X., Holtzman D.M., and Paul S.M. 2002. Immunization reverses memory deficits without reducing brain Abeta burden in Alzheimer's disease model. *Nat. Neurosci.* **5:** 452–457.

Fagan A.M., Mintun M.A., Mach R.H., Lee S., Dence C.S., Shah A.R., LaRossa G.N., Spinner M.L., Klunk W.E., Mathis C.A., DeKosky S.T., Morris J.C., and Holtzman D.M. 2005. Inverse relation between in vivo amyloid imaging load and cerebrospinal fluid Abeta42 in humans. *Ann. Neurol.* **59:** 512–519.

Farris W., Mansourian S., Chang Y., Lindsley L., Eckman E.A., Frosch M.P., Eckman C.B., Tanzi R.E., Selkoe D.J., and Guenette S. 2003. Insulin-degrading enzyme regulates the levels of insulin, amyloid beta-protein, and the beta-amyloid precursor protein intracellular domain in vivo. *Proc. Natl. Acad. Sci.* **100:** 4162–4167.

Farzan M., Schnitzler C.E., Vasilieva N., Leung D., and Choe H. 2000. BACE2, a β-secretase homolog, cleaves at the β site and within the amyloid-β region of the amyloid-β precursor protein. *Proc. Natl. Acad. Sci.* **97:** 9712–9717.

Federoff H.J. and Bowers W.J. 2005. Immune shaping and the development of Alzheimer's disease vaccines. *Sci. Aging Knowledge Environ.* **2005:** e35.

Francis P.T. and Bowen D.M. 1985. Relevance of reduced concentrations of somatostatin in Alzheimer's disease. *Biochem. Soc. Trans.* **13:** 170–171.

Francis P.T., Palmer A.M., Sims N.R., Bowen D.M., Davison A.N., Esiri M.M., Neary D., Snowden J.S., and Wilcock G.K. 1985. Neurochemical studies of early-onset Alzheimer's disease. Possible influence on treatment. *N. Engl. J. Med.* **313:** 7–11.

Gandy S. and Walker L. 2004. Toward modeling hemorrhagic and encephalitic complications of Alzheimer amyloid-beta vaccination in nonhuman primates. *Curr. Opin. Immunol.* **16:** 607–615.

Gatz M., Reynolds C.A., Fratiglioni L., Johansson B., Mortimer J.A., Berg S., Fiske A., and Pedersen N.L. 2006. Role of genes and environments for explaining Alzheimer disease. *Arch. Gen. Psychiatry* **63:** 168–174.

Ghiso J. and Wisniewski T. 2004. An animal model of vascular amyloidosis. *Nat. Neurosci.* **7:** 902–904.

Goedert M. and Spillantini M.G. 2006a. A century of Alzheimer's disease. *Science* **314:** 777–781.

———. 2006b. Neurodegenerative alpha-synucleinopathies and tauopathies. In *Basic neurochemistry: Molecular, cellular, and medical aspects* (ed. G.J. Siegel et al.), pp. 745–759. Elsevier, Boston, Massachusetts.

Gong Y., Chang L., Viola K.L., Lacor P.N., Lambert M.P., Finch C.E., Krafft G.A., and Klein W.L. 2003. Alzheimer's disease-affected brain: Presence of oligomeric A beta ligands (ADDLs) suggests a molecular basis for reversible memory loss. *Proc. Natl. Acad. Sci.* **100:** 10417–10422.

Gotz J., Chen F., Van Dorpe J., and Nitsch R.M. 2001. Formation of neurofibrillary tangles in P3011 tau transgenic mice induced by Abeta fibrils. *Science* **293:** 1491–1495.

Goutte C., Tsunozaki M., Hale V.A., and Priess J.R. 2002. APH-1 is a multipass membrane protein essential for the Notch signaling pathway in *Caenorhabditis elegans* embryos. *Proc. Natl. Acad. Sci.* **99:** 775–779.

Hardy J. 2006a. A hundred years of Alzheimer's disease research. *Neuron* **52:** 3–13.

Hardy J. 2006b. Amyloid double trouble. *Nat. Genet.* **38:** 11–12.

Hartley D.M., Walsh D.M., Ye C.P., Diehl T., Vasquez S., Vassilev P.M., Teplow D.B., and Selkoe D.J. 1999. Protofibrillar intermediates of amyloid beta-protein induce acute electrophysiological changes and progressive neurotoxicity in cortical neurons. *J. Neurosci.* **19:** 8876–8884.

Herget T., Specht H., Esdar C., Oehrlein S.A., and Maelicke A. 1998. Retinoic acid induces apoptosis-associated neural differentiation of a murine teratocarcinoma cell line. *J. Neurochem.* **70:** 47–58.

Herzig M.C., Winkler D.T., Burgermeister P., Pfeifer M., Kohler E., Schmidt S.D., Danner S., Abramowski D., Sturchler-Pierrat C., Burki K., van Duinen S.G., Maat-Schieman M.L., Staufenbiel M., Mathews P.M., and Jucker M. 2004. Abeta is targeted to the vasculature in a mouse model of hereditary cerebral hemorrhage with amyloidosis. *Nat. Neurosci.* **7:** 954–960.

Hock C., Konietzko U., Papassotiropoulos A., Wollmer A., Streffer J., von Rotz R.C., Davey G., Moritz E., and Nitsch R.M. 2002. Generation of antibodies specific for beta-amyloid by vaccination of patients with Alzheimer disease. *Nat. Med.* **8:** 1270–1275.

Hock C., Konietzko U., Streffer J.R., Tracy J., Signorell A., Muller-Tillmanns B., Lemke U., Henke K., Moritz E., Garcia E., Wollmer M.A., Umbricht D., de Quervain D.J., Hofmann M., Maddalena A., Papassotiropoulos A., and Nitsch R.M. 2003. Antibodies against beta-amyloid slow cognitive decline in Alzheimer's disease. *Neuron* **38:** 547–554.

Hulette C., Mirra S., Wilkinson W., Heyman A., Fillenbaum G., and Clark C. 1995. The consortium to establish a registry for Alzheimer's disease (CERAD). Part IX. A prospective cliniconeuropathologic study of Parkinson's features in Alzheimer's disease. *Neurology* **45:** 1991–1995.

Hutton M. and McGowan E. 2004. Clearing tau pathology with abeta immunotherapy-reversible and irreversible stages revealed. *Neuron* **43:** 293–294.

Hyman B.T., Van Hoesen G.W., Damasio A.R., and Barnes C.L. 1984. Alzheimer's disease: Cell-specific pathology isolates the hippocampal formation. *Science* **225:** 1168–1170.

Iwata N., Mizukami H., Shirotani K., Takaki Y., Muramatsu S., Lu B., Gerard N.P., Gerard C., Ozawa K., and Saido T.C. 2004. Presynaptic localization of neprilysin contributes to efficient clearance of amyloid-beta peptide in mouse brain. *J. Neurosci.* **24:** 991–998.

Iwata N., Tsubuki S., Takaki Y., Shirotani K., Lu B., Gerard N.P., Gerard C., Hama E., Lee H.J., and Saido T.C. 2001. Metabolic regulation of brain Abeta by neprilysin, *Science* **292:** 1550–1552.

Iwata N., Tsubuki S., Takaki Y., Watanabe K., Sekiguchi M., Hosoki E., Kawashima-Morishima M., Lee H.J., Hama E., Sekine-Aizawa Y., and Saido T.C. 2000. Identification of the major Abeta1-42-degrading catabolic pathway in brain parenchyma: Suppression leads to biochemical and pathological deposition. *Nat. Med.* **2:** 143–150.

Iwatsubo T. 2004. The gamma-secretase complex: Machinery for intramembrane proteolysis. *Curr. Opin. Neurobiol.* **14:** 379–383.

Jankowsky J.L., Fadale D.J., Anderson J., Xu G.M., Gonzales V., Jenkins N.A., Copeland N.G., Lee M.K., Younkin L.H., Wagner S.L., Younkin S.G., and Borchelt D.R. 2004. Mutant presenilins specifically elevate the levels of the 42 residue β-amyloid peptide in vivo: Evidence for augmentation of a 42-specific γ-secretase. *Hum. Mol. Genet.* **13:** 159–170.

Jicha G.A., Parisi J.F., Dickson D.W., Johnson K., Cha R., Ivnik R.J., Tangalos E.G., Boeve B.F., Knopman D.S., Braak H., and Petersen R.C. 2006. Neuropathologic outcome of mild cognitive impairment following progression to clinical dementia. *Arch. Neurol.* **63:** 674–681.

Kawarabayashi T., Shoji M., Younkin L.H., Wen-Lang L., Dickson D.W., Murakami T., Matsubara E., Abe K., Ashe K.H., and Younkin S.G. 2004. Dimeric amyloid beta protein rapidly accumulates in lipid rafts followed by apolipoprotein E and phosphorylated tau accumulation in the Tg2576 mouse model of Alzheimer's disease. *J. Neurosci.* **24:** 3801–3809.

Kimberly W.T., LaVoie M.J., Ostaszewski B.L., Ye W., Wolfe M.S., and Selkoe D.J. 2003. Gamma-secretase is a membrane protein complex comprised of presenilin, nicastrin, Aph-1, and Pen-2. *Proc. Natl. Acad. Sci.* **100:** 6382–6387.

Kitt C.A., Price D.L., Struble R.G., Cork L.C., Wainer B.H., Becher M.W., and Mobley W.C. 1984. Evidence for cholinergic neurites in senile plaques. *Science* **226:** 1443–1445.

Kitt C.A., Struble R.G., Cork L.C., Mobley W.C., Walker L.C., Joh T.H., and Price D.L. 1985. Catecholaminergic neurites in senile plaques in prefrontal cortex of aged non-human primates. *Neuroscience* 16: 691–699.

Klein W.L., Krafft G.A., and Finch C.E. 2001. Targeting small Aβ oligomers: The solution to an Alzheimer's disease conundrum? *Trends Neurosci.* **24:** 219–223.

Klunk W.E., Engler H., Nordberg A., Wang Y., Blomqvist G., Holt D.P., Bergstrom M., Savitcheva I., Huang G.F., Estrada S., Ausen B., Debnath M.L., Barletta J., Price J.C., Sandell J., Lopresti B.J., Wall A., Koivisto P., Antoni G., Mathis C.A., and Langstrom B. 2004. Imaging brain amyloid in Alzheimer's disease with Pittsburgh Compound-B. *Ann. Neurol.* **55:** 306–319.

Klyubin I., Walsh D.M., Lemere C.A., Cullen W.K., Shankar G.M., Betts V., Spooner E.T., Jiang L., Anwyl R., Selkoe D.J., and Rowan M.J. 2005. Amyloid beta protein immunotherapy neutralizes Abeta oligomers that disrupt synaptic plasticity in vivo. *Nat. Med.* **11:** 556–561.

Kobayashi D., Zeller M., Cole T., Buttini M., McConlogue L., Sinha S., Freedman S., Morris R.G.M., and Chen K.S. 2007. *BACE1* gene deletion: Impact on behavioral function in a model of Alzheimer's disease. *Neurobiol. Aging* (in press).

Koo E.H., Sisodia S.S., Archer D.R., Martin L.J., Beyreuther K.T., Weidemann A., and Price D.L. 1989. Amyloid precursor protein (APP) undergoes fast anterograde transport. *Soc. Neurosci. Abstr.* **15:** 23.

Kotilinek L.A., Bacskai B.J., Westerman M., Kawarabayashi T., Younkin L., Hyman B.T., Younkin S., and Ashe K.H. 2002. Reversible memory loss in a mouse transgenic model of Alzheimer's disease. *J. Neurosci.* **22:** 6331–6335.

Laird F.M., Cai H., Savonenko A.V., Farah M.H., He K., Melnikova T., Wen H., Chiang, H.C., Xu G., Koliatsos V.E., Borchelt D.R., Price D.L., Lee H.K., and Wong P.C. 2005. BACE1, a major determinant of selective vulnerability of the brain to Aβ amyloidogenesis is essential for cognitive, emotional and synaptic functions. *J. Neurosci.* **25:** 11693–11709.

Lambert M.P., Barlow A.K., Chromy B.A., Edwards C., Freed R., Liosatos M., Morgan T.E., Rozovsky I., Trommer B., Viola K.L., Wals P., Zhang C., Finch C.E., Krafft G.A., and Klein W.L. 1998. Diffusible, nonfibrillar ligands derived from Abeta1-42 are potent central nervous system neurotoxins. *Proc. Natl. Acad. Sci.* **95:** 6448–6453.

Lauritzen M. and Gold L. 2003. Brain function and neurophysiological correlates of signals used in functional neuroimaging. *J. Neurosci.* **23:** 3972–3980.

Lazarov O., Lee M., Peterson D.A., and Sisodia S.S. 2002. Evidence that synaptically released beta-amyloid accumulates as extracellular deposits in the hippocampus of transgenic mice. *J. Neurosci.* **22:** 9785–9793.

Lazarov O., Robinson J., Tang Y.P., Hairston I.S., Korade-Mirnics Z., Lee V.M., Hersh L.B., Sapolsky R.M., Mirnics K., and Sisodia S.S. 2005a. Environmental enrichment reduces Abeta levels and amyloid deposition in transgenic mice. *Cell* **120:** 701–713.

Lazarov O., Morfini G.A., Lee E.B., Farah M.H., Szodorai A., Deboer S.R., Koliatsos V.E., Kins S., Lee V.M., Wong P.C., Price D.L., Brady S.T., and Sisodia S.S. 2005b. Axonal transport, amyloid precursor protein, kinesin-1, and the processing apparatus: Revisited. *J. Neurosci.* **25:** 2386–2395.

Lee V.M., Goedert M., and Trojanowski J.Q. 2001. Neurodegenerative tauopathies. *Annu. Rev. Neurosci.* **24:** 1121–1159.

Leissring M.A., Farris W., Chang A.Y., Walsh D.M., Wu X., Sun X., Frosch M.P., and Selkoe D.J. 2003. Enhanced proteolysis of beta-amyloid in APP transgenic mice prevents plaque formation, secondary pathology, and premature death. *Neuron* **40:** 1087–1093.

Lesne S., Koh M.T., Kotilinek L., Kayed R., Glabe C.G., Yang A., Gallagher M., and Ashe K.H. 2006. A specific amyloid-β protein assembly in the brain impairs memory. *Nature* **440:** 352–357.

Lewis J., Dickson D.W., Lin W.L., Chisholm L., Corral A., Jones G., Yen S.H., Sahara N., Skipper L., Yager D., Eckman C., Hardy J., Hutton M., and McGowan E. 2001. Enhanced neurofibrillary degeneration in transgenic mice expressing mutant tau and APP. *Science* **293:** 1487–1491.

Li T., Ma G., Cai H., Price D.L., and Wong P.C. 2003. Nicastrin is required for assembly of presenilin/gamma-secretase complexes to mediate Notch signaling and for processing and trafficking of beta-amyloid precursor protein in mammals. *J. Neurosci.* **23:** 3272–3277.

Li T., Wen H., Brayton C., Laird F.M., Ma G., Peng S., Placanica L., Wu T.-C., Crain B.J., Price D.L., et al. 2007. Moderate reduction of gamma-secretase attenuates amyloid burden and limits mechanism-based liabilities. *J. Neurosci.* (in press).

Lim G.P., Yang F., Chu T., Chen P., Beech W., Teter B., Tran T., Ubeda O., Ashe K.H., Frautschy S.A., and Cole G.M. 2000. Ibuprofen suppresses plaque pathology and inflammation in a mouse model for Alzheimer's disease. *J. Neurosci.* **20:** 5709–5714.

Lipton S. 2004. Paradigm shift in NMDA receptor antagonist drug development: Molecular mechanism of uncompetitive inhibition by memantine in the treatment of Alzheimer's disease and other neurologic disorders. *J. Alzheimer's Dis.* (suppl.) **6:** S61–S74.

Luo Y., Bolon B., Kahn S., Bennett B.D., Babu-Khan S., Denis P., Fan W., Kha H., Zhang J., Gong Y., Martin L., Louis J.C., Yan Q., Richards W.G., Citron M., and Vassar R. 2001. Mice deficient in BACE1, the Alzheimer's beta-secretase, have normal phenotype and abolished beta-amyloid generation. *Nat. Neurosci.* **4:** 231–232.

Ma G., Li T., Price D.L., and Wong P.C. 2005. APH-1a is the principal mammalian APH-1 isoform present in γ-secretase complexes during embryonic development. *J. Neurosci.* **25:** 192–198.

Mahley R.W., Weisgraber K.H., and Huang Y. 2006. Apolipoprotein E4: A causative factor and therapeutic target in neuropathology, including Alzheimer's disease. *Proc. Natl. Acad. Sci.* **103:** 5644–5651.

Mandelkow, E.-M., Thies, E., and Mandelkow E. 2007. Tau and axonal transport. In *Alzheimer's disease: Advances in genetics, molecular and cellular biology* (ed. S.S. Sisodia and R.E. Tanzi), pp. 237–256. Springer, New York.

Markesbery W.R., Schmitt F.A., Kryscio R.J., Davis D.G., Smith C.D., and Wekstein D.R. 2006. Neuropathologic substrate of mild cognitive impairment. *Arch. Neurol.* **63:** 38–46.

Marr R.A., Rockenstein E., Mukherjee A., Kindy M.S., Hersh L.B., Gage F.H., Verma I.M., and Masliah E. 2003. Neprilysin gene transfer reduces human amyloid pathology in transgenic mice. *J. Neurosci.* **23:** 1992–1996.

Martin L.J., Pardo C.A., Cork L.C., and Price D.L. 1994. Synaptic pathology and glial responses to neuronal injury precede the formation of senile plaques and amyloid deposits in the aging cerebral cortex. *Am. J. Pathol.* **145:** 1358–1381.

Masliah E., Hansen L., Adame A., Crews L., Bard F., Lee C., Seubert P., Games D., Kirby L., and Schenk D. 2005. Aβ vaccination effects on plaque pathology in the absence of encephalitis in Alzheimer disease. *Neurology* **64:** 129–131.

Mayeux R. 2003. Epidemiology of neurodegeneration. *Annu. Rev. Neurosci.* **26:** 81–104.

McGowan E., Eriksen J., and Hutton M. 2006. A decade of modeling Alzheimer's disease in transgenic mice. *Trends Genet.* **22:** 281–289.

McKhann G., Drachman D., Folstein M., Katzman R., Price D., and Stadlan E.M. 1984. Clinical diagnosis of Alzheimer's disease: Report of the NINCDS-ADRDA Work

Group under the auspices of the Department of Health and Human Services Task Force on Alzheimer's disease. *Neurology* **34:** 939–944.

McLean C.A., Cherny R.A., Fraser F.W., Fuller S.J., Smith M.J., Beyreuther K.T., Bush A.I., and Masters C.L. 1999. Soluble pool of Abeta amyloid as a determinant of severity of neurodegeneration in Alzheimer's disease. *Ann. Neurol.* **46:** 860–866.

Meyer M.R., Tschanz J.T., Norton M.C., Welsh-Bohmer K.A., Steffens D.C., Wyse B.W., and Breitner J.C. 1998. APOE genotype predicts when—not whether—one is predisposed to develop Alzheimer disease. *Nat. Genet.* **19:** 321–322.

Milano J., McKay J., Dagenais C., Foster-Brown L., Pognan F., Gadient R., Jacobs R.T., Zacco A., Greenberg B., and Ciaccio P.J. 2004, Modulation of Notch processing by γ-secretase inhibitors causes intestinal goblet cell metaplasia and induction of genes known to specify gut secretory lineage differentiation. *Toxicol. Sci.* **82:** 341–358.

Miller B.C., Eckman E.A., Sambamurti K., Dobbs N., Chow K.M., Eckman C.B., Hersh L.B., and Thiele D.L. 2003. Amyloid-beta peptide levels in brain are inversely correlated with insulysin activity levels in vivo. *Proc. Natl. Acad. Sci.* **100:** 6221–6226.

Monsonego A. and Weiner H.L. 2003. Immunotherapeutic approaches to Alzheimer's disease. *Science* **302:** 834–838.

Morgan D., Diamond D.M., Gottschall P.E., Ugen K.E., Dickey C., Hardy J., Duff K., Jantzen P., DiCarlo G., Wilcock D., Connor K., Hatcher J., Hope C., Gordon M., and Arendash G.W. 2000. Aβ peptide vaccination prevents memory loss in an animal model of Alzheimer's disease. *Nature* **408:** 982–985.

Morris J.C. and Price J.L. 2001. Pathologic correlates of nondemented aging, mild cognitive impairment, and early-stage Alzheimer's disease. *J. Mol. Neurosci.* **17:** 101–118.

Morris J.C., Storandt M., Miller J.P., McKeel D.W., Price J.L., Rubin E.H., and Berg L. 2001. Mild cognitive impairment represents early-stage Alzheimer disease. *Arch. Neurol.* **58:** 397–405.

Mucke L., Masliah E., Yu G.Q., Mallory M., Rockenstein E.M., Tatsuno G., Hu K., Kholodenko D., Johnson-Wood K., and McConlogue L. 2000. High-level neuronal expression of $A\beta_{1-42}$ in wild-type human amyloid protein precursor transgenic mice: Synaptotoxicity without plaque formation. *J. Neurosci.* **20:** 4050–4058.

Nestor P.J., Scheltens P., and Hodges J.R. 2004. Advances in the early detection of Alzheimer's disease. *Nat. Med.* (suppl.) **10:** S34–S41.

Nicolas G., Bennoun M., Porteu A., Mativet S., Beaumont C., Grandchamp B., Sirito M., Sawadogo M., Kahn A., and Vaulont S. 2002. Severe iron deficiency anemia in transgenic mice expressing liver hepcidin. *Proc. Natl. Acad. Sci.* **99:** 4596–4601.

Nicolas M., Wolfer A., Raj K., Kummer J.A., Mill P., van Noort M., Hui C.C., Clevers H., Dotto G.P., and Radtke F. 2003. Notch1 functions as a tumor suppressor in mouse skin. *Nat. Genet.* **33:** 416–421.

Nicoll J.A., Wilkinson D., Holmes C., Steart P., Markham H., and Weller R.O. 2003. Neuropathology of human Alzheimer disease after immunization with amyloid-beta peptide: A case report. *Nat. Med.* **9:** 448–452.

Nishizaki T., Matsuoka T., Nomura T., Kondoh T., Watabe S., Shiotani T., and Yoshii M. 2000. Presynaptic nicotinic acetylcholine receptors as a functional target of nefiracetam in inducing a long-lasting facilitation of hippocampal neurotransmission. *Alzheimer Dis. Assoc. Disord.* **14:** S82–S94.

Oddo S., Billings L., Kesslak J.P., Cribbs D.H., and LaFerla F.M. 2004. Abeta immunotherapy leads to clearance of early, but not late, hyperphosphorylated tau aggregates via the proteasome. *Neuron* **43:** 321–332.

Oddo S., Caccamo A., Shepherd J.D., Murphy M.P., Golde T.E., Kayed R., Metherate R., Mattson M.P., Akbari Y., and LaFerla F.M. 2003 Triple-transgenic model of Alzheimer's disease with plaques and tangles: Intracellular Abeta and synaptic dysfunction. *Neuron* **39:** 409–421.

Ohno M., Sametsky E.A., Younkin L.H., Oakley H., Younkin S.G., Citron M., Vassar R., and Disterhoft J.F. 2004. BACE1 deficiency rescues memory deficits and cholinergic dysfunction in a mouse model of Alzheimer's disease. *Neuron* **41:** 27–33.

Perry E.K., Perry R.H., Blessed G., and Tomlinson B.E. 1978a. Changes in brain cholinesterases in senile dementia of Alzheimer type. *Neuropathol. Appl. Neurobiol.* **4:** 273–277.

Perry E.K., Gibson P.H., Blessed G., Perry R.H., and Tomlinson B.E. 1977. Neurotransmitter enzyme abnormalities in senile dementia. *J. Neurol. Sci.* **34:** 247–265.

Perry E.K., Tomlinson B.E., Blessed G., Bergmann K., Gibson P.H., and Perry R.H. 1978b. Correlation of cholinergic abnormalities with senile plaques and mental test scores in senile dementia. *Br. Med. J.* **2:** 1457–1459.

Perry T.L., Hansen S., Currier R.D., and Berry K. 1978. Abnormalities in neurotransmitter amino acids in dominantly inherited cerebellar disorders. In *The inherited ataxias: Biochemical, viral, and pathological studies* (ed. R.A.P. Kark et al.), pp. 303–314. Raven Press, New York.

Petersen R.C. 2003. Mild cognitive impairment clinical trials. *Nat. Rev. Drug Discov.* **2:** 646–653.

Petersen R.C., Doody R., Kurz A., Mohs R.C., Morris J.C., Rabins P.V., Ritchie K., Rossor M., Thal L., and Winblad B. 2001. Current concepts in mild cognitive impairment. *Arch. Neurol.* **58:** 1985–1992.

Petersen R.C., Parisi J.E., Dickson D.W., Johnson K.A., Knopman D.S., Boeve B.F., Jicha G.A., Ivnik R.J., Smith G.E., Tangalos E.G., Braak H., and Kokmen E. 2006. Neuropathologic features of amnestic mild cognitive impairment. *Arch. Neurol.* **63:** 665–672.

Pfeifer M., Boncristiano S., Bondolfi L., Stalder A., Deller T., Staufenbiel M., Mathews P.M., and Jucker M. 2002. Cerebral hemorrhage after passive anti-Abeta immunotherapy. *Science* **298:** 1379.

Prasher V.P., Farrer M.J., Kessling A.M., Fisher E.M., West R.J., Barber P.C., and Butler A.C. 1998. Molecular mapping of Alzheimer-type dementia in Down's syndrome. *Ann. Neurol.* **43:** 380–383.

Price D.L. and Sisodia S.S. 1998. Mutant genes in familial Alzheimer's disease and transgenic models. *Annu. Rev. Neurosci.* **21:** 479–505.

Price D.L., Tanzi R.E., Borchelt D.R., and Sisodia. S.S. 1998. Alzheimer's disease: Genetic studies and transgenic models. *Annu. Rev. Genet.* **32:** 461–493.

Reisberg B., Doody R., Stoffler A., Schmitt F., Ferris S., and Mobius H.J. 2003. Memantine in moderate-to-severe Alzheimer's disease. *N. Engl. J. Med.* **348:** 1333–1341.

Ritchie C.W., Bush A.I., Mackinnon A., Macfarlane S., Mastwyk M., MacGregor L., Kiers L., Cherny R., Li Q., Tammer A., Carrington D., Mavros C., Volitakis I., Xilinas M., Ames D., Davis S., Beyreuther K., Tanzi R.E., and Masters C.L. 2003. Metal-protein attenuation with iodochlorhydroxyquin (olioquinol) targeting $A\beta$ amyloid deposition and toxicity in Alzheimer disease: A pilot phase 2 clinical trial. *Arch. Neurol.* **60:** 1685–1691.

Roberson E.D. and Mucke L. 2006. 100 years and counting: Prospects for defeating Alzheimer's disease. *Science* **314:** 781–784.

Rovelet-Lecrux A., Hannequin D., Raux G., Meur N.L., Laquerriere A., Vital A., Dumanchin C., Feuillette S., Brice A., Vercelletto M., Dubas F., Frebourg T., and Campion D. 2006. APP locus duplication causes autosomal dominant early-onset Alzheimer disease with cerebral amyloid angiopathy. *Nat. Genet.* **38:** 24–26.

Saunders A.M., Strittmatter W.J., Schmechel D., St. George-Hyslop P.H., Pericak-Vance M.A., Joo S.H., Rosi B.L., Gusella J.F., Crapper-MacLachlan D.R., Alberts M.J., Hulette C., Crain B., Goldgaber D., and Roses A.D. 1993. Association of apolipoprotein E allele ε4 with late-onset familial and sporadic Alzheimer's disease. *Neurology* **43:** 1467–1472.

Saura C.A., Choi S.Y., Beglopoulos V., Malkani S., Zhang D., Rao B.S., Chattarji S., Kelleher R.J., III, Kandel E.R., Duff K., Kirkwood A., and Shen J. 2004. Loss of presenilin function causes impairments of memory and synaptic plasticity followed by age-dependent neurodegeneration. *Neuron* **42:** 23–36.

Savonenko A.V., Laird F.M., Troncoso J.C., Wong P.C., and Price D.L. 2006. Role of Alzheimer's disease models in designing and testing experimental therapeutics. *Drug Discov. Today* **2:** 305–312.

Savonenko A.V., Xu G.M., Price D.L., Borchelt D.R., and Markowska A.L. 2003. Normal cognitive behavior in two distinct congenic lines of transgenic mice hyperexpressing mutant APP_{swe}. *Neurobiol. Dis.* **12:** 194–211.

Savonenko A., Xu G.M., Melnikova T., Morton J.L., Gonzales V., Wong M.P.F., Price D.L., Tang F., Markowska A.L., and Borchelt D.R. 2005. Episodic-like memory deficits in the APPswe/PS1dE9 mouse model of Alzheimer's disease: Relationships to β-amyloid deposition and neurotransmitter abnormalities. *Neurobiol. Dis.* **18:** 602–617.

Schenk D., Hagen M., and Seubert P. 2004. Current progress in beta-amyloid immunotherapy. *Curr. Opin. Immunol.* **16:** 599–606.

Schenk D., Barbour R., Dunn W., Gordon G., Grajeda H., Guido T., Hu K., Huang J.P., Johnson-Wood K., Khan K., Kholodenko D., Lee M., Liao Z.M., Lieberburg I., Motter R., Mutter L., Soriano F., Shopp G., Vasquez N., Vandevert C., Walker S., Wogulis M., Yednock T., Games D., and Seubert P. 1999. Immunization with amyloid-beta attenuates Alzheimer disease-like pathology in the PDAPP mouse *Nature* **400:** 173–177.

Schmitt F.A., Davis D.G., Wekstein D.R., Smith C.D., Ashford J.W., and Markesbery W.R. 2000. Preclinical AD revisited. *Neurology* **55:** 370–376.

Selkoe D. and Kopan R. 2003. Notch and presenilin: Regulated intramembrane proteolysis links development and degeneration. *Annu. Rev. Neurosci.* **26:** 565–597.

Selkoe D.J. and Schenk D. 2003. Alzheimer's disease: Molecular understanding predicts amyloid-based therapeutics. *Annu. Rev. Pharmacol. Toxicol.* **43:** 545–584.

Selkoe D.J., Bell D.S., Podlisny M.B., Price D.L., and Cork L.C. 1987. Conservation of brain amyloid proteins in aged mammals and humans with Alzheimer's disease. *Science* **235:** 873–877.

Serneels L., Dejaegere T., Craessaerts K., Horre K., Jorissen E., Tousseyn T., Hebert S., Coolen M., Martens G., Zwijsen A., Annaert W., Hartmann D., and De Strooper B. 2005. Differential contribution of the three Aph1 genes to γ-secretase activity in vivo. *Proc. Natl. Acad. Sci.* **102:** 1719–1724.

Shah S., Lee S.F., Tabuchi K., Hao Y.H., Yu C., LaPlant Q., Ball H., Dann C.E., III, Sudhof T., and Yu G. 2005. Nicastrin functions as a gamma-secretase-substrate receptor. *Cell* **122:** 435–447.

Shen J., Bronson R.T., Chen D.F., Xia W., Selkoe D.J., and Tonegawa S. 1997. Skeletal and CNS defects in presenilin-1-deficient mice. *Cell* **89:** 629–639.

Sheng J.G., Price D.L., and Koliatsos V.E. 2002. Disruption of corticocortical connections ameliorates amyloid burden in terminal fields in a transgenic model of Abeta amyloidosis. *J. Neurosci.* **22:** 9794–9799.

————. 2003. The beta-amyloid-related proteins presenilin 1 and BACE1 are axonally transported to nerve terminals in the brain. *Exp. Neurol.* **184:** 1053–1057.

Sherrington R., Rogaev E.I., Liang Y., Rogaeva E.A., Levesque G., Ikeda M., Chi H., Lin C., Li G., Holman K., Tsuda T., Mar L., Foncin J., Bruni A.C., Montesi M.P., Sorbi S., Rainero I., Pinessi L., Nee L., Chumakov I., Pollen D., Brookes A., Sanseau P., Polinsky R.J., and Wasco W., et al. 1995. Cloning of a gene bearing missense mutations in early-onset familial Alzheimer's disease. *Nature* **375:** 754–760.

Sigurdsson E.M., Knudsen E., Asuni A., Fitzer-Attas C., Sage D., Quartermain D., Goni F., Frangione B., and Wisniewski T. 2004. An attenuated immune response is sufficient to enhance cognition in an Alzheimer's disease mouse model immunized with amyloid-beta derivatives. *J. Neurosci.* **24:** 6277–6282.

Singer O., Marr R.A., Rockenstein E., Crews L., Coufal N.G., Gage F.H., Verma I.M., and Masliah E. 2005. Targeting BACE1 with siRNAs ameliorates Alzheimer disease neuropathology in a transgenic model. *Nat. Neurosci.* **8:** 1343–1349.

Sisodia S.S., Koo E.H., Beyreuther K.T., Unterbeck A., and Price D.L. 1990. Evidence that β-amyloid protein in Alzheimer's disease is not derived by normal processing. *Science* **248:** 492–495.

Sisodia S.S., Koo E.H., Hoffman P.N., Perry G., and Price D.L. 1993. Identification and transport of full-length amyloid precursor proteins in rat peripheral nervous system. *J. Neurosci.* **13:** 3136–3142.

Small S.A. and Gandy S. 2006. Sorting through the cell biology of Alzheimer's disease: Intracellular pathways to pathogenesis. *Neuron* **52:** 15–31.

Solitare G.B. and Lamarche J.B. 1966. Alzheimer's disease and senile dementia as seen in mongoloids: Neuropathological observations. *Am. J. Ment. Defic.* **70:** 840–848.

Stokin G.B., Lillo C., Falzone T.L., Brusch R.G., Rockenstein E., Mount S.L., Raman R., Davies P., Masliah E., Williams D.S., and Goldstein L.S. 2005. Axonopathy and transport deficits early in the pathogenesis of Alzheimer's disease. *Science* **307:** 1282–1288.

Strittmatter W.J., Saunders A.M., Schmechel D., Pericak-Vance M., Enghild J., Salvesen G.S., and Roses A.D. 1993. Apolipoprotein E: High-avidity binding to β-amyloid and increased frequency of type 4 allele in late-onset familial Alzheimer disease. *Proc. Natl. Acad. Sci.* **90:** 1977–1981.

Sunderland T., Linker G., Mirza N., Putnam K.T., Friedman D.L., Kimmel L.H., Bergeson J., Manetti G.J., Zimmermann M., Tang B., Bartko J.J., and Cohen R.M. 2003. Decreased beta-amyloid1-42 and increased tau levels in cerebrospinal fluid of patients with Alzheimer disease. *J. Am. Med. Assoc.* **289:** 2094–2103.

Sze C.I., Troncoso J.C., Kawas C.H., Mouton P.R., Price D.L., and Martin L.J. 1997. Loss of the presynaptic vesicle protein synaptophysin in hippocampus correlates with early cognitive decline in aged humans. *J. Neuropathol. Exp. Neurol.* **56:** 933–944.

Tanzi R.E. 2005. The synaptic Abeta hypothesis of Alzheimer disease. *Nat. Neurosci.* **8:** 977–979.

Tanzi R.E. and Bertram L. 2001. New frontiers in Alzheimer's disease genetics. *Neuron* **32:** 181–184.

Tanzi R.E., Moir R.D., and Wagner S.L. 2004. Clearance of Alzheimer's Abeta peptide: The many roads to perdition. *Neuron* **43:** 605–608.

Thinakaran G., Harris C.L., Ratovitski T., Davenport F., Slunt H.H., Price D.L., Borchelt D.R., and Sisodia S.S. 1997. Evidence that levels of presenilins (PS1 and PS2) are coordinately regulated by competition for limiting cellular factors. *J. Biol. Chem.* **272:** 28415–28422.

Troncoso J.C., Martin L.J., Dal Forno G., and Kawas C.H. 1996. Neuropathology in controls and demented subjects from the Baltimore Longitudinal Study of Aging. *Neurobiol. Aging* **17:** 365–371.

Vassar R., Bennett B.D., Babu-Khan S., Kahn S., Mendiaz E.A., Denis P., Teplow D.B., Ross S., Amarante P., Loeloff R., Luo L., Fisher S., Fuller J., Edenson S., Lile J., Jarosinski M.A., Biere A.L., Curran E., Burgess T., Louis J.C., Collins F., Treanor J., Rogers G., and Citron M. 1999. β-secretase cleavage of Alzheimer's amyloid precursor protein by the transmembrane aspartic protease BACE. *Science* **286:** 735–741.

Vekrellis K., Ye Z., Qiu W.Q., Walsh D., Hartley D., Chesneau V., Rosner M.R., and Selkoe D.J. 2000. Neurons regulate extracellular levels of amyloid beta-protein via proteolysis by insulin-degrading enzyme. *J. Neurosci.* **20:** 1657–1665.

Walsh D.M. and Selkoe D.J. 2004. Deciphering the molecular basis of memory failure in Alzheimer's disease. *Neuron* **44:** 181–193.

Walsh D.M., Hartley D.M., Kusumoto Y., Fezoui Y., Condron M.M., Lomakin A., Benedek G.B., Selkoe D.J., and Teplow D.B. 1999. Amyloid beta-protein fibrillogenesis. Structure and biological activity of protofibrillar intermediates. *J. Biol. Chem.* **274:** 25945–25952.

Walsh D.M., Klyubin I., Faden A.I., Fadeeva J.V., Cullen W.K., Anwyl R., Wolfe M.S., Rowan M.J., and Selkoe D.J. 2002. Naturally secreted oligomers of amyloid β-protein potently inhibit hippocampal LTP *in vivo*. *Nature* **416:** 535–539.

Wang H.W., Pasternak J.F., Kuo H., Ristic H., Lambert M.P., Chromy B., Viola K.L., Klein W.L., Stine W.B., Krafft G.A., and Trommer B.L. 2002. Soluble oligomers of beta amyloid (1–42) inhibit long-term potentiation but not long-term depression in rat dentate gyrus. *Brain Res.* **924:** 133–140.

Weggen S., Eriksen J.L., Das P., Sagl S.A., Wang R., Pietrzik C.U., Findlay K.A., Smith T.E., Murphy M.P., Bulter T., Kang D.E., Marquez-Sterling N., Golde T.E., and Koo E.H. 2001. A subset of NSAIDs lower amyloidogenic Aβ42 independently of cyclooxygenase activity. *Nature* **414:** 212–216.

Weninger S.C. and Yankner B.A. 2001. Inflammation and Alzheimer disease: The good, the bad, and the ugly. *Nat. Med.* **7:** 527–528.

West M.J., Coleman P.D., Flood D.G., and Troncoso J.C. 1994. Differences in the pattern of hippocampal neuronal loss in normal ageing and Alzheimer's disease. *Lancet* **344:** 769–772.

West M.J., Kawas C.H., Martin L.J., and Troncoso J.C. 2000. The CA1 region of the human hippocampus is a hot spot in Alzheimer's disease. *Ann. N.Y. Acad. Sci.* **908:** 255–259.

West M.J., Kawas C.H., Stewart W.F., Rudow G., and Troncoso J.C. 2004. Hippocampal neurons in pre-clinical Alzheimer's disease. *Neurobiol. Aging* **25:** 1205–1212.

Whitehouse P.J., Price D.L., Clark A.W., Coyle J.T., and DeLong M.R. 1981. Alzheimer disease: Evidence for selective loss of cholinergic neurons in the nucleus basalis. *Ann. Neurol.* **10:** 122–126.

Whitehouse P.J., Price D.L., Struble R.G., Clark A.W., Coyle J.T., and DeLong M.R. 1982. Alzheimer's disease and senile dementia: Loss of neurons in the basal forebrain. *Science* **215:** 1237–1239.

Wilcock D.M., Munireddy S.K., Rosenthal A., Ugen K.E., Gordon M.N., and Morgan D. 2004a. Microglial activation facilitates Abeta plaque removal following intracranial anti-Abeta antibody administration. *Neurobiol. Dis.* **15:** 11–20.

Wilcock D.M., DiCarlo G., Henderson D., Jackson J., Clarke K., Ugen K.E., Gordon M.N., and Morgan D. 2003. Intracranially administered anti-Abeta antibodies reduce beta-amyloid deposition by mechanisms both independent of and associated with microglial activation. *J. Neurosci.* **23:** 3745–3751.

Wilcock D.M., Rojiani A., Rosenthal A., Levkowitz G., Subbarao S., Alamed J., Wilson D., Wilson N., Freeman M.J., Gordon M.N., and Morgan D. 2004b. Passive amyloid immunotherapy clears amyloid and transiently activates microglia in a transgenic mouse model of amyloid deposition. *J. Neurosci.* **24:** 6144–6151.

Winblad B., Kilander L., Eriksson S., Minthon L., Batsman S., Wetterholm A.L., Jansson-Blixt C., and Haglund A. 2006. Donepezil in patients with severe Alzheimer's disease: Double-blind, parallel-group, placebo-controlled study. *Lancet* **367:** 1057–1065.

Winkler D.T., Bondolfi L., Herzig M.C., Jann L., Calhoun M.E., Wiederhold K.H., Tolnay, M., Staufenbiel M., and Jucker M. 2001. Spontaneous hemorrhagic stroke in a mouse model of cerebral amyloid angiopathy. *J. Neurosci.* **21:** 1619–1627.

Wolfe M.S. and Kopan R. 2004. Intramembrane proteolysis: Theme and variations. *Science* **305:** 1119–1123.

Wolfe M.S., Xia W., Ostaszewski B.L., Diehl T.S., Kimberly W.T., and Selkoe D.J. 1999. Two transmembrane aspartates in presenilin-1 required for presenilin endoproteolysis and gamma-secretase activity. *Nature* **398:** 513–517.

Wong G.T., Manfra D., Poulet F.M., Zhang Q., Josien H., Bara T., Engstrom L., Pinzon-Ortiz M., Fine J.S., Lee H.J., Zhang L., Higgins G.A., and Parker E.M. 2004. Chronic treatment with the gamma-secretase inhibitor LY-411,575 inhibits beta-amyloid peptide production and alters lymphopoiesis and intestinal cell differentiation. *J. Biol. Chem.* **279:** 12876–12882.

Wong P.C., Li T., and Price D.L. 2006. Neurobiology of Alzheimer's disease. In *Basic neurochemistry: Molecular, cellular, and medical aspects,* 7th edition (ed. G.J. Siegel et al.), pp. 781–790. Elsevier, Boston, Massachusetts.

Wong P.C., Price D.L., and Cai H. 2001. The brain's susceptibility to amyloid plaques. *Science* **293:** 1434–1435.

Wong P.C., Cai H., Borchelt D.R., and Price D.L. 2002. Genetically engineered mouse models of neurodegenerative diseases. *Nat. Neurosci.* **5:** 633–639.

Wong P.C., Zheng H., Chen H., Becher M.W., Sirinathsinghji D.J., Trumbauer M.E., Chen H.Y., Price D.L., Van der Ploeg L.H., and Sisodia S.S. 1997. Presenilin 1 is required for Notch1 and Dll1 expression in the paraxial mesoderm. *Nature* **387:** 288–292.

Xia X., Qian S., Soriano S., Wu Y., Fletcher A.M., Wang X.J., Koo E.H., Wu X., and Zheng H. 2001. Loss of presenilin 1 is associated with enhanced beta-catenin signaling and skin tumorigenesis. *Proc. Natl. Acad. Sci.* **98:** 10863–10868.

Xu G., Gonzales V., and Borchelt D.R. 2002. Abeta deposition does not cause the aggregation of endogenous tau in transgenic mice. *Alzheimer Dis. Assoc. Disord.* **16:** 196–201.

Zamora E., Handisurya A., Shafti-Keramat S., Borchelt D., Rudow G., Conant K., Cox C., Troncoso J.C., and Kirnbauer R. 2006. Papillomavirus-like particles are an effective platform for amyloid-beta immunization in rabbits and transgenic mice. *J. Immunol.* **177:** 2662–2670.

Zhao X., Kuryatov A., Lindstrom J.M., Yeh J.Z., and Natahashi T. 2001. Nootropic drug modulation of neuronal nicotinic acetylcholine receptors in rat cortical neurons. *Mol. Pharmacol.* **59:** 674–683.

15

How Does Caloric Restriction Increase the Longevity of Mammals?

Richard Weindruch
Department of Medicine, University of Wisconsin
Madison, Wisconsin 53706
Geriatric Research, Education, and Clinical Center
William S. Middleton Memorial Veterans Hospital
Madison, Wisconsin 53705

Ricki J. Colman
Wisconsin National Primate Research Center
Madison, Wisconsin 53715

Viviana Pérez[1,2] and Arlan G. Richardson[1,2,3]
[1] Department of Cellular & Structural Biology and
[2] Barshop Institute for Longevity and Aging Studies
   at the University of Texas Health Science Center
   at San Antonio, Texas 78245-3207
[3] Geriatric Research Education and Clinical Center
   of the South Texas Veterans Health Care System
   San Antonio, Texas 78245

THE CLASSIC STUDY BY McCAY ET AL. IN 1935 showed that one could increase the life span of rats by reducing their food consumption. Since this initial observation, numerous laboratories have confirmed these results and have shown that reducing food consumption 30–50% (without malnutrition) consistently increases both the mean and maximum life spans of laboratory rodents (Weindruch and Walford 1988; Masoro 2005). Caloric restriction is also able to oppose the development of diverse age-associated diseases arising in laboratory rodents, including many types of cancer, diabetes, and renal disease (Weindruch and Walford 1988). This paradigm has been termed caloric restriction, dietary restriction, or food restriction. In this chapter, we use the term caloric restriction (CR) because the decreased intake of total calories appears to be responsible for the increased life span of rodents (Masoro 2005), rather than the reduction in a specific nutrient, such as dietary protein or fat (Iwasaki et al. 1988; Masoro et al. 1989). It is important to note that the effect of CR on longevity is not limited to rodents, as it increases the life span of a variety of invertebrates, e.g., yeast, *Caenorhabditis elegans,* and *Drosophila* (Min and Tatar 2006), as well as of dogs (Kealy et al. 2002). In this review chapter, we focus on what currently is known of the biological mechanism responsible for the life-extending action of CR in mammals, specifically laboratory rodents and nonhuman primates.

LABORATORY RODENTS

Since the seminal observation by McCay et al. in 1935, CR has been shown to significantly increase the mean and maximum life spans of a variety of strains of both male and female laboratory rats and mice (Weindruch and Walford 1988; Tuturro et al. 1999). CR is even shown to increase the life span (~15%) of long-lived Ames dwarf mice (Bartke et al. 2001), which live 50–60% longer than their normal, wild-type littermates. A large amount of research during the past four decades has shown that CR retards/reduces age-related declines in most physiological functions and age-related increases in pathology and disease (Weindruch and Walford 1988; Masoro 2005). These observations have led to the view that CR increases the life span of laboratory rodents by retarding/slowing down aging.

   A widely accepted explanation for how CR extends the life span of rodents is through alterations in oxidative damage/oxidative stress. During the past three decades, a large number of studies have shown that CR retards/reduces age-related increases in the levels of oxidative damage to lipid, protein, and DNA in a variety of tissues of laboratory rats and mice (for review, see Bokov et al. 2004). In addition, CR mice have been shown to be more resistant to oxidative stress (Sun et al. 2001; Richardson et al. 2004). Thus, it has been argued that the anti-aging action of CR arises from a reduction of oxidative damage and/or enhanced resistance to oxidative stress. Although the activity of one or more of the major antioxidant enzymes has been reported to be increased significantly in tissues of CR mice or rats, there is no clear, overall trend in the expression of antioxidant enzymes with CR (Bokov et al. 2004). However, CR has been observed consistently to reduce the production of reactive oxygen species (ROS) in mitochondria isolated from various tissues of mice (Sohal et al. 1994) and rats (Gredilla et al. 2001; Lopez-Torres et al. 2002;

Lambert and Merry 2004). The decrease in ROS production by isolated mitochondria from tissues of rodents on CR appears to be due to an up-regulation of uncoupling proteins (UCPs) (Cadenas et al. 1999; Lee et al. 2002; Bevilacqua et al. 2004, 2005; Xiao et al. 2004), which is believed to lead to a more rapid electron flux through the respiratory chain, leading to energy dissipation, a decrease in proton leak, and reduced ROS production. Several investigators have shown a reduced proton leak in mitochondria isolated from tissues of CR rats (Vidal-Puig et al. 2000; Cline et al. 2001; Krauss et al. 2002; Bevilacqua et al. 2004, 2005; Hagopian et al. 2005).

What molecular pathways/processes are responsible for CR's ability to slow the aging process? Data from genetic manipulations of invertebrate models and research with laboratory rodents suggest that changes in signaling pathways have a role in the life-extending action of CR. Figure 1 shows schematically the four signaling pathways that are likely to have an important role in the reduction in oxidative damage/oxidative stress in tissues of CR rodents: insulin/insulin-like growth factor-1 (IGF-1), sirtuin, redox, and TOR (target of rapamycin) signaling pathways.

Insulin/IGF-1 Signaling Pathway

Studies in *C. elegans* during the past 5–10 years have been the major driving force behind the view that a reduction in insulin/IGF-1 signaling retards aging. A subset of *daf* genes, which are associated with the insulin-like signaling pathway, dramatically increases life span, and the main target of *daf* signaling is *daf-16*, which encodes a fork-head transcription factor that regulates downstream target genes (Guarente and Kenyon 2000). FOXO, a family of mammalian fork-head transcription factors, is similar in sequence to *daf-16*. In the presence of insulin and growth factors, FOXO is localized to the cytoplasm and degraded via the ubiquitin-proteasome pathway. A reduction in insulin and growth factors results in the translocation of FOXO proteins to the nucleus, resulting in the up-regulation of a series of target genes promoting cell cycle arrest, stress resistance, and apoptosis (Greer and Brunet 2005). Thus, the up-regulation of this pathway could have an important role in the increased stress resistance and reduced levels of oxidative damage observed in tissues of CR rodents. Data from CR studies in the 1980s and 1990s show that CR alters insulin and IGF-1 levels in rodents. For example, Masoro et al. (1989, 1992) observed that CR resulted in reduced (~50%) serum levels of insulin throughout most of the day over the life span of CR rats compared to rats fed ad libitum. Other groups also found that CR reduced

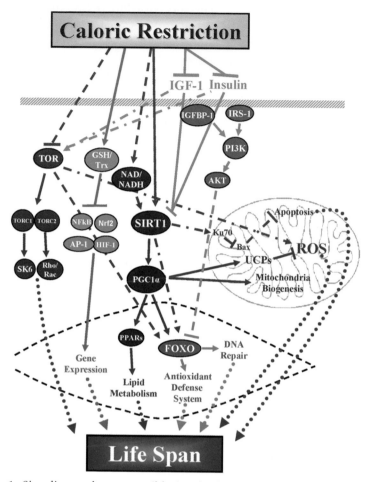

Figure 1. Signaling pathways possibly involved in life span extension by caloric restriction. The figure shows a schematic representation of potential mechanisms of how CR increases the life span of mammals through four signaling pathways: the insulin/IGF-1 pathway (*blue*), the sirtuin pathway (*purple*), the redox-signaling pathway (*green*), and the TOR pathway (*red*). The solid lines indicate biochemical/molecular processes known to be altered by CR in mammals. The dashed lines indicate biochemical/molecular processes shown to be altered by CR in invertebrates; however, these processes have not been studied in mammals. The dotted-dashed lines indicate biochemical/molecular processes that have been shown to occur in mammalian cells in the TOR pathway. The dotted lines indicate potential biochemical/molecular processes that could increase mammalian life span.

serum levels of insulin in rats (Kalant et al. 1988; Wetter et al. 1999) and mice (Dhahbi et al. 2001). CR also reduces plasma levels of IGF-1 in rats and mice (Breese et al. 1991; Sonntag et al. 1999). However, the relative reduction in IGF-1 levels is generally not as marked as the reduction in serum insulin; e.g., CR results in a 15–22% reduction in plasma IGF-1 levels. Kenyon (2001) has argued that reduced circulating levels of insulin and IGF-I in CR rodents led to increased life span through a mechanism of reduced insulin/IGF-1 signaling, as has been observed in *C. elegans.* However, there is limited evidence for CR reducing downstream signaling molecules in tissues of rodents, even though CR clearly reduces plasma levels of insulin and IGF-1.

Sirtuin Signaling Pathway

Studies by Guarente's laboratory in yeast produced the first data suggesting that the *sir2* gene has an important role in extension of life span by CR (for review, see Bordone and Guarente 2005). SIR2 is a member of the conserved sirtuin family of nicotinamide adenine dinucleotide (NAD)-dependent deacetylases (Hekimi and Guarente 2003). Wood et al. (2004) demonstrated that the increase in life span in *Drosophila* by CR also requires the SIR2 ortholog. On the basis of these studies in invertebrates, investigators have studied the effect of CR on SIRT1, the mammalian SIR2 ortholog, in rodents. In 2004, Cohen et al. (2004b) reported that CR induced the expression of SIRT1 protein in liver, fat, brain, and kidney of rats. Nisoli et al. (2005) also reported that CR increased SIRT1 protein levels in white adipose tissue (WAT) of mice. Using 3T3-L1 fibroblasts, Picard et al. (2004) showed that activation of SIRT1 induces lipolysis of triglycerides and the release of free fatty acids by inhibiting the nuclear receptor peroxisome proliferator-activated receptor-γ. Because CR results in a major decrease in the amount of WAT, they also studied the effect of fasting on fatty acid release from mice heterozygous for SIRT1. The mobilization of fatty acids from WAT upon fasting was compromised in the SIRT1$^{+/-}$ mice, suggesting a role of SIRT1 in the reduction in WAT by CR. Subsequently, Chen et al. (2005) studied the effect of CR in SIRT1 knockout mice. They observed that the up-regulation of physical activity, which is observed in CR rodents, does not occur with CR in the knockout mice, indicating that the gene coding for SIRT1 is required for increased physical activity observed with CR.

Bordone and Guarente (2005) proposed that SIR2/SIRT1 has a role in the mechanism of life extension by CR in yeast by connecting changes in cellular energetics to increased life span. An increase in the NAD/NADH ratio is proposed to lead to an increase in the activity of SIR2/SIRT1, which results in the deacetylation of key substrates, e.g., histones in chromatin or transcription factors, which alter gene expression and, thereby, cellular functions. Kobayashi et al. (2005) showed that SIRT1 binds to FOXO4, catalyzing its deacetylation and increasing its *trans*-activation activity. Cohen et al. (2004a) showed that SIRT1 catalyzes the deacetylation of Ku70, a DNA-repair factor that becomes tightly associated with Bax and suppresses Bax-mediated apoptosis when deacetylated. Several studies show that various parameters of the intrinsic (mitochondrial) pathway of apoptosis are reduced in tissues of CR rats (Selman et al. 2003; Shelke and Leeuwenburg 2003; Dirks and Leeuwenburg 2004; Lee et al. 2004) and mice (Avula and Fernandes 2002). The peroxisome proliferator-activated receptor-γ coactivator-1α (PGC-1α) appears to have a particularly important role in the effect of SIRT1 on CR. SIRT1 has been shown to physically interact with and deacetylate PGC-1α at multiple lysine sites, leading to increased PGC-1α activity (Nemoto et al. 2005). PGC-1α is a transcriptional coactivator that increases the rate of transcription by interacting with transcription factors and regulates a wide range of biological responses, e.g., mitochondrial biogenesis, energy metabolism, thermal regulation, glucose metabolism, fat metabolism, and muscle fiber-type switching (Puigserver and Spiegelman 2003).

Recently, Lagouge et al. (2006) showed that mice fed resveratrol, which induces SIRT1 activity, had increased aerobic capacity, as evidenced by their increased running time and consumption of oxygen in muscle fibers, and an induction of genes for oxidative phosphorylation and mitochondrial biogenesis. These changes were largely explained by a decrease in PGC-1α acetylation and an increase in PGC-1α activity, which is consistent with resveratrol activating SIRT1. Baur et al. (2006) also observed that feeding resveratrol to mice induced PGC-1α activity, which was correlated with increased mitochondrial number, increased insulin sensitivity, reduced IGF-1 levels, and improved motor function. Using cells treated with plasma from rats fed either ad libitum or CR, Cohen et al. (2004b) reported that cells treated with CR plasma showed increased expression of SIRT1, which was inhibited when insulin and/or IGF-1 was added to the CR serum. Subsequently, de Cabo's laboratory (Lopez-Lluch et al. 2006) showed that CR plasma reduced oxidative stress in the cells at the same time that it stimulated the proliferation of mitochondria through the PGC-1α signaling pathway. The mitochondria from the cells treated with the CR plasma generated less ROS than mitochondria in cells treated with plasma from ad libitum rats, which is consistent with the observation that PGC-1α induces mitochondrial uncoupling (Scarpulla 2002). Nisoli et al. (2005) also showed that the increase in SIRT1 expression by CR in WAT of mice was correlated to increased expression of PGC-1α and mitochondrial biogenesis. Thus, it is inviting to speculate that many of the effects of CR on mitochondrial function arise through the SIRT1 activation of the PGC-1α pathway.

Redox Signaling Pathway

Lavrovsky et al. (2000) and Merry (2004) argue that changes in redox signaling by CR might have a greater impact on cell function rather than changes in oxidative damage. The redox status of a cell is normally maintained in a reducing, thiol-rich, intracellular state through thiol-disulfide exchange reactions, which involve the thioredoxin and the glutathione systems. It is generally believed that the cellular environment of the cell becomes more oxidative with age (Jones 2006), which would lead to changes in the activities of redox-sensitive transcription factors, such as NF-κB, AP-1, Nrf2, and HIF-1 (Hansen et al. 2006). In 2002, Kim et al. reported that CR attenuates the activation and translocation of the redox-sensitive transcription factor, NF-κB, in kidneys of rats. Cho et al. (2003) showed that both the thioredoxin and glutathione systems are negatively modulated by aging in kidney and that CR suppresses the age-related decrease in GSH and thioredoxin levels. They also showed that the nuclear translocation of thioredoxin increases with age and that this effect is blunted by CR. This is an important observation, because it is known that the nuclear import of thioredoxin leads it to interact with Ref-1 to form a dimer and facilitate the transcriptional activities of redox-sensitive transcription factors, such as NF-κB, AP-1, and HIF-1 by keeping cysteine residues reduced (Demple 1998). It is argued that alterations in the activation of these transcription factors by CR lead to changes in gene expression profiles and cell function (Merry 2004). During the past 6 years, several groups have compared the RNA expression profiles of tissues from rats and mice fed ad libitum and CR using microarrays (Lee et al. 1999, 2002; Han et al. 2000; Spindler 2001; Weindruch et al. 2001; Fu et al. 2006). These studies demonstrate that gene expression profiles are altered in tissues of CR rodents and that many age-related changes in gene expression are prevented/suppressed by CR, resulting in a more "youthful" expression profile.

Target of Rapamycin Signaling Pathway

The TOR pathway is an evolutionarily conserved signaling pathway that regulates cell growth (accumulation of mass) in response to a variety of environmental cues, such as nutrient availability, growth factors, hypoxia, DNA damage, and osmotic stress. TOR controls cell growth by regulating translation, transcription, ribosome biogenesis, nutrient transport, and autophagy (Reiling and Sabatini 2006). Recent studies in yeast (Kaeberlein et al. 2005), *C. elegans* (Vellai et al. 2003; Jia et al. 2004), and *Drosophila* (Kapahi et al. 2004) show that the targeted deletions of genes in the TOR pathway increase life span. In *C. elegans,* the increase in life span arising from the down-regulation of the TOR pathway is *daf-16*-dependent; i.e., it involves the activation of FOXO. Kaeberlein et al. (2005) found that CR fails to increase life span further in yeast; therefore, life extension by CR in yeast appears to involve the down-regulation of TOR signaling. Currently, there are no mammalian data on the influence of CR on TOR signaling pathways. However, studies using cultures of mammalian cells show that the TOR pathway can be activated by insulin and other growth factors (Sarbassov et al. 2005), and a reduction of TOR signaling results in reduced ROS production by mitochondria (Kim et al. 2005). Recently, Sharp and Bartke (2005) showed that the TOR pathway is reduced in Ames dwarf mice compared to control, wild-type mice. Therefore, as researchers study the effect of CR on the nutrient-sensing TOR pathway in rodents, it is likely that CR will be observed to alter the TOR signaling pathway.

NONHUMAN PRIMATES

Nonhuman primates are often used to model human health and disease. Among the most commonly used nonhuman primate species is the Old World rhesus monkey (*Macaca mulatta*). The vast array of information available for this species, particularly related to the aging process, has made it an obvious choice for studying the efficacy of CR in nonhuman primates. Three such long-term studies of CR in rhesus monkeys exist. Each study uses a slightly different form of CR; however, as detailed below, many outcomes from the studies are similar.

The NIA Study

In 1987, the first long-term, broad-based, controlled study of CR in a nonhuman primate species was initiated at the National Institute on Aging. This study began with a group of 30 male rhesus monkeys and was doubled to 60 males in 1988. In 1992, 60 females were added. Animals ranged in age from 1 to 23 years of age when they were initiated into the study; thus, some monkeys were prepubertal, whereas others were considered old. All animals in this study were fed a semisynthetic diet. Individual food allotments were based on estimates of ad libitum food intake according to National Research Guidelines. Regular monitoring of food intake over the study confirmed that the control animals' intakes approximated ad libitum feeding, whereas experimental animals took in approximately 70–75% as many calories as age- and weight-matched controls (Mattison et al. 2003). Details of the experimental design have been described previously (Ingram et al. 1990; Lane et al. 1992; Mattison et al. 2003).

Among the most consistent findings of this study is the total body and fat tissue mass reduction in experimental animals (Lane et al. 1999). Furthermore, the rate of body weight gain in the growing animals in this study was significantly slower in experimental animals compared to controls, suggesting that physical development was slowed by CR (Lane et al. 1992; Weindruch et al. 1995). Skeletal maturation, as measured by crown–rump length, total body bone mineral content, and total serum alkaline phosphatase levels (Lane et al. 1995b), and sexual maturation, as measured by maturational increases in serum testosterone levels (Roth et al. 1993), showed further evidence of a slowed, but normal maturation rate.

It is not surprising, given the general improvements in body composition, that experimental animals are better able than controls to regulate glucose. Experimental animals in this study had reduced circulating glucose levels and acute insulin response to a glucose load compared to controls (Lane et al. 1995a). Other findings of this study include, but are not limited to, reduction in body temperature with CR, lack of difference in lean body mass corrected energy expenditure, and improvements in cardiovascular disease profile (see Mattison et al. 2003).

The University of Wisconsin Study

Investigators at the Wisconsin National Primate Research Center located at the University of Wisconsin, Madison, began a study of CR in nonhuman primates in 1989 (Kemnitz et al. 1993). Similar to the NIA study, this study chose a paradigm of an approximately 30% CR in rhesus macaques. In 1989, 30 adult males (8–14 years of age) were entered into the study and randomized into control and experimental groups. After randomization, animals in the experimental group were reduced in food intake, based on their individually quantified baseline values, by 10% per month over a 3-month period to attain the desired level of 30% CR. In 1994, an additional 30 adult females and 16 adult males were added to the study and treated similarly to the initial group (Colman et al. 1998). Animals in this study were fed a pelleted, semipurified diet, with experimental animals receiving a diet enhanced by 30% in vitamins and minerals to assure that both groups received a similar level of micronutrients. Details of the experimental design have been published previously (Kemnitz et al. 1993; Ramsey et al. 2000).

Unlike the NIA study, all animals in this study had reached full adult size at entry into the study. Not unexpectedly, rapidly following restriction, experimental animals had lower body mass and fat mass, as well as decreased levels of fat localized to the abdomen when compared to controls (Colman et al. 1999). Total body bone mass is also lower in experimental animals compared to controls; however, this likely reflects the decreased mechanical load in these animals.

Among the most striking findings to date in this study is the improvement in glucoregulatory function, specifically insulin sensitivity, conferred by CR (Gresl et al. 2001). At entry into the study, two control and two experimental animals had glucoregulatory profiles suggestive of a progression to diabetes. Since that time, all experimental animals have healthy glucoregulatory profiles, whereas four controls have progressed to diabetes and six more are classified as at risk of developing diabetes. Other findings of this study include but are not limited to decreased levels of osteoarthritis, improved cardiovascular profile including reduced C-reactive protein (CRP) levels and number of triglyceride and phospholipid molecules in low-density lipoproteins (LDL) (Edwards et al. 2001), altered skeletal muscle transcriptional profiles (Kayo et al. 2001), lowering of oxidative damage in skeletal muscle (Zainal et al. 2000), and possibly delayed presbycusis (Fowler et al. 2002).

The University of Maryland Study

The third nonhuman primate study of CR and aging has been controversial due to the aspects of study design, subject characteristics, and statistical methodology (see Lane et al. 2004). This study, performed at the University of Maryland, was designed to prevent obesity by maintaining adult male rhesus monkeys at a stable body weight (~10–12 kg) as opposed to testing CR per se. This weight clamp regimen resulted in a reduction in caloric intake of approximately 35%, similar to that seen in studies specifically designed to test the effectiveness of CR. As opposed to the other two studies, this study focuses almost entirely on the effect of the diet regimen on carbohydrate metabolism. Similar to the NIA and UW studies, experimental animals in this study were leaner and showed positive alterations in carbohydrate metabolism. This study is the first to claim that CR leads to an increased average age of death in primates, which they believe is associated with the prevention of hyperinsulinemia and the mitigation of age-related disease (Bodkin et al. 2003, 2005; see also Lane et al. 2004). Details on experimental design and results for this study have been published previously (Bodkin et al. 1995, 2003; DeLany et al. 1999).

Conclusions from Nonhuman Primate Studies

The effectiveness of CR in humans is challenging to study. These nonhuman primate studies allow us to ask whether CR can be safely and effectively carried out in a primate species while still maintaining the strict control of laboratory research. These three studies have shown that CR can be safely and effectively carried out in a nonhuman primate. The three studies have important differences in experimental design; however, all show improvements in body composition and glucoregulatory function with CR. It is too early to know whether CR will extend life span in a nonhuman primate model, but there are currently clear indications of improved span of good health with this experimental paradigm. The same can be said of people on long-term CR who display many of the core features of CR, such as reductions in circulating insulin, glucose, and CRP levels as well as other overt signs of protection from cardiovascular disease (Fontana et al. 2004) and age-associated changes in diastolic function (Meyer et al. 2006).

SUMMARY

The effect of CR on the longevity of rodents has been studied for more than 50 years; therefore, it is the first and best-characterized experimental manipulation known to increase life span. CR has been shown to be effective in a wide range of organisms, ranging from invertebrates to dogs, and current studies suggest that CR will be effective in rhesus monkeys. Studies with rodents demonstrate that the increase in longevity observed with CR arises from retarding aging; i.e., the old CR rodents appear more youthful than the control-fed rodents throughout their life span. The general consensus is that the anti-aging action of CR is at least partially due to reduced oxidative damage and increased resistance to oxidative stress in tissues of CR rodents. Research over the past decade suggests that changes in the signaling pathways shown in Figure 1 have a central role in the increased resistance of CR rodents to oxidative stress and reduced oxidative damage, and in the shifts in energy metabolism that likely underlie the actions of CR.

ACKNOWLEDGMENTS

This work was supported partially by Merit Review grants (A.G.R.), from the Department of Veteran Affairs, and National Institutes of Health grants R01 AG23843 (A.G.R.), R37 AG26557 (A.G.R.), P01 AG11915 (R.W.), and P20 CA103697 (R.W.).

REFERENCES

Avula C.P. and Fernandes G. 2002. Inhibition of H_2O_2-induced apoptosis of lymphocytes by calorie restriction during aging. *Microsc. Res. Tech.* **59:** 282–292.

Bartke A., Wright J.C., Mattison J.A., Ingram D.K., Miller R.A., and Roth G.S. 2001. Extending the life span of long-lived mice. *Nature* **414:** 412.

Baur J.A., Pearson K.J., Price N.L., Jamieson H.A., Lerin C., Kalra A., Prabhu V.V., Allard J. S., Lopez-Lluch G., Lewis K., et al. 2006. Resveratrol improves health and survival of mice on a high-calorie diet. *Nature* **444:** 337–342.

Bevilacqua L., Ramsey J.J., Hagopian K., Weindruch R., and Harper M.E. 2004. Effects of short- and medium-term calorie restriction on muscle mitochondrial proton leak and reactive oxygen species production. *Am. J. Physiol. Endocrinol. Metab.* **286:** E852–E861.

———. 2005. Long-term caloric restriction increases UCP3 content but decreases proton leak and reactive oxygen species production in rat skeletal muscle mitochondria. *Am. J. Physiol. Endocrinol. Metab.* **289:** E429–E438.

Bodkin N.L., Ortmeyer H.K., and Hansen B.C. 1995. Long-term dietary restriction in older-aged rhesus monkeys: Effects on insulin resistance. *J. Gerontol. A Biol. Sci. Med. Sci.* **50A:** B142–B147.

———. 2005. A comment on the comment: Relevance of nonhuman primate dietary restriction to aging in humans. *J. Gerontol. A Biol. Sci. Med. Sci.* **60A:** 951–952.

Bodkin N.L., Alexander T.M., Ortmeyer H.K., Johnson E., and Hansen B.C. 2003. Mortality and morbidity in laboratory-maintained rhesus monkeys and effects of long-term dietary restriction. *J. Gerontol. A Biol. Sci. Med. Sci.* **58A:** 212–219.

Bokov A., Chaudhuri A., and Richardson A. 2004. The role of oxidative damage and stress in aging. *Mech. Ageing Dev.* **125:** 811–826.

Bordone L. and Guarente L. 2005. Calorie restriction, SIRT1 and metabolism: Understanding longevity. *Nat. Rev. Mol. Cell Biol.* **6:** 298–305.

Breese C.R., Ingram R.L., and Sonntag W.E. 1991. Influence of age and long-term dietary restriction on plasma insulin-like growth factor-1 (IGF-1), IGF-1 gene expression, and IGF-1 binding proteins. *J. Gerontol.* **46:** B180–B187.

Cadenas S., Buckingham J.A., Samec S., Seydoux J., Din N., Dulloo A.G., and Brand M.D. 1999. UCP2 and UCP3 rise in starved rat skeletal muscle but mitochondrial proton conductance is unchanged. *FEBS Lett.* **462:** 257–260.

Chen D., Steele A.D., Lindquist S., and Guarente L. 2005. Increase in activity during calorie restriction requires Sirt1. *Science* **310:** 1641.

Cho C.G., Kim H.J., Chung S.W., Jung K.J., Shim K.H., Yu B.P., Yodoi J., and Chung H.Y. 2003. Modulation of glutathione and thioredoxin systems by calorie restriction during the aging process. *Exp. Gerontol.* **38:** 539–548.

Cline G.W., Vidal-Puig A.J., Dufour S., Cadman K.S., Lowell B.B., and Shulman G.I. 2001. In vivo effects of uncoupling protein-3 gene disruption on mitochondrial energy metabolism. *J. Biol. Chem.* **276:** 20240–20244.

Cohen H.Y., Lavu S., Bitterman K.J., Hekking B., Imahiyerobo T.A., Miller C., Frye R., Ploegh H., Kessler B.M., and Sinclair D.A. 2004a. Acetylation of the C terminus of Ku70 by CBP and PCAF controls Bax-mediated apoptosis. *Mol. Cell* **13:** 627–638.

Cohen H.Y., Miller C., Bitterman K.J., Wall N.R., Hekking B., Kessler B., Howitz K.T., Gorospe M., de Cabo R., and Sinclair D.A. 2004b. Calorie restriction promotes mammalian cell survival by inducing the SIRT1 deacetylase. *Science* **305:** 390–392.

Colman R.J., Roecker E.B., Ramsey J.J., and Kemnitz J.W. 1998. The effect of dietary restriction on body composition in adult male and female rhesus macaques. *Aging Clin. Exp. Res.* **10:** 83–92.

Colman R.J., Ramsey J.J., Roecker E.B., Havighurst T., Hudson J.C., and Kemnitz J.W. 1999. Body fat distribution with long-term dietary restriction in adult male rhesus macaques. *J. Gerontol. A Biol. Sci. Med. Sci.* **54A:** B283–B290.

DeLany J.P., Hansen B.C., Bodkin N.L., Hannah J., and Bray G.A. 1999. Long-term calorie restriction reduces energy expenditure in aging monkeys. *J. Gerontol. A Biol. Sci. Med. Sci.* **54A:** B5–11.

Demple B. 1998. A bridge to control. *Science* **279:** 1655–1656.

Dhahbi J.M., Mote P.L., Wingo J., Rowley B.C., Cao S.X., Walford R.L., and Spindler S.R. 2001. Caloric restriction alters the feeding response of key metabolic enzyme genes. *Mech. Ageing Dev.* **122:** 1033–1048.

Dirks A.J. and Leeuwenburgh C. 2004. Aging and lifelong calorie restriction result in adaptations of skeletal muscle apoptosis repressor, apoptosis-inducing factor, X-linked inhibitor of apoptosis, caspase-3, and caspase-12. *Free Radic. Biol. Med.* **36:** 27–39.

Edwards I.J., Rudel L.L., Terry J.G., Kemnitz J.W., Weindruch R., Zaccaro D.J., and Cefalu W.T. 2001. Caloric restriction lowers plasma lipoprotein (a) in male but not female rhesus monkeys. *Exp. Gerontol.* **36:** 1413–1418.

Fontana L., Meyer T.E., Klein S.L., and Holloszy J.O. 2004. Long-term calorie restriction is highly effective in reducing the risk for atherosclerosis in humans. *Proc. Natl. Acad. Sci.* **101:** 6659–6663.

Fowler C.G., Torre P., III, and Kemnitz J.W. 2002. Effects of caloric restriction and aging on the auditory function of rhesus monkeys *(Macaca mulatta):* The University of Wisconsin study. *Hear. Res.* **169:** 24–35.

Fu C., Hickey M., Morrison M., McCarter R., and Han E.S. 2006. Tissue specific and non-specific changes in gene expression by aging and by early stage CR. *Mech. Ageing Dev.* **127:** 905–916.

Gredilla R., Barja G., and Lopez-Torres M. 2001. Effect of short-term caloric restriction on H2O2 production and oxidative DNA damage in rat liver mitochondria and location of the free radical source. *J. Bioenerg. Biomembr.* **33:** 279–287.

Greer E.L. and Brunet A. 2005. FOXO transcription factors at the interface between longevity and tumor suppression. *Oncogene* **24:** 7410–7425.

Gresl T.A., Colman R.J., Roecker E.B., Havighurst T.C., Huang Z., Allison D.B., Bergman R.N., and Kemnitz J.W. 2001. Dietary restriction and glucose regulation in aging rhesus monkeys: A follow-up report at 8.5 years. *Am. J. Physiol.* **281:** E757–E765.

Guarente L. and Kenyon C. 2000. Genetic pathways that regulate ageing in model organisms. *Nature* **408:** 255–262.

Hagopian K., Harper M.E., Ram J.J., Humble S.J., Weindruch R., and Ramsey J.J. 2005. Long-term calorie restriction reduces proton leak and hydrogen peroxide production in liver mitochondria. *Am. J. Physiol. Endocrinol. Metab.* **288:** E674–E684.

Han E., Hilsenbeck S.G., Richardson A., and Nelson J.F. 2000. cDNA expression arrays reveal incomplete reversal of age-related changes in gene expression by calorie restriction. *Mech. Ageing Dev.* **115:** 157–174.

Hansen J.M., Go Y.M., and Jones D.P. 2006. Nuclear and mitochondrial compartmentation of oxidative stress and redox signaling. *Annu. Rev. Pharmacol. Toxicol.* **46:** 215–234.

Hekimi S. and Guarente L. 2003. Genetics and the specificity of the aging process. *Science* **299:** 1351–1354.

Ingram D.K., Cutler R.G., Weindruch R., Renquist D.M., Knapka J.J., Milton A., Belcher C.T., Clark M.A., Hatcherson C.D., Marriott B.M., and Roth G.S. 1990. Dietary restriction and aging: The initiation of a primate study. *J. Gerontol.* **45:** B148–B163.

Iwasaki K., Gleiser C.A., Masoro E.J., McMahan C.A., Seo E.J., and Yu B.P. 1988. The influence of dietary protein source on longevity and age-related disease processes of Fischer rats. *J. Gerontol.* **43:** B5–B12.

Jia K., Chen D., and Riddle D.L. 2004. The TOR pathway interacts with the insulin signaling pathway to regulate *C. elegans* larval development, metabolism and lifespan. *Development* **131:** 3897–3906.

Jones D.P. 2006. Extracellular redox state: Refining the definition of oxidative stress in aging. *Rejuvenation Res.* **9:** 169–181.

Kaeberlein M., Powers R.W., Steffen K.K., Westman E.A., Hu D., Dang N. Kerr E.O., Kirkland K.T., Fields S., and Kennedy B.K. 2005. Regulation of yeast replicative lifespan by TOR and Sch9 in response to nutrients. *Science* **310:** 1193–1196.

Kalant N., Stewart J., and Kaplan R. 1988. Effect of diet restriction on glucose metabolism and insulin responsiveness in aging rats. *Mech. Ageing Dev.* **46:** 89–104.

Kapahi P., Zid B.M., Harper T., Koslover D., Sapin V., and Benzer S. 2004. Regulation of lifespan in *Drosophila* by modulation of genes in the TOR signaling pathway. *Curr. Biol.* **14:** 885–890.

Kayo T., Allison D.B., Weindruch R., and Prolla T.A. 2001. Influences of aging and caloric restriction on the transcriptional profile of skeletal muscle from rhesus monkeys. *Proc. Natl. Acad. Sci.* **98:** 5093–5098.

Kealy R.D., Lawler D.F., Ballam J.M., Mantz S.L., Biery D.N., Greeley E.H., Lust G., Segre M., Smith G.K., and Stowe H.D. 2002. Effects of diet restriction on lifespan and age-related changes in dogs. *J. Am. Vet. Med. Assoc.* **220:** 1315–1320.

Kemnitz J.W., Weindruch R., Roecker E.B., Crawford K., Kaufman P.L., and Ershler W.B. 1993. Dietary restriction of adult male rhesus monkeys: Design, methodology, and preliminary findings from the first year of study. *J. Gerontol.* **48:** B17–B26.

Kenyon C. 2001. A conserved regulatory system for aging. *Cell* **105:** 165–168.

Kim H.J., Yu B.P., and Chung H.Y. 2002. Molecular exploration of age-related NF-kappaB/IKK downregulation by calorie restriction in rat kidney. *Free Radic. Biol. Med.* **32:** 991–1005.

Kim J.H., Chu S.C., Gramlich J.L., Pride Y.B., Babendreier E., Chauhan D., Salgia R., Podar K., Griffin J.D., and Sattler M. 2005. Activation of the PI3K/mTOR pathway by BCR-ABL contributes to increased production of reactive oxygen species. *Blood* **105:** 1717–1723.

Kobayashi Y., Furukawa-Hibi Y., Chen C., Horio Y., Isobe K., Ikeda K., and Motoyama N. 2005. SIRT1 is critical regulator of FOXO-mediated transcription in response to oxidative stress. *Int. J. Mol. Med.* **16:** 237–243.

Krauss S., Zhang C.Y., and Lowell B.B. 2002. A significant portion of mitochondrial proton leak in intact thymocytes depends on expression of UCP2. *Proc. Natl. Acad. Sci.* **99:** 118–122.

Lagouge M., Argmann C., Gerhart-Hines Z., Meziane H., Lerin C., Daussin F., Messadeq N., Milne J., Lambert P., Elliott P., et al. 2006. Resveratrol improves mitochondrial function and protects against metabolic disease by activating SIRT1 and PGC-1alpha. *Cell* **127:** 1109–1122.

Lambert A.J. and Merry B.J. 2004. Effect of caloric restriction on mitochondrial reactive oxygen species production and bioenergetics: Reversal by insulin. *Am. J. Physiol. Regul. Integr. Comp. Physiol.* **286:** R71–R79.

Lane M.A., Ingram D.K., and Roth G.S. 1999. Calorie restriction in nonhuman primates: Effects on diabetes and cardiovascular disease risk. *Toxicol. Sci.* **52:** 41–48.

Lane M.A., Mattison J.A., Roth G.S., Brant L.J., and Ingram D.K. 2004. Effects of long-term diet restriction on aging and longevity in primates remain uncertain. *J. Gerontol. A Biol. Sci. Med. Sci.* **59A:** 405–407.

Lane M.A., Ingram D.K., Cutler R.G., Knapka J.J., Bernard D.E., and Roth G.S. 1992. Dietary restriction in nonhuman primates: Progress report on the NIA study. *Ann. N.Y. Acad. Sci.* **673:** 36–45.

Lane M.A., Ball S.S., Ingram D.K., Cutler R.G., Engel J., Read V., and Roth G.S. 1995a. Diet restriction in rhesus monkeys lowers fasting and glucose-stimulated glucoregulatory endpoints. *Am. J. Physiol.* **268:** E941–E948.

Lane M.A., Reznick A.Z., Tilmont E.M., Lanir A., Ball S.S., Read V., Ingram D.K., Cutler R.G., and Roth G.S. 1995b. Aging and food restriction alter some indices of bone metabolism in male rhesus monkeys *(Macaca mulatta). J. Nutr.* **125:** 1600–1610.

Lavrovsky Y., Chatterjee B., Clark R.A, and Roy A.K. 2000. Role of redox-regulated transcription factors in inflammation, aging and age-related diseases. *Exp. Gerontol.* **35:** 521–532.

Lee C.K., Klopp R.G., Weindruch R., and Prolla T.A. 1999. Gene expression profile of aging and its retardation by caloric restriction. *Science* **285:** 1390–1393.

Lee C.K., Allison D.B., Brand J., Weindruch R., and Prolla T.A. 2002. Transcriptional profiles associated with aging and middle age-onset caloric restriction in mouse hearts. *Proc. Natl. Acad. Sci.* **99:** 14988–14993.

Lee J.H., Jung K.J., Kim J.W., Kim H.J., Yu B.P., and Chung H.Y. 2004. Suppression of apoptosis by calorie restriction in aged kidney. *Exp. Gerontol.* **39:** 1361–1368.

Lopez-Lluch G., Hunt N., Jones B., Zhu M., Jamieson H., Hilmer S., Cascajo M.V., Allard J., Ingram D.K., Navas P., and de Cabo R. 2006. Calorie restriction induces mitochondrial biogenesis and bioenergetic efficiency. *Proc. Natl. Acad. Sci.* **103:** 1768–1773.

Lopez-Torres M., Gredilla R., Sanz A., and Barja G. 2002. Influence of aging and long-term caloric restriction on oxygen radical generation and oxidative DNA damage in rat liver mitochondria. *Free Radic. Biol. Med.* **32:** 882–889.

Masoro E.J. 2005. Overview of caloric restriction and ageing. *Mech. Ageing Dev.* **126:** 913–922.

Masoro E.J., McCarter R.J., Katz M.S., and McMahan C.A. 1992. Dietary restriction alters characteristics of glucose fuel use. *J. Gerontol.* **47:** B202–B208.

Masoro E.J., Iwasaki K., Gleiser C.A., McMahan C.A., Seo E.J., and Yu B.P. 1989. Dietary modulation of the progression of nephropathy in aging rats: An evaluation of the importance of protein. *Am. J. Clin. Nutr.* **49:** 1217–1227.

Mattison J.A., Lane M.A., Roth G.S., and Ingram D.K. 2003. Calorie restriction in rhesus monkeys. *Exp. Gerontol.* **38:** 35–46.

McCay C.M., Crowell M.F., and Maynard L.A. 1935. The effect of retarded growth upon the length of lifespan and upon the ultimate body size. *J. Nutr.* **10:** 63–79.

Merry B.J. 2004. Oxidative stress and mitochondrial function with aging—The effects of calorie restriction. *Aging Cell* **3:** 7–12.

Meyer T.E., Kovacs S.J., Ehsani A.A., Klein S., Holloszy J.O., and Fontana L. 2006. Long-term caloric restriction ameliorates the decline in diastolic function in humans. *J. Am. Coll. Cardiol.* **47:** 398–402.

Min K.J. and Tatar M. 2006. Restriction of amino acids extends lifespan in *Drosophila melanogaster*. *Mech. Ageing Dev.* **127:** 643–646.

Nemoto S., Fergusson M.M., and Finkel T. 2005. SIRT1 functionally interacts with the metabolic regulator and transcriptional coactivator PGC-1α. *J. Biol. Chem.* **280:** 16456–16460.

Nisoli E., Tonello C., Cardile A., Cozzi V., Bracale R., Tedesco L., Falcone S., Valerio A., Cantoni O., Clementi E., et al. 2005. Calorie restriction promotes mitochondrial biogenesis by inducing the expression of eNOS. *Science* **310:** 314–317.

Picard F., Kurtev M., Chung N., Topark-Ngarm A., Senawong T., Machado De Oliveira R., Leid M., McBurney M.W., and Guarente L. 2004. Sirt1 promotes fat mobilization in white adipocytes by repressing PPAR-gamma. *Nature* **429:** 771–776.

Puigserver P. and Spiegelman B.M. 2003. Peroxisome proliferator-activated receptor-gamma coactivator 1α (PGC-1α): Transcriptional coactivator and metabolic regulator. *Endocr. Rev.* **24:** 78–90.

Ramsey J.J., Colman R.J., Binkley N.C., Christensen J.D., Gresl T.A., Kemnitz J.W., and Weindruch R. 2000. Dietary restriction and aging in rhesus monkeys: The University of Wisconsin study. *Exp. Gerontol.* **35:** 1131–1149.

Reiling J.H. and Sabatini D.M. 2006. Stress and mTORture signaling. *Oncogene* **25:** 6373–6383.

Richardson A., Liu F., Adamo M.L., Van Remmen H., and Nelson J.F. 2004. The role of insulin and insulin-like growth factor-I in mammalian ageing. *Res. Clin. Endocrinol. Metab.* **18:** 393–406.

Roth G.S., Blackman M.R., Ingram D.K., Lane M.A., Ball S.S., and Cutler R.G. 1993. Age-related changes in androgen levels of rhesus monkeys subjected to diet restriction. *Endocr. J.* **1:** 227–234.

Sarbassov D.D., Ali S.M., and Sabatini D.M. 2005. Growing roles for the mTOR pathway. *Curr. Opin. Cell Biol.* **17:** 596–603.

Scarpulla R.C. 2002. Transcriptional activators and coactivators in the nuclear control of mitochondrial function in mammalian cells. *Gene* **286:** 81–89.

Selman C., Gredilla R., Phaneuf S., Kendaiah S., Barja G., and Leeuwenburgh C. 2003. Short-term caloric restriction and regulatory proteins of apoptosis in heart, skeletal muscle and kidney of Fischer 344 rats. *Biogerontology* **4:** 141–147.

Sharp Z.D. and Bartke A. 2005. Evidence for down-regulation of phosphoinositide 3-kinase/Akt/mammalian target of rapamycin (PI3K/Akt/mTOR)-dependent translation regulatory signaling pathways in Ames dwarf mice. *J. Gerontol.* **60:** 293–300.

Shelke R.R. and Leeuwenburgh C. 2003. Lifelong caloric restriction increases expression of apoptosis repressor with a caspase recruitment domain (ARC) in the brain. *FASEB J.* **17:** 494–496.

Sohal R.S., Ku H.H., Agarwal S., Forster M.J., and Lal H. 1994. Oxidative damage, mitochondrial oxidant generation and antioxidant defenses during aging and in response to food restriction in the mouse. *Mech. Ageing Dev.* **74:** 121–133.

Sonntag W.E., Lynch C.D., Cefalu W.T., Ingram R.L., Bennett S.A., Thornton P. L., and Khan A.S. 1999. Pleiotropic effects of growth hormone and insulin-like growth factor (IGF)-1 on biological aging: Inferences from moderate caloric-restricted animals. *J. Gerontol.* **54:** B521–B538.

Spindler S.R. 2001. Calorie restriction enhances the expression of key metabolic enzymes associated with protein renewal during aging. *Ann. N.Y. Acad. Sci.* **928:** 296–304.

Sun D., Muthukumar A.R., Lawrence R.A., and Fernandes G. 2001. Effects of calorie restriction on polymicrobial peritonitis induced by cecum ligation and puncture in young C57BL/6 mice. *Clin. Diagn. Lab. Immunol.* **8:** 1003–1011.

Turturro A., Witt W.W., Lewis S., Hass B.S., Lipman R.D., and Hart R.W. 1999. Growth curves and survival characteristics of the animals used in the Biomarkers of Aging Program. *J. Gerontol. A Biol. Sci. Med. Sci.* **54:** B492–B501.

Vellai T., Takacs-Vellai K., Zhang Y., Kovacs A.L., Orosz L., and Muller F. 2003. Genetics: Influence of TOR kinase on lifespan in *C. elegans. Nature* **426:** 620.

Vidal-Puig A.J., Grujic D., Zhang C.Y., Hagen T., Boss O., Ido Y., Szczepanik A., Wade J., Mootha V., Cortright R., et al. 2000. Energy metabolism in uncoupling protein 3 gene knockout mice. *J. Biol. Chem.* **275:** 16258–16266.

Weindruch R. and Walford R. 1988. *The retardation of aging and disease by dietary restriction.* C.C. Thomas, Springfield, Illinois.

Weindruch R., Kayo T., Lee C.K., and Prolla T.A. 2001. Microarray profiling of gene expression in aging and its alteration by caloric restriction in mice. *J. Nutr.* **131:** 918S–923S.

Weindruch R., Marriott B.M., Conway J., Knapka J.J., Lane M.A., Cutler R.G., Roth G.S., and Ingram D.K. 1995. Measures of body size and growth in rhesus and squirrel monkeys subjected to long-term dietary restriction. *Am. J. Primatol.* **35:** 207–228.

Wetter T.J., Gazdag A.C., Dean D.J., and Cartee G.D. 1999. Effect of calorie restriction on in vivo glucose metabolism by individual tissues in rats. *Am. J. Physiol.* **276:** E728–E738.

Wood J.G., Rogina B., Lavu S., Howitz K., Helfand S.L., Tatar M., and Sinclair D. 2004. Sirtuin activators mimic caloric restriction and delay ageing in metazoans. *Nature* **430:** 686–689.

Xiao H., Massaro D., Massaro G.D., and Clerch L.B. 2004. Expression of lung uncoupling protein-2 mRNA is modulated developmentally and by caloric intake. *Exp. Biol. Med.* **229:** 479–485.

Zainal T.A., Oberley T.D., Allison D.B., Szweda L.I., and Weindruch R. 2000. Caloric restriction of rhesus monkeys lowers oxidative damage in skeletal muscle. *FASEB J.* **14:** 1825–1836.

16

Determination of Aging Rate by Coordinated Resistance to Multiple Forms of Stress

Gordon J. Lithgow
Buck Institute for Age Research
Novato, California 94945

Richard A. Miller
Department of Pathology and Geriatrics Center
University of Michigan School of Medicine
and Ann Arbor VA Medical Center
Ann Arbor, Michigan 48109-2200

THIS CHAPTER EXPLORES THE HYPOTHESIS THAT DECLINES in aging rate, whether produced by evolutionary adaptations, single gene mutations, or dietary interventions, reflect alterations in a stress resistance pathway that increases cellular resistance to multiple forms of stress. We argue that such a stress resistance pathway evolved early in the eukaryotic lineage to allow small short-lived organisms to adjust their life history styles to intermittent environmental fluctuations and that as multicellular organisms evolved, they linked regulated stress-response mechanisms to a variety of hormonal and nutritional triggers specific for their own environmental niche. We believe that this model, although surely oversimplified, helps to explain much of the experimental data on stress and aging. We will try to show that the model provides a helpful heuristic framework for developing new experimental approaches to learn about the connections linking stress resistance, developmental biology, and endocrine controls to the aging process and, ultimately, to modulation of life span and most if not all of the diseases of aging.

We must start with working definitions of two key terms: aging and stress. By "aging" we mean the process that gradually transforms healthy and vigorous adults into older adults with diminished ability to meet a wide range of physiological challenges and, concomitantly, increased susceptibility to multiple forms of illness, injury, and death. This definition emphasizes the process of aging, rather than its outcome, the aged individual, in order to highlight the changes that occur, even in young and middle-aged adults, and lead eventually to infirmity. In this view, aging is occurring in 4-day-old worms, 3-year-old dogs, and 25-year-old people: Even if their performance on most age-sensitive tests is not yet seriously impaired, the amount of additional time that will elapse before functional decline has been shortened, through pathways still obscure. We deliberately refer to the aging process in the singular, rather than to aging processes. We recognize, of course, that aging impairs many distinct processes, ranging (in mammals) from lymphocyte activation to reflex speed to cardiovascular resilience to cognition to lens turbidity. But we are impressed with the now overwhelming evidence that low-calorie diets and single gene mutations can produce parallel retardation in nearly all of the age-sensitive processes, and we infer from this that the many effects of aging, however distinct in their underlying pathophysiology, must also be modulated at least in part by a common mechanism. The details of this mechanism, and its linkage to age-related declines in multiple cells and tissues, are still unknown; one goal of this chapter is to explore the idea that tunable stress resistance is indeed this "missing link." We will try to distinguish carefully between aging and longevity: Although diminution in the rate of aging typically produces exceptionally long-lived individuals, the reverse inference—short life span as a sign of accelerated aging—is very risky, because there are many ways to cause illness and shorten life span that do not work by an increase in the rate of aging.

The term "stress" is used to refer, in both casual and scientific conversations, to a wide range of overlapping concepts that must be carefully operationalized for an article on stress and aging to be intelligible. Cells can be stressed by exposing them to conditions that threaten to perturb their homeostatic state, for example, by damaging DNA, throttling protein translation, altering intracellular ion concentrations, reducing the levels of available ATP, or denaturing key proteins. Cells can respond to stress by repairing the damage, by succumbing to apoptotic or nonapoptotic cell death pathways, or by differentiation to an alternate state, which may involve neoplastic transformation or other changes that put tissue or organismic function at risk. Organisms, too, can be stressed, by conditions as diverse as exposure to heat or cold, infection, nutrient deprivation, blood loss or mechanical trauma, or psychological factors from sleep deprivation to approaching deadlines. In simple microscopic animals, it is clear that organismic stress rapidly leads to loss of homeostasis at the cellular level. In more complex animals, it is not clear whether or to what extent stress at the organismic level leads to stress responses that perturb homeostasis of specific cell types. Multicellular organisms have evolved complex, interacting networks of stress-response pathways, often based on neuroendocrine and immune control circuits, for tuning their responses to stressful conditions whether acute, intermittent, or enduring. Mammals respond to a highly diverse range of agents, including cold, heat, radiation, exercise, and psychological stress, by production of the pituitary hormone adrenocorticotrophic hormone (ACTH), which in turn leads to enhanced release of glucocorticoids from the adrenal cortex. Glucocorticoid effects can be beneficial or detrimental to cell function, depending on the amount of hormone, length of exposure, cell type, and exposure to other neuroendocrine factors. It is likely that these intercellular messages may induce, in individual cells of multiple tissues, some of the same circuitry that helps to protect the cell from homeostatic threats. Our understanding of the factors that mediate stress resistance at the cellular level, and how these contribute to defenses against organismic stress, although improving, is still lamentably far from complete.

Many of the cellular responses to stress are conserved throughout evolution. Exposed to heat or other sources of damage, cells often alter gene expression, protein synthesis, and metabolism in ways that both promote survival of the ongoing stress and improve the chances of surviving a recurrence. After thermal stress, for example, normal protein synthesis is essentially suspended, and a specialized set of proteins, the heat shock proteins (HSPs), are synthesized through alterations in both transcription and translation. Many HSPs are expressed, at lower levels, even in unstressed cells, and they influence protein folding, protein degradation, and the assembly of multisubunit protein complexes. These "molecular chaperones" bind to unfolded proteins and either promote folding in an energy-dependent fashion or target the protein for degradation.

Our stated hypothesis contains a third key word: We are alleging that anti-aging maneuvers act via a pathway that regulates resistance to *multiple* forms of stress at the same time. The hypothesis is thus different, in a critical way, from models that purport to tie the aging process to DNA damage, or to accumulations of irreversibly glycated proteins, or to oxidative destruction of proteins or lipids, or to telomere-dependent proliferative blockade, or indeed to any other single form of cell and tissue injury. We are proposing, instead, that early in evolutionary history, eukaryotic cells, already blessed with several systems for mitigating and repairing several forms of cellular injuries, further developed a control pathway that could induce the coordinated up-regulation of these multiple stress resistance systems at the same time. Hypothetically, an organism (perhaps a single-celled organism) with coordinated inducible defense pathways may have had higher Darwinian fitness than its cousins whose defenses were always "on," and also in comparison to cousins whose various protective mechanisms were under unlinked independent control. Over time, such a control circuit could have evolved further, in different organisms, to acquire links to additional forms of cellular repair mechanisms, different triggering signals (e.g., hormones of the insulin family), and a range of species-specific developmental switches, such as dauer formation in worms, estivation in flies, perhaps caste commitment in some insects, and perhaps hibernation in some vertebrate species. It is easy to imagine how such a system, originally devoted to protecting cells from stressful milieus, might be drafted to coordinate responses to nutrient deprivation, eventually including cues from hormonal indicators of environmental richness. It is also easy to imagine how environmental niches that reward extended life histories and long-lasting tissues, i.e., niches that permit and support evolution of long-lived species from shorter-lived progenitors, might usefully have called upon preexisting pathways for multiplex stress resistance to stave off cellular injury over prolonged time periods, i.e., over intervals longer than were helpful to their shorter-lived ancestors.

This hypothesis—*multiplex* cellular stress resistance as the key substrate for induced and evolved postponement of aging—although speculative and sketchy, has emerged from a striking confluence of new discoveries, initially and still mostly from studies of worms, flies, and yeast, but gradually coming to include evidence from vertebrates as well. This chapter starts with a summary of the evidence from the invertebrate systems, presents relevant evidence from studies of mammals and from cross-species comparisons, and then concludes with a prospectus for future work. To avoid duplicate coverage of material presented by experts elsewhere in this volume, we skip lightly over many subjects of obvious relevance, including considerations of the biochemistry of cell damage and repair, signal transduction machinery connecting insulinoid receptors to gene expression, and the detailed physiology of slow-aging animals from dwarf mice to dauer-constituent worms and many varieties of calorie-deprived creatures. We also avoid detailed presentation of work on resistance to single forms of stress—the "oxidative stress" theory of aging, for aging, for example, or the "somatic mutation" theory of aging—except insofar as data generated in pursuit of these ideas are relevant to the idea that coordinated resistance to multiple forms of stress is the key anti-aging mechanism triggered by experimental manipulation, intraspecies genetic variants, and evolutionary opportunities to exploit niches suited to slow-aging species.

INVERTEBRATE LIFE SPAN IS MODULATED BY STRESS

Temperature Effects on Life Span

Some of the earliest insights into the relationship of stress to aging came from studies in which transient exposure of flies to higher temperatures was found to increase their life expectancy, perhaps by induction of a long-lasting increase in their stress resistance: the "hormesis" effect. Temperature modulation is a major source of environmental stress to organisms, especially poikilotherms, such as fish and insects, that have limited mechanisms to control internal body temperature. Many mechanisms have evolved to minimize the detrimental effects of temperature change, including behavioral adjustments, physiological adjustments, or changes at the cellular level such as those mediated by repression of protein synthesis and accumulation of HSPs. Early studies of temperature modulation in aging studies were motivated by the "rate-of-living" hypothesis (Pearl 1923), the idea that life span is determined by the rate at which a series of developmental or metabolic steps proceeds. To evaluate this idea, early studies made use of poikilotherms where high ambient temperature can lead to increased metabolic rate. As predicted by the "rate of living" idea, high temperature leads to shorter life span in poikilotherms kept at constant temperature throughout their life span. Loeb and Northrop (1916, 1917), for example, in early studies of the fruit fly *Drosophila melanogaster*, observed a tenfold increase in life span for a temperature decrease of 20°C. This suggested to some researchers that aging was driven by simple rate-limiting reactions that can be described in terms of a chemical reaction.

Thermal Stress Alters Invertebrate Life Span

The initial observations prompted a series of experiments on the temperature dependence of life span conducted by Maynard Smith (1958a,b) with the fruit fly *D. subobscura*, which documented more complex effects. The protocol involved temperature-shift experiments between 20°C and 30.5°C. Female flies were maintained at 20°C until the sixth day of adulthood; subgroups were then shifted to 30.5°C for 5, 8, or 12 days and then returned to 20°C for the remainder of the life span. The initial prediction was that the flies shifted to the higher temperature would exhibit a shorter life span. Surprisingly, for male flies, life expectancy at 17 days of age was unaltered for the shifted subgroups compared to flies maintained at 20°C throughout. Female flies shifted to 30.5°C for 5 days showed a 47% increase in life expectancy compared to unshifted controls. These data, and follow up experiments (Lamb 1968; Hollingsworth 1969), led Maynard Smith to propose that life span was linked to reproduction in *Drosophila*. To test this hypothesis, he measured survival of flies in which reproductive effort was curtailed by a mutation in the *grandchildless* gene and observed a highly significant increase in life span compared to mated control flies and a slight increase over virgin controls, suggesting that life span extension in the temperature-shift protocol resulted from reduction in reproductive effort (Maynard Smith 1958a). This early prescient manuscript thus not only documented the first case of life span extension by a single gene mutation, but also provided the first clear instance of life span extension by manipulation of stress responses. This interpretation of the temperature-shift experiments was developed prior to an understanding of the response of biological systems to stress. A modern interpretation would consider the possibility that the induction of HSPs or other stress response factors could contribute to the life span increase.

Maynard Smith's discoveries lay in abeyance for almost three decades, but his experimental approaches have been critical to the development of experimental biogerontology. They resurfaced, for example, in the work of Khazaeli et al. (1995), who documented a decrease in age-specific mortality of *D. melanogaster* after exposure to desiccation stress. This group went on to demonstrate a reduction of age-specific mortality following transient exposure of fly cohorts to periods at 36°C (Khazaeli et al. 1997). Interestingly, they found no inverse correlation between life span and the number of eggs lain for this species, at odds with Maynard Smith's proposal that life span is prolonged at the expense of reproductive effort. This suggested that life span extension was not simply the result of shifting metabolic resources from the expensive business of producing eggs to maintenance but rather that the stress was inducing beneficial responses.

Other invertebrates also exhibit an increase in life span after stress. Johnson and Hartman (1988) published occasional life span increases in the microscopic nematode *Caenorhabditis elegans* when these worms were exposed to γ-radiation. This result was reminiscent of reported beneficial effects of low-dose radiation in humans (radiation *hormesis;* from the

Greek *hormaein* which means "to excite"). In the mid 1990s, a clear hormetic effect of thermal stress on life span was reported; transient exposure of synchronously aging cultures of *C. elegans* to "heat shock" resulted in significant increases in subsequent survival at normal temperatures (Lithgow et al. 1995). *C. elegans* lives for approximately 16–20 days at the optimal reproductive temperature of 20°C. When cohorts of worms were shifted to 35°C for periods of 3–12 hours, life span increases of up to 30% were observed, with a mean increase of 14%. Repeated heat shocks throughout life can enhance life span further in both *C. elegans* (Olsen et al. 2006b) and *Drosophila* (Hercus et al. 2003). Such heat shocks were also shown to alter tolerance to subsequent lethal thermal stress; heat shock confers increases in both longevity and thermotolerance, suggesting that a common mechanism could be at play (see below) (Lithgow et al. 1995; A.A. Khazaeli et al. 1997). Since the role of induced HSPs in thermotolerance was already under investigation in a number of laboratories (Parsell et al. 1993), HSPs became candidates as mediators of life span extension (Lithgow et al. 1995; Lithgow 1996).

The relevance of hormesis to human aging is unclear, but some investigators have speculated on the possible therapeutic avenues against age-related pathologies (Rattan 2001). The notion that small doses of otherwise toxic substances or low doses of radiation can be of benefit is not surprisingly a highly contentious public health issue. Research with human cell culture demonstrates strong hormesis effects across a wide range of circumstances (Rattan 1998; Fonager et al. 2002). However, as discussed later, the relationship between cellular and organismal stress resistance is not clear, and therefore the use of hormesis in slowing mammalian aging awaits investigation.

Heterogeneity in the Hormetic Response Predicts Life Span

Life span and age-dependent phenotypes vary greatly among individuals even in genetically uniform cohorts. The basis for this nongenetic heterogeneity is not fully understood, but variation in stress resistance may have an important role in that the magnitude of response to a stressful event can predict longevity outcomes. This has been studied in populations of *C. elegans* carrying a green fluorescent protein (GFP) reporter transcription unit controlled by an HSP transcriptional promoter (*hsp-16.2*) (Rea et al. 2005). The worms were exposed to a mild heat shock on the first day of adulthood and subsequently sorted, using an automated worm-handling system, into subpopulations based on expression levels of the *hsp-16::GFP* reporter. In general, worms that displayed high levels of *hsp-16::GFP* expression exhibited both higher thermotolerance and longer life span than worms that scored low for expression. At a minimum, the result suggests that the expression of the HSP genes can identify worms with higher life expectancies; more provocatively, the result implies that heterogeneity in stress response underpins heterogeneity in survival (Rea et al. 2005).

GENETIC MODULATION OF LIFE SPAN: A ROLE FOR STRESS RESISTANCE

The most comprehensive evidence for a mechanistic relationship between multiplex stress resistance and aging comes from genetic studies of life span in invertebrate model organisms. With few exceptions, single gene mutations that extend life span in worms and flies also result in resistance to a range of extrinsic stresses. Mutations in similar mammalian genes also confer stress resistance, suggesting that the relationship between longevity and stress resistance is conserved across diverse phyla (see further below). Remarkably, stress resistance is detectable even early in the life of long-lived mutants and is thus plausibly a cause, rather than an effect, of the delayed aging process. Resistance to multiple stresses, including oxidative agents, heat, and heavy metals, provided one of the first indicators that long-lived mutants were different from wild types in early adult life. The stresses administered in such studies are generally acute and cause death within hours or days, thus providing rapid assays and selection schemes for genetic screens for life span extension mutations.

Few studies have addressed the relationship between aging and the stress resistance of older animals. This is an important omission; response to stress in late life is likely to be critical to immediate survival, thus influencing the life span of a given strain. Stress resistance in early life is likely to predict stress resistance in late life (Lithgow et al. 1994), but stress resistance could simply develop late in life; studies that test stress resistance only in young adults might therefore present an incomplete picture.

Despite this limitation, there is still significant supporting evidence for a mechanistic relationship between stress resistance and life span when genes are manipulated in invertebrates and in mammals. There are two general nonexclusive explanations for such a stress/aging genetic connection. First, the process of aging impairs health by inducing changes similar to those that result from experimentally induced stress; the factors that confer stress resistance are induced by age-related changes and sustain life. The second explanation is that the longer-lived animals are intrinsically different from their wild-type counterparts even in the absence of stress, but this difference is beneficial either during aging or during stress. In *C. elegans*, there is evidence that long-lived and control strains differ both in their intrinsic resistance to stress and in the extent to which stress resistance pathways can be induced.

Signaling Pathways That Regulate Stress Resistance and Modulate Invertebrate Life Span

Many genes that modulate life span encode components of intracellular signaling pathways (Kenyon 2005), some of which have long been known to alter gene expression in ways that affect responses to cytotoxic or geno-toxic stress, and others of which have been shown more recently to affect stress resistance. Many of these signal pathways have well-documented homologs in mammalian cells, providing encouragement that manipulation of these pathways may result in mammalian life span extension. We review pathways that provide evidence that multiplex stress resistance has a significant role on life span determination.

Insulin Signaling Axis

Insulin/insulin-like growth factor (IGF) signaling (IIS) pathways have a large influence on stress resistance and life span in a range of diverse species including budding yeast, the nematode, the fruit fly, and the mouse. There is also considerable interest in whether IIS pathways modulate life span or aging phenotypes in human populations (Rincon et al. 2005; van Heemst et al. 2005). Manipulation of this pathway has profound effects on the ability of *C. elegans* to survive episodes of acute stress, and there is evidence that stress resistance is a feature of long-lived *Drosophila* and mouse mutants that affect components of this pathway.

The effect of IIS on survival and stress resistance is evident during worm development. In *C. elegans*, hypomorphic mutations in the DAF-2 insulin/IGF receptor-like protein (Daf, dauer formation) causes conditional formation of a diapause larval form called the dauer (Riddle et al. 1981; Kimura et al. 1997). Under conditions of sufficient food and low population density, *C. elegans* has a 3-day life cycle composed of four larval stages (L1 to L4), before the final molt into a reproducing adult (Riddle et al. 1997). However, poor nutritional conditions or overcrowding channel development toward an alternate larval stage called the dauer larva, in which the animal is nonfeeding, nonreproducing, and resistant to thermal and oxidative stress (Anderson 1978, 1982). The dauers enter reproductive development when food is reintroduced. Mutations affecting this process may be dauer-formation-constitutive (Daf-c), such that dauers are formed in the presence of food, or dauer-formation-defective (Daf-d), where true dauers fail to form even in the absence of food. Hypomorphic mutants at DAF-2, the worm receptor for IISs, are Daf-c at 25°C, but when cultured at 20°C, they develop into reproducing adults that are highly stress-resistant and exhibit a 100% increase in life span (Kenyon et al. 1993; Lithgow et al. 1995). Life span and stress resistance are also increased by mutation of *age-1* (Friedman and Johnson 1988a,b; Johnson 1990; Larsen 1993; Vanfleteren 1993; Lithgow et al. 1994, 1995), which encodes a p110 subunit of phosphatidylinositol-3 kinase (PI3K) (Morris et al. 1996) required for response to IIS receptor signals. Life span extension and stress resistance by either *daf-2* or *age-1* mutations requires the activity of DAF-16 (Kenyon et al. 1993; Tissenbaum and Ruvkun 1998; Hsu et al. 2003), a member of the FOXO family of transcription factors (Lin et al. 1997; Ogg et al. 1997) and DAF-18, a homolog of the PTEN phosphatase, which is a negative regulator of PI3K signaling (Dorman et al. 1995; Gil et al. 1999). Under conditions of abundant food, the wild-type versions of the DAF-2 and AGE-1 proteins act indirectly to suppress the activity of DAF-16. Mutant *daf-16* worms have a reduced life span (Lin et al. 2001). Expression of a DAF-16::GFP fusion protein in wild-type worms confers a slight life span increase and high levels of thermotolerance in young animals (Henderson and Johnson 2001). Moreover, stress events result in translocation of cytoplasmic DAF-16 to the nucleus (Henderson and Johnson 2001; Lin et al. 2001).

Comparison of the worm IIS pathway to its mammalian counterpart leads to the identification of additional components such as *pdk-1* (Paradis et al. 1999), *akt-1, akt-2* (Paradis and Ruvkun 1998), and *sgk-1* (Hertweck et al. 2004). SGK-1 acts in parallel to the AKTs to regulate DAF-16 by phosphorylation. DAF-16 is also regulated by ubiquitination. The E3 ubiquitin ligase RLE-1 reduces levels of DAF-16 leading to proteosomal degradation. Mutation of the *rle-1* gene results in elevation of DAF-16 levels and life span extension (Li et al. 2007). Moreover, *rle-1* mutants are highly resistant to thermal stress and UV radiation and to the pathogenic bacterium *Pseudomonas aeruginosa* (Li et al. 2007).

DAF-16 is key to the effects of a number of signaling pathways. Other regulators of DAF-16, including SMK-1 and SIR-2, also have roles in stress resistance (Tissenbaum and Guarente 2001; Wolff et al. 2006). There is also evidence that the *wnt* signaling mediator, β-catenin (encoded by the *bar-1* gene in *C. elegans*), directly interacts with DAF-16 to suppress expression of at least *sod-3* and potentially other stress response genes (Essers et al. 2005). The *bar-1* mutant worms are short-lived and sensitive to oxidative stress. Since mammalian β-catenin appears to interact with both FOXO4 and FOXO3a, there may be conservation of the effects of β-catenin on stress and life span (Essers et al. 2005). Life span extension by overexpressing a tyrosine-kinase-like gene, *old-1*, depends on the wild-type allele of *daf-16* (Murakami and Johnson 2001), suggesting it may also be a regulator of DAF-16. OLD-1 provides another link between life span and multiplex stress resistance; expression of OLD-1 is induced by heat, UV light, and starvation, and its overexpression leads to stress resistance.

Taken together, the genetics suggests that worm cells use intracellular circuits, similar to those that respond to insulin and IGF-I signals in mammals, to regulate stress resistance in ways that also affect life span. More specifically, the data show that these circuits, in normal worms, actively suppress the function of the FOXO family protein DAF-16, and thereby diminish life span and stress tolerance to levels well below those seen when DAF-16 is rendered fully active by mutations in upstream regulatory pathways.

Endocrine factors also have major effects on the rate of fly aging, but the case for multiplex stress resistance in IIS mutants extending life span is somewhat less convincing than it is for *C. elegans*. In *D. melanogaster*, mutations in *Inr* (insulin-like receptor) and *chico* (the insulin receptor substrate homolog) extend life span (Clancy et al. 2001; Tatar et al. 2001). Tissue-specific expression of the FOXO-like transcription factor dFOXO results in extended life span, consistent with the *C. elegans* findings (Giannakou et al. 2004; Hwangbo et al. 2004). Expression of dFOXO in the head fat body also results in resistance to oxidative stress (Giannakou et al. 2004) and decreases the levels of *Drosophila* insulin-like peptides (dilp) (Hwangbo et al. 2004). Unlike the components of the intracellular IIS pathways, which are generally encoded by single genes, insulin-like peptides are encoded by multiple genes in invertebrate genomes: 7 in the *Drosophila* genome and 38 in the *C. elegans* genome. Three dilps are produced in median neurosecretary cells (mNSC) in the adult *Drosophila* brain. Flies in which these cells have been eliminated by apoptosis exhibit increased storage of lipid and carbohydrate, reduced fecundity, and (unexpectedly) *reduced* tolerance of heat and cold (Broughton et al. 2005). Although they are thermosensitive, these flies exhibit an extension of median and maximal life span and increased resistance to oxidative stress and starvation. Mutation of *chico* does lead to stress resistance, but the effects are much less striking than in the worm (Clancy et al. 2001). Moreover, there is no effect of IIS on metabolic rate (Bohni et al. 1999). It would be informative to investigate whether older long-lived flies develop higher stress resistance with increasing age.

These disparities between worms and flies raise the question of whether IISs alter life span by modulation of stress resistance per se. There is certainly a remarkable correlation between longevity and stress resistance among mutants of *C. elegans* (Fig. 1). For example, there is a strong correlation between thermotolerance and life span across 15 alleles of the *daf-2* gene (Gems et al. 1998). Long-lived mutant worms are also better able to survive other stresses such as oxidative stress induced by paraquat (Vanfleteren 1993) or hydrogen peroxide (Larsen 1993), UV radiation (Murakami and Johnson 1996), heavy metal toxicity (Barsyte et al. 2001), hypertonic stress (Lamitina and Strange 2005), and bacterial infection (Garsin et al. 2003). Resistance to such a wide range of agents presumably reflects changes in expression of a diverse group of genes encoding stress response functions, metabolic functions, and others. Indeed, long-lived mutants exhibit an up-regulation of superoxide dismutase (SOD) activity (Vanfleteren 1993), mitochondrial SOD mRNA levels (Honda and Honda 1999; Murphy et al. 2003), catalase activity (Larsen 1993), catalase mRNAs (Murphy et al. 2003), metallothionein RNA (Barsyte et al. 2001; Murphy et al. 2003), small HSP (sHSP) levels following stress (Walker et al. 2001), sHSP mRNA levels (Hsu et al. 2003; Halaschek-Wiener et al. 2005), and glutathione-*S*-transferase (GST) mRNA (Murphy et al. 2003). Microarray analysis of mRNA abundance demonstrates that stress genes are a major class that distinguishes long-lived mutant worms from controls early in adult life (McElwee et al. 2003; Murphy et al. 2003; Fisher and Lithgow 2006). Some of these genes are known to be direct targets of the transcription factor DAF-16 (Oh et al. 2006).

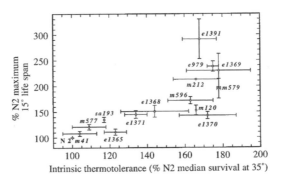

Figure 1. Plot of thermotolerance and life span of an allelic series of *daf-2* compared to wild type (N2 strain). Bars represent standard errors. (Reprinted, with permission, from Gems et al. 1998 [© Genetics Society of America].)

From this, it is clear why long-lived mutants are resistant to multiple forms of stress, but additional work will be needed to elucidate the contribution that stress resistance makes to life span extension in IIS mutants and whether some stress resistance factors are more important than others. These issues are far from being resolved, but if we turn our attention to experiments in which the expression of IIS-regulated genes have been manipulated, there is indeed evidence that IIS affects life span through stress gene expression. For example, knocking down the expression of individual HSPs decreases the life span and stress resistance of long-lived *daf-2* (Hsu et al. 2003) and *age-1* mutants (Morley and Morimoto 2004; Lamitina and Strange 2005) but not in wild-type worms (Morley and Morimoto 2004), and overexpressing HSP genes is sufficient to extend life span (see Fig. 2) (Yokoyama et al. 2002; Walker and Lithgow 2003). That said, knocking down genes that have no obvious role in stress response, such as isocitrate lyase/malate synthase, similarly reduces *daf-2* life span (Murphy et al. 2003), suggesting that stress response is perhaps not the exclusive mechanism of life span extension. Neither is stress resistance in young animals sufficient to confer extended life span. The notable example is the worm transforming growth factor-β (TGF-β)-like signaling pathway that influences dauer formation (Estevez et al. 1993) and regulates stress resistance (Lithgow et al. 1995) but has no apparent effect on life span. One possible explanation is that stress resistance is lost in these mutants as they age, but this has not been tested.

In summary, IIS regulates stress resistance through the expression of stress response genes, some of which have been shown to be necessary for life span extension and some where overexpression is sufficient for life span extension. Although other factors may be required for life span extension, the evidence to date supports the idea that stress resistance is a major contributing factor to the longevity of IIS mutant worms. Whether this is also true for IIS mutant *Drosophila* awaits further investigation.

The Lipophilic Endocrine Systems in Invertebrates

Downstream from both insulin signaling and TGF-β pathways in dauer formation in *C. elegans* is a secondary endocrine pathway that also influences stress resistance and life span. The pathway contains *daf-12*, a member of the nuclear hormone receptor superfamily (Antebi et al. 1998, 2000; Snow and Larsen 2000), and *daf-9*, a cytochrome P450 which is likely to be involved in the synthesis of a lipophilic hormone that acts through DAF-12 (Albert and Riddle 1988; Gerisch et al. 2001; Jia et al. 2002). DAF-12, which is expressed by many cells throughout the worm, shares homology with the vertebrate liver X receptor (LXR), the thyroid hormone receptor, and the pregnane X and vitamin D receptors. Interestingly, naturally occurring polymorphic variants of the LXR gene in human populations may affect life span (Mooijaart et al. 2005).

The role of DAF-12 in determining life span is complex, and DAF-12 is required for life span extension under some circumstances but inhibits life span extension in others. Loss-of-function mutations in *daf-9* are Daf-c and have been shown to increase life span, consistent with models in which the DAF-12 ligand limits life span (Gerisch et al. 2001; Jia et al. 2002). DAF-16 is not required for life span extension in *daf-9* mutants, showing that *daf-9* acts on life span downstream from insulin signaling. One Daf-c *daf-12* mutant lives 10% longer than wild-type worms, but another allele results in a shorter life span (Fisher and Lithgow 2006). The long-lived allele of *daf-12* also exhibits thermotolerance and paraquat resistance, whereas worms with the short-lived allele are sensitive to both stresses compared to wild types (Fisher and Lithgow 2006). *daf-9* and *daf-12* are also required for the life span extension observed in animals lacking cycling germ-line stem cells (Hsin and Kenyon 1999). The complexity of the relationship between *daf-12* and life span is illustrated by the fact that the extended life span of weak *daf-2* alleles is suppressed by some *daf-12* Daf-d alleles, but in contrast, strong *daf-2* alleles have synergistic effects on life span with *daf-12* (Larsen et al. 1995). Interestingly, one long-lived *daf-2;daf-12* double mutant exhibited lower accumulation of protein carbonyl damage than wild type or the *daf-2* mutant alone, suggesting that proteins associated with oxygen radical generation or protein repair or turnover could be involved. Perhaps the identification of sterol ligands for DAF-12 (Held et al. 2006; Motola et al. 2006) will result in a better understanding of the effects of this hormone axis on stress and aging.

Lipophilic hormone signaling is also a feature of life span determination in *Drosophila*. When long-lived *InR* mutants were treated with juvenile hormone (JH), life span was restored to normal, suggesting that IIS influences JH levels and that JH is part of a downstream endocrine system (Tatar et al. 2001). Indeed, *InR* mutants have reduced JH synthesis (particularly the major subtype, JH III-bisepoxide) and *chico* mutants reduce JH levels to a lesser extent (Tu et al. 2005). Another developmentally important hormone, ecdysone, is also reduced in *InR* mutants. There is evidence that JH is a regulator of reproductive responses to thermal stress in flies (Gruntenko et al. 2000, 2003; Flatt et al. 2005), and flies heterozygous for the ecdysone receptor (EcR) are resistant to oxidative stress and are also long-lived, suggesting that there may be parallels to *C. elegans.*

Cell Cycle Checkpoint Signaling and p53

DNA-damage cell cycle checkpoint proteins participate in a complex pathway that arrest cell division in response to stalled replication forks and other forms of genotoxic stress. The worm checkpoint proteins CID-1 (the homolog of caffeine-induced death protein-1 [spCid1] in the fission yeast *Schizosaccharomyces pombe*) (Wang et al. 2000; Saitoh et al. 2002), CHK-1, and CDC-25 all affect survival after exposure to stress; worms treated with RNA interference (RNAi) sequences corresponding to these genes become highly resistant to a lethal thermal stress (35°C) and also have increased normal life span (Olsen et al. 2006a). Cell division is confined to the germ line in adult worms, suggesting that these checkpoint proteins can influence worm survival through actions in postmitotic cells. This signaling pathway appears to act independently, at least in part, from the insulin signaling pathway in that the life span extension and stress resistance do not require DAF-16 or DAF-12. Independence of DAF-16 and DAF-12 also suggests that the *C. elegans* checkpoint proteins regulate somatic maintenance through processes different from those regulated by the germ-line signal and the insulin signaling pathways, but resemble the IIS pathway in the association with increased stress resistance (Olsen et al. 2006a).

The Ku protein also regulates checkpoint responses to genotoxic stress and cytotoxic stress responses and influences aging (McColl et al. 2005). Mammalian Ku, a heterodimer of Ku70 and Ku80, has a role in DNA double-strand break repair, telomere maintenance, checkpoint function, tumor suppression, and cellular stress resistance. RNAi of the *C. elegans* homolog of Ku70 (*cku-70*), in contrast to genes in the *chk-1* pathway, extended life span in an insulin-signaling-dependent manner.

Invertebrate proteins with similarity to the mammalian tumor suppressor p53 have also been implicated in life span determination. Expression in *Drosophila* of a dominant-negative p53 in neurons resulted in a significant life span extension and an increase in resistance to genotoxic stress (Bauer et al. 2005). This effect may be related to a calorie-restricted mechanism, because expression of this protein does not further increase the life span of calorie-restricted flies. Whether this is related to the observed accelerated aging phenotypes associated with overexpression of forms of p53 in rodents (Tyner et al. 2002; Maier et al. 2004) remains to be seen.

c-Jun Amino-terminal Kinase and p38 Kinase Signaling in Invertebrates

JNK, also known as "stress-activated protein kinase," has a key role in helping many mammalian cell types respond to many forms of cellular injury. JNK signaling also appears to have a conserved role in stress response and significantly effects life span. JNK signaling activity has been identified as a key regulator of the response to oxygen radicals in various models (Nemoto et al. 2000; Wang et al. 2003). Mutant *Drosophila* flies with enhanced JNK signaling accumulate less oxidative damage and live longer than wild-type flies (Wang et al. 2003). JNK signaling has also been shown to affect life span in *C. elegans,* where it appears to act in parallel with insulin signaling but requires the FOXO transcription factor DAF-16 (Oh et al. 2005). It directly interacts with and phosphorylates DAF-16 and promotes translocation of DAF-16 into the nucleus following heat shock (Oh et al. 2005). Thus, we may expect JNK to influence stress and aging by altering some of the downstream stress response genes described above for IIS.

The conserved p38 mitogen-activated protein kinase (MAPK) pathway has been intensely studied in mammalian cells, and this pathway, like the one involving JNK, responds to stress (Guyton et al. 1996; Liu et al. 1996). In addition, p38 regulates in vitro cellular senescence and apoptosis in a variety of circumstances. In *C. elegans,* there are three genes with similarity to mammalian p38 (*pmk-1 -3*) (Berman et al. 2001); *pmk-1* is required for *daf-2* mutant-associated resistance to pathogenic bacteria (Kim et al. 2002; Aballay et al. 2003; Sifri et al. 2003). There are a number of similarities between resistance to bacterial infection and resistance to other stresses (Lithgow 2003). The *pmk-1* gene is also required for life span extension by mutation of *daf-2* but has no effect on wild-type life span (Troemel et al. 2006), suggesting that in this circumstance, MAPK and JNK signaling have opposite effects on stress resistance and life span in the worm.

The Influence of Mitochondrial Function on Life Span and Stress Resistance

It is clear that mitochondrial oxidative stress and the production of reactive oxygen species (ROS) are associated with many age-related changes. However, the notion that ROS is the single *cause* of aging is perhaps an oversimplification that is not supported by a large number of studies. This topic is discussed in detail elsewhere (Chapter 1), but it is worth noting here that alterations in mitochondrial function seem to have variable effects on invertebrate stress resistance and life span. The invertebrate models provide some of the best direct evidence that ROS damage contributes to age-related decline, but there are many paradoxical and inconsistent findings. For example, mutations and RNAi treatments that alter expression of genes in the mitochondrial electron transport chain (ETC) have had unpredictable results, sometimes leading to shorter life span and in other cases to extended life span. For instance, a mutation in the *C. elegans* iron sulphur protein (*isp-1*) of mitochondrial complex III results in low oxygen consumption, resistance to paraquat (oxidative stress), and an increased life span (Feng et al. 2001). In a similar vein, mutation of the *clk-1* gene, which encodes a di-iron carboxylate enzyme responsible for the final hydroxylase step in the synthesis of ubiquinone (Q), results in extended life span when grown on bacteria that provide a source of Q (Wong et al. 1995; Ewbank et al. 1997; Jonassen et al. 1998, 2001; Felkai et al. 1999; Vajo et al. 1999; Stenmark et al. 2001). *clk-1* mutants are resistant to UV radiation (Murakami and Johnson 1996), and so these results, taken together with the *isp-1* data, suggest that reducing mitochondrial electron transport increases life span and stress resistance. However, mutations in genes encoding other components of the ETC result in shortened life span; *mev-1* encodes the cytochrome *b* large subunit (Cyt-1/ceSDHC) in complex II and displays hypersensitivity to elevated oxygen levels, a more rapid accumulation of fluorescent material and protein carbonyls, and a shortened life span (Ishii et al. 1990, 1998; Hosokawa et al. 1994; Adachi et al. 1998; Senoo-Matsuda et al. 2003). Mutations in *gas-1*, which encodes a 49-kD iron protein subunit of the complex I NADH-ubiquinone-oxidoreductase (Kayser et al. 1999, 2001, 2003), are also short lived. Both of these short-lived mutants are sensitive to oxidative stress (Ishii et al. 1990; Hartman et al. 2001; Kayser et al. 2001); there is a correlation between stress resistance and life span but not electron transport function and life span.

A wider survey of mitochondrial perturbations suggests that the picture could be more complex. In systematic RNAi screens for longevity genes, mitochondrial functions feature prominently (Dillin et al. 2002; Lee et al. 2003). Nine genes that exhibit increased life span upon RNAi have also been surveyed for resistance to oxidative stress and thermotolerance. Some exhibited mild increases in thermotolerance and resistance to hydrogen peroxide, but no resistance to paraquat was observed (Lee et al. 2003). We can conclude that stress resistance may have a role on the effects of mitochondrial ETC activity on life span in *C. elegans,* but whether multiplex stress resistance is important is not yet clear.

Despite the fact that no correlation exists between mitochondrial ROS production and life span in *Drosophila* (Miwa et al. 2004), genetic studies in this species generally support the notion that stress resistance has a role in mitochondrial life span determination. For example, A mutant with a defect in subunit b of succinate dehydrogenase (SDH; mitochondrial complex II) is hypersensitive to oxygen, is short-lived, and exhibits signs of accelerated aging (Walker et al. 2006). Mutation of another mitochondrial protein, with homology to acyl-CoA dehydrogenases, in female flies results in extended life span and resistance to paraquat (Mourikis et al. 2006). In another example, when a mitochondrial antioxidant function is enhanced, life span is increased; high-level overexpression of the mitochondrial SOD gene increased life span by 37%, with lower levels of expression resulting in smaller life span increases (Sun et al. 2002).

Genetic Manipulation of Stress Resistance Modulates Invertebrate Life Span

The observation that anti-aging mutations typically render invertebrates resistant to multiple forms of stress has prompted experiments to determine whether genetic manipulation of stress resistance per se leads to increased longevity. Such experiments have inspired parallel experiments in the mouse (Schriner et al. 2005). The strategy has also been exploited as a useful strategy for the identification of new aging genes (Walker et al. 1998). In one screen, researchers subjected mutagenized first larval stage *C. elegans* to heat stress (30°C for 7 days) and then allowed survivors to recover and grow at 15°C. This resulted in the identification of 49 long-lived alleles, most of which were also involved in dauer formation, but some of which had no other observable phenotype (Munoz and Riddle 2003). The effect of checkpoint proteins in nematode aging was also uncovered using a similar screen, but in this case, the selection was performed on adult worms, rather than larvae (Olsen et al. 2006a).

Wang et al. (2004) used a different approach by identifying genes in *Drosophila* that were up-regulated by heat, oxidative stress, and starvation by subtractive hybridization. Two of the genes revealed by this method encoded small heat shock proteins (Hsp26 and Hsp27), and when these genes were overexpressed, the flies exhibited resistance to paraquat, thermal stress, and starvation (Fig. 2).

This study is one of many demonstrating that HSPs have potent effects on life span in multiple invertebrate species (Fig. 2). No mouse studies have addressed the affects of chaperones on life span, but there is some evidence that naturally occurring polymorphic variants of chaperone genes affect human life span (Altomare et al. 2003). The effects of chaperones on organismal survival were first observed when the Lindquist lab began engineering *Drosophila* for enhanced thermotolerance by overexpression of Hsp70 (Welte et al. 1993). Tatar demonstrated that these thermotolerant flies exhibited reduced mortality during normal aging (Tatar et al. 1997; Silbermann and Tatar 2000). Later, a related HSP-70 was shown to prolong life span in *C. elegans* (Fig. 2) (Yokoyama et al. 2002). Transgenic lines carrying extra copies of the *hsp-70F* gene had a mean life span 43% longer than controls when the gene was expressed predominantly in muscle tissue.

There have been no comprehensive studies of the tissue or organelle specificity of the chaperone effect on life span. However, overexpression of another small HSP gene, *hsp-22*, also extended life span in *Drosophila*, even when the overexpression was limited to motor neurons (Morrow et al. 2004). The resulting flies were resistant to both oxidative and thermal stress. This HSP is a mitochondrial matrix protein that exhibits chaperone activity (Morrow et al. 2000). Decreased expression of Hsp22 makes flies very sensitive to stress and short-lived. Elevated Hsp22 levels were also observed in lines of flies genetically selected for increased longevity (Kurapati et al. 2000).

It is likely that overexpression of HSPs also explains, in part, the increased life span of *C. elegans* IIS mutants (Lithgow 1996). Overexpression of the gene encoding HSP-16, which is up-regulated in *daf-2* and *age-1* mutants (Walker et al. 2001; Hsu et al. 2003), results in resistance to thermal stress and an extended life span (Fig. 2) (Walker and Lithgow 2003). The expression of the *hsp-16* transgenes in these worm lines is dependent on IIS such that loss of *daf-16* results in reduced HSP-16 levels and shorter life span. Consistent with this report, knocking down expression of individual HSP genes, including those encoding small HSPs, can prevent extended life span in long-lived IIS mutants (Hsu et al. 2003; Morley and Morimoto 2004). In addition, there are genetic interactions between IIS and HSF (Hsu et al. 2003; Walker et al. 2003; Morley and Morimoto 2004), and overexpression of HSF is sufficient to extend *C. elegans* life span (Hsu et al. 2003; Morley and Morimoto 2004) presumably due to enhanced HSP expression during aging.

The role of chaperones as inhibitors of both aging and stress has implications for pathogenesis and perhaps treatment of age-related diseases, particularly those thought to be the result of toxic protein aggregation such Huntington's disease (HD), Alzheimer's disease, and

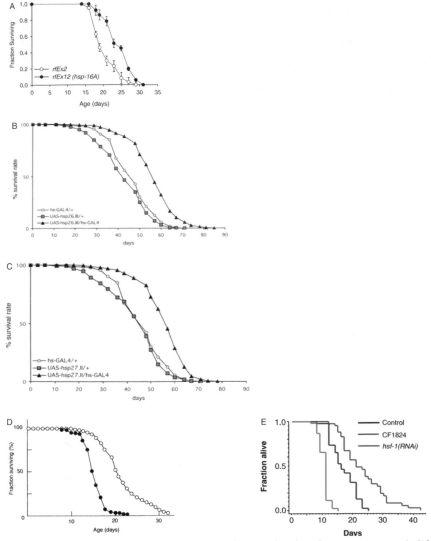

Figure 2. The overexpression of genes encoding molecular chaperones extends life span. Many molecular chaperone genes are up-regulated to a wide variety of stresses and most were initially characterized as heat shock proteins (HSPs). Inducible HSP genes are subject to positive transcriptional regulation by the DNA-binding protein, heat shock factor (HSF). (*A*) Survivorship of a *C. elegans* strain carrying extra copies of *hsp-16A* rfEx12(hsp-16A)] and its transgenic control rfEx2. Life span is significantly extended by the overexpression of this single molecular chaperone. (*B*) Survivorship of *Drosophila* strains overexpressing hsp26 (UAS-hsp26.III/hs-GAL4) and two control strains. Overexpression of the chaperone-encoding gene extends life span. (*C*) Survivorship of *Drosophila* strains overexpressing hsp27 (UAS-hsp27.II/ hs-GAL4) and two control strains. Overexpression of the chaperone-encoding gene

Parkinson's Disease. Worms engineered to express proteins with polyglutamine (poly-Q) expansions characteristic of HD show interesting interactions with both chaperones and insulin signaling (Morley et al. 2002; Hsu et al. 2003). Expression of poly-Q proteins results in the induction of molecular chaperone genes, and aggregate formation is retarded in long-lived worms carrying a mutation in *age-1* (Morley et al. 2002). Aggregation of these proteins is accelerated by down-regulation of HSPs known to determine life span (Hsu et al. 2003). Similarly, DAF-16 and HSF have distinctive roles in the toxicity of the AD-associated Aβ(1–42) peptide (Cohen et al. 2006). These results in invertebrate models of neurodegenerative illnesses suggest that molecular chaperones may provide a mechanistic connection between at least some age-related diseases and the aging process. The extent to which protein aggregation or other consequences of altered protein structure contribute to age-related dysfunction is still uncertain, and it is possible that chaperone function may guard against cellular damage through a variety of pathways, including regulation of translation and autophagy of damaged intracellular structures. Further studies of the involvement of HSPs and related proteins as modulators of resistance to cellular injury are likely to shed new light on these relationships.

PHARMACOLOGICAL LIFE SPAN EXTENSION BY ENHANCING STRESS RESISTANCE

Many of the chemical compounds that extend invertebrate life span also enhance stress resistance or at least are associated with stress resistance in other systems. As a class, these compounds can be termed "stress response mimetics" (SRMs). For example, many antioxidants have been tested and positive effects have been observed. Exposure of *D. melanogaster* to *N*-acetylcysteine (NAC), an antioxidant and precursor to glutathione, resulted

---

significantly extends life span. (*D*) Survivorship of control and hsp70F overexpressing *C. elegans*. (control; *filled circles*, hsp70F overexpression strain; *open circles*). (*E*) Survivorship of strain carrying extra copies of the gene encoding heat shock factor (HSF; CF1824) compared to a control strain. Life span is significantly increased by HSF overexpression. RNAi knockdown of HSF reduces life span. (*A*, Reprinted, with permission, from Walker and Lithgow 2003 [© Blackwell Publishing]; *C*, reprinted, with permission, from Wang et al. 2004 [© National Academy of Sciences]; *D*, reprinted, with permission, from Yokoyama et al. 2002 [© Elsevier]; *E*, reprinted, with permission, from Hsu et al. 2003 [© AAAS].)

in a dose-dependent increase in median and maximum life span of about 30% (Brack et al. 1997). In addition, life span extension was achieved with 4-phenylbutyrate (PBA) administration in *Drosophila* (Kang et al. 2002). Critically, PBA also enhanced survival to starvation and oxidative stress. Another group has reported that longevity can be significantly increased in *C. elegans* by treatment with catalytic compounds EUK-134 and EUK-8, which exhibit SOD and catalase activities (Melov et al. 2000). Life span extension was achieved without effects on development rate or fertility. This result appears to be dependent on subtle variations in culture conditions in that other investigators have failed to observe a life span increase or have instead observed toxic effects (Keaney and Gems 2003; Keaney et al. 2004). These SRMs reduce oxidative damage in a range of disease models in rodents (Musleh et al. 1994; Baker et al. 1998; Melov et al. 2001, 2005) and confer resistance to oxidative stress and thermal stress in the worm (Keaney et al. 2004; Sampayo et al. 2003). Whether they act by directly reducing reactive oxygen species or indirectly by altering intrinsic stress resistance is unknown. The compounds appear to have no effects on insect life span (Bayne and Sohal 2002; Magwere et al. 2006).

A number of life span extension studies have focused on the effects of natural products that may also act by a multiplex stress resistance mechanism. These include, among others, studies of Vitamin E, coenzyme Q, and complex extracts from ginkgo biloba and blueberries. Some of these studies also evaluated effects on resistance to stress in the treated worms. For example, the ginkgo biloba extract, EGb 761, has been shown to increase resistance to stress and extend life span in the worm (Wu et al. 2002; Strayer et al. 2003), and similar results were recently observed for blueberry phenols (Wilson et al. 2006). Resveratrol, found in many botanical preparations, has also been shown to extend life span of yeast, flies, and worms and may have beneficial effects on mice subjected to the perils of a high-fat diet (Howitz et al. 2003; Wood et al. 2004). There is some evidence, still controversial, that resveratrol acts through stimulation of Sir2, a modulator of several stress pathways including those coordinated by *daf-16* and other members of the FOXO family (Sinclair 2005).

As would be predicted from genetic studies, some compounds extend life span without any apparent effects on stress resistance, including anticonvulsants in *C. elegans* (Evason et al. 2005; Collins et al. 2006).

STUDIES IN VERTEBRATES: OVERVIEW AND PROSPECTUS

Studies of stress resistance in long-lived mutant worms provided the initial evidence for the idea that multicellular animals retain an evolutionarily ancient control pathway that can coordinately activate multiple defense mechanisms and that this pathway is critical to deceleration of aging both within species and across evolutionary time. Laboratory investigation of invertebrates is proceeding rapidly to test the generality of this assertion, to define how the pathway works by altering expression of specific sets of genes, and to delineate how intercellular signals connect this hypothesized defense system with environmental signals and developmental events. Watching mouse researchers starting to wrestle with these themes and pathways is like watching a team of highly trained elephants impersonating a ballet troupe, a site to inspire varying mixtures of admiration and pity. Many studies focus on single specific aspects of stress resistance, typically resistance to oxidative injury, in calorically restricted rodents or in comparisons across species, but there are few studies that evaluate multiple forms of stress resistance in the same set of animal models. Thus, an assessment of current data on the "multiplex stress resistance" idea as a central theme for slow aging in vertebrates has a patchwork quality, providing a basis for further investigations, rather than a conclusive, air-tight argument.

Several lines of evidence are available with varying levels of relevance to our theme. These include studies of resistance to specific forms of stress in long-lived mutant mice or in calorically restricted vertebrates, evaluation of cellular and hormonal pathways that mediate anti-aging effects in worms and flies, and analyses of the distribution of stress resistance among species with varying longevity.

CELLULAR STRESS VS. ORGANISMIC STRESS IN MAMMALS

The distinction between defense mechanisms that operate at the organismic level and those that reflect cell-autonomous defense mechanisms becomes progressively more salient in animals larger and more complex than flies and worms. Exposing a fly or worm to a warm environment quickly increases the temperature of all cells in the animal's body; however, mammals have complex controls over heat distribution, involving central and local adjustments of heat production, blood flow, and evaporative loss that give different organs and tissues differential protection against the ill effects of overheating. Similarly, specific cell types in the body of mammals enjoy variable protection against direct exposure to UV light, the toxic effects of atmospheric oxygen, and ingested toxins, agents that might lead to relative uniform access to most cells in a smaller, less-opaque organism. Thus, experimental protocols that simply expose the entire organism to a source of injury, such as a heavy metal, UV irradiation, or an oxidizing agent, may provide useful information about cellular resistance to injury in worms or flies, but not be applicable for studies of cell biology in rodents or humans. There is unfortunately a paucity of well-developed methods for analysis of the relative vulnerability of mammalian endothelial cells, neurons, cardiomyocytes, kidney cells, etc., to different forms of cell injury. Those methods that have been developed, typically for investigation of clinical models such as ischemia/reperfusion injury, can be difficult to interpret in that the extent of injury and recovery can depend not simply on cellular stress resistance, but on details of vascularization, recruitment, and activation of inflammatory cells, and the dynamics of sequestration of the agent by binding proteins, lipid depots, and other local sinks. Even an apparently straightforward approach, such as intravenous injection of an oxidative liver toxin such as acetaminophen (Harper et al. 2006), is complicated by differences between animals in rates of metabolic conversion of the administered agent, rates that can be influenced by both dietary and hormonal factors. Reductionist approaches, in which stress resistance of mammalian cells is evaluated in short- or long-term cell cultures, pose their own problems, including the difficulty of culturing most adult cell types, and both selective and adaptive pressures for cell growth and survival that may quickly alter the cellular properties of interest. These difficulties greatly complicate attempts to evaluate hypotheses about the possible role of stress resistance as a modulator of aging rate in mammals and must be taken into account when interpreting many of the studies presented below.

STRESS RESISTANCE IN LONG-LIVED RODENTS

Caloric Restriction

There is strong and consistent evidence that oxidative damage is diminished in rats and mice whose life span has been extended by a calorie-restricted (CR) diet; classical and more recent reviews (Yu and Masoro 1995; Bokov et al. 2004) list and discuss reports of decreased lipid peroxidation, protein carbonyl levels, and DNA oxidation damage and cite numerous reports of CR-induced increases in enzymes, including catalase, both SODs, and glutathione peroxidase that could contribute to protection against oxidative damage. There is some uncertainty as to whether the protective effects involve diminished production of damaged macromolecules by free radicals (Gredilla et al. 2001) or augmented removal of damaged molecules once damage has occurred (Pamplona et al. 2002), but it seems likely that both mechanisms, coordinately induced, may contribute to the net effect. Rhesus monkeys receiving a CR diet also accumulate lower levels of oxidatively damaged proteins and appear to have lower levels of lipid peroxidation as well (Zainal et al. 2000). These studies were in part motivated by the notion that diminished food intake might lead to a decline in cellular metabolic activity and thus to diminished rates of production of free radicals. Rodents on CR diets, however, reduce their body weight, and their utilization of both glucose and oxygen per gram of lean or metabolic mass is indistinguishable from levels in control animals (McCarter et al. 1985; Masoro et al. 1992). Thus, the pathways that connect the CR diet to diminution of oxidant damage are not yet well understood. Could these connections involve circuits that also increase resistance to other forms of stress?

A good deal of evidence suggests that caloric restriction reduces the rate of somatic mutation, thought to contribute to many forms of late-life pathology. In some cases, such as the induction of mutations in rat T cells by bleomycin (Aidoo et al. 1999), the mutagenic effect depends at least partly on oxidative modification of DNA, making it difficult to decide if the protective effect of the CR diet represents changes in antioxidant defenses, more rapid repair of DNA lesions, or both. The effect of the CR diet seems to be quite rapid, with significant effects seen as early as 4 weeks after transition to the fully restricted diet (Aidoo et al. 1999), arguing that the differences between control and restricted rats do not simply reflect a deceleration by CR of age-related mutant accumulation. Other studies (Casciano et al. 1996) have hinted that resistance of CR rodents to chemical mutagenesis can reflect diminished bioactivation, by cytochrome P450s, of procarcinogenic agents. Two studies have evaluated spontaneous mutation frequency in T cells of CR rats and mice that had not been deliberately exposed to mutagens, and both found a lower frequency in the CR animals (Dempsey et al. 1993; Aidoo et al. 2003) when tested as early as 3 or 6 months after imposition of the CR diet. Aidoo et al. (2003) noted that the CR effect was relatively specific for base substitutions and frame-shift mutations but had less impact on large-scale changes, such as deletions. They interpret this observation as consistent with the hypothesis that CR is acting by an effect on antioxidant defenses, but it seems hard to exclude models that feature augmented repair instead.

Indeed, several groups have shown an increase in enzymes responsible for repair of damaged DNA in CR rodents. One study, for example, evaluated removal of cyclopyrimidine dimers from UV-irradiated short-term cultures of rat hepatocytes as a measure of nucleotide excision repair and documented (Guo et al. 1998) more rapid repair of these lesions using a test of global repair rate, as well as for repair of damage to an actively transcribed gene and two genes known not to be active in hepatocytes. This study evaluated only aged animals, however, making it impossible to know if the observed CR effect was merely a consequence of retardation of age effects or instead a rapid effect of the CR diet and thus potentially a contributor to the delayed pace of aging. A second group (Cabelof et al. 2003) evaluated CR effects on base excision repair and found that even 2.5 to 4.5 months of exposure to the CR diet leads to improved base excision repair in rat liver; these authors showed improved repair in older rats for brain, spleen, or testes. This group also showed that short-term CR induces an increase in protein and mRNA for β-polymerase, the rate-limiting enzyme in the base excision repair response. In a parallel study, mice bearing a *lacI* transgenic sequence as a mutation target were exposed either to dimethylsulfate or to 2-nitropropane; the former induces DNA alkylation directly, which can be removed by the base excision repair pathway, and the latter induces oxidative damage to DNA. Both varieties of induced mutation were diminished, in young mice, by the CR diet.

Calorically restricted rodents have also been used to test ideas about the role performed by translational fidelity and protein conformation in aging. One group found that long-term CR prevented the decrease with age in Hsp70 expression in rat hepatocytes exposed to heat in primary culture (Heydari et al. 1993). The experimental design did not include young animals on short-term CR, and therefore it is not possible to decide if the higher Hsp70 levels in the old CR rats might have contributed to, or instead simply be a consequence of, delayed aging. A similar approach (Selsby et al. 2005) revealed a similar decline with age in HSP72 and HSP25 levels in rat soleus muscle, which was prevented by CR diet, but again, the design omitted animals tested only shortly after initial CR exposure.

Methionine Restriction

Although low-calorie diets have long been a mainstay of mammalian aging research, it is now clear that diets low in methionine also extend life span in both rats (Orentreich et al. 1993; Zimmerman et al. 2003) and mice (Miller et al. 2005). Methionine-restricted rodents differ from controls in glutathione distribution (Richie et al. 2004) and levels of thyroid hormones, glucose, insulin, and IGF-I (Miller et al. 2005). The longevity effect can be seen in multiple rat strains, is not attributable to reduced energy intake (Zimmerman et al. 2003), inhibits at least some stages of carcinogenesis (Komninou et al. 2006), and retards the development of cataracts and immune system senescence (Miller et al. 2005), suggesting that like CR, the methionine-deficient diet slows many different aspects of aging in addition to increasing maximal life span. Rats placed on a low-methionine diet for 6 weeks show changes in heart and liver mitochondria, including diminished levels of Complex I and Complex IV of the electron transport chain, diminished production of H_2O_2, lower electron leak rate, diminished accumulation of oxidative damage to mitochondrial DNA, and lower levels of protein oxidation (Sanz et al. 2006). These authors also noted changes in the fatty acid composition of mitochondrial membranes, similar to those noted in long-lived species (see below). Methionine-restricted mice are resistant to the oxidative hepatotoxicity caused by injection of acetaminophen (Miller et al. 2005). The mechanisms that connect a low-methionine diet to extended rodent longevity are unclear, with plausible arguments involving alterations in DNA methylation, glutathione metabolism, transcription rate and protein turnover, and hormone levels. It is also noteworthy that a comparison of amino acid abundance among eight mammalian species found only one—methionine—that showed a dramatic correlation with species life span ($R^2 = 0.93$, $p < 0.001$), longer-lived species having relatively low levels of methionine in total heart protein mixtures (Ruiz et al. 2005).

Long-lived Mutants: p66Shc

The *Shc* locus encodes proteins of 46 kD and 52 kD that are known for their role in signal transduction, as well as a 66-kD splice variant, p66^{shc}, with an altered amino-terminal sequence. Mice with a mutant form of *Shc* that prevents formation of the p66^{shc} form have been shown to live 30% longer than controls (Migliaccio et al. 1999), and the limited data suggest that heterozygotes may also be somewhat longer-lived than controls. The original report showed that fibroblasts from p66^{shc}-deficient embryos were resistant, in culture, to death induced either by UV light or by H_2O_2 and that these mouse embryonic fibroblast (MEF) cells were deficient in a p53-independent pathway that induced the cyclin-dependent kinase inhibitor p21 after exposure to UV or H_2O_2. The p66^{shc}-defective mice were also resistant to the lethal effects of paraquat. Further work in several laboratories has shown that these mutant mice are resistant to a number of forms of late-life pathology, including the kidney damage produced by streptozotocin diabetes (Menini et al. 2006), aortic injury induced by a high-fat diet (Napoli et al. 2003), and age-dependent defects in acetylcholine-induced relaxation of vascular endothelium in an aortic ring preparation (Francia et al. 2004). In the diabetes model, cell cultures grown from renal mesangial tissue of p66^{shc} KO mice were found to be resistant to the effects of high glucose levels on cell survival and induction of reactive oxygen species in an NF-κB-dependent stress pathway. In the high-fat diet model, evidence of systemic oxidation damage, including oxidation of low-density lipoprotein and accumulation of isoprostanes, was diminished in mice carrying the KO mutation. All of these data thus support the hypothesis that the wild-type version of p66^{shc}, whatever its (unknown) selective benefits to wild-type mice, also renders them more susceptible to multiple forms of late-life pathology and to oxidation damage. It would be of considerable interest to determine if cells (fibroblasts, endothelial cells, mesangial cells, etc.) of p66^{shc} KO mice were resistant to multiple forms of stress; the data on UV resistance are interpreted in the original report as further evidence of resistance to oxidative damage, but there is at least some evidence (Salmon et al. 2005) that UV-mediated damage to fibroblasts is independent of oxidative pathways.

Altered stimulation of p66^{shc} might contribute to the stress resistance and anti-aging effects in other varieties of long-lived mutant mice. Embryonic fibroblasts from mice hemizygous for the IGF-I receptor, for example, show lower levels of tyrosine-specific phosphorylation of p66^{shc} after stimulation by IGF-I (Holzenberger et al. 2003). p66^{shc} also regulates the phosphorylation state, and thus activity, of the mammalian Forkhead protein FRKHL1 (Nemoto and Finkel 2002), the analog of the DAF-16 transcription factor so central to anti-aging mutations in *C. elegans*, in mammalian cells exposed to H_2O_2, although in this system, it is unclear if H_2O_2 is acting as a source of oxidative stress or as a mitotic stimulus. Because in nematodes the Forkhead protein DAF-16 can induce a wide range of protective cellular responses, more work on p66^{shc}-mediated changes in cell defense to both oxidant and nonoxidant stresses is likely to be productive.

Cellular metabolic pathways also seem to be altered by the p66^{shc} KO mutation (Nemoto et al. 2006). Fibroblasts from these mice use less O_2, both in their basal state and when uncoupling agents have been used to stimulate maximal respiration. Conversely, the KO cells require higher levels of glucose to maintain ATP levels and produce more lactate, suggesting a shift to a greater reliance on glycolytic metabolism. Finally, this group noted that the p66^{shc} KO cells showed a relative poor ability to regenerate NADH levels after laser-induced NADH photolysis and that permeabilized mitochondria from these cells were relatively slow to take up exogenous NADH from solution. These data thus suggest that the wild-type version of the p66^{shc} protein may regulate the balance between oxidative and glycolytic metabolism in at least some cell types, through an alteration in one or more NADH-dependent mitochondrial reactions. The relationship of these metabolic changes to stress resistance is uncertain, but it is noteworthy that fibroblasts from adult Snell dwarf mice are comparatively resistant to the effects of low-glucose medium on NADH-dependent reduction of an extracellular electron acceptor and that cell lines from normal mice show correlations between their resistance to this effect of glucose deprivation and their resistance to death from H_2O_2 or cadmium toxicity (Leiser et al. 2006). Similarly, fibroblasts derived from the skin of long-lived species of rodents are, like Snell dwarf cells, resistant to the inhibitory effects of low-glucose medium, as well as to the lethal effects of H_2O_2 and cadmium (Harper et al. 2007).

Long-lived Mutants: Endocrine Mutations in Mice

The first published observation of a single gene mutation that extended life span in a mammal was the report of exceptional longevity in the Ames dwarf mouse (Prop-1$^{df/df}$), in which a defect in the development of the embryonic pituitary leads to low levels of serum growth hormone (GH), thyrotropin (TSH), and prolactin, with secondary deficits in IGF-I and thyroid hormones (Brown-Borg et al. 1996). As young adults, Ames dwarf mice are about one third the size of normal littermates and live approximately 40% longer lives. The Snell dwarf mutation (Pit-1$^{dw/dw}$) leads to a very similar set of endocrine abnormalities and also produces life span extension (Flurkey et al. 2001), as well as delays in age-related changes in the eye lens, joints, immune system, collagen, and kidneys (Silberberg 1972; Flurkey et al. 2001; Vergara et al. 2004). Although the Snell and Ames dwarf mice exhibit abnormalities in multiple hormones, three other mutations whose primary effects are limited to the GH/IGF-I axis are also long-lived, suggesting that diminished IGF-I levels contribute to the delayed aging in these mutant mice (Flurkey et al. 2001; Coschigano et al. 2003; Holzenberger et al. 2003). The biology of these long-lived mutants is discussed in more detail elsewhere in this book (Chapter 13); here, we present studies relevant to questions of cellular stress resistance.

Fibroblast cultures derived from the skin of young adult Snell dwarf mice have been found (Murakami et al. 2003; Salmon et al. 2005) to be resistant to the lethal effects of six different agents: H_2O_2, paraquat, cadmium, heat, UV irradiation, and the DNA alkylating agent MMS (methylmethane sulfonate). The differential stress resistance is detectable in cells grown in culture for several weeks through multiple cell divisions and thus seems to represent a stable epigenetic trait. Cells taken from mice less than 1 week old do not show differential stress resistance (Salmon et al. 2005), suggesting that the stress resistance phenotype emerges in juvenile or young animals when precursors of fibroblasts differentiate in the hormonally abnormal environment of the Snell dwarf mouse. Fibroblasts from Ames dwarf mice show a similar pattern of in vitro resistance to these stresses, as do cells from the long-lived GH receptor KO mutant (Salmon et al. 2005), suggesting that the stress resistance trait is robust across several background genotypes and housing environments. Entry of serially passaged cells into "growth crisis," characterized by a failure of logarithmic expansion and the emergence of transformed aneuploid variants, is also delayed in dwarf-derived cells (Maynard and Miller 2006), suggesting that they are resistant to the oxygen-dependent accumulation of mutations and chromosomal rearrangements characteristic of mouse (but not human) cells grown under 20% O_2 concentrations (Busuttil et al. 2003; Parrinello et al. 2003). In this system, cell death induced by UV irradiation is not impaired by the inclusion of antioxidants (N-acetylcysteine or ascorbic acid) in the culture medium, suggesting that the cells are resistant to both oxidant and nonoxidant injury (Salmon et al. 2005). The resistance to death induced by MMS (Salmon et al. 2005) and evidence that Snell dwarf cells are superior to control cells at repair of UV-induced DNA damage (A.B. Salmon et al., unpubl.) are consistent with the idea that fibroblasts from these mice, like many long-lived forms of mutant nematode, exhibit multiplex stress resistance.

Mice that are heterozygous for a null allele of the IGF-I receptor (IGF-IR) (Holzenberger et al. 2003) are also relatively long-lived, although the life span effect is significant only in females. Control mice in this experiment are rather short-lived (19 months), and survival in the mutants (25 months) is lower than mean survival in most other laboratory mouse stocks. In addition, the colony does not appear to be specific-pathogen-free, and the very low tumor incidence (5%) suggests that the life table may reflect, in part, differences in the vigor of response to infectious agents. Fibroblasts derived from these mice are, like those from Snell and Ames dwarf mice, relatively resistant to death induced by H_2O_2. In contrast to the Snell dwarf data, however, cells from the IGF-IR mutant mice are peroxide-resistant even when derived from embryonic tissues. It seems likely that the stress resistance in the IGF-IR model reflects alterations of IGF-I-dependent signals in the tested fibroblasts themselves, rather than adaptive changes to maturation in a low-hormone environment thought to be involved in differentiation to stress resistance in the Snell dwarf cell lines. Interestingly, female mice heterozygous for the IGF-IR mutation are slightly but significantly more resistant than controls to the lethal effects of injected paraquat, an oxidative toxin.

Similarly, mice of a related stock, a compound heterozygote for a null allele and a hypomorphic allele of IGF-IR, show unusual resistance to the pulmonary damage caused by 72-hour exposure to hyperbaric (90%) oxygen (Ahamed et al. 2005), consistent with the notion that lower levels of IGF-IR help protect cells from oxidant damage. Evaluation of the resistance of cells and tissues from these IGF-I-hyporesponsive mice to other sources of cellular stress would be of great interest.

Data about stress resistance of other cell types in long-lived endocrine mutant mice are, alas, sparse. Contrary to hypothesis, Snell dwarf mice are more sensitive than controls to liver injury induced by the oxidative toxin acetaminophen (Harper et al. 2006), possibly because of augmented conversion of this drug to its more toxic metabolite. Male mice of the GHR-KO stock have been tested for sensitivity to paraquat, and again contrary to hypothesis, they were found to be more sensitive than controls (Hauck et al. 2002) (there was no genotype effect among females). Similarly, GHR-KO mice are more sensitive than controls in the acetaminophen toxicity test (Harper et al. 2006). In contrast, calorically restricted mice, and mice whose life span has been extended by a diet low in methionine, are much more resistant than normal mice to this form of acetaminophen toxicity (Miller et al. 2005; Harper et al. 2006). A systematic survey of multiple cell types and tissues for resistance to a wider range of cellular insults will be needed to develop a comprehensive test of whether the stress resistance seen in skin-derived fibroblasts from Ames, Snell, and GHR-KO mice is an oddity of a single cell type in vitro or a reflection of in vivo stress resistance of multiple cell types in intact mice.

Studies of stress resistance signaling pathways in long-lived mutant mice are also in their infancy. One group (Madsen et al. 2004) has evaluated the response of Snell dwarf mice to 3-nitropropionic acid (3-NPA), which induces liver cell death by free radical toxicity. These authors report that the activation of stress-sensitive kinases ERK and MEK is induced more strongly by 3-NPA in normal mice than in dwarf mice, consistent with the idea that initial cellular injury, to which MEK and ERK respond, may for unknown reasons be lower in hepatocytes from dwarf mice. The finding that Snell dwarf fibroblasts are slow to induce HSP-70 protein when subjected to the stress of serum deprivation (Maynard and Miller 2006) is also consistent with this idea. The limited data (Madsen et al. 2004) suggest that prior to 3-NPA exposure, dwarf hepatocytes may have higher baseline levels of phosphorylated (activated) MEK and ERK, lower levels of the c-Jun protein that is a target of JNK, and much lower levels of Elk-1, a substrate of ERK and JNK. Thus, the molecular basis for diminished activation of ERK and MEK in liver of Snell dwarf mice is at this point unclear, but likely to repay attention.

COMPARISONS ACROSS SPECIES

Throughout the mammalian evolutionary tree—among primates, rodents, bats, whales—evolutionary pressures have been able, given suitable ecological niches, to develop long-lived species from progenitors of shorter life span. Very little is known so far about the cellular and physiological adjustments that can lead to the delay of aging across species. It is an open question as to whether each such evolutionary transition represents a unique solution to postpone age-dependent change, or whether the same set of physiological alterations is called upon repeatedly when a species evolves to fit a niche that supports postponement of aging at the expense of immediate reproduction. If, as we postulate, eukaryotes early evolved a mechanism for coordinated augmentation of cellular resistance to multiple forms of injury, then deployment of this multiple stress resistance system might serve as a critical element in the evolution of exceptional longevity.

Assessment of this kind of evidence faces four challenges, three of them straightforward, and one more controversial. The first problem is that we have very little data about resistance to any kind of stress in most cell types of most mammals; more data of this kind are urgently needed. The second difficulty is that we still know very little about the molecular circuitry that coordinates multiplex stress resistance in mammalian cells, although studies of these pathways, catalyzed by analysis of invertebrate genetics and gene expression patterns, are accumulating rapidly. A third complication is the likelihood that each long-lived species may evolve an idiosyncratic set of adaptations, over and above the hypothesized up-regulation of shared stress resistance strategies. For example, long-lived species, such as most birds, that require high glucose levels and high metabolic rates to support the demands of flight may make use of different anti-aging adaptations than species, like porcupines and naked mole rats, for whom rapid acceleration and sustained bouts of aerial locomotion would be atypical. This complexity of niche-specific adaptation will certainly complicate, but perhaps not prevent, a search for common threads.

The fourth difficulty deserves more detailed discussion: the problem of body size as a confounder for cross-species analyses. Among mammals larger than approximately 1 kg (Austad and Fischer 1991), there is a trend for larger species to have longer maximal life span, a plausible surrogate for slower aging rates. For this reason, a regression analysis that tests the association of a measured trait against species-specific life span is likely to show a strong positive relationship for any trait that varies with body size. Basal metabolic rate (per gram mass or per gram metabolic tissue), for example, is one such trait, because the relatively high ratio of skin area to body volume in smaller animals requires greater metabolic activity to maintain a steady body temperature. Many aspects of cellular physiology, including rates of oxidative phosphorylation and free radical production, vary with metabolic rates and would thus show a similar association with species-specific longevity. Because each of these traits—and many others, such as ankle thickness—vary with body size, each will show a strong association with life span when compared across species. In some cases, the temptation to leap to a causal inference is easily avoided: The idea that thick ankles cause slow aging does not seem biologically plausible. In other cases, however, the association has a surface plausibility: Perhaps the evolution of long-lived creatures does involve modulation of rate of oxygen consumption, or cellular levels of lactate dehydrogenase, or number of heart beats, or basal metabolic rate?

The fallacy of inferring that a trait modulates aging rate from the observation that it varies with longevity across mammals has long been appreciated, and in particular, various strategies have been proposed for adjusting regression analyses to remove the portion of the association that reflects body weight. For example, each species can be assigned a "longevity quotient," calculated as the ratio of its maximal life span to the average life span of mammals of the same size, and subsequent analyses can relate physiological characteristics of interest (metabolic rate, levels of a given enzyme, etc.) to longevity quotient rather than to species life span per se (Austad and Fischer 1991). A related approach (Speakman 2005) calculates, for each species and each trait of interest, the residual for that species derived from a regression of the trait against body weight across species and looks for associations with the residual of longevity against body weight for the same species. The goal in each case is to find traits that find species, or groups of species, that are relatively longer lived than species of similar body size. Such discoveries can be of great interest: These approaches can help focus attention, for example, on groups of animals (like bats and birds) that are likely to be longer lived than nonflying mammals of the same size and prompt investigation of hypotheses to explain the disparity (Austad and Fischer 1991).

Requiring that all cross-species comparisons begin by factoring out differences in body size, although recommended by some experts (Speakman 2005), may have the unfortunate consequence of distracting attention from (hypothetical) traits that in fact do modulate aging rate and do vary with body size. If, for example, a specific enzyme or chaperone protein or membrane lipid was a critical component of an anti-aging pathway and was invariably higher in long-lived species than in short-lived species, it would show a strong positive association with body size in mammals, because the long-lived mammals are in general of larger size. Statistical adjustments that remove the portion of association attributable to body size would greatly diminish the strength of the association between the trait of interest and longevity and thereby obscure an association that does (in this hypothetical example) help to explain the basis for slow aging among the species studied. Statistical adjustments for body size therefore have the great benefit of reducing false-positive inferences, which are expected to be common, but the great weakness of obscuring true cause-and-effect relationships, rare though they are. The relationship between body size and life span across species now seems almost certain to reflect evolutionary, rather than metabolic, effects: Larger size animals must postpone age-related diseases long enough to grow to full size for reproductive maturity and, in addition, can afford to postpone reproduction because of the relative immunity to predation conferred by body size. The association of small body size with longer life span within dogs (Li et al. 1996; Patronek et al. 1997; Michell 1999), mice (Miller et al. 2000, 2002), horses (Brosnahan and Paradis 2003), and, for neoplastic diseases, humans (Miller and Austad 2006) argues cogently against the idea that large body size or traits such as metabolic rate that are a consequence of body size differences (Speakman 2005) cause slow aging. Thus, from this perspective, evidence that a specific stress resistance pathway varies among species might in principle reflect an artifact of association with body size or metabolic rate or might reflect an important element in the evolution of slower aging. Making this discrimination will require experimental designs that increase life span by augmenting stress resistance, rather than by statistical adjustments to observational data sets.

There has been substantial interest in the idea that longer-lived species have higher rates of repair of DNA damage. Much of the work published prior to 1998 has been reviewed by Perez-Campo et al. (1998). A correlation between rates of repair of UV-induced DNA damage was first studied systematically by Hart and Setlow (1974) and confirmed for UV-induced excision repair in a separate study of fibroblasts derived from 18 mammalian species (Francis et al. 1981). A reanalysis of data from multiple groups using rat cells as a shared standard also found a consistent relationship between species life span and DNA-repair rate (Cortopassi and Wang 1996). In a thoughtful review, Hanawalt (2001) has noted two important complications: (1) changes in repair of UV-damaged DNA that occur when cells adapt to short-term in vitro cultures, which make it hazardous to use the in vitro data as an index of the UV sensitivity of cells in the donor animal, and (2) the lack of any correspondence between species longevity and the ability to resist UV-induced cell death in tissue culture. This author points out that the mechanisms responsible for repair of UV lesions in actively transcribed areas of the genome ("transcription-coupled repair" or TCR) are distinct from those responsible for removal of the bulk of UV-induced lesions ("global genome repair" or GGR), in part due to the requirement for p53 function for the GGR pathway. He notes that cell survival after UV irradiation depends more on TCR and speculates that this functionally important aspect of repair may remain at similar levels in long- and short-lived species. Although as Hanawalt points out, GGR activity is typically low for rodent cell lines compared to human cells, a comparison among eight species of primates found a positive correlation between life span and repair of UV-induced damage in both fibroblasts and lymphocytes (Hall et al. 1984), suggesting that this association may be detectable within as well as across orders of mammals.

Other contributors to genomic stability also vary consistently across mammalian species. DNA strand breaks lead to addition of poly-ADP-ribosyl moieties to proteins catalyzed by the enzyme PARP-1 (poly[ADP-ribose] polymerase-1). A study of blood mononuclear cells from 132 individuals of 13 species showed a strong positive correlation between PARP function and species maximal longevity (Grube and Bürkle 1992). Differences among species did not appear to reflect variation in the levels of PARP-1 itself, but they may reflect changes in the kinetic parameters of enzyme action (Beneke et al. 2000) or, potentially, differences in factors that activate or recruit PARP-1 at the site of DNA strand breaks. Differences in the ability to convert procarcinogens to mutagenic derivatives may also help to protect long-lived species against environmental sources of DNA damage (Moore and Schwartz 1978).

We have noted above the strong correlation, across mammalian species, of low methionine abundance to species life span (Ruiz et al. 2005). These authors also documented strong correlations ($R^2 > 0.96$, $p < 0.001$), across these eight species, in accumulation of glutamic semialdehyde and aminoadipic semialdehyde, both indices of protein carbonylation, with higher levels of these products in the longer-lived species. In contrast, a marker of lipoxidation-dependent protein modification, malondialdehydelysine, showed a strong negative correlation, as did a calculated index of fatty acid saturation. These authors argue that resistance to both methionine oxidation and protein lipoxidation are important to the evolution of longer-lived species of mammals. It would be of interest to evaluate these indices of cellular damage in rodents subjected to low-calorie or low-methionine diets and in a variety of cell types from mice carrying antiaging mutants.

Several groups have undertaken surveys of the resistance of cultured cells to various forms of lethal stress in tissue culture. One report compared responses of fibroblasts from eight species of mammals to hydrogen peroxide, paraquat (which increases intracellular superoxide generation), sodium arsenite (also a catalyst for free radical generation), tert-butyl hydroperoxide (which induces lipid peroxidation), and sodium hydroxide. For each agent, there was a significant positive correlation between species maximum life span and the LD_{90}, i.e., dose of stress that resulted in 90% reduction in cell viability (Kapahi et al. 1999). A second group (Harper et al. 2007) evaluated skin-derived fibroblasts from nine species (eight rodents and one species of bat) and found that cells from longer-lived species were significantly more resistant to cell death induced by cadmium or by hydrogen peroxide, with a similar trend for death induced by heat or by the DNA alkylating agent MMS. Adjustment of the regression for differences in body weight did not change the significance of the association for cadmium or H_2O_2 and increased to statistical significance the association with resistance to death induced by heat stress. Interestingly, there was within this group of species no evidence for an association between maximal life span and resistance to the lethal effects of UV irradiation or paraquat (see Fig. 3). The differences in the results for paraquat and hydrogen peroxide are striking and may reflect differences in the mode of cell death; paraquat, for which no species association was seen, is thought to increase intracellular levels of reactive oxygen species, whereas H_2O_2 may kill cultured cells by oxidation of plasma membrane components.

There have been several attempts to explain the basis for the greater longevity of bird species compared to mammals of similar size (Holmes et al. 2001). The relatively high levels of glucose and of O_2 usage in birds have long stood as challenges to models in which the pace of aging is attributed simply to differences in oxidative damage. There is some evidence to suggest (Herrero and Barja 1998) that bird mitochondria may have adaptations that limit production of H_2O_2, and more generally to suggest that the rate of free radical production by mitochondria correlates better than metabolic rate with life span among species. Comparisons of birds to mammals have also suggested (Pamplona et al. 1999a,b) that a relatively low proportion of double bonds in plasma membrane lipids of bird cells may help to protect them from lipid per-

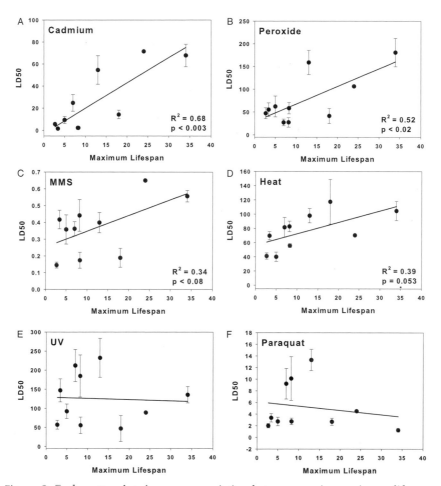

Figure 3. Each scatterplot shows an association between species maximum life span and mean LD_{50} value for each of ten species (treating laboratory mice and wild-trapped mice as separate species). From left to right, points represent laboratory mouse, wild-trapped mouse, rat, red squirrel, white-footed mouse, deer mouse, fox squirrel, porcupine, beaver, and little brown bat. Error bars show standard errors of the mean. The line shows the outcome of a least squares regression. Pearson R2 and *P* values (quoted only where *P* < 0.1) reflect standard linear regression of maximum life span against mean LD_{50} values. Units for LD_{50} are in micromoles (cadmium and H_2O_2), millimoles (MMS and paraquat), J m^{-2} (UV light), or min at 42°C (heat). (Reprinted, with permission, from Harper et al. 2007 [© Blackwell Publishing].)

oxidation. Among mammals, there is also a strong correlation of maximum life span and mitochondrial lipid composition, with longer-lived mammals typically resembling birds in their low levels of unsaturation and thus higher resistance to peroxidation (Pamplona et al. 1998).

Consistent with these studies, an independent analysis of multiple species of both birds and mammals (Hulbert 2005) has shown that in both classes, longer-lived species have plasma membrane lipids that are relatively resistant to peroxidation because of a lower ratio of $n - 3$ to $n - 6$ polyunsaturated fatty acids. This relationship was notable both in liver and in skeletal muscle. Interestingly, although regression lines for plots of metabolic rate (or size) against life span are dramatically different between birds and mammals, with birds showing longer life span than mammals of the same size or metabolic rate, a plot of life span against an index of lipid composition is very similar for both birds and mammalian species (Hulbert 2005). Data comparing breeds or stocks that differ in life span within a species are limited, but it is noteworthy that caloric restriction also changes lipid composition of plasma membranes toward decreased susceptibility to peroxidation (for references, see Hulbert 2005) and that mouse stocks recently derived from wild-trapped animals are longer lived than laboratory-adapted stocks and also have membranes relatively resistant to peroxidation (Hulbert et al. 2006b). Similarly, a comparison of mitochondrial lipid composition in the muscle and liver of mice to those from the naked mole rat, a rodent of similar size but much greater longevity, showed the mole rat membranes to have far lower levels of docosahexaenoic acid, presumably rendering them comparatively resistant to peroxidation damage (Hulbert et al. 2006a). Studies that explicitly compare resistance to agents that induce damage by lipid peroxidation, either in vitro or in multiple tissues of long- or short-lived animals, in long-lived mutants, or in animals on anti-aging diets, would help to show whether such changes in plasma membrane and/or mitochondrial lipid composition contribute to resistance to oxidative stress or instead to multiple forms of cellular injury.

It is possible, as noted by Speakman (2005), that such relationships among species might reflect connections between lipid properties and the metabolic demands imposed by differences in body size, itself strongly correlated with life span, rather than a causal chain in which altered lipids lead to longer life span by inhibition of peroxidation. The hypothesis that the association reflects an underlying causal connection is strengthened by demonstrations within a species that small body size is associated with both high metabolic rate and longer life span (Speakman et al. 2003; Miller and Austad 2006) and that interclass differences (e.g., birds vs. mammals) in the relationship between metabolic rate and longevity can be reconciled by evaluation of mitochondrial free radical leak and membrane lipid composition.

AN ALTERNATE THEORY: FREE RADICAL DAMAGE AS THE CRITICAL CONTROLLER OF AGING RATE

The idea that damage due to free radicals is the principal cause of aging and that variations in oxidative damage are responsible for differences in the rate of aging within and among species has earned wide contemporary support and is now easily the most popular molecular framework for discussion of cellular biogerontology. We have presented a number of results from invertebrate studies that generally support the theory, but most studies lack biochemical or bioenergetic analysis. A recent review (Bokov et al. 2004) has presented a sophisticated and well-documented summary of the evidence for and against the free radical theory. This attractive idea has in recent years begun to show signs of wear and tear, as accumulating evidence has failed to confirm some of its key predictions. One concern, noted as early as 25 years ago (Comfort 1979), and still true today, is the absence of any compelling evidence that adding antioxidants to the diet of mammals can slow aging or increase maximal life span. A recent critical review (Howes 2006) has tabulated 15 such studies in humans, involving long-term treatment of more than 550,000 volunteers with a variety of antioxidants, and concluded that in none of the studies was there evidence for a positive effect of antioxidant agents on human health. Genetically modified mice constructed so as to have diminished levels of the mitochondrial manganese superoxide dismutase (Mn-SOD) have been shown, as anticipated, to have ample evidence of increased oxidative damage in multiple tissues, but have a normal life span, and also show no acceleration of age-related changes in cataract formation and immune responsiveness. These results thus suggest that accumulation of oxidant damage is not the limiting factor on longevity (or the other age-sensitive traits evaluated) at least in this strain (Van Remmen et al. 2003). Conversely, there is little or no evidence that genetic manipulations that improve defenses against oxidant damage can extend maximum life span in mammals. A paper reporting that catalase targeted to mitochondria could extend life span in transgenic mice (Schriner et al. 2005) may represent the first step in such a direction. However, this requires replication, in part because the effect is noted to depend on interaction with background genes, diminishing by an unstated amount when the transgene is evaluated on a C57BL/6 background.

New results from studies of the naked mole rat also call into some question the primacy of oxidative defenses as the key modulator of aging rates in mammals. Naked mole rats are perhaps the longest-lived rodents, with a mean body weight similar to that of mice but with a maximal life span approximately six times longer than that of mice (Buffenstein 2005); at least one individual has exceeded the age of 28 years. Extensive life table and necropsy data have shown no evidence for spontaneous malignancy in this species, nor any evidence for the age-related increase in mortality risk characteristic of nearly every other vertebrate species that has been carefully evaluated. Despite this evidence of an exceptionally slow rate of aging, young adult naked mole rats have higher levels of oxidation damage to proteins, lipids, and DNA compared to young adult mice (Andziak et al. 2006), tenfold higher levels of lipid peroxidation, and a significantly lower ratio of reduced to oxidized glutathione, consistent with the idea of a relatively oxidizing intracellular milieu. This species has evolved to fit an ecological niche that is unusual, among mammals, in many respects, including constant temperature, low exposure to sunlight, and unusually how oxygen tension; these factors, together with its eusocial breeding scheme and poikilothermy, may well have led to highly idiosyncratic changes in cell biology and physiology. Nonetheless, it is difficult to reconcile the high levels of oxidative damage in this species with models in which exceptional longevity requires exceptionally strong defenses against oxidative damage.

A model based on the idea of a pathway for coordinated up-regulation of resistance to many different kinds of cellular injuries may help to reconcile the evidence for oxidative injury as an important contributor to age-related pathology with the remarkable lack of evidence that altering resistance to oxidation damage can retard aging or extend maximal longevity. In this view, the synchronous decline in tissue and cellular and organismic integrity seen in 2-year-old mice, 20-year-old horses, and 80-year-old people may reflect the compounded and reinforcing effects of many forms of damage to many kinds of cells, some perhaps particularly vulnerable to protein misfolding, others to somatic mutations, still others to oxidative alterations of lipid or proteins. Just as substituting a single young midfielder into an over-50 football squad would have little effect on its chances for success in the World Cup, interventions that merely improve the cellular defenses against oxidation damage might be expected, from our perspective, to have little or no leverage in retarding the overall process of aging. And yet it is clear that the overall rate of aging can indeed be diminished, by relatively simple changes in diet, by a few hundreds or perhaps a few dozens of generations of selective breeding, by mutations at each of at least a dozen genes in mice and at least 100 mutations in nematodes, and, repeatedly, by evolutionary adaptation to specific niches in many distantly related clades.

SUMMARY AND PROSPECTUS

We suspect that a search for the "cause" of aging, when it takes the form of arguments about the relative importance of somatic mutations, oxidative stress, telomere-triggered clonal exhaustion, misfolded proteins, uncontrolled inflammatory responses, or similar threats, may be off-target in a fundamental way. We see the key implication of recent breakthroughs in the understanding of anti-aging mutations in nematodes as the idea that ancient conserved cellular pathways may have developed to augment multiple forms of cellular defenses in parallel, to deal with frequent but unpredictable intervals of resource shortage, and then been coopted to postpone aging in larger animals to fit ecological niches in which more leisurely maturation and extended period of reproduction were advantageous. Further progress in working out ... fundamental control of aging rate—its synchrony among cells and tissues in any single species, its dramatic variation among species—is likely to emerge not so much from investigation of any one form of cellular injury, but from elucidation of the underlying process that coordinates multiple forms of cellular defenses in response to signals from the ancient resource-sensing system. Worm and fly biologists are making rapid progress using genetic tricks to work out the relevant cellular biochemistry, and researchers who are beginning to translate these findings into the molecular language used by mammalian cells are also making important discoveries. The next stage in the pathway, the trek from cell biology to rodent (and eventually human) pathophysiology, is still an uphill climb and one likely to reveal unexpected twists en route.

ACKNOWLEDGMENTS

R.A.M. acknowledges support from National Institutes of Health (NIH) grants AG023122 and AG024824. G.J.L. is supported by NIH grants AG21069, AG22868, NS050789-01, the Ellison Medical Foundation, the Glenn Foundation for Medical Research, the Herbert Simon Family Medical Foundation, and the Larry L. Hillblom Foundation.

REFERENCES

Aballay A., Drenkard E., Hilbun L.R., and Ausubel F.M. 2003. *Caenorhabditis elegans* innate immune response triggered by *Salmonella enterica* requires intact LPS and is mediated by a MAPK signaling pathway. *Curr. Biol.* **13:** 47–52.

Adachi H., Fujiwara Y., and Ishii N. 1998. Effects of oxygen on protein carbonyl and aging in *Caenorhabditis elegans* mutants with long (age-1) and short (mev-1) life spans. *J. Gerontol. A Biol. Sci. Med. Sci.* **53A:** B240–B244.

Ahamed K., Epaud R., Holzenberger M., Bonora M., Flejou J.F., Puard J., Clement A., and Henrion-Caude A. 2005. Deficiency in type 1 insulin-like growth factor receptor in mice protects against oxygen-induced lung injury. *Respir. Res.* **6:** 31.

Aidoo A., Desai V.G., Lyn-Cook L.E., Chen J.J., Feuers R.J., and Casciano D.A. 1999. Attenuation of bleomycin-induced Hprt mutant frequency in female and male rats by calorie restriction. *Mutat. Res.* **430:** 155–163.

Aidoo A., Mittelstaedt R.A., Bishop M.E., Lyn-Cook L.E., Chen Y.J., Duffy P., and Heflich R.H. 2003. Effect of caloric restriction on Hprt lymphocyte mutation in aging rats. *Mutat. Res.* **527:** 57–66.

Albert P.S. and Riddle D.L. 1988. Mutants of *Caenorhabditis elegans* that form dauer-like larvae. *Dev. Biol.* **126:** 270–293.

Altomare K., Greco V., Bellizzi D., Berardelli M., Dato S., DeRango F., Garasto S., Rose G., Feraco E., Mari V., Passarino G., Franceschi C., and De Benedictis G. 2003. The allele (A)(-110) in the promoter region of the HSP70-1 gene is unfavorable to longevity in women. *Biogerontology* **4:** 215–220.

Anderson G.L. 1978. Responses of dauer larvae of *Caenorhabditis elegans* (Nematoda: Rhabditidae) to thermal stress and oxygen deprivation. *Can. J. Zool.* **56:** 1786–1791.

———. 1982. Superoxide dismutase activity in dauer larvae of *Caenorhabditis elegans* (Nematoda: Rhabditidae). *Can. J. Zool.* **60:** 288–291.

Andziak B., O'Connor T.P., Qi W., DeWaal E.M., Pierce A., Chaudhuri A.R., Van Remmen H., and Buffenstein R. 2006. High oxidative damage levels in the longest-living rodent, the naked mole-rat. *Aging Cell* **5:** 463–471.

Antebi A., Culotti J.G., and Hedgecock E.M. 1998. *daf-12* regulates developmental age and the dauer alternative in *Caenorhabditis elegans*. *Development* **125:** 1191–1205.

Antebi A., Yeh W.H., Tait D., Hedgecock E.M., and Riddle D.L. 2000. *daf-12* encodes a nuclear receptor that regulates the dauer diapause and developmental age in *C. elegans*. *Genes Dev.* **14:** 1512–1527.

Austad S.N. and Fischer K.E. 1991. Mammalian aging, metabolism, and ecology: Evidence from the bats and marsupials. *J. Gerontol.* **46:** B47–B53.

Baker K., Marcus C.B., Huffman K., Kruk H., Malfroy B., and Doctrow S.R. 1998. Synthetic combined superoxide dismutase/catalase mimetics are protective as a delayed treatment in a rat stroke model: A key role for reactive oxygen species in ischemic brain injury. *J. Pharmacol. Exp. Ther.* **284:** 215–221.

Barsyte D., Lovejoy D.A., and Lithgow G.J. 2001. Longevity and heavy metal resistance in *daf-2* and *age-1* long-lived mutants of *Caenorhabditis elegans*. *FASEB J.,* **15:** 627–634.

Bauer J.H., Poon P.C., Glatt-Deeley H., Abrams J.M., and Helfand S.L. 2005. Neuronal expression of p53 dominant-negative proteins in adult *Drosophila melanogaster* extends life span. *Curr. Biol.* **15:** 2063–2068.

Bayne A.C. and Sohal R.S. 2002. Effects of superoxide dismutase/catalase mimetics on life span and oxidative stress resistance in the housefly, *Musca domestica*. *Free Radic. Biol. Med.* **32:** 1229–1234.

Beneke S., Alvarez-Gonzalez R. and Bürkle A. 2000. Comparative characterisation of poly(ADP-ribose) polymerase-1 from two mammalian species with different life span. *Exp. Gerontol.* **35:** 989–1002.

Berman K., McKay J., Avery L., and Cobb M. 2001. Isolation and characterization of pmk-(1–3): Three p38 homologs in *Caenorhabditis elegans*. *Mol. Cell Biol. Res. Commun.* **4:** 337–344.

Bohni R., Riesgo-Escovar J., Oldham S., Brogiolo W., Stocker H., Andruss B.F., Beckingham K., and Hafen E. 1999. Autonomous control of cell and organ size by CHICO, a *Drosophila* homolog of vertebrate IRS1-4. *Cell* **97:** 865–875.

Bokov A., Chaudhuri A., and Richardson A. 2004. The role of oxidative damage and stress in aging. *Mech. Ageing Dev.* **125:** 811–826.

Brack C., Bechter-Thuring E., and Labuhn M. 1997. N-acetylcysteine slows down ageing and increases the life span of *Drosophila melanogaster. Cell. Mol. Life Sci.* **53:** 960–966.

Brosnahan M.M. and Paradis M.R. 2003. Demographic and clinical characteristics of geriatric horses: 467 cases (1989–1999). *J. Am. Vet. Med. Assoc.* **223:** 93–98.

Broughton S.J., Piper M.D., Ikeya T., Bass T.M., Jacobson J., Driege Y., Martinez P., Hafen E., Withers D.J., Leevers S.J., and Partridge L. 2005. Longer lifespan, altered metabolism, and stress resistance in *Drosophila* from ablation of cells making insulin-like ligands. *Proc. Natl. Acad. Sci.* **102:** 3105–3110.

Brown-Borg H.M., Borg K.E., Meliska C.J., and Bartke A. 1996. Dwarf mice and the ageing process. *Nature* **384:** 33.

Buffenstein R. 2005. The naked mole-rat: A new long-living model for human aging research. *J. Gerontol. A Biol. Sci. Med. Sci.* **60:** 1369–1377.

Busuttil R.A., Rubio M., Dolle M.E., Campisi J., and Vijg J. 2003. Oxygen accelerates the accumulation of mutations during the senescence and immortalization of murine cells in culture. *Aging Cell* **2:** 287–294.

Cabelof D.C., Yanamadala S., Raffoul J.J., Guo Z., Soofi A., and Heydari A.R. 2003. Caloric restriction promotes genomic stability by induction of base excision repair and reversal of its age-related decline. *DNA Repair* **2:** 295–307.

Casciano D.A., Chou M., Lyn-Cook L.E., and Aidoo A. 1996. Calorie restriction modulates chemically induced in vivo somatic mutation frequency. *Environ. Mol. Mutagen.* **27:** 162–164.

Clancy D.J., Gems D., Harshman L.G., Oldham S., Stocker H., Hafen E., Leevers S.J., and Partridge L. 2001. Extension of life-span by loss of CHICO, a *Drosophila* insulin receptor substrate protein. *Science* **292:** 104–106.

Cohen E., Bieschke J., Perciavalle R.M., Kelly J.W., and Dillin A. 2006. Opposing activities protect against age-onset proteotoxicity. *Science* **313:** 1604–1610.

Collins J.J., Evason K., and Kornfeld K. 2006. Pharmacology of delayed aging and extended lifespan of *Caenorhabditis elegans. Exp. Gerontol.* **41:** 1032–1039.

Comfort A. 1979. *The biology of senescence.* Elsevier, New York.

Cortopassi G.A. and Wang E. 1996. There is substantial agreement among interspecies estimates of DNA repair activity. *Mech. Ageing Dev.* **91:** 211–218.

Coschigano K.T., Holland A.N., Riders M.E., List E.O., Flyvbjerg A., and Kopchick J.J. 2003. Deletion, but not antagonism, of the mouse growth hormone receptor results in severely decreased body weights, insulin, and insulin-like growth factor I levels and increased life span. *Endocrinology* **144:** 3799–3810.

Dempsey J.L., Pfeiffer M., and Morley A.A. 1993. Effect of dietary restriction on in vivo somatic mutation in mice. *Mutat. Res.* **291:** 141–145.

Dillin A., Hsu A.L., Arantes-Oliveira N., Lehrer-Graiwer J., Hsin H., Fraser A.G., Kamath R.S., Ahringer J., and Kenyon C. 2002. Rates of behavior and aging specified by mitochondrial function during development. *Science* **298:** 2398–2401.

Dorman J.B., Albinder B., Shroyer T., and Kenyon C. 1995. The *age-1* and *daf-2* genes function in a common pathway to control the lifespan of *Caenorhabditis elegans. Genetics* **141:** 1399–1406.

Essers M.A., De Vries-Smits L.M., Barker N., Polderman P.E., Burgering B.M., and Korswagen H.C. 2005. Functional interaction between beta-catenin and FOXO in oxidative stress signaling. *Science* **308:** 1181–1184.

Estevez M., Attisano L., Wrana J.L., Albert P.S., Massague J., and Riddle D.L. 1993. The *daf-4* gene encodes a bone morphogenetic protein receptor controlling *C. elegans* dauer larva development. *Nature* **365:** 644–649.

Evason K., Huang C., Yamben I., Covey D.F., and Kornfeld K. 2005. Anticonvulsant medications extend worm life-span. *Science* **307:** 258–262.

Ewbank J.J., Barnes T.M., Lakowski B., Lussier M., Bussey H., and Hekimi S. 1997. Structural and functional conservation of the *Caenorhabditis elegans* timing gene *clk-1*. *Science* **275:** 980–983.

Felkai S., Ewbank J.J., Lemieux J., Labbe J.C., Brown G.G., and Hekimi S. 1999. CLK-1 controls respiration, behavior and aging in the nematode *Caenorhabditis elegans*. *EMBO J.* **18:** 1783–1792.

Feng J., Bussiere F., and Hekimi S. 2001. Mitochondrial electron transport is a key determinant of life span in *Caenorhabditis elegans*. *Dev. Cell* **1:** 633–644.

Fisher A.L. and Lithgow G.J. 2006. The nuclear hormone receptor DAF-12 has opposing effects on *Caenorhabditis elegans* lifespan and regulates genes repressed in multiple long-lived worms. *Aging Cell* **5:** 127–138.

Flatt T., Tu M.P., and Tatar M. 2005. Hormonal pleiotropy and the juvenile hormone regulation of *Drosophila* development and life history. *Bioessays* **27:** 999–1010.

Flurkey K., Papaconstantinou J., Miller R.A., and Harrison D.E. 2001. Lifespan extension and delayed immune and collagen aging in mutant mice with defects in growth hormone production. *Proc. Natl. Acad. Sci.* **98:** 6736–6741.

Fonager J., Beedholm R., Clark B.F., and Rattan S.I. 2002. Mild stress-induced stimulation of heat-shock protein synthesis and improved functional ability of human fibroblasts undergoing aging in vitro. *Exp. Gerontol.* **37:** 1223–1228.

Francia P., delli Gatti C., Bachschmid M., Martin-Padura I., Savoia C., Migliaccio E., Pelicci P.G., Schiavoni M., Luscher T.F., Volpe M., and Cosentino F. 2004. Deletion of p66shc gene protects against age-related endothelial dysfunction. *Circulation* **110:** 2889–2895.

Francis A.A., Lee W.H., and Regan J.D. 1981. The relationship of DNA excision repair of ultraviolet-induced lesions to the maximum life span of mammals. *Mech. Ageing Dev.* **16:** 181–189.

Friedman D.B. and Johnson T.E. 1988a. A mutation in the *age-1* gene in *Caenorhabditis elegans* lengthens life and reduces hermaphrodite fertility. *Genetics* **118:** 75–86.

———. 1988b. Three mutants that extend both mean and maximum life span of the nematode, *Caenorhabditis elegans*, define the *age-1* gene. *J. Gerontol.* **43:** B102–B109.

Garsin D.A., Villanueva J.M., Begun J., Kim D.H., Sifri C.D., Calderwood S.B., Ruvkun G., and Ausubel F.M. 2003. Long-lived *C. elegans daf-2* mutants are resistant to bacterial pathogens. *Science* **300:** 1921.

Gems D., Sutton A.J., Sundermeyer M.L., Albert P.S., King K.V., Edgley M.L., Larsen P.L., and Riddle D.L. 1998. Two pleiotropic classes of *daf-2* mutation affect larval arrest, adult behavior, reproduction and longevity in *Caenorhabditis elegans*. *Genetics* **150:** 129–155.

Gerisch B., Weitzel C., Kober-Eisermann C., Rottiers V., and Antebi A. 2001. A hormonal signaling pathway influencing *C. elegans* metabolism, reproductive development, and life span. *Dev. Cell* **1:** 841–851.

Giannakou M.E., Goss M., Junger M.A., Hafen E., Leevers S.J., and Partridge L. 2004. Long-lived *Drosophila* with overexpressed dFOXO in adult fat body. *Science* **305:** 361.

Gil E.B., Malone L.E., Liu L.X., Johnson C.D., and Lees J.A. 1999. Regulation of the insulin-like developmental pathway of *Caenorhabditis elegans* by a homolog of the PTEN tumor suppressor gene. *Proc. Natl. Acad. Sci.* **96:** 2925–2930.

Gredilla R., Sanz A., Lopez-Torres M., and Barja G. 2001. Caloric restriction decreases mitochondrial free radical generation at complex I and lowers oxidative damage to mitochondrial DNA in the rat heart. *FASEB J.* **15:** 1589–1591.

Grube K. and Bürkle A. 1992. Poly(ADP-ribose) polymerase activity in mononuclear leukocytes of 13 mammalian species correlates with species-specific life span. *Proc. Natl. Acad. Sci.* **89:** 11759–11763.

Gruntenko N.E., Khlebodarova T.M., Vasenkova I.A., Sukhanova M.J., Wilson T.G., and Rauschenbach I.Y. 2000. Stress-reactivity of a *Drosophila melanogaster* strain with impaired juvenile hormone action. *J. Insect Physiol.* **46:** 451–456.

Gruntenko N.E., Chentsova N.A., Andreenkova E.V., Bownes M., Segal D., Adonyeva N.V., and Rauschenbach I.Y. 2003. Stress response in a juvenile hormone-deficient *Drosophila melanogaster* mutant apterous56f. *Insect Mol. Biol.* **12:** 353–363.

Guo Z., Heydari A., and Richardson A. 1998. Nucleotide excision repair of actively transcribed versus nontranscribed DNA in rat hepatocytes: Effect of age and dietary restriction. *Exp. Cell Res.* **245:** 228–238.

Guyton K.Z., Liu Y., Gorospe M., Xu Q., and Holbrook N.J. 1996. Activation of mitogen-activated protein kinase by H_2O_2. Role in cell survival following oxidant injury. *J. Biol. Chem.* **271:** 4138–4142.

Halaschek-Wiener J., Khattra J.S., McKay S., Pouzyrev A., Stott J.M., Yang G.S., Holt R.A., Jones S.J., Marra M.A., Brooks-Wilson A.R., and Riddle D.L. 2005. Analysis of long-lived *C. elegans daf-2* mutants using serial analysis of gene expression. *Genome Res.* **15:** 603–615.

Hall K.Y., Hart R.W., Benirschke A.K., and Walford R.L. 1984. Correlation between ultraviolet-induced DNA repair in primate lymphocytes and fibroblasts and species maximum achievable life span. *Mech. Ageing Dev.* **24:** 163–173.

Hanawalt P.C. 2001. Revisiting the rodent repairadox. *Environ. Mol. Mutagen.* **38:** 89–96.

Harper J.M., Salmon A.B., Leiser S.F., Galecki A.T., and Miller R.A. 2007. Skin-derived fibroblasts from long-lived species are resistant to some, but not all, lethal stresses and to the mitochondrial inhibitor rotenone. *Aging Cell* **6:** 1–13.

Harper J.M., Salmon A.B., Chang Y., Bonkowski M., Bartke A., and Miller R.A. 2006. Stress resistance and aging: Influence of genes and nutrition. *Mech. Ageing Dev.* **127:** 687–694.

Hart R.W. and Setlow R.B. 1974. Correlation between deoxyribonucleic acid excision repair and lifespan in a number of mammalian species. *Proc. Natl. Acad. Sci.* **71:** 2169–2173.

Hartman P.S., Ishii N., Kayser E.B., Morgan P.G., and Sedensky M.M. 2001. Mitochondrial mutations differentially affect aging, mutability and anesthetic sensitivity in *Caenorhabditis elegans*. *Mech. Ageing Dev.* **122:** 1187–1201.

Hauck S.J., Aaron J.M., Wright C., Kopchick J.J., and Bartke A. 2002. Antioxidant enzymes, free-radical damage, and response to paraquat in liver and kidney of long-living growth hormone receptor/binding protein gene-disrupted mice. *Horm. Metab. Res.* **34:** 481–486.

Held J.M., White M.P., Fisher A.L., Gibson B.W., Lithgow G.J., and Gill M.S. 2006. DAF-12-dependent rescue of dauer formation in *Caenorhabditis elegans* by (25S)-cholestenoic acid. *Aging Cell* **5:** 283–291.

Henderson S.T. and Johnson T.E. 2001. *daf-16* integrates developmental and environmental inputs to mediate aging in the nematode *Caenorhabditis elegans*. *Curr. Biol.* **11:** 1975–1980.

Hercus M.J., Loeschcke V., and Rattan S.I. 2003. Lifespan extension of *Drosophila melanogaster* through hormesis by repeated mild heat stress. *Biogerontology* **4:** 149–156.

Herrero A. and Barja G. 1998. H_2O_2 production of heart mitochondria and aging rate are slower in canaries and parakeets than in mice: Sites of free radical generation and mechanisms involved. *Mech. Ageing Dev.* **103:** 133–146.

Hertweck M., Gobel C., and Baumeister R. 2004. *C. elegans* SGK-1 is the critical component in the Akt/PKB kinase complex to control stress response and life span. *Dev. Cell* **6:** 577–588.

Heydari A.R., Wu B., Takahashi R., Strong R., and Richardson A. 1993. Expression of heat shock protein 70 is altered by age and diet at the level of transcription. *Mol. Cell. Biol.* **13:** 2909–2918.

Hollingsworth M.J. 1969. Fluctuating temperatures and the length of life in *Drosophila*. *Nature* **221:** 857–858.

Holmes D.J., Fluckiger R., and Austad S.N. 2001. Comparative biology of aging in birds: An update. *Exp. Gerontol.* **36:** 869–883.

Holzenberger M., Dupont J., Ducos B., Leneuve P., Geloen A., Even P.C., Cervera P., and Le Bouc Y. 2003. IGF-1 receptor regulates lifespan and resistance to oxidative stress in mice. *Nature* **421:** 182–187.

Honda Y. and Honda S. 1999. The *daf-2* gene network for longevity regulates oxidative stress resistance and Mn-superoxide dismutase gene expression in *Caenorhabditis elegans*. *FASEB J.* **13:** 1385–1393.

Hosokawa H., Ishii N., Ishida H., Ichimori K., Nakazawa H., and Suzuki K. 1994. Rapid accumulation of fluorescent material with aging in an oxygen-sensitive mutant *mev-1* of *Caenorhabditis elegans*. *Mech. Ageing Dev.* **74:** 161–170.

Howes R.M. 2006. The free radical fantasy: A panoply of paradoxes. *Ann. N.Y. Acad. Sci.* **1067:** 22–26.

Howitz K.T., Bitterman K.J., Cohen H.Y., Lamming D.W., Lavu S., Wood J.G., Zipkin R.E., Chung P., Kisielewski A., Zhang L.L., Scherer B., and Sinclair D.A. 2003. Small molecule activators of sirtuins extend *Saccharomyces cerevisiae* lifespan. *Nature* **425:** 191–196.

Hsin H. and Kenyon C. 1999. Signals from the reproductive system regulate the lifespan of *C. elegans*. *Nature* **399:** 362–366.

Hsu A.L., Murphy C.T., and Kenyon C. 2003. Regulation of aging and age-related disease by DAF-16 and heat-shock factor. *Science* **300:** 1142–1145.

Hulbert A.J. 2005 On the importance of fatty acid composition of membranes for aging. *J. Theor. Biol.* **234:** 277–288.

Hulbert A.J., Faulks S.C., and Buffenstein R. 2006a. Oxidation-resistant membrane phospholipids can explain longevity differences among the longest-living rodents and similarly-sized mice. *J. Gerontol. A Biol. Sci. Med. Sci.* **61:** 1009–1018.

Hulbert A.J., Faulks S.C., Harper J.M., Miller R.A., and Buffenstein R. 2006b. Extended longevity of wild-derived mice is associated with peroxidation-resistant membranes. *Mech. Ageing Dev.* **127:** 653–657.

Hwangbo D.S., Gershman B., Tu M.P., Palmer M., and Tatar M. 2004. *Drosophila* dFOXO controls lifespan and regulates insulin signalling in brain and fat body. *Nature* **429:** 562–566.

Ishii N., Takahashi K., Tomita S., Keino T., Honda S., Yoshino K., and Suzuki K. 1990. A methyl viologen-sensitive mutant of the nematode *Caenorhabditis elegans*. *Mutat. Res.* **237:** 165–171.

Ishii N., Fujii M., Hartman P.S., Tsuda M., Yasuda K., Senoo-Matsuda N., Yanase S., Ayusawa D., and Suzuki K. 1998. A mutation in succinate dehydrogenase cytochrome b causes oxidative stress and ageing in nematodes. *Nature* **394:** 694–697.

Jia K., Albert P.S., and Riddle D.L. 2002. DAF-9, a cytochrome P450 regulating *C. elegans* larval development and adult longevity. *Development* **129:** 221–231.

Johnson T.E. 1990. Increased life-span of *age-1* mutants in *Caenorhabditis elegans* and lower Gompertz rate of aging. *Science* **249:** 908–912.

Johnson T.E. and Hartman P.S. 1988. Radiation effects on life span in *Caenorhabditis elegans*. *J. Gerontol.* **43:** B137–B141.

Jonassen T., Larsen P.L., and Clarke C.F. 2001. A dietary source of coenzyme Q is essential for growth of long-lived *Caenorhabditis elegans clk-1* mutants. *Proc. Natl. Acad. Sci.* **98:** 421–426.

Jonassen T., Proft M., Randez-Gil F., Schultz J.R., Marbois B.N., Entian K.D., and Clarke C.F. 1998. Yeast Clk-1 homologue (Coq7/Cat5) is a mitochondrial protein in coenzyme Q synthesis. *J. Biol. Chem.* **273:** 3351–3357.

Kang H.L., Benzer S., and Min K.T. 2002. Life extension in *Drosophila* by feeding a drug. *Proc. Natl. Acad. Sci.* **99:** 838–843.

Kapahi P., Boulton M.E., and Kirkwood T.B. 1999. Positive correlation between mammalian life span and cellular resistance to stress. *Free Radic. Biol. Med.* **26:** 495–500.

Kayser E.B., Morgan P.G., and Sedensky M.M. 1999. GAS-1: A mitochondrial protein controls sensitivity to volatile anesthetics in the nematode *Caenorhabditis elegans*. *Anesthesiology* **90:** 545–554.

Kayser E.B., Hoppel C.L., Morgan P.G., and Sedensky M.M. 2003. A mutation in mitochondrial complex I increases ethanol sensitivity in *Caenorhabditis elegans*. *Alcohol Clin. Exp. Res.* **27:** 584–592.

Kayser E.B., Morgan P.G., Hoppel C.L., and Sedensky M.M. 2001. Mitochondrial expression and function of GAS-1 in *Caenorhabditis elegans*. *J. Biol. Chem.* **276:** 20551–20558.

Keaney M. and Gems D. 2003. No increase in lifespan in *Caenorhabditis elegans* upon treatment with the superoxide dismutase mimetic EUK-8. *Free Radic. Biol. Med.* **34:** 277–282.

Keaney M., Matthijssens F., Sharpe M., Vanfleteren J., and Gems D. 2004. Superoxide dismutase mimetics elevate superoxide dismutase activity in vivo but do not retard aging in the nematode *Caenorhabditis elegans*. *Free Radic. Biol. Med.* **37:** 239–250.

Kenyon C. 2005. The plasticity of aging: Insights from long-lived mutants. *Cell* **120:** 449–460.

Kenyon C., Chang J., Gensch E., Rudner A., and Tabtiang R. 1993. A *C. elegans* mutant that lives twice as long as wild type. *Nature* **366:** 461–464.

Khazaeli A.A., Xiu L., and Curtsinger J.W. 1995. Stress experiments as a means of investigating age-specific mortality in *Drosophila melanogaster*. *Exp. Gerontol.* **30:** 177–184.

Khazaeli A.A, Tatar M., Pletcher S.D., and Curtsinger J.W. 1997. Heat-induced longevity extension in *Drosophila*. I. Heat treatment, mortality, and thermotolerance. *J. Gerontol. A Biol. Sci. Med. Sci.* **52:** B48–B52.

Kim D.H., Feinbaum R., Alloing G., Emerson F.E., Garsin D.A., Inoue H., Tanaka-Hino M., Hisamoto N., Matsumoto K., Tan M.W., and Ausubel F.M. 2002. A conserved p38 MAP kinase pathway in *Caenorhabditis elegans* innate immunity. *Science* **297:** 623–626.

Kimura K.D., Tissenbaum H.A., Liu Y., and Ruvkun G. 1997. daf-2, an insulin receptor-like gene that regulates longevity and diapause in *Caenorhabditis elegans*. *Science* **277:** 942–946.

Komninou D., Leutzinger Y., Reddy B.S., and Richie J.P., Jr. 2006. Methionine restriction inhibits colon carcinogenesis. *Nutr. Cancer* **54:** 202–208.

Kurapati R., Passananti H.B., Rose M.R., and Tower J. 2000. Increased hsp22 RNA levels in *Drosophila* lines genetically selected for increased longevity. *J. Gerontol. A Biol. Sci. Med. Sci.* **55:** B552–B559.

Lamb M.J. 1968. Temperature and lifespan in *Drosophila. Nature* **220:** 808–809.

Lamitina S.T. and Strange K. 2005. Transcriptional targets of DAF-16 insulin signaling pathway protect *C. elegans* from extreme hypertonic stress. *Am. J. Physiol. Cell Physiol.* **288:** C467–C474.

Larsen P.L. 1993. Aging and resistance to oxidative damage in *Caenorhabditis elegans. Proc. Natl. Acad. Sci.* **90:** 8905–8909.

Larsen P.L., Albert P.S., and Riddle D.L. 1995. Genes that regulate both development and longevity in *Caenorhabditis elegans. Genetics* **139:** 1567–1583.

Lee S.S., Lee R.Y., Fraser A.G., Kamath R.S., Ahringer J., and Ruvkun G. 2003. A systematic RNAi screen identifies a critical role for mitochondria in *C. elegans* longevity. *Nat. Genet.* **33:** 40–48.

Leiser S.F., Salmon A.B., and Miller R.A. 2006. Correlated resistance to glucose deprivation and cytotoxic agents in fibroblast cell lines from long-lived pituitary dwarf mice. *Mech. Ageing Dev.* **127:** 821–829.

Li W., Gao B., Lee S.M., Bennett K., and Fang D. 2007. RLE-1, an E3 ubiquitin ligase, regulates *C. elegans* aging by catalyzing DAF-16 polyubiquitination. *Dev. Cell* **12:** 235–246.

Li Y., Deeb B., Pendergrass W., and Wolf N. 1996. Cellular proliferative capacity and life span in small and large dogs. *J. Gerontol. A Biol. Sci. Med. Sci.* **51:** B403–B408.

Lin K., Dorman J.B., Rodan A., and Kenyon C. 1997. daf-16: An HNF-3/forkhead family member that can function to double the life-span of *Caenorhabditis elegans. Science* **278:** 1319–1322.

Lin K., Hsin H., Libina N., and Kenyon C. 2001. Regulation of the *Caenorhabditis elegans* longevity protein DAF-16 by insulin/IGF-1 and germline signaling. *Nat. Genet.* **28:** 139–145.

Lithgow G.J. 1996. Invertebrate gerontology: The *age* mutations of *Caenorhabditis elegans. Bioessays* **18:** 809–815.

———. 2003. Does anti-aging equal anti-microbial? *Sci. Aging Knowledge Environ.* **2003:** E16.

Lithgow G.J., White T.M., Hinerfeld D.A., and Johnson T.E. 1994. Thermotolerance of a long-lived mutant of *Caenorhabditis elegans. J. Gerontol.* **49:** B270–B276.

Lithgow G.J., White T.M., Melov S., and Johnson T.E. 1995. Thermotolerance and extended life-span conferred by single-gene mutations and induced by thermal stress. *Proc. Natl. Acad. Sci.* **92:** 7540–7544.

Liu Y., Guyton K.Z., Gorospe M., Xu Q., Kokkonen G.C., Mock Y.D., Roth G.S., and Holbrook N.J. 1996. Age-related decline in mitogen-activated protein kinase activity in epidermal growth factor-stimulated rat hepatocytes. *J. Biol. Chem.* **271:** 3604–3607.

Loeb J. and Northrop J.H. 1916. Is there a temperature coefficient for the duration of life? *Proc. Natl. Acad. Sci.* **2:** 456–457.

———. 1917. What determines the duration of life in metazoa? *Proc. Natl. Acad. Sci.* **3:** 382–386.

Madsen M.A., Hsieh C.C., Boylston W.H., Flurkey K., Harrison D., and Papaconstantinou J. 2004. Altered oxidative stress response of the long-lived Snell dwarf mouse. *Biochem. Biophys. Res. Commun.* **318:** 998–1005.

Magwere T., West M., Riyahi K., Murphy M.P., Smith R.A., and Partridge L. 2006. The effects of exogenous antioxidants on lifespan and oxidative stress resistance in *Drosophila melanogaster. Mech. Ageing Dev.* **127:** 356–370.

Maier B., Gluba W., Bernier B., Turner T., Mohammad K., Guise T., Sutherland A., Thorner M., and Scrable H. 2004. Modulation of mammalian life span by the short isoform of p53. *Genes Dev.* **18:** 306–319.

Masoro E.J., McCarter R.J., Katz M.S., and McMahan C.A. 1992. Dietary restriction alters characteristics of glucose fuel use. *J. Gerontol.* **47:** B202–B208.

Maynard Smith J. 1958a. The effects of temperature and of egg-laying on the longevity of *Drosophila subobscura*. *J. Exp. Biol.* **35:** 832–843.

———. 1958b. Prolongation of the life of *Drosophila subobscura* by brief exposure of adults to a high temperature. *Nature* **181:** 496–497.

Maynard S.P. and Miller R.A. 2006. Fibroblasts from long-lived Snell dwarf mice are resistant to oxygen-induced in vitro growth arrest. *Aging Cell* **5:** 89–96.

McCarter R., Masoro E.J., and Yu B.P. 1985. Does food restriction retard aging by reducing the metabolic rate? *Am. J. Physiol.* **248:** E488–E490.

McColl G., Vantipalli M.C., and Lithgow G.J. 2005. The *C. elegans* ortholog of mammalian Ku70 interacts with insulin-like signaling to modulate stress resistance and life span. *FASEB J.* **19:** 1716–1718.

McElwee J., Bubb K., and Thomas J.H. 2003. Transcriptional outputs of the *Caenorhabditis elegans* forkhead protein DAF-16. *Aging Cell* **2:** 111–121.

Melov S., Wolf N., Strozyk D., Doctrow S.R., and Bush A.I. 2005. Mice transgenic for Alzheimer disease beta-amyloid develop lens cataracts that are rescued by antioxidant treatment. *Free Radic. Biol. Med.* **38:** 258–261.

Melov S., Doctrow S.R., Schneider J.A., Haberson J., Patel M., Coskun P.E., Huffman K., Wallace D.C., and Malfroy B. 2001. Lifespan extension and rescue of spongiform encephalopathy in superoxide dismutase 2 nullizygous mice treated with superoxide dismutase-catalase mimetics. *J. Neurosci.* **21:** 8348–8353.

Melov S., Ravenscroft J., Malik S., Gill M.S., Walker D.W., Clayton P.E., Wallace D.C., Malfroy B., Doctrow S.R., and Lithgow G.J. 2000. Extension of life-span with superoxide dismutase/catalase mimetics. *Science* **289:** 1567–1569.

Menini S., Amadio L., Oddi G., Ricci C., Pesce C., Pugliese F., Giorgio M., Migliaccio E., Pelicci P., Iacobini C., and Pugliese G. 2006. Deletion of p66Shc longevity gene protects against experimental diabetic glomerulopathy by preventing diabetes-induced oxidative stress. *Diabetes* **55:** 1642–1650.

Michell A.R. 1999. Longevity of British breeds of dog and its relationships with sex, size, cardiovascular variables and disease. *Vet. Rec.* **145:** 625–629.

Migliaccio E., Giorgio M., Mele S., Pelicci G., Reboldi P., Pandolfi P.P., Lanfrancone L., and Pelicci P.G. 1999. The p66shc adaptor protein controls oxidative stress response and life span in mammals. *Nature* **402:** 309–313.

Miller R.A. and Austad S.N. 2006. Growth and aging: Why do big dogs die young? In *Handbook of the biology of aging*, 6th edition (ed. E.J. Masoro and S.N. Austed), ch. 19, pp. 512–533. Academic Press, New York.

Miller R.A., Chrisp C., and Atchley W.R. 2000. Differential longevity in mouse stocks selected for early life growth trajectory. *J. Gerontol. Biol. Sci.* **55A:** B455–B461.

Miller R.A., Harper J.M., Galecki A., and Burke D.T. 2002. Big mice die young: Early-life body weight predicts longevity in genetically heterogeneous mice. *Aging Cell* **1:** 22–29.

Miller R.A., Buehner G., Chang Y., Harper J.M., Sigler R., and Smith-Wheelock M. 2005. Methionine-deficient diet extends mouse lifespan, slows immune and lens aging, alters glucose, T4, IGF-I and insulin levels, and increases hepatocyte MIF levels and stress resistance. *Aging Cell* **4:** 119–125.

Miwa S., Riyahi K., Partridge L., and Brand M.D. 2004. Lack of correlation between mito-chondrial reactive oxygen species production and life span in *Drosophila. Ann. N.Y. Acad. Sci.* **1019:** 388–391.

Mooijaart S.P., Brandt B.W., Baldal E.A., Pijpe J., Kuningas M., Beekman M., Zwaan B.J., Slagboom P.E., Westendorp R.G., and van Heemst D. 2005. *C. elegans* DAF-12, Nuclear Hormone Receptors and human longevity and disease at old age. *Ageing Res. Rev.* **4:** 351–371.

Moore C.J. and Schwartz A.G. 1978. Inverse correlation between species lifespan and capacity of cultured fibroblasts to convert benzo(a)pyrene to water-soluble metabo-lites. *Exp. Cell Res.* **116:** 359–364.

Morley J.F. and Morimoto R.I. 2004. Regulation of longevity in *Caenorhabditis elegans* by heat shock factor and molecular chaperones. *Mol. Biol. Cell* **15:** 657–664.

Morley J.F., Brignull H.R., Weyers J.J., and Morimoto R.I. 2002. The threshold for polyglutamine-expansion protein aggregation and cellular toxicity is dynamic and influ-enced by aging in *Caenorhabditis elegans. Proc. Natl. Acad. Sci.* **99:** 10417–10422.

Morris J.Z., Tissenbaum H.A., and Ruvkun G. 1996. A phosphatidylinositol-3-OH kinase family member regulating longevity and diapause in *Caenorhabditis elegans. Nature* **382:** 536–539.

Morrow G., Inaguma Y., Kato K., and Tanguay R.M. 2000. The small heat shock protein Hsp22 of *Drosophila melanogaster* is a mitochondrial protein displaying oligomeric organization. *J. Biol. Chem.* **275:** 31204–31210.

Morrow G., Samson M., Michaud S., and Tanguay R.M. 2004. Overexpression of the small mitochondrial Hsp22 extends *Drosophila* life span and increases resistance to oxida-tive stress. *FASEB J.* **18:** 598–599.

Motola D.L., Cummins C.L., Rottiers V., Sharma K.K., Li T., Li Y., Suino-Powell K., Xu H.E., Auchus R.J., Antebi A., and Mangelsdorf D.J. 2006. Identification of ligands for DAF-12 that govern dauer formation and reproduction in *C. elegans. Cell* **124:** 1209–1223.

Mourikis P., Hurlbut G.D., and Artavanis-Tsakonas S. 2006. Enigma, a mitochondrial pro-tein affecting lifespan and oxidative stress response in *Drosophila. Proc. Natl. Acad. Sci.* **103:** 1307–1312.

Munoz M.J. and Riddle D.L. 2003. Positive selection of *Caenorhabditis elegans* mutants with increased stress resistance and longevity. *Genetics* **163:** 171–180.

Murakami S. and Johnson T.E. 1996. A genetic pathway conferring life extension and resistance to UV stress in *Caenorhabditis elegans. Genetics* **143:** 1207–1218.

———. 2001. The OLD-1 positive regulator of longevity and stress resistance is under DAF-16 regulation in *Caenorhabditis elegans. Curr. Biol.* **11:** 1517–1523.

Murakami S., Salmon A., and Miller R.A. 2003. Multiplex stress resistance in cells from long-lived dwarf mice. *FASEB J.* **17:** 1565–1566.

Murphy C.T., McCarroll S.A., Bargmann C.I., Fraser A., Kamath R.S., Ahringer J., Li H., and Kenyon C. 2003. Genes that act downstream of DAF-16 to influence the lifespan of *Caenorhabditis elegans. Nature* **424:** 277–283.

Musleh W., Bruce A., Malfroy B., and Baudry M. 1994. Effects of EUK-8, a synthetic cat-alytic superoxide scavenger, on hypoxia- and acidosis-induced damage in hippocam-pal slices. *Neuropharmacology* **33:** 929–934.

Napoli C., Martin-Padura I., de Nigris F., Giorgio M., Mansueto G., Somma P., Condorelli M., Sica G., De Rosa G., and Pelicci P. 2003. Deletion of the p66Shc longevity gene reduces systemic and tissue oxidative stress, vascular cell apoptosis, and early atherogenesis in mice fed a high-fat diet. *Proc. Natl. Acad. Sci.* **100:** 2112–2116.

Nemoto S. and Finkel T. 2002. Redox regulation of forkhead proteins through a p66shc-dependent signaling pathway. *Science* **295:** 2450–2452.

Nemoto S., Takeda K., Yu Z.X., Ferrans V.J., and Finkel T. 2000. Role for mitochondrial oxidants as regulators of cellular metabolism. *Mol. Cell. Biol.* **20:** 7311–7318.

Nemoto S., Combs C.A., French S., Ahn B.H., Fergusson M.M., Balaban R.S., and Finkel T. 2006. The mammalian longevity-associated gene product p66shc regulates mitochondrial metabolism. *J. Biol. Chem.* **281:** 10555–10560.

Ogg S., Paradis S., Gottlieb S., Patterson G.I., Lee L., Tissenbaum H.A., and Ruvkun G. 1997. The Fork head transcription factor DAF-16 transduces insulin-like metabolic and longevity signals in *C. elegans*. *Nature* **389:** 994–999.

Oh S.W., Mukhopadhyay A., Dixit B.L., Raha T., Green M.R., and Tissenbaum H.A. 2006. Identification of direct DAF-16 targets controlling longevity, metabolism and diapause by chromatin immunoprecipitation. *Nat. Genet.* **38:** 251–257.

Oh S.W., Mukhopadhyay A., Svrzikapa N., Jiang F., Davis R.J., and Tissenbaum H.A. 2005. JNK regulates lifespan in *Caenorhabditis elegans* by modulating nuclear translocation of forkhead transcription factor/DAF-16. *Proc. Natl. Acad. Sci.* **102:** 4494–4499.

Olsen A., Vantipalli M.C., and Lithgow G.J. 2006a. Checkpoint proteins control survival of the postmitotic cells in *Caenorhabditis elegans*. *Science* **312:** 1381–1385.

———. 2006b. Lifespan extension of *Caenorhabditis elegans* following repeated mild hormetic heat treatments. *Biogerontology* **7:** 221–230.

Orentreich N., Matias J.R., DeFelice A., and Zimmerman J.A. 1993. Low methionine ingestion by rats extends life span. *J. Nutr.* **123:** 269–274.

Pamplona R., Portero-Otin M., Requena J., Gredilla R., and Barja G. 2002. Oxidative, glycoxidative and lipoxidative damage to rat heart mitochondrial proteins is lower after 4 months of caloric restriction than in age-matched controls. *Mech. Ageing Dev.* **123:** 1437–1446.

Pamplona R., Portero-Otin M., Requena J.R., Thorpe S.R., Herrero A., and Barja G. 1999a. A low degree of fatty acid unsaturation leads to lower lipid peroxidation and lipoxidation-derived protein modification in heart mitochondria of the longevous pigeon than in the short-lived rat. *Mech. Ageing Dev.* **106:** 283–296.

Pamplona R., Portero-Otin M., Riba D., Ledo F., Gredilla R., Herrero A., and Barja G. 1999b. Heart fatty acid unsaturation and lipid peroxidation, and aging rate, are lower in the canary and the parakeet than in the mouse. *Aging* **11:** 44–49.

Pamplona R., Portero-Otin M., Riba D., Ruiz C., Prat J., Bellmunt M.J., and Barja G. 1998. Mitochondrial membrane peroxidizability index is inversely related to maximum life span in mammals. *J. Lipid Res.* **39:** 1989–1994.

Paradis S. and Ruvkun G. 1998. *Caenorhabditis elegans* Akt/PKB transduces insulin receptor-like signals from AGE-1 PI3 kinase to the DAF-16 transcription factor. *Genes Dev.* **12:** 2488–2498.

Paradis S., Ailion M., Toker A., Thomas J.H., and Ruvkun G. 1999. A PDK1 homolog is necessary and sufficient to transduce AGE-1 PI3 kinase signals that regulate diapause in *Caenorhabditis elegans*. *Genes Dev.* **13:** 1438–1452.

Parrinello S., Samper E., Krtolica A., Goldstein J., Melov S., and Campisi J. 2003. Oxygen sensitivity severely limits the replicative lifespan of murine fibroblasts. *Nat. Cell Biol.* **5:** 741–747.

Parsell D.A., Taulien J., and Lindquist S. 1993. The role of heat-shock proteins in thermotolerance. *Philos. Trans. R. Soc. Lond. B Biol. Sci.* **339:** 279–285.

Patronek G.J., Waters D.J., and Glickman L.T. 1997. Comparative longevity of pet dogs and humans: Implications for gerontology research. *J. Gerontol. A Biol. Sci. Med. Sci.* **52:** B171–B178.

Pearl R. 1923. *The rate of living.* Knopf, New York.

Perez-Campo R., Lopez-Torres M., Cadenas S., Rojas C., and Barja G. 1998. The rate of free radical production as a determinant of the rate of aging: Evidence from the comparative approach. *J. Comp. Physiol. B* **168:** 149–158.

Rattan S.I. 1998. Repeated mild heat shock delays ageing in cultured human skin fibroblasts. *Biochem. Mol. Biol. Int.* **45:** 753–759.

———. 2001. Applying hormesis in aging research and therapy (see discussion on pp. 293–294). *Hum. Exp. Toxicol.* **20:** 281–285.

Rea S.L., Wu D., Cypser J.R., Vaupel J.W., and Johnson T.E. 2005. A stress-sensitive reporter predicts longevity in isogenic populations of *Caenorhabditis elegans. Nat. Genet.* **37:** 894–898.

Richie J.P., Jr., Komninou D., Leutzinger Y., Kleinman W., Orentreich N., Malloy V., and Zimmerman J.A. 2004. Tissue glutathione and cysteine levels in methionine-restricted rats. *Nutrition* **20:** 800–805.

Riddle D.L., Swanson M.M., and Albert P.S. 1981. Interacting genes in nematode dauer larva formation. *Nature* **290:** 668–671.

Riddle D.R., Blumenthal T., Meyer B.J., and Priess J.R., Eds. 1997. Introduction to *C. elegans.* In C. elegans II, pp. 1–22. Cold Spring Harbor Laboratory Press, Cold Spring Harbor, New York.

Rincon M., Rudin E., and Barzilai N. 2005. The insulin/IGF-1 signaling in mammals and its relevance to human longevity. *Exp. Gerontol.* **40:** 873–877.

Ruiz M.C., Ayala V., Portero-Otin M., Requena J.R., Barja G., and Pamplona R. 2005. Protein methionine content and MDA-lysine adducts are inversely related to maximum life span in the heart of mammals. *Mech. Ageing Dev.* **126:** 1106–1114.

Saitoh S., Chabes A., McDonald W.H., Thelander L., Yates J.R., and Russell P. 2002. Cid13 is a cytoplasmic poly(A) polymerase that regulates ribonucleotide reductase mRNA. *Cell* **109:** 563–573.

Salmon A.B., Murakami S., Bartke A., Kopchick J., Yasumura K., and Miller R.A. 2005. Fibroblast cell lines from young adult mice of long-lived mutant strains are resistant to multiple forms of stress. *Am. J. Physiol. Endocrinol. Metab.* **289:** E23–E29.

Sampayo J.N., Olsen A., and Lithgow G.J. 2003. Oxidative stress in *Caenorhabditis elegans:* Protective effects of superoxide dismutase/catalase mimetics. *Aging Cell* **2:** 319–326.

Sanz A., Caro P., Ayala V., Portero-Otin M., Pamplona R., and Barja G. 2006. Methionine restriction decreases mitochondrial oxygen radical generation and leak as well as oxidative damage to mitochondrial DNA and proteins. *FASEB J.* **20:** 1064–1073.

Schriner S.E., Linford N.J., Martin G.M., Treuting P., Ogburn C.E., Emond M., Coskun P.E., Ladiges W., Wolf N., Van Remmen H., Wallace D.C., and Rabinovitch P.S. 2005. Extension of murine lifespan by overexpression of catalase targeted to mitochondria. *Science* **308:** 1875–1876.

Selsby J.T., Judge A.R., Yimlamai T., Leeuwenburgh C., and Dodd S.L. 2005. Life long calorie restriction increases heat shock proteins and proteasome activity in soleus muscles of Fisher 344 rats. *Exp. Gerontol.* **40:** 37–42.

Senoo-Matsuda N., Hartman P.S., Akatsuka A., Yoshimura S., and Ishii N. 2003. A complex II defect affects mitochondrial structure, leading to ced-3- and ced-4-dependent apoptosis and aging. *J. Biol. Chem.* **278:** 22031–22036.

Sifri C.D., Begun J., Ausubel F.M., and Calderwood S.B. 2003. *Caenorhabditis elegans* as a model host for *Staphylococcus aureus* pathogenesis. *Infect. Immun.* **71:** 2208–2217.

Silberberg R. 1972. Articular aging and osteoarthritis in dwarf mice. *Pathol. Microbiol.* **38:** 417–430.

Silbermann R. and Tatar M. 2000. Reproductive costs of heat shock protein in transgenic *Drosophila melanogaster. Evolution Int. J. Org. Evolution* **54:** 2038–2045.

Sinclair D.A. 2005. Toward a unified theory of caloric restriction and longevity regulation. *Mech. Ageing Dev.* **126:** 987–1002.

Snow M.I. and Larsen P.L. 2000. Structure and expression of *daf-12:* A nuclear hormone receptor with three isoforms that are involved in development and aging in *Caenorhabditis elegans. Biochim. Biophys. Acta* **1494:** 104–116.

Speakman J.R. 2005. Correlations between physiology and lifespan: Two widely ignored problems with comparative studies. *Aging Cell* **4:** 167–175.

Speakman J.R., van Acker A., and Harper E.J. 2003. Age-related changes in the metabolism and body composition of three dog breeds and their relationship to life expectancy. *Aging Cell* **2:** 265–275.

Stenmark P., Grunler J., Mattsson J., Sindelar P.J., Nordlund P., and Berthold D.A. 2001. A new member of the family of di-iron carboxylate proteins. Coq7 (clk-1), a membrane-bound hydroxylase involved in ubiquinone biosynthesis. *J. Biol. Chem.* **276:** 33297–33300.

Strayer A., Wu Z., Christen Y., Link C.D., and Luo Y. 2003. Expression of the small heat-shock protein Hsp16-2 in *Caenorhabditis elegans* is suppressed by Ginkgo biloba extract EGb 761. *FASEB J.* **17:** 2305–2307.

Sun J., Folk D., Bradley T.J., and Tower J. 2002. Induced overexpression of mitochondrial Mn-superoxide dismutase extends the life span of adult *Drosophila melanogaster. Genetics* **161:** 661–672.

Tatar M., Khazaeli A.A., and Curtsinger J.W. 1997. Chaperoning extended life (letter). *Nature* **390:** 30.

Tatar M., Kopelman A., Epstein D., Tu M.P., Yin C.M., and Garofalo R.S. 2001. A mutant *Drosophila* insulin receptor homolog that extends life-span and impairs neuroendocrine function. *Science* **292:** 107–110.

Tissenbaum H.A. and Guarente L. 2001. Increased dosage of a *sir-2* gene extends lifespan in *Caenorhabditis elegans. Nature* **410:** 227–230.

Tissenbaum H.A. and Ruvkun G. 1998. An insulin-like signaling pathway affects both longevity and reproduction in *Caenorhabditis elegans. Genetics* **148:** 703–717.

Troemel E.R., Chu S.W., Reinke V., Lee S.S., Ausubel F.M., and Kim D.H. 2006. p38 MAPK regulates expression of immune response genes and contributes to longevity in *C. elegans. PLoS Genet.* **2:** e183.

Tu M.P., Yin C.M., and Tatar M. 2005. Mutations in insulin signaling pathway alter juvenile hormone synthesis in *Drosophila melanogaster. Gen. Comp. Endocrinol.* **142:** 347–356.

Tyner S.D., Venkatachalam S., Choi J., Jones S., Ghebranious N., Igelmann H., Lu X., Soron G., Cooper B., Brayton C., Hee P.S., Thompson T., Karsenty G., Bradley A., and Donehower L.A. 2002. p53 mutant mice that display early ageing-associated phenotypes. *Nature* **415:** 45–53.

Vajo Z., King L.M., Jonassen T., Wilkin D.J., Ho N., Munnich A., Clarke C.F., and Francomano C.A. 1999. Conservation of the *Caenorhabditis elegans* timing gene *clk-1* from yeast to human: A gene required for ubiquinone biosynthesis with potential implications for aging. *Mamm. Genome* **10:** 1000–1004.

Vanfleteren J.R. 1993. Oxidative stress and ageing in *Caenorhabditis elegans. Biochem. J.* **292:** 605–608.

van Heemst D., Beekman M., Mooijaart S.P., Heijmans B.T., Brandt B.W., Zwaan B.J., Slagboom P.E., and Westendorp R.G. 2005. Reduced insulin/IGF-1 signalling and human longevity. *Aging Cell* **4:** 79–85.

Van Remmen H., Ikeno Y., Hamilton M., Pahlavani M., Wolf N., Thorpe S.R., Alderson N.L., Baynes J.W., Epstein C.J., Huang T.T., Nelson J., Strong R., and Richardson A. 2003. Life-long reduction in MnSOD activity results in increased DNA damage and higher incidence of cancer but does not accelerate aging. *Physiol. Genomics* **16:** 29–37.

Vergara M., Smith-Wheelock M., Harper J.M., Sigler R., and Miller R.A. 2004. Hormone-treated Snell dwarf mice regain fertility but remain long-lived and disease resistant. *J. Gerontol. A Biol. Sci. Med. Sci.* **59:** 1244–1250.

Walker D.W., Hajek P., Muffat J., Knoepfle D., Cornelison S., Attardi G., and Benzer S. 2006. Hypersensitivity to oxygen and shortened lifespan in a *Drosophila* mitochondrial complex II mutant. *Proc. Natl. Acad. Sci.* **103:** 16382–16387.

Walker G.A. and Lithgow G.J. 2003. Lifespan extension in *C. elegans* by a molecular chaperone dependent upon insulin-like signals. *Aging Cell* **2:** 131–139.

Walker G.A., Walker D.W., and Lithgow G.J. 1998. Genes that determine both thermotolerance and rate of aging in *Caenorhabditis elegans. Ann. N.Y. Acad. Sci.* **851:** 444–449.

Walker G.A., Thompson F.J., Brawley A., Scanlon T., and Devaney E. 2003. Heat shock factor functions at the convergence of the stress response and developmental pathways in *Caenorhabditis elegans. FASEB J.* **17:** 1960–1962.

Walker G.A., White T.M., McColl G., Jenkins N.L., Babich S., Candido E.P., Johnson T.E., and Lithgow G.J. 2001. Heat shock protein accumulation is upregulated in a long-lived mutant of *Caenorhabditis elegans. J. Gerontol. A Biol. Sci. Med. Sci.* **56:** B281–B287.

Wang H.D., Kazemi-Esfarjani P., and Benzer S. 2004. Multiple-stress analysis for isolation of *Drosophila* longevity genes. *Proc. Natl. Acad. Sci.* **101:** 12610–12615.

Wang M.C., Bohmann D., and Jasper H. 2003. JNK signaling confers tolerance to oxidative stress and extends lifespan in *Drosophila. Dev. Cell* **5:** 811–816.

Wang S.W., Toda T., MacCallum R., Harris A.L., and Norbury C. 2000. Cid1, a fission yeast protein required for S-M checkpoint control when DNA polymerase delta or epsilon is inactivated. *Mol. Cell. Biol.* **20:** 3234–3244.

Welte M.A., Tetrault J.M., Dellavalle R.P., and Lindquist S.L. 1993. A new method for manipulating transgenes: Engineering heat tolerance in a complex, multicellular organism. *Curr. Biol.* **3:** 842–853.

Wilson M.A., Shukitt-Hale B., Kalt W., Ingram D.K., Joseph J.A., and Wolkow C.A. 2006. Blueberry polyphenols increase lifespan and thermotolerance in *Caenorhabditis elegans. Aging Cell* **5:** 59–68.

Wolff S., Ma H., Burch D., Maciel G.A., Hunter T., and Dillin A. 2006. SMK-1, an essential regulator of DAF-16-mediated longevity. *Cell* **124:** 1039–1053.

Wong A., Boutis P., and Hekimi S. 1995. Mutations in the *clk-1* gene of *Caenorhabditis elegans* affect developmental and behavioral timing. *Genetics* **139:** 1247–1259.

Wood J.G., Rogina B., Lavu S., Howitz K., Helfand S.L., Tatar M., and Sinclair D. 2004. Sirtuin activators mimic caloric restriction and delay ageing in metazoans. *Nature* **430:** 686–689.

Wu Z., Smith J.V., Paramasivam V., Butko P., Khan I., Cypser J.R., and Luo Y. 2002. Ginkgo biloba extract EGb 761 increases stress resistance and extends life span of *Caenorhabditis elegans. Cell. Mol. Biol.* **48:** 725–731.

Yokoyama K., Fukumoto K., Murakami T., Harada S., Hosono R., Wadhwa R., Mitsui Y., and Ohkuma S. 2002. Extended longevity of *Caenorhabditis elegans* by knocking in extra copies of hsp70F, a homolog of mot-2 (mortalin)/mthsp70/Grp75. *FEBS Lett.* **516:** 53–57.

Yu B.P. and Masoro E.J. 1995. Putative interventions. In *Handbook of physiology: Aging* (ed. E.J. Masoro), Sect. 11, pp. 613–631. Oxford University Press, New York.

Zainal T.A., Oberley T.D., Allison D.B., Szweda L.I., and Weindruch R. 2000. Caloric restriction of rhesus monkeys lowers oxidative damage in skeletal muscle. *FASEB J.* **14:** 1825–1836.

Zimmerman J.A., Malloy V., Krajcik R., and Orentreich N. 2003. Nutritional control of aging. *Exp. Gerontol.* **38:** 47–52.

17

Molecular Mechanisms of Aging: Insights from Budding Yeast

Su-Ju Lin

Section of Microbiology, College of Biological Sciences
University of California, Davis, California 95616

David Sinclair

Paul F. Glenn Laboratories for the Biological Mechanisms of Aging
Department of Pathology, Harvard Medical School
Boston, Massachusetts 02115

UNTIL THE LATE 1980S, THE PREVAILING VIEW among researchers was that life span of any organism, even yeast, could not be regulated, let alone by just a few genes. The view was based on the fact that aging is an incredibly complex process that is affected by thousands of genes. Then, in just a few years, genetic studies in model organisms such as *Saccharomyces cerevisiae* and *Caenorhabditis elegans* uncovered numerous single-gene mutations that extend life span (Jazwinski et al. 1993; Kenyon et al. 1993; Kennedy et al. 1995). What had researchers overlooked prior to 1990? The major oversight appears to have been the failure to foresee that organisms have evolved to promote their survival, and hence longevity, during times of adversity. Longevity regulation, as it has come to be known, is now thought of as a highly adaptive biological trait that is conserved all the way from yeast to mammals (Kirkwood and Holliday 1979; Kenyon 2001).

When Andrew Barton first proposed in 1950 that *S. cerevisiae* might serve as a model for aging, he was met with considerable skepticism (Barton 1950). It was difficult for most researchers to accept that a simple unicellular organism could provide any information about aging. But we have since learned never to underestimate a fungus. Today, *S. cerevisiae* is one of the most highly utilized models for aging, and dozens of longevity genes have been identified. Translating these findings to mammals is one of the major challenges for researchers during the next decade.

BIOLOGY OF YEAST AGING

Replicative versus Chronological Yeast Aging

Aging in *S. cerevisiae* is measured two ways. Yeast "replicative life span" is defined as the number of divisions an individual yeast cell undergoes before dying. One attractive feature of *S. cerevisiae*, as opposed to many other simple eukaryotes, is that the progenitor cell is easily distinguished from its descendants because cell division is asymmetric: A newly formed "daughter" cell is almost always smaller than the "mother" cell that gave rise to it. Yeast mother cells divide about 20 times before dying and, as described below, undergo characteristic structural and metabolic changes as they age.

The alternative measure, "chronological life span," also referred to as "post-diauxic survival," is the length of time a population of yeast cells remains viable in a nondividing state following nutrient deprivation (Longo and Fabrizio 2002). Yeast cells grown in a nutrient-rich medium multiply until all readily utilizable nutrients are exhausted. At this point, cells cease dividing and enter a post-diauxic, hypometabolic state known as stationary phase, where they can remain viable for weeks. In synthetic medium, cells will deplete the medium and cease dividing, yet retain relatively high metabolism (Longo and Fabrizio 2002). Such cells have a short chronological life span relative to cells in rich medium and are thought to more closely resemble postmitotic cells in multicellular organisms (Longo et al. 1996).

Basic Biology of Yeast Aging

Replicative Aging

Because many mutations can shorten life span, it is necessary to have a set of characteristics that distinguish accelerated aging from cell sickness. When a bud separates from a mother cell, the boundary between the two cells constricts, leaving behind on the mother cell's surface a circular chitin-containing remnant termed the bud scar. Bud scars remain permanently deposited on the surface of the mother cell. As a mother cell goes through successive rounds of cell division, bud scars accumulate on the cell surface, serving as a convenient marker for the number of divisions realized by a single cell. Although it has been hypothesized that the accumulation of bud scars may impose a theoretical upper limit on a cell's replicative potential (Cabib et al. 1974), evidence indicates that it does not result in senescence of wild-type strains.

After three or four divisions, mother cells are easily distinguished from daughter cells due to their increased size. It has also been proposed that one cause of aging might be that the cells have an upper size limit (Mortimer and Johnston 1959). Although this could be true for very long-lived mutant strains, this is not a cause of aging for most wild-type laboratory strains (Kennedy et al. 1994).

Yeast cells continue to give rise to small daughter cells throughout most of their life span. However, very old mother cells tend to produce large, short-lived daughter cells (Johnston 1966; Jazwinski et al. 1989; Kennedy et al. 1994) that often do not separate from the mother until both cells are similar in size (Johnston 1966; Jazwinski et al. 1989). Young yeast cells are fertile because the two silent repositories of mating-type information, *HMR* and *HML*, remain in a transcriptionally inert state (Loo and Rine 1995). This is due to the activity of the Sir2/3/4p silencing complex. When yeast cells grow, the Sir complex relocalizes from *HM* loci (and telomeres) to the rDNA locus (Kennedy et al. 1997), and old yeast cells become sterile (Muller 1985; Smeal et al. 1996). This phenotype remains one of the most reliable markers of yeast aging.

The nucleolus is a nuclear structure containing the ribosomal DNA (rDNA) genes and other components required for ribosome assembly (Melese and Xue 1995; Cockell and Gasser 1999). Yeast rDNA, located on chromosome XII, contains 100–200 tandemly repeated copies of a 9.1-kb unit. In young yeast cells, the nucleolus forms a crescent-shaped structure retained near the nuclear periphery (Schimmang et al. 1989). In old cells, however, the nucleolus becomes enlarged and fragmented into multiple, rounded structures (Sinclair et al. 1997). Fragmentation of the nucleolus and relocalization of Sir3 may be a response to the accumulation of extrachromosomal rDNA (Sinclair and Guarente 1997), as described below.

As yeast age, they also undergo metabolic changes. There is a progressive shift away from glycolysis toward gluconeogenesis and energy storage, which is associated with the induction of genes involved in glycogen production, fatty acid degradation, gluconeogenesis, and the glyoxylate cycle. Much of the up-regulation of genes involved in gluconeogenesis appears to be due to the translocation of the Mig1 transcriptional repressor from the nucleus to the cytoplasm as cells age (Ashrafi et al. 2000). ATP levels do not decline with age, but there is an approximately 30% drop in NAD between ages 0–1 and 7–8. Interestingly, many of these changes are recapitulated in short-lived *sip2* cells, which lack the putative repressor of Snf1, a kinase that regulates cellular responses to glucose deprivation. The role of the Snf1/Sip2 pathway in longevity is described below.

Numerous hypotheses have been proposed to explain why dividing yeast cells grow old, and all but one have been disproved. Despite some heated debates, researchers now agree that the primary cause of yeast replicative aging stems from changes within the nucleolus, the distinct nuclear region responsible for ribosomal RNA (rRNA) transcription and ribosome assembly. In 1997, Sinclair and Guarente (1997) proposed that yeast replicative aging stems from genomic instability at the rDNA locus, *RDN1*. The yeast rDNA locus is inherently recombinagenic due to its repetitive nature and unidirectional mode of DNA replication. Homologous recombination between adjacent repeats is known to result in the excision of extrachromosomal circular forms of rDNA, known as ERCs (Fig. 1). The important aspect of the aging mechanism is that ERCs replicate during S phase but are inefficiently segregated to daughter cells along with linear chromosomes. Because ERCs can double in copy number every S phase, their abundance increases exponentially in mother cells at a rate timed by cell division (Sinclair and Guarente 1997). ERCs accumulate to more than 1000 copies per old cell, which totals more DNA than the rest of the yeast genome, and all the ERCs in old cells are likely derived from a single initial recombination event. The mechanism by which ERCs cause death is not known, but, given their abundance, it is thought that they titrate away vital transcription and/or replication factors (Sinclair et al. 1998).

A major mechanism by which yeast cells suppress ERC formation is the packaging of DNA and histones into "silent" heterochromatin. Heterochromatin in yeast occurs at telomeres, *HM* loci, and the rDNA (Moazed 2001). The formation of heterochromatin at *HM* loci and telomeres is mediated by the silent information regulatory complex Sir2/3/4 (Hecht et al. 1996; Strahl-Bolsinger et al. 1997). Alternatively, heterochromatin at the rDNA locus is catalyzed by the RENT (regulator of nucleolar silencing and telophase exit) complex, which includes Sir2, Net1, and Cdc14 (Ghidelli 2001; Shou et al. 2001). Of these proteins, Sir2 is the only factor that is indispensable for silencing at all three silent regions (Gottschling et al. 1990; Bryk et al. 1997; Smith and Boeke 1997). To be precise, "rDNA silencing" is actually a misnomer. Although pol-II-transcribed marker genes integrated at the rDNA are transcriptionally silenced, transcription of the native rDNA is seemingly unaf-

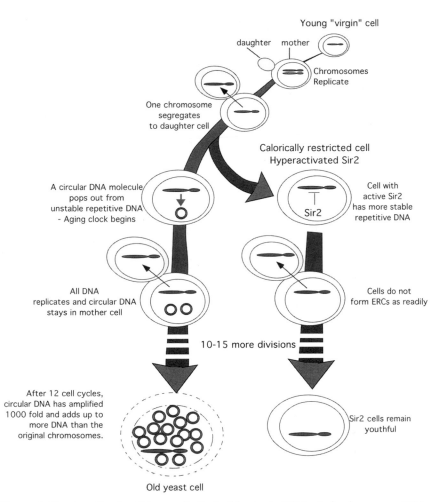

Young "virgin" cell

daughter   mother

Chromosomes
Replicate

One chromosome
segregates
to daughter cell

Calorically restricted cell
Hyperactivated Sir2

A circular DNA molecule
pops out from
unstable repetitive DNA
- Aging clock begins

Sir2

Cell with
active Sir2
has more stable
repetitive DNA

All DNA
replicates and circular DNA
stays in mother cell

Cells do not
form ERCs as readily

10-15 more divisions

After 12 cell cycles,
circular DNA has amplified
1000 fold and adds up to
more DNA than the
original chromosomes.

Sir2 cells remain
youthful

Old yeast cell

Figure 1. Yeast replicative life span is limited by toxic DNA circles known as ERCs. A major cause of replicative aging in *S. cerevisiae* stems from genomic instability. Like all eukaryotes, budding yeast possesses a highly repetitive locus which encodes the rRNAs that are assembled into ribosomes to facilitate translation. Recombination between rDNA repeats can lead to a circular form of rDNA known as an ERC. ERCs have the propensity to be replicated and to stay within the mother cell, resulting in their exponential accumulation to a quantity of DNA that exceeds the yeast genome. Genes such as *SIR2* suppress rDNA recombination and extend replicative life span. The cause of death due to ERC accumulation is not known, but it seems likely that they titrate essential transcription or replication factors away from the genome.

fected by deletion of *SIR2* (Smith and Boeke 1997). This suggests that the primary function of heterochromatin at the rDNA locus is to suppress recombination rather than to silence transcription.

More than 40 mutations are known to extend yeast life span (Table 1). Most do so by increasing the extent of heterochromatin at the rDNA, which seems to inhibit rDNA recombination. Examples of this class of longevity genes include *TPK1-3*, *SIR2*, and *PNC1*. The other class includes those that directly suppress homologous recombination between rDNA repeats such as *FOB1* (Defossez et al. 1998). *FOB1* encodes a nucleolar protein that is required for a DNA replication fork-block immediately downstream from rDNA origins (Kobayashi et al. 1998; Johzuka and Horiuchi 2002). Strains lacking *FOB1* have a 100-fold lower rate of rDNA recombination (Lin and Keil 1991) and have manyfold fewer ERCs than age-matched wild-type cells (Defossez et al. 1999). As predicted by the ERC model, *fob1* loss-of-function mutants live almost twice as long as wild-type controls.

It would be a mistake to assume that there is only one cause of replicative aging, but as yet, there is no other known cause of replicative aging (McMurray and Gottschling 2003), although some strains have been found to "age" independently of ERC formation (Kim et al. 1999; Ray et al. 2003). One proposed cause of aging is disregulation of rDNA transcription during the life span of a cell (Jazwinski 2000b), and another appears to be governed by the Slt2 kinase via phosphorylation of Sir3, although the details of how this mechanism influences life span are not yet known.

Chronological Aging

A population of yeast cells that enters stationary phase undergoes a number of dramatic physiological and biochemical transformations. These cells accumulate glycogen and trehalose and develop thick cell walls (Werner-Washburne et al. 1996). Cells are also significantly more thermotolerant and resistant to various forms of oxidative damage (Werner-Washburne et al. 1993; Stephen et al. 1995). As may be expected, protein synthesis rates drop dramatically; although interestingly, the number and identity of different proteins being synthesized are similar to those of exponentially growing cells (Werner-Washburne et al. 1996). As Longo and colleagues point out, although the chronological life span of yeast may appear to be a starvation phase distinct from the aging of higher eukaryotic postmitotic cells, nondividing yeast are not starving (Longo and Fabrizio 2002). During

Table 1. Longevity genes in yeast *S. cerevisiae*

Pathway	Gene/Allele	Replicative LS		Chronological LS		Function	References
		Δ	OE	Δ	OE		
Glucose/RAS-cAMP-PKA	cdc25*	+				Ras GEF	Lin et al. (2000)
	cdc35*			+		adenylate cyclase	Fabrizio et al. (2001); Lin et al. (2000)
	tpk1-3*	+				PKA catalytic subunit	Lin et al. (2000)
	bcy1	–				PKA regulatory subunit	Sun et al. (1994)
	hxk2	+				hexokinase	Lin et al. (2000)
	gpr1	+				GPCR	Lin et al. (2000)
	gpa2	+				GTPase	Lin et al. (2000)
	HAP4		+			transcription factor	Lin et al. (2002)
	ras1	+	N.E.			GTPase	Sun et al. (1994)
	RAS2	–	+	+		GTPase	Sun et al. (1994)
SNF1/AMP-kinase	SIP2	–				protein kinase activator	Ashrafi et al. (2000)
	snf4	+				protein kinase activator	Ashrafi et al. (2000)
NAD salvage	NPT1	N.E.	+			nicotinic acid phosphoribosyl-transferases	Anderson et al. (2002); Lin et al. (2000)
	PNC1	–	+			nicotinamide deamidase	Anderson et al. (2002, 2003)
Genome stability/ rDNA recombination	UTH1	–				silencing	Kennedy et al. (1995)
	Sir4-42	–	+			DNA repair/silencing	Kennedy et al. (1995, 1997)
	UTH4	–	+			silencing	Kennedy et al. (1995)
	SGS1	–	+			genome stability	Sinclair et al. (1997)

(continued)

Table 1. (*continued*)

Pathway	Gene/Allele	Replicative LS		Chronological LS		Function	References
		Δ	OE	Δ	OE		
	rpd3	+				histone deacetylase	Kim et al. (1999)
	fob1	+				replication fork block protein	Defossez et al. (1999)
	zds1	+				silencing	Roy and Runge (2000)
	zds2	–				silencing	Roy and Runge (2000)
	SIR3	–				silencing	Kaeberlein et al. (1999); Kim et al. (1999)
	SIR3^P	+				silencing	Ray et al. (2003)
	SIR2	–	+	+		histone deacetylase	Fabrizio et al. (2005); Kaeberlein et al. (1999); Kim et al. (1999)
	HST2	–	+			histone deacetylase	Lamming et al. (2005)
	cdc6*	+				ATPase/replication	Sinclair and Guarente (1997)
	LAG1/2	–	+			ceramide synthesis	Childress et al. (1996); D'Mello et al. (1994)
	gcn5	+				histone acetyltransferase	Kim et al. (2004)

Category	Gene	Description			Reference
Nitrogen sensing	gln3	transcription factor	+		Powers et al. (2006)
	lys12	lysine synthesis	+		Powers et al. (2006)
	mep2-3	ammonium transporter	+		Powers et al. (2006)
	agp1	amino acid permease	+		Powers et al. (2006)
	sch9	Serine/threonine kinase	+	+	Fabrizio et al. (2001, 2004); Kaeberlein et al. (2005b)
	tor1	target of rapamycin kinase	+	+	Kaeberlein et al. (2005b); Powers et al. (2006)
	rpl31a	ribosomal protein		+	Kaeberlein et al. (2005b)
	rpl6b	ribosomal protein		+	Kaeberlein et al. (2005b)
Stress response	SOD2	mitochondrial superoxide dismutase	–	+	Longo et al. (1999)
	hsc82	chaperone protein		N.E.	Harris et al. (2001)
	hsp82	chaperone protein	+	–	Harris et al. (2001)
Retrograde response	rtg2	HSP70-like ATPase	–	–	Barros et al. (2004); Borghouts et al. (2004); Jiang et al. (2000)
	rtg3	transcription factor	–		Borghouts et al. (2004); Jiang et al. (2000)

(N.E.) No effect; (*) point mutation; (LS) life span; (Δ) deletion; (OE) overexpression; (–) decreased life span; (+) increased life span; (^{P}) altered phosphorylation sites.

these post-diauxic phases, yeast are breaking down glycogen and utilizing other stored nutrients, similar to hibernating animals and the diapause state of other metazoans. Respiration is the primary source of energy in these cells, and the limited resources appear to be directed toward resisting cellular damage and stress (Longo and Fabrizio 2002).

LONGEVITY GENES

Replicative Aging

Longevity Genes Regulating Replicative Life Span

Identifications and characterizations of longevity genes in yeast have provided insight into the molecular mechanisms underlying longevity regulation (summarized in Table 1). Many of these longevity genes are highly conserved among species and have been shown to increase longevity in higher eukaryotes (Kim et al. 1999; Ashrafi et al. 2000; Jazwinski 2000a; Lin et al. 2000; Fabrizio et al. 2001; Kenyon 2001; Tissenbaum and Guarente 2001; Rogina et al. 2002; Bitterman et al. 2003; Rogina and Helfand 2004). The major pathways, and arguably the best understood, are shown in Figure 2.

SGS1. SGS1 encodes a RecQ-like DNA helicase that functions in maintaining genomic integrity. Sgs1 forms a complex with the type-1 DNA topoisomerase Top3, which stabilizes the replication fork block (Gangloff et al. 1994; Bjergbaek et al. 2005). Sgs1 also has important roles during double-strand break repair (Ira et al. 2003), meiotic crossing-over (Rockmill et al. 2003), and recombination-mediated telomere maintenance/elongation in cells lacking telomerase (Cohen and Sinclair 2001; Huang et al. 2001; Johnson et al. 2001). The *sgs1* mutants exhibit shorter life span and a higher recombination rate at rDNA (Sinclair et al. 1997). It is suggested that *sgs1* mutant cells have a short replicative life span because of accelerated aging. Consistent with this, *sgs1* mutants show signs of premature aging such as sterility and nucleolar fragmentation (Sinclair et al. 1997) and have increased numbers of ERCs over age-matched wild-type cells (Sinclair and Guarente 1997). Cells lacking *sgs1* have also been shown to possess a defect in telomere maintenance and to lose telomeric DNA at a faster rate than wild-type cells (Cohen and Sinclair 2001; Huang et al. 2001; Johnson et al. 2001), although the telomere defect is not thought to limit replicative life span.

The human SGS1 homologs are genes for Bloom syndrome (BLM) and Werner's syndrome (WRN). Werner's syndrome is a disease with manifes-

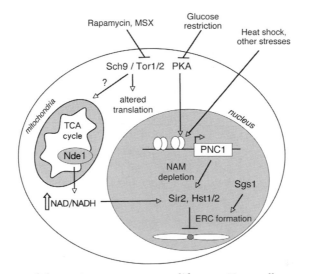

Figure 2. Impact of the environment on yeast life span. Yeast cells possess an elaborate system to sense the environment and, when times become harsh, increase defenses. These defenses against adversity are thought to underlie the life span extension seen during nutrient restriction, not just in yeast, but in most eukaryotes, including mammals. There are two main pathways that regulate yeast replicative life span: *TOR/SCH9* signaling and *PNC1/SIR2*. The extent of overlap between these pathways is the subject of ongoing research. Whether *TOR* and *SCH9* regulate life span by altering mitochondrial function, the NAD/NADH ratio, the expression of *PNC1*, a regulator of *SIR2*-mediated ERC suppression, or by other means, remains to be determined.

tations resembling premature aging (Lebel 2001). Cells cultured from Werner's patients exhibit reduced division potential (Salk et al. 1985), suggesting a possible link between replicative potential and organismal life span. Interestingly, cells from WRN knockout mice and WRN patients have recently been shown to possess telomere maintenance defects, paralleling the yeast *sgs1* mutants (Du et al. 2004; Crabbe et al. 2007). The relevance of this finding to the human condition is not yet known, but it is tempting to speculate that it underlies aspects of their premature aging.

The LAG genes. *LAG1* and *LAG2* (longevity assurance gene) were identified as genes that are preferentially expressed in young cells (Egilmez et al. 1989) and have been shown to have a role in determining yeast longevity (D'Mello et al. 1994; Childress et al. 1996). Deletion of *LAG1* or *LAG2* shortens life span, whereas overexpression of *LAG1* or *LAG2* extends life span. Lag1 and its close homolog Lac1 have been found to be essential components of ceramide synthase (Guillas et al. 2001; Schorling et al. 2001), suggesting that ceramide signaling/production has a role in longevity regulation (Jiang et al. 2004). Human *LAG1* homologs have been identified and shown to also function in ceramide synthesis (Guillas et al. 2003; Riebeling et al. 2003). It is currently unknown whether *LAG1* or ceramide signaling affects longevity in mammals. The mechanism by which *LAG1/LAG2*/ceramide signaling extends yeast life span also remains unclear.

The cAMP/protein kinase A pathway. One of the first signaling pathways implicated in regulation of longevity in yeast is the Ras/cAMP/protein kinase A (cAMP/PKA) pathway (Sun et al. 1994; Lin et al. 2000; Fabrizio et al. 2001) that links glucose availability with the control of growth, proliferation, metabolism, stress resistance, and longevity (Thevelein and de Winde 1999; Guarente and Kenyon 2000). Multiple signals are required for the activation of the PKA pathway: The signal for glucose availability transduced through the G-protein-coupled receptor Gpr1 and the Gα protein Gpa2 and a glucose phosphorylation event by glucokinase (Glk1) or hexokinase (Hxk1 or 2) are required for adenylate cyclase Cyr1 (Cdc35) activation. Upon stimulation by the GTP/GDP-binding Ras proteins (Ras1, Ras2), Cyr1 produces cAMP, which activates PKA by promoting the dissociation of the regulatory subunits (Bcy1) from the catalytic subunits Tpk1, 2, and 3 (Broach 1991; Thevelein and de Winde 1999; Rolland et al. 2002). Mutations in components of this pathway, which decrease the PKA activity, extend yeast life span in multiple strain backgrounds (Sun et al. 1994; Lin et al. 2000, 2002, 2004; Fabrizio et al. 2001; Kaeberlein et al. 2004; Lamming et al. 2005).

Sch9 is a serine/threonine protein kinase that has homology with yeast PKA and mammalian protein kinase B (PKB) (the Akt kinase family) (Toda et al. 1988; Kandel and Hay 1999). Deletion of *SCH9* leads to life span extension (Kaeberlein et al. 2005b). In addition to longevity regulation, Sch9 is involved in the control of cell size and oxidative stress resistance (Fabrizio et al. 2001; Jorgensen et al. 2002, 2004). Sch9 is also a central component of the "fermentable growth medium" (FGM)-induced pathway, which monitors the presence of fermentable sugars and essential nutrients in the growth media (Crauwels et al. 1997; Thevelein et al. 2000). The Sch9 homolog, PKB, also regulates life span as part of the insulin/PI3K (phosphatidylinositol-3 kinase) signaling pathway in multicellular organisms (Barbieri et al. 2003; Bartke et al. 2003; Fulop et al. 2003; Longo and Finch 2003; Kenyon 2005).

Detailed mechanisms underlying life span extension in *sch9Δ* and low PKA activity mutants remain to be elucidated. One mechanism suggested is that increased rDNA stability leads to life span extension. Many of these mutants show a lower rate of rDNA recombination and rDNA circle accumulation (Defossez et al. 1999; Lin et al. 2000; Prusty and Keil 2004). In support for the rDNA circle model, deletion of the *FOB1* gene, which encodes an rDNA-specific replication fork-block protein, decreases rDNA recombination and extends life span (Defossez et al. 1999).

The RTG genes. The retrograde response, a signaling pathway respond- ing to the functional state of mitochondria through changes in nuclear gene expression (Butow and Avadhani 2004; Jazwinski 2005), also affects life span (Kirchman et al. 1999; Jiang et al. 2000; Borghouts et al. 2004). Key components of the retrograde response are the transcription factors, Rtg1 and Rtg3, and their upstream activator Rtg2, a cytoplasmic Hsp70- type ATP-binding protein (Butow and Avadhani 2004; Jazwinski 2005). The precise role for RTG genes in yeast longevity is somewhat elusive— on the one hand, functional retrograde response is required for increased life span in respiratory-deficient petite cells. On the other hand, deletion of *RTG3* leads to an increased life span (Kirchman et al. 1999; Jiang et al. 2000; Borghouts et al. 2004). Deletion of the histone acetyltransferase gene *GCN5* suppresses life span extension by induction of the retrograde response (Kim et al. 2004). Rtg2 and Gcn5 are components of the SAGA- like (SLIK) histone acetyltransferase, transcriptional coactivator complex that potentiates the activation of retrograde responsive genes. It is there- fore suggested that the retrograde response through SLIK links metabo- lism, gene regulation, and genome stability in yeast aging (Jazwinski 2005). Paradoxically, induction of the retrograde response leads to pro- duction of rDNA circles. One explanation put forward is that although induction of the retrograde response compensates for age-associated mitochondrial dysfunction, it also progressively monopolizes the activity of Rtg2 protein, and thus Rtg2 is not available to repress genome insta- bility at the rDNA repeats (Borghouts et al. 2004; Jazwinski 2005).

The Sir (silencing regulator) genes and genes regulating rDNA silencing. In yeast, members of the Sir family of proteins mediate transcriptional silencing at telomeres, at the two silent mating-type loci (*HML* and *HMR*) and at the rDNA (*RDN1*) (Loo and Rine 1995; Moazed 2001). Each locus has a unique set of DNA-binding proteins that recruit spe- cific Sir proteins to mediate transcriptional silencing. *SIR4* was first iden- tified as a longevity factor in a genetic screen for starvation-resistant mutants (Kennedy et al. 1995). The *SIR4-42* semidominant mutant allele encodes truncated Sir4 proteins, leading to increased rDNA silencing and life span extension (Kennedy et al. 1995, 1997). Two additional longevity genes, *UTH1* and *UTH4*, were identified in the same screen. *UTH1* encodes a SUN domain protein whose deletion results in a global increase in silencing and life span extension (Austriaco and Guarente 1997). *UTH4* encodes a *Drosophila* Pumilio homolog (Edwards et al. 2000) that affects the distribution of the Sir complex between rDNA and telomere (Kennedy et al. 1997). Deleting *UTH4* decreases rDNA silencing and life span, whereas overexpressing *UTH4* increases rDNA silencing and life span (Kennedy et al. 1997).

Sir2 exhibits an NAD-dependent histone deacetylase activity that is conserved among the Sir2 family members including Hst2 (Homolog of Sir Two) in yeast and sirtuins (silent mating-type information regulation 2 homolog) in higher eukaryotes (Imai et al. 2000; Landry et al. 2000; Smith et al. 2000). Overexpressing *SIR2* extends life span, whereas deleting *SIR2* reduces life span (Kaeberlein et al. 1999). *HST2* also regulates yeast life span in a dose-dependent manner (Lamming et al. 2005). The longevity regulating activity of Sir2/Hst2 involves suppressing recombination among the rDNA repeats, thereby decreasing the levels of toxic rDNA circles (Sinclair and Guarente 1997; Guarente 2000; Lamming et al. 2005). Sir2 may also extend life span by increasing the fitness of newborn cells (Aguilaniu et al. 2003). It has been shown that oxidatively damaged proteins were retained in mother cells and not inherited by daughter cells during cytokinesis. Interestingly, this asymmetric partitioning of oxidatively damaged proteins is lost in the *sir2Δ* mutants, suggesting that a Sir2-dependent free-radical defense mechanism may also contribute to life span regulation. Overexpressing Sir2 homologs in worms (*sir-2.1*) (Tissenbaum and Guarente 2001) and flies (*dSir2*) also extends life span (Rogina and Helfand 2004). Although it is still unknown whether the Sir2 family has an important role in mammalian aging, studies reveal that the targets of mammalian Sir2 proteins are involved in regulating cell survival under stress through controlling the acetylation status of both the tumor suppressor gene p53 (Luo et al. 2001; Vaziri et al. 2001) and the FOXO transcription factors (Brunet et al. 2004; Daitoku et al. 2004; Motta et al. 2004).

Genes involved in NAD biosynthesis also affect life span. Overexpressing components of the NAD salvage pathway such as Npt1 (nicotinamide phosphoribosyl transferase) and Pnc1 (nicotinamide deamidase) extend life span (Anderson et al. 2002). It is suggested that these genes extend life span by increasing Sir2 activity, thereby increasing rDNA silencing and stability (Anderson et al. 2002).

Deleting the gene encoding Rpd3, a histone deacetylase whose activity is independent of NAD, causes an increase in rDNA silencing and extends life span (Kim et al. 1999). The *rpd3*Δ mutation rescues the silencing defect of the *sir2*Δ mutant, yet its life span extension requires a functional Sir2. It is suggested that Rpd3 and Sir2 regulate life span by different mechanisms (Kim et al. 1999; Jiang et al. 2002); however, the precise role for Rpd3 in life span regulation remains unclear. In flies, deleting the Rpd3 homolog also extends life span, and this life span extension requires dSir2, placing Rpd3 upstream of dSir2.

The phosphorylation status of Sir3 is regulated by the Slt2/MAPK (mitogen-activated protein) pathway, which affects the distribution of silencing factors at the telomere, mating-type loci, and rDNA (Ai et al. 2002; Ray et al. 2003). Elimination of the Sir3 phosphorylation site at Ser-275 extends life span (Ray et al. 2003). The silencing factors Zds1 and Zds2 have been reported to affect Sir3 phosphorylation and life span (Roy and Runge 2000). Deletion of *ZDS1* results in an increase in rDNA silencing, Sir3 phosphorylation, and life span, whereas deletion of *ZDS2* causes opposite effects (Roy and Runge 2000). Interestingly, Sir3 appears to regulate life span independent of rDNA silencing/recombination (Ray et al. 2003). The details of the mechanism by which Sir3 affects life span remain to be determined.

The Snf1 complex. Snf1 is a heterotrimeric complex composed of a catalytic subunit (Snf1) that phosphorylates target proteins at serine/threonine residues, an activating subunit (Snf4), and an inhibitory subunit (Sip1, Sip2, or Gal83) (Hardie et al. 1998; Ashrafi et al. 2000). Snf1 is the homolog of mammalian AMP-activated protein kinase (AMPK) that is involved in regulating cellular stress responses and energy sensing (Hardie et al. 1998). In yeast, the Snf1 complex responds to glucose starvation by catalyzing phosphorylation of a number of target proteins, including transcriptional regulators of genes involved in alternative carbon source utilization, gluconeogenesis, and respiration (Hardie et al. 1998). Forced expression of *SNF1* or deletion of *SIP2* shortens life span, whereas deletion of *SNF4* extends life span (Ashrafi et al. 2000). Snf1 has also been shown to exhibit an age-associated increase in histone H3 kinase activity, which is associated with increased rDNA recombination (Lin et al. 2003). Interestingly, overexpressing the Snf1 homolog, AAK-2, in worms extends life span (Apfeld et al. 2004), raising the possibility that AMPKs may regulate life span through a different mechanism in higher eukaryotes.

Tor1. Tor1 is a highly conserved serine/threonine protein kinase belonging to the phosphatidylinositol kinase-related kinase family (PIKK)

(Schmelzle and Hall 2000; Jacinto and Hall 2003). Tor1 is also the specific cellular target of rapamycin, an immunosuppressant and antiproliferative drug (Schmelzle and Hall 2000). Reducing the TOR signaling activity extends life span in yeast (Kaeberlein et al. 2005b) and higher eukaryotes (Vellai et al. 2003; Jia et al. 2004; Kapahi et al. 2004; Meissner et al. 2004; Martin and Hall 2005). In yeast, treating cells with rapamycin, deleting *TOR1* or the downstream effectors/transcriptional targets involved in the utilization of alternative nitrogen sources, increases life span (Table 1) (Kaeberlein et al. 2005b). Although it is suggested that the TOR signaling affects life span by regulating ribosome biogenesis (Kaeberlein et al. 2005b), the precise role for the TOR signaling in longevity regulation is yet to be determined.

Chronological Aging

Among the first genes to be implicated in the chronological aging of yeast were *SOD1* and *SOD2*, which encode cytoplasmic and mitochondrial dismutases, respectively (Longo et al. 1996). It was found that long-term survival in stationary phase required the expression of each of these genes, sparking speculation that chronological death is the result of extensive free-radical damage. Indeed, it has been further demonstrated that overexpression of both *SOD1* and *SOD2* can extend chronological survival by 30% and that loss of mitochondrial function occurs just before the death of both wild type and *sod2* mutants (Longo et al. 1999; Fabrizio et al. 2001). Analysis of the *sod2* mutant led to the identification of aconitase as one of the primary mitochondrial targets of oxidative damage (Longo et al. 1999). Interestingly, aging fruit flies also show increased oxidation and inactivation of aconitase and, in *sod2* knockout mice, mitochondrial aconitase activity is 67% that of wild type in some tissues (Yan et al. 1997; Melov et al. 1998).

These results have given support to the free radical theory of aging, first proposed half a century ago by Harman (1956). Subsequent identification of other yeast genes involved in chronological aging, however, has clearly shown that survival in stationary phase requires more than just antioxidant protection. For example, whereas overexpression of *SOD1* and *SOD2* extends life span by 30%, mutation of specific signaling proteins discussed below can increase longevity by as much as threefold. There is no question, however, that free radicals have an important role in this process (see below). At the very least, chronologically aging yeast have proven to be a valuable model for the study of oxidative damage in the postmitotic tissues of higher eukaryotes (Longo et al. 1997; Matsuyama et al. 1998; Shaham et al. 1998).

The most notable of the above-mentioned discoveries came from a transposon-mediated mutagenesis screen performed by Fabrizio et al. (2001). In a search for mutants that were both long-lived and stress-resistant, the researchers isolated strains with transposon-insertions in two genes, SCH9 and CYR1 (the transposon was in the promoter region of SCH9). Loss-of-function mutations in these genes had dramatic effects on chronological longevity. Mutant sch9 and cyr1 cells survived approximately threefold and twofold longer than wild type, respectively. Interestingly, these two mutants not only had the greatest extension in chronological life span when compared with wild-type cells, but were also the only two genes isolated independently from both heat and oxidative stress selections. This supports a model whereby survival in stationary phase and extension of chronological life span depend on development of resistance to multiple forms of stress.

SCH9 encodes a serine/threonine protein kinase, whereas CYR1 encodes adenylate cyclase, required for stimulation of cAMP-dependent protein kinase (PKA) activity. These proteins appear to function in separate, yet parallel, signaling pathways, both of which mediate glucose/nutrient signaling, stimulate growth and glycolysis, and function to down-regulate stress resistance, glycogen accumulation, and gluconeogenesis (Lillie and Pringle 1980; Longo and Fabrizio 2002). Consistent with a role of cAMP signaling in regulation of chronological aging, deletion of the gene encoding the GTP-binding protein Ras2, an upstream regulator of Cyr1, doubles chronological life span (Fabrizio et al. 2003). Similar to the above mutants, long-lived ras2 cells also display multiple stress resistances, as they are thermotolerant and more resistant to oxidative damage. Furthermore, the life span extension of both cyr1 and ras2 strains requires the stress-response transcription factors Msn2 and Msn4. The life span extension in sch9 strains does not require these factors, however, and may act via the protein kinase Rim15 (Fabrizio et al. 2001). Rim15 in turn functions via the stress-response transcription factor Gis1 (Pedruzzi et al. 2000), which binds post-diauxic shift (PDS) elements found in the promoters of such genes as HSP26, HSP12, and SOD2. The above results again highlight the close association between stress-response pathways and chronological longevity.

In accordance with previous observations, it was also shown that the age-dependent inactivation of aconitase was significantly lower in both sch9 and cyr1 mutants (Fabrizio et al. 2001). This suggests that increased survival of the strains is due, at least in part, to increased protection from oxidative damage. This was further supported by the recent confirmation that Sod2 functions downstream from Sch9 (Fabrizio et al. 2003). Deletion of *SOD2* abolishes the life span extension of an *sch9* mutant, and furthermore, expression of endogenous *SOD2* was shown to be elevated in *sch9* cells. As was mentioned above, however, the limited survival obtained by overexpression of *SOD1/SOD2* indicates that these pathways likely regulate many downstream genes. Identification of these targets will be a major step toward understanding the relationship between stress resistance and chronological aging in this organism.

The kinase domain of Sch9 is 47% and 49% identical to that of *C. elegans* AKT-2 and AKT-1, respectively. AKT-1 and AKT-2 function in a longevity and diapause regulatory pathway downstream from the insulin receptor homolog DAF-2 (Guarente and Kenyon 2000; Kenyon 2001). Loss-of-function mutations in this insulin/insulin-like growth factor-1 (IGF-1) signaling pathway cause worms to enter a state of diapause called dauer, a state normally initiated in response to nutrient limitation or crowding. Weak mutations in this pathway, however, can extend the life span of the adult worm by as much as twofold (Johnson 1990; Kenyon et al. 1993). Similar to the yeast Ras/Cyr1/PKA and Sch9 pathways, the insulin/IGF-1 pathway of worms also acts to down-regulate stress resistance and the storage of reserve nutrients (Kenyon et al. 1993; Morris et al. 1996; Kimura et al. 1997). Thus, not only is conservation between specific factors involved, but a similar strategy for the regulation of longevity and stress resistance may be conserved as well.

These results have been extended to *Drosophila*, where it has been shown that mutation of components of the fly insulin/IGF-1 pathway can extend life span by 85% (Clancy et al. 2001; Tatar et al. 2001). These mutants also up-regulate nutrient storage and SOD expression, similar to yeast and worms. The similarity in strategies, pathways, and factors involved in such distantly related organisms has led to the proposal that this common longevity regulatory system arose early in evolution as a way to delay reproduction and increase chances of survival during times of nutrient limitation (Kenyon 2001). This degree of conservation ensures that the study of chronological aging in yeast will continue to shed light on the regulation of, and mechanisms underlying, numerous critical processes in higher eukaryotes.

In 2006, Kennedy and colleagues published the results of a genome-wide screen (~4800 mutants) for mutations that extend chronological life span (Powers et al. 2006). Of the approximately 90 longest-lived strains, 16 had mutations in genes that were implicated in TOR signaling and nutrient acquisition. These included *GLN3*, encoding a TOR-regulated transcription factor of nitrogen acquisition genes; *LYS12*, encoding an enzyme in lysine biosynthesis; *MEP3*, encoding an ammonium permease; and *AGP1*, encoding an amino acid permease. Deletion of the TOR-regulated transcription factor *GLN3* conferred the largest magnitude of life span increase (about twofold). Both deletion of *TOR1* and treatment of cells with the TOR inhibitors rapamycin or MSX extend chronological life span up to twofold (Powers et al. 2006), consistent with TOR signaling being a major antagonist of starvation-induced life span extension.

Interestingly, mutation of the PKA and TOR pathways extends both the chronological and replicative life spans of yeast (Lin et al. 2000), arguing for some degree of overlap between these two distinct measures of aging. It is worth noting that replicative life span extension can be observed in the absence of Msn2/Msn4 (Lin et al. 2000), so although there may be some commonality in the pathways involved, the downstream mechanisms regulating these two processes may be distinct.

Although it is known that inhibition of TOR by rapamycin leads to a transcriptional response reminiscent of nutrient deprivation, the precise mechanisms by which TOR inhibition extends replicative and chronological life span are unknown (Fig. 2). It may be due to increased respiration or to an alteration in protein synthesis. A recent study showed that treatment with rapamycin causes Tor1 to exit from the nucleus (Li et al. 2006), suggesting that TOR may be more intimately involved in nuclear processes than first thought. Clearly, further work will be required to elucidate the mechanisms by which TOR signaling limits life span.

IMPACT OF THE ENVIRONMENT

In addition to genetic manipulations, several environmental factors have been reported to extend life span in yeast. Mild heat stress extends yeast replicative life span (Shama et al. 1998; Anderson et al. 2003), and this life span extension requires the Ras proteins, the heat shock protein HSP104 (Shama et al. 1998), and Pnc1, a nicotinamide deamidase functioning in NAD biosynthesis (Anderson et al. 2003). It has also been shown that yeast cells grown in the presence of a high concentration of external osmolytes, such as sorbitol, have a longer life span. It is suggested that increased osmolarity extends life span by activating Hog1, leading to an increase in the biosynthesis of glycerol from glycolytic intermediates (Kaeberlein et al. 2002). Both heat- and high-osmolarity-induced longevity have been associated with an increase in NAD or the $NAD^+/NADH$ ratio (Kaeberlein et al. 2002; Anderson et al. 2003).

Calorie restriction (CR, a moderate reduction in calorie intake), also called dietary restriction (DR), is the most effective intervention known to extend life span in a variety of species, including mammals (for review, see Weindruch and Walford 1988; Roth et al. 2001). CR has also been shown to delay the onset or reduce the incidence of many age-related diseases, including cancer, diabetes, and cardiovascular disorders (Weindruch and Walford 1988; Roth et al. 2001; Bordone and Guarente 2005). Moderate CR can be imposed in yeast by reducing the glucose concentration from 2% to 0.5% in rich media (Lin et al. 2000, 2002, 2004; Kaeberlein et al. 2002; Anderson et al. 2003; Lamming et al. 2005). Under this CR condition, the growth rate remains robust, and yeast mother cells show an extended replicative life span of about 20–30%. Variations in CR protocols have been described where a further reduction in the glucose concentration and limitation of amino acids and other nutrients are utilized (Jiang et al. 2000, 2002; Kaeberlein et al. 2004). These CR regimens are more stringent and may also induce pathways that function in parallel to 0.5% glucose-mediated CR.

Genetic models of CR have also been identified and studied in multiple strain backgrounds (Lin et al. 2000, 2002, 2004; Fabrizio et al. 2001; Kaeberlein et al. 2004; Lamming et al. 2005). These CR mimics include a hexokinase mutant (*hxk2Δ*) and mutations that down-regulate the glucose-sensing cAMP/PKA pathway, and deletions of the glucose-sensing protein genes *Gpa2* and *Gpr1*. Additional CR genetic models, the *tor1Δ* and *sch9Δ* mutants, have recently been reported to extend yeast life span (Fabrizio et al. 2001; Kaeberlein et al. 2005b).

In yeast, CR requires NAD and Sir2 to extend life span (Lin et al. 2000). It is suggested that CR activates Sir2 to extend life span by increasing rDNA silencing, thereby decreasing the production of rDNA circles (Guarente 2000). The Sir2 family in higher eukaryotes also has an important role in CR-induced life-span extension (Lin et al. 2000, 2002; Rogina and Helfand 2004; Wood et al. 2004; Chen et al. 2005; Nisoli et al. 2005; Wang and Tissenbaum 2006). Two models for Sir2 activation have been proposed. One model suggests that CR activates Sir2 by increasing the $NAD^+/NADH$ ratio (Fig. 2). In yeast, CR induces an increase in intracellular $NAD^+/NADH$, ratios by decreasing the NADH levels (Lin et al. 2004). CR also induces a metabolic shift from fermentation to mitochondrial respiration (Lin et al. 2002). The fact that respiration produces NAD^+ from NADH, as well as the finding that NADH can function as a competitive inhibitor of Sir2 activity (Lin et al. 2004), suggests that an increase in the $NAD^+/NADH$, ratio activates Sir2 during CR. In support for this model, genetic manipulations that cause a decrease in

NADH levels are shown to increase Sir2 activity and extend life span (Lin et al. 2002, 2004). This model, however, is challenged by recent studies. One study suggests that CR does not require functional mitochondria to extend life span (Kaeberlein et al. 2005a). It is noteworthy that this study utilizes 0.05% glucose rather than a moderate CR condition—0.5% glucose. It is very likely that the discrepancies are due to different CR conditions (Kaeberlein et al. 2006a; Lin and Guarente 2006). Another study indicates that reported in vivo NADH levels are too low to inhibit Sir2 activity (Schmidt et al. 2004), suggesting that CR is unlikely to activate Sir2 by decreasing NADH levels. It is possible that intracellular compartmentalization of NAD^+ and NADH and specific protein–protein interactions create local high NAD^+/NADH ratios, thereby activating Sir2 in vivo. It is suggested that the affinity/sensitivity of Sir2 toward its substrates and inhibitors varied when Sir2 was in complex with different interacting partners (Tanny et al. 2004).

In addition to regulation by NAD^+/NADH, Sir2 activity is regulated by the concentration of a noncompetitive inhibitor, nicotinamide (NAM), a by-product in the Sir2-mediated deacetylation reaction (Bitterman et al. 2002). It is therefore suggested that CR extends life span by decreasing the level of NAM (Anderson et al. 2003). This model is further supported by the evidence that overexpressing nicotinamidase, Pnc1, increases yeast life span and suppresses the inhibitory effect of NAM on Sir2 deacetylase activity in vivo and in vitro (Anderson et al. 2003; Gallo et al. 2004). Pnc1 is essential for CR-mediated life span extension and is induced by CR and multiple stress conditions that extend longevity (Anderson et al. 2003). It is therefore suggested that Pnc1 is a master longevity regulator that responds to CR and multiple environmental stresses to activate Sir2, thereby extending life span (Anderson et al. 2003). These studies have supported the "Hormesis" theory of CR. In this model, CR acts as a nutritional stress and mobilizes various defense mechanisms to elicit a well-coordinated multilevel protection (Yu and Chung 2001; Masoro 2005; Sinclair 2005). Therefore, low-intensity stress induced by CR can evoke metabolic changes that lead to a stronger resistance to various forms of stress (Yu and Chung 2001; Masoro 2005; Sinclair 2005).

In yeast, a Sir2-independent CR pathway has recently been described (Kaeberlein et al. 2004). In this model, CR appears to extend life span in certain sir2Δ mutants when a second gene, FOB1, is deleted (Kaeberlein et al. 2004). Deletion of the FOB1 gene, which encodes an rDNA-specific replication fork-block protein, decreases rDNA recombination, thereby rescuing the short life span of the sir2Δ mutants to wild-type level

(Defossez et al. 1999). A *fob1Δ* mutation has been routinely introduced into the *sir2Δ* mutants to help reveal the true requirement of Sir2 in CR (Lin et al. 2000, 2002, 2004; Kaeberlein et al. 2004; Lamming et al. 2005). Although two Sir2 family members, Hst2 and Hst1, have been suggested to have a role in this Sir2-independent pathway under 0.5% glucose CR (Lamming et al. 2005), certain CR conditions such as 0.05% glucose appear to be totally independent of the Sir2 family (Kaeberlein et al. 2006b; Lamming et al. 2006). Therefore, different CR conditions are likely to activate different yet overlapping downstream targets to extend life span. Further studies are required to elucidate the components and interactions of the multiple pathways of CR.

SMALL MOLECULES THAT EXTEND LIFE SPAN

The existence of longevity pathways makes it feasible that small-molecule modulators that extend life span can be discovered in yeast. The fact that the longevity pathways are conserved also means that they might extend life span in higher organisms, even mammals. The number of compounds that extend yeast life span are surprisingly few, although this number is expected to grow substantially in coming years. Small molecules that extend yeast life span currently fall into two categories, TOR inhibitors and the sirtuin-activating compounds (STACs).

TOR inhibitors are interesting because they are the only compounds known to extend both replicative and chronological life span (Kaeberlein et al. 2005b; Powers et al. 2006). The most commonly used inhibitor of TOR is rapamycin, also known as sirolimus. Rapamycin is a small molecule (m.w. = 914) that was isolated in 1975 from *Streptomyces hygroscopicus*, a bacterial isolate from soil on Easter Island. In mammals, a rapamycin–FKBP12 complex inhibits the TOR pathway through direct binding to the mTOR Complex 1 (mTORC1). In yeast, TOR activity can also be decreased by treating cells with methionine sulfoximine (MSX), an inhibitor of glutamine synthetase. Direct inhibition of the TOR pathway by low doses of rapamycin and MSX extends chronological life span up to about 2-fold in a dose-responsive manner. Similarly, treatment of cells with MSX also increases chronological life span in a dose-dependent manner up to about 1.5-fold. Replicative life span extension by MSX is on the order of 20% (Kaeberlein et al. 2005b).

In yeast and mammals, rapamycin can be toxic at moderate doses, so it must be titrated to levels that provide the health benefits without causing untoward toxicity. In yeast replicative life span experiments, for example, care must be taken not to block cell division; therefore, doses of these compounds must be given in a narrow range (100 pg/ml to 1 ng/ml for rapamycin). Whether or not rapamycin or MSX can extend the life span of higher organisms is not known, but inhibition of TOR by genetic means does extend life span in both worms and flies (Vellai et al. 2003; Kapahi et al. 2004). Thus, if inhibition of TOR in mammals could be achieved without toxicity, there is a chance that life span could be extended.

The other classes of life-span-extending compounds, STACs, were first discovered in a screen of various small-molecule libraries for activators of SIRT1 deacetylase activity. The screen identified two polyphenolic compounds: piceatannol, a stilbene, and quercetin, a flavone (Howitz et al. 2003). Screening of additional polyphenols identified 15 additional SIRT1 activators, the most potent of which was resveratrol (3,5,4'-trihydroxystilbene). A second class of STACs, which includes the compound isonicotinamide, was developed by rational design by Sauve and Schramm (2003). These compounds work by interfering with NAM inhibition. These two classes of STACs have different structures, seem to bind to different places on the sirtuin surface, and alter activity via two distinct mechanisms. Polyphenolic STACs lower the K_m for the substrate, whereas NAM inhibitors increase V_{max}.

Structural features common to the polyphenolic STACs include two coplanar aromatic rings in a *trans* arrangement, with differing bridges connecting the rings (Fig. 3). STACs activate sirtuins by lowering the K_m values for the acetylated peptide substrate while having little or no effect on V_{max} (Howitz et al. 2003). This result suggests that resveratrol is a K-type allosteric effector of SIRT1 (Monod et al. 1965). Qualitatively similar effects on the K_m values of the yeast sirtuin, Hst2, can be achieved by deletion of autoinhibitory sequences in the amino terminus (Zhao et al. 2004). On the basis of these Hst2 structural and kinetic studies, Zhao et al. (2004) proposed that a polyphenol-binding-induced reconfiguration of the conserved β1-α2 loop and/or zinc-binding domain could act to enhance sirtuin substrate binding.

STACs have thus far been shown to extend life span in *S. cerevisiae*, *D. melanogaster*, and *C. elegans* (Howitz et al. 2003; Wood et al. 2004). In each of these model systems, the STAC-induced life span extension required the presence of a functional Sir2/SIRT1 ortholog, and in each case, recombinant preparations of these enzymes were activated by STACs in vitro (Howitz et al. 2003; Wood et al. 2004). In the two organisms for which this was tested, *S. cerevisiae* and *Drosophila*, STACs did not increase the life span extension provided by CR (Howitz et al. 2003; Wood et al. 2004), which is consistent with the idea that STACs act to extend life span

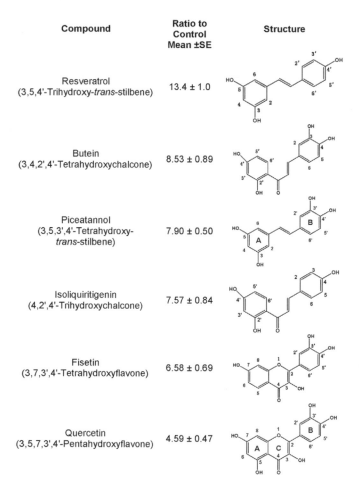

Compound	Ratio to Control Mean ±SE	Structure
Resveratrol (3,5,4'-Trihydroxy-*trans*-stilbene)	13.4 ± 1.0	
Butein (3,4,2',4'-Tetrahydroxychalcone)	8.53 ± 0.89	
Piceatannol (3,5,3',4'-Tetrahydroxy-*trans*-stilbene)	7.90 ± 0.50	
Isoliquiritigenin (4,2',4'-Trihydroxychalcone)	7.57 ± 0.84	
Fisetin (3,7,3',4'-Tetrahydroxyflavone)	6.58 ± 0.69	
Quercetin (3,5,7,3',4'-Pentahydroxyflavone)	4.59 ± 0.47	

All ratio data were calculated from experiments in which the total deacetylation in the control

reaction was 0.25-1.25 µM peptide or 1-5% of the initial concentration of acetylated peptide.

Figure 3. Stimulation of SIRT1 catalytic rate by plant polyphenols (100 µM). Structures of polyphenolic sirtuin-activating compounds (STACs). There are two families of compounds that extend yeast life span: TOR inhibitors and STACs. Shown are representatives of STACs from the silbene, chalcone, and flavone class, which are produced by plants in response to stress and may modulate plant defenses. The compounds were discovered in a screen of molecules that could activate the human SIRT1 enzyme. Resveratrol, a potent natural STAC, extends the life span of yeast, worms, and flies in a *SIR2*-dependent manner. (SE) Standard error of the mean. All ratio data were calculated from experiments in which the total deacetylation in the control reaction was 0.25–1.25 µM peptide or 1–5% of the initial concentration of acetylated peptide. (Reprinted, with permission, from Howitz et al. 2003 [Nature Publishing Group].)

by the same pathway as CR, with direct sirtuin stimulation providing the most straightforward mechanistic explanation.

Two recent papers reported that resveratrol shifts the metabolism of mice toward those on a lower-calorie diet (Baur et al. 2006; Lagouge et al. 2006). Resveratrol-fed mice had a gene expression pattern closely resembling that of a mouse on a lean or CR diet, along with decreased acetylation of PGC-1α, indicating increased SIRT1 activity. The mice not only showed altered physiology, but were also able to perform physical tasks better than the obese controls, including a two-fold increase in endurance on a treadmill. In cell culture experiments, the effects of resveratrol are diminished by knocking down SIRT1, indicating that at least some of the metabolic changes were mediated by this pathway. A more convincing series of experiments requires that SIRT1 knockout mice be treated with resveratrol.

PERSPECTIVE

Although our understanding of yeast aging has come far, there are still many unresolved questions. Are the major longevity pathways linked or separate? How much overlap is there between pathways that regulate replicative and chronological life span? How does TOR signaling extend life span? By altering protein synthesis? By reducing ERCs? How many of the 40 known longevity genes will extend life span in mammals? While we ponder such questions, it is worth considering how far the field has come in the past 10 years. It has moved from knowing almost nothing about aging in this organism and facing skepticism, to finding dozens of longevity genes and small molecules, many of which extend life span in yeast and higher organisms, possibly even mammals. The reason that researchers are able to make such broadly applicable discoveries in yeast seems to date back a billion years, when the first "longevity genes" evolved in eukaryotes to promote survival during harsh times. Perhaps we will learn to harness these pathways to treat diseases of aging such as diabetes and atherosclerosis. If that occurs, it would be the greatest gift from a fungus to humanity since antibiotics.

ACKNOWLEDGMENTS

We are grateful to the researchers whose work provided the basis for this review. We also thank members of the Lin laboratory and the Sinclair laboratory for discussions and suggestions. D.S. is a consultant and board member of Sirtris Pharmaceuticals.

REFERENCES

Aguilaniu H., Gustafsson L., Rigoulet M., and Nystrom T. 2003. Asymmetric inheritance of oxidatively damaged proteins during cytokinesis. *Science* **299:** 1751–1753.

Ai W., Bertram P.G., Tsang C.K., Chan T.F., and Zheng X.F. 2002. Regulation of subtelomeric silencing during stress response. *Mol. Cell* **10:** 1295–1305.

Anderson R.M., Bitterman K.J., Wood J.G., Medvedik O., and Sinclair D.A. 2003. Nicotinamide and Pnc1 govern life span extension by calorie restriction in *S. cerevisiae*. *Nature* **423:** 181–185.

Anderson R.M., Bitterman K.J., Wood J.G., Medvedik O., Cohen H., Lin S.S., Manchester J.K., Gordon J.I., and Sinclair D.A. 2002. Manipulation of a nuclear NAD+ salvage pathway delays aging without altering steady-state NAD+ levels. *J. Biol. Chem.* **277:** 18881–18890.

Apfeld J., O'Connor G., McDonagh T., DiStefano P.S., and Curtis R. 2004. The AMP-activated protein kinase AAK-2 links energy levels and insulin-like signals to lifespan in *C. elegans*. *Genes Dev.* **18:** 3004–3009.

Ashrafi K., Lin S.S., Manchester J.K., and Gordon J.I. 2000. Sip2p and its partner snf1p kinase affect aging in *S. cerevisiae*. *Genes Dev.* **14:** 1872–1885.

Austriaco N.R., Jr. and Guarente L.P. 1997. Changes of telomere length cause reciprocal changes in the lifespan of mother cells in *Saccharomyces cerevisiae*. *Proc. Natl. Acad. Sci.* **94:** 9768–9772.

Barbieri M., Bonafe M., Franceschi C., and Paolisso G. 2003. Insulin/IGF-I-signaling pathway: An evolutionarily conserved mechanism of longevity from yeast to humans. *Am. J. Physiol. Endocrinol. Metab.* **285:** E1064–E1071.

Barros M.H., Bandy B., Tahara E.B., and Kowaltowski A.J. 2004. Higher respiratory activity decreases mitochondrial reactive oxygen release and increases life span in *Saccharomyces cerevisiae*. *J. Biol. Chem.* **279:** 49883–49888.

Barton A. 1950. Some aspects of cell division in *Saccharomyces cerevisiae*. *J. Gen. Microbiol.* **4:** 84–86.

Bartke A., Chandrashekar V., Dominici F., Turyn D., Kinney B., Steger R., and Kopchick J.J. 2003. Insulin-like growth factor 1 (IGF-1) and aging: Controversies and new insights. *Biogerontology* **4:** 1–8.

Baur J.A., Pearson K.J., Price N.L., Jamieson H.A., Lerin C., Kalra A., Prabhu V.V., Allard J.S., Lopez-Lluch G., Lewis K., et al. 2006. Resveratrol improves health and survival of mice on a high-calorie diet. *Nature* **444:** 337–342.

Bitterman K.J., Medvedik O., and Sinclair D.A. 2003. Longevity regulation in *Saccharomyces cerevisiae*: Linking metabolism, genome stability, and heterochromatin. *Microbiol. Mol. Biol. Rev.* **67:** 376–399.

Bitterman K.J., Anderson R.M., Cohen H.Y., Latorre-Esteves M., and Sinclair D.A. 2002. Inhibition of silencing and accelerated aging by nicotinamide, a putative negative regulator of yeast sir2 and human SIRT1. *J. Biol. Chem.* **277:** 45099–45107.

Bjergbaek L., Cobb J.A., Tsai-Pflugfelder M., and Gasser S.M. 2005. Mechanistically distinct roles for Sgs1p in checkpoint activation and replication fork maintenance. *EMBO J.* **24:** 405–417.

Bordone L. and Guarente L. 2005. Calorie restriction, SIRT1 and metabolism: Understanding longevity. *Nat. Rev. Mol. Cell Biol.* **6:** 298–305.

Borghouts C., Benguria A., Wawryn J., and Jazwinski S.M. 2004. Rtg2 protein links metabolism and genome stability in yeast longevity. *Genetics* **166:** 765–777.

Broach J.R. 1991. Ras-regulated signaling processes in *Saccharomyces cerevisiae*. *Curr. Opin. Genet. Dev.* **1**: 370–377.

Brunet A., Sweeney L.B., Sturgill J.F., Chua K.F., Greer P.L., Lin Y., Tran H., Ross S.E., Mostoslavsky R., Cohen H.Y., et al. 2004. Stress-dependent regulation of FOXO transcription factors by the SIRT1 deacetylase. *Science* **303**: 2011–2015.

Bryk M., Banerjee M., Murphy M., Knudsen K.E., Garfinkel D.J., and Curcio M.J. 1997. Transcriptional silencing of Ty1 elements in the RDN1 locus of yeast. *Genes Dev.* **11**: 255–269.

Butow R.A. and Avadhani N.G. 2004. Mitochondrial signaling: The retrograde response. *Mol. Cell* **14**: 1–15.

Cabib E., Ulane R., and Bowers B. 1974. A molecular model for morphogenesis: The primary septum of yeast. *Curr. Top. Cell. Regul.* **8**: 1–32.

Chen D., Steele A.D., Lindquist S., and Guarente L. 2005. Increase in activity during calorie restriction requires Sirt1. *Science* **310**: 1641.

Childress A.M., Franklin D.S., Pinswasdi C., and Kale S. 1996. LAG2, a gene that determines yeast longevity. *Microbiology* **142**: 2289–2297.

Clancy D.J., Gems D., Harshman L.G., Oldham S., Stocker H., Hafen E., Leevers S.J., and Partridge L. 2001. Extension of life-span by loss of CHICO, a *Drosophila* insulin receptor substrate protein. *Science* **292**: 104–106.

Cockell M.M. and Gasser S.M. 1999. The nucleolus: Nucleolar space for RENT. *Curr. Biol.* **9**: R575–R576.

Cohen H. and Sinclair D.A. 2001. Recombination-mediated lengthening of terminal telomeric repeats requires the Sgs1 DNA helicase. *Proc. Natl. Acad. Sci.* **98**: 3174–3179.

Crabbe L., Jauch A., Naeger C.M., Holtgreve-Grez H., and Karlseder J. 2007. Telomere dysfunction as a cause of genomic instability in Werner syndrome. *Proc. Natl. Acad. Sci.* **104**: 2205–2210.

Crauwels M., Donaton M.C., Pernambuco M.B., Winderickx J., de Winde J.H., and Thevelein J.M. 1997. The Sch9 protein kinase in the yeast *Saccharomyces cerevisiae* controls cAPK activity and is required for nitrogen activation of the fermentable-growth-medium-induced (FGM) pathway. *Microbiology* **143**: 2627–2637.

Daitoku H., Hatta M., Matsuzaki H., Aratani S., Ohshima T., Miyagishi M., Nakajima T., and Fukamizu A. 2004. Silent information regulator 2 potentiates Foxo1-mediated transcription through its deacetylase activity. *Proc. Natl. Acad. Sci.* **101**: 10042–10047.

Defossez P.A., Park P.U., and Guarente L. 1998. Vicious circles: A mechanism for yeast aging. *Curr. Opin. Microbiol.* **1**: 707–711.

Defossez P.A., Prusty R., Kaeberlein M., Lin S.J., Ferrigno P., Silver P.A., Keil R.L., and Guarente L. 1999. Elimination of replication block protein Fob1 extends the life span of yeast mother cells. *Mol. Cell* **3**: 447–455.

D'Mello N.P., Childress A.M., Franklin D.S., Kale S.P., Pinswasdi C., and Jazwinski S.M. 1994. Cloning and characterization of LAG1, a longevity-assurance gene in yeast. *J. Biol. Chem.* **269**: 15451–15459.

Du X., Shen J., Kugan N., Furth E.E., Lombard D.B., Cheung C., Pak S., Luo G., Pignolo R.J., DePinho R.A., et al. 2004. Telomere shortening exposes functions for the mouse Werner and Bloom syndrome genes. *Mol. Cell. Biol.* **24**: 8437–8446.

Edwards T.A., Trincao J., Escalante C.R., Wharton R.P., and Aggarwal A.K. 2000. Crystallization and characterization of Pumilo: A novel RNA binding protein. *J. Struct. Biol.* **132**: 251–254.

Egilmez N.K., Chen J.B., and Jazwinski S.M. 1989. Specific alterations in transcript prevalence during the yeast life span. *J. Biol. Chem.* **264:** 14312–14317.

Fabrizio P., Pletcher S.D., Minois N., Vaupel J.W., and Longo V.D. 2004. Chronological aging-independent replicative life span regulation by Msn2/Msn4 and Sod2 in *Saccharomyces cerevisiae. FEBS Lett.* **557:** 136–142.

Fabrizio P., Gattazzo C., Battistella L., Wei M., Cheng C., McGrew K., and Longo V.D. 2005. Sir2 blocks extreme life-span extension. *Cell* **123:** 655–667.

Fabrizio P., Liou L.L., Moy V.N., Diaspro A., SelverstoneValentine J., Gralla E.B., and Longo V.D. 2003. SOD2 functions downstream of Sch9 to extend longevity in yeast. *Genetics* **163:** 35–46.

Fulop T., Larbi A., and Douziech N. 2003. Insulin receptor and ageing. *Pathol. Biol.* **51:** 574–580.

Gallo C.M., Smith D.L., Jr., and Smith J.S. 2004. Nicotinamide clearance by Pnc1 directly regulates Sir2-mediated silencing and longevity. *Mol. Cell. Biol.* **24:** 1301–1312.

Gangloff S., McDonald J.P., Bendixen C., Arthur L., and Rothstein R. 1994. The yeast type I topoisomerase Top3 interacts with Sgs1, a DNA helicase homolog: A potential eukaryotic reverse gyrase. *Mol. Cell. Biol.* **14:** 8391–8398.

Ghidelli S.D., Donze D., Dhillon N., and Kamakaka R.T. 2001. Sir2p exists in two nucleosome-binding complexes with distinct deacetylase activities. *EMBO J.* **20:** 4522–4535.

Gottschling D.E., Aparicio O.M., Billington B.L., and Zakian V.A. 1990. Position effect at *S. cerevisiae* telomeres: Reversible repression of Pol II transcription. *Cell* **63:** 751–762.

Guarente L. 2000. Sir2 links chromatin silencing, metabolism, and aging. *Genes Dev.* **14:** 1021–1026.

Guarente L. and Kenyon C. 2000. Genetic pathways that regulate ageing in model organisms. *Nature* **408:** 255–262.

Guillas I., Kirchman P.A., Chuard R., Pfefferli M., Jiang J.C., Jazwinski S.M., and Conzelmann A. 2001. C26-CoA-dependent ceramide synthesis of *Saccharomyces cerevisiae* is operated by Lag1p and Lac1p. *EMBO J.* **20:** 2655–2665.

Guillas I., Jiang J.C., Vionnet C., Roubaty C., Uldry D., Chuard R., Wang J., Jazwinski S.M., and Conzelmann A. 2003. Human homologues of LAG1 reconstitute Acyl-CoA-dependent ceramide synthesis in yeast. *J. Biol. Chem.* **278:** 37083–37091.

Hardie D.G., Carling D., and Carlson M. 1998. The AMP-activated/SNF1 protein kinase subfamily: Metabolic sensors of the eukaryotic cell? *Annu. Rev. Biochem.* **67:** 821–855.

Harman D. 1956. Aging: A theory based on free radical and radiation chemistry. *J. Gerontol.* **11:** 298–304.

Harris N., MacLean M., Hatzianthis K., Panaretou B., and Piper P.W. 2001. Increasing *Saccharomyces cerevisiae* stress resistance, through the overactivation of the heat shock response resulting from defects in the Hsp90 chaperone, does not extend replicative life span but can be associated with slower chronological ageing of nondividing cells. *Mol. Genet. Genomics* **265:** 258–263.

Hecht A., Strahl-Bolsinger S., and Grunstein M. 1996. Spreading of transcriptional repressor SIR3 from telomeric heterochromatin. *Nature* **383:** 92–96.

Howitz K.T., Bitterman K.J., Cohen H.Y., Lamming D.W., Lavu S., Wood J.G., Zipkin R.E., Chung P., Kisielewski A., Zhang L.L., et al. 2003. Small molecule activators of sirtuins extend *Saccharomyces cerevisiae* lifespan. *Nature* **425:** 191–196.

Huang P., Pryde F.E., Lester D., Maddison R.L., Borts R.H., Hickson I.D., and Louis E.J. 2001. SGS1 is required for telomere elongation in the absence of telomerase. *Curr. Biol.* **11:** 125–129.

Imai S., Armstrong C.M., Kaeberlein M., and Guarente L. 2000. Transcriptional silencing and longevity protein Sir2 is an NAD-dependent histone deacetylase. *Nature* **403:** 795–800.

Ira G., Malkova A., Liberi G., Foiani M., and Haber J.E. 2003. Srs2 and Sgs1-Top3 suppress crossovers during double-strand break repair in yeast. *Cell* **115:** 401–411.

Jacinto E. and Hall M.N. 2003. Tor signalling in bugs, brain and brawn. *Nat. Rev. Mol. Cell Biol.* **4:** 117–126.

Jazwinski S.M. 2000a. Metabolic control and ageing. *Trends Genet.* **16:** 506–511.

———. 2000b. Metabolic control and gene dysregulation in yeast aging. *Ann. N.Y. Acad. Sci.* **908:** 21–30.

———. 2005. The retrograde response links metabolism with stress responses, chromatin-dependent gene activation, and genome stability in yeast aging. *Gene* **354:** 22–27.

Jazwinski S.M., Chen J.B., and Sun J. 1993. A single gene change can extend yeast life span: The role of Ras in cellular senescence. *Adv. Exp. Med. Biol.* **330:** 45–53.

Jazwinski S.M., Egilmez N.K., and Chen J.B. 1989. Replication control and cellular life span. *Exp. Gerontol.* **24:** 423–436.

Jia K., Chen D., and Riddle D.L. 2004. The TOR pathway interacts with the insulin signaling pathway to regulate *C. elegans* larval development, metabolism and life span. *Development* **131:** 3897–3906.

Jiang J.C., Jaruga E., Repnevskaya M.V., and Jazwinski S.M. 2000. An intervention resembling caloric restriction prolongs life span and retards aging in yeast. *FASEB J.* **14:** 2135–2137.

Jiang J.C., Kirchman P.A., Allen M., and Jazwinski S.M. 2004. Suppressor analysis points to the subtle role of the LAG1 ceramide synthase gene in determining yeast longevity. *Exp. Gerontol.* **39:** 999–1009.

Jiang J.C., Wawryn J., Shantha Kumara H.M., and Jazwinski S.M. 2002. Distinct roles of processes modulated by histone deacetylases Rpd3p, Hda1p, and Sir2p in life extension by caloric restriction in yeast. *Exp. Gerontol.* **37:** 1023–1030.

Johnson F.B., Marciniak R.A., McVey M., Stewart S.A., Hahn W.C., and Guarente L. 2001. The *Saccharomyces cerevisiae* WRN homolog Sgs1p participates in telomere maintenance in cells lacking telomerase. *EMBO J.* **20:** 905–913.

Johnson T.E. 1990. Increased life-span of age-1 mutants in *Caenorhabditis elegans* and lower Gompertz rate of aging. *Science* **249:** 908–912.

Johnston J.R. 1966. Reproductive capacity and mode of death of yeast cells. *Antonie van Leeuwenhoek* **32:** 94–98.

Johzuka K. and Horiuchi T. 2002. Replication fork block protein, Fob1, acts as an rDNA region specific recombinator in *S. cerevisiae*. *Genes Cells* **7:** 99–113.

Jorgensen P., Nishikawa J.L., Breitkreutz B.J., and Tyers M. 2002. Systematic identification of pathways that couple cell growth and division in yeast. *Science* **297:** 395–400.

Jorgensen P., Rupes I., Sharom J.R., Schneper L., Broach J.R., and Tyers M. 2004. A dynamic transcriptional network communicates growth potential to ribosome synthesis and critical cell size. *Genes Dev.* **18:** 2491–2505.

Kaeberlein M., McVey M., and Guarente L. 1999. The SIR2/3/4 complex and SIR2 alone promote longevity in *Saccharomyces cerevisiae* by two different mechanisms. *Genes Dev.* **13:** 2570–2580.

Kaeberlein M., Andalis A.A., Fink G.R., and Guarente L. 2002. High osmolarity extends life span in *Saccharomyces cerevisiae* by a mechanism related to calorie restriction. *Mol. Cell. Biol.* **22:** 8056–8066.

Kaeberlein M., Kirkland K.T., Fields S., and Kennedy B.K. 2004. Sir2-independent life span extension by calorie restriction in yeast. *PLoS Biol.* **2:** E296.

Kaeberlein M., Hu D., Kerr E.O., Tsuchiya M., Westman E.A., Dang N., Fields S., and Kennedy B.K. 2005a. Increased life span due to calorie restriction in respiratory-deficient yeast. *PLoS Genet.* **1:** e69.

———. 2006a. Author's reply. *PLoS Genet.* **2:** e34.

Kaeberlein M., Steffen K.K., Hu D., Dang N., Kerr E.O., Tsuchiya M., Fields S., and Kennedy B.K. 2006b. Comment on "HST2 mediates SIR2-independent life-span extension by calorie restriction." *Science* **312:** 1312.

Kaeberlein M., Powers R.W., III, Steffen K.K., Westman E.A., Hu D., Dang N., Kerr E.O., Kirkland K.T., Fields S., and Kennedy B.K. 2005b. Regulation of yeast replicative life span by TOR and Sch9 in response to nutrients. *Science* **310:** 1193–1196.

Kandel E.S. and Hay N. 1999. The regulation and activities of the multifunctional serine/threonine kinase Akt/PKB. *Exp. Cell Res.* **253:** 210–229.

Kapahi P., Zid B.M., Harper T., Koslover D., Sapin V., and Benzer S. 2004. Regulation of lifespan in *Drosophila* by modulation of genes in the TOR signaling pathway. *Curr. Biol.* **14:** 885–890.

Kennedy B.K., Austriaco N.R., Jr., and Guarente L. 1994. Daughter cells of *Saccharomyces cerevisiae* from old mothers display a reduced life span. *J. Cell Biol.* **127:** 1985–1993.

Kennedy B.K., Austriaco N.R., Jr., Zhang J., and Guarente L. 1995. Mutation in the silencing gene SIR4 can delay aging in *S. cerevisiae*. *Cell* **80:** 485–496.

Kennedy B.K., Gotta M., Sinclair D.A., Mills K., McNabb D.S., Murthy M., Pak S.M., Laroche T., Gasser S.M., and Guarente L. 1997. Redistribution of silencing proteins from telomeres to the nucleolus is associated with extension of life span in *S. cerevisiae*. *Cell* **89:** 381–391.

Kenyon C. 2001. A conserved regulatory system for aging. *Cell* **105:** 165–168.

———. 2005. The plasticity of aging: Insights from long-lived mutants. *Cell* **120:** 449–460.

Kenyon C., Chang J., Gensch E., Rudner A., and Tabtiang R. 1993. A *C. elegans* mutant that lives twice as long as wild type. *Nature* **366:** 461–464.

Kim S., Benguria A., Lai C.Y., and Jazwinski S.M. 1999. Modulation of life-span by histone deacetylase genes in *Saccharomyces cerevisiae*. *Mol. Biol. Cell* **10:** 3125–3136.

Kim S., Ohkuni K., Couplan E., and Jazwinski S.M. 2004. The histone acetyltransferase GCN5 modulates the retrograde response and genome stability determining yeast longevity. *Biogerontology* **5:** 305–316.

Kimura K.D., Tissenbaum H.A., Liu Y., and Ruvkun G. 1997. daf-2, an insulin receptor-like gene that regulates longevity and diapause in *Caenorhabditis elegans*. *Science* **277:** 942–946.

Kirchman P.A., Kim S., Lai C.Y., and Jazwinski S.M. 1999. Interorganelle signaling is a determinant of longevity in *Saccharomyces cerevisiae*. *Genetics* **152:** 179–190.

Kirkwood T.B. and Holliday R. 1979. The evolution of ageing and longevity. *Proc. R. Soc. Lond. B Biol. Sci.* **205:** 531–546.

Kobayashi T., Heck D.J., Nomura M., and Horiuchi T. 1998. Expansion and contraction of ribosomal DNA repeats in *Saccharomyces cerevisiae*: Requirement of replication fork blocking (Fob1) protein and the role of RNA polymerase I. *Genes Dev.* **12:** 3821–3830.

Lagouge M., Argmann C., Gerhart-Hines Z., Meziane H., Lerin C., Daussin F., Messadeq N., Milne J., Lambert P., Elliott P., et al. 2006. Resveratrol improves mitochondrial function and protects against metabolic disease by activating SIRT1 and PGC–1alpha. *Cell* **127:** 1109–1122.

Lamming D.W., Latorre-Esteves M., Medvedik O., Wong S.N., Tsang F.A., Wang C., Lin S.J., and Sinclair D.A. 2005. HST2 mediates SIR2-independent life-span extension by calorie restriction. *Science* **309**: 1861–1864.

———. 2006. Response to comment on "HST2 mediates SIR2-independent life-span extension by calorie restriction". *Science* **312**: 1312c.

Landry J., Sutton A., Tafrov S.T., Heller R.C., Stebbins J., Pillus L., and Sternglanz R. 2000. The silencing protein SIR2 and its homologs are NAD-dependent protein deacetylases. *Proc. Natl. Acad. Sci.* **97**: 5807–5811.

Lebel M. 2001. Werner syndrome: Genetic and molecular basis of a premature aging disorder. *Cell. Mol. Life Sci.* **58**: 857–867.

Li H., Tsang C.K., Watkins M., Bertram P.G., and Zheng X.F. 2006. Nutrient regulates Tor1 nuclear localization and association with rDNA promoter. *Nature* **442**: 1058–1061.

Lillie S.H. and Pringle J.R. 1980. Reserve carbohydrate metabolism in *Saccharomyces cerevisiae*: Responses to nutrient limitation. *J. Bacteriol.* **143**: 1384–1394.

Lin S.J. and Guarente L. 2006. Increased life span due to calorie restriction in respiratory-deficient yeast. *PLoS Genet.* **2**: e33.

Lin S.J., Defossez P.A., and Guarente L. 2000. Requirement of NAD and SIR2 for life-span extension by calorie restriction in *Saccharomyces cerevisiae*. *Science* **289**: 2126–2128.

Lin S.J., Ford E., Haigis M., Liszt G., and Guarente L. 2004. Calorie restriction extends yeast life span by lowering the level of NADH. *Genes Dev.* **18**: 12–16.

Lin S.J., Kaeberlein M., Andalis A.A., Sturtz L.A., Defossez P.-A., Culotta V.C., Fink G.R., and Guarente L. 2002. Calorie restriction extends life span by shifting carbon toward respiration. *Nature* **418**: 344–348.

Lin S.S., Manchester J.K., and Gordon J.I. 2003. Sip2, an N-myristoylated beta subunit of Snf1 kinase, regulates aging in *Saccharomyces cerevisiae* by affecting cellular histone kinase activity, recombination at rDNA loci, and silencing. *J. Biol. Chem.* **278**: 13390–13397.

Lin Y.H. and Keil R.L. 1991. Mutations affecting RNA polymerase I-stimulated exchange and rDNA recombination in yeast. *Genetics* **127**: 31–38.

Longo V.D. and Fabrizio P. 2002. Regulation of longevity and stress resistance: A molecular strategy conserved from yeast to humans? *Cell. Mol. Life Sci.* **59**: 903–908.

Longo V.D. and Finch C.E. 2003. Evolutionary medicine: From dwarf model systems to healthy centenarians? *Science* **299**: 1342–1346.

Longo V.D., Gralla E.B., and Valentine J.S. 1996. Superoxide dismutase activity is essential for stationary phase survival in *Saccharomyces cerevisiae*. Mitochondrial production of toxic oxygen species in vivo. *J. Biol. Chem.* **271**: 12275–12280.

Longo V.D., Liou L.L., Valentine J.S., and Gralla E.B. 1999. Mitochondrial superoxide decreases yeast survival in stationary phase. *Arch. Biochem. Biophys.* **365**: 131–142.

Longo V.D., Ellerby L.M., Bredesen D.E., Valentine J.S., and Gralla E.B. 1997. Human Bcl-2 reverses survival defects in yeast lacking superoxide dismutase and delays death of wild-type yeast. *J. Cell Biol.* **137**: 1581–1588.

Loo S. and Rine J. 1995. Silencing and heritable domains of gene expression. *Annu. Rev. Cell Dev. Biol.* **11**: 519–548.

Luo J., Nikolaev A.Y., Imai S., Chen D., Su F., Shiloh A., Guarente L., and Gu W. 2001. Negative control of p53 by Sir2alpha promotes cell survival under stress. *Cell* **107**: 137–148.

Martin D.E. and Hall M.N. 2005. The expanding TOR signaling network. *Curr. Opin. Cell Biol.* **17**: 158–166.

Masoro E.J. 2005. Overview of caloric restriction and ageing. *Mech. Ageing Dev.* **126:** 913–922.

Matsuyama S., Xu Q., Velours J., and Reed J.C. 1998. The Mitochondrial F0F1-ATPase proton pump is required for function of the proapoptotic protein Bax in yeast and mammalian cells. *Mol. Cell* **1:** 327–336.

McMurray M.A. and Gottschling D.E. 2003. An age-induced switch to a hyper-recombinational state. *Science* **301:** 1908–1911.

Meissner B., Boll M., Daniel H., and Baumeister R. 2004. Deletion of the intestinal peptide transporter affects insulin and TOR signaling in *Caenorhabditis elegans. J. Biol. Chem.* **279:** 36739–36745.

Melese T. and Xue Z. 1995. The nucleolus: An organelle formed by the act of building a ribosome. *Curr. Opin. Cell Biol.* **7:** 319–324.

Melov S., Schneider J.A., Day B.J., Hinerfeld D., Coskun P., Mirra S.S., Crapo J.D., and Wallace D.C. 1998. A novel neurological phenotype in mice lacking mitochondrial manganese superoxide dismutase. *Nat. Genet.* **18:** 159–163.

Moazed D. 2001. Common themes in mechanisms of gene silencing. *Mol. Cell* **8:** 489–498.

Monod J., Wyman J., and Changeux J.-P. 1965. On the nature of allosteric transitions. *J. Mol. Biol.* **12:** 88–118.

Morris J.Z., Tissenbaum H.A., and Ruvkun G. 1996. A phosphatidylinositol-3-OH kinase family member regulating longevity and diapause in *Caenorhabditis elegans. Nature* **382:** 536–539.

Mortimer R.K. and Johnston J.R. 1959. Life span of individual yeast cells. *Nature* **183:** 1751–1752.

Motta M.C., Divecha N., Lemieux M., Kamel C., Chen D., Gu W., Bultsma Y., McBurney M., and Guarente L. 2004. Mammalian SIRT1 represses forkhead transcription factors. *Cell* **116:** 551–563.

Muller I. 1985. Parental age and the life-span of zygotes of *Saccharomyces cerevisiae. Antonie van Leeuwenhoek* **51:** 1–10.

Nisoli E., Tonello C., Cardile A., Cozzi V., Bracale R., Tedesco L., Falcone S., Valerio A., Cantoni O., Clementi E., et al. 2005. Calorie restriction promotes mitochondrial biogenesis by inducing the expression of eNOS. *Science* **310:** 314–317.

Pedruzzi I., Burckert N., Egger P., and De Virgilio C. 2000. *Saccharomyces cerevisiae* Ras/cAMP pathway controls post-diauxic shift element-dependent transcription through the zinc finger protein Gis1. *EMBO J.* **19:** 2569–2579.

Powers R.W., III, Kaeberlein M., Caldwell S.D., Kennedy B.K., and Fields S. 2006. Extension of chronological life span in yeast by decreased TOR pathway signaling. *Genes Dev.* **20:** 174–184.

Prusty R. and Keil R.L. 2004. SCH9, a putative protein kinase from *Saccharomyces cerevisiae*, affects HOT1-stimulated recombination. *Mol. Genet. Genomics* **272:** 264–274.

Ray A., Hector R.E., Roy N., Song J.H., Berkner K.L., and Runge K.W. 2003. Sir3p phosphorylation by the Slt2p pathway effects redistribution of silencing function and shortened lifespan. *Nat. Genet.* **33:** 522–526.

Riebeling C., Allegood J.C., Wang E., Merrill A.H., Jr., and Futerman A.H. 2003. Two mammalian longevity assurance gene (LAG1) family members, trh1 and trh4, regulate dihydroceramide synthesis using different fatty acyl–CoA donors. *J. Biol. Chem.* **278:** 43452–43459.

Rockmill B., Fung J.C., Branda S.S., and Roeder G.S. 2003. The Sgs1 helicase regulates chromosome synapsis and meiotic crossing over. *Curr. Biol.* **13:** 1954–1962.

Rogina B. and Helfand S.L. 2004. Sir2 mediates longevity in the fly through a pathway related to calorie restriction. *Proc. Natl. Acad. Sci.* **101:** 15998–16003.

Rogina B., Helfand S.L., and Frankel S. 2002. Longevity regulation by *Drosophila* Rpd3 deacetylase and caloric restriction. *Science* **298:** 1745.

Rolland F., Winderickx J., and Thevelein J.M. 2002. Glucose-sensing and -signalling mechanisms in yeast. *FEMS Yeast Res.* **2:** 183–201.

Roth G.S., Ingram D.K., and Lane M.A. 2001. Caloric restriction in primates and relevance to humans. *Ann. N.Y. Acad. Sci.* **928:** 305–315.

Roy N. and Runge K.W. 2000. Two paralogs involved in transcriptional silencing that antagonistically control yeast life span. *Curr. Biol.* **10:** 111–114.

Salk D., Bryant E., Hoehn H., Johnston P., and Martin G.M. 1985. Growth characteristics of Werner syndrome cells in vitro. *Adv. Exp. Med. Biol.* **190:** 305–311.

Sauve A.A. and Schramm V.L. 2003. Sir2 regulation by nicotinamide results from switching between base exchange and deacetylation chemistry. *Biochemistry* **42:** 9249–9256.

Schimmang T., Tollervey D., Kern H., Frank R., and Hurt E.C. 1989. A yeast nucleolar protein related to mammalian fibrillarin is associated with small nucleolar RNA and is essential for viability. *EMBO J.* **8:** 4015–4024.

Schmelzle T. and Hall M.N. 2000. TOR, a central controller of cell growth. *Cell* **103:** 253–262.

Schmidt M.T., Smith B.C., Jackson M.D., and Denu J.M. 2004. Coenzyme specificity of Sir2 protein deacetylases: Implications for physiological regulation. *J. Biol. Chem.* **279:** 40122–40129.

Schorling S., Vallee B., Barz W.P., Riezman H., and Oesterhelt D. 2001. Lag1p and Lac1p are essential for the Acyl-CoA-dependent ceramide synthase reaction in *Saccharomyces cerevisae*. *Mol. Biol. Cell* **12:** 3417–3427.

Shaham S., Shuman M.A., and Herskowitz I. 1998. Death-defying yeast identify novel apoptosis genes. *Cell* **92:** 425–427.

Shama S., Lai C.Y., Antoniazzi J.M., Jiang J.C., and Jazwinski S.M. 1998. Heat stress–induced life span extension in yeast. *Exp. Cell Res.* **245:** 379–388.

Shou W., Sakamoto K.M., Keener J., Morimoto K.W., Traverso E.E., Azzam R., Hoppe G.J., Feldman R.M.R., DeModena J., Moazed D., et al. 2001. Net1 stimulates RNA polymerase I transcription and regulates nucleolar structure independently of controlling mitotic exit. *Mol. Cell* **8:** 45–55.

Sinclair D.A. 2005. Toward a unified theory of caloric restriction and longevity regulation. *Mech. Ageing Dev.* **126:** 987–1002.

Sinclair D.A. and Guarente L. 1997. Extrachromosomal rDNA circles—A cause of aging in yeast. *Cell* **91:** 1033–1042.

Sinclair D.A., Mills K., and Guarente L. 1997. Accelerated aging and nucleolar fragmentation in yeast *sgs1* mutants. *Science* **277:** 1313–1316.

———. 1998. Molecular mechanisms of yeast aging. *Trends Biochem. Sci.* **23:** 131–134.

Smeal T., Claus J., Kennedy B., Cole F., and Guarente L. 1996. Loss of transcriptional silencing causes sterility in old mother cells of *S. cerevisiae*. *Cell* **84:** 633–642.

Smith J.S. and Boeke J.D. 1997. An unusual form of transcriptional silencing in yeast ribosomal DNA. *Genes Dev.* **11:** 241–254.

Smith J.S., Brachmann C.B., Celic I., Kenna M.A., Muhammad S., Starai V.J., Avalos J.L., Escalante-Semerena J.C., Grubmeyer C., Wolberger C., and Boeke J.D. 2000. A phylogenetically conserved NAD+-dependent protein deacetylase activity in the Sir2 protein family. *Proc. Natl. Acad. Sci.* **97:** 6658–6663.

Stephen D.W., Rivers S.L., and Jamieson D.J. 1995. The role of the YAP1 and YAP2 genes in the regulation of the adaptive oxidative stress responses of *Saccharomyces cerevisiae*. *Mol. Microbiol.* **16:** 415–423.

Strahl-Bolsinger S., Hecht A., Luo K., and Grunstein M. 1997. SIR2 and SIR4 interactions differ in core and extended telomeric heterochromatin in yeast. *Genes Dev.* **11:** 83–93.

Sun J., Kale S.P., Childress A.M., Pinswasdi C., and Jazwinski S.M. 1994. Divergent roles of RAS1 and RAS2 in yeast longevity. *J. Biol. Chem.* **269:** 18638–18645.

Tanny J.C., Kirkpatrick D.S., Gerber S.A., Gygi S.P., and Moazed D. 2004. Budding yeast silencing complexes and regulation of Sir2 activity by protein-protein interactions. *Mol. Cell. Biol.* **24:** 6931–6946.

Tatar M., Kopelman A., Epstein D., Tu M.P., Yin C.M., and Garofalo R.S. 2001. A mutant *Drosophila* insulin receptor homolog that extends life-span and impairs neuroen-docrine function. *Science* **292:** 107–110.

Thevelein J.M. and de Winde J.H. 1999. Novel sensing mechanisms and targets for the cAMP-protein kinase A pathway in the yeast *Saccharomyces cerevisiae*. *Mol. Microbiol.* **33:** 904–918.

Thevelein J.M., Cauwenberg L., Colombo S., de Winde J.H., Donation M., Dumortier F., Kraakman L., Lemaire K., Ma P., Nauwelaers D., et al. 2000. Nutrient-induced signal transduction through the protein kinase A pathway and its role in the control of metabolism, stress resistance, and growth in yeast. *Enzyme Microb. Technol.* **26:** 819–825.

Tissenbaum H.A. and Guarente L. 2001. Increased dosage of a sir-2 gene extends lifespan in *Caenorhabditis elegans*. *Nature* **410:** 227–230.

Toda T., Cameron S., Sass P., and Wigler M. 1988. SCH9, a gene of *Saccharomyces cerevisiae* that encodes a protein distinct from, but functionally and structurally related to, cAMP-dependent protein kinase catalytic subunits. *Genes Dev.* **2:** 517–527.

Vaziri H., Dessain S.K., Eaton E.N., Imai S.I., Frye R.A., Pandita T.K., Guarente L., and Weinberg R.A. 2001. hSIR2(SIRT1) functions as an NAD-dependent p53 deacetylase. *Cell* **107:** 149–159.

Vellai T., Takacs-Vellai K., Zhang Y., Kovacs A.L., Orosz L., and Muller F. 2003. Genetics: Influence of TOR kinase on lifespan in *C. elegans*. *Nature* **426:** 620.

Wang Y. and Tissenbaum H.A. 2006. Overlapping and distinct functions for a *Caenorhabditis elegans* SIR2 and DAF-16/FOXO. *Mech. Ageing Dev.* **127:** 48–56.

Weindruch R. and Walford R.L. 1988. *The retardation of aging and disease by dietary restriction*. C.C. Thomas, Springfield, Illinois.

Werner-Washburne M., Braun E.L., Crawford M.E., and Peck V.M. 1996. Stationary phase in *Saccharomyces cerevisiae*. *Mol. Microbiol.* **19:** 1159–1166.

Werner-Washburne M., Braun E., Johnston G.C., and Singer R.A. 1993. Stationary phase in the yeast *Saccharomyces cerevisiae*. *Microbiol. Rev.* **57:** 383–401.

Wood J.G., Rogina B., Lavu S., Howitz K., Helfand S.L., Tatar M., and Sinclair D. 2004. Sirtuin activators mimic caloric restriction and delay ageing in metazoans. *Nature* **430:** 686–689.

Yan L.J., Levine R.L., and Sohal R.S. 1997. Oxidative damage during aging targets mito-chondrial aconitase. *Proc. Natl. Acad. Sci.* **94:** 11168–11172.

Yu B.P. and Chung H.Y. 2001. Stress resistance by caloric restriction for longevity. *Ann. N.Y. Acad. Sci.* **928:** 39–47.

Zhao K., Chai X., and Marmorstein R. 2004. Structure and substrate binding properties of cobB, a Sir2 homolog protein deacetylase from *Escherichia coli*. *J. Mol. Biol.* **337:** 731–741.

18

Genetics of Exceptional Longevity

Thomas T. Perls

New England Centenarian Study
Geriatrics Section, Department of Medicine
Boston University School of Medicine
Boston Medical Center, Boston, Massachusetts 02118

Paola Sebastiani

Department of Biostatistics
Boston University School of Public Health
New England Centenarian Study
Boston, Massachusetts 02118

CENTENARIANS MAINTAIN THEIR GOOD HEALTH AND ABILITY to function independently often until the last few years of their lives (Hitt et al. 1999). Most subjects experience a decline in their cognitive function only in the last 3–5 years of their lifetime (Silver et al. 2001). Thus, these individuals are a human model not only of longevity, but also of the delay or escape of age-related disability and various age-related diseases.

THE INFLUENCE OF DEMOGRAPHIC SELECTION UPON THE EXCEPTIONAL LONGEVITY PHENOTYPE

The compression of disability of centenarians toward the end of their very long lives is likely due to the phenomenon of demographic selection. Demographic selection describes the dying off of individuals prone to diseases that typically peak in incidence in the sexagenarian

An expanded version of this chapter by Thomas T. Perls and Paola Sebastiani will be published in *Exceptional Longevity* (ed: T. Perls) by Johns Hopkins University Press, Baltimore, Maryland in 2008.

An abbreviated version of this chapter by Thomas T. Perls and Paola Sebastiani will be published in *The Molecular Biology of Aging* (ed. L. Guarente, L. Partridge, and D.C. Wallace) by the Cold Spring Harbor Laboratory Press, Cold Spring Harbor, New York (© 2008).

through octogenarian years, thus leaving behind a cohort of select survivors. The observation by some research groups that the incidence of Alzheimer's disease (AD) plateaus at very old ages is consistent with this phenomenon. Ritchie and Kildea (1995) performed a meta-analysis of nine epidemiological studies, finding that the rate of increase in dementia prevalence fell off among octogenarians and plateaued at approximately 40% at age 95. In a longitudinal study of older people living in Cache County, Utah in the United States, the incidence of both dementia and AD increased almost exponentially until ages 85–90 but declined after age 93 for men and age 97 for women (Tschanz et al. 2005). A genetic example of demographic selection is the decreased frequency of the apolipoprotein E-ε4 (ApoE-ε4) allotype in the oldest old. The allelic frequency of ApoE-ε4 drops off dramatically in the oldest age groups, presumably because of its association with AD and vascular disease (Kervinen et al. 1994; Louhija et al. 1994; Rebeck et al. 1994; Schachter et al. 1994; Sobel et al. 1995). Interestingly, the effect of the ApoE allotype upon AD incidence appears to decrease with age at these very old ages at least in Caucasians (Rebeck et al. 1994; Sobel et al. 1995).

EXCEPTIONAL LONGEVITY RUNS STRONGLY IN FAMILIES

A number of centenarian studies support the possibility of a substantial familial (genetic *and* exposure factors in common among family members) component to exceptional longevity (EL). In an analysis of 444 centenarian families in the United States that included 2092 siblings of centenarians, Perls et al. (2002) found, after accounting for race and education, that the net survival advantage of siblings of centenarians was 16 years greater than the average for their birth cohort. Figure 1 shows that from age 20 until age 100 years, the siblings of centenarians generally maintained half the mortality risk of their birth cohort. This year-to-year survival advantage translated into very high relative survival probabilities of living to age 100. Ultimately, the male siblings of centenarians had an 18 times greater risk of surviving to age 100 and the female siblings had an 8.5 times greater risk, compared with the general population born in 1900. A similar finding of reduced mortality among siblings of centenarians was found in a study of 348 Okinawan families (Willcox et al. 2006). From age 20, offspring of centenarians also appear to have a significant survival advantage compared with others in their birth cohort (Perls et al. 2000; Perls and Terry 2003a,b; Atzmon et al. 2004; Kemkes-Grottenthaler 2004; Terry et al. 2004).

The New England Centenarian Study (NECS) identified and enrolled members of six families who demonstrated exceptional clustering for

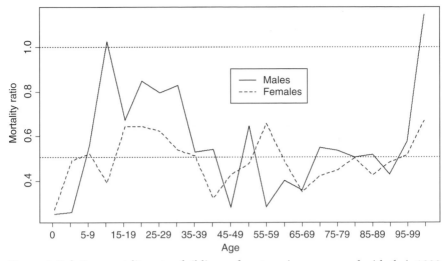

Figure 1. Relative mortality rate of siblings of centenarians compared with their 1900 birth cohort for ages 20 to 100 (Perls et al. 2002).

extreme old age. In one family, 5 of 7 siblings attained age 100 or older and 21 of 40 cousins were nonagenarians or older (Perls et al. 2000). The probability of encountering families such as these by chance is fewer than one per all the families that exist today. Thus, these family members must have factors in common with one another that facilitate such EL.

This observed familial aggregation for EL is likely due to both important environmental and genetic factors that family members have in common. These results do not help us discern what proportion of this aggregation is genetic versus environmental and behavioral, although as noted below, recent findings from the Scandinavian twin studies indicate a greater genetic role with extreme old age.

PHENOTYPIC AND GENETIC COMPLEXITY OF EXCEPTIONAL LONGEVITY

The search for the genetic and exposure bases of EL faces two levels of complexity: the definition of the phenotype and the many different paths to EL. EL can generally be defined as survival beyond a certain age or percentile of survival. There are likely many paths to achieving EL that differ according to secular trends in life expectancy, gender, social, environmental and behavioral factors, and their interactions with a spectrum of genetic variations. Such genetic variations might be grossly categorized

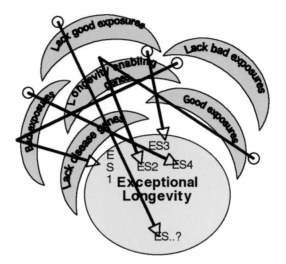

Figure 2. There are multiple paths (exceptional survival subphenotypes, ES1 . . . ES?) composed of different combinations of deleterious, neutral, and beneficial exposures and genetic variations leading to EL.

into those that facilitate longer survival, "longevity enabling genes," and those that predispose to disease, or "disease genes."

These paths can be categorized as exceptional survival (ES) phenotypes (Perls 2005). Definitions of ES might include, for example, disability-free survival past a specified age or disease-free survival past a specified age. Figure 2 illustrates in albeit a reductionist manner the complex interaction of disease predisposing and longevity-enabling exposures and genetic variations that lead to different patterns of ES and subsequently, EL.

For example, based on the occurrence of the most common diseases of aging, Evert et al. (2003) noted that centenarians in the NECS sample fit into one of three categories: survivors, delayers, or escapers. About 43% of the centenarians in the sample were survivors or those who lived with an age-related disease for 20 years or more. About 42% were delayers or those who delayed the onset of age-related disease until after age 80. The remaining 15% were escapers or those without demonstrable disease at age 100 (32% of the male and 15% of the female centenarians). These categories likely differ for a wide range of underlying factors including varying degrees of functional reserve or adaptive capacity. This underscores the need for well-characterized phenotypes that go beyond simple survival. In Figure 2, the categories of survivor, delayer, and escaper would be examples of ES phenotypes.

ROLES OF ENVIRONMENT AND GENES IN ACHIEVING EXTREME OLD AGE

There is substantial heterogeneity in how people age. Some individuals age relatively quickly and develop age-related illnesses such as heart disease or stroke in their 40s and 50s, whereas others appear to age very slowly, and if they do develop age-related illnesses, they might only do so toward the end of their very long lives. Within these two extremes are the majority of people who, at least in industrialized nations, have an average life expectancy in their late 70s. Much of the variation around this average appears to be due to differences in environment, behavior, and stochastic factors.

One approach to the question of the impact of environment and behavior on life expectancy is to study populations with relatively high average life expectancies and determine how they are different from other groups. Seventh Day Adventists have the highest mean life expectancy, of 88 years, in the United States. Given that the racial, ethnic, and geographic makeup of followers of this faith is heterogeneous, it appears that "much" of their increased life expectancy, over that of the general population, appears to be due to the healthy behaviors dictated by their religion: frugal vegetarian diet, regular daily exercise, no tobacco or alcohol use, and substantial time devoted to family and church activities (which might have roles in managing stress). On the basis of the Seventh Day Adventist data, it would seem that most Americans, if they were to emulate the habits of Adventists, should be able to achieve an average life expectancy of 88 years (Fraser and Shavlik 2001).

In another approach to the nature versus nurture question, twin studies have estimated the heritability of longevity to be from 15% to 35% (Herskind et al. 1996; Ljungquist et al. 1998; Gudmundsson et al. 2000). However, the oldest subjects in these studies were in their mid to late 80s and very few if any lived to extreme old age. Thus, about 25% of the variation in *average* life expectancy can be explained by genetic variation and the remainder (~75%), explained by variation in environment (Perls and Terry 2003a). More specifically addressing the genetic contribution to EL though, Kaare Christensen and colleagues recently analyzed survival data from Danish, Finnish, and Swedish twins born between 1870 and 1910 and found that the relative recurrence risk of reaching age 92 was 4.8 for monozygotic males compared to 1.8 for dizygotic males and 2.5 for monozygotic females versus 1.6 for dizygotic females (Hjelmborg et al. 2006). From these findings, the authors concluded that there may be an increasingly greater genetic component to survival at extreme ages.

MULTIFACTORIAL MODEL FOR EXCEPTIONAL LONGEVITY AND EXCEPTIONAL SURVIVAL PHENOTYPES

The fact that siblings maintain half the mortality risk of their birth cohort from age 20 to extreme age (see Fig. 2, above) supports a multifactorial model for achieving EL. For a simplistic example, sociodemographic advantages may have key roles at younger ages, whereas genetic advantages may distinguish the ability to go from old age to extreme old age. Undoubtedly, EL is much more complicated, with temporally overlapping roles for major genes, polygenic, environmental, and stochastic components. Such a scenario would be consistent with a *threshold model*, where predisposition for the EL trait (and ES phenotypes) can be measured on a quantitative scale. Figure 3 illustrates the generic threshold model proposed by Falconer (1965), where it is predicted that the proportion of affected relatives will be highest among the most severely affected individuals. In the case of EL, perhaps severity could be measured by additional years beyond a certain age (threshold) or in further years of delay in the case of ES phenotypes such as age of onset of disability or certain diseases.

Examples of phenotypes fitting the threshold model are early-onset breast cancer or AD, where relatives of patients who develop these diseases at unusually young ages are themselves at increased risk or liability. Thus, a 104-year-old's "liability" or predisposition for EL is further beyond the

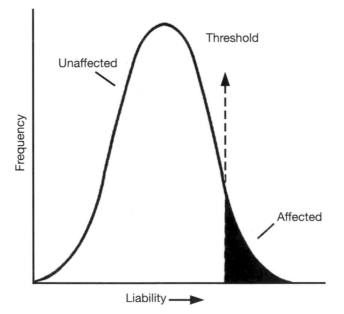

Figure 3. Threshold model of a multifactorial trait (Falconer 1965).

threshold than someone more mildly affected, e.g., a person who died at age 99 years. One interpretation of data indicating the higher relative survival probability of male siblings of centenarians compared to female siblings is that the males carry a higher liability for the trait given the presence of the requisite traits. The model predicts that if a multifactorial trait is more frequent in one sex (as is the case with EL which is predominantly represented by females), the liability will be higher for relatives of the less-"susceptible" sex (males, in the case of EL) (Farrer and Cupples 1998). Although the NECS has not yet looked at relative survival probability of siblings of male versus female probands (something that certainly needs to be done), these elevated risks for male versus female siblings are interesting in this context. The model also predicts that the risk for EL will be sharply lower for second-degree relatives compared to first-degree relatives, another observation we hope to test by having access to many expanded pedigrees. The ramifications of this model holding true for EL (and/or ES phenotypes) include the following.

- The older the subject, the better for discovering traits predisposing for EL.
- Gender-related differences exist in both relatives and probands in "liability" for EL given the presence of specific traits conducive to EL.

DISCOVERING GENETIC VARIATIONS THAT INFLUENCE EXCEPTIONAL LONGEVITY

In humans, the search for genes that influence EL have been for the most part based on a candidate gene approach, comparing frequencies of specific genetic variations or markers (single-nucleotide polymorphisms, SNPs) linked to those variations between long-lived individuals and various types of controls. Most positive findings have demonstrated only modest differences, thus supporting the "common diseases, common variants" hypothesis.

A SAMPLING OF GENES OF INTEREST

In studying genes that impact upon longevity, we can grossly divide these into genes found to influence the basic biology of aging, such as those that influence DNA repair and free-radical production, and genes that influence susceptibility to age-related diseases. However, age-related diseases are likely related to aging because the functional changes associated with aging at the DNA and cellular levels certainly have important roles in their pathogenesis. Thus, genes in the "basic biology of aging genes" are likely critical players in both aging and age-related diseases.

Although a number of basic biology of aging genes have been noted to have roles in human aging, few genes have been reported in the literature that substantially differentiate the ability to achieve exceptional old age. For many of us in the field, it is these longevity-enabling genes that are the Holy Grail.

Genes Influencing the Basic Biology of Aging

DNA Stability and Repair

Poly(ADP-ribose) polymerase (PARP): Oxidative and other sources of accumulated DNA damage likely contribute to aging. Defenses against this damage are obvious candidates for the genetic study of longevity and life span. As shown in Figure 4, Grube and Bürkle (1992) noted a strong correlation between species-specific life spans and their poly(ADP-ribose) polymerase (PARP) activities. This enzyme catalyzes poly(ADP-ribosyl)ation of nuclear proteins, thus assisting with DNA base-excision repair in response to DNA strand breaks, and has a role in DNA repair. For example, PARP activity was five times greater in human cells compared to rat cells in response to DNA damage. Interestingly, *PARP-1*, which accounts for the vast majority of cellular poly(ADP-ribose), interacts with WRN, the protein deficient in the progeroid syndrome, Werner syndrome, to facilitate its DNA helicase activity (Li et al. 1994) and to prevent

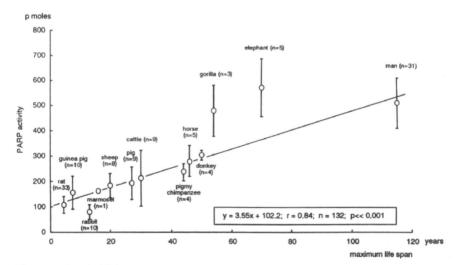

Figure 4. Species' life spans correlate with poly(ADP-ribose) polymerase-1 (PARP-1) (DNA-repair capacity) activity (Grube and Bürkle 1992).

carcinogenesis (von Kobbe et al. 2003). *PARP-1* and *PARP-2* have also been shown to have roles in telomere length regulation (Bürkle et al. 2005).

Investigating this finding further, a French Centenarian Study assessed PARP activity in 49 centenarians and 51 young controls. Muiras et al. (1998) found the median enzyme activity to be 1.6-fold higher in the centenarians than in the controls. However, in this early (1998) genetic association study, no difference was noted for a single polymorphism of the PARP promoter gene. A subsequent study in 2000 of four SNPs of *PARP-1* again revealed no significant differences between an even larger comparison of centenarians and young controls (Cottet et al. 2000).

Werner Syndrome gene (WRN): Werner Syndrome is a progeroid syndrome that is usually autosomal recessive. Patients have, at a young age, many signs of advanced aging, including cataracts, osteoporosis, coronary artery disease, and insulin resistance. *WRN* encodes a RecQ DNA helicase with several mutations having been found responsible for Werner Syndrome (Yu et al. 1996; Gray et al. 1997).

The population-based Leiden 85-plus Study assessed among its sample ($n = 1245$, age ≥ 85 years) the frequencies of three *WRN* variants according to the presence of cardiovascular disease, cognitive function, and overall mortality (Kuningas et al. 2006). In this study, no significant association was observed except for a minor frequency difference for those with and without cognitive impairment.

XPF-ERCC1 endonuclease: Another recently discovered progeroid syndrome has as its basis a severe mutation of the *XPF* gene. Some milder mutations of this gene lead to xeroderma pigmentosum and others have been linked to increased susceptibility to certain cancers such as melanoma. This endonuclease assists in the repair of damage to the tertiary and secondary structures of DNA. Niedernhofer et al. (2006) created an *XPF-ERCC1*-deficient mouse model of the progeroid syndrome exhibited in a 15-year-old boy. These mice exhibited reduced antioxidant defenses and reduced insulin-like growth factor-1 (IGF-1) signaling. Otherwise normal old mice were found to have liver mRNA (the product of DNA expression) molecules similar to those of the progeroid mouse model. Thus, the authors concluded that unrepaired DNA damage is a key contributor to aging and functional decline, whereas variations in genes that defend against this damage can explain to a substantial extent the variability in rates of DNA damage and functional decline.

Sirtuins: The protein Sir2 has been shown to extend life span in yeast (Lin et al. 2000). In mammals, there are seven genes homologous for Sir2, called sirtuins (*SIRT*). *SIRT1*, like Sir2, removes acetyl groups from other proteins. However, unlike Sir2, *SIRT1* does not stabilize DNA, but rather interacts with proteins such as the tumor suppressor p53 and proteins in insulin-signaling pathways. In a German sample comparing 1026 very old individuals (ages 95–109) and 547 controls (ages 60–75), no frequency differences were noted for five SNPs distributed across the *SIRT1* gene, including its promoter region (Flachsbart et al. 2006). However, another sirtuin, *SIRT6*, like Sir2, has been more directly linked to basic mechanisms of aging and the stabilization of DNA. In a mouse model, *SIRT6*-deficient cells were found to be defective in base-excision repair, and the mice displayed signs of premature aging and metabolic defects (Mostoslavsky et al. 2006).

Defenses against Reactive Oxygen Species

Various theories abound regarding the underlying causes of aging and how and why aging increases susceptibility to various diseases. Free-radical generation as a normal consequence of cellular metabolism has been implicated as a potent contributor to aging (Harman 1993; Sohal et al. 1993, 1994; Butterfield et al. 1995; Stadtman 2001) as well as to the pathogenesis of common age-related diseases including cardiovascular disease (CVD) (Salonen et al. 1994; Ascherio et al. 2001), Alzheimer's disease (Grundke-Iqbal et al. 1990; Connor et al. 1992; Behl et al. 1994; Hensley et al. 1994; Benzi and Moretti 1995; Butterfield et al. 1995; Schubert and Chevion 1995; Kala et al. 1996; Smith et al. 1997, 1998; Markesbery and Carney 1999; Bishop et al. 2002), diabetes mellitus (Medalie et al. 1975; Wilson et al. 1981; Catalano et al. 1996; Wannamethee et al. 1996; Tuomainen et al. 1997; Facchini 1998; Fernandez-Real et al. 1998, 2002; Salonen et al. 1998; Ford and Cogswell 1999; Bertelsen et al. 2001; Hua et al. 2001; Lao et al. 2001; Wilson et al. 2003; Jiang et al. 2004), and cancer (Stevens et al. 1988).

p66Shc (SHC1): The p66shc protein has a signal transduction role in cellular responses to oxidative stress. Mice lacking p66shc (p66shc$^{-/-}$) have increased resistance to reactive oxygen species (ROS) and a 30% prolonged life span (Migliaccio et al. 1999). Because ROS has important roles in endothelial dysfunction and vascular disease, Francia et al. (2004) investigated the vascular endothelium from wild-type and p66shc$^{-/-}$ mice and found that although present in the wild-type mice, there was no age-related decline in nitric oxide release, increased oxide production, and nitrotyrosine in the p66shc$^{-/-}$ mice. The investigators concluded that p66shc has an important role in endothelial integrity and vascular aging. A recent study has shown that p66shc regulates mitochondrial oxidative capacity in mice and suggests that this protein may extend life span by repartitioning metabolic energy conversion away from oxidative and toward glycolytic pathways (Nemoto et al. 2006).

Few human studies of *p66Shc* have been reported. One negative finding study compared the frequencies of two SNPs (one in the regulatory region) among 78 subjects with early-onset coronary artery disease (CAD) (mean age 49 years) and 93 long-living control subjects (mean age 89 years). No substantial difference was found, although it would seem that the authors' conclusion, based on this small and limited study, that the *p66SHc* gene has no role in susceptibility to CAD is premature (Sentinelli et al. 2006). The Italian Centenarian Study found that *p66Shc* was highly expressed in fibroblasts from centenarians after exposure to oxidative stress, and the pattern of expression was associated with a polymorphism of p53 (Pandolfi et al. 2005). This finding requires replication in other studies.

Paraoxonase 1 (PON1): PON1 is an arylesterase that has been noted to protect low-density lipoprotein (LDL) against oxidative damage. Peroxidation of LDL is a key step in the production of atherosclerotic plaques. The Italian Centenarian Study assessed the frequencies of two common *PON1* polymorphisms in 308 centenarians and 579 young controls (Rea et al. 2004). A difference was observed for one of the SNPs, but this finding was not reproduced in a separate comparison group of Irish octogenarians versus controls.

Klotho (KL): The *Klotho* gene encodes a β-glucuronidase. *Klotho*-deficient mice were found to exhibit signs of accelerated aging, especially premature atherosclerosis, impaired endothelial function, and angiogenesis (Shimada et al. 2004). Some findings support the possibility that *Klotho* has a role in the production of nitric oxide by the endothelium. Overexpression of *Klotho* in mice increased their life span from 19% to 31% (Kurosu et al. 2005). In a study with human fibroblasts, down-regulation of *Klotho* induced premature cell senescence via its influence on the p53/p21 pathway and ultimately the cell cycle (de Oliveira 2006). In humans, the VS allele of the *Klotho* gene was found to confer an increased risk for CAD and that this risk was modified by other well-known risk factors such as hypertension, high-density lipoprotein (HDL) level, and smoking (Arking et al. 2005; Rhee et al. 2006).

Superoxide dismutase (SOD): SOD enzymes (Sod: copper/zinc enzyme; Sod2: manganese enzyme based in mitochondria; and Sod3: extracellular-based enzyme) catalyze superoxide, one of the most prevalent ROSs, into water and hydrogen peroxide. In *Drosophila*, overexpression of *SOD* extends life span (Orr and Sohal 1994). In 1998, the Italian Centenarian Study found no significant frequency difference between centenarians and controls for one Sod2 polymorphism (De Benedictis et al. 1998). Other studies in humans have not been reported.

Catalase (CAT): Catalase protects cells from hydrogen peroxide. Overexpression of *CAT* extends the life span of short-lived strains of fruit flies but not in long-lived strains (Orr et al. 2003) and in mice, where overexpression of *CAT* in mitochondria is induced. However, it is thought that the survival advantage is cardiac-specific, rather than a basic biology of aging (Schriner et al. 2005).

Hemochromatosis gene (HFE): HFE assists in the regulation of the interactions between the transferrin receptor with transferrin and therefore in iron absorption. *HFE* also has a role in modulating levels of another iron regulatory protein, hepcidin. Approximately 20 mutations of the *HFE* gene are known to cause type-1 hemochromatosis. The result of such mutations is a variably impaired interaction with the transferring receptor that leads to too much iron entering cells, most importantly, the cells lining the small intestine. Thus, particularly when the patient is homozygous for the mutation, the body absorbs too much iron, leading to iron overload. Iron overload has been associated with diabetes, AD, CVD, and cancer.

The mechanism by which iron affects aging and age-related diseases is likely related to its essential role in mitochondrial free-radical generation (Sohal et al. 1985; Sullivan 1989, 1991, 2003a,b). Some data support the potentiating role of iron in lipid peroxidation, the first step in the formation of atherosclerotic lesions. Macrophages and endothelial cells are involved in this process, but the exact mechanism and the sites of the interactions between these cells, iron, and LDL are still unclear. The available epidemiological evidence suggests that elevated iron levels are involved in the pathogenesis of atherosclerosis.

Heterozygotes might also absorb too much iron, although lesser amounts. It was with this hypothesis in mind that Coppin et al. (2003) studied 492 centenarians from the French Centenarian Study with the hypothesis that the C282Y mutation would be underrepresented in the oldest old compared to younger subjects (mean age of 51 years). In this study, there was no significant difference between the comparison groups. In fact, two centenarians were found to be homozygous for the mutation.

Mitochondrial Genes

A discussion of free radicals and energy dynamics with respect to longevity must include mitochondrial genes. Sohal and Weindruch (1996), in their *Science* article, provided compelling arguments for ROS as important facilitators of aging. Observations of their own and of other investigators supporting a key role include the following: In mouse studies, measures of oxidative stress are lower in the caloric-restricted state (see also López-Lluch et al. 1996); differences in life spans for different species correlate with rates of free-radical production; and increased production of free-radical scavenging enzymes increases life span in the *Drosophila melanogaster* model. Trifunovic et al. (2004) produced a progeroid mouse by engineering a mutation in a DNA polymerase gene, resulting in impaired proofreading and a greater accumulation of mitochondrial DNA mutations. Their findings support the mitochondrial theory of aging stating that accumulation of such mutations at least contribute to aging (Wallace 1997).

Several mitochondrial gene studies of oldest old humans have been published. Tanaka et al. (1998) received significant attention for their sequencing of all of the coding regions of mitochondrial DNA in 11 centenarians. Several nucleotide substitutions were noted to be more frequent in the centenarians. Most notably, 9 of 11 centenarians demonstrated a substitution at nucleotide position 5178, within the NADH dehydrogenase subunit 2 gene (*ND2*), which was observed in only a quarter of 43 young controls. A 2003 Finnish study of three age groups ranging from infants (n = 257) to middle age (n = 400) to old (aged 90–91, n = 225) individuals noted some haplogroup and haplogroup cluster frequency differences between the old and the younger groups (Niemi et al. 2003). Haplogroup H and cluster HV were significantly less frequent, and haplogroups U and J and clusters UK and WIX were more frequent. The Finnish researchers along with the Tokyo Centenarian Study subsequently studied the 150T SNP among 321 very old and 489 middle-aged subjects from Finland and Japan (Niemi et al. 2005). Although the 150T polymorphism was higher in frequency in the very old subjects of both samples, these frequencies were significantly influenced by two other SNPs: 489C and 10398G. As noted above, the *p66Shc* gene has garnered interest as a longevity determinant. Protein kinase C β and prolylisomerase 1 have been found to be key players in the effect of *p66Shc* upon mitochondrial function (Pinton et al. 2007).

Genes Impacting Age-related Diseases

Considering that CVD is significantly delayed among the offspring of centenarians and that 88% of centenarians either delay or escape CVD and stroke beyond the age of 80, it follows that the frequency of genetic polymorphisms that have a role in the risk for such diseases would be differentiated between centenarians and the general population (Evert et al. 2003; Terry et al. 2003). However, it is also apparent that in association studies comparing centenarians against controls, the differences in frequencies for various CVD-related alleles are relatively modest.

Apolipoprotein E

We have previously observed that CVD is markedly delayed in both centenarians and their children; 88% of centenarians either delay or escape CVD and stroke beyond the age of 80 years. The children of centenarians exhibit a 60% reduced relative risk for cardiovascular disease compared to children of the same birth cohort who had at least one parent die prior to the age of 77 years (Terry et al. 2004). Thus, it makes sense that the frequency of genetic polymorphisms that have a role in the risk for CVD would be differentiated between centenarians and the general population. Schachter et al. (1994), of the French Centenarian Study, noted that the frequency of the ApoE-ε4 allele was very low among centenarians, presumably because of its association with AD and vascular disease (Kervinen et al. 1994; Louhja et al. 1994; Rebeck et al. 1994; Schachter et al. 1994; Sobel et al. 1995). Interestingly, the effect of the ApoE allotype on AD incidence appears to decrease with age at very old ages at least in Caucasians (Rebeck et al. 1994; Sobel et al. 1995).

Cholesteryl Ester Transfer Protein

A study of Ashkenazi Jewish centenarians and their families identified another cardiovascular pathway and gene that is differentiated between centenarians and controls (Barzilai et al. 2003). These authors noted that HDL and LDL particles were significantly larger among centenarians and their offspring and that particle size differentiated subjects with and without CVD, hypertension, and metabolic syndrome. In a candidate gene approach, the researchers then searched the literature for genes that affect HDL and LDL particle size. The genes for hepatic lipase and cholesteryl ester transfer protein (CETP) emerged as candidates. Genotyping the common alleles for these genes, the VV variant of CETP emerged as significantly increased among the centenarians and unrelated centenarian offspring compared to controls. The individuals with the VV variation also had lower levels of CETP, higher HDL levels, and larger lipoprotein particles.

In a subsequent study, Barzilai's group segregated subjects according to cognitive function. They found that subjects with good function (Mini-Mental Status Examination scores >25) had the CETP VV genotype more frequently compared to those with poorer function (29% vs. 14%,

$p = 0.02$) (Barzilai et al. 2006). In addition, as in the earlier more general study, these subjects had a more advantageous chemistry profile (lower CETP, higher HDL, and larger lipoprotein particles). In a separate sample (the Einstein Aging Study) of longitudinally followed older subjects, this profile was associated with a significantly lower prevalence of dementia.

Apolipoprotein C3 (ApoC3)

The Ashkenazi Jewish Centenarian Study also noted an age-dependent frequency pattern for the ApoC3 promoter. The CC polymorphism occurred in 25% of centenarians, 20% in their offspring, and 10% in controls. This variation is associated with lower cholesterol and higher HDL levels, as well as larger particle size.

Microsomal Transfer Protein (MTP)

Puca et al. (2001) conducted a genome-wide sibling pair linkage study to discover polymorphisms linked to EL; 137 centenarians and their 171 siblings were included in a genome-wide concordant sib-pair study. Using nonparametric analysis, significant evidence for linkage was noted for a locus on chromosome 4 at D4S1564 with a maximum logarithm of odds score of 3.65 ($P = 0.044$) (Puca et al. 2001). A detailed haplotype map was created of the chromosome 4 locus that extended over 12 million base pairs and involved the genotyping of more than 1000 SNP markers in 200 centenarians and 200 controls. The resulting genetic association study identified a haplotype marker within the microsomal transfer protein (*MTP*) gene as a modifier of human life span (Geesaman et al. 2003). All known SNPs for *MTP* and its promoter were genotyped in 200 centenarians and 200 young controls. After haplotype reconstruction of the area was completed, a single variant—the −493 G/T variant in the promoter of MTP—was underrepresented in the long-lived individuals, accounting for the majority of the statistical variability at the locus (~15% among the subjects vs. 23% in the controls). The chromosome 4 linkage result was replicated in a study of clinically healthy dizygotic twins older than 70 (Reed et al. 2004), but the MTP finding was not replicated by European studies (Nebel et al. 2005; Beekman et al. 2006).

Prolyl Isomerase (PIN1)

Pin1 is a protein-folding chaperone and has been found to have an important role in AD. It acts to restore misshapen Tau and amyloid precursor proteins to their original healthy shape, possibly preventing the onset of neurodegeneration and development of dementia. This enzyme helps prevent abnormal cleavage of the amyloid precursor protein into the toxic amyloid-β peptides Aβ40 and Aβ42 and the subsequent formation of amyloid plaques (Pastorino et al. 2007). Certain genetic variations of the *PIN1* gene are associated with reduced Pin1 protein levels and subsequently increased risk for AD. Pin1 also binds with SHC1, leading to conformational changes in that protein and its accumulation within mitochondria, a pathway thought to be conducive to longevity (Pinton et al. 2007).

It may be that centenarians are rare not because of a few rare factors that have a dramatic impact, but rather because of rare combinations of genetic and environmental exposures (Perls and Terry 2003a). Complicating matters, many claimed associations appear to vary substantially depending on the population sample and/or control being studied. For example, as indicated above, the MTP finding was not replicated in several European studies. Even in the case of ApoE, an exception to the rule was noted in a large Korean centenarian cohort (Choi et al. 2003). Thus, not only are rare combinations likely important, but their significances also vary across populations. The difficulty in reproducing these findings speaks to the importance of ethnicity and environmental factors that can result in very different survival-related outcomes for specific genotypes (Panza et al. 2004).

ANALYTIC CHALLENGES TO THE STUDY OF EXCEPTIONAL LONGEVITY

The Choice of Controls

Ethnicity and other demographic-specific genetic background differences can lead to false-positive results where a difference associated with longevity is actually due to population stratification. Thus, the choice of controls becomes very important. This is a particularly dicey issue when it comes to centenarian studies. To date, numerous centenarian studies have relied upon the premise that younger controls can be used because, given the rarity of centenarians, it is unlikely that a younger control will go on to survive to extreme old age. However, given that centenarians are the fastest-growing segment of our population, this is really no longer a safe assumption. Furthermore, Lewis and Brunner (2004), in using ApoE as an example, proposed that environmental and behavioral differences that exist between different birth cohorts of a specific population and ethnicity can lead to vastly different effects of genotypes upon survival and risk for specific diseases.

The ideal control for EL-related studies would thus be individuals belonging to the same birth cohort as the experimentals, but who died at a substantially younger age due to nontraumatic causes. Longitudinal studies, such as the Framingham Heart Study, Normative Study on Aging, and the Baltimore Longitudinal Study on Aging may be valuable resources for obtaining phenotypic data and genetic material for such deceased controls.

Accounting for Multiple Determinants and Interactions

EL is a complex trait characterized by multiple stochastic, environmental/ behavioral, and genetic factors interacting with one another. Although the relative risk of achieving exceptional old age might be attributed to a single factor, one even as potent as, say, smoking, obesity, or the ApoE-ε4 allele, it would be overly optimistic to believe that other factors do not exist which can markedly modify these attributable risks.

Several analytic approaches exist to account for such complex systems. One procedure is based on cluster analysis and aims at grouping those subjects whose data are similar to each other and differ from those of the subjects in other groups. The result of the analysis is the assignment of each individual to one cluster, and this cluster membership can be taken to represent a data-driven phenotype to be used in the genetic association analysis. This procedure offers the advantage of summarizing complex phenotypes determined by many interacting variables into a single categorical variable, but it has some problematic aspects. The cluster membership is determined from a sample of subjects and is therefore a stochastic rather than a deterministic variable, and the uncertainty of the cluster membership should be taken into account. Another disadvantage is that clustering techniques do not usually produce predictive models. An alternative solution is to use model-based clustering (Fraley and Raftery 2002) or rely on more sophisticated statistical methods.

Bayesian network (BN) modeling is a significant improvement upon cluster analysis in its ability to take into account not only the interaction of many variables at once, but also generate the set of variables that can predict risk for the outcome of interest with a high degree of reproducibility. A BN is described by a graph in which the nodes represent the variables of the model and the directed arcs represent stochastic dependencies between primary (also called "parent") and secondary ("child") nodes that are quantified by conditional probability distributions. For example, in Figure 5, the node P (phenotype) is a child of the nodes G (genotypes of a SNP), meaning that the two variables are associated. The dependency is quantified by the conditional probability table of the node P given the possible configurations of the parent nodes. These conditional probability distributions—one for each node in the network—

represent the uncertainty in making predictions about individual subjects. The conditional probability distributions can be multiplied together to obtain the joint probability distribution of the variables in the network, which can be used to show how one or more events effect each other in the production of one or more outcomes.

The algorithm to make these calculations is based on Bayes' theorem—hence, the name Bayesian networks—that updates the prior ("marginal") probability of each event or node into the posterior ("conditional") probability of the event, which is influenced by other events. Both the structure of dependency between the nodes and the conditional probability tables can be estimated from a database of cases using a standard Bayesian methodology that is described, for example, by Cowell et al. (1999) in their text, *Probabilistic Networks and Expert Systems.* Figure 5 illustrates the basic reasoning behind the calculation of a joint probability generated by a number of factors for a particular outcome.

The BN model developed by Sebastiani et al. (2005) for the genetic dissection of stroke in sickle cell disease (HbSS) is a small-scale example of the successful and powerful use of this analytic approach. This model was based on a BN that captures the interplay between genetic and clinical variables to modulate the risk for stroke in HbSS. The model was inferred from data using a Bayesian model search strategy imple-

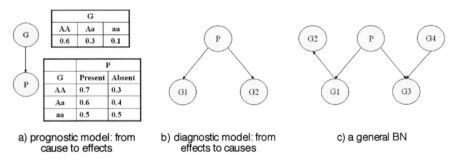

a) prognostic model: from cause to effects   b) diagnostic model: from effects to causes   c) a general BN

Figure 5. Examples of BN structures. (*a*) A simple BN with two nodes representing an SNP (G) and a phenotype (P). The probability distribution of G represents the genotype distribution in the population, and the conditional probability distribution of P describes the distribution of the phenotype given each genotype. (*b*) Association between SNPs and phenotype can be reversed using Bayes' theorem. (*c*) A BN linking four SNPs (G_1–G_4) to a phenotype P. The phenotype is independent of G2, once we know the SNPs G1, G3, and G4. The joint probability distribution of the network is fully specified by the five distributions representing the distribution of G1 given P, of G2 given G1, of G3 given P and G4 (6 parameters), of G4, and of P. By Bayes' theorem, one can compute the probability distribution of the phenotype, given a genetic profile specified by the SNPs G1, G3, and G4 that are the necessary and sufficient predictors of the phenotypes.

mented in the program Discoverer (http://www.bayesware.com/), and it can be used as a diagnostic model to identify those genetic profiles that increase the risk for stroke, as well as prognostic model for risk prediction based on different clinical presentations. In particular, the network describes the interplay between 31 SNPs in 12 genes that, together with hemoglobin F (HbF), modulate the risk for stroke. This network of interactions includes three genes in the transforming growth factor-β (TGF-β) pathway and the gene *SELP* that was already associated with stroke in the general population. Sebastiani et al. (2005) validated this model in a different population by predicting the occurrence of stroke in 114 subjects with 98.2% accuracy and showed that his approach outperformed standard statistical models based on logistic regression. Figure 6 illustrates the use of the model to predict the risk for stroke of a subject given his genetic and clinical profile.

In a preliminary study, we used a BN to describe interrelationships among age, sociodemographic characteristics including birth cohort, sex, education, and the health history of diseases and conditions in approximately 600 NECS subjects. Figure 7 displays the computer-program (Bayesware Discoverer)-generated network of associations among the variables that are the nodes of the directed graph.

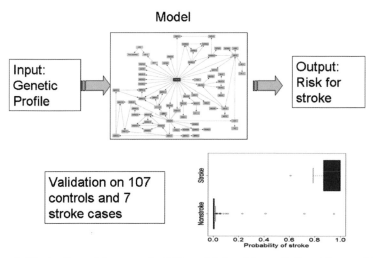

Figure 6. Illustration of the use of a BN model for risk prediction. The model in the middle is the BN that describes the interrelations between genetic variants, clinical variables, and stroke in sickle cell anemia (Sebastiani et al. 2005). Given the genetic profile of a patient with sickle cell anemia, the network can be used to compute the risk for stroke. We applied this model to compute the risk for stroke in 114 subjects and the box plots show the predicted risk of the 107 controls (*blue*) and the predicted risk of the 7 stroke cases (*red*).

The network describes expected associations such as those between cancer, years of education, and age at death, or the association between HTN and cardiovascular and cerebrovascular accidents (nodes stroke and circulatory disease). Inspection of the probability distributions associated with these nodes shows, for example, that centenarians with both circulatory problems and diabetes have 2.4 times the odds for stroke compared to centenarians with diabetes alone. These associations change when we consider gender and history of HTN and show that female centenarians with a history of HTN have a higher prevalence of stroke than male centenarians, whereas no history of HTN makes the prevalence of stroke in male centenarians higher compared to female centenarians. When we consider profiles of centenarians by gender, male centenarians present with a healthier profile characterized by a smaller prevalence of hypertension and congestive heart failure and a much smaller prevalence of dementia (21% compared to 34% in females) and osteoporosis (8% compared to 34%).

The multivariate model described by the BN provides a powerful approach in which, rather than "labeling" centenarians by some particular pattern of ES that summarizes the information in many variables, we use this information in a model that can be expanded by adding genetic information.

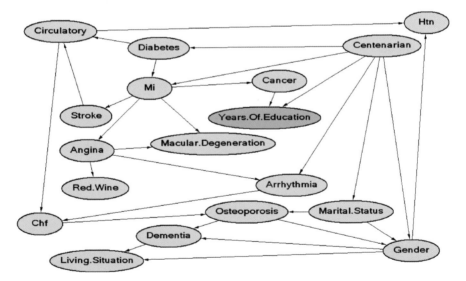

Figure 7. The BN that summarizes the mutual relationships between socioeconomic, demographic, and health variables in a samples of 600 subjects of the NECS.

FUTURE DIRECTIONS

Despite extensive research on longevity in animal models, we continue to have a poor understanding of the genetic basis of human longevity. Insulin and growth factor, immune and inflammatory, free-radical and DNA-repair-related networks appear to be the hottest areas of interest right now in the study of longevity. However, except for the case of ApoE, the candidate gene approach in centenarian versus control studies has generally not been fruitful (Kojima et al. 2004; Flachsbart et al. 2006) probably because of (1) inappropriate controls, (2) inadequate power in terms of either sample size and/or the effect of the genotype upon longevity, (3) selection of candidate genes that are biased toward age-related disease rather than longevity, and (4) the inability of traditional statistical analyses to detect and quantify multiple interactions between genes and environmental/behavioral factors. All of these shortcomings speak for the need to perform an unbiased approach to gene discovery (e.g., GW analysis) in highly powered studies and to replicate the findings in different populations in order to identify the necessary genetic components of exceptional longevity.

Many humans begin to experience significant health decline from age-related diseases when they reach their seventh decade. With optimal environments and behaviors, an average person lives to approximately age 85 (Fraser and Shavlik 2001). Centenarians live 15–25 years beyond what is average and markedly delay disability and age-related lethal diseases (Hitt et al. 1999; Evert et al. 2003). To live to such an old age, centenarians are likely to have relatively unique genetic and environmental exposures and interactions.

GW association studies could uncover the complex relationships among genes, environment, and phenotype and permit physicians to identify patients who are at risk, based on their genetic profile, for premature mortality and allow them to advise patients how best to modify risk behaviors and environmental exposures. Such studies might also assist in identifying therapeutic targets that could potentially extend life span.

ACKNOWLEDGMENTS

This work is supported by a National Institute on Aging grant awarded to T.P. (K24)AG025727 and by a National Heart Lung Blood Institute grant awarded to P.S. (R21)HL080463.

REFERENCES

Arking D.E., Atzmon G., Arking A.A., Barzilai N., and Dietz H.C. 2005. Association between a functional variant of the KLOTHO gene and high-density lipoprotein cholesterol, blood pressure, stroke, and longevity. *Circ. Res.* **96:** 412–418.

Ascherio A., Rimm E.B., Giovannucci E., Willett W.C., and Stampfer M.J. 2001. Blood donations and risk of coronary heart disease in men. *Circulation* **103:** 52–57.

Atzmon G., Schechter C., Greiner W., Davidson D., Rennert G., and Barzilai N. 2004. Clinical phenotype of families with longevity. *J. Am. Geriatr. Soc.* **52:** 274–277.

Atzmon G., Rincon M., Schechter C.B., Shuldiner A.R., Lipton R.B., Bergman A., and Barzilai N. 2006. Lipoprotein genotype and conserved pathway for exceptional longevity in humans. *PLoS Biol.* **4:** e113.

Barzilai N., Atzmon G., Derby C.A., Bauman J.M., Lipton R.B. 2006. A genotype of exceptional longevity is associated with preservation of cognitive function. *Neurology* **67:** 2170–2175.

Barzilai N., Atzmon G., Schechter C., Schaefer E.J., Cupples A.L., Lipton R., Cheng S., and Shuldiner. 2003. Unique lipoprotein phenotype and genotype associated with exceptional longevity. *J. Am. Med. Assoc.* **290:** 2030–2040.

Beekman M., Blauw G.J., Houwing-Duistermaat J.J., Brandt B.W., Westendorp R.G., and Slagboom P.E. 2006. Chromosome 4q25, microsomal transfer protein gene, and human longevity: Novel data and a meta-analysis of association studies. *J. Gerontol. A Biol. Sci. Med. Sci.* **61:** 355–362.

Behl C., Davis J.B., Lesley R., and Schubert D. 1994. Hydrogren peroxide mediates amyloid beta protein toxicity. *Cell* **75:** 817–827.

Benzi G. and Moretti A. 1995. Are reactive oxygen species involved in Alzheimer's disease? *Neurobiol. Aging.* **16:** 661–674.

Bertelsen M., Anggard E.E., and Carrier M.J. 2001. Oxidative stress impairs insulin internalization in endothelial cells in vitro. *Diabetologia* **44:** 605–613.

Bishop G.M., Robinson S.R., Liu Q., Perry G., Atwood C.S., and Smith M.A. 2002. Iron: A pathological mediator of Alzheimer disease? *Dev. Neurosci.* **24:** 184–187.

Bürkle A., Brabeck C., Diefenbach J., and Beneke S. 2005. The emerging role of poly(ADP-ribose) polymerase-1 in longevity. *Int. J. Biochem. Cell Biol.* **37:** 1043–1053.

Butterfield D.A., Howard B., Yatin S., Koppal T., Drake J., Hensley K., Aksenov J.M., Aksenova M., Subramanain R., Varadarajan S., Harris-White M.E., Pedigo N.W., Jr., and Carney J.M. 1995. Elevated oxidative stress in models of normal brain aging and Alzheimer's disease. *Life Sci.* **65:** 193–210.

Catalano C., Muscelli E., Quinones A., Baldi S., Ciorciaro D., Seghieri G., and Ferrannini E. 1996. Reciprocal association between insulin sensitivity and the hematocrit in man. *Diabetes Care* **45:** 323. (Abstr.)

Choi Y.H., Kim J.H., Kim D.K., Kim J.W., Kim D.K., Lee M.S., Kim C.H., and Park S.C. 2003. Distributions of ACE and APOE polymorphisms and their relations with dementia status in Korean centenarians. *J. Gerontol. A Biol. Sci. Med. Sci.* **58:** 225–226.

Connor J.R., Snyder B.S., Beard J.L., Fine R.E., and Mufson E.J. 1992. Regional distribution of iron and iron-regulatory proteins in the brain in aging and Alzheimer's disease. *J. Neurosci. Res.* **31:** 327–335.

Coppin H., Bensaid M., Fruchon S., Borot N., Blanche H., Roth M.P. 2003. Longevity and carrying the C282Y mutation for haemochromatosis on the HFE gene: Case control study of 492 French centenarians. *Brit. Med. J.* **327:** 132–133.

Cottet F., Blanche H., Verasdonck P., Le Gall I., Schachter F., Bürkle A., and Muiras M.L. 2000. New polymorphisms in the human poly(ADP-ribose) polymerase-1 coding sequence: Lack of association with longevity or with increased cellular poly(ADP-ribosyl)ation capacity. *J. Mol. Med.* **78:** 431–440.

Cowell R.G., Dawid A.P., Lauritzen S.L., and Spiegelhalter D.J. 1999. *Probabilistic networks and expert systems.* Springer Verlag, New York.

De Benedictis G., Carotenuto L., Carrieri G., De Luca M., Falcone E., Rose G., Cavalcanti S., Corsonello F., Feraco E., Baggio G., Bertolini S., Mari D., Mattace R., Yashin A.I., Bonafe M., and Franceschi C. 1998. Gene/longevity association studies at four autosomal loci (REN, THO, PARP, SOD2). *Eur. J. Hum. Genet.* **6:** 534–541.

de Oliveira R. 2006. Klotho RNAi induces premature senescence of human cells via a p53/p21 dependent pathway. *FEBS Lett.* **580:** 5753–5758.

Evert J., Lawler E., Bogan H., and Perls T. 2003. Morbidity profiles of centenarians: Survivors, delayers, and escapers. *J. Gerontol. A Biol. Sci. Med. Sci.* **58:** 232–237.

Facchini F.S. 1998. Effect of phlebotomy on plasma glucose and insulin concentrations. *Diabetes Care* **21:** 2190.

Falconer D. 1965. The inheritance and liability to certain disease estimated from the incidence among relatives. *Ann. Hum. Genet.* **29:** 51–76.

Farrer L. and Cupples A. 1998. Determining the genetic component of a disease. In *Approaches to gene mapping in complex human diseases* (ed. J. Haines and M.A. Pericak-Vance). pp. 93–130. John Wiley & Sons, New York.

Fernandez-Real J.M., Lopez-Bermejo A., and Ricart W. 2002. Cross-talk between iron metabolism and diabetes. *Diabetes Care* **51:** 2348–2354.

Fernandez-Real J.M., Ricart W., Arroyo E., Balanca R., Casamitjana R., and Cabrero D. 1998. Serum ferritin as a component of the insulin resistance syndrome. *Diabetes Care* **21:** 62–68.

Flachsbart F., Croucher P.J., Nikolaus S., Hampe J., Cordes C., Schreiber S., and Nebel A. 2006. Sirtuin 1 (SIRT1) sequence variation is not associated with exceptional human longevity. *Exp. Gerontol.* **41:** 98–102.

Ford E.S. and Cogswell M.E. 1999. Diabetes and serum ferritin concentration among U.S. adults. *Diabetes Care* **22:** 1978–1983.

Fraley C. and Raftery A.E. 2002. Model-based clustering, discriminant analysis, and density estimation. *J. Am. Stat. Assoc.* **97:** 611–631.

Francia P., delli Gatti C., Bachschmid M., Martin-Padura I., Savoia C., Migliaccio E., Pelicci P.G., Schiavoni M., Luscher T.F., Volpe M., and Cosentino F. 2004. Deletion of p66shc gene protects against age-related endothelial dysfunction. *Circulation* **110:** 2889–2895.

Fraser G.E. and Shavlik D.J. 2001. Ten years of life: Is it a matter of choice? *Arch. Intern. Med.* **161:** 1645–1652.

Geesaman B.J., Benson E., Brewster S.J., Kunkel L.M., Blanche H., Thomas G., Perls T.T., Daly M.J., and Puca A.A. 2003. Haplotype-based identification of a microsomal transfer protein marker associated with the human lifespan. *Proc. Natl. Acad. Sci.* **100:** 14115–14120.

Gray M.D., Shen J.C., Kamath-Loeb A.S., Blank A., Sopher B.L., Martin G.M., Oshima J., and Loeb L.A. 1997. The Werner syndrome protein is a DNA helicase. *Nat. Genet.* **17:** 100–103.

Grube K. and Bürkle A. 1992. Poly(ADP-ribose) polymerase activity in mononuclear leukocytes of 13 mammalian species correlates with species-specific life span. *Proc. Natl. Acad. Sci.* **89:** 11759–11763.

Grundke-Iqbal I., Fleming J., Tung Y.C., Lassmann H., Iqbal K., and Joshi J.G. 1990. Ferritin is a component of the neuritic (senile) plaque in Alzheimer dementia. *Acta Neuropathol.* **81:** 105–110.

Gudmundsson H., Gudbjartsson D.F., Frigge M., Gulcher J.R., and Stefansson K. 2000. Inheritance of human longevity in Iceland. *Eur. J. Hum. Genet.* **8:** 743–749.

Harman D. 1993. Free radical involvement in aging. Pathophysiology and therapeutic implications. *Drugs Aging* **3:** 60–80.

Hensley K., Carney J.M., Mattson M.P., Aksenova M., Harris M., Wu J.F., Floyd R.A., and Butterfield D.A. 1994. A model for β-amyloid aggregation and neurotoxicity based on free radical generation by the peptide: Relevance to Alzheimer's disease. *Proc. Natl. Acad. Sci.* **91:** 3270–3274.

Herskind A.M., McGue M., Holm N.V., Sorensen T.I., Harvald B., and Vaupel J.W. 1996. The heritability of human longevity: A population-based study of 2872 Danish twin pairs born 1870-1900. *Hum. Genet.* **97:** 319–323.

Hitt R., Young-Xu Y., Silver M., and Perls T. 1999. Centenarians: The older you get, the healthier you have been. *Lancet* **354:** 652.

Hjelmborg J., Iachine I., Skytthe A., Vaupel J.W., McGue M., Koskenvuo M., Kaprio M.J., Pedersen N.L., and Christensen K. 2006. Genetic influence on human lifespan and longevity. *Hum. Genet.* **119:** 312–321.

Hua N.W., Stoohs R.A., and Facchini F.S. 2001. Low iron status and enhanced insulin sensitivity in lacto-ovo vegetarians. *Br. J. Nutr.* **86:** 515–519.

Jiang R., Manson J.E., Meigs J.B., Ma J., Rifai N., and Hu F.B. 2004. Body iron stores in relation to risk of type 2 diabetes in apparently healthy women. *J. Am. Med. Assoc.* **291:** 711–717.

Kala S.V., Hasinoff B.B., and Richardson J.S. 1996. Brain samples from Alzheimer's patients have elevated levels of loosely bound iron. *Int. J. Neurosci.* **86:** 263–269.

Kemkes-Grottenthaler A. 2004. Parental effects on offspring longevity—Evidence from 17th to 19th century reproductive histories. *Ann. Hum. Biol.* **31:** 139–158.

Kervinen K., Savolainen M.J., Salokannel J., Hynninen A., Heikkinen J., Ehnholm C., Koistinen M.J., and Kesaniemi Y.A. 1994. Apolipoprotein E and B polymorphisms—Longevity factors assessed in nonagenarians. *Atherosclerosis* **105:** 89–95.

Kojima T., Kamei H., Aizu T., Arai Y., Takayama M., Nakazawa S., Ebihara Y., Inagaki H., Masui Y., Gondo Y., Sakaki Y., and Hirose N. 2004. Association analysis between longevity in the Japanese population and polymorphic variants of genes involved in insulin and insulin-like growth factor 1 signaling pathways. *Exp. Gerontol.* **39:** 1595–1598.

Kuningas M., Slagboom P.E., Westendorp R.G., and van Heemst D. 2006. Impact of genetic variations in the WRN gene on age related pathologies and mortality. *Mech. Ageing Dev.* **127:** 307–313.

Kurosu H., Yamamoto M., Clark J.D., Pastor J.V., Nandi A., Gurnani P., McGuinness O.P., Chikuda H., Yamaguchi M., Kawaguchi H., Shimomura I., Takayama Y., Herz J., Kahn C.R., Rosenblatt K.P., and Kuro-o M. 2005. Suppression of aging in mice by the hormone Klotho. *Science* **309:** 1829–1833.

Lao T.T., Chan P.L., and Tam K.F. 2001. Gestational diabetes mellitus in the last trimester: A feature of maternal iron excess? *Diabet. Med.* **18:** 218–223.

Lewis S.J. and Brunner E.J. 2004. Methodological problems in genetic association studies of longevity—The apolipoprotein E gene as an example. *Int. J. Epidemiol.* **33:** 962–970.

Li B., Navarro S., Kasahara N., and Comai L. 2004. Identification and biochemical characterization of a Werner's syndrome protein complex with Ku70/80 and poly(ADP-ribose) polymerase-1. *J. Biol. Chem.* **279:** 13659–13667.

Lin S.-J., Defossez P.-A., and Guarente L. 2000. Requirement of NAD and SIR2 for life-span extension by calorie restriction in *Saccharomyces cerevisiae*. *Science* **289:** 2126–2128.

Ljungquist B., Berg S., Lanke J., McClearn G.E., and Pedersen N.L. 1998. The effect of genetic factors for longevity: A comparison of identical and fraternal twins in the Swedish Twin Registry. *J. Gerontol. A Biol. Sci. Med. Sci.* **53:** M441–M446.

López-Lluch G., Hunt N., Jones B., Zhu M., Jamieson H., Hilmer S., Cascajo M.V., Allard J., Ingram D.K., Navas P., and de Cabo R. 1996. Calorie restriction induces mitochondrial biogenesis and bioenergetic efficiency. *Proc. Natl. Acad. Sci.* **103:** 1768–1773.

Louhija J., Miettinen H.E., Kontula K., Tikkanen M.J., Miettinen T.A., and Tilvis R.S. 1994. Aging and genetic variation of plasma apolipoproteins: Relative loss of apolipoprotein E4 phenotype in centenarians. *Arterioscler. Thromb.* **14:** 1084–1089.

Markesbery W.R. and Carney J.M. 1999. Oxidative alterations in Alzheimer's disease. *Brain Pathol.* **9:** 133–146.

Medalie J.H., Papier C.M., Goldburt U., and Herman J.B. 1975. Major factors in the development of diabetes mellitus in 10,000 men. *Arch. Intern. Med.* **135:** 811–817.

Migliaccio E., Giorgio M., Mele S., Pelicci G., Reboldi P., Pandolfi P.P., Lanfrancone L., and Pelicci P.G. 1999. The p66shc adaptor protein controls oxidative stress response and life span in mammals. *Nature* **402:** 309–313.

Mostoslavsky R., Chua K.F., Lombard D.B., Pang W.W., Fischer M.R., Gellon L., Liu P., Mostoslavsky G., Franco S., Murphy M.M., Mills K.D., Patel P., Hsu J.T., Hong A.L., Ford E., Cheng H.L., Kennedy C., Nunez N., Bronson R., Frendewey D., Auerbach W., Valenzuela D., Karow M., Hottiger M.O., and Hursting S., et al. 2006. Genomic instability and aging-like phenotype in the absence of mammalian SIRT6. *Cell* **124:** 315–329.

Muiras M.L., Muller M., Schachter F., and Bürkle A. 1998. Increased poly(ADP-ribose) polymerase activity in lymphoblastoid cell lines from centenarians. *J. Mol. Med.* **76:** 346.

Nebel A., Croucher P.J., Stiegeler R., Nikolaus S., Krawczak M., and Schreiber S. 2005. No association between microsomal triglyceride transfer protein (MTP) haplotype and longevity in humans. *Proc. Natl. Acad. Sci.* **102:** 7906–7909.

Nemoto S., Combs C.A., French S., Ahn B.H., Fergusson M.M., Balaban R.S., and Finkel T. 2006. The mammalian longevity-associated gene product p66shc regulates mitochondrial metabolism. *J. Biol. Chem.* **281:** 10555–10560.

Niedernhofer L.J., Garinis G.A., Raams A., Lalai A.S., Robinson A.R., Appeldoorn E., Odijk H., Oostendorp R., Ahmad A., van Leeuwen W., Theil A.F., Vermeulen W., van der Horst G.T., Meinecke P., Kleijer W.J., Vijg J., Jaspers N.G., and Hoeijmakers J.H. 2006. A new progeroid syndrome reveals that genotoxic stress suppresses the somatotroph axis. *Nature* **444:** 1038–1043.

Niemi A.K., Hervonen A., Hurme M., Karhunen P.J., Jylhä M., and Majamaa K. 2003. Mitochondrial DNA polymorphisms associated with longevity in a Finnish population. *Hum. Genet.* **112:** 29–33.

Niemi A.K., Moilanen J.S., Tanaka M., Hervonen A., Hurme M., Lehtimäki T., Arai Y., Hirose N., and Majamaa K. 2005. A combination of three common inherited mitochondrial DNA polymorphisms promotes longevity in Finnish and Japanese subjects. *Eur. J. Hum. Genet.* **13:** 166–170.

Orr W.C. and Sohal R.S. 1994. Extension of life-span by overexpression of superoxide dismutase and catalase in *Drosophila melanogaster*. *Science* **263:** 1128–1130.

Orr W.C., Mockett R.J., Benes J.J., and Sohal R.S. 2003. Effects of overexpression of copper-zinc and manganese superoxide dismutases, catalase, and thioredoxin reductase genes on longevity in *Drosophila melanogaster. J. Biol. Chem.* **278**: 26418–26422.

Pandolfi S., Bonafe M., Di Tella L., Tiberi L., Salvioli S., Monti D., Sorbi S., and Franceschi C. 2005. p66(shc) is highly expressed in fibroblasts from centenarians. *Mech. Ageing Dev.* **126**: 839–844.

Panza F., D'Introno A., Colacicco A.M., Capurso C., Capurso S., Kehoe P.G., Capurso A., and Solfrizzi V. 2004. Vascular genetic factors and human longevity. *Mech. Ageing Dev.* **125**: 169–178.

Pastorino L., Sun A., Lu P.J., Zhou X.Z., Balastik M., Finn G., Wulf G., Lim J., Li S.H., Li X., Xia W., Nicholson L.K., and Lu K.P. 2007. The prolyl isomerase Pin1 regulates amyloid precursor protein processing and amyloid-beta production. *Nature* **446**: 342.

Perls T. 2005. The different paths to age one hundred. *Ann. N.Y. Acad. Sci.* **1055**: 13–25.

Perls T. and Terry D. 2003a. Understanding the determinants of exceptional longevity. *Ann. Intern. Med.* **139**: 445–449.

———. 2003b. Genetics of exceptional longevity. *Exp. Gerontol.* **38**: 725–730.

Perls T., Shea-Drinkwater M., Bowen-Flynn J., Ridge S.B., Kang S., Joyce E., Daly M., Brewster S.J., Kunkel L., and Puca A.A. 2000. Exceptional familial clustering for extreme longevity in humans. *J. Am. Geriatr. Soc.* **48**: 1483–1485.

Perls T.T., Wilmoth J., Levenson R., Drinkwater M., Cohen M., Bogan H., Joyce E., Brewster S., Kunkel L., and Puca A. 2002. Life-long sustained mortality advantage of siblings of centenarians. *Proc. Natl. Acad. Sci.* **99**: 8442–8447.

Pinton P., Rimessi A., Marchi S., Orsini F., Migliaccio E., Giorgio G., Contursi C., Minucci S., Mantovani F., Wieckowski M.R., Del Sal G., Pelicci P.G., and Rizzuto R. 2007. Protein kinase C β and prolyl isomerase 1 regulate mitochondrial effects of the life-span determinant p66Shc. *Science* **315**: 659–663.

Puca A.A., Daly M.J., Brewster S.J., Matise T.C., Barrett J., Shea-Drinkwater M., Kang S., Joyce E., Nicoli J., Benson E., Kunkel L.M., and Perls T. 2001. A genome-wide scan for linkage to human exceptional longevity identifies a locus on chromosome 4. *Proc. Natl. Acad. Sci.* **98**: 10505–10508.

Rea I.M., McKeown P.P., McMaster D., Young I.S., Patterson C., Savage M.J., Belton C., Marchegiani F., Olivieri F., Bonafe M., and Franceschi C. 2004. Paraoxonase polymorphisms PON1 192 and 55 and longevity in Italian centenarians and Irish nonagenarians. A pooled analysis. *Exp. Gerontol.* **39**: 629–635.

Rebeck G.W., Perls T.T., West H.L., Sodhi P., Lipsitz L.A., and Hyman B.T. 1994. Reduced apolipoprotein epsilon 4 allele frequency in the oldest old Alzheimer's patients and cognitively normal individuals. *Neurology* **44**: 1513–1516.

Reed T., Dick D.M., Uniacke S.K., Foroud T., and Nichols W.C. 2004. Genome-wide scan for a healthy aging phenotype provides support for a locus near D4S1564 promoting healthy aging. *J. Gerontol. A Biol. Sci. Med. Sci.* **59**: 227–232.

Rhee E.J., Oh K.W., Lee W.Y., Kim S.Y., Jung C.H., Kim B.J., Sung K.C., Kim B.S., Kang J.H., Lee M.H., Kim S.W., and Park J.R. 2006. The differential effects of age on the association of KLOTHO gene polymorphisms with coronary artery disease. *Metabolism* **55**: 1344–1351.

Ritchie K. and Kildea D. 1995. Is senile dementia "age-related" or "ageing-related"?—Evidence from meta-analysis of dementia prevalence in the oldest old. *Lancet* **346**: 931–934.

Salonen J.T., Tuomainen T.P., Nyyssonen K., Lakka H.M., and Punnonen K. 1998. Relation between iron stores and non-insulin dependent diabetes in men: Case-control study. *Brit. Med. J.* **317:** 727–730.

Salonen J.T., Yla-Herttuala S., Yamamoto R., Butler S., Korpela H., Salonen R., Nyyssonen K., Palinski W., and Witztum J.L. 1992. Autoantibody against oxidised LDL and progression of carotid atherosclerosis. *Lancet* **339:** 883–887.

Schachter F., Faure-Delanef L., Guenot F., Rouger H., Froguel P., Lesueur-Ginot L., and Cohen D. 1994. Genetic associations with human longevity at the APOE and ACE loci. *Nat. Genet.* **6:** 29–32.

Schriner S.E., Linford N.J., Martin G.M., Treuting P., Ogburn C.E., Emond M., Coskun P.E., Ladiges W., Wolf N., Van Remmen H., Wallace D.C., and Rabinovitch P.S. 2005. Extension of murine life span by overexpression of catalase targeted to mitochondria. *Science* **308:** 1909–1911.

Schubert D. and Chevion M. 1995. The role of iron in beta amyloid toxicity. *Biochem. Biophys. Res. Commun.* **216:** 702–707.

Sebastiani P., Ramoni M.F., Nolan V., Baldwin C.T., and Steinberg M.H. 2005. Genetic dissection and prognostic modeling of overt stroke in sickle cell anemia. *Nat. Genet.* **37:** 435–440.

Sentinelli F., Romeo S., Barbetti F., Berni A., Filippi E., Fanelli M., Fallarino M., and Baroni M.G. 2006. Search for genetic variants in the p66Shc longevity gene by PCR-single strand conformational polymorphism in patients with early-onset cardiovascular disease. *BMC Genet.* **7:** 14.

Shimada T., Takeshita Y., Murohara T., Sasaki K., Egami K., Shintani S., Katsuda Y., Ikeda H., Nabeshima Y., and Imaizumi T. 2004. Angiogenesis and vasculogenesis are impaired in the precocious-aging klotho mouse. *Circulation* **110:** 1148–1155.

Silver M.H., Jilinskaia E., and Perls T.T. 2001. Cognitive functional status of age-confirmed centenarians in a population-based study. *J. Gerontol. B Psychol. Sci. Soc. Sci.* **56:** P134–P140.

Smith M.A., Harris P.L., Sayre L.M., and Perry G. 1997. Iron accumulation in Alzheimer disease is a source of redox generated free radical. *Proc. Natl. Acad. Sci.* **94:** 9866–9868.

Smith M.A., Wehr K., Harris P.L., Siedlak S.L., Connor J.R., and Perry G. 1998. Abnormal localization of iron regulatory protein in Alzheimer's disease. *Brain Res.* **788:** 232–236.

Sobel E., Louhija J., Sulkava R., Davavipour Z., Kontula K., Miettinen H., Tikkanen M., Kainulainen K., and Tilvis R. 1995. Lack of association of apolipoprotein E allele epsilon 4 with late-onset Alzheimer's disease among Finnish centenarians. *Neurology* **45:** 903–907.

Sohal R.S. and Weindruch R. 1996. Oxidative stress, caloric restriction, and aging. *Science* **273:** 59–63.

Sohal R.S., Agarwal S., Dubey A., and Orr W.C. 1993. Protein oxidative damage is associated with life expectancy of houseflies. *Proc. Natl. Acad. Sci.* **90:** 7255–7259.

Sohal R.S., Allen R.G., Farmer K.J., and Newton R.K. 1985. Iron induces oxidative stress and may alter the rate of aging in the housefly, *Musca domestica. Mech. Aging Dev.* **32:** 33–38.

Sohal R.S., Ku H.H., Agarwal S., Forster M.J., and Lal H. 1994. Oxidative damage, mitochondrial oxidant generation and antioxidant defenses during aging and in response to food restriction in the mouse. *Mech. Ageing Dev.* **74:** 121–133.

Stadtman E. 2001. Protein oxidation in aging and age-related diseases. *Ann. N.Y. Acad. Sci.* **928:** 22–23.

Stevens R.G., Jones D.Y., Micozzi M.S., and Taylor P.R. 1988. Body iron stores and the risk of cancer. *N. Engl. J. Med.* **319:** 1047–1052.

Sullivan J.L. 1989. The iron paradigm of ischemic heart disease. *Am. Heart J.* **117:** 1177–1188.

———. 1991. Antioxidants and coronary heart disease. *Lancet* **337:** 432–433.

———. 2003a. Are menstruating women protected from heart disease because of, or in spite of, estrogen? Relevance to the iron hypothesis. *Am. Heart J.* **145:** 190–194.

———. 2003b. Interactions of stored iron with traditional and inflammatory cardiovascular risk factors. *Atherosclerosis* **167:** 169.

Tanaka M., Gong J.S., Zhang J., Yoneda M., and Yagi K. 1998. Mitochondrial genotype associated with longevity. *Lancet* **351:** 185–186.

Terry D.F., Wilcox M.A., McCormick M.A., and Perls T.T. 2004. Cardiovascular disease delay in centenarian offspring. *J. Gerontol. A Biol. Sci. Med. Sci.* **59:** M385–M389.

Terry D.F., Wilcox M., McCormick M.A., Lawler E., and Perls T.T. 2003. Cardiovascular advantages among the offspring of centenarians. *J. Gerontol. A Biol. Sci. Med. Sci.* **58:** M425–M431.

Trifunovic A., Wredenberg A., Falkenberg M., Spelbrink J.N., Rovio A.T., Bruder C.E., Bohlooly-Y M., Gidlöf S., Oldfors A., Wibom R., Törnell J., Jacobs H.T., and Larsson N.G. Premature ageing in mice expressing defective mitochondrial DNA polymerase. *Nature* **429:** 417–423.

Tschanz J.T., Treiber K., Norton M.C., Welsh-Bohmer K.A., Toone L., Zandi P.P., Szekely C.A., Lyketsos C., and Breitner J.C. (Cache County Study on Memory, Health, and Aging). 2005. A population study of Alzheimer's disease: Findings from the Cache County Study on Memory, Health, and Aging. *Care Manag. J.* **6:** 107–114.

Tuomainen T.P., Nyyssonen K., Salonen R., Tervahauta A., Korpela H., Lakka T., Kaplan G.A., and Salonen J.T. 1997. Body iron stores are associated with serum insulin and blood glucose concentrations: Population study in 1,013 eastern Finnish men. *Diabetes Care* **20:** 426–428.

von Kobbe C., Harrigan J.A., May A., Opresko P.L., Dawut L., Cheng W.H., and Bohr V.A. 2003. Central role for the Werner syndrome protein/poly(ADP-ribose) polymerase 1 complex in the poly(ADP-ribosyl)ation pathway after DNA damage. *Mol. Cell. Biol.* **23:** 8601–8613.

Wallace D. 1997. Mitochondrial DNA in aging and disease. *Sci. Am.* **277:** 40–47.

Wannamethee S.G., Perry I.J., and Shaper A.G. 1996. Hematocrit and risk of NIDDM. *Diabetes Care* **45:** 576–579.

Willcox B.J., Willcox D.C., He Q., Curb J.D., and Suzuki M. 2006. Siblings of Okinawan centenarians share lifelong mortality advantages. *J. Gerontol. A Biol. Sci. Med. Sci.* **61:** 345–354.

Wilson J.G., Hoff Lindquist J., Grambow S.C., Crook E.D., and Maher J.F. 2003. Potential role of increased iron stores in diabetes. *Am. J. Med. Sci.* **325:** 332–339.

Wilson P.W., McGee D.L., and Kannel W.B. 1981. Obesity, very low density lipoproteins and glucose intolerance over fourteen years: The Framingham Study. *Am. J. Epidemiol.* **114:** 697–704.

Yu C.E., Oshima J., Fu Y.H., Wijsman E.M., Hisama F., Alisch R., Matthews S., Nakura J., Miki T., Ouais S., Martin G.M., Mulligan J., and Schellenberg G.D. 1996. Positional cloning of the Werner's syndrome gene. *Science* **272:** 258–262.

19

Mammalian Metabolism in Aging

Pere Puigserver
Dana-Farber Cancer Institute and Department of Cell Biology
Harvard Medical School, Boston, Massachusetts 02215

C. Ronald Kahn
Joslin Diabetes Center and Department of Medicine
Harvard Medical School, Boston, Massachusetts 02215

LIFE SPAN IN ORGANISMS IS CONTROLLED BY A COMPLEX interaction of genetic and metabolic factors. Although in lower organisms, some of these factors that determine longevity have been identified, in mammals, they are largely unknown. In this chapter, we focus on the metabolic factors associated with aging. Perhaps the most striking evidence that the aging process is under strict metabolic control comes from the highly investigated effects of caloric restriction in extending life span. There are several theories of aging which support the concept that metabolic stability is a major factor that determines life span. We concentrate on the main metabolic/signaling pathways that favor anabolic or catabolic responses and are altered as mammals age.

INSULIN ACTION IN AGING

It is well established that food restriction reduces age-dependent mortality and prolongs life span of animal species ranging in complexity from single-cell organisms to mammals (Masoro 2005). Insulin signaling was first linked to life span when it was shown that mutations in *daf-2* dramatically lengthen the life span of *Caenorhabditis elegans* (Kenyon et al. 1993), and *daf-2* was found to encode the *C. elegans* homolog of the insulin and insulin-like growth factor-1 (IGF-1) receptors (Kimura et al. 1997). During the past 15 years, an extensive body of work has shown that mutations in many genes in pathways homologous to the insulin and IGF-1 signaling pathways of mammals can prolong the life span of the nematode worm *C. elegans* and the fruit fly *Drosophila melanogaster* (for review, see Tatar et al. 2003; Kenyon 2005). The life span effects of the insulin/IGF-1-like pathway appear to occur primarily through the phosphatidylinositol-3-kinase (PI3K) pathway and a forkhead transcription factor of the FoxO subgroup. Similarly, down-regulation of the insulin/IGF-1 pathways has also been linked to longevity in mice, suggesting the generalization of these findings to higher species and providing a link between the role of metabolic regulation and life span across a wide range of species.

The Insulin/IGF-1-like Signaling Pathway in *C. elegans* and *D. melanogaster*

Figure 1 illustrates the remarkable similarity of the insulin and IGF-1 signaling pathways in worms, flies, and mammals. Both worms and flies have a single ortholog of the mammalian insulin and IGF-1 receptor tyrosine kinases called *daf-2* and *InR,* respectively (Kimura et al. 1997; Tatar et al. 2001). Upon binding of one of a family of insulin-like ligands (Pierce et al. 2001; Li et al. 2003; Murphy et al. 2003), the tyrosine kinase activities of DAF-2 and InR are activated and either phosphorylate directly the enzyme PI3K or, in flies, act indirectly through an insulin receptor substrate (CHICO) that is analogous to the IRS proteins of mammals that act to facilitate the InR PI3K activation (Bohni et al. 1999; Clancy et al. 2001; Tatar et al. 2001; Tu et al. 2002). The activated PI3K catalyzes the synthesis of phosphatidylinositol-3,4,5-triphosphate (PIP_3) (Friedman and Johnson 1988; Morris et al. 1996), which in turn activates 3-phosphoinositide-dependent protein kinase-1 (PDK-1). PDK-1 then phosphorylates and activates AKT/protein kinase B (PKB) and the serum/glucocorticoid-regulated kinase (SGK) (Hertweck et al. 2004). The most important AKT/PKB/SGK substrates from the standpoint of aging appear to be FoxO family forkhead transcription factors (Lin et al. 1997; Ogg et al. 1997). When the AKT/PKB/SGK kinases phosphorylate the forkhead transcription factors (DAF-16 in worms and dFOXO in flies), they are sequestered in the cytoplasm, reducing their effects on a variety of positively and negatively regulated genes.

Decreased signaling through the insulin/IGF-1-like pathways can increase the life span of worms by up to 250% (Arantes-Oliveira et al. 2003) and that of flies by up to 85% (Tatar et al. 2001). Mutations that decrease signaling through the insulin/IGF-1-like signaling pathway

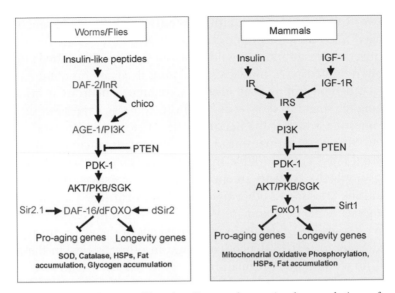

Figure 1. Insulin and IGF-1-like signaling pathways in the regulation of aging. Parallel pathways in *C. elegans* (worms), *D. melanogaster* (flies), and *M. musculus* (mice) and other mammals regulate the activity of forkhead transcription factors (DAF-16, dFOXO, or FoxO1) through a conserved signaling pathway. The action of insulin-like peptides (in worms and flies) or IGF-1 and insulin (in mice) ultimately leads to phosphorylation of the forkhead transcription factor, blocking its transcriptional activity. Inhibition of the insulin/IGF-1 pathways allows forkhead transcription factors to positively regulate the expression of anti-aging genes and negatively regulate the expression of proaging genes. Some of the DAF-16 targets identified in worms are listed. (SOD) Superoxide dismutase; (GST) glutathione-*S*-transferase; (HSP) heat shock protein (S.S. Lee et al. 2003; Murphy et al. 2003).

inhibit phosphorylation of forkhead transcription factors, allowing translocation to the nucleus (Henderson and Johnson 2001; Lin et al. 2001), where they modulate the expression of many genes that may work to increase life span (S. S. Lee et al. 2003; Murphy et al. 2003; Oh et al. 2006). Life span extension by signaling through the insulin/IGF-1-like signaling pathway requires *daf-16* in worms (Kenyon et al. 1993). Likewise, overexpression of dFOXO is sufficient to increase life span in flies (Hwangbo et al. 2004).

One of the most intriguing aspects of the insulin/IGF-1-related longevity pathways in worms and flies is cell nonautonomy, i.e., genetic alterations made in a small subset of cells or tissues increase the life span of the whole organism. Cell nonautonomous regulation of aging in the *daf-2* pathway in *C. elegans* was first demonstrated using mosaic analysis.

Thus, worms with a *daf-2* mutation in only a subset of neurons have a prolonged life span (Apfeld and Kenyon 1998).

Conversely, restoration of wild-type *daf-2* to specific subsets of neurons, but not intestine or muscle cells, in long-lived *daf-2* mutants reverses longevity, demonstrating that insulin signaling in neurons is critical in regulating life span (Wolkow et al. 2000). In contrast, expression of *daf-16* in the intestine, which is the major site of fat storage in the worm (analogous to the fat cell in mammals), can substantially increase life span of short-lived *daf-2 daf-16* double-null mutants (Libina et al. 2003). These findings imply the existence of endocrine longevity signals from the neurons and the intestine to other tissues of the worm. The exact identity of these signals remains unknown. What is known, however, is that somehow, overexpression of *daf-16* in intestinal cells of the worm up-regulates the expression of *daf-16*-dependent genes in other tissues. In flies, similar cell nonautonomous regulation of aging has been demonstrated by the finding that transgenic overexpression of wild-type dFOXO or a dFOXO mutant not responsive to insulin/IGF-1-like signaling in the cerebral fat body is sufficient to prolong life span (Giannakou et al. 2004; Hwangbo et al. 2004). As in the worm, in some way, overexpression of dFOXO in the cerebral fat body increases the amount of dFOXO in the nuclei of peripheral fat body cells, consistent with the existence of an endocrine signal or signals regulated by dFOXO in the cerebral fat body that can activate dFOXO in peripheral fat.

Regulation of Life Span by Insulin and IGF-1 in Mammals

Mammals possess distinct insulin and IGF-1 receptors that are homologous to the single insulin/IGF-1-like receptor present in worms and flies. The first evidence that they might be involved in control of longevity came from studies of several strains of mice with genetic lesions in the growth hormone (GH)–IGF-1 axis leading to IGF-1 deficiency (Liang et al. 2003). Thus, the Ames dwarf mouse, which has a mutation in a transcription factor Prop1 that is required for normal pituitary development, has reduced prolactin, TSH, and growth hormone and, consequently, low levels of circulating IGF-1. This is associated with an increase in mean life span by 49% for males and 68% for females compared to wild-type controls (Brown-Borg et al. 1996). The Snell dwarf mouse has a homozygous deletion of Pit1, another transcription factor involved in pituitary development, and also lives longer than wild-type controls by 29% for males and 51% for females (Flurkey et al. 2002). Similarly, "Little" mice have a mutation in the GH-releasing hormone receptor and consequently low GH and IGF-1 levels and live longer than controls by 23–25% (Flurkey et al. 2001). Finally, mice with a homozygous deletion of growth hormone receptor/growth-hormone-binding protein (GHR/BP) have dramatically lower levels of IGF-1 than controls and are also long-lived (Coschigano et al. 2000, 2003).

More recently, these findings have been extended by creation of genetically engineered mice with alterations in the IGF-1 or insulin signaling pathway that also exhibit increased longevity. Thus, whereas homozygous IGF-1 receptor deletion is lethal, heterozygous IGF-1 receptor knockout mice have increased life span (Holzenberger et al. 2003). The phenotype is most marked in female IGF-1 receptor knockout mice that live 35% longer than wild-type controls. This is interesting in the context that food restriction, which also increases life span, is associated with a decrease in IGF-1 levels and hence decreased IGF-1 signaling. In addition, wild-caught mice have lower IGF-1 levels than mixed-background laboratory mice and, when brought into the laboratory, are longer-lived (Harper et al. 2006). Taken together, these findings suggest that IGF-1 may be an important regulator of longevity.

The relationship between insulin signaling and aging in mammals is more complicated because of the many metabolic effects of raising and lowering insulin levels. Caloric restriction, which increases longevity, is well known to be associated with decreased insulin levels and increased insulin sensitivity (Masoro 2005). Under most circumstances, however, insulin deficiency or insulin resistance is associated with metabolic disorders, such as diabetes and obesity, and these are clearly linked to shortened, not lengthened, life span. On the other hand, using the Cre-*lox* system to selectively reduce insulin responsiveness in a single tissue has uncovered an interesting form of cell nonautonomous control of aging. This model is the fat-specific IR knockout (FIRKO) mouse.

FIRKO mice have a selective loss of insulin receptors in adipose tissue, whereas in other tissues, the receptors are intact. Although adipocytes from FIRKO animals are highly resistant to the effects of insulin, the animals are lean and are protected from insulin resistance and diabetes both on a high-fat diet and during normal aging (Fig. 2) (Bluher et al. 2002). FIRKO animals are protected from age-associated obesity that occurs despite the fact that food intake is equivalent on a per-mouse basis and increased on a weight-adjusted basis. The finding of leanness despite normal or increased food intake allowed one to ask the question: What would happen to longevity in a situation where there was no caloric restriction, but where mice were lean and had selective alterations in the insulin signaling system? When this was studied, it became clear that in

FIRKO Mice Are Lean Despite Normal Food Intake

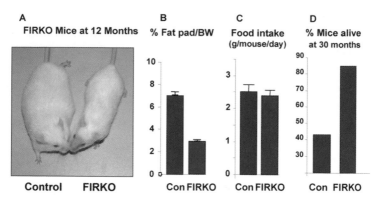

Figure 2. Effects of insulin receptor knockout in adipose tissue of mice metabolism and aging. (*A*) Picture of control and FIRKO mouse at 12 months of age showing obvious differences in body fat. (*B*) Epididymal fat pad weight of 12-month control and FIRKO mice expressed as percentage of body weight. (*C*) Food intake in grams per mouse per day for control and FIRKO mice. (*D*) Percentage of mice alive at 30 months. At this age, less than 50% of the control male mice remain alive, whereas more than 80% of male FIRKO mice are still alive. The data are similar for females.

addition to the metabolic phenotype, FIRKO mice have an increase in median (18%) and maximum (14%) life span compared to littermate controls (Bluher et al. 2002).

Although there has been some controversy over whether food restriction might increase longevity by decreasing metabolic rate, the discordance between food intake and weight gain in the FIRKO mouse suggested an increase in metabolic rate, and this was confirmed by studies of basal metabolic rate (M. Katic and C.R. Kahn, unpubl.). Furthermore, gene expression studies using adipose tissue taken from these mice as they aged revealed that they have higher levels of expression of the genes of mitochondrial electron transport and fatty acid oxidation than control mice, and an increase in mitochondrial DNA content. Therefore, in this model, a selective decrease in insulin signaling in fat produced increased life span of the animal as a whole, demonstrating that insulin signaling can regulate aging in mammals, i.e., the FIRKO mouse, in a noncell autonomous way. In addition, this model allowed us to separate the effects of leanness from reduced food intake in producing the long-lived phenotype. The FIRKO mouse also points to increased, rather than decreased, mitochondrial activity as beneficial for this phenotype.

The cell nonautonomy of insulin effects on life span implies the existence of humoral signals produced in fat that regulates the rate of aging in other tissues critical for longevity. In this case, it is important to note that IGF-1 levels are not different between FIRKO and control animals, so the longevity phenotype is not due to a secondary change in IGF-1 levels. There are some hormonal changes that could potentially have a role. For example, leptin levels in FIRKO mice are similar to those in controls despite decreased adipose mass, and this may be significant because in F_2 hybrids between long-lived wild-derived and shorter-lived mixed-background laboratory mice, life span correlated positively with leptin level but negatively with body weight (Harper et al. 2006). Adiponectin expression is also increased in the fat of FIRKO mice (Bluher et al. 2002), and increased adiponectin is a feature of calorie-restricted rodents and humans (Zhu et al. 2004; Weiss et al. 2006). The recent identification of steroid hormones as mediators of life span in worms (Broue et al. 2007) raises the interesting possibility that insulin signaling may also modulate steroid hormone metabolism by fat. Genes involved in the synthesis or metabolism of steroid hormones are expressed in mammalian fat (Belanger et al. 2002), and targets of activated DAF-16 in worms include steroid-metabolizing enzymes (Murphy et al. 2003). In contrast to the studies in worms and flies, at present, a link between FoxO signaling and longevity has not yet been made in mammals. There are several reasons for this. First, in parallel with the increased complexity of the insulin/IGF-1 pathway in mammals, there are four FoxO proteins (FoxO1, FoxO3, FoxO4, and FoxO6) that are orthologous to *daf-16* and dFOXO. All are regulated by Akt/PKB phosphorylation. Of these, FoxO1 is most closely related to *daf-16* and dFOXO and is the most studied (Wijchers et al. 2006). The effects of FoxO1 activity vary with the tissue, further increasing the complexity. As noted in the section above, in the liver, FoxO1, in collaboration with its coactivator PGC-1α, promotes glucose production by activating transcription of the gluconeogenic enzymes glucose-6-phosphatase (G6P) and phosphoenolpyruvate carboxy kinase (PEPCK) (Nakae et al. 2001; Puigserver et al. 2003). In skeletal muscle, insulin represses FoxO1, leading to reduced expression of PGC-1α and decreased expression of oxidative phosphorylation genes. In adipocytes, FoxO1 inhibits adipocyte differentiation, whereas dominant-negative FoxO1 restores differentiation to adipocytes lacking the insulin receptor. Consistent with these effects, FoxO1 haploinsufficiency lowers hepatic glucose production (Nakae et al. 2002; Samuel et al. 2006) and inhibits adipocyte hypertrophy, increasing adipocyte insulin sensitivity (Nakae et al. 2003). However, effects of altered FoxO1 levels or activity on life span have not been reported.

The Interaction of Sirt1 with Insulin/IGF-1 Signaling

In addition to regulation by the insulin/IGF-1 signaling pathway, the activities of DAF-16, dFOXO, and FoxO1 are also modulated by acetylation. Sirtuins are NAD^+-dependent class III protein/histone deacetylases (Imai et al. 2000). The founding member of this family, Sir2, was first identified in yeast, where its overexpression was found to extend life span. Overexpression of the homologous Sir2.1 in worms and dSir2 in flies also prolongs life span of those organisms (for review, see Haigis and Guarente 2006; Longo and Kennedy 2006). The effect of Sir2.1 overexpression in worms requires *daf-16*, but Sir2.1 is not required for life span extension by the insulin/IGF-1-like signaling pathway (Berdichevsky et al. 2006). Therefore, it appears that signals from the insulin/IGF-1-like signaling pathway and from Sir2.1 are integrated at the level of DAF-16 in worms. In mammals, Sirt1 is a member of a family of seven sirtuins and the closest homolog of Sir2. Sirt1 has a complex relationship with insulin signaling and FoxO1 in multiple tissues (Fig. 3). Whole-body deletion of Sirt1 leads to perinatal lethality, but surviving animals overexpress IGF-binding protein-1 (IGFBP-1), resulting in lower free IGF-1 levels and a

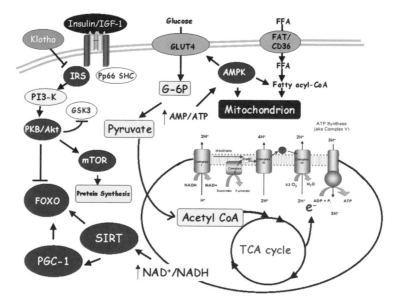

Figure 3. Relationship of insulin/IGF-1 and AMPK signaling to mitochondrial metabolism and oxidative phosphorylation. Insulin triggers a signaling cascade that promotes an anabolic response increasing glucose uptake, glycogen synthesis, fatty acid and triglyceride synthesis, and protein synthesis. On the opposite side, a catabolic response is induced through the AMPK pathway. For further details, see the text.

dwarf phenotype (McBurney et al. 2003; Lemieux et al. 2005). In the liver, Sirt1 deacetylates PGC-1α, leading to induction of gluconeogenic genes, hepatic glucose output, and the inhibition of glycolytic genes (Rodgers et al. 2005). In adipose tissue, Sirt1 represses the peroxisome proliferator activated receptor-γ (PPARγ) by recruiting a corepressor to PPARγ target genes. This leads to inhibition of fat storage and lipid mobilization via lipolysis (Picard et al. 2004). In pancreatic β cells, Sirt1 enhances glucose-stimulated insulin secretion by suppressing uncoupling protein 2 (UCP2) and thereby increasing available ATP levels (Moynihan et al. 2005; Bordone et al. 2006). Sirt1 activity is also modulated by the NAD^+/NADH ratio or the nicotinamide concentration, and these may change in different ways in different tissues (Haigis and Guarente 2006; Longo and Kennedy 2006).

Resveratrol is a small-molecule activator of Sirt1 activity, and treatment of mice on a high-fat diet with resveratrol has been shown to limit the accumulation of fat without reducing food intake (Lagouge et al. 2006). Resveratrol-treated animals also had increased exercise tolerance and an increase in resting energy expenditure. This is associated with an increase in expression of genes of the electron transport chain and oxidative phosphorylation in muscle, corresponding to decreased acetylation of PGC-1α and increased expression of PGC-1α, estrogen-related receptor-α (ERRα), nuclear respiratory factor (NRF-1), mitochondrial transcription factor A (Tfam), and uncoupling proteins 1 and 3 (UCP-1 and UCP-3). These gene expression changes are expected to increase fatty acid oxidation and heat production. Increased insulin sensitivity and glucose tolerance were also significantly increased by resveratrol treatment (Lagouge et al. 2006). Long-term (6 months) treatment with resveratrol also prevented the development of fatty liver and reduced the death rate of high-fat-diet mice to the level of chow-fed controls (Baur et al. 2006). The livers of resveratrol-fed mice had increased mitochondria, but in contrast to the findings in muscle, oxidative phosphorylation and many mitochondrial genes, including electron transport genes, were down-regulated in the liver.

It is not clear that all of the effects of resveratrol feeding are mediated through Sirt1, but it is intriguing to note that the metabolic phenotype is very similar to that of the FIRKO mouse and that the gene expression changes are diametrically opposed to those seen in the muscle of insulin-resistant and diabetic subjects. The gene expression changes seen in the muscle of resveratrol-fed mice are similar to those seen in the fat of FIRKO mice. However, there was no change in the expression levels of Sirt1 or FoxO proteins in the fat of FIRKO animals (M. Katic and C.R. Kahn, unpubl.). Further study will be needed to define the interrelationships between these phenotypes and insulin/IGF-1 signaling.

Klotho

An unexpected connection between insulin signaling and aging has come through the discovery of the *klotho* gene. *klotho* was identified serendipitously in an insertional mutagenesis screen as a hypomorphic allele resulting in severe early degenerative changes and short life span (Kuro-o et al. 1997). The homozygous mutant animals appear to be normal until 3 weeks of age and then exhibit severe growth retardation, osteoporosis, ectopic calcifications, skin atrophy, arteriosclerosis, emphysema, and premature atrophy of the thymus, testes, and ovaries. Mice homozygous for *klotho* die at an average age of 61 days. Analysis revealed that this phenotype was due to a single insertion of the transgene into the 5′-untranslated region of what was subsequently named the *klotho* gene, severely reducing the levels of *klotho* expression in homozygous mutant mice.

The *klotho* gene encodes a protein with a single transmembrane domain and an extracellular domain made of up two repeats (termed KL1 and KL2), which have some homology with the β-glycosidase family. *klotho* is expressed predominantly in choroid plexus of the brain and distal convoluted tubule of the kidney (Kuro-o et al. 1997; Tohyama et al. 2004). An alternatively spliced transcript is also made that lacks the transmembrane domain (Shiraki-Iida et al. 1998). The extracellular domain of *klotho* is detectable in the blood and cerebrospinal fluid of mice and humans (Imura et al. 2004), and a circulating form of *klotho* is present at a concentration of 100 pM in wild-type mice, suggesting that *klotho* might act as a hormone (Kurosu et al. 2005). Consistent with this hypothesis, transgenic overexpression of the *Klotho* gene is able to rescue the phenotypes of the *klotho* mouse. Indeed, expression of the *Klotho* transgene limited to the brain and testis is sufficient for rescue of the accelerated aging phenotype, indicating that *Klotho* can act in a cell nonautonomous manner (Kuro-o et al. 1997; Kurosu et al. 2005). Wild-type animals overexpressing *Klotho* also live longer than congenic controls (31% in males and 19% in females) (Kurosu et al. 2005).

Determining the exact mechanism of insulin action of *Klotho* has been difficult and somewhat controversial. In contrast to findings in food-restricted animals and the FIRKO mouse, the longevity associated with Klotho administration is associated with insulin resistance, as demonstrated by elevated insulin levels, despite normal blood glucose. Conversely, male Klotho-deficient mice are hypoglycemic, hypoinsulinemic, and more sensitive to the hypoglycemic effects of insulin and IGF-1 than wild-type controls (Mori et al. 2000; Utsugi et al. 2000). At a physiological and molecular level, intraperitoneal administration of the extracellular domain of Klotho causes insulin resistance in both male and female mice. This effect is not due to competition with insulin for binding to its receptor, but to inhibition of autophosphorylation and activation of the insulin and IGF-1 receptor tyrosine kinases leading to decreases in downstream signaling events such as phosphorylation of insulin receptor substrate-1 (IRS-1) and IRS-2 (Kurosu et al. 2005). Consistent with the hypothesis that Klotho prolongs life span at least in part by inhibiting insulin and/or IGF-1 signaling, heterozygous deletion of the IRS-1 gene partially rescues the reduced life span of male *klotho* animals and the histologic findings reminiscent of aging (Kurosu et al. 2005).

More recently, Klotho has been shown to have a major role in calcium and phosphate homeostasis. Levels of both calcium and phosphate are elevated in *klotho* mice (Kuro-o et al. 1997; Yoshida et al. 2002) due to elevated $1,25\text{-}(OH)_2$-vitamin D levels, and normalization of $1,25\text{-}(OH)_2$-vitamin D levels by placing mice on a vitamin-D-deficient diet at least partially reverses some of the *klotho* phenotypes, including slow growth, ectopic calcification, and early death (Tsujikawa et al. 2003). Insight into the mechanism of this altered vitamin D metabolism came from the observation that the mice with a deletion of the FGF-23 gene share many features with the *klotho* mouse, including a very short life span (Shimada et al. 2004; Nabeshima 2006). Fibroblast growth factor-23 (FGF-23) is a circulating growth factor produced in bone that inhibits phosphate transport in renal proximal tubular cells. Deficiency in FGF-23 results in phosphate retention and constitutively elevated expression of 1α-hydroxylase producing elevated $1,25\text{-}(OH)_2$-vitamin D levels and hypercalcemia (Shimada et al. 2004). Surprisingly, multiple FGF receptor isoforms can be immunoprecipitated with Klotho from tissue culture cells, and Klotho potentiates the interaction between FGF-23 and these receptors, thus enhancing FGF-23 signaling (Kurosu et al. 2006). Defective FGF-23 signaling could explain many phenotypes of the *klotho*-deficient mouse, but it is unclear how this signaling pathway contributes to the increased life span of mice overexpressing Klotho. FGF-23 knockout mice are hypoglycemic relative to littermates, suggesting that FGF-23 signaling might antagonize the action of insulin. FGF-23 is also known to activate the mitogen-activated protein kinases ERK1/2 and p38 (Yamashita et al. 2002). ERK is capable of serine phosphorylating IRS-1, thereby negatively regulating insulin signaling. Therefore, FGF-23 may prolong life span at least in part by inhibiting insulin and/or IGF-1 signaling at the level of the IRS proteins. This is consistent with the finding that IRS-1 deletion can partially suppress the short life span phenotype of *klotho* mice (Kurosu et al. 2005). Which of these actions of Klotho is most relevant for aging is not yet clear. A human homolog of *klotho* has been identified (Kuro-o et al. 1997). Interestingly, a polymorphism in the *klotho* gene that results in amino acid substitutions has been associated with longevity in some human populations (Arking et al. 2002).

mTOR

Another key nutrient pathway that in lower organisms has been negatively associated with aging is TOR (target of rapamycin). The TOR pathway is active under high-nutrient conditions, and it is a downstream component of the insulin signaling (Wullschleger et al. 2006). In *C. elegans,* loss-of-function mutations in *daf-15* (ortholog of the mammalian gene raptor that is part of the mTOR protein complex TORC1) or *let-363* (ortholog of the mammalian gene, mTOR) prolongs life span (Vellai et al. 2003; Jia et al. 2004). A TOR target gene that seems to be important for TOR-dependent longevity effects is *bec-1* (ortholog of the mammalian gene, beclin 1) that promotes autophagy. Interestingly, *bec-1* is required for increases of life span in *daf2* mutants (Melendez et al. 2003). In flies, dTOR also extends life span and specific reductions of the dTOR pathway in the fat body are sufficient to increase longevity (Kapahi et al. 2004). Taken together, in these organisms, TOR is connected to the nutrient- and insulin-signaling effects as they relate to life span.

In mammals, the mTOR pathway metabolic effects in whole-body physiology are complex and not completely understood. In addition, there are no data as to its involvement in mammalian aging. On the one hand, part of the metabolic effects of insulin are mediated through mTOR, but at the same time, the negative feedback loop generated by S6K provides uncertainty about the final outcomes of constitutive activation or inhibition of mTOR activity (Shah et al. 2004). In fact, knockout mice of two downstream targets of mTOR, S6K1 and 4EBP1, display opposite phenotypes. In the case of S6K1 KO, mice are resistant to diet-induced and age-related obesity and are insulin-sensitive (Um et al. 2004). In contrast, mice null for 4E-BP1 or 4E-BP2 have different phenotypes depending on the genetic background, C57BL/6J KO mice have a lean phenotype but BALB/c KO mice have increased body weight and adiposity (Tsukiyama-Kohara et al. 2001; Le Bacquer et al. 2007). Future studies with mutant mice for different components of the mTOR pathway will provide valuable information about the effects of mTOR in whole-body nutrient and energy metabolism and its connection to aging.

AMPK

The AMPK (AMP-activated protein kinase) pathway is activated under low nutrient or energy conditions. Recent studies in *C. elegans* have shown that it has a role in longevity. Overexpression of *aak-2,* the mammalian ortholog of the AMPKα subunit, increases life span (Apfeld et al. 2004). In addition, *aak-2* is required for the life span extension of *daf-2* mutants or Sir2-overexpressing worms (Curtis et al. 2006). In mammals, multiple types of experiments also implicate an important role for AMPK in aging processes. For example, recent work from Shulman and colleagues has shown that the ability to activate AMPK in response to AMPK activators or exercise is blunted in old compared to young rodents (Reznick et al. 2007).

AMPK is a component of a signaling pathway that functions as a cellular sensor for the energy status (Carling 2004; Hardie and Sakamoto 2006). It was first identified as a protein kinase that phosphorylates and inactivates enzymes involved in cholesterol and fatty acid synthesis, HMG-CoA reductase, and the acetyl-CoA carboxylases, ACC1 and ACC2 (Carling et al. 1987). AMPK is a heterotrimer formed by a catalytic α subunit and β and γ regulatory subunits. This kinase complex is activated by increases of AMP:ATP ratios that occur during nutrient deprivation through different molecular mechanisms. Interestingly, naturally occurring mutations in AMPK in the γ regulatory subunit of AMPK, which affects AMP binding, causes hereditary heart disease in humans (Scott et al. 2004).

This highly conserved nutrient pathway has a central role in the control of energy and nutrient homeostasis. This control is mainly achieved through specific metabolic effects in central and peripheral tissues. In the hypothalamus, fasting activates AMPK, whereas refeeding inhibits it (Andersson et al. 2004; Minokoshi et al. 2004). Among the possible regulators, leptin decreases AMPK activity, in part, through the melanocortin-4 receptor, and AMPK is required for the anorexigenic effects of leptin (Minokoshi et al. 2004). Hypothalamic activation of AMPK correlates with increases of food intake and body weight (Kim et al. 2004). In peripheral tissues, AMPK directly activates catabolic pathways to maintain energy homeostasis and allows cells to function in this physiological context. Physical exercise is perhaps one of the most potent physiological activators of AMPK in skeletal muscle (Mu et al. 2001; Hardie and Sakamoto 2006). During this highly energetically demanding situation, AMPK increases glycogen breakdown, glucose uptake and glycolysis, and oxidation of fatty acids. Although the presumed key regulator of AMPK in response to exercise is AMP, it is possible that other factors or pathways can also lead to increased AMPK activity. For instance, changes in interleukin-6 (IL-6) or adipokines, such as adiponectin or leptin, also regulate AMPK activity (Kahn et al. 2005).

Recent studies in yeast and in mammalian cells have placed the LKB1 kinase as an upstream activator of AMPK that mediates its effects in glucose and fatty acid metabolism (Hawley et al. 2003; Hong et al. 2003; Shaw et al. 2004). In response to exercise, LKB1 is required for activation of AMPK (Hawley et al. 2003). Metformin, an antidiabetic drug that mainly targets hepatic glucose output through the AMPK pathway, also requires LKB1 kinase to decrease blood glucose levels (Shaw et al. 2005).

The effects of AMPK in peripheral tissues are important for the dysregulation of metabolic pathways that occurs in aging. For example, AMPK activation controls oxidation of fatty acids acutely through direct phosphorylation of ACC, a key enzyme in fatty acid synthesis that catalyzes conversion of acetyl-CoA to malonyl-CoA. Malonyl-CoA is a key intracellular metabolite that potently inhibits CPT1 (carnitine:palmitoyl-CoA transferase-1), a rate-limiting enzyme in the fatty acid oxidation pathway that facilitates entrance of long-chain fatty acyl-CoA into mitochondria for complete oxidation. As a consequence of these metabolic effects, AMPK activation will decrease cellular malonyl-CoA levels and result in fine-tuning regulation of fatty acid oxidation. AMPK rapidly regulates metabolic changes through direct phosphorylation of enzymes and indirectly by controlling expression of metabolic genes (Kahn et al. 2005; Hardie and Sakamoto 2006). Among the transcriptional downstream targets of AMPK is the transcriptional coactivator PGC-1α that controls mitochondrial and fatty acid oxidation. Indeed, stimuli such as exercise or the AMPK activator AICAR induces PGC-1α gene expression (Suwa et al. 2003; Lee et al. 2006). However, how mechanistically AMPK controls this coactivator is unknown. Notably, with regard to aging, blunted AMPK activation in aged rodents also correlates with a lack of PGC-1α gene expression induction.

PGC-1 AND SIRTUINS

In lower organisms, Sir2—the homolog of the mammalian Sirtuin-1 (SIRT1)—regulates life span (for more detailed information, see the above and Chapter 2). The transcriptional coactivator PGC-1α is another component of the mammalian metabolic machinery that regulates different programs linked to glucose and lipid mitochondrial oxidation (Finck and Kelly 2006; Handschin and Spiegelman 2006). Recent studies have shown that SIRT1 deacetylates and regulates PGC-1α, functionally linking these two pathways (Nemoto et al. 2005; Rodgers et al. 2005). Fasting is a physiological situation in which PGC-1α is deacetylated, an effect that can be mimicked by low concentrations of glucose. Under these conditions of nutrient restriction, PGC-1α is deacetylated and turns on genes of mitochondrial and fatty acid oxidation to allow a nutrient switch from glucose to fatty acid oxidation. Importantly, SIRT1 is required for this switch in nutrient utilization, indicating that low concentrations of glucose trigger a pathway that is sensed by SIRT1 (Gerhart-Hines et al. 2007). A similar regulation occurs in liver during food deprivation. Again, SIRT1 activity is increased and deacetylates PGC-1α, which increases hepatic glucose production and β-oxidation of fatty acids (Rodgers et al. 2005). A crucial question is whether these constant fluctuations and switches in response to nutrients that are controlled by these proteins have a role in the aging process. Recent studies have shown that under conditions of high caloric intake, rodents treated with resveratrol, a plant polyphenol that activates SIRT1 deacetylase, display a longer life span (Baur et al. 2006). Parallel studies also indicated that an important target of resveratrol is the protein complex formed by PGC-1α and SIRT1. In this context, resveratrol induces PGC-1α deacetylation in tissues such as skeletal muscle, and this is correlated with improved mitochondrial function and resistance to diet-induced obesity (Fig. 4) (Lagouge et al. 2006).

With respect to age-associated diseases, it is clear that dysregulation of these nutrient switches can be related to some of the metabolic inflexibilities observed in type 2 diabetes. Furthermore, it is important to note that dysregulation of these glucose and lipid pathways is also connected to oxidative stress (discussed in Chapters 1 and 3). These increases in free radical production have been causatively associated with insulin resistance (Houstis et al. 2006). The control of ROS production is complex, and many different pathways and regulators are involved in both acute and chronic responses. In this context, PGC-1α (St-Pierre et al. 2006), SIRT1, and other SIRT1 targets, such as FoxO proteins (Brunet et al. 2004) and p53 (Luo et al. 2001; Vaziri et al. 2001), control ROS production in response to stress conditions and are important to maintain survival. In fact, one of the main theories explaining the biological consequences of the aging process is oxidative stress. A key tissue vulnerable to oxidative stress is the brain, and recent work from the Spiegelman and other labs has provided evidence that PGC-1α has an important role to protect neurons in response to oxidative damage (Cui et al. 2006; St-Pierre et al. 2006; Weydt et al. 2006). Since neurodegeneration is one of the main pathologies associated with aging, it is conceivable that the control of

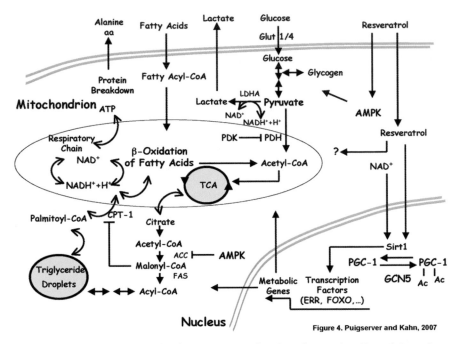

Figure 4. Metabolic control of nutrient-regulated pathways implicated in aging. Cellular metabolism is controlled through a number of hormonal and nutrient pathways. These metabolic pathways are controlled either by regulation of key enzymes or at the level of activity and gene expression. Aging or associated diseases display metabolic dysregulation such as resistance to AMPK activation or accumulation of triglycerides linked to insulin resistance. A natural compound, resveratrol targets several of these nutrient pathways to induce mitochondrial oxidative activity through activation of Sirt1/PGC-1α. For further details, see the text.

PGC-1α might serve as a drug target (McGill and Beal 2006). A key mechanistic and biological question is what are the targets of PGC-1α and SIRT1 that execute the protective program in response to oxidative stress? Although different enzymes involved in ROS detoxification have been proposed, it is not clear whether they are sufficient to maintain cellular integrity and survival in response to oxidative stress.

GLUCOSE METABOLISM

Glucose is the most basic energetic molecule, and certain cells such as neurons almost entirely depend on glucose to maintain their energetic status. Mammalian cells have different glucose transporters that control glucose entrance in response to different nutrient or stress stimuli

(Shepherd and Kahn 1999; Uldry and Thorens 2004). Once transported into the cell, glucose is metabolized through different biochemical pathways. Under high-nutrient conditions, glucose is stored through synthesis of glycogen, particularly in liver and muscle. A percentage of glucose is used through glycolysis to lactate or by full mitochondrial oxidation (Fig. 4). An interesting theory has been developed and supports the notion that rates of glycolysis might contribute to the aging process (Hipkiss 2006). The basis is that glycolytic intermediates, for example, methyglyoxal—a precursor of dihydroxyacetone and glyceraldehyde-3-phosphates—can glycate proteins and increase free radical production. Interestingly, as glycolysis is a key insulin metabolic target, it opens the question as to what extent the effects of the insulin pathway on aging are affected by the glycolytic rates. Under conditions of high glucose and insulin, increased uptake of glucose can be routed to glycogen synthesis. This occurs by an effect of insulin to promote dephosphorylation of glycogen synthase through activation of PP1 (protein phosphatase 1) and by inhibition of cAMP-activated PKA, which stimulates phosphorylation of glycogen synthase (Cross et al. 1995). Insulin does not directly regulate the PP1 enzymatic activity, rather, it controls the localization of PP1 to glycogen through PTG (protein targeting to glycogen) (Brady et al. 1997). Heterozygous PTG mice display glucose intolerance and insulin resistance with age, indicating the importance of maintaining tight regulation of glycogen synthesis and breakdown to maintain nutrient homeostasis (Crosson et al. 2003; Meynial-Denis et al. 2005).

Hepatic glucose production from different precursors—mainly lactate and alanine from skeletal muscle and glycerol from adipose tissue—maintains blood glucose levels in narrow limits during fasting or prolonged exercise (Pilkis and Granner 1992; Gurney et al. 1994). Gluconeogenesis is tightly controlled negatively by insulin and positively by counterregulatory hormones, such as glucagon, glucocorticoids, and catecholamines. Hepatic glucose metabolism is also controlled indirectly by the hypothalamus through effects of both nutrients (Lam et al. 2005) and insulin (Fisher and Kahn 2003). The key enzymes of this pathway, PEPCK and glucose-6-phosphatase, are transcriptionally regulated during nutrient deprivation by a complex transcriptional network (Hanson and Reshef 1997). Among components of this network are the transcription factors CREB and its coactivator TORC1 (Herzig et al. 2001; Koo et al. 2005), and HNF-4α and FoxO1, which is controlled through the transcriptional coactivator PGC-1α (Yoon et al. 2001; Puigserver et al. 2003). In addition to the hormonal control, fasting triggers a hepatic nutrient pathway that also activates SIRT1 and deacetylates PGC-1α, and likely other unidentified targets, to induce gluconeogenic expression and repress glycolytic gene expression (Rodgers et al. 2005). AMPK is increased in nutrient deprivation states and negatively regulates gluconeogenesis (Lochhead et al. 2000). Although the physiological relevance of AMPK to repress glucose production is not clear, it may act as a safety mechanism to maintain hepatocyte ATP levels in prolonged fasted states. In summary, hepatic glucose production is controlled at different levels and through different mechanisms that involve posttranslational modifications such as phosphorylation and acetylation.

LIPID METABOLISM

Similar to carbohydrate metabolism, lipid homeostasis is under tight control by hormonal and nutrient inputs. The effects of these inputs in lipid metabolic pathways are controlled through several kinases, such as AKT, AMPK, and PKA, that directly phosphorylate key metabolic enzymes. In addition, the same network of kinases also targets transcriptional components that regulate gene expression of lipid metabolism enzymes. Two main classes of transcription factors control lipid metabolism: (1) the basic helix-loop-helix class SREBPs (Horton et al. 2002; Goldstein et al. 2006) and ChREBP and (2) nuclear hormone receptors such as PPARs and LXRs (C. H. Lee et al. 2003; Beaven and Tontonoz 2006; Kalaany and Mangelsdorf 2006). In each case, an important role of this regulation also occurs at the transcriptional coactivator level (Spiegelman and Heinrich 2004; Lin et al. 2005; Yang et al. 2006). Control of lipid homeostasis is achieved through metabolic pathways of synthesis and degradation. For example, high nutrient concentrations favor fatty acid synthesis by effects on LXR and SREBP transcriptional pathways. In contrast, nutrient deprivation activates fatty acid oxidation in peripheral tissues than can be completed through OXPHOS activity or partial oxidation through hepatic ketonegenesis. In aging, there is a clear tendency to dysregulation of lipid metabolic pathways that leads to accumulation of triglycerides and deficits in oxidation of fatty acids (Petersen et al. 2004). The inability to efficiently catabolize these substrates could lead to important oxidative damage in different tissues and to the accumulation of intracellular free fatty acids and other lipids that may also have a role in inducing insulin resistance.

In humans, alleles of apolipoproteins have been genetically associated with longevity. In particular, the ApoE4 allele that is linked to Alzheimer's disease and cardiovascular disease has been correlated with aging. Recent studies have confirmed a significantly lower ApoE4 allele frequency in human centenarians (Capri et al. 2006). However, it is currently unknown how this specific apolipoprotein might influence biological processes that affect aging.

PROTEIN METABOLISM

Protein and amino acid metabolism is under continuous nutrient control and is integrated into glucose and lipid pathways. One of important consequences of dysregulated protein metabolism at both the synthesis and degradation level is muscle loss that occurs in aging-associated diseases (Bales and Ritchie 2002; Greenlund and Nair 2003). A major pathway controlling protein synthesis is mTOR, which in lower organisms has been shown to negatively influence life span (Syntichaki et al. 2007). Interestingly, a diet deficient in methionine increases longevity in rodents (Orentreich et al. 1993). To what extent this effect is through reduced levels of protein synthesis is unknown; however, methionine is the first amino acid to initiate polypeptides. In contrast, protein degradation through the proteosomal and lysosomal pathways is the major determinant of skeletal muscle mass (Jagoe and Goldberg 2001). Little is known about the transcriptional regulatory pathways that control protein synthesis and degradation. Recently, the forkhead transcription factor FOXO1, a target of growth factors and insulin, has been shown to control protein degradation in situations of nutrient deprivation (Sandri et al. 2004). How protein metabolism controlled by FoxO transcription factors might mediate part of the effects on life span is unknown.

CONCLUSIONS AND PERSPECTIVES

The insulin/IGF-1-like signaling pathways of worms and flies are the best-studied models of endocrine regulation in life span, identifying insulin-like peptides as mediators of longevity in these organisms. In mammals, IGF-1 is the best correlated endocrine regulator of aging, and insulin appears to have a role in specific tissues, such as fat, to regulate life span. In both the lower species and mammals, some of these effects occur in a cell nonautonomous fashion, i.e., a genetic alteration of a subset of cells or tissues can alter the life span of the organisms through endocrine signals. These actions may be further modified by a number of proteins, including extracellular signaling proteins, such as Klotho, and intracellular modifiers of insulin/IGF-1 action, such as the sirtuins. In addition, these endocrine effects may contribute to the well-known effects of caloric restriction to prolong life span. The fact that some of these mechanisms function in a cell nonautonomous manner in both lower species and mammals suggests that circulating mediators of aging exist. Almost certainly, these also are circulating mediators of metabolism. Defining these mediators may offer new therapeutic approaches to control metabolic disease and improve the quantity and quality of life span.

Two other unexplained aspects regarding the role of insulin/IGF-1 signaling and longevity are worth further consideration. One is the major difference between insulin/IGF-1 signaling in worms and mammals exemplified by the dauer state. Dauer is a developmental stage between the larval stage and adulthood that occurs when worms are raised under unfavorable conditions, such as food deprivation or crowding. The dauer form is highly stress-resistant and very long-lived, as time spent as a dauer does not reduce the post-dauer life span. In fact, the *daf* genes, which include most of the components of the insulin/IGF-1 signaling system in the worm, received their names because of their effects on dauer formation. What is poorly understood about dauer, however, is that in contrast to higher animals exposed to starvation or dietary restriction that break down fat stores, worms in dauer accumulate fat in intestinal cells. Strong mutations in *daf* genes lead to dauer formation even under favorable conditions, whereas some weak mutations in *daf-2* prolong life span without significantly altering development. In this regard, the metabolic physiology of the worm with respect to fat storage seems to be diametrically opposite the effects of decreased insulin/IGF-1 signaling in mammals.

A second consideration about this system is why insulin/IGF-1 deficiency or resistance in mammals appears to be, on the one hand, beneficial for longevity and, on the other hand, linked to disease states such as diabetes and growth retardation. It is worth noting that although insulin resistance is a major risk factor for type 2 diabetes and the metabolic syndrome, about 25% of normal adults are as insulin-resistant as individuals with type 2 diabetes, and this alteration in insulin sensitivity appears to be genetically programmed. Perhaps the persistence of this abnormality in such an important metabolic pathway has persisted through evolution because of its beneficial effects on longevity. This should eventually become clear as we dissect the complex interaction between insulin signaling, metabolism, and control of life span (Fig. 5).

Finally, understanding the mechanisms of AMPK function will be also crucial in identifying how AMPK activity becomes insensitive to different stimuli as mammals age. Although these effects on AMPK "resistance" were reported in skeletal muscle, other tissues might share this lack of stimulation. Consequently, in tissues from old mammals, downstream mediators of AMPK, including PGC-1 and possibly sirtuins, might not

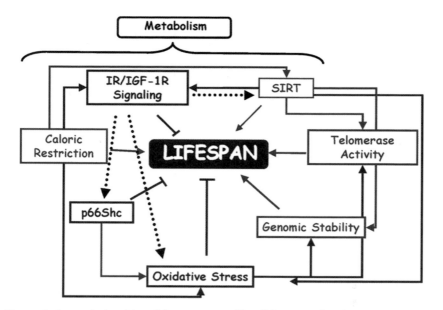

Figure 5. Interrelationship of factors controlling life span related to insulin action and metabolism. Life span is regulated by metabolic and genetic factors that control activities of different biological processes. Among these processes, the mechanisms that maintain the capacity to respond to nutrient and hormonal signals will determine organismal age.

be able to reprogram cellular metabolism and maintain the energy status compromising whole-body activity. Thus, aging-linked decreases in AMPK activity might be an important factor that can contribute to a reduced mitochondrial function associated with intracellular lipid accumulation in nonadipose tissues. It is even plausible that this dysregulation might be closely connected to aging-associated pathologies.

REFERENCES

Andersson U., Filipsson K., Abbott C.R., Woods A., Smith K., Bloom S.R., Carling D., and Small C.J. 2004. AMP-activated protein kinase plays a role in the control of food intake. *J. Biol. Chem.* **279:** 12005–12008.

Apfeld J. and Kenyon C. 1998. Cell nonautonomy of *C. elegans daf-2* function in the regulation of diapause and life span. *Cell* **95:** 199–210.

Apfeld J., O'Connor G., McDonagh T., DiStefano P.S., and Curtis R. 2004. The AMP-activated protein kinase AAK-2 links energy levels and insulin-like signals to life span in *C. elegans. Genes Dev.* **18:** 3004–3009.

Arantes-Oliveira N., Berman J.R., and Kenyon C. 2003. Healthy animals with extreme longevity. *Science* **302:** 611.

Arking D.E., Krebsova A., Macek M., Sr., Macek M., Jr., Arking A., Mian I.S., Fried L., Hamosh A., Dey S., McIntosh I., and Dietz H.C. 2002. Association of human aging with a functional variant of klotho. *Proc. Natl. Acad. Sci.* **99:** 856–861.

Bales C.W. and Ritchie C.S. 2002. Sarcopenia, weight loss, and nutritional frailty in the elderly. *Annu. Rev. Nutr.* **22:** 309–323.

Baur J.A., Pearson K.J., Price N.L., Jamieson H.A., Lerin C., Kalra A., Prabhu V.V., Allard J.S., Lopez-Lluch G., Lewis K., Pistell P.J., Poosala S., Becker K.G., Boss O., Gwinn D., Wang M., Ramaswamy S., Fishbein K.W., Spencer R.G., Lakatta E.G., Le Couteur D., Shaw R.J., Navas P., Puigserver P., Ingram D.K., de Cabo R., and Sinclair D.A. 2006. Resveratrol improves health and survival of mice on a high-calorie diet. *Nature* **444:** 337–342.

Beaven S.W. and Tontonoz P. 2006. Nuclear receptors in lipid metabolism: Targeting the heart of dyslipidemia. *Annu. Rev. Med.* **57:** 313–329.

Belanger C., Luu-The V., Dupont P., and Tchernof A. 2002. Adipose tissue intracrinology: Potential importance of local androgen/estrogen metabolism in the regulation of adiposity. *Horm. Metab. Res.* **34:** 737–745.

Berdichevsky A., Viswanathan M., Horvitz H.R., and Guarente L. 2006. *C. elegans* SIR-2.1 interacts with 14-3-3 proteins to activate DAF-16 and extend life span. *Cell* **125:** 1165–1177.

Bluher M., Michael M.D., Peroni O.D., Ueki K., Carter N., Kahn B.B., and Kahn C.R. 2002. Adipose tissue selective insulin receptor knockout protects against obesity and obesity-related glucose intolerance. *Dev. Cell* **3:** 25–38.

Bohni R., Riesgo-Escovar J., Oldham S., Brogiolo W., Stocker H., Andruss B.F., Beckingham K., and Hafen E. 1999. Autonomous control of cell and organ size by CHICO, a *Drosophila* homolog of vertebrate IRS1-4. *Cell* **97:** 865–875.

Bordone L., Motta M.C., Picard F., Robinson A., Jhala U.S., Apfeld J., McDonagh T., Lemieux M., McBurney M., Szilvasi A., Easlon E.J., Lin S.J., and Guarente L. 2006. Sirt1 regulates insulin secretion by repressing UCP2 in pancreatic beta cells. *PLoS Biol.* **4:** e31.

Brady M.J., Printen J.A., Mastick C.C., and Saltiel A.R. 1997. Role of protein targeting to glycogen (PTG) in the regulation of protein phosphatase-1 activity. *J. Biol. Chem.* **272:** 20198–20204.

Broue F., Liere P., Kenyon C., and Baulieu E.E. 2007. A steroid hormone that extends the lifespan of *Caenorhabditis elegans*. *Aging Cell* **6:** 87–94.

Brown-Borg H.M., Borg K.E., Meliska C.J., and Bartke A. 1996. Dwarf mice and the ageing process. *Nature* **384:** 33.

Brunet A., Sweeney L.B., Sturgill J.F., Chua K.F., Greer P.L., Lin Y., Tran H., Ross S.E., Mostoslavsky R., Cohen H.Y., Hu L.S., Cheng H.L., Jedrychowski M.P., Gygi S.P., Sinclair D.A., Alt F.W., and Greenberg M.E. 2004. Stress-dependent regulation of FOXO transcription factors by the SIRT1 deacetylase. *Science* **303:** 2011–2015.

Capri M., Salvioli S., Sevini F., Valensin S., Celani L., Monti D., Pawelec G., De Benedictis G., Gonos E.S., and Franceschi C. 2006. The genetics of human longevity. *Ann. N.Y. Acad. Sci.* **1067:** 252–263.

Carling D. 2004. The AMP-activated protein kinase cascade: A unifying system for energy control. *Trends Biochem. Sci.* **29:** 18–24.

Carling D., Zammit V.A., and Hardie D.G. 1987. A common bicyclic protein kinase cascade inactivates the regulatory enzymes of fatty acid and cholesterol biosynthesis. *FEBS Lett.* **223:** 217–222.

Clancy D.J., Gems D., Harshman L.G., Oldham S., Stocker H., Hafen E., Leevers S.J., and Partridge L. 2001. Extension of life-span by loss of CHICO, a *Drosophila* insulin receptor substrate protein. *Science* **292:** 104–106.

Coschigano K.T., Clemmons D., Bellush L.L., and Kopchick J.J. 2000. Assessment of growth parameters and life span of GHR/BP gene-disrupted mice. *Endocrinology* **141:** 2608–2613.

Coschigano K.T., Holland A.N., Riders M.E., List E.O., Flyvbjerg A., and Kopchick J.J. 2003. Deletion, but not antagonism, of the mouse growth hormone receptor results in severely decreased body weights, insulin, and insulin-like growth factor I levels and increased life span. *Endocrinology* **144:** 3799–3810.

Cross D.A., Alessi D.R., Cohen P., Andjelkovich M., and Hemmings B.A. 1995. Inhibition of glycogen synthase kinase-3 by insulin mediated by protein kinase B. *Nature* **378:** 785–789.

Crosson S.M., Khan A., Printen J., Pessin J.E., and Saltiel A.R. 2003. PTG gene deletion causes impaired glycogen synthesis and developmental insulin resistance. *J. Clin. Invest.* **111:** 1423–1432.

Cui L., Jeong H., Borovecki F., Parkhurst C.N., Tanese N., and Krainc D. 2006. Transcriptional repression of PGC-1alpha by mutant huntingtin leads to mitochondrial dysfunction and neurodegeneration. *Cell* **127:** 59–69.

Curtis R., O'Connor G., and DiStefano P.S. 2006. Aging networks in *Caenorhabditis elegans:* AMP-activated protein kinase (aak-2) links multiple aging and metabolism pathways. *Aging Cell* **5:** 119–126.

Finck B.N. and Kelly D.P. 2006. PGC-1 coactivators: Inducible regulators of energy metabolism in health and disease. *J. Clin. Invest.* **116:** 615–622.

Fisher S.J. and Kahn C.R. 2003. Insulin signaling is required for insulin's direct and indirect action on hepatic glucose production. *J. Clin. Invest.* **111:** 463–468.

Flurkey K., Papaconstantinou J., and Harrison D.E. 2002. The Snell dwarf mutation Pit1(dw) can increase life span in mice. *Mech. Ageing Dev.* **123:** 121–130.

Flurkey K., Papaconstantinou J., Miller R.A., and Harrison D.E. 2001. Lifespan extension and delayed immune and collagen aging in mutant mice with defects in growth hormone production. *Proc. Natl. Acad. Sci.* **98:** 6736–6741.

Friedman D.B. and Johnson T.E. 1988. A mutation in the *age-1* gene in *Caenorhabditis elegans* lengthens life and reduces hermaphrodite fertility. *Genetics* **118:** 75–86.

Gerhart-Hines Z., Rodgers J.T., Bare O., Lerin C., Kim S.H., Mostoslavsky R., Alt F.W., Wu Z., and Puigserver P. 2007. Metabolic control of mitochondrial function and fatty acid oxidation through SIRT1/PGC-1alpha. *EMBO J.* **26:** 1913–1923.

Giannakou M.E., Goss M., Junger M.A., Hafen E., Leevers S.J., and Partridge L. 2004. Long-lived *Drosophila* with overexpressed dFOXO in adult fat body. *Science* **305:** 361.

Goldstein J.L., DeBose-Boyd R.A., and Brown M.S. 2006. Protein sensors for membrane sterols. *Cell* **124:** 35–46.

Greenlund L.J. and Nair K.S. 2003. Sarcopenia: Consequences, mechanisms, and potential therapies. *Mech. Ageing Dev.* **124:** 287–299.

Gurney A.L., Park E.A., Liu J., Giralt M., McGrane M.M., Patel Y.M., Crawford D.R., Nizielski S.E., Savon S., and Hanson R.W. 1994. Metabolic regulation of gene transcription. *J. Nutr.* (suppl. 8) **124:** 1533S–1539S.

Haigis M.C. and Guarente L.P. 2006. Mammalian sirtuins: Emerging roles in physiology, aging, and calorie restriction. *Genes Dev.* **20:** 2913–2921.

Handschin C. and Spiegelman B.M. 2006. Peroxisome proliferator-activated receptor gamma coactivator 1 coactivators, energy homeostasis, and metabolism. *Endocr. Rev.* **27:** 728–735.

Hanson R.W. and Reshef L. 1997. Regulation of phosphoenolpyruvate carboxykinase (GTP) gene expression. *Annu. Rev. Biochem.* **66:** 581–611.

Hardie D.G. and Sakamoto K. 2006. AMPK: A key sensor of fuel and energy status in skeletal muscle. *Physiology* **21:** 48–60.

Harper J.M., Durkee S.J., Dysko R.C., Austad S.N., and Miller R.A. 2006. Genetic modulation of hormone levels and life span in hybrids between laboratory and wild-derived mice. *J. Gerontol. A Biol. Sci. Med. Sci.* **61:** 1019–1029.

Hawley S.A., Boudeau J., Reid J.L., Mustard K.J., Udd L., Makela T.P., Alessi D.R., and Hardie D.G. 2003. Complexes between the LKB1 tumor suppressor, STRAD alpha/beta and MO25 alpha/beta are upstream kinases in the AMP-activated protein kinase cascade. *J. Biol.* **2:** 28.

Henderson S.T. and Johnson T.E. 2001. *daf-16* integrates developmental and environmental inputs to mediate aging in the nematode *Caenorhabditis elegans. Curr. Biol.* **11:** 1975–1980.

Hertweck M., Gobel C., and Baumeister R. 2004. *C. elegans* SGK-1 is the critical component in the Akt/PKB kinase complex to control stress response and life span. *Dev. Cell* **6:** 577–588.

Herzig S., Long F., Jhala U.S., Hedrick S., Quinn R., Bauer A., Rudolph D., Schutz G., Yoon C., Puigserver P., Spiegelman B., and Montminy M. 2001. CREB regulates hepatic gluconeogenesis through the coactivator PGC-1. *Nature* **413:** 179–183.

Hipkiss A.R. 2006. Does chronic glycolysis accelerate aging? Could this explain how dietary restriction works? *Ann. N.Y. Acad. Sci.* **1067:** 361–368.

Holzenberger M., Dupont J., Ducos B., Leneuve P., Geloen A., Even P.C., Cervera P., and Le Bouc Y. 2003. IGF-1 receptor regulates lifespan and resistance to oxidative stress in mice. *Nature* **421:** 182–187.

Hong S.P., Leiper F.C., Woods A., Carling D., and Carlson M. 2003. Activation of yeast Snf1 and mammalian AMP-activated protein kinase by upstream kinases. *Proc. Natl. Acad. Sci.* **100:** 8839–8843.

Horton J.D., Goldstein J.L., and Brown M.S. 2002. SREBPs: Activators of the complete program of cholesterol and fatty acid synthesis in the liver. *J. Clin. Invest.* **109:** 1125–1131.

Houstis N., Rosen E.D., and Lander E.S. 2006. Reactive oxygen species have a causal role in multiple forms of insulin resistance. *Nature* **440:** 944–948.

Hwangbo D.S., Gershman B., Tu M.P., Palmer M., and Tatar M. 2004. *Drosophila* dFOXO controls lifespan and regulates insulin signalling in brain and fat body. *Nature* **429:** 562–566.

Imai S., Armstrong C.M., Kaeberlein M., and Guarente L. 2000. Transcriptional silencing and longevity protein Sir2 is an NAD-dependent histone deacetylase. *Nature* **403:** 795–800.

Imura A., Iwano A., Tohyama O., Tsuji Y., Nozaki K., Hashimoto N., Fujimori T., and Nabeshima Y. 2004. Secreted Klotho protein in sera and CSF: Implication for post-translational cleavage in release of Klotho protein from cell membrane. *FEBS Lett.* **565:** 143–147.

Jagoe R.T. and Goldberg A.L. 2001. What do we really know about the ubiquitin-proteasome pathway in muscle atrophy? *Curr. Opin. Clin. Nutr. Metab. Care* **4:** 183–190.

Jia K., Chen D., and Riddle D.L. 2004. The TOR pathway interacts with the insulin signaling pathway to regulate *C. elegans* larval development, metabolism and life span. *Development* **131:** 3897–3906.

Kahn B.B., Alquier T., Carling D., and Hardie D.G. 2005. AMP-activated protein kinase: Ancient energy gauge provides clues to modern understanding of metabolism. *Cell Metab.* **1:** 15–25.

Kalaany N.Y. and Mangelsdorf D.J. 2006. LXRS and FXR: The yin and yang of cholesterol and fat metabolism. *Annu. Rev. Physiol.* **68:** 159–191.

Kapahi P., Zid B.M., Harper T., Koslover D., Sapin V., and Benzer S. 2004. Regulation of lifespan in *Drosophila* by modulation of genes in the TOR signaling pathway. *Curr. Biol.* **14:** 885–890.

Kenyon C. 2005. The plasticity of aging: Insights from long-lived mutants. *Cell* **120:** 449–460.

Kenyon C., Chang J., Gensch E., Rudner A., and Tabtiang R. 1993. A *C. elegans* mutant that lives twice as long as wild type. *Nature* **366:** 461–464.

Kim E.K., Miller I., Aja S., Landree L.E., Pinn M., McFadden J., Kuhajda F.P., Moran T.H., and Ronnett G.V. 2004. C75, a fatty acid synthase inhibitor, reduces food intake via hypothalamic AMP-activated protein kinase. *J. Biol. Chem.* **279:** 19970–19976.

Kimura K.D., Tissenbaum H.A., Liu Y., and Ruvkun G. 1997. daf-2, an insulin receptor-like gene that regulates longevity and diapause in *Caenorhabditis elegans*. *Science* **277:** 942–946.

Koo S.H., Flechner L., Qi L., Zhang X., Screaton R.A., Jeffries S., Hedrick S., Xu W., Boussouar F., Brindle P., Takemori H., and Montminy M. 2005. The CREB coactivator TORC2 is a key regulator of fasting glucose metabolism. *Nature* **437:** 1109–1111.

Kuro-o M., Matsumura Y., Aizawa H., Kawaguchi H., Suga T., Utsugi T., Ohyama Y., Kurabayashi M., Kaname T., Kume E., Iwasaki H., Iida A., Shiraki-Iida T., Nishikawa S., Nagai R., and Nabeshima Y.I. 1997. Mutation of the mouse *klotho* gene leads to a syndrome resembling ageing. *Nature* **390:** 45–51.

Kurosu H., Ogawa Y., Miyoshi M., Yamamoto M., Nandi A., Rosenblatt K.P., Baum M.G., Schiavi S., Hu M.C., Moe O.W., and Kuro-o M. 2006. Regulation of fibroblast growth factor-23 signaling by klotho. *J. Biol. Chem.* **281:** 6120–6123.

Kurosu H., Yamamoto M., Clark J.D., Pastor J.V., Nandi A., Gurnani P., McGuinness O.P., Chikuda H., Yamaguchi M., Kawaguchi H., Shimomura I., Takayama Y., Herz J., Kahn C.R., Rosenblatt K.P., and Kuro-o M. 2005. Suppression of aging in mice by the hormone Klotho. *Science* **309:** 1829–1833.

Lagouge M., Argmann C., Gerhart-Hines Z., Meziane H., Lerin C., Daussin F., Messadeq N., Milne J., Lambert P., Elliott P., Geny B., Laakso M., Puigserver P., and Auwerx J. 2006. Resveratrol improves mitochondrial function and protects against metabolic disease by activating SIRT1 and PGC-1alpha. *Cell* **127:** 1109–1122.

Lam T.K., Pocai A., Gutierrez-Juarez R., Obici S., Bryan J., Aguilar-Bryan L., Schwartz G.J., and Rossetti L. 2005. Hypothalamic sensing of circulating fatty acids is required for glucose homeostasis. *Nat. Med.* **11:** 320–327.

Le Bacquer O., Petroulakis E., Paglialunga S., Poulin F., Richard D., Cianflone K., and Sonenberg N. 2007. Elevated sensitivity to diet-induced obesity and insulin resistance in mice lacking 4E-BP1 and 4E-BP2. *J. Clin. Invest.* **117:** 387–396.

Lee C.H., Olson P., and Evans R.M. 2003. Minireview: Lipid metabolism, metabolic diseases, and peroxisome proliferator-activated receptors. *Endocrinology* **144:** 2201–2207.

Lee S.S., Lee R.Y., Fraser A.G., Kamath R.S., Ahringer J., and Ruvkun G. 2003. A systematic RNAi screen identifies a critical role for mitochondria in *C. elegans* longevity. *Nat. Genet.* **33:** 40–48.

Lee W.J., Kim M., Park H.S., Kim H.S., Jeon M.J., Oh K.S., Koh E.H., Won J.C., Kim M.S., Oh G.T., Yoon M., Lee K.U., and Park J.Y. 2006. AMPK activation increases fatty acid oxidation in skeletal muscle by activating PPARalpha and PGC-1. *Biochem. Biophys. Res. Commun.* **340:** 291–295.

Lemieux M.E., Yang X., Jardine K., He X., Jacobsen K.X., Staines W.A., Harper M.E., and McBurney M.W. 2005. The Sirt1 deacetylase modulates the insulin-like growth factor signaling pathway in mammals. *Mech. Ageing Dev.* **126:** 1097–1105.

Li W., Kennedy S.G., and Ruvkun G. 2003. *daf-28* encodes a *C. elegans* insulin superfamily member that is regulated by environmental cues and acts in the DAF-2 signaling pathway. *Genes Dev.* **17:** 844–858.

Liang H., Masoro E.J., Nelson J.F., Strong R., McMahan C.A., and Richardson A. 2003. Genetic mouse models of extended lifespan. *Exp. Gerontol.* **38:** 1353–1364.

Libina N., Berman J.R., and Kenyon C. 2003. Tissue-specific activities of *C. elegans* DAF-16 in the regulation of lifespan. *Cell* **115:** 489–502.

Lin J., Yang R., Tarr P.T., Wu P.H., Handschin C., Li S., Yang W., Pei L., Uldry M., Tontonoz P., Newgard C.B., and Spiegelman B.M. 2005. Hyperlipidemic effects of dietary saturated fats mediated through PGC-1beta coactivation of SREBP. *Cell* **120:** 261–273.

Lin K., Dorman J.B., Rodan A., and Kenyon C. 1997. *daf-16:* An HNF-3/forkhead family member that can function to double the life-span of *Caenorhabditis elegans. Science* **278:** 1319–1322.

Lin K., Hsin H., Libina N., and Kenyon C. 2001. Regulation of the *Caenorhabditis elegans* longevity protein DAF-16 by insulin/IGF-1 and germline signaling. *Nat. Genet.* **28:** 139–145.

Lochhead P.A., Salt I.P., Walker K.S., Hardie D.G., and Sutherland C. 2000. 5-Aminoimidazole-4-carboxamide riboside mimics the effects of insulin on the expression of the 2 key gluconeogenic genes PEPCK and glucose-6-phosphatase. *Diabetes* **49:** 896–903.

Longo V.D. and Kennedy B.K. 2006. Sirtuins in aging and age-related disease. *Cell* **126:** 257–268.

Luo J., Nikolaev A.Y., Imai S., Chen D., Su F., Shiloh A., Guarente L., and Gu W. 2001. Negative control of p53 by Sir2alpha promotes cell survival under stress. *Cell* **107:** 137–148.

Masoro E.J. 2005. Overview of caloric restriction and ageing. *Mech. Ageing Dev.* **126:** 913–922.

McBurney M.W., Yang X., Jardine K., Bieman M., Th'ng J., and Lemieux M. 2003. The absence of SIR2alpha protein has no effect on global gene silencing in mouse embryonic stem cells. *Mol. Cancer Res.* **1:** 402–409.

McGill J.K. and Beal M.F. 2006. PGC-1alpha, a new therapeutic target in Huntington's disease? *Cell* **127:** 465–468.

Melendez A., Talloczy Z., Seaman M., Eskelinen E.L., Hall D.H., and Levine B. 2003. Autophagy genes are essential for dauer development and life-span extension in *C. elegans. Science* **301:** 1387–1391.

Meynial-Denis D., Miri A., Bielicki G., Mignon M., Renou J.P., and Grizard J. 2005. Insulin-dependent glycogen synthesis is delayed in onset in the skeletal muscle of food-deprived aged rats. *J. Nutr. Biochem.* **16:** 150–154.

Minokoshi Y., Alquier T., Furukawa N., Kim Y.B., Lee A., Xue B., Mu J., Foufelle F., Ferre P., Birnbaum M.J., Stuck B.J., and Kahn B.B. 2004. AMP-kinase regulates food intake by responding to hormonal and nutrient signals in the hypothalamus. *Nature* **428:** 569–574.

Mori K., Yahata K., Mukoyama M., Suganami T., Makino H., Nagae T., Masuzaki H., Ogawa Y., Sugawara A., Nabeshima Y., and Nakao K. 2000. Disruption of *klotho* gene causes an abnormal energy homeostasis in mice. *Biochem. Biophys. Res. Commun.* **278:** 665–670.

Morris J.Z., Tissenbaum H.A., and Ruvkun G. 1996. A phosphatidylinositol-3-OH kinase family member regulating longevity and diapause in *Caenorhabditis elegans*. *Nature* **382**: 536–539.

Moynihan K.A., Grimm A.A., Plueger M.M., Bernal-Mizrachi E., Ford E., Cras-Meneur C., Permutt M.A., and Imai S. 2005. Increased dosage of mammalian Sir2 in pancreatic beta cells enhances glucose-stimulated insulin secretion in mice. *Cell Metab.* **2**: 105–117.

Mu J., Brozinick J.T., Jr., Valladares O., Bucan M., and Birnbaum M.J. 2001. A role for AMP-activated protein kinase in contraction- and hypoxia-regulated glucose transport in skeletal muscle. *Mol. Cell* **7**: 1085–1094.

Murphy C.T., McCarroll S.A., Bargmann C.I., Fraser A., Kamath R.S., Ahringer J., Li H., and Kenyon C. 2003. Genes that act downstream of DAF-16 to influence the lifespan of *Caenorhabditis elegans*. *Nature* **424**: 277–283.

Nabeshima Y. 2006. Toward a better understanding of Klotho. *Sci. Aging Knowledge Environ.* **2006**: pe11.

Nakae J., Kitamura T., Silver D.L., and Accili D. 2001. The forkhead transcription factor Foxo1 (Fkhr) confers insulin sensitivity onto glucose-6-phosphatase expression. *J. Clin. Invest.* **108**: 1359–1367.

Nakae J., Kitamura T., Kitamura Y., Biggs W.H., III, Arden K.C., and Accili D. 2003. The forkhead transcription factor Foxo1 regulates adipocyte differentiation. *Dev. Cell* **4**: 119–129.

Nakae J., Biggs W.H., III, Kitamura T., Cavenee W.K., Wright C.V., Arden K.C., and Accili D. 2002. Regulation of insulin action and pancreatic beta-cell function by mutated alleles of the gene encoding forkhead transcription factor Foxo1. *Nat. Genet.* **32**: 245–253.

Nemoto S., Fergusson M.M., and Finkel T. 2005. SIRT1 functionally interacts with the metabolic regulator and transcriptional coactivator PGC-1(alpha). *J. Biol. Chem.* **280**: 16456–16460.

Ogg S., Paradis S., Gottlieb S., Patterson G.I., Lee L., Tissenbaum H.A., and Ruvkun G. 1997. The Fork head transcription factor DAF-16 transduces insulin-like metabolic and longevity signals in *C. elegans*. *Nature* **389**: 994–999.

Oh S.W., Mukhopadhyay A., Dixit B.L., Raha T., Green M.R., and Tissenbaum H.A. 2006. Identification of direct DAF-16 targets controlling longevity, metabolism and diapause by chromatin immunoprecipitation. *Nat. Genet.* **38**: 251–257.

Orentreich N., Matias J.R., DeFelice A., and Zimmerman J.A. 1993. Low methionine ingestion by rats extends life span. *J. Nutr.* **123**: 269–274.

Petersen K.F., Dufour S., Befroy D., Garcia R., and Shulman G.I. 2004. Impaired mitochondrial activity in the insulin-resistant offspring of patients with type 2 diabetes. *N. Engl. J. Med.* **350**: 664–671.

Picard F., Kurtev M., Chung N., Topark-Ngarm A., Senawong T., Machado De Oliveira R., Leid M., McBurney M.W., and Guarente L. 2004. Sirt1 promotes fat mobilization in white adipocytes by repressing PPAR-gamma. *Nature* **429**: 771–776.

Pierce S.B., Costa M., Wisotzkey R., Devadhar S., Homburger S.A., Buchman A.R., Ferguson K.C., Heller J., Platt D.M., Pasquinelli A.A., Liu L.X., Doberstein S.K., and Ruvkun G. 2001. Regulation of DAF-2 receptor signaling by human insulin and *ins-1*, a member of the unusually large and diverse *C. elegans* insulin gene family. *Genes Dev.* **15**: 672–686.

Pilkis S.J. and Granner D.K. 1992. Molecular physiology of the regulation of hepatic gluconeogenesis and glycolysis. *Annu. Rev. Physiol.* **54**: 885–909.

Puigserver P., Rhee J., Donovan J., Walkey C.J., Yoon J.C., Oriente F., Kitamura Y., Altomonte J., Dong H., Accili D., and Spiegelman B.M. 2003. Insulin-regulated hepatic gluconeogenesis through FOXO1-PGC-1alpha interaction. *Nature* **423:** 550–555.

Reznick R.M., Zong H., Li J., Morino K., Moore I.K., Yu H.J., Liu Z.X., Dong J., Mustard K.J., Hawley S.A., Befroy D., Pypaert M., Hardie D.G., Young L.H., and Shulman G.I. 2007. Aging-associated reductions in AMP-activated protein kinase activity and mitochondrial biogenesis. *Cell Metab.* **5:** 151–156.

Rodgers J.T., Lerin C., Haas W., Gygi S.P., Spiegelman B.M., and Puigserver P. 2005. Nutrient control of glucose homeostasis through a complex of PGC-1alpha and SIRT1. *Nature* **434:** 113–118.

Samuel V.T., Choi C.S., Phillips T.G., Romanelli A.J., Geisler J.G., Bhanot S., McKay R., Monia B., Shutter J.R., Lindberg R.A., Shulman G.I., and Veniant M.M. 2006. Targeting foxo1 in mice using antisense oligonucleotide improves hepatic and peripheral insulin action. *Diabetes* **55:** 2042–2050.

Sandri M., Sandri C., Gilbert A., Skurk C., Calabria E., Picard A., Walsh K., Schiaffino S., Lecker S.H., and Goldberg A.L. 2004. Foxo transcription factors induce the atrophy-related ubiquitin ligase atrogin-1 and cause skeletal muscle atrophy. *Cell* **117:** 399–412.

Scott J.W., Hawley S.A., Green K.A., Anis M., Stewart G., Scullion G.A., Norman D.G., and Hardie D.G. 2004. CBS domains form energy-sensing modules whose binding of adenosine ligands is disrupted by disease mutations. *J. Clin. Invest.* **113:** 274–284.

Shah O.J., Wang Z., and Hunter T. 2004. Inappropriate activation of the TSC/Rheb/mTOR/S6K cassette induces IRS1/2 depletion, insulin resistance, and cell survival deficiencies. *Curr. Biol.* **14:** 1650–1656.

Shaw R.J., Kosmatka M., Bardeesy N., Hurley R.L., Witters L.A., DePinho R.A., and Cantley L.C. 2004. The tumor suppressor LKB1 kinase directly activates AMP-activated kinase and regulates apoptosis in response to energy stress. *Proc. Natl. Acad. Sci.* **101:** 3329–3335.

Shaw R.J., Lamia K.A., Vasquez D., Koo S.H., Bardeesy N., DePinho R.A., Montminy M., and Cantley L.C. 2005. The kinase LKB1 mediates glucose homeostasis in liver and therapeutic effects of metformin. *Science* **310:** 1642–1646.

Shepherd P.R. and Kahn B.B. 1999. Glucose transporters and insulin action: Implications for insulin resistance and diabetes mellitus. *N. Engl. J. Med.* **341:** 248–257.

Shimada T., Takeshita Y., Murohara T., Sasaki K., Egami K., Shintani S., Katsuda Y., Ikeda H., Nabeshima Y., and Imaizumi T. 2004. Angiogenesis and vasculogenesis are impaired in the precocious-aging *klotho* mouse. *Circulation* **110:** 1148–1155.

Shiraki-Iida T., Aizawa H., Matsumura Y., Sekine S., Iida A., Anazawa H., Nagai R., Kuro-o M., and Nabeshima Y. 1998. Structure of the mouse *klotho* gene and its two transcripts encoding membrane and secreted protein. *FEBS Lett.* **424:** 6–10.

Spiegelman B.M. and Heinrich R. 2004. Biological control through regulated transcriptional coactivators. *Cell* **119:** 157–167.

St-Pierre J., Drori S., Uldry M., Silvaggi J.M., Rhee J., Jager S., Handschin C., Zheng K., Lin J., Yang W., Simon D.K., Bachoo R., and Spiegelman B.M. 2006. Suppression of reactive oxygen species and neurodegeneration by the PGC-1 transcriptional coactivators. *Cell* **127:** 397–408.

Suwa M., Nakano H., and Kumagai S. 2003. Effects of chronic AICAR treatment on fiber composition, enzyme activity, UCP3, and PGC-1 in rat muscles. *J. Appl. Physiol.* **95:** 960–968.

Syntichaki P., Troulinaki K., and Tavernarakis N. 2007. eIF4E function in somatic cells modulates ageing in *Caenorhabditis elegans*. *Nature* **445:** 922–926.

Tatar M., Bartke A., and Antebi A. 2003. The endocrine regulation of aging by insulin-like signals. *Science* **299:** 1346–1351.

Tatar M., Kopelman A., Epstein D., Tu M.P., Yin C.M., and Garofalo R.S. 2001. A mutant *Drosophila* insulin receptor homolog that extends life-span and impairs neuroendocrine function. *Science* **292:** 107–110.

Tohyama O., Imura A., Iwano A., Freund J.N., Henrissat B., Fujimori T., and Nabeshima Y. 2004. Klotho is a novel beta-glucuronidase capable of hydrolyzing steroid beta-glucuronides. *J. Biol. Chem.* **279:** 9777–9784.

Tsujikawa H., Kurotaki Y., Fujimori T., Fukuda K., and Nabeshima Y. 2003. *Klotho*, a gene related to a syndrome resembling human premature aging, functions in a negative regulatory circuit of vitamin D endocrine system. *Mol. Endocrinol.* **17:** 2393–2403.

Tsukiyama-Kohara K., Poulin F., Kohara M., DeMaria C.T., Cheng A., Wu Z., Gingras A.C., Katsume A., Elchebly M., Spiegelman B.M., Harper M.E., Tremblay M.L., and Sonenberg N. 2001. Adipose tissue reduction in mice lacking the translational inhibitor 4E-BP1. *Nat. Med.* **7:** 1128–1132.

Tu M.P., Epstein D., and Tatar M. 2002. The demography of slow aging in male and female *Drosophila* mutant for the insulin-receptor substrate homologue chico. *Aging Cell* **1:** 75–80.

Uldry M. and Thorens B. 2004. The SLC2 family of facilitated hexose and polyol transporters. *Pflugers Arch.* **447:** 480–489.

Um S.H., Frigerio F., Watanabe M., Picard F., Joaquin M., Sticker M., Fumagalli S., Allegrini P.R., Kozma S.C., Auwerx J., and Thomas G. 2004. Absence of S6K1 protects against age- and diet-induced obesity while enhancing insulin sensitivity. *Nature* **431:** 200–205.

Utsugi T., Ohno T., Ohyama Y., Uchiyama T., Saito Y., Matsumura Y., Aizawa H., Itoh H., Kurabayashi M., Kawazu S., Tomono S., Oka Y., Suga T., Kuro-o M., Nabeshima Y., and Nagai R. 2000. Decreased insulin production and increased insulin sensitivity in the *klotho* mutant mouse, a novel animal model for human aging. *Metabolism* **49:** 1118–1123.

Vaziri H., Dessain S.K., Ng Eaton E., Imai S.I., Frye R.A., Pandita T.K., Guarente L., and Weinberg R.A. 2001. hSIR2(SIRT1) functions as an NAD-dependent p53 deacetylase. *Cell* **107:** 149–159.

Vellai T., Takacs-Vellai K., Zhang Y., Kovacs A.L., Orosz L., and Muller F. 2003. Genetics: Influence of TOR kinase on lifespan in *C. elegans. Nature* **426:** 620.

Weiss E.P., Racette S.B., Villareal D.T., Fontana L., Steger-May K., Schechtman K.B., Klein S., and Holloszy J.O. 2006. Improvements in glucose tolerance and insulin action induced by increasing energy expenditure or decreasing energy intake: A randomized controlled trial. *Am. J. Clin. Nutr.* **84:** 1033–1042.

Weydt P., Pineda V.V., Torrence A.E., Libby R.T., Satterfield T.F., Lazarowski E.R., Gilbert M.L., Morton G.J., Bammler T.K., Strand A.D., Cui L., Beyer R.P., Easley C.N., Smith A.C., Krainc D., Luquet S., Sweet I.R., Schwartz M.W., and La Spada A.R. 2006. Thermoregulatory and metabolic defects in Huntington's disease transgenic mice implicate PGC-1alpha in Huntington's disease neurodegeneration. *Cell Metab.* **4:** 349–362.

Wijchers P.J., Burbach J.P., and Smidt M.P. 2006. In control of biology: Of mice, men and foxes. *Biochem. J.* **397:** 233–246.

Wolkow C.A., Kimura K.D., Lee M.S., and Ruvkun G. 2000. Regulation of *C. elegans* lifespan by insulinlike signaling in the nervous system. *Science* **290:** 147–150.

Wullschleger S., Loewith R., and Hall M.N. 2006. TOR signaling in growth and metabolism. *Cell* **124:** 471–484.

Yamashita T., Konishi M., Miyake A., Inui K., and Itoh N. 2002. Fibroblast growth factor (FGF)-23 inhibits renal phosphate reabsorption by activation of the mitogen-activated protein kinase pathway. *J. Biol. Chem.* **277:** 28265–28270.

Yang F., Vought B.W., Satterlee J.S., Walker A.K., Jim Sun Z.Y., Watts J.L., DeBeaumont R., Saito R.M., Hyberts S.G., Yang S., Macol C., Iyer L., Tjian R., van den Heuvel S., Hart A.C., Wagner G., and Naar A.M. 2006. An ARC/Mediator subunit required for SREBP control of cholesterol and lipid homeostasis. *Nature* **442:** 700–704.

Yoon J.C., Puigserver P., Chen G., Donovan J., Wu Z., Rhee J., Adelmant G., Stafford J., Kahn C.R., Granner D.K., Newgard C.B., and Spiegelman B.M. 2001. Control of hepatic gluconeogenesis through the transcriptional coactivator PGC-1. *Nature* **413:** 131–138.

Yoshida T., Fujimori T., and Nabeshima Y. 2002. Mediation of unusually high concentrations of 1,25-dihydroxyvitamin D in homozygous *klotho* mutant mice by increased expression of renal 1alpha-hydroxylase gene. *Endocrinology* **143:** 683–689.

Zhu M., Miura J., Lu L.X., Bernier M., DeCabo R., Lane M.A., Roth G.S., and Ingram D.K. 2004. Circulating adiponectin levels increase in rats on caloric restriction: The potential for insulin sensitization. *Exp. Gerontol.* **39:** 1049–1059.

20

Telomeres and Telomerase in Aging and Cancer

Jerry W. Shay and Woodring E. Wright
The University of Texas Southwestern Medical Center
Department of Cell Biology
Dallas, Texas 75390-9039

THE ROLE OF TELOMERES IN MAINTAINING CHROMOSOMAL integrity was proposed by Barbara McClintock (for review, see Blackburn 2006). Studying telomeres in maize chromosomes, McClintock observed that if not capped by telomeres, the ends of chromosomes had a tendency to fuse. Her observations were confirmed 50 years later in yeast and mice when it was demonstrated that without telomeric ends, chromosomes undergo aberrant end-to-end fusions, forming multicentric chromosomes with a propensity to break during mitosis, activating DNA-damage checkpoints and, in some cases, leading to widespread cell death (Zakian 1989). It is now known that the shortening of telomeres due to cell divisions forms the basis of replicative aging, the growth arrest originally described by Hayflick and Moorhead (1961).

Aging is associated with the gradual decline in the performance of organ systems, resulting in the loss of reserve capacity, leading to an increased chance of death (Gompertz 1825). In some organ systems, this loss of reserve capacity with increasing age can be attributed to the loss of cell function (Martin et al. 1970). Chronic localized stress to specific tissues/cell types may result in increased cell turnover, and it has been hypothesized that this may lead to focal areas of replicative senescence (Hayflick and Moorhead 1961), followed by alterations in patterns of gene expression (West 1994; West et al. 1996). This could result in reduced tissue regeneration, culminating in some of the clinical pathologies that are often associated with increased age.

In addition to replicative aging, a variety of mechanisms can induce an irreversible growth arrest that has also been called senescence not involving critically shortened telomeres. This chapter reviews some of the aspects of telomere biology that relate to these different processes, the roles these may have in human longevity, as well as how they can be exploited to increase our understanding of the mechanisms underlying certain aspects of human aging and how they relate to our understanding of cancer and the potential to treat cancer.

TELOMERES AND TELOMERASE

Telomeres and the End-replication Problem

In vertebrates, telomeres are long stretches of simple repetitive noncoding DNA located at the ends of all eukaryotic chromosomes. Human telomeres contain the 6-bp sequence TTAGGG, repeated many thousands of times (for review, see deLange 2006). Evidence suggests that as chromosome "caps," telomeres have at least three critical functions related to understanding the role of telomeres in aging: (1) to protect chromosome ends from being recognized as DNA damage, thus leading to protection from enzymatic degradation and abnormal fusion reactions, (2) to serve as a buffer zone to protect coding genes against the "end-replication" problem, and (3) to serve as a gauge for mitotic age (the divisional clock [Fig. 1] or replicometer).

The first of these critical functions is provided by a combination of special structures and proteins that prevent the telomeres from being seen as double-strand breaks. Without a masking mechanism, the ends of linear chromosome would be ligated together by the process of nonhomologous end-joining (NHEJ). This would then induce a mitotic catastrophe when the cell attempted to divide. One of the major functions of telomeres is thus to hide the ends of the chromosomes. One structure believed to contribute to this end protection is the t-loop, in which the single-strand G-rich 3′overhang at the end of the telomere is inserted as a "D-loop" into the double-strand telomeric sequences, forming an overall structure called a t-loop (Griffith et al. 1999). The proteins and potential mechanisms involved in this process have been reviewed recently (de Lange 2006).

The second major critical function of telomeres relates to the process of semiconservative DNA replication. During each round of cell division, 50–200 bp are lost from the ends of linear human chromosomes (Hastie

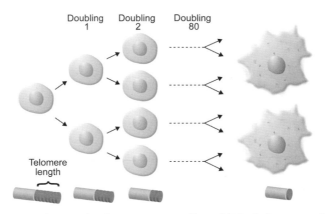

Figure 1. Aging of normal cells. It is now well established that normal cells, even those that have some detectable telomerase activity, are not immortal. After a finite number of divisions (population doublings), cells will undergo replicative senescence. Progressive telomere shortening correlates with each doubling until a few short telomeres become uncapped, leading to double-stranded DNA-damage signals. In normal cells, the DNA damage cannot be repaired, and such cells can remain in a metabolically active but nondividing state for years. In cells with abrogated cell cycle check points, the DNA damage is ignored, cells continue to divide, and telomeres continue to shorten, eventually leading to genomic instability and crisis (a period of apoptosis being balanced by cell growth). Only a rare cell escapes crisis, and this is almost universally due to the up-regulation of telomerase.

et al. 1990; Lindsey et al. 1991; Allsopp et al. 1992, 1995; Shay and Wright 2004a). This "end-replication" problem (Fig. 2) occurs because conventional DNA replication machinery is unable to replicate completely the 3′ ends of chromosomal DNA during the S phase of each cell cycle. The polymerases that copy parental DNA strands prior to cell division synthesize DNA only in the 5′ to 3′ direction and require a short RNA primer to begin. These primers are then degraded and filled in by DNA synthesis extending from the upstream primer. However, at the end of a linear chromosome, there is no "upstream" DNA synthesis to fill in the gap between the final RNA priming event and the end of the chromosome.

Although the daughter telomere that is the product of lagging-strand synthesis could have an overhang without further processing, the leading daughter would initially be blunt-ended. The formation of the t-loop described above requires a 3′ overhang, so the leading-strand daughter must be processed to generate this overhang if it is to be packaged in a t-loop. There is good evidence for such processing events, although the

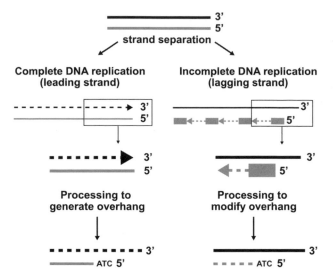

Figure 2. DNA replication at telomeres. Our current understanding of telomere replication is that the leading G-rich strand is able to replicate all the way to the end of
its template strand, which may then be processed to leave a small 3′ G-rich overhang
to which end-protecting proteins can bind. A much larger G-rich overhang is left
following synthesis of the C-rich lagging strand, since there is no DNA beyond the
end of the chromosome that could serve as a priming event to fill in the gap between
the last Okazaki priming event and the end of the chromosome.

sizes of the overhangs at leading and lagging daughter telomeres are different (Wright et al. 1997; Sfeir et al. 2005; Chai et al. 2006a,b).

This replication strategy predicts that with each round of cell division, there will be progressive shortening of the 3′ end of chromosomal
DNA. Telomeric DNA therefore provides a cushion of expendable noncoding sequences to protect against the potentially catastrophic attrition
of important chromosomal material.

In the absence of any compensating mechanism, we would be extinct
since the progressive shortening described above would eventually exhaust
the telomeric buffer zone. However, our germ line (and certain renewal
tissues and stem cells) expresses telomerase, an enzyme that can add
telomeric repeats to the end of the chromosome to compensate for the
end-replication problem. Telomerase is a ribonucleoprotein that has reverse
transcriptase activity. The RNA component, hTERC (or hTR), contains
the complement of the telomeric TTAGGG sequence, and the protein
component (hTERT) contains the reverse transcription activity (Chen and
Greider 2006; Cristofari and Lingner 2006). The enzyme is processive, so

following the addition of one telomeric sequence to the end of the telomeric G-strand, the enzyme pauses and repositions the template (hTR) RNA for the addition of the next telomeric repeat. Telomerase can thus add sufficient repeats to compensate for the end-replication problem and maintain telomere length in some but not all cell types.

In humans, telomerase is turned off or down-regulated in somatic tissues during development (Shay and Wright 2006a), and from that point on, telomeres shorten with each cell division, even in cells that have some detectable enzyme activity. This forms the basis of the counting mechanism that limits the maximal number of divisions a cell can undergo (e.g., replicative aging; see Fig. 1). There are mechanisms other than the end-replication problem such as oxidative damage and perhaps other end-processing events that can cause telomere shortening, and thus complicate the simple interpretation of the telomere shortening observed in a variety of aging studies.

Setting the Number of Doublings

A primary purpose of replicative aging may be to serve as a barrier for the formation of tumors. In most cases, it takes at least four to six mutations (if not more) to form a cancer cell (Vogelstein and Kinzler 1993). Each mutation occurs in a single cell (e.g., clonal origin of tumors), which has to expand considerably before there is a reasonable probability for a rare additional mutation to occur. Furthermore, most cancer mutations are recessive. This means that once the original mutant cell expands, one of those cells needs to eliminate the remaining wild-type normal copy. This daughter then needs to expand again to form a sufficient population size to permit additional mutations to occur. Thus, limiting the total number of times a cell could divide could prevent premalignant cells with one or two mutations from being able to divide and accumulate additional mutations (Fig. 3). Senescence (M1) and even crisis (M2) would then form a powerful protection mechanism limiting cancer formation (Wright and Shay 1995).

The amount of energy devoted to the maintenance and repair processes that keep our bodies healthy can be understood from theories of the evolution of aging (see Chapter 4, Evolutionary Theory). The *Disposable Soma Theory* (Kirkwood 1996) postulates that there is a trade-off between the amount of energy invested in the repair of the soma (body) and reproductive fitness. The balance between energy devoted to reproduction versus somatic repair may be an important key to understanding the rate at which species age. Species that have a high rate of

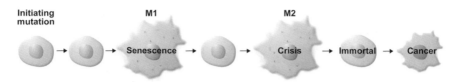

Figure 3. Senescence and cancer: Are telomeres the connection? It requires many cell divisions for a normal cell to become a cancer cell. Two telomere-based growth check points (M1, senescence and M2, crisis) help protect our cells from obtaining the unlimited proliferative potential that is often a hallmark of cancer cells. These blockades to immortalization may be thought of tumor suppressor pathways that help protect our cells from the early development of cancer.

annual mortality (and thus are unlikely to survive very long anyway) must invest most of their energy in early reproduction and relatively little in somatic maintenance and repair. A mouse that repaired itself sufficiently to live for 20 years would be making a bad investment if most mice are eaten by foxes and owls within 3 months. The mouse would be better off investing more energy in early reproduction and less in maintenance and repair. Humans, whose average survival is much longer than a field mouse, would be selected for devoting much more energy toward tissue maintenance and repair compared to mice. A whole variety of maintenance and repair processes, such as the efficiency of DNA repair, protection against oxidative damage, the rate of protein turnover, and the efficiency of the immune system, likely contribute to the genetic component of aging. It is important in this context to recognize that limits in the amount of energy devoted to many different maintenance and repair processes may be contributing to aging and that the role of replicative aging discussed below would be only one (and perhaps a minor one) of a large variety of aging mechanisms.

An efficient way of keeping cells healthy is triggering programmed cell death (apoptosis) of damaged cells and replacing them with new healthy ones. Replacing a dying cell with a freshly divided cell also dilutes the buildup of "unrepairable and indigestible" products that could contribute to aging. However, using cell turnover to repair tissues carries risks as well as benefits. Mistakes in copying the DNA during cell division can lead to harmful mutations, so more divisions can lead to a greater risk of cancer. In contrast, limiting the number of divisions provides an independent way of preventing malignancy (but may promote or lead to the development of chronic diseases often associated with aging). Even though it is believed that the primary reason for counting and limiting the number of cell divisions is to form a barrier against the formation of cancer, these limits are likely to contribute to the physiology of aging for the following reason. Very few of our Stone Age ancestors lived beyond 30–40 years of age, largely due to mortality from predators and infections. Evolution would have sought a balance between the advantages of cell turnover for maintaining healthy tissues and the advantages of limiting the total number of times a cell divided to prevent cancer. Human beings must have a rate of cell division that is sufficient to keep them relatively fit during the reproductive years, but they should not have an enormous reservoir of unused cell divisions. As long as the probability of surviving beyond age 40 was historically very small, having sufficient cell divisions to allow us to be extraordinarily fit until age 120 would have carried an increased risk of cancer without any benefit. The number of permitted cell divisions would thus represent the balance between reducing the number as much as possible to increase the effectiveness as a brake against cancer formation while permitting enough divisions for adequate maintenance and repair until about age 40. Modern sanitation, antibiotics, vaccines, and other disease prevention interventions as well as better treatment of established diseases have now resulted in most of us living well beyond age 40. We believe that the restrictions on cell turnover imposed by replicative aging, which would have little effect on physiology before age 40, are now contributing to part of the decline in tissue function that we call aging as we grow older.

Senescence

The term senescence has been used to describe any irreversible growth arrest of cells in culture, accompanied by a distinct set of cellular phenotypic changes. A variety of mechanisms can trigger cessation of cell proliferation (Fig. 4) and all may be part of potent anticarcinogenic programs (Shay et al. 1991; Shay and Roninson 2004; Campisi 2005). In addition to senescence being initiated by the shortening of telomeres, other endogenous and exogenous acute and chronic stress signals that can also lead to irreversible cell cycle arrest include oxidative damage, overexpression of oncoproteins, chromatin changes, and DNA damage (Fig. 4). The process of neoplastic transformation involves a series of events that allow cells to bypass or overcome these senescence pathways. For additional information on nontelomere-based initiators of senescence, see Chapter 8 (Senescence) and several recent reviews (Shay and Wright 2002b; Shay and Roninson 2004; Campisi 2005). We now know that replicative senescence is caused by a DNA-damage signal initiated by the "uncapping" of critically shortened telomeres (Zou et al. 2004).

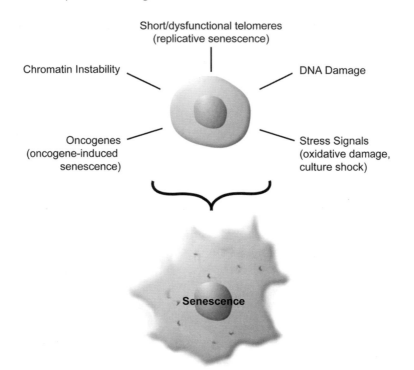

Figure 4. Does cellular aging antagonize or promote cancer? In addition to progressive telomere shortening leading to irreversible growth arrest, other pathways include DNA damage, chromatin instability, overexpression of oncoproteins, and a variety of stress signals (including oxidative damage). Although morphologically, the cells appear to be similar (greatly enlarge size), only replicative senescence shows shortened telomeres. The other mechanisms occur with normal length telomeres. However, all of these mechanisms may be potent tumor suppressor pathways, initially limiting the growth of potentially malignant cells.

In the absence of telomerase or other mechanisms to restore the end-protection property of telomeres, these senescent cells can remain viable for years.

Species That Do Not Use Replicative Aging

Mouse telomeres are extremely long in comparison to human telomeres, so that the shortest mouse telomere may be longer than the longest human telomere. Mouse cells stop dividing after 10–15 doublings in culture, and "spontaneously immortalize" so readily that it was initially debated whether or not they exhibited senescence at all. In contrast, the frequency of spontaneous immortalization in human fibroblasts in cell culture is essentially zero, and even after blocking the activity of the retinoblastoma protein (Rb) and p53 with SV40 large T antigen, the frequency is only 10^{-7} (Shay et al. 1999). This suggests that there might be fundamental differences between what was described as senescence in mouse fibroblasts and what was described as replicative aging in human fibroblasts.

When the mouse telomerase template RNA (mTERC) was knocked out to eliminate telomerase activity, mice survived for many generations. Their telomeres became progressively shorter in successive generations as expected. Importantly, the senescence that occurred in culture after 10–15 doublings in mouse embryo fibroblasts happened at the same number of doublings regardless of whether first-generation or third-generation mice were used and was thus independent of the initial telomere length (Blasco et al. 1997). This growth arrest thus could not be short-telomere-initiated replicative aging, since it was not caused by progressive replication-dependent telomere shortening.

Tissue oxygen concentrations are normally approximately 1–6% (Guyton and Hall 1966) so that the 21% oxygen of the typical sea-level culture room actually represents a hyperoxic environment. The hypothesis that the senescence of mouse embryo fibroblasts represented a general DNA-damage response from oxidative damage was confirmed when it was shown that they showed no evidence of senescence if grown in low oxygen (Parrinello et al. 2003). Thus, there is no evidence for replicative aging in mouse cells obtained from inbred strains, since the only senescence seen is the growth arrest due to other causes.

The study of the mTERC knockout mice showed that telomerase was not needed for mouse tumor formation (Blasco et al. 1997). This is in marked contrast to the behavior of human tumors, where many studies have shown that the ongoing proliferation of most human cancer cells (which usually have very short telomeres) is dependent on telomerase being able to maintain telomere length (Shay and Bachetti 1997). The concept that replicative aging is an anti-cancer mechanism is based on progressive telomere shortening limiting the number of available cells divisions during which oncogenic mutations could accumulate. The normal frequency of tumor formation in the telomerase knockout mice implies that replicative aging is not used as an anticancer mechanism in mice, consistent with the size of their telomeres being so large that progressive shortening could not be used to count cell divisions in one generation.

Replicative aging has been shown to be used in humans, primates (Steinert et al. 2002), sheep (Cui et al. 2002), deer (Zou et al. 2002), and cows (Lanza et al. 2000), thus it is not unique to humans. The presence of extremely long telomeres and the failure to exhibit replicative aging are not restricted to inbred laboratory mice but are also present in rabbits (Forsyth et al. 2005) and many other wild species (unpublished results). A working hypothesis to explain this in the larger context of aging is that there may be trade-offs between the advantages of having short telomeres/using replicative aging as an anticancer mechanism and the advantages having very long telomeres/not using replicative aging, so that different evolutionary choices can be made.

One possible, but as yet unproven, source of this trade-off might be the amount of energy invested in protecting telomeres from oxidative damage. Oxidative intermediates can travel along the DNA and preferentially produce damage at triplet GGG sequences (Hall et al. 1996; Oikawa and Kawanishi 1999). Not only are telomeres thus more sensitive to oxidative damage, but DNA damage at telomeres is repaired with less efficiency due to the fundamental nature of telomeric proteins hiding telomeres from being recognized as damaged DNA needing repair (Kruk et al. 1995; Petersen et al. 1998). Species occupying a niche in which their annual mortality is so high that they are overinvesting in oxidative protection mechanisms (as a general repair/longevity assurance mechanism) would be selected to reduce that investment. One consequence of this would be increased telomeric damage and the production of cells with a rapid growth arrest due to one or a few severely shortened telomeres. If the annual mortality of this species was sufficiently high so that they very rarely got cancer, it might thus be advantageous to compensate by abandoning the anticancer benefits of short telomeres and replicative aging in favor of extremely long telomeres that could tolerate greater levels of oxidative breakage as well as not suppressing telomerase in most tissues during development, so that broken telomeres could be elongated rather than stimulating a growth arrest. In fact, telomerase expression is found in several adult mouse tissues (e.g., liver) in which it is suppressed in humans. Given that some species age perfectly well without appearing to use replicative aging, it is important not to overemphasize the role of replicative aging in overall aging but to focus on areas in which human aging may differ from aging in the mouse (Wright and Shay 2000).

Telomere Length Changes in Human Aging

There are many correlative studies demonstrating a link between telomere length and aging and there is a heritable component to telomere length (Slagboom et al. 1994; Wu et al. 2003). In newborn humans, telomeres are approximately 15–20 kb in length and shorten gradually throughout life, suggesting that telomere length may serve as a surrogate marker for aging, reflecting a variety of processes that could include both causes (such as replicative aging) or consequences (such as oxidative damage) to telomeres due to underlying pathologies. It is believed that when a subset of cells in a tissue reaches a critically shortened telomere length, the senescent cells may produce a different constellation of proteins compared to those that are nonsenescent but quiescent adjacent cells. When cells become senescent in a tissue, this could change the homeostasis of that tissue leading to aging phenotypes.

These observations have led to a vast array of epidemiological studies in which telomere lengths have been measured in different tissues. For example, a number of recent reports examining telomere length in peripheral blood mononuclear cells (PBMCs) have reported correlations between shortened telomeres and a wide variety of age-related diseases such as early myocardial infarction, vascular dementia, atherosclerosis, and Alzheimer's disease (Okuda et al. 2000; von Zglinicki et al. 2000; Samani et al. 2001; Brouilette et al. 2003; Panossian et al. 2003; Benetos et al. 2004). Shortened telomeres from cells in affected tissues have been reported in patients with liver cirrhosis, Barrett's esophagus, ulcerative colitis, and myeloproliferative disorders (Rudolph et al. 2001; O'Sullivan et al. 2002; Wiemann et al. 2002; Farazi et al. 2003; Meeker et al. 2004). Other precancerous lesions such as ductal carcinoma in situ for breast cancer and prostatic and cervical intraepithelial neoplasias have been shown to have critically shortened telomeres in situ (Meeker et al. 2004). A correlation between shortened PBMCs and increased risk of death has even been found in subgroups of patients (Cawthon et al. 2003). It is thus fair to ask do these short telomeres promote development of advanced disease or do the shorten telomeres and subsequent senescence actually act as potent anticancer protection mechanisms?

There is considerable interindividual variability in telomere length; thus, adequate sample sizes are critical in making firm cause and effect conclusions (Jeanclos et al. 2000; Benetos et al. 2001; Aviv 2004; Epel et al. 2004; Nawrot et al. 2004; Gardner et al. 2005; Unryn et al. 2005; Valdes et al. 2005; Martin-Ruiz et al. 2005, 2006; Andrew et al. 2006; Aviv et al. 2006; Bischoff et al. 2006; Honig et al. 2006; Rando 2006). There are issues of survivor bias, as well as environmental factors such as inflammatory disease/chronic infections, that could lead to misinterpretation of results. Importantly, the various techniques for measuring telomere length may not be reliable or reproducible within small ranges. Thus, measurements of telomere lengths in PBMCs are at best correlative, and the utility of measuring telomeres as a surrogate for increased risk for age-related morbidity or for predicting risk for the development of disease that has been reported in the literature may in some instances be misleading (Rando 2006). It is important to reemphasize that there is significant variability of telomere length within the cells and tissues of an individual and among groups of otherwise similar individuals based on health status. Because there is a large amount overlap in the range of telomere length in the control and disease groups studied, the numbers of patients required to make meaningful interpretations is very high (in the hundreds if not thousands of patients), and few published studies have achieved these large data sets. Given the absence of a narrow "normal" telomere range for a given chronological age, it is difficult to imagine the utility of a single measurement of telomere length in an individual for prognostic purposes. Nevertheless, measurements of telomere length and studies of the mechanisms of telomere shortening may offer some insights into the mechanisms of aging (Rando 2006).

Diseases Caused by Short Telomeres

Several human diseases of telomere dysfunction have been discovered (Heiss et al. 1998; Mitchell et al. 1999; Shay and Wright 1999, 2004b; Vulliamy et al. 2001, 2004; Fogarty et al. 2003; Mason et al. 2005; Yamaguchi et al. 2005; Vulliamy and Dokal 2006), and individuals born with reduced levels of telomerase have short telomeres. This leads to telomere dysfunction in highly proliferative cells such as the bone marrow resulting in diseases such as aplastic anemia and in some instances (likely in combination with additional genetic and epigenetic changes) increased risk for the development of leukemia or other tumors (Heiss et al. 1998; Mitchell et al. 1999; Shay and Wright 1999, 2004b; Vulliamy et al. 2001, 2004; Fogarty et al. 2003; Mason et al. 2005; Yamaguchi et al. 2005; Vulliamy and Dokal 2006). This suggests that a more detailed knowledge of telomerase and telomere function may provide insights into human diseases. Since most of these individuals with genetically inherited short telomeres are haploinsuffcent for telomerase and have shortened life spans, it also suggests that humans may need full telomerase activity during development or in adult tissues that have high rates of proliferation to have a full human life span.

There is good evidence that in the absence of p53 checkpoint activity, short telomeres can lead to genomic instability and that this genomic instability can contribute to cancer formation (e.g., in DKC [dyskeratosis congenita], ulcerative colitis, and Barrett's esophagus). Telomere shortening has thus been described as a two-edged sword, both preventing and contributing to tumorigenesis, which calls into question the interpretation that replicative aging serves as an anticancer mechanism. However, it is important to remember that the frequency of cancer in DKC patients, although much higher than in the general population, is nonetheless low and still requires many years to develop. This is in the context of a whole-body-wide reduction in telomere length, and thus a huge increase in the population size of cells with very short telomeres that are "at risk" for genomic instability. In the absence of this genetic defect, the balance between preventing cancer and contributing to tumorigenesis could be heavily in favor of prevention. Similarly, diseases such as ulcerative colitis and Barrett's esophagus are both generally of late onset and affect small fractions of the population, and thus the adverse effects on genomic stability of short telomeres in these diseases of local high cell turnover (and/or oxidative damage, etc.) would be offset by the generally protective effect in the general population. Since replicative aging protects against the *clonal* expansion of malignant precursors, it can be viewed as advantageous in situations where short telomeres result from clonal processes but less protective when other conditions produce much larger populations of nonclonal cells with short telomeres.

Telomerase Therapy for Age-associated Disease

We and other investigators have shown that the expression of the catalytic subunit of human telomerase (hTERT) reconstitutes telomerase activity and circumvents the induction of senescence (Bodnar et al. 1998; Shay and Wright 2000; Thomas et al. 2000; Vaughan et al. 2004, 2006; Poh et al. 2005; Robertson et al. 2005). We have used hTERT to extend the life span of a variety of human cell types including skin and oral keratinocytes, dermal fibroblasts, muscle satellite (stem) cells, endothelial cells, retinal-pigmented epithelial cells, breast epithelial cells, and both corneal fibroblasts (keratocytes) and corneal epithelial cells. Such hTERT-immortalized cells have normal cell cycle controls, functional p53, $p21^{Cip1}$, and $p16^{Ink4a}$/pRb checkpoints, are contact-inhibited, anchorage-dependent, require growth factors for proliferation, and possess a normal or close to normal karyotype (Morales et al. 1999). Telomerase itself thus does not contribute to tumorigenic properties but is primarily permissive by abolishing the counting mechanism that normally limits ongoing proliferation. Human corneal fibroblasts/epithelial cells, skin keratinocytes/dermal fibroblasts, and human bronchial epithelial cells and lung fibroblasts expressing hTERT have been used to form three-dimensional organotypic cultures (Vaughan et al. 2004, 2006; Robertson et al. 2005). Such organotypic cultures express differentiation-specific proteins, suggesting that hTERT does not inhibit normal differentiation functions of cells.

Because hTERT preferentially elongates the shortest telomeres (Steinert et al. 2000; Hemann et al. 2001), the transient expression of telomerase could have profound effects on cell life span (Fig. 5). Since telomerase in the absence of other alterations is not oncogenic, the production of transiently or reversibly immortalized engineered cells may offer the possibility of treating a variety of chronic diseases and aged-related medical conditions that are due to telomere-based replicative senescence. From a theoretical standpoint, this is an appealing approach,

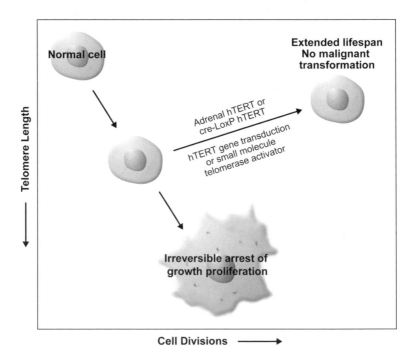

Figure 5. Telomerase activation as a target of age-associated disease. Normal cells show progressive telomere shortening leading to a telomere-induced growth arrest. Almost all cancer cells overcome replicative senescence by reactivating or up-regulating the endogenous hTERT gene resulting in telomerase expression and telomere maintenance. In diseases associated with telomere decline, transiently rejuvenating cells using hTERT transduction or a small-molecule telomerase activator may permit cells to continue to divide. In cell culture, this has been shown to extend cellular life span without malignant transformation. In addition, these cells do not show the hallmarks of cancer cells and are fully capable of normal differentiation.

but from a practical point of view, there are many regulatory safety issues that have to first be addressed. The challenge is how to rapidly progress these basic research advances into translational approaches to treat age-associated diseases that are due to telomere decline.

So where do we start? Since telomerase RNA is limiting in vivo in DKC and telomeres are not maintained to age-appropriate lengths, is DKC the perfect disease to test the therapeutic value of telomere rejuvenation? Showing cause and effect will require demonstrating that slowing down the rate of telomere loss or resetting the telomere clock reverses or delays the onset of disease. To test if telomere rejuvenation could impact on the progression of disease in DKC, one approach would be to isolate hematopoietic stem cells (CD34$^+$) from DKC patients, expand them in the laboratory in the presence of transiently expressed hTERT (perhaps using adenoviral hTERT that does not integrate into nuclear DNA) until telomeres become sufficiently elongated, and then return the rejuvenated stem-like cells to the patient. The obvious advantage of this approach is that these are the patient's own cells, which would avoid problems of rejection. In addition, this could be done without ablating the patients own bone marrow cells. Clearly, safety and efficacy standards, as well as quality and control assurances, will need to be carefully considered prior to initiating these studies. However, if this strategy for engineering rejuvenated DKC cells improves the health and longevity of DKC patients, it could then be considered for treatment of other telomere-based proliferative deficiencies produced by disease or aging.

Targeting Telomerase for Cancer Therapeutics

Advances in chromosome dynamics have increased our understanding of the significant role of telomeres and telomerase in cancer. Telomerase activity is present in most cancers and often correlates with the acquisition of a more malignant phenotype (Kim et al. 1994). Telomere length is also typically shorter in tumor cells when compared to adjacent non-cancerous cells. Thus, a therapeutic window exists in which tumor cells may be efficiently targeted, whereas normal telomerase-expressing cells, such as proliferating stem cells, may remain unaffected as a result of their longer telomere lengths (Fig. 6).

Telomerase is unusual among cancer molecular targets because a large body of outstanding basic science in telomere biology has preceded development of effective lead compounds, allowing potential problems to be anticipated before evidence of efficacy in model systems is in hand, exactly the opposite of the situation faced during most drug development.

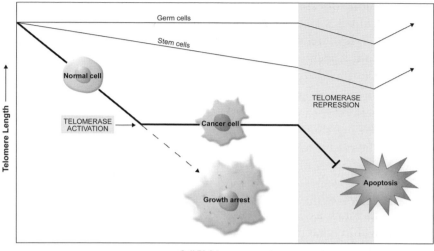

Figure 6. Effects of telomerase inhibition on normal and cancer cells. Normal cells progressively shorten telomeres until they undergo senescence or become immortal. To become immortal, cells almost universally reactivate telomerase and together with additional genetic and epigenetic changes can become tumorigenic. Telomerase inhibitors, such as GRN163L, that are in clinical trials would be predicted to lead the already shortened cancer telomeres to get even shorter, eventually leading to apoptotic cell death. Germ-line and normal stem cells that have regulated telomerase have longer telomeres compared to most cancer cells. Thus, there may be a therapeutic window to drive cancer cells into apoptosis without major side effects on normal cells.

Even though telomerase does not cause cancer, and its role in cancer is most probably permissive, cancer therapy directed at telomerase has in some instances already advanced to clinical trials to validate safety and specificity (Shay et al. 1997; Shay 1998, 2003; White et al. 2001; Shay and Wright 2002a, 2005; Gellert et al. 2005).

Numerous telomerase inhibitor strategies have been developed that are in preclinical or various stages of clinical trials. These have been reviewed in detail recently (Shay and Wright 2006a,b). Perhaps one of the most rapidly advancing areas of telomerase therapy involves immunotherapy or vaccines targeting telomerase. Telomerase may be processed differently in cancers, and it has been determined that hTERT-specific epitopes are expressed on cancer cells but not on normal cells. Clinical trials involving patients with lung, breast, prostate, and pancreatic cancers have been conducted with excellent and encouraging results (Vonderheide et al. 2004; Su et al. 2005; Brunsvig et al. 2006; Carpenter and Vonderheide 2006). None of the patients have had treatment-related side effects (such as autoimmune disease or bone marrow stem cell depression). Thus, late-stage clinical trials are now being initiated.

Another lead telomerase inhibitor, GRN163L, is currently being tested in both leukemia and solid tumors (Dikmen et al. 2005; Shay and Wright 2006b). GRN163L is an oligonucleotide that targets the template region of hTERC and thus is a telomerase antagonist. The rationale for developing GRN163L is that hTERC must be exposed in order to add new telomeric repeats onto the chromosome ends, thus making it an accessible target for oligonucleotides. Rather than acting by "antisense" mechanisms to degrade mRNA or inhibit translation, oligonucleotides targeting the hTERC template region at the active site functions as classical enzymatic inhibitors of telomerase activity. With typical mRNAs, targeting as much as 90% or more must occur to be effective. However, targeting hTERC may be effective at considerably lower percentages driving telomere shortening and cancer cell death.

Telomerase inhibitors not only might directly limit or stop the growth of human tumors, but also might act in a synergistic fashion with existing therapeutic modalities and amplify their effectiveness. After initial chemotherapy or surgery, telomerase inhibitors might be used to inhibit the recovery of residual cancer cells, making them more susceptible to killing by existing chemotherapeutic agents or other novel therapies.

SUMMARY

Aging is complex and there are many theories that attempt to describe organismal aging in terms of single highly conserved mechanisms that are likely to be partially or totally wrong. The role of telomeres in aging has often been greatly exaggerated, and although most scientists accept the hypothesis that telomere shortening can function as a tumor-protection mechanism, there is great skepticism about replicative aging having any role in organismal aging. Many species, including mice, appear to have abandoned replicative aging entirely in favor of telomeres which are so long that they could not function effectively to count cell divisions. Aging mechanisms shared between mice and humans cannot depend on replicative aging, and thus the contributions to human aging from replicative aging must lie in aspects of aging that are not shared between these species. However, since telomeres can shorten due to other causes, such as oxidative damage, telomere shortening could nonetheless contribute to some shared mechanisms. There are conceptual reasons for believing that replicative aging should nonetheless contribute to some aspects of human aging, and one challenge for the future is identifying if, where, and how human aging differs from mouse aging due to telomere shortening from counting cell divisions. In addition, the use of telomerase for cell and tissue engineering and inhibiting telomerase as a promising and novel cancer therapy offer a glimpse for a future of this area of research.

ACKNOWLEDGMENTS

This work was supported by Lung Cancer SPORE P50 CA75907, NSCOR NNJ05HD36G, and the Ted Nash Foundation. We also thank Angela Diehl for the illustrations. Additional information can be obtained at the following Web site: http://www4.utsouthwestern.edu/ cellbio/shay-wright/.

REFERENCES

Allsopp R.C., Chang E., Kashefi-Aazam M., Rogaev E.I., Piatyszek M.A., Shay J.W., and Harley C.B. 1995. Telomere shortening is associated with cell division *in vitro* and *in vivo*. *Exp. Cell Res.* **220:** 194–200.

Allsopp R.C., Vaziri H., Patterson C., Goldstein S., Younglai E.V., Futcher A.B., Greider C.W., and Harley C.B. 1992. Telomere length predicts the replicative capacity of human fibroblasts. *Proc. Natl. Acad. Sci.* **85:** 10114–10118.

Aviv A. 2004. Telomeres and human aging: Facts and fibs. *Sci. Aging Knowledge Environ.* **1:** 43–45.

Aviv A., Valdes A., Gardner J.P., Swaminathan R., Kimura M., and Spector T.D. 2006. Menopause modifies the association of leukocyte telomere length with insulin resistance and inflammation. *J. Clin. Endocrinol. Metab.* **91:** 635–640.

Andrew T., Aviv A., Falchi M., Surdulescu G.L., Gardner J.P., Lu X., Kimura M., Kato B.S., Valdes A.M., and Spector T.D. 2006. Mapping genetic loci that determine leukocyte telomere length in a large sample of unselected, female sibling pairs. *Am. J. Hum. Genet.* **78:** 480–486.

Benetos A., Gardner J.P., and Zureik M. 2004. Short telomeres are associated with increased carotid atherosclerosis in hypertensive subjects. *Hypertension* **43:** 182–185.

Benetos A., Okuda K., Lajemi M., Kimura M., Thomas F., Skurnick J., Labat C., Bean K., and Aviv A. 2001. Telomere length as an indicator of biological aging: The gender effect and relation with pulse pressure and pulse wave velocity. *Hypertension* **37:** 381–385.

Bischoff C., Petersen H.C., Graakjaer J., Andersen-Ranberg K., Vaupel J.W., Bohr V.A., Kolvraa S., and Christensen K. 2006. No association between telomere length and survival among the elderly and oldest old. *Epidemiology* **17:** 190–194.

Blackburn E.H. 2006. The history of telomere biology. In *Telomeres*, 2nd edition (ed. T. de Lange et al.), pp. 1–19. Cold Spring Harbor Laboratory Press, Cold Spring Harbor, New York.

Blasco M.A., Lee H.W., Hande M.P., Samper E., Lansdorp P.M., DePinho R.A., and Greider C.W. 1997. Telomere shortening and tumor formation by mouse cells lacking telomerase RNA. *Cell* **91:** 25–34.

Bodnar A.G., Ouellette M., Frolkis M., Holt S.E., Chiu C.P., Morin G.B., Harley C.B., Shay J.W., Lichtsteiner S., and Wright W.E. 1998. Extension of lifespan by introduction of telomerase in normal human cells. *Science* **279:** 349–352.

Brouilette S., Singh R.K., Thompson J.R., Goodall A.H., and Samani N.J. 2003. White cell telomere length and risk of premature myocardial infarction. *Arterioscler. Thromb. Vasc. Biol.* **23:** 842–846.

Brunsvig P.F., Aamdal S., Gjertsen M.K., Kvalheim G., Markowski-Grimsrud C.J., Sve I., Dyrhaug M., Trachsel S., Moller M., Eriksen J.A., and Gaudernack G. 2006. Telomerase peptide vaccination: A phase I/II study in patients with non-small cell lung cancer. *Cancer Immunol. Immunother.* **55:** 1553–1564.

Campisi J. 2005. Senescent cells, tumor suppression, and organismal aging: Good citizens, bad neighbors. *Cell* **120:** 513–522.

Carpenter E.L. and Vonderheide R.H. 2006. Telomerase-based immunotherapy of cancer. *Expert Opin. Biol. Ther.* **6:** 1031–1039.

Cawthon R.M., Smith K.R., O'Brien E., Sivatchenko A., and Kerber R.A. 2003. Association between telomere length in blood and mortality in people aged 60 years or older. *Lancet* **361:** 393–395.

Chai W., Du Q., Shay J.W., and Wright W.E. 2006a. Human telomeres have different overhang sizes at leading versus lagging strands. *Mol. Cell* **21:** 427–435.

Chai W., Sfeir A.J., Hoshiyama H., Shay J.W., and Wright W.E. 2006b. The MRN complex is required for the generation of proper G-overhangs at human telomeres. *EMBO Rep.* **7:** 225–230.

Chen J.-L. and Greider C.W. 2006. Telomerase biochemistry and biogenesis. In *Telomeres*, 2nd edition (ed. T. de Lange et al.), pp. 49–80. Cold Spring Harbor Laboratory Press, Cold Spring Harbor, New York.

Cristofari G. and Lingner J. 2006. The telomerase ribonucleoprotein particle. In *Telomeres*, 2nd edition (ed. T. de Lange et al.), pp. 21–48. Cold Spring Harbor Laboratory Press, Cold Spring Harbor, New York.

Cui W., Aslam S., Fletcher J., Wylie D., Clinton M., and Clark A.J. 2002. Stabilization of telomere length and karyotypic stability are directly correlated with the level of hTERT gene expression in primary fibroblasts. *J. Biol. Chem.* **277:** 38531–38539.

de Lange T. 2006. Mammalian telomeres. In *Telomeres*, 2nd edition (ed. T. deLange et al.), pp. 387–431. Cold Spring Harbor Laboratory Press, Cold Spring Harbor, New York.

Dikmen Z.G., Gellert G.C., Jackson S., Gryaznov S., Tressler R., Dogan P., Wright W.E., and Shay J.W. 2005. *In vivo* inhibition of lung cancer by GRN163L—A novel human telomerase inhibitor. *Cancer Res.* **65:** 7866–7873.

Epel E.S., Blackburn E.H., Lin J., Dhabhar F.S., Adler N.E., Morrow J.D., and Cawthon R.M. 2004. Accelerated telomere shortening in response to life stress. *Proc. Natl. Acad. Sci.* **101:** 17312–17315.

Farazi P.A., Glickman J., Jiang S., Yu A., Rudolph K.L., and DePinho R.A. 2003. Differential impact of telomere dysfunction on initiation and progression of hepatocellular carcinoma. *Cancer Res.* **63:** 5021–5027.

Fogarty P.F., Yamaguchi H., Wiestner A., Baerlocher G.M., Sloand E., Zeng W.S., Read E.J., Lansdorp P.M., and Young N.S. 2003. Late presentation of dyskeratosis congenital as apparently acquired aplastic anemia due to mutations in telomerase RNA. *Lancet* **362:** 1628–1630.

Forsyth N.R., Elder F.E., Shay J.W., and Wright W.E. 2005. Lagomorphs (rabbits, pikas and hares) do not use telomere-directed replicative aging *in vitro. Mech. Ageing Dev.* **126:** 685–691.

Gardner J.P., Li S., Srinivasan S.R., Chen W., Kimura M., Lu X., Berenson G.S., and Aviv A. 2005. Rise in insulin resistance is associated with escalated telomere attrition. *Circulation* **111:** 2171–2177.

Gellert G.C., Jackson S.R., Dikmen G., Wright W.E., and Shay J.W. 2005. Telomerase as a therapeutic target in cancer. *Drug Discov. Today* **2:** 159–164.

Gompertz B. 1825. On the nature and function expressivity of the law of human mortality and on a new mode of determining life contingencies. *Philos. Trans. R. Soc. Lond.* **115:** 513–585.

Griffith J.D., Corneau L., Rosenfield S., Stansel R.M., Bianchi A., Moss H., and de Lange T. 1999. Mammalian telomeres end in a large duplex loop. *Cell* **97:** 503–514.

Guyton A.C. and Hall J.E. 1966. Transport of oxygen and carbon dioxide in the blood and bodily fluids. In *Textbook of medical physiology*, 3rd edition, pp. 513–523. Saunders, Philadelphia, Pennsylvania.

Hall D.B., Holmlin D.B., and Barton J.K. 1996. Oxidative DNA damage through long-range electron transfer. *Nature* **382:** 731–735.

Hastie N.D., Dempster M., Dunlop M.G., Thompson A.M., Green D.K., and Allshire R.C. 1990. Telomere reduction in human colorectal carcinoma and with ageing. *Nature* **346:** 866–868.

Hayflick L. and Moorhead P.S. 1961. The limited in vitro lifetime of human diploid cell strains. *Exp. Cell Res.* **25:** 585–621.

Hemann M.T., Strong M.A., Hao L.Y., and Greider C.W. 2001. The shortest telomere, not average telomere length, is critical for cell viability and chromosome stability. *Cell* **107:** 67–77.

Heiss N.S., Knight S.W., Vulliamy T.J., Klauck S.M., Wiemann S., Mason P.J., Poustka A., and Dokal I. 1998. X-linked dyskeratosis congenita is caused by mutations in a highly conserved gene with putative nucleolar functions. *Nat. Genet.* **19:** 32–38.

Honig L.S., Schupf N., Lee J.H., Tang M.X., and Mayeux R. 2006. Shorter telomeres are associated with mortality in those with APOE E4 and dementia. *Ann. Neurol.* **60:** 181–187.

Jeanclos E., Schork N.J., Kyvik K.O., Kimura M., Skurnick J.H., and Aviv A. 2000. Telomere length inversely correlates with pulse pressure and is highly familial. *Hypertension* **36:** 195–200.

Kim N.W., Piatyszek M.A., Prowse K.R., Harley C.B., West M.D., Ho P.L., Coviello G.M., Wright W.E., Weinrich S.L., and Shay J.W. 1994. Specific association of human telomerase activity with immortal cells and cancer. *Science* **266:** 2011–2015.

Kirkwood T.B.L. 1996. Human senescence. *Bioessays* **19:** 1009–1016.

Kruk P.A., Rampino N.J., and Bohr V.A. 1995. DNA damage and repair in telomeres: Relation to aging. *Proc. Natl. Acad. Sci.* **92:** 258–262.

Lanza R.P., Cibelli J.B., Diaz F., Moraes C.T., Farin P.W., Farin C.E., Hammer C.J., West M.D., and Damiani P. 2000. Cloning of an endangered species (*Bos gaurus*) using interspecies nuclear transfer. *Cloning* **2:** 79–90.

Lindsey J., McGill N.I., Lindsey L.A., Green D.K., and Cooke H.J. 1991. In vivo loss of telomeric repeats with age in humans. *Mutat. Res.* **256:** 45–48.

Martin-Ruiz C.M., Gussekloo J., van Heemst D., von Zglinicki T., and Westendorp R.G. 2005. Telomere length in white blood cells is not associated with morbidity or mortality in the oldest old: A population-based study. *Aging Cell* **4:** 287–290.

Martin-Ruiz C., Dickinson H.O., Key B., Rowan E., Kenny R.A, and von Zgliniki T. 2006. Telomere length predicts poststroke mortality, dementia, and cognitive decline. *Ann. Neurol.* **60:** 174–180.

Martin G.M., Sprague C.A., and Epstein C.J. 1970. Replicative lifespan of cultivated human cells: Effect of donors age, tissue and genotype. *Lab. Invest.* **23:** 86–92.

Mason P.J., Wilson D.B., and Bessler M. 2005. Dyskeratosis congenita—A disease of dysfunctional telomere maintenance. *Curr. Mol. Med.* **5:** 159–170.

Meeker A.K., Hicks J.L., Iacobuzio-Donahue C.A., Montgomery E.A., Westra W.H., Chan T.Y., Ronnett B.M., and De Marzo A.M. 2004. Telomere length abnormalities occur early in the initiation of epithelial carcinogenesis. *Clin. Cancer Res.* **10:** 317–326.

Mitchell J.R., Wood E., and Collins K.A. 1999. A telomerase component is defective in the human disease dyskeratosis congenita. *Nature* **402:** 551–555.

Morales C.P., Holt S.E., Ouellette M., Kaur K.J., Yan Y., Wilson K.S., White M.A., Wright W.E., and Shay J.W. 1999. Absence of cancer-associated changes in human fibroblasts immortalized with telomerase. *Nat. Genet.* **21:** 115–118.

Nawrot T.S., Staessen J.A., Gardner J.P., and Aviv A. 2004. Telomere length and possible link to X chromosome. *Lancet* **363:** 507–510.

Oikawa S. and Kawanishi S. 1999. Site-specific DNA damage at GGG sequence by oxidative stress may accelerate telomere shortening. *FEBS Lett.* **453:** 365–368.

Okuda K., Khan M.Y., Skurnick J. Kimura M., Aviv H., and Aviv A. 2000. Telomere attrition of the human abdominal aorta: Relationships with age and atherosclerosis. *Atherosclerosis* **152:** 391–398.

O'Sullivan J.N., Bronner M.P., Brentnall T.A., Finley J.C., Shen W.T., Emerson S., Emond M.J., Gollahon K.A., Moskovitz A.H., Crispin D.A., Potter J.D., and Rabinovitch P.S. 2002. Chromosomal instability in ulcerative colitis is related to telomere shortening. *Nat. Genet.* **32:** 280–284.

Panossian L.A., Porter V.R., Valenzuela H.F., Zhu X., Reback E., Materman D., Cumming J.L., and Effros R.B. 2003. Telomere shortening in T cells correlate with Alzheimer's disease status. *Neurobiol. Aging* **24:** 77–84.

Parrinello S., Samper E., Krtolica A., Goldstein J., Melov S., and Campisi J. 2003. Oxygen sensitivity severely limits the replicative lifespan of murine fibroblasts. *Nat. Cell Biol.* **5:** 741–747.

Petersen S., Saretzki G., and von Zglinicki T. 1998. Preferential accumulation of single-stranded regions in telomeres of human fibroblasts. *Exp. Cell Res.* **239:** 152–160.

Poh M., Boyer M., Solan A., Dahl S.L., Pedrotty D., Banik S.S., McKee J.A., Klinger R.Y., Counter C.M., and Niklason L.E. 2005. Blood vessels engineered from human cells. *Lancet* **365:** 2122–2124.

Rando T. 2006. Prognostic value of telomere length: The long and short of it. *Ann. Neurol.* **60:** 155–157.

Robertson D.M., Li L., Fisher S., Pearce V.P., Shay J.W., Wright W.E., Cavanagh H.D., and Jester J.V. 2005. Characterization of growth and differentiation in a telomerase-immortalized human corneal epithelial cell line. *Invest. Ophthalmol. Vis. Sci.* **46:** 470–478.

Rudolph K.L., Millard M., Bosenberg M.W., and DePinhho R.A. 2001. Telomere dysfunction and evolution of intestinal carcinoma in mice and humans. *Nat. Genet.* **28:** 155–159.

Samani N.J., Boultby R., Butler R., Thompson J.R., and Goodall A.H. 2001. Telomere shortening in atherosclerosis. *Lancet* **358:** 472–473.

Sfeir A.J., Chai W., Shay J.W., and Wright W.E. 2005. Telomere-end processing; the terminal nucleotides of human chromosome. *Mol. Cell* **18:** 131–138.

Shay J.W. 1998. Telomerase in cancer: Diagnostic, prognostic and therapeutic implications. *Cancer J. Sci. Am.* **4:** 26–34.

———. 2003. Telomerase therapeutics: Telomeres recognized as a DNA damage signal. *Clin. Cancer Res.* **9:** 3521–3525.

Shay J.W. and Bacchetti S. 1997. A survey of telomerase in human cancer. *Eur. J. Cancer* **33:** 787–791.

Shay J.W. and Roninson I.B. 2004. Hallmarks of senescence in carcinogenesis and cancer therapy. *Oncogene* **23:** 2919–2933.

Shay J.W. and Wright W.E. 1999. Telomeres in dyskeratosis congenita. *Nat. Genet.* **36:** 437–438.

———. 2000. The use of "telomerized" cells for tissue engineering. *Nat. Biotechnol.* **18:** 22–23.

———. 2002a. Telomerase: A target for cancer therapy. *Cancer Cell* **2:** 257–265.

———. 2002b. Historical claims and current interpretations of replicative aging. *Nat. Biotechnol.* **20:** 682–688.

———. 2004a. Senescence and immortalization: Role of telomeres and telomerase. *Carcinogenesis* **25:** 1–8.

———. 2004b. Mutant dyskerin ends relationship with telomerase. *Science* **286:** 2284–2285.

———. 2005. Mechanism-based combination telomerase inhibition therapy. *Cancer Cell* **7:** 1–2.

———. 2006a. Telomerase and human cancer. In *Telomeres,* 2nd edition (ed. T. de Lange et al.), pp. 81–108. Cold Spring Harbor Laboratory Press, Cold Spring Harbor, New York.

———. 2006b. Telomerase therapeutics for cancer: Challenges and new directions. *Nat. Rev. Drug Discov.* **5:** 577–584.

Shay J.W., Werbin H., and Wright W.E. 1997. Telomerase assays in the diagnosis and prognosis of cancer. *Ciba Found. Symp.* **211:** 148–160.

Shay J.W., Wright W.E., and Werbin H. 1991. Defining the molecular mechanism of human cell immortalization. *Biochim. Biophys. Acta* **1072:** 1–7.

Shay J.W., Van der Haegen B.A., Ying Y., and Wright W.E. 1999. The frequency of immortalization of human fibroblast and mammary epithelial cells transfected with SV40 large T-antigen. *Exp. Cell Res.* **209:** 45–52.

Slagboom P.E., Droog S., and Boomsma D.I. 1994. Genetic determination of telomere size in humans: A twin study of three age groups. *Am. J. Hum. Genet.* **55:** 876–882.

Steinert S., Shay J.W., and Wright W.E. 2000. Transient expression of human telomerase (hTERT) extends the life span of normal human fibroblasts. *Biochem. Biophys. Res. Commun.* **273:** 1095–1098.

Steinert S., White D.M., Zou Y., Shay J.W., and Wright W.E. 2002. Telomere biology and cellular aging in non-human primate cells. *Exp. Cell Res.* **272:** 146–152.

Su Z., Dannull J., Yang B.K., Dahm P., Coleman D., Yancey D., Sichi S., Niedzwiecki D., Boczkowski D., Gilboa E., and Vieweg J. 2005. Telomerase mRNA-tranfected dendritic cells stimulate antigen-specific CD8+ and CD4+ cell responses in patients with metastatic prostate cancer. *J. Immunol.* **174:** 3798–3807.

Thomas M., Yang L., and Hornsby P.J. 2000. Formation of normal functional tissue from transplanted adrenocortical cells expressing telomerase reverse transcriptase (TERT). *Nat. Biotechnol.* **18:** 39–42.

Unryn B.M., Cook L.S., and Riabowol K.T. 2005. Paternal age is positively linked to telomere length of children. *Aging Cell* **4:** 97–101.

Valdes A.M., Andrew T., Gardner J.P., Kimura M., Oelsner E., Cherkas L.F., Aviv A., and Spector T.D. 2005. Obesity, cigarette smoking, and telomere length in women. *Lancet* **366:** 662–664.

Vaughan M.B., Ramirez R.D., Wright W.E., Minna J.D., and Shay J.W. 2006. A three-dimensional model of differentiation of immortalized human bronchial epithelial cells. *Differentiation* **74:** 141–148.

Vaughan M.B., Ramirez R.D., Brown S.A., Yang J.C., Wright W.E., and Shay J.W. 2004. A reproducible laser-wounded skin equivalent model to study the effect of aging *in vitro*. *Regen. Med.* **2:** 99–110.

Vogelstein B. and Kinzler K.W. 1993. The multistep nature of cancer. *Trends Genet.* **9:** 138–141.

Vonderheide R.H., Domchek S.M., Schultze J.L., George D.J., Hoar K.M., Chen D.Y., Stephans K.F., Masutomi K., Loda M., Xia Z., Anderson K.S., Hahn W.C., and Nadler L.M. 2004. Vaccination of cancer patients against telomerase induces functional anti-tumor CD8+ T lymphocytes. *Clin. Cancer Res.* **10:** 828–839.

von Zglinicki T., Serra V., and Lorenz M. 2000. Short telomeres in patients with vascular dementia: An indicator of low antioxidative capacity and a possible risk factor? *Lab. Invest.* **80:** 1739–1747.

Vulliamy T. and Dokal I. 2006. Dyskeratosis congenita. *Semin. Hematol.* **43:** 157–166.

Vulliamy T., Marrone A., Szydlo R., Walne A., Mason P.J., and Dokal I. 2004. Disease anticipation is associated with progressive telomere shortening in families with dyskeratosis congenita due to mutations in TERC. *Nat. Genet.* **36:** 447–449.

Vulliamy T., Marrone A., Goldman F., Dearlove A., Bessler M., Mason P.J., and Dokal I. 2001. The RNA component of telomerase is mutated in autosomal dominant dyskeratosis congenita. *Nature* **413:** 432–435.

Wiemann S.U., Satyanarayana A., Tsahuridu M., Tillmann H.L., Zender L., Klempnauer J., Flemming P., Franco S., Blasco M.A., Manns M.P., and Rudolph K.L. 2002. Hepatocyte telomere shortening and senescence are general markers of human liver cirrhosis. *FASEB J.* **16:** 935–942.

West M.D. 1994. The cellular and molecular biology of skin aging. *Arch. Dermatol.* **130:** 87–95.

West M.D., Shay J.W., Wright W.E., and Linksken M.H.K. 1996. Altered expression of plasminogen activator and plasminogen activator inhibitor during cellular senescence. *Exp. Gerontol.* **31:** 175–193.

White L., Wright W.E., and Shay J.W. 2001. Telomerase inhibitors. *Trends Biotechnol.* **19:** 114–120.

Wright W.E. and Shay J.W. 1995. Time, telomeres and tumors: Is cellular senescence more than an anticancer mechanism. *Trends Cell Biol.* **5:** 293–296.

———. 2000. Fundamental differences in human and mouse telomere biology. *Nat. Med.* **6:** 849–851.

Wright W.E., Tesmer V.M., Hoffman K.E., Levene S.D., and Shay J.W. 1997. Normal human chromosomes have long G-rich telomeric overhangs at one end. *Genes Dev.* **11:** 2801–2809.

Wu X., Amos C.I., Zhu Y., Zhao H., Grossman B.H., Shay J.W., Swan G.E., Benowitz N.L., Luo S., and Spitz M.R. 2003. Telomere dysfunction: A potential cancer predisposition factor. *J. Natl. Cancer Inst.* **95:** 1211–1218.

Yamaguchi H., Calado R.T., Ly H., Kajigaya S., Baerlocher G.M., Chanock S.J., Lansdorp P.M., and Young N.S. 2005. Mutations in TERT, the gene for telomerase reverse transcriptase, in aplastic anemia. *N. Engl. J. Med.* **352:** 1413–1424.

Zakian V.A. 1989. The structure and function of telomeres. *Annu. Rev. Genet.* **23:** 579–604.

Zou Y., Sfeir A., Shay J.W., and Wright W.E. 2004. Does a sentinel or groups of short telomeres determine replicative senescence? *Mol. Biol. Cell* **15:** 3709–3718.

Zou Y., Yi X., Wright W.E., and Shay J.W. 2002. Human telomerase can immortalize Indian muntjac cells. *Exp. Cell Res.* **281:** 63–76.

Index